Student Solutions Manual
Katy Murphy
Michael J. Sullivan, III & Michael Sullivan

Algebra & Trigonometry
ENHANCED WITH GRAPHING UTILITIES

Precalculus
ENHANCED WITH GRAPHING UTILITIES

MICHAEL SULLIVAN

MICHAEL SULLIVAN, III

PRENTICE HALL Upper Saddle River NJ 07458

For Pat and Yola

Acquisitions Editor: **Sally Denlow**
Supplement Editor: **Audra Walsh**
Production Editor: **Barbara Kraemer**
Production Supervisor: **Barbara Murray**
Production Coordinator: **Alan Fischer**

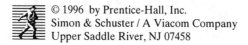

© 1996 by Prentice-Hall, Inc.
Simon & Schuster / A Viacom Company
Upper Saddle River, NJ 07458

Printed in the United States of America

10 9 8 7 6 5 4 3

ISBN 0-13-238700-X

Prentice-Hall International (UK) Limited, *London*
Prentice-Hall of Australia Pty. Limited, *Sydney*
Prentice-Hall Canada, Inc., *Toronto*
Prentice-Hall Hispanoamericana, S.A., *Mexico*
Prentice-Hall of India Private Limited, *New Delhi*
Prentice-Hall of Japan, Inc., *Tokyo*
Simon & Schuster Asia Pte. Ltd., *Singapore*
Editora Prentice-Hall do Brasil, Ltda., *Rio de Janeiro*

This manual contains detailed solutions to all the odd-numbered problems in *Algebra & Trigonometry Enhanced with Graphing Utilities* and *Precalculus Enhanced with Graphing Utilities*, by Michael Sullivan and Michael Sullivan, III. Chapter 6 for *Precalculus* is found at the end of the manual and begins on page 649. Preceding the solutions to some of the groups of problems we have listed step-by-step procedures which may be applied to each exercise in the group. Hopefully, these will enable you to develop a systematic approach to solving certain types of exercises. Our desire is that after seeing several examples, you will be able to solve problems without referring to this manual.

We wish to thank everyone who has helped with this project. The enormous job of typing was done by Brenda Dobson, and the art work was prepared by Kelly Evans. A special thank you to our families for their support during this project.

Finally, we would be grateful to hear from readers who discover any errors in this solutions manual.

Michael Sullivan, III
Katy Murphy
and
Michael Sullivan

Contents

Contents

Chapter 1

GRAPHS

1.1 Rectangular Coordinates; Graphing Utilities

1. (a) Quadrant II
    (b) Positive x-axis
    (c) Quadrant III
    (d) Quadrant I
    (e) Negative y-axis
    (f) Quadrant IV

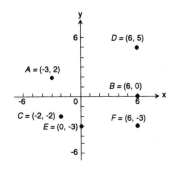

3. The points will be on a vertical line that is two units to the right of the y-axis.

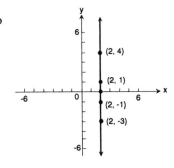

5. $(-1, 4)$

7. $(3, 1)$

9. $x\text{min} = -11$
    $x\text{max} = 6$
    $x\text{scl} = 1$
    $y\text{min} = -3$
    $y\text{max} = 6$
    $y\text{scl} = 1$

11. $x\text{min} = -30$
    $x\text{max} = 50$
    $x\text{scl} = 10$
    $y\text{min} = -90$
    $y\text{max} = 50$
    $y\text{scl} = 10$

13. $x\text{min} = -10$
    $x\text{max} = 110$
    $x\text{scl} = 10$
    $y\text{min} = -10$
    $y\text{max} = 160$
    $y\text{scl} = 10$

15. $x\text{min} = -6$
    $x\text{max} = 6$
    $x\text{scl} = 2$
    $y\text{min} = -4$
    $y\text{max} = 4$
    $y\text{scl} = 2$

17. $x\text{min} = -9$
    $x\text{max} = 9$
    $x\text{scl} = 3$
    $y\text{min} = -4$
    $y\text{max} = 4$
    $y\text{scl} = 2$

19. $x\text{min} = -6$
    $x\text{max} = 6$
    $x\text{scl} = 1$
    $y\text{min} = -8$
    $y\text{max} = 8$
    $y\text{scl} = 2$

21. $x\text{min} = -6$
$x\text{max} = 6$
$x\text{scl} = 2$
$y\text{min} = -1$
$y\text{max} = 3$
$y\text{scl} = 1$

23. $x\text{min} = 3$
$x\text{max} = 9$
$x\text{scl} = 1$
$y\text{min} = 2$
$y\text{max} = 10$
$y\text{scl} = 2$

25. $d(P_1, P_2) = \sqrt{(2 - 0)^2 + (1 - 0)^2} = \sqrt{4 + 1} = \sqrt{5}$

27. $d(P_1, P_2) = \sqrt{(-1 - 1)^2 + (3 - 1)^2} = \sqrt{4 + 4} = \sqrt{8} = 2\sqrt{2}$

29. $P_1 = (3, -4), P_2 = (5, 4)$
$d(P_1, P_2) = \sqrt{(5 - 3)^2 + [4 - (-4)]^2}$
$d(P_1, P_2) = \sqrt{(2)^2 + (8)^2}$
$d(P_1, P_2) = \sqrt{4 + 64}$
$d(P_1, P_2) = \sqrt{68} = 2\sqrt{17}$

31. $P_1 = (-3, 2), P_2 = (6, 0)$
$d(P_1, P_2) = \sqrt{[6 - (-3)]^2 + (0 - 2)^2}$
$d(P_1, P_2) = \sqrt{(9)^2 + (-2)^2}$
$d(P_1, P_2) = \sqrt{81 + 4}$
$d(P_1, P_2) = \sqrt{85}$

33. $P_1 = (4, -3), P_2 = (6, 4)$
$d(P_1, P_2) = \sqrt{(6 - 4)^2 + [4 - (-3)]^2}$
$d(P_1, P_2) = \sqrt{(2)^2 + (7)^2}$
$d(P_1, P_2) = \sqrt{4 + 49}$
$d(P_1, P_2) = \sqrt{53}$

35. $P_1 = (-0.2, 0.3), P_2 = (2.3, 1.1)$
$P_1, P_2) = \sqrt{[2.3 - (-0.2)]^2 + (1.1 - 0.3}$
$P_1, P_2) = \sqrt{(2.5)^2 + (.8)^2}$
$P_1, P_2) = \sqrt{6.25 + .64}$
$P_1, P_2) = \sqrt{6.89}$
$P_1, P_2) = 2.625$ (with calculator)

37. $P_1 = (a, b), P_2 = (0, 0)$
$d(P_1, P_2) = \sqrt{0 - a)^2 + (0 - b)^2}$
$d(P_1, P_2) = \sqrt{a^2 + b^2}$

39. $P_1 = (1, 3), P_2 = (5, 15)$
$d(P_1, P_2) = \sqrt{(5 - 1)^2 + (15 - 3)^2}$
$d(P_1, P_2) = \sqrt{(4)^2 + (12)^2}$
$d(P_1, P_2) = \sqrt{16 + 144}$
$d(P_1, P_2) = \sqrt{160}$
$d(P_1, P_2) = 4\sqrt{10}$

41. $P_1 = (-4, 6), P_2 = (4, 8)$
$d(P_1, P_2) = \sqrt{(4 - (-4))^2 + (8 - 6)^2}$
$d(P_1, P_2) = \sqrt{(8)^2 + (2)^2}$
$d(P_1, P_2) = \sqrt{64 + 4}$
$d(P_1, P_2) = \sqrt{68}$
$d(P_1, P_2) = 2\sqrt{17}$

43. $A = (-2, 5), B = (1, 3), C = (-1, 0)$

$$d(A, B) = \sqrt{[1 - (-2)^2 + (3 - 5)^2]}$$
$$d(A, B) = \sqrt{3^2 + (-2)^2}$$
$$d(A, B) = \sqrt{9 + 4}$$
$$d(A, B) = \sqrt{13}$$
$$d(B, C) = \sqrt{(-1 - 1)^2 + (0 - 3)^2}$$
$$d(B, C) = \sqrt{(-2)^2 + (-3)^2}$$
$$d(B, C) = \sqrt{4 + 9}$$
$$d(B, C) = \sqrt{13}$$
$$d(A, C) = \sqrt{[-1 - (-2)]^2 + (0 - 5)^2}$$
$$d(A, C) = \sqrt{(-1)^2 + (-5)^2}$$
$$d(A, C) = \sqrt{1 + 25}$$
$$d(A, C) = \sqrt{26}$$

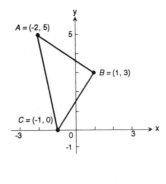

Verify that ABC is a right triangle by the Pythagorean Theorem:

$$[d(A, B)]^2 + (d(B, C)]^2 = [d(A, C)]^2$$
$$(\sqrt{13})^2 + (\sqrt{13})^2 = (\sqrt{26})^2$$
$$13 + 13 = 26$$
$$26 = 26$$

Area of a triangle is $A = \dfrac{1}{2}bh$. In this problem,

$$A = \frac{1}{2}[d(B, C)][d(A, B)]$$

$$A = \frac{1}{2}(\sqrt{13})(\sqrt{13})$$

$$A = \frac{1}{2}(13)$$

$$A = \frac{13}{2} \text{ square units}$$

45. $A = (-5, 3), B = (6, 0), C = (5, 5)$

$$d(A, B) = \sqrt{[6 - (-5)^2 + (0 - 3)^2]}$$
$$d(A, B) = \sqrt{(11)^2 + (-3)^2}$$
$$d(A, B) = \sqrt{121 + 9}$$
$$d(A, B) = \sqrt{130}$$
$$d(B, C) = \sqrt{(5 - 6)^2 + (5 - 0)^2}$$
$$d(B, C) = \sqrt{(-1)^2 + (5)^2}$$
$$d(B, C) = \sqrt{1 + 25}$$
$$d(B, C) = \sqrt{26}$$
$$d(A, C) = \sqrt{[5 - (-5)]^2 + (5 - 3)^2}$$
$$d(A, C) = \sqrt{(10)^2 + (2)^2}$$
$$d(A, C) = \sqrt{100 + 4}$$
$$d(A, C) = \sqrt{104}$$

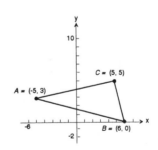

Verify that ABC is a right triangle by the Pythagorean Theorem:

$$[d(A,\ C)]^2 + (d(B,\ C)]^2 = [d(A,\ B)]^2$$
$$\left(\sqrt{104}\right)^2 + \left(\sqrt{26}\right)^2 = \left(\sqrt{130}\right)^2$$
$$104 + 26 = 130$$
$$130 = 130$$

Area of a triangle is $A = \dfrac{1}{2}bh$. In this problem,

$$A = \frac{1}{2}[d(A,\ C)][d(B,\ C)]$$

$$A = \frac{1}{2}\left(\sqrt{104}\right)\left(\sqrt{26}\right)$$

$$A = \frac{1}{2}\left(\sqrt{2704}\right)$$

$$A = \frac{1}{2}(52)$$

$$A = 26 \text{ square units}$$

47. $A = (4,\ -3),\ B = (0,\ -3),\ C = (4,\ 2)$

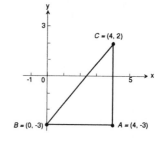

$d(A,\ B) = \sqrt{(0 - 4)^2 + [-3 - (-3)]^2}$
$d(A,\ B) = \sqrt{(-4)^2 + (0)^2}$
$d(A,\ B) = \sqrt{16 + 0}$
$d(A,\ B) = 4$
$d(B,\ C) = \sqrt{(4 - 0)^2 + [(2 - (-3)]^2}$
$d(B,\ C) = \sqrt{(4)^2 + (5)^2}$
$d(B,\ C) = \sqrt{16 + 25}$
$d(B,\ C) = \sqrt{41}$
$d(A,\ C) = \sqrt{(4 - 4)^2 + [2 - (-3)]^2}$
$d(A,\ C) = \sqrt{(0)^2 + (5)^2}$
$d(A,\ C) = \sqrt{0 + 25}$
$d(A,\ C) = 5$

Verify that ABC is a right triangle by the Pythagorean Theorem:

$$[d(A,\ C)]^2 + (d(A,\ B)]^2 = [d(B,\ C)]^2$$
$$5^2 + 4^2 = \left(\sqrt{41}\right)^2$$
$$25 + 16 = 41$$
$$41 = 41$$

Area of a triangle is $A = \dfrac{1}{2}bh$. In this problem,

$$A = \frac{1}{2}[d(A,\ B)][d(A,\ C)]$$

$$A = \frac{1}{2}(4)(5)$$

$$A = 10 \text{ square units}$$

49. All points having an x-coordinate of 2 would be of the form $(2, y)$. Those which are 5 units from $(-2, -1)$ would be:

$$\sqrt{[2 - (-2)]^2 + [y - (-1)]^2} = 5$$
$$\sqrt{(4)^2 + (y + 1)^2} = 5$$

Square both sides:
$$4^2 + (y + 1)^2 = 25$$
$$16 + y^2 + 2y + 1 = 25$$
$$y^2 + 2y + 17 = 25$$
$$y^2 + 2y - 8 = 0$$
$$(y + 4)(y - 2) = 0$$
$$y + 4 = 0 \quad \text{or} \quad y - 2 = 0$$
$$y = -4 \quad \text{or} \quad y = 2$$

Therefore, the points are $(2, -4), (2, 2)$.

51. All points on the x-axis would be of the form $(x, 0)$. Those which are 5 units from $(4, -3)$ would be:

$$\sqrt{(x - 4)^2 + [0 - (-3)]^2} = 5$$
$$\sqrt{(x - 4)^2 + (3)^2} = 5$$

Square both sides:
$$(x - 4)^2 + 9 = 25$$
$$x^2 - 8x + 16 + 9 = 25$$
$$x^2 - 8x + 25 = 25$$
$$x^2 - 8x = 0$$
$$x(x - 8) = 0$$
$$x = 0 \quad \text{or} \quad x = 8$$

Therefore, the points are $(0, 0)$ and $(8, 0)$.

53. $P_1 = (5, -4), P_2 = (3, 2)$

Let $\quad x_1 = 5 \quad y_1 = -4$
$\quad\quad\quad x_2 = 3 \quad y_2 = 2$

Then, the coordinates (x, y) of the midpoint are:

$$x = \frac{x_1 + x_2}{2} = \frac{5 + 3}{2} = \frac{8}{2} = 4$$

$$y = \frac{y_1 + y_2}{2} = \frac{-4 + 2}{2} = -\frac{2}{2} = -1$$

Midpoint $= (4, -1)$

55. $P_1 = (-3, 2), P_2 = (6, 0)$

Let $\quad x_1 = -3 \quad y_1 = 2$
$\quad\quad\quad x_2 = 6 \quad y_2 = 0$

Then, the coordinates (x, y) of the midpoint are:

$$x = \frac{x_1 + x_2}{2} = \frac{-3 + 6}{2} = \frac{3}{2}$$

$$y = \frac{y_1 + y_2}{2} = \frac{2 + 0}{2} = 1$$

Midpoint $= \left(\frac{3}{2}, 1\right)$

57. $P_1 = (4, -3), P_2 = (6, 1)$

Let $\quad x_1 = 4 \quad y_1 = -3$
$\quad\quad\quad x_2 = 6 \quad y_2 = 1$

Then, the coordinates (x, y) of the midpoint are:

$$x = \frac{x_1 + x_2}{2} = \frac{4 + 6}{2} = \frac{10}{2} = 5$$

$$y = \frac{y_1 + y_2}{2} = \frac{-3 + 1}{2} = \frac{-2}{2} = -1$$

Midpoint $= (5, -1)$

59. $P_1 = (-0.2, 0.3), P_2 = (2.3, 1.1)$

Let $\quad x_1 = -0.2 \quad\quad y_1 = 0.3$
$\quad\quad\quad x_2 = 2.3 \quad\quad y_2 = 1.1$

Then, the coordinates (x, y) of the midpoint are:

$$x = \frac{x_1 + x_2}{2} = \frac{-0.2 + 2.3}{2} = \frac{2.1}{2} = 1.05$$

$$y = \frac{y_1 + y_2}{2} = \frac{0.3 + 1.1}{2} = \frac{1.4}{2} = .7$$

Midpoint $= (1.05, 0.7)$

61. $P_1 = (a, b)$, $P_2 = (0, 0)$

Let $x_1 = a$ $y_1 = b$
             $x_2 = 0$ $y_2 = 0$

Then, the coordinates (x, y) of the midpoint are:

$$x = \frac{x_1 + x_2}{2} = \frac{a + 0}{2} = \frac{a}{2}$$

$$y = \frac{y_1 + y_2}{2} = \frac{b + 0}{2} = \frac{b}{2}$$

Midpoint $= \left(\dfrac{a}{2}, \dfrac{b}{2} \right)$

63.   Let's label the points:

     $A = (0, 0)$
     $B = (0, 6)$
     $C = (8, 0)$

Then find each midpoint of each side of triangle ABC and label it.

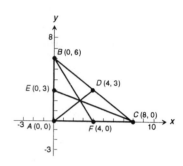

$$D = \text{midpoint of } BC = \frac{8 + 0}{2}, \frac{0 + 6}{2} = (4, 3)$$

$$E = \text{midpoint of } AB = \frac{0 + 0}{2}, \frac{0 + 6}{2} = (0, 3)$$

$$F = \text{midpoint of } AC = \frac{0 + 8}{2}, \frac{0 + 0}{2} = (4, 0)$$

The medians are AD, CE, and BF. Using the distance formula, find the length of each median.

$d(A, D) = \sqrt{(4 - 0)^2 + (3, 0)^2} = \sqrt{16 + 9} = \sqrt{25} = 5$

$d(C, E) = \sqrt{(8 - 0)^2 + (0 - 3)^2} = \sqrt{64 + 9} = \sqrt{73}$

$d(B, F) = \sqrt{(0 - 4)^2 + (6 - 0)^2} = \sqrt{16 + 36} = \sqrt{52} = 2\sqrt{13}$

65. $P_1 = (2, 1)$, $P_2 = (-4, 1)$, $P_3 = (-4, -3)$

$d(P_1, P_2) = \sqrt{(-4 - 2)^2 + (1 - 1)^2} = \sqrt{(-6)^2 + (0)^2} = \sqrt{36} = 6$

$d(P_2, P_3) = \sqrt{[-4 - (-4)]^2 + (-3 - 1)^2} = \sqrt{(0)^2 + (-4)^2} = \sqrt{16} = 4$

$d(P_1, P_3) = \sqrt{(-4 - 2)^2 + (3 - 1)^2} = \sqrt{(-6)^2 + (-4)^2} = \sqrt{36 + 16} = \sqrt{52} = 2\sqrt{13}$

Since no two sides are of equal length, the triangle is not isosceles.

By the Pythagorean Theorem,

$$\left[d(P_1, P_2) \right]^2 + \left[d(P_2, P_3) \right]^2 = \left[d(P_1, P_3) \right]^2$$
$$6 + 4^2 = \left(2\sqrt{13} \right)^2$$
$$36 + 16 = 52$$
$$52 = 52$$

Therefore, triangle $P_1 P_2 P_3$ is a right triangle.

67. $P_1 = (-2, -1)$, $P_2 = (0, 7)$, $P_3 = (3, 2)$

$d(P_1, P_2) = \sqrt{[0 - (-2)]^2 + [7 - (-1)]^2}$

            $= \sqrt{2^2 + 8^2} = \sqrt{4 + 64} = \sqrt{68}$

$d(P_2, P_3) = \sqrt{(3 - 0)^2 + (2 - 7)^2}$

            $= \sqrt{3^2 + (-5)^2} = \sqrt{9 + 25} = \sqrt{34}$

$d(P_1, P_3) = \sqrt{[3 - (-2)]^2 + [2 - (-1)]^2}$

            $= \sqrt{5^2 + 3^2} = \sqrt{25 + 9} = \sqrt{34}$

Since $d(P_2, P_3) = d(P_1, P_3)$, the triangle is isosceles.

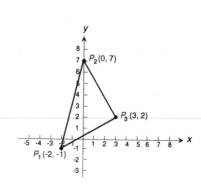

By the Pythagorean Theorem,

$$\left[d(P_2, P_3)\right]^2 + \left[d(P_1, P_3)\right]^2 = \left[d(P_1, P_2)\right]^2$$
$$\left(\sqrt{34}\right)^2 + \left(\sqrt{34}\right)^2 = \left(\sqrt{68}\right)^2$$
$$34 + 34 = 68$$
$$68 = 68$$

Therefore, triangle is an isosceles right triangle.

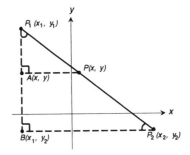

69. Given: $\dfrac{d(P_1, P)}{d(P_1, P_2)} = r$

Prove: $P(x, y)$ has coordinates $x = (1 - r)x_1 + rx_2$
and $y = (1 - r)y_1 + ry_2$

Proof: $\triangle APP_1 \cong \triangle BP_2P_1$, so corresponding sides are proportional.

$$\frac{d(A, P)}{d(B, P_2)} = \frac{d(P_1P)}{d(P_1, P_2)} = r \text{ and } \frac{d(P_1, A)}{d(P_1, B)} = \frac{d(P_1P)}{d(P_1, P_2)} = r$$

$$\frac{\sqrt{(x - x_1)^2 + (y - y)^2}}{\sqrt{(x_2 - x_1)^2 + (y_2 - y_2)^2}} = r \qquad \frac{\sqrt{(x_1 - x_1)^2 + (y - y_1)^2}}{\sqrt{(x_1 - x_1)^2 + (y_2 - y_1)^2}} = r$$

$$\frac{x - x_1}{x_2 - x_1} = r \qquad\qquad \frac{y - y_1}{y_2 - y_1} = r$$

$$x - x_1 = rx_2 - rx_1 \qquad\qquad y - y_1 = ry_2 - ry_1$$
$$x = x_1 - rx_1 + rx_2 \qquad\qquad y = y_1 - ry_1 + ry_2$$
$$x = x_1 + r(x_2 - x_1) \qquad\qquad y = y_1 + r(y_2 - y_1)$$

71. If $r = 1$, then

$$x = (1 - 1)x_1 + (1)x_2 \qquad y = (1 - 1)y_1 + (1)y_2$$
$$x = 0 + x_2 \qquad\qquad y = 0 + y_2$$
$$x = x_2 \qquad\qquad y = y_2$$
$$(x_2, y_2) = P_2$$

73. $d(P_1, P) = 2d(P_1, P_2)$

$$\frac{d(P_1, P)}{d(P_1, P_2)} = 2 = r$$

$$x = x_1 + r(x_2 - x_1) = 2 + 2(5 - 2) = 8$$
$$y = y_1 + r(y_2 - y_1) = 4 + 2(6 - 4) = 8$$

The point P we seek is $(8, 8)$.

75.
$$90^2 + 90^2 = d^2$$
$$8100 + 8100 = d^2$$
$$16200 = d^2$$
$$\sqrt{16200} = d$$
$$90\sqrt{2} = d$$
$$127.28 = d$$
127.28 ft.

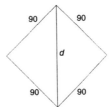

77. **(a)**

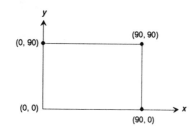

(b) $d = \sqrt{(310 - 90)^2 + (15 - 90)^2}$
$= \sqrt{(220)^2 + (75)^2}$
$= 232.4$ feet

(c) $d = \sqrt{(300 - 0)^2 + (300 - 90)^2}$
$= \sqrt{(300)^2 + (210)^2}$
$= 366.2$ feet

79. Midpoint of $AD = M_1 = \left(\dfrac{s + 0}{2}, \dfrac{s + 0}{2} \right) = \left(\dfrac{s}{2}, \dfrac{s}{2} \right)$

Midpoint of $BC = M_2 = \left(\dfrac{0 + s}{2}, \dfrac{s + 0}{2} \right) = \left(\dfrac{s}{2}, \dfrac{s}{2} \right)$

Therefore, the diagonals of a square intersect at their midpoints.

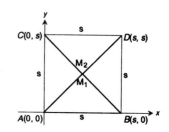

81. The automobile heading east moves a distance of $40t$ after t hours. The truck heading south moves a distance of $30t$ after t hours. Their distance apart after 1 hour is:

$d = \sqrt{(30)^2 + (40)^2}$
$= \sqrt{900 + 1600}$
$= \sqrt{2500} = 50$

$d = 50t$ is the expression for their distance apart after t hours.

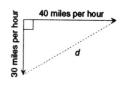

1.2 Graphs of Equations

1.

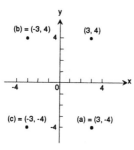

3.

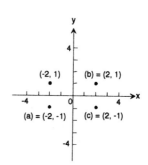

5.

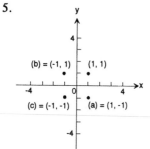

7.

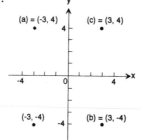

9.

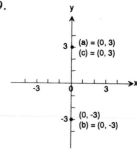

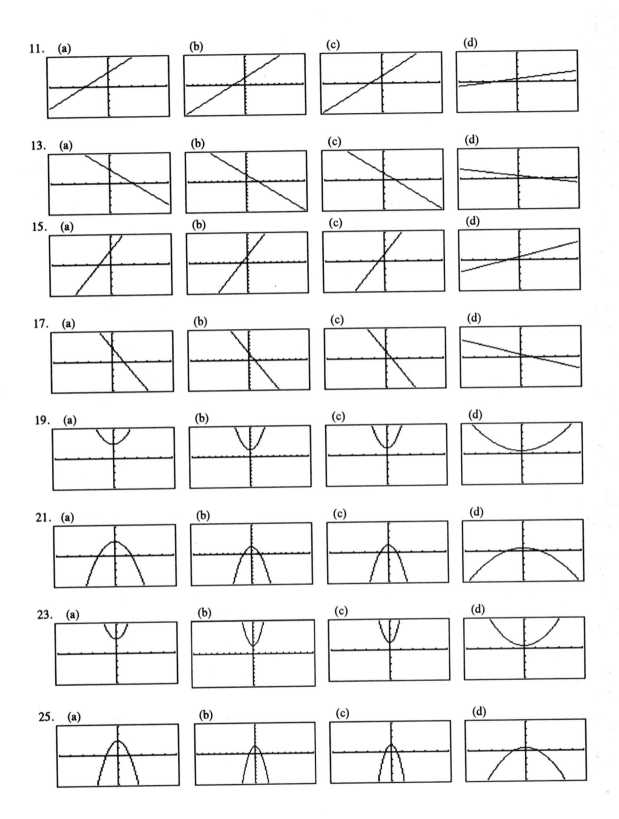

11. (a)   (b)   (c)   (d)

13. (a)   (b)   (c)   (d)

15. (a)   (b)   (c)   (d)

17. (a)   (b)   (c)   (d)

19. (a)   (b)   (c)   (d)

21. (a)   (b)   (c)   (d)

23. (a)   (b)   (c)   (d)

25. (a)   (b)   (c)   (d)

27. (a) 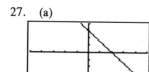 (b) 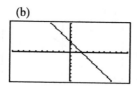 (c) 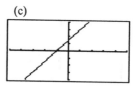 (d)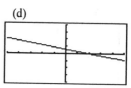

29. (a) 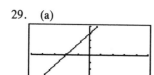 (b) 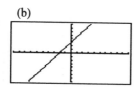 (c) 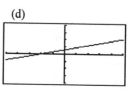 (d)

31. (a) $(-1, 0); (1, 0)$
    (b) x-axis, y-axis, origin

33. (a) $\left(-\dfrac{\pi}{2}, 0\right), \left(\dfrac{\pi}{2}, 0\right), (0, 1)$
    (b) y-axis

35. (a) $(0, 0)$
    (b) x-axis

37. (a) $(1, 0)$
    (b) none

39. (a) $(-3, 0), (0, 2), (3, 0)$
    (b) y-axis

41. (a) $(x, 0), 0 \le x < 1$
    (b) none

43. (a) $(-1.5, 0), (0, -2),$
        $(1.5, 0)$
    (b) y-axis

45. (a) none
    (b) origin

47. $y = x^4 + \sqrt{x}$
    $0 = 0^4 + \sqrt{0}$
    $0 = 0$
    $1 = 1^4 + \sqrt{1}$
    $1 \ne 2$
    $0 = (-1)^4 + \sqrt{-1}$
    $0 \ne 1 + \sqrt{-1}$
    $(0, 0)$ is on the graph.

49. $y^2 = x^2 + 4$
    $2^2 = 0^2 + 4$
    $4 = 4$
    $0^2 = 2^2 + 4$
    $0 \ne 8$
    $0^2 = (-2)^2 + 4$
    $0 \ne 4$
    $(0, 2)$ is on the graph.

51. $x^2 + y^2 = 4$
    $0^2 + 2^2 = 4$
    $4 = 4$
    $(-2)^2 + (2)^2 = 4$
    $8 \ne 4$
    $\left(\sqrt{2}\right)^2 + \left(\sqrt{2}\right)^2 = 4$
    $4 = 4$
    $(0, 2)$ and $\left(\sqrt{2}, \sqrt{2}\right)$ are on the graph.

53. $y = 5x + 4$
    $2 = 5a + 4$
    $5a = -2$
    $a = \dfrac{-2}{5}$

55. $2x + 3y = 6$
    $2a + 3b = 6$

57.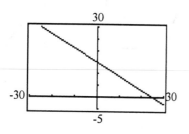

59.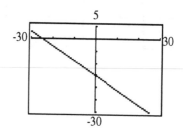

61.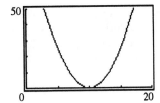

63.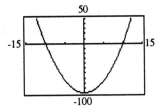

65.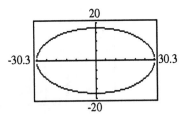

67.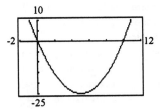

69.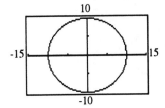

71.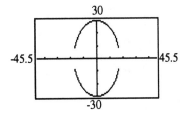

73.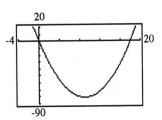

75.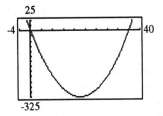

77. Symmetry with respect to the *x*-axis means
$(x, -y)$ is on the graph for every (x, y) on the
graph. Therefore, given $(-4, 1)$, $(-2, 1)$,
$(2, -1)$, $(4, 1)$; plot $(-4, -1)$, $(-2, -1)$,
$(2, 1)$, $(4, -1)$.

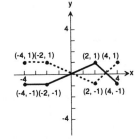

79. Symmetry with respect to the origin means
$(-x, -y)$ is on the graph for every (x, y) on
the graph. Therefore, given $(-4, 1)$, $(-2, 1)$,
$(2, -1)$, $(4, 1)$; plot $(4, -1)$, $(2, -1)$, $(-2, 1)$,
$(-4, -1)$.

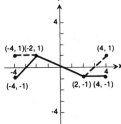

81. Symmetry with respect to the x-axis means $(x, -y)$ is on the graph for every (x, y) on the graph. Therefore, given $(-3, 1)$, $(0, 0)$, $(1, 2)$, $(4, 2)$; plot $(-3, -1)$, $(0, 0)$, $(1, -2)$, $(4, -2)$.

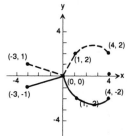

83. Symmetry with respect to the origin means $(-x, -y)$ is on the graph for every (x, y) on the graph. Therefore, given $(-3, 1)$, $(0, 0)$, $(1, 2)$, $(4, 2)$; plot $(3, -1)$, $(0, 0)$, $(-1, -2)$, $(-4, -2)$.

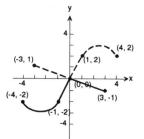

85. $x^2 = y$

y-intercept: Let $x = 0$ so $y = 0$ $(0, 0)$
x-intercept: Let $y = 0$ so $x = 0$ $(0, 0)$

Test for symmetry:

     x-axis: Replace y by $-y$ so $x^2 = -y$, which is not equivalent to $x^2 = y$.

     y-axis: Replace x by $-x$ so $(-x)^2 = y$ or $x^2 = y$ is equivalent to $x^2 = y$.

     Origin: Replace x by $-x$ and y by $-y$ so $(-x)^2 = -y$ or $x^2 = -y$ is not equivalent to $x^2 = y$.

Therefore, symmetric with respect to the y-axis.

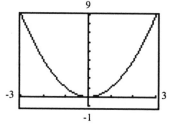

87. $y = 3x$

y-intercept: Let $x = 0$ so $y = 0$ $(0, 0)$
x-intercept: Let $y = 0$ so $x = 0$ $(0, 0)$

Test for symmetry:

     x-axis: Replace y by $-y$ so $-y = 3x$ is not equivalent to $y = 3x$.

     y-axis: Replace x by $-x$ so $y = -3x$ is not equivalent to $y = 3x$.

     Origin: Replace x by $-x$ and y by $-y$ so $-y = -3x$ is $y = 3x$ which is equivalent to $y = 3x$.

Therefore, symmetric with respect to the origin.

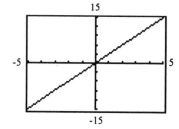

89. $x^2 + y - 9 = 0$

y-intercept: Let $x = 0$ so $y = 9$ $(0, 9)$
x-intercept: Let $y = 0$ so $x^2 = 9$ $(-3, 0)$
                              $x = \pm 3$ $(3, 0)$

Test for symmetry:

     x-axis: Replace y by $-y$ so $x^2 - y - 9 = 0$ is not equivalent to $x^2 + y - 9 = 0$.

     y-axis: Replace x by $-x$ so $(-x)^2 + y - 9 = 0$ is $x^2 + y - 9 = 0$, which is equivalent to $x^2 + y - 9 = 0$.

     Origin: Replace x by $-x$ and y by $-y$ so $(-x)^2 - y - 9 = 0$ is $x^2 - y - 9 = 0$, which is not equivalent to $x^2 + y - 9 = 0$.

Therefore, symmetric with respect to the y-axis.

91. $9x^2 + 4y^2 = 36$

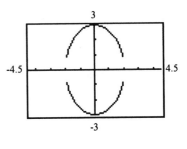

      y-intercept: Let $x = 0$ so $4y^2 = 36$ $(0, -3)$
                                       $y^2 = 9$ $(0, 3)$
                                         $y = \pm 3$
      x-intercept: Let $y = 0$ so $9x^2 = 36$ $(-2, 0)$
                                      $x^2 = 4$ $(2, 0)$
                                      $x = \pm 2$

Test for symmetry:
    x-axis: Replace y by $-y$ so $4x^2 + 9(-y)^2 = 36$
                                 $4x^2 + 9y^2 = 36$
               is equivalent to $4x^2 + 9y^2 = 36$
    y-axis: Replace x by $-x$ so $4(-x)^2 + 9y^2 = 36$
                                 $4x^2 + 9y^2 = 36$
               is equivalent to $4x^2 + 9y^2 = 36$
    Origin: Replace x by $-x$ and y by $-y$ so $4(-x)^2 + 9(-y)^2 = 36$
                                       $4x^2 + 9y^2 = 36$
               is equivalent to $4x^2 + 9y^2 = 36$

Therefore, symmetric with respect to the x-axis, y-axis, and origin.

93. $y = x^3 - 27$

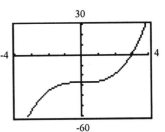

      y-intercept: Let $x = 0$ so $y = -27$ $(0, -27)$
      x-intercept: Let $y = 0$ so $0 = x^3 - 27$
                                     $27 = x^3$
                                     $3 = x$ $(3, 0)$

Test for symmetry:
    x-axis: Replace y by $-y$ so $-y = x^3 - 27$ is not equivalent to $y = x^3 - 27$
    y-axis: Replace x by $-x$ so $y = (-x)^3 - 27$
               $y = -x^3 - 27$ is not equivalent to $y = x^3 - 27$
    Origin: Replace x by $-x$ and y by $-y$ so $-y = (-x)^3 - 27$
                                       $-y = -x^3 - 27$
              $y = x^3 + 27$ is not equivalent to $y = x^3 - 27$

Therefore, no symmetry.

95. $y = x^2 - 3x - 4$

      y-intercept: Let $x = 0$ so $y = -4$ $(0, -4)$
      x-intercept: Let $y = 0$ so $0 = x^2 - 3x - 4$ $(4, 0)$
                                 $0 = (x - 4)(x + 1)$ $(-1, 0)$
                        $x - 4 = 0$ or $x + 1 = 0$
                           $x = 4$ or $x = -1$

Test for symmetry:
    x-axis: Replace y by $-y$ so $-y = x^2 - 3x - 4$ is not equivalent to $y = x^2 - 3x - 4$.

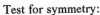

    y-axis: Replace x by $-x$ so $y = (-x)^2 - 3(-x) - 4$
              $y = x^2 + 3x - 4$ is not equivalent to $y = x^2 - 3x - 4$.
    Origin: Replace x by $-x$ and y by $-y$ so
                                     $-y = (-x)^2 - 3(-x) - 4$
                                     $-y = x^2 + 3x - 4$
              $y = -x^2 - 3x + 4$ is not equivalent to $y = x^2 - 3x - 4$

Therefore, no symmetry.

97. $y = \dfrac{x}{x^2 + 9}$

        y-intercept: Let $x = 0$ so $y = 0$ (0, 0)

        x-intercept: Let $y = 0$ so $0 = \dfrac{x}{x^2 + 9}$

                              $0 = x$ (0, 0)

Test for symmetry:

      x-axis: Replace y by $-y$ so $-y = \dfrac{x}{x^2 + 9}$ is not equivalent

                  to $y = \dfrac{x}{x^2 + 9}$

      y-axis: Replace x by $-x$ so $y = \dfrac{-x}{(-x)^2 + 9}$

             $y = \dfrac{-x}{x^2 + 9}$ is not equivalent to $y = \dfrac{x}{x^2 + 9}$

      Origin: Replace x by $-x$ and y by $-x$ so $-y = \dfrac{-x}{(-x)^2 + 9}$

                        $-y = \dfrac{-x}{x^2 + 9}$

             $y = \dfrac{x}{x^2 + 9}$ is not equivalent to $y = \dfrac{x}{x^2 + 9}$

Therefore, symmetric with respect to the origin.

99.   (a)

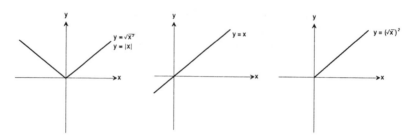

    (b)   Since $\sqrt{x^2} = |x|$, then for all x, the graphs of $y = \sqrt{x^2}$ and $y = |x|$ are the same.

    (c)   For $y = \left(\sqrt{x}\right)^2$, the domain of the variable x is $x \geq 0$; for $y = x$, the domain of the variable x is all real numbers. Thus, $\left(\sqrt{x}\right)^2 = x$ only for $x \geq 0$.

    (d)   For $y = \sqrt{x^2}$, the domain of the variable y is $y \geq 0$; for $y = x$, the domain of the variable y is all real numbers. Also, $\sqrt{x^2} = |x|$ equals x only if $x \geq 0$.

1.3 The Straight Line

1.   Slope $= \dfrac{1 - 0}{2 - 0} = \dfrac{1}{2}$ **3.**   Slope $= \dfrac{3 - 1}{-1 - 1} = \dfrac{2}{-2} = -1$

5. (x_1, y_1) (x_2, y_2)
    $(2, 3)$ $(4, 0)$

   $\text{Slope} = \dfrac{y_2 - y_1}{x_2 - x_1} = \dfrac{0 - 3}{4 - 2} = \dfrac{-3}{2}$

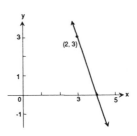

7. (x_1, y_1) (x_2, y_2)
    $(-2, 3)$ $(2, 1)$

   $\text{Slope} = \dfrac{y_2 - y_1}{x_2 - x_1} = \dfrac{1 - 3}{2 - (-2)} = \dfrac{-2}{4} = -\dfrac{1}{2}$

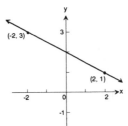

9. (x_1, y_1) (x_2, y_2)
    $(-3, -1)$ $(2, -1)$

   $\text{Slope} = \dfrac{y_2 - y_1}{x_2 - x_1} = \dfrac{-1 - (-1)}{2 - (-3)} = \dfrac{0}{5} = 0$

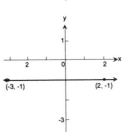

11. (x_1, y_1) (x_2, y_2)
    $(-1, 2)$ $(-1, -2)$

   $\text{Slope} = \dfrac{y_2 - y_1}{x_2 - x_1} = \dfrac{-2 - 2}{1 - (-1)} = \dfrac{-4}{0}$

   (undefined)

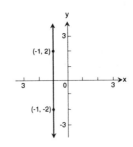

13. (x_1, y_1) (x_2, y_2)
    $\left(\sqrt{2}, 3\right)$ $\left(1, \sqrt{3}\right)$

   $\text{Slope} = \dfrac{y_2 - y_1}{x_2 - x_1} = \dfrac{\sqrt{3} - 3}{1 - \sqrt{2}}$

   $= \dfrac{1.732 - 3}{1 - 1.414} = \dfrac{-1.268}{-.414} = 3.06$

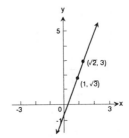

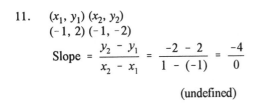

15.

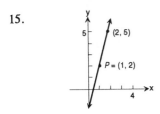

17.

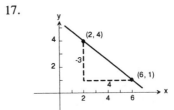

19.

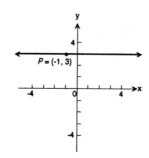

21.

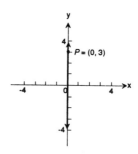

23. (0, 0) and (2, 1) are points on the line.

Slope $= \dfrac{1 - 0}{2 - 0} = \dfrac{1}{2}$

y-intercept is 0; so using $y = mx + b$, we get

$$y = \dfrac{1}{2}x + 0$$
$$2y = x$$
$$0 = x - 2y$$
$$x - 2y = 0 \quad \text{or} \quad y = \dfrac{1}{2}x$$

25. (−1, 3) and (1, 1) are points on the line.

Slope $= \dfrac{1 - 3}{1 - (-1)} = \dfrac{-2}{2} = -1$

Use $y - y_1 = m(x - x_1)$ with (x_1, y_1) being either point on the line.

$$y - 3 = -1[x - (-1)]$$
$$y - 3 = -(x + 1)$$
$$y - 3 = -x - 1$$
$$x + y - 2 = 0 \text{ or } y = -x + 2$$

27. $(x_1, y_1) = (-2, 3)$
Slope = 3
$y - y_1 = m(x - x_1)$

$y - 3 = 2[x - (-2)]$	$y - 3 = (x - (-2))$
$y - 3 = 2(x + 2)$	$y - 3 = 3(x + 2)$
$y - 3 = 2x + 4$	$y - 3 = 3x + 6$
$0 = 2x - y + 7$	$0 = 3x - y + 9$
$3x - y + 9 = 0 \quad \text{or}$	$y = 3x + 9$

29. $(x_1, y_1) = (1, -1)$

Slope $= -\dfrac{2}{3}$

$$y - y_1 = m(x - x_1)$$
$$y - (-1) = -\dfrac{2}{3}(x - 1)$$
$$y + 1 = -\dfrac{2}{3}x + \dfrac{2}{3}$$

Multiply both sides by 3:
$$3y + 3 = -2x + 2$$
$$2x + 3y + 1 = 0 \text{ or } y = \dfrac{-2}{3}x - \dfrac{1}{3}$$

31. Passing through (1, 3) and (−1, 2).

Slope $= \dfrac{2 - 3}{-1 - 1} = \dfrac{-1}{-2} = \dfrac{1}{2}$

Use either point and slope.

$$y - 3 = \dfrac{1}{2}(x - 1)$$
$$y - 3 = \dfrac{1}{2}x - \dfrac{1}{2}$$

Multiply both sides by 2:
$$2y - 6 = x - 1$$
$$0 = x - 2y + 5$$
$$x - 2y + 5 = 0 \text{ or } y = \dfrac{1}{2}x + \dfrac{5}{2}$$

33. $m = -3$
$b = y$-intercept $= 3$
Use $y = mx + b$
$$y = (-3)x + 3$$
$$y = -3x + 3$$
$$3x + y - 3 = 0 \text{ or } y = -3x + 3$$

35. x-intercept $= 2$
y-intercept $= -1$

Using the intercept form   or   use points $(2, 0)$ and $(0, -1)$

$$\frac{x}{a} + \frac{y}{b} = 1 \qquad\qquad m = \frac{-1 - 0}{0 - 2} = \frac{1}{2}$$

$$\frac{x}{2} + \frac{y}{-1} = 1 \qquad\qquad b = -1$$

Multiply both sides by 2: $\qquad\qquad y = mx + b$

$$x - 2y = 2 \qquad\qquad\qquad y = \frac{1}{2}x - 1$$

$$x - 2y - 2 = 0 \text{ or } y = \frac{1}{2}x - 1 \qquad 2y = x - 2$$

$$0 = x - 2y - 2$$
$$x - 2y - 2 = 0$$

37. Slope undefined; passing through $(2, 4)$.
Vertical line $x = a$ has undefined slope.
Thus, $\qquad\qquad x = 2$
$\qquad\qquad x - 2 = 0$; No slope intercept form

39.
$$y = 2x + 3$$
$$y = mx + b$$
$$m: \text{ slope} = 2$$
$$b: \ y\text{-intercept} = 3$$
Using intercepts, draw graph.
$$\left(-\frac{3}{2}, 0\right)$$

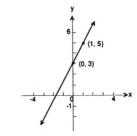

41.
$$\frac{1}{2}y = x - 1$$
$$y = 2x - 2$$
$$y = mx + b$$
$$m: \text{ slope} = 2$$
$$b: \ y\text{-intercept} = -2$$
Using intercepts, draw graph.
$$(0, -2)$$
$$(1, 0)$$

43.
$$y = \frac{1}{2}x + 2$$
$$m: \text{ slope} = \frac{1}{2}$$
$$b: \ y\text{-intercept} = 2$$

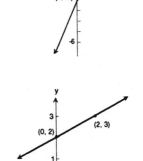

45.
$$x + 2y = 4$$
$$2y = -x + 4$$
$$y = -\frac{1}{2}x + 2$$

m: slope $= -\frac{1}{2}$

b: y-intercept $= 2$

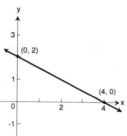

47.
$$2x - 3y = 6$$
$$-3y = -2x + 6$$
$$y = \frac{2}{3}x - 2$$
$$y = mx + b$$

m: slope $= \frac{2}{3}$

b: y-intercept $= -2$

Using intercepts, draw graph.

(0, -2)

(3, 0)

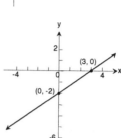

49.
$$x + y = 1$$
$$y = -x + 1$$
m: slope $= -1$

b: y-intercept $= 1$

Using intercepts, draw graph.

(0, 1)

(1, 0)

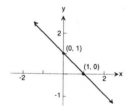

51. $x = -4$

of the form $x = a$, a vertical line slope is undefined; no y-intercept.

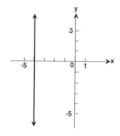

53. $y = 5$

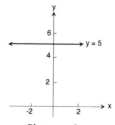

Slope $= 0$

y-intercept $= 5$

55. $y - x = 0$

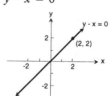

Slope $= 1$

y-intercept $= 0$

57.
$$2y - 3x = 0$$
$$2y = 3x$$
$$y = \frac{3}{2}x$$
$$y = mx + b$$
m: slope $= \dfrac{3}{2}$

b: y-intercept $= 0$

Using intercept and another point, draw graph.

(0, 0)

(2, 3)

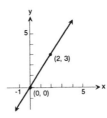

59.   The general equation of the x-axis is
$y = 0$ because the slope $= 0$ and x can be
any real number on the number line but
y must equal zero.

$$y = 0x + 0$$
$$y = 0$$

61.
$$(^\circ C, {}^\circ F) = (0, 32)$$
$$(^\circ C, {}^\circ F) = (100, 212)$$
$$\text{slope} = \frac{212 - 32}{100 - 0} = \frac{180}{100} = 1.8 \text{ or } \frac{9}{5}$$
$$y - 32 = \frac{9}{5}(x - 0)$$
$$^\circ F - 32 = \frac{9}{5}(^\circ C)$$
$$^\circ C = \frac{5}{9}(^\circ F - 32)$$

If $^\circ F = 70$, then
$$^\circ C = \frac{5}{9}(70 - 32) = \frac{5}{9}(38)$$
$$^\circ C \approx 21^\circ$$

63.   (a)   Since there is only a profit of $0.50 per copy and the expense of $100 must
be deducted, then the profit is
$$P = (0.50)x - 100$$
$$\text{or } P = 0.5x - 100$$
Plot points: (0, -100)

(200, 0)

(b) $P = 0.5(1000) - 100 = 500 - 100 = \$400$

(c) $P = 0.5(5000) - 100 = 2500 - 100 = \$2400$

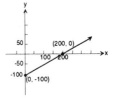

65. $C = 0.10819x + 9.06$
For 300 kWh, $C = 0.10819(300) + 9.06$
$$= \$41.52$$
For 900 kWh, $C = 0.10819(900) + 9.06$
$$= \$106.43$$

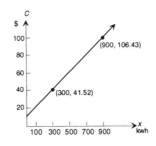

67.   (b)

69.   (d)

71.
$$y - 2 = x$$
$$y = x + 2$$
or
$$x - y + 2 = 0$$

73.
$$y - 0 = -\frac{1}{3}(x - 3)$$
$$y = -\frac{1}{3}x + 1$$
$$3y = -x + 3$$
$$x + 3y - 3 = 0 \text{ or } y = -\frac{1}{3}x + 1$$

75.
$$y - -2 = -\frac{2}{3}(x - 3)$$
$$y + 2 = -\frac{2}{3}x + 2$$
$$3y + 6 = -2x + 6$$
$$2x + 3y = 0 \text{ or } y = -\frac{2}{3}x$$

81. Its slope is -1.

83. Yes, if the y-intercept $= 0$.

1.4 Parallel and Perpendicular Lines; Circles

1. $y = 4x$
(a) 4
(b) $-\frac{1}{4}$

3. $y = \frac{-1}{2}x + 2$
(a) $\frac{-1}{2}$
(b) 2

5. $2x - 4y + 5 = 0$
$$4y = 2x + 5$$
$$y = \frac{1}{2}x + \frac{5}{4}$$
(a) $\frac{1}{2}$
(b) -2

7. $3x + 5y - 10 = 0$
$$5y = -3x + 10$$
$$y = \frac{-3}{5}x + 2$$
(a) $\frac{-3}{5}$
(b) $\frac{5}{3}$

9. $x = 4$ is a vertical line; slope is undefined.
(a) Slope is undefined.
(b) Since $x = 4$ is vertical, a line perpendicular would be horizontal; slope $= 0$.

11.
$$y - y_1 = m(x - x_1), \ m = 2$$
$$y - 3 = 2(x - 3)$$
$$y - 3 = 2x - 6$$
$$2x - y - 3 = 0 \ \text{ or } \ y = 2x - 3$$

13. $y - y_1 = m(x - x_1), \ m = -\frac{1}{2}$
$$y - 2 = -\frac{1}{2}(x - 1)$$
$$2y - 4 = -x + 1$$
$$x + 2y - 5 = 0 \text{ or } y = -\frac{1}{2}x + \frac{5}{2}$$

15. Parallel to $y = 2x$;
passing through $(-1, 2)$.
Slope $= 2$ and parallel line has same slope.
$$y - 2 = 2(x - (-1))$$
$$y - 2 = 2(x + 1)$$
$$y - 2 = 2x + 2$$
$$0 = 2x - y + 4$$
$$2x - y + 4 = 0 \ \text{ or } \ y = 2x + 4$$

17. Parallel to $2x - y + 2 = 0$; passing through $(0, 0)$.
$$2x + 2 = y$$
$$y = 2x + 2$$
Slope $= 2$ and parallel line has same slope.
$$y - 0 = 2(x - 0)$$
$$y = 2x$$
$$0 = 2x - y$$
$$2x - y = 0 \text{ or } y = 2x$$

19. Parallel to $x = 5$; passing through $(4, 2)$.
$x = 5$ is a vertical line; slope is undefined.
Therefore, $x = 4$ is parallel to $x = 5$.
$x - 4 = 0$; no y-intercept form

21. Perpendicular to $y = \dfrac{1}{2}x + 4$; passing through $(1, -2)$.
Slope of given line $= \dfrac{1}{2}$
Slope of perpendicular line $= -2$
$$[y - (-2)] = -2(x - 1)$$
$$y + 2 = -2x + 2$$
$$2x + y = 0 \text{ or } y = -2x$$

23. Perpendicular to $2x + y - 2 = 0$; passing through $(-3, 0)$.
$$y = -2x + 2$$
Slope of given line $= -2$
Slope of perpendicular line $= \dfrac{1}{2}$
$$y - 0 = \dfrac{1}{2}[x - (-3)]$$
$$y = \dfrac{1}{2}(x + 3)$$
$$y = \dfrac{1}{2}x + \dfrac{3}{2}$$
$$2y = x + 3$$
$$0 = x - 2y + 3$$
$$x - 2y + 3 = 0 \text{ or } y = \dfrac{1}{2}x + \dfrac{3}{2}$$

25. Perpendicular to $x = 8$; passing through $(3, 4)$.
Since $x = 8$ is vertical, a line perpendicular would be horizontal of the form $y = b$.
Therefore, $y = 4$ is the line perpendicular to $x = 8$.
$$y - 4 = 0 \text{ or } y = 4$$

27. Center $= (2, 1)$
Radius $=$ Distance from $(0, 1)$ to $(2, 1)$
$$= \sqrt{(2 - 0)^2 + (1 - 1)^2}$$
$$= \sqrt{4} = 2$$
$$(x - 2)^2 + (y - 1)^2 = 4$$

29. Center $=$ Midpoint of $(1, 2)$ and $(4, 2)$
$$= \left(\dfrac{1 + 4}{2}, \dfrac{2 + 2}{2} \right)$$
$$= \left(\dfrac{5}{2}, 2 \right)$$
Radius $=$ Distance from $\left(\dfrac{5}{2}, 2 \right)$ to $(4, 2)$
$$= \sqrt{\left(4 - \dfrac{5}{2} \right)^2 + (2 - 2)^2}$$
$$= \sqrt{\dfrac{9}{4}} = \dfrac{3}{2}$$
$$\left(x - \dfrac{5}{2} \right)^2 + (y - 2)^2 = \dfrac{9}{4}$$

31. Use $(x - h)^2 + (y - k)^2 = r^2$, where
$r = 1$; $(h, k) = (1, -1)$

$$(x - 1)^2 + [y - (-1)]^2 = 1^2$$
$$(x - 1)^2 + (y + 1)^2 = 1$$

Square each part and gather up terms:
$$(x^2 - 2x + 1) + (y^2 + 2y + 1) = 1$$
$$x^2 + y^2 - 2x + 2y + 2 = 1$$
$$x^2 + y^2 - 2x + 2y + 1 = 0$$

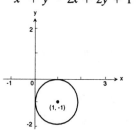

33. Use $(x - h)^2 + (y - k)^2 = r^2$, where
$r = 2$; $(h, k) = (0, 2)$

$$(x - 0)^2 + (y - 2)^2 = 2^2$$
$$x^2 + (y - 2)^2 = 4$$
$$x^2 + y^2 - 4y + 4 = 4$$
$$x^2 + y^2 - 4y = 0$$

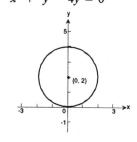

35. Use $(x - h)^2 + (y - k)^2 = r^2$, where
$r = 5$; $(h, k) = (4, -3)$

$$(x - 4)^2 + [y - (-3)]^2 = (5)^2$$
$$(x - 4)^2 + (y + 3)^2 = 25$$

Square each part and gather up terms:
$$(x^2 - 8x + 16) + (y^2 + 6y + 9) = 25$$
$$x^2 + y^2 - 8x + 6y + 25 = 25$$
$$x^2 + y^2 - 8x + 6y = 0$$

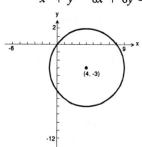

37. Since $(h, k) = (0, 0)$, the center is at the
origin so use:

$$x^2 + y^2 = r^2 \text{ when } r = 2$$
$$x^2 + y^2 = 2^2$$
$$x^2 + y^2 = 4$$
$$x^2 + y^2 - 4 = 0$$

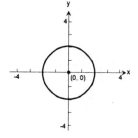

39. $r = \dfrac{1}{2}$; $(h, k) = \left(\dfrac{1}{2}, 0\right)$

$$\left(x - \dfrac{1}{2}\right)^2 + (y - 0)^2 = \left(\dfrac{1}{2}\right)^2$$
$$\left(x - \dfrac{1}{2}\right)^2 + y^2 = \dfrac{1}{4} \quad \text{standard form}$$
$$x^2 - x + \dfrac{1}{4} + y^2 = \dfrac{1}{4}$$
$$x^2 + y^2 - x = 0 \quad \text{general form}$$

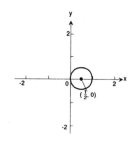

41. Since $x^2 + y^2 = 4$ is of the form $x^2 + y^2 = r^2$, it can be
    written as $x^2 + y^2 = (2)^2$, so $(h, k) = (0, 0)$ and $r = 2$.

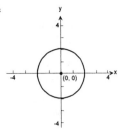

43. $(x - 3)^2 + y^2 = 4$ can be written as:
    $$(x - 3)^2 + (y - 0)^2 = 2^2,$$
    so compare to standard form
    $(x - h)^2 + (y - k)^2 = r^2$ and get
    $(h, k) = (3, 0)$ and $r = 2$.

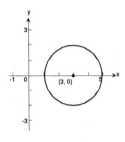

45. $x^2 + y^2 + 4x - 4y - 1 = 0$
    Group together the x-terms and the y-terms and rearrange constant to right side:
    $$(x^2 + 4x) + (y^2 - 4y) = 1$$
    Complete the square of each expression in parentheses by taking $\dfrac{1}{2}$ (coefficient

    of variable to first degree) and then squaring it and adding the number to the left
    side and the right side.

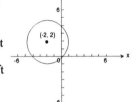

$$\text{Add } \left[\frac{1}{2}(4)\right]^2 = 4 \quad \text{Add } \left[\frac{1}{2}(-4)\right]^2 = 4$$
$$(x^2 + 4x + 4) + (y^2 - 4y + 4) = 1 + 4 + 4$$
$$(x + 2)^2 + (y - 2)^2 = 9$$
$$[x-(-2)]^2 + (y - 2)^2 = 3^2$$
$$(h, k) = (-2, 2)$$
$$r = 3$$

47. $$x^2 + y^2 - x + 2y + 1 = 0$$
    Proceed as in 45:
    $$(x^2 - x) + (y^2 + 2y) = -1$$
    $$\text{Add } \left[\frac{1}{2}(-1)\right]^2 = \frac{1}{4} \quad \text{Add } \left[\frac{1}{2}(2)\right]^2 = 1$$
    $$\left(x^2 - x + \frac{1}{4}\right) + (y^2 + 2y + 1) = -1 + \frac{1}{4} + 1$$
    $$\left(x - \frac{1}{2}\right)^2 + (y + 1)^2 = \frac{1}{4}$$
    $$\left(x - \frac{1}{2}\right)^2 + [y - (-1)]^2 = \left(\frac{1}{2}\right)^2$$
    $$(h, k) = \left(\frac{1}{2}, -1\right)$$
    $$r = \frac{1}{2}$$

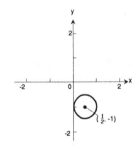

49. $2x^2 + 2y^2 - 12x + 8y - 24 = 0$

The coefficients of x^2 and y^2 should be 1 in order to put the equation in standard form, so divide each term by 2.

$$x^2 + y^2 - 6x + 4y - 12 = 0$$
$$(x^2 - 6x) + (y^2 + 4y) = 12$$

$$\text{Add}\left[\frac{1}{2}(-6)\right]^2 = 9 \quad \text{Add}\left[\frac{1}{2}(4)\right]^2 = 4$$

$$(x^2 - 6x + 9) + (y^2 + 4y + 4) = 12 + 9 + 4$$
$$(x - 3)^2 + (y + 2)^2 = 25$$
$$(x - 3)^2 + [y - (-2)]^2 = 5^2$$
$$(h, k) = (3, -2)$$
$$r = 5$$

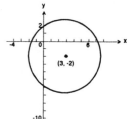

51. Center at origin and containing the point $(-2, 3)$.
$$r = d(C, P) = \sqrt{(-2 - 0)^2 + (3 - 0)^2}$$
$$= \sqrt{4 + 9}$$
$$= \sqrt{13}$$

Use $x^2 + y^2 = r^2$ since center is at origin.
$$x^2 + y^2 = \left(\sqrt{13}\right)^2$$
$$x^2 + y^2 = 13$$
$$x^2 + y^2 - 13 = 0$$

53. Center at $(2, 3)$ and touching x-axis.
$$r = d(C, P) = \sqrt{(2 - 2)^2 + (3 - 0)^2} = \sqrt{9} = 3$$
[r could have been obtained by carefully examining the graph.]
$$(x - 2)^2 + (y - 3)^2 = 3^2$$
$$x^2 - 4x + 4 + y^2 - 6y + 9 = 9$$
$$x^2 + y^2 - 4x - 6y + 4 = 0$$

r is perpendicular to x-axis at point of tangency so P has coordinate $(2, 0)$.

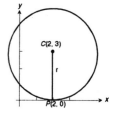

55. Endpoints of diameter $(1, 4)$ and $(-3, 2)$; midpoint of diameter is center of circle.
$$C = \left(\frac{1 + (-3)}{2}, \frac{4 + 2}{2}\right)$$
$$C = (-1, 3)$$

The distance from the center to either endpoint is the radius.
$$r = \sqrt{(-1 - 1)^2 + (3 - 4)^2}$$
$$r = \sqrt{4 + 1}$$
$$r = \sqrt{5}$$
$$[x - (-1)]^2 + (y - 3)^2 = \left(\sqrt{5}\right)^2$$
$$(x + 1)^2 + (y - 3)^2 = 5$$
$$x^2 + 2x + 1 + 2y^2 - 6y + 9 = 5$$
$$x^2 + y^2 + 2x - 6y + 5 = 0$$

57. Plot points $P_1(-2, 5)$, $P_2(1, 3)$, and $P_3(-1, 0)$.

Slope $P_1P_2 = \dfrac{3 - 5}{1 + 2} = \dfrac{-2}{3} = m_1$

Slope $P_2P_3 = \dfrac{3 - 0}{1 - (-1)} = \dfrac{3}{2} = m_2$

Since $m_1m_2 = \left(-\dfrac{2}{3}\right)\left(\dfrac{3}{2}\right) = -1$, the lines are perpendicular so points P_1, P_2, and

P_3 form a right triangle.

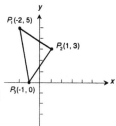

59. Plot points $P_1(-1, 0)$, $P_2(2, 3)$, and $P_3(1, -2)$, and $P_4(4, 1)$.

Slope $P_1P_2 = m_1 = \dfrac{3 - 0}{2 - (-1)} = \dfrac{3}{3} = 1$

Slope $P_3P_4 = m_2 = \dfrac{1 - (-2)}{4 - 1} = \dfrac{3}{3} = 1$

Slope $P_1P_3 = m_3 = \dfrac{-2 - 0}{1 - (-1)} = \dfrac{-2}{2} = -1$

Slope $P_2P_4 = m_4 = \dfrac{1 - 3}{4 - 2} = \dfrac{-2}{2} = -1$

Since the opposite sides are parallel and the adjacent sides are perpendicular, the points form a rectangle.

61. (c)   63. (b)

65. $(x + 3)^2 + (y - 1)^2 = 16$ 67. $(x - 2)^2 + (y - 2)^2 = 9$

69. Plot $(0, 0)$, $(4, 0)$, $(1, 3)$, and $(5, 3)$.

Area = bh $b = 4$, $h = 3$

Area = $(4)(3)$

Area = 12 sq units

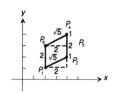

71. Plot $P_1(2, 1)$, $P_2(4, 2)$, $P_3(2, 3)$, and $P_4(4, 4)$.

The base P_2P_4 is 2.

The altitude to the base P_3P_5 is 2.

The area $= 2 \cdot 2 = 4$ sq. units.

73. Plot $P_1(1, 1)$, $P_2(3, 2)$, $P_3(2, -1)$, and $P_4(4, 0)$.

Slope $P_1P_2 = \dfrac{2 - 1}{3 - 1} = \dfrac{1}{2}$

Slope $P_1P_3 = \dfrac{-1 - 1}{2 - 1} = -2$

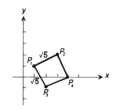

(Since they are perpendicular, P_1P_3, is the altitude to P_1P_2.)

Area $= \left(\sqrt{5}\right)\left(\sqrt{5}\right) = 5$ sq. units

$d(P_1, P_2) = \sqrt{(3 - 1)^2 + (2 - 1)^2} = \sqrt{4 + 1} = \sqrt{5}$

$d(P_1, P_3) = \sqrt{(2 - 1)^2 + (-1 - 1)^2} = \sqrt{1 + 4} = \sqrt{5}$

75. Refer to Fig. 81 in the text.

If $m_1m_2 = -1$, then

$$d(A, B) = \sqrt{(1 - 1)^2 + (m_2 - m_1)^2} = \sqrt{(m_2 - m_1)^2}$$

$$d(O, A) = \sqrt{(1 - 0)^2 + (m_2 - 0)^2} = \sqrt{1 + m_2^2}$$

$$d(O, B) = \sqrt{(1 - 0)^2 + (m_1 - 0)^2} = \sqrt{1 + m_1^2}$$

Show: $[d(O, B)]^2 + [d(O, A)]^2 = [d(A, B)]^2$

$$\left(\sqrt{1 + m_1^2}\right)^2 + \left(\sqrt{1 + m_2^2}\right)^2 = \left(\sqrt{(m_2 - m_1)^2}\right)^2$$

$$1 + m_1^2 + 1 + m_2^2 = (m_2 - m_1)^2$$

$$m_1^2 + m_2^2 + 2 = m_2^2 - 2m_1m_2 + m_1^2$$

Since $m_1m_2 = -1$, $m_1^2 + m_2^2 + 2 = m_2^2 - 2(-1) + m_1^2$

$$m_1^2 + m_2^2 + 2 = m_1^2 + m_2^2 + 2$$

77. (a)

$$x^2 + (mx + b)^2 = r^2$$

$$(1 + m^2)x^2 + 2mbx + b^2 - r^2 = 0$$

One solution if and only if discriminant $= 0$

$$(2mb)^2 - 4(1 + m^2)(b^2 - r^2) = 0$$

$$-4b^2 + 4r^2 + 4m^2r^2 = 0$$

$$r^2(1 + m^2) = b^2$$

(b) $x = \dfrac{-2mb}{2(1 + m^2)} = \dfrac{-2mb}{2b^2/r^2} = \dfrac{-r^2m}{b}$

$y = m\left(\dfrac{-r^2m}{b}\right) + b = \dfrac{-r^2m^2}{b} + b = \dfrac{-r^2m^2 + b^2}{b} = \dfrac{r^2}{b}$

(c) Slope of tangent line $= m$

Slope of line joining center to point of tangency $= \dfrac{r^2/b}{-r^2m/b} = \dfrac{-1}{m}$

79.

$$x^2 - 4x + y^2 + 6y = -4$$

$$(x - 2)^2 + (y + 3)^2 = 9$$

Center $(2, -3)$

Slope from center to $\left(3, 2\sqrt{2} - 3\right)$ is $\dfrac{2\sqrt{2} - 3 + 3}{3 - 2} = 2\sqrt{2}$

Slope of tangent line is $\dfrac{-1}{2\sqrt{2}} = \dfrac{-2\sqrt{2}}{4}$

$$y - \left(2\sqrt{2} - 3\right) = \dfrac{-\sqrt{2}}{4}(x - 3)$$

$$\sqrt{2}x + 4y - 11\sqrt{2} + 12 = 0$$

81. $x^2 + y^2 - 4x + 6y + 4 = 0$ $\qquad$ $x^2 + y^2 + 6x + 4y + 9 = 0$

$\quad\;\;\; x^2 - 4x + y^2 + 6y = -4$ $\qquad\;\;\;$ $x^2 + 6x + y^2 + 4y = -9$

$\quad\;\;\; (x - 2)^2 + (y + 3)^2 = 9$ $\qquad$ $(x + 3)^2 + (y + 2)^2 = 4$

$\qquad\qquad$ Center $(2, -3)$ $\qquad\qquad\qquad$ Center $(-3, -2)$

Slope of line joining centers is $\dfrac{-2 + 3}{-3 - 2} = \dfrac{1}{-5}$

$$y + 3 = \frac{-1}{5}(x - 2)$$

$x + 5y + 13 = 0$

83. $2x - y + c = 0$

$\qquad$ If $c = 4$, $2x - y + 4 = 0$

$\qquad\qquad$ Plot points $\quad (0, -4)$
$\qquad\qquad\qquad\qquad\quad\;\; (1, -2)$
$\qquad\qquad\qquad\qquad\quad\;\; (2, 0)$

$\qquad$ If $c = 0$, $2x - y + 0 = 0$ or $2x = y$

$\qquad\qquad$ Plot points $\quad (0, 0)$
$\qquad\qquad\qquad\qquad\quad\;\; (1, 2)$

$\qquad$ If $c = 2$, $2x - y + 2 = 0$

$\qquad\qquad$ Plot points $\quad (0, 2)$
$\qquad\qquad\qquad\qquad\quad\;\; (1, 4)$

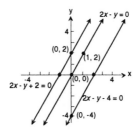

All have the same slope, 2. The lines are parallel.

85. $y = 2$

1.5 Linear Curve Fitting

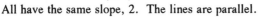

1. Strong linear relation $\qquad$ **3.** Strong linear relation $\qquad$ **5.** Nonlinear relation

7. (a) $\qquad\qquad\qquad\qquad$ **9.** (a) $\qquad\qquad\qquad\qquad$ **11.** (a)

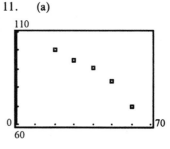

$\qquad$ (b) $\quad y = 2.0357x - 2.3571$ $\qquad$ (b) $\quad y = 2.2x + 1.2$ $\qquad$ (b) $\quad y = -0.72x + 116.6$

13. (a)

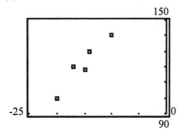

(b) $y = 3.861313869x + 180.2919708$

15. (a)

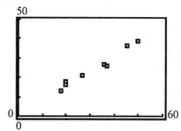

(b) $I = 0.7548890222C + 0.6266346731$
(c) As disposable income increases by \$1, consumption increases by about \$0.75.
(d) \$32,000

17. (a)

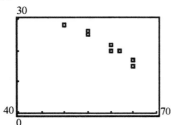

(b) $y = -0.8265306122x + 70.39030612$
(c) As speed increases by 1 mph, mpg decreases by 0.8265.
(d) 20 mpg

19. (a)

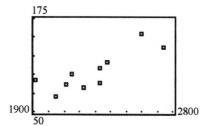

(b) NIBT $= 0.084195641$ sales $- 88.57761368$
(c) As sales increase by \$1, NIBT increases by \$0.08.
(d) \$118.2 billion

1.6 Variation

1. $y = kx$, $y = 2$ when $x = 10$
   $2 = k(10)$
   $\dfrac{1}{5} = k$
   $y = \dfrac{1}{5}x$

3. $A = kx^2$, $A = 4\pi$ when $x = 2$
   $4\pi = k(2)^2$
   $4\pi = 4k$
   $\pi = k$
   $A = \pi x^2$

5. $F = \dfrac{K}{d^2}$, $F = 10$ when $d = 5$
   $10 = \dfrac{k}{25}$
   $250 = k$
   $F = \dfrac{250}{d^2}$

7. $z = k(x^2 + y^2)$, $z = 5$
   when $x = 3$, $y = 4$
   $5 = k(9 + 16)$
   $5 = k(25)$
   $\dfrac{1}{5} = k$
   $z = \dfrac{1}{5}(x^2 + y^2)$

9. $M = k\dfrac{d^2}{\sqrt{x}}, M = 24$

when $x = 9, d = 4$

$24 = k\left(\dfrac{16}{\sqrt{9}}\right)$

$24 = k\left(\dfrac{16}{3}\right)$

$\dfrac{9}{2} = k$

$M = \dfrac{9}{2}\left(\dfrac{d^2}{\sqrt{x}}\right) = \dfrac{9d^2}{2\sqrt{x}}$

11. $T^2 = k\dfrac{a^3}{d^2}$, $T = 2$ when $a = 2$, $d = 4$

$4 = k\dfrac{8}{16}$

$8 = k$

$T^2 = 8\dfrac{a^3}{d^2}$

13. $V = kr^3, k = \dfrac{4\pi}{3}$

$V = \dfrac{4\pi}{3}r^3$

15. $A = k(bh), k = \dfrac{1}{2}$

$A = \dfrac{1}{2}bh$

17. $V = k(r^2h), k = \pi$
$V = \pi r^2h$

19. $F = G\left(\dfrac{mM}{d^2}\right), G = 6.67 \times 10^{-11}$

$F = 6.67 \times 10^{-11}\left(\dfrac{mM}{d^2}\right)$

21. $s = kt^2$ $64 = 16t^2$
$16 = k(1^2)$ $4 = t^2$
$16 = k$ $\pm 2 = t$
$s = 16t^2$ time is positive, so
$s = 16(3)^2$ $t = 2$ seconds
$s = 16(9)$
$s = 144$ ft.

23. $E = kW, E = 3$ when $W = 20$

$3 = k(20)$

$\dfrac{3}{20} = k$

$E = \dfrac{3}{20}W$, if $W = 15$

$E = \dfrac{3}{20}(15)$

$E = \dfrac{9}{4} = 2\dfrac{1}{4} = 2.25$

25. $W = \dfrac{k}{d^2}, W = 55$ when $d = 4 \times 10^3$

$55 = \dfrac{k}{(4 \times 10^3)^2}$

$(4 \times 10^3)^2(55) = k$

$(16 \times 10^6)(55) = k$

$(880)(10^6) = k$

$(8.8)(10^8) = k$

$W = \dfrac{8.8 \times 10^8}{(4.4 \times 10^3)^2}$

when $d = 4.4 \times 10^3$

$W = \dfrac{8.8 \times 10^8}{19.36 \times 10^6}$

$W = 0.4545 \times 10^2$

$W = 45.45$ pounds

27. $h = k(sd^3), d = 2$
$$h = 36$$
$$s = 75$$
$$36 = k(75)(8)$$
$$36 = 600k$$
$$\frac{36}{600} = k$$
$$0.06 = k$$
$h = 0.06(sd^3),$ $d = ?$ when $h = 45, s = 125$
$$45 = 0.06(125)d^3$$
$$45 = 7.5d^3$$
$$6 = d^3$$
$$\sqrt[3]{6} = d$$
$$1.8171 = d$$
$$1.82 \text{ inches} \approx d$$

29. $K = k(mv^2)$ $m = 25$ pounds, $v = 100$ ft./sec., $K = 400$ foot-pounds
$$400 = k(25)(100)^2$$
$$400 = 250,000k$$
$$0.0016 = k$$
$K = 0.0016mv^2,$ $m = 25$ pounds
   $v = 150$ ft./sec.
   $K = ?$
$$K = 0.0016(25)(150)^2$$
$$K = 900 \text{ ft.-lb.}$$

31. $S = k\dfrac{(pd)}{t},$ $S = 100, d = 5, t = .75, p = 25$
$$100 = k\frac{(25)(5)}{(.75)}$$
$$100 = 166.6667k$$
$$0.5999 = k$$
$S = (0.5999)\dfrac{pd}{t}$ $S = ?, p = 48, d = 8, t = 0.5$
$$S = (0.5999)\frac{(40)(8)}{0.5}$$
$$S = 383.999$$
$$S = 384 \text{ psi}$$

33. $R = \dfrac{k\ell}{r^2}$ $\ell = 50, r = 6 \times 10^{-3}, R = 10$ $R = \dfrac{(7.2 \times 10^{-6}\ell)}{r^2}$ $R = ?, \ell = 100, r = 7 \times 10^{-3}$

$$10 = \frac{k(50)}{(6 \times 10^{-3})}$$
   $$R = \frac{(7.2 \times 10^{-6})(100)}{(7 \times 10^{-3})^2}$$
$$10 = \frac{50}{36 \times 10^{-6}}k$$
   $$R = \frac{7.2 \times 10^{-4}}{49 \times 10^{-6}}$$
$$10 = (1.3888 \times 10^6)k$$
   $$R = 0.1469388 \times 10^2$$
$$7.3 \times 10^{-6} = k$$
   $$R = 14.69 \text{ ohms}$$

35. $v = k\sqrt{r} = \sqrt{g}$

$v = \sqrt{g} \cdot \sqrt{r}$

$v = \sqrt{gr}$

37. $v = \sqrt{g}\sqrt{(4000 + r)}$

$v = \sqrt{g(4000 + r)}$

$r = 100 \qquad g = 79,036$

$v = \sqrt{79,036(4000 + 100)} = 18,001$ mph

39. $v = \sqrt{gr}$

diameter $= 2(4000 + r)$, distance $d = $ diameter $\cdot \pi$

$v = \dfrac{d}{t} = \dfrac{2(4000 + r)\pi}{1.5} = 4.19(4000 + r)$

$v = 4.19(4000 + r) = \sqrt{(79,036)(4000 + r)}$ (from Problem 37)

$[4.19(4000 + r)]^2 = 79,036(4000 + r)$

$17.56(4000 + r) = 79,036$

$4000 + r = \dfrac{79,036}{17.56}$

$4000 + r = 4504.5$

$r = 504.5$ miles

≈ 505 miles

41. $F = \dfrac{mv^2}{r}$

43. $F = \dfrac{m(1.1v)^2}{r} = \dfrac{(1.21)mv^2}{r}$

21% increase

45. $F = \dfrac{m(3v)^2}{r} = \dfrac{9mv^2}{r}$

9 times

1 Chapter Review

1. Slope $= -2$; passing through $(3, -1)$

Use
$$y - y_1 = m(x - x_1)$$
$$y - (-1) = -2(x - 3)$$
$$y + 1 = -2x + 6$$
$$2x + y - 5 = 0 \text{ or } y = -2x + 5$$

3. Slope undefined; passing through $(-3, 4)$.

Use $x = a$ so $x = -3$ or $x + 3 = 0$;
no y-intercept form

5. y-intercept $= -2$; passing through $(5, -3)$

y-intercept: $(0, -2)$

$m = \dfrac{-3 - (-2)}{5 - 0} = \dfrac{-3 + 2}{5} = -\dfrac{1}{5}$

$y = mx + b$

$y = -\dfrac{1}{5}x - 2$

$5y = -x - 10$

$x + 5y + 10 = 0 \text{ or } y = -\dfrac{1}{5}x - 2$

7. Parallel to $2x - 3y + 4 = 0$; passing through $(-5, 3)$.

$$-3y = -2x - 4$$
$$y = \frac{2}{3}x + \frac{4}{3}$$
$$m = \frac{2}{3}$$

Line parallel has same slope.

Use $y - y_1 = m(x - x_1)$

$$y - 3 = \frac{2}{3}[x - (-5)]$$
$$y - 3 = \frac{2}{3}(x + 5)$$
$$y - 3 = \frac{2}{3}x + \frac{10}{3}$$
$$3y - 9 = 2x + 10$$
$$0 = 2x - 3y + 19$$
$$2x - 3y + 19 = 0 \text{ or } y = \frac{2}{3}x + \frac{19}{3}$$

9. Perpendicular to $x + y - 2 = 0$; passing through $(4, -3)$.

$$x + y - 2 = 0$$
$$y = -x + 2$$
$$m = -1$$
$$m_{\perp} = 1$$
$$y - (-3) = 1(x - 4)$$
$$y + 3 = x - 4$$
$$0 = x - y - 7$$
$$x - y - 7 = 0$$
$$\text{or } -x + y + 7 = 0 \text{ or } y = x - 7$$

11. $4x - 5y + 20 = 0$
$$-5y = -4x - 20$$
$$y = \frac{4}{5}x + 4$$

x-intercept: $(-5, 0)$
y-intercept: $(0, 4)$

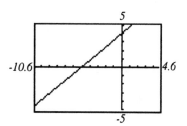

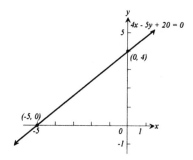

13. $\frac{1}{2}x - \frac{1}{3}y + \frac{1}{6} = 0$

$$-\frac{1}{3}y = -\frac{1}{2}x - \frac{1}{6}$$
$$y = \frac{3}{2}x + \frac{1}{2}$$

Let $y = 0$, so x-intercept: $\left(-\frac{1}{3}, 0\right)$

Let $x = 0$, so y-intercept: $\left(0, \frac{1}{2}\right)$

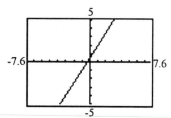

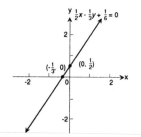

15. $\sqrt{2}x + \sqrt{3}y = \sqrt{6}$

$\sqrt{3}y = -\sqrt{2}x + \sqrt{6}$

$y = \dfrac{-\sqrt{6}}{3}x + \sqrt{2}$

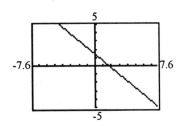

Let $y = 0$, so $\sqrt{2}x = \sqrt{6}$

$x = \dfrac{\sqrt{6}}{\sqrt{2}}$

$= \sqrt{3}$

x-intercept: $\left(\sqrt{3}, 0\right)$

Let $x = 0$, so $\sqrt{3}y = \sqrt{6}$

$x = \dfrac{\sqrt{6}}{\sqrt{3}}$

$y = \sqrt{2}$

y-intercept: $\left(0, \sqrt{2}\right)$

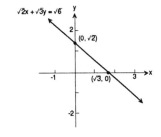

17. $x^2 + y^2 - 2x + 4y - 4 = 0$

$(x^2 - 2x \quad) + (y^2 + 4y \quad) = 4$

Complete the square twice:

$(x^2 - 2x + 1) + (y^2 + 4y + 4) = 4 + 1 + 4$

$(x - 1)^2 + (y + 2)^2 = 9$

Center $(1, -2)$

Radius $= \sqrt{9} = 3$

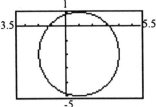

19. $3x^2 + 3y^2 - 6x + 12y = 0$

Divide by 3:

$x^2 + y^2 - 2x + 4y = 0$

$(x^2 - 2x \quad) + (y^2 + 4y \quad) = 0$

Complete the square twice:

$(x^2 - 2x + 1) + (y^2 + 4y + 4) = 1 + 4$

$(x - 1)^2 + (y + 2)^2 = 5$

Center $(1, -2)$

Radius $= \sqrt{5}$

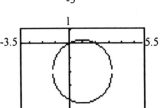

21. $(7, 4)$ and $(-3, 2)$

Slope $= \dfrac{2 - 4}{-3 - 7} = \dfrac{-2}{-10} = \dfrac{1}{5}$

Distance $= \sqrt{(-3 - 7)^2 + (2 - 4)^2} = \sqrt{(-10)^2 + (-2)^2} = \sqrt{100 + 4} = \sqrt{104} = 2\sqrt{26}$

Midpoint: $x = \dfrac{7 - 3}{2} = \dfrac{4}{2} = 2$

$y = \dfrac{4 + 2}{2} = \dfrac{6}{2} = 3$

Midpoint $= (2, 3)$

23. The lines shown are parallel, so they have the same slope. This slope must be a positive number. The y-intercepts of one is positive and the other one is negative.

(a) These lines have unequal slopes.

(b) These lines have the same slope, but it is negative (-1).

(c) These lines might be the ones graphed; each has slope 1; one has a positive y-intercept (2); the other y-intercept is -1.

(d) These lines have the same slope (1), but also have the same y-intercept (2).

(e) These lines have equal slopes, but it is negative $\left(-\dfrac{1}{2}\right)$.

The only possibility is (c).

25. $2x = 3y^2$

x-intercept: $(0, 0)$

y-intercept: $(0, 0)$

Replace y by $(-y)$.

$2x = 3(-y)^2$

$2x = 3y^2$ so symmetric with respect to x-axis.

Replace x by $(-x)$.

$2(-x) = 3y^2$

$-2x = 3y^2$, *not* symmetric with respect to y-axis.

Replace x by $(-x)$ and y by $(-y)$.

$2(-x) = 3(-y)^2$

$-2x = 3y^2$, *not* symmetric with respect to origin.

27. $x^2 + 4y^2 = 16$

To find x-intercepts, let $y = 0$.

$x^2 = 16$

$x = \pm 4$ \qquad (4, 0) and (-4, 0)

To find y-intercepts, let $x = 0$.

$4y^2 = 16$

$y^2 = 4$

$y = \pm 2$ \qquad (0, 2) and (0, -2)

Replace y by $(-y)$.

$x^2 + 4(-y)^2 = 16$

$x^2 + 4y^2 = 16$ so symmetric with respect to x-axis

Replace x by $(-x)$.

$(-x)^2 + 4y^2 = 16$

$x^2 + 4y^2 = 16$ so symmetric with respect to y-axis

Replace x by $(-x)$ and y by $(-y)$.

$(-x)^2 + 4(-y)^2 = 16$

$x^2 + 4y^2 = 16$ so symmetric with respect to the origin

29. $y = x^4 + 2x^2 + 1$

To find x-intercept, let $y = 0$.

$x^4 + 2x^2 + 1 = 0$

$(x^2 + 1)^2 = 0$

$x^2 = -1$ \; no x-intercept

To find y-intercept, let $x = 0$.

$y = 0 + 0 + 1$

$y = 1$ \qquad (0, 1)

Replace y by $-y$.

$-y = x^4 + 2x^2 + 1$ so **not** symmetric with respect to x-axis

Replace x by $-x$.

$y = (-x)^4 + 2(-x)^2 + 1$

$y = x^4 + 2x^2 + 1$ so symmetric with respect to y-axis

Replace x by $(-x)$ and y by $(-y)$.

$-y = (-x)^4 + 2(-x^2) + 1$

$-y = x^4 + 2x^2 + 1$ so **not** symmetric with respect to the origin

31. $x^2 + x + y^2 + 2y = 0$

To find x-intercepts, let $y = 0$.

$x^2 + x = 0$

$x(x + 1) = 0$

$x = 0$ or $\qquad\qquad x + 1 = 0$

$\qquad\qquad\qquad\qquad x = -1 \qquad$ (0, 0) and (-1,0)

To find y-intercept, let $x = 0$

$\qquad\qquad$ $(-1, 0)(0, 0)(0, -2)$

$y^2 + 2y = 0$

$y(y + 2) = 0$

$y = 0 \quad$ or $\quad y = -2 \qquad$ (0, 0) and (0, -2)

Replace x by $-x$.

$(-x)^2 - x + y^2 + 2y = 0$

$x^2 - x + y^2 + 2y = 0$ so **not** symmetric with respect to x-axis

Replace y by $-y$.

$x^2 + x + (-y)^2 + 2(-y) = 0$

$x^2 + x + y^2 - 2y = 0$ so **not** symmetric with respect to y-axis

Replace x by $-x$ and y by $-y$.

$(-x)^2 + (-x) + (-y)^2 + 2(-y) = 0$

$x^2 - x + y^2 - 2y = 0$ so **not** symmetric with respect to the origin

33. (a)

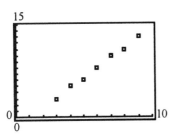

(b) $\quad y = 1.642857143x - 1.857142857$

(c) As x increases by 1, y increases about 1.64.

35. (a)

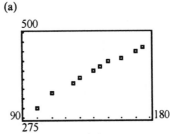

(b) $\quad y = 2.018273184x + 112.4326176$

(c) As x increases by 1, y increases about 2.02.

(b) Calculus $= 1.018354237 \cdot$ Algebra $+ 52.6898134$

(c) As the algebra score increases by 1 point, the calculus score increases by about 1.02 points.

(d) 73

37. (a)

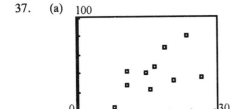

39. $A = ks^2$, $A = \dfrac{\sqrt{3}}{4}$ if $s = 1$

$$\frac{\sqrt{3}}{4} = k(1)^2$$

$$k = \frac{\sqrt{3}}{4}$$

$$A = \frac{\sqrt{3}}{4}s^2, \; A = 16, \; s = ?$$

$$16 = \frac{\sqrt{3}}{4}s^2$$

$$64 = \sqrt{3}s^2$$

$$\frac{64}{\sqrt{3}} = s^2$$

$$\sqrt{\frac{64}{\sqrt{3}}} = s$$

$$\frac{\sqrt{64}}{\sqrt[4]{3}} = s$$

$$\frac{8}{\sqrt[4]{3}} = s$$

$6.08 \; cm \approx s$ (with calculator)

41. $T^2 = ka^3$

for Earth: $\quad a = 93 \times 10^6,$
$\qquad\qquad\quad T = 365$ days

$$k = \frac{T^2}{a^3} = \frac{(365)^2}{\left(93 \times 10^6\right)^3}$$

for Mercury: $\; T = 88$ days

$$a^3 = \frac{T^2}{k} = \frac{(88)^2}{\dfrac{(365)^2}{(93 \times 10^6)^3}}$$

$$a = \sqrt[3]{\frac{(88)^2(93 \times 10^6)^3}{(365)^2}}$$

$$= 36{,}025{,}449 \text{ miles}$$
$$\approx 36 \text{ million miles}$$

43.
$$M = \left(\frac{a + 0}{2}, \; \frac{0 + b}{2}\right)$$

$$M = \left(\frac{a}{2}, \; \frac{b}{2}\right)$$

$$D(A, M) = \sqrt{\left(a - \frac{a}{2}\right)^2 + \left(0 - \frac{b}{2}\right)^2}$$

$$= \sqrt{\frac{a^2}{4} + \frac{b^2}{4}} = \frac{1}{2}\sqrt{a^2 + b^2}$$

$$D(B, M) = \sqrt{\left(0 - \frac{a}{2}\right)^2 + \left(b - \frac{b}{2}\right)^2}$$

$$= \sqrt{\frac{a^2}{4} + \frac{b^2}{4}} = \frac{1}{2}\sqrt{a^2 + b^2}$$

$$D(O, M) = \sqrt{\left(\frac{a}{2} - 0\right)^2 + \left(\frac{b}{2} - 0\right)^2}$$

$$= \sqrt{\frac{a^2}{4} + \frac{b^2}{4}} = \frac{1}{2}\sqrt{a^2 + b^2}$$

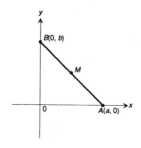

45. $A = (3, 4)$ $B = (1, 1)$ $C = (-2, 3)$

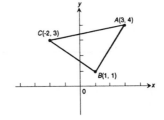

$$d(B, C) = \sqrt{(-2 - 1)^2 + (3 - 1)^2}$$
$$= \sqrt{9 + 4} = \sqrt{13}$$
$$d(A, B) = \sqrt{(3 - 1)^2 + (4 - 1)^2}$$
$$= \sqrt{4 + 9} = \sqrt{13}$$

Since $BC = AB$, Triangle ABC is isosceles.

47. $A = (2, 5)$ $B = (6, 1)$ $C = (8, -1)$

$$m_{AB} = \frac{1 - 5}{6 - 2} = \frac{-4}{4} = -1$$

$$m_{BC} = \frac{-1 - 1}{8 - 6} = \frac{-2}{2} = -1$$

49. Endpoints of diameter: $A(-3, 2)$ and $B(5, -6)$.

Midpoint of diameter $= C = \left(\dfrac{-3 + 5}{2}, \dfrac{2 - 6}{2} \right) = (1, -2)$

radius $= d(A, C) = \sqrt{(-3 - 1)^2 + (2 - (-2)^2} = \sqrt{16 + 16} = 4\sqrt{2}$

equation:
$$(x - 1)^2 + (y + 2)^2 = \left(4\sqrt{2}\right)^2$$
$$x^2 - 2x + 1 + y^2 + 4y + 4 = 32$$
$$x^2 + y^2 - 2x + 4y - 27 = 0$$

Chapter 2

FUNCTIONS AND THEIR GRAPHS

2.1 Functions

1. $f(x) = -3x^2 + 2x - 4$

   (a) $f(0) = -3(0)^2 + 2(0) - 4 = -4$

   (b) $f(1) = -3(1)^2 + 2(1) - 4$
   $= -3 + 2 - 4 = -5$

   (c) $f(-1) = -3(-1)^2 + 2(-1) - 4$
   $= -3 - 2 - 4 = -9$

   (d) $f(3) = -3(3)^2 + 2(3) - 4$
   $= -27 + 6 - 4 = -25$

3. $f(x) = \dfrac{x}{x^2 + 1}$

   (a) $f(0) = \dfrac{0}{0^2 + 1} = 0$

   (b) $f(1) = \dfrac{1}{1^2 + 1} = \dfrac{1}{2}$

   (c) $f(-1) = \dfrac{-1}{(-1)^2 + 1} = -\dfrac{1}{2}$

   (d) $f(3) = \dfrac{3}{3^2 + 1} = \dfrac{3}{10}$

5. $f(x) = |x| + 4$

   (a) $f(0) = |0| + 4 = 4$

   (b) $f(1) = |1| + 4 = 5$

   (c) $f(-1) = |-1| + 4 = 1 + 4 = 5$

   (d) $f(3) = |3| + 4 = 3 + 4 = 7$

7. $f(x) = \dfrac{2x + 1}{3x - 5}$

   (a) $f(0) = \dfrac{2(0) + 1}{3(0) - 5} = \dfrac{1}{-5} = -\dfrac{1}{5}$

   (b) $f(1) = \dfrac{2(1) + 1}{3(1) - 5} = \dfrac{3}{-2} = -\dfrac{3}{2}$

   (c) $f(-1) = \dfrac{2(-1) + 1}{3(-1) - 5} = \dfrac{-2 + 1}{-3 - 5} = \dfrac{-1}{-8} = \dfrac{1}{8}$

   (d) $f(3) = \dfrac{2(3) + 1}{3(3) - 5} = \dfrac{7}{9 - 5} = \dfrac{7}{4}$

9. $f(0) = 3$ since $(0, 3)$ is on graph
   $f(-6) = -3$ since $(-6, -3)$ is on graph

11. $f(2) =$ is positive since $f(2) = 4$

13. $f(x) = 0$ when $x = -3$
    $x = 6$
    $x = 10$

15. Domain: $[-6, 11]$ or
    $\{x \mid -6 \le x \le 11\}$

17. The x-intercepts are -3, 6, 10.

19. 3 times

21. $f(x) = \dfrac{x + 2}{x - 6}$

    (a) $14 \overset{?}{=} \dfrac{3 + 2}{3 - 6}$

         $14 \overset{?}{=} \dfrac{5}{-3}$

         No, $(3, 14)$ is not on the graph of f.

    (b) $f(4) = \dfrac{4 + 2}{4 - 6} = \dfrac{6}{-2} = -3$

    (c) $2 = \dfrac{x + 2}{x - 6}$

        $2x - 12 = x + 2$

             $x = 14$

    (d)   Domain of $f = \{x \mid x \neq 6\}$

23. $f(x) = \dfrac{2x^2}{x^4 + 1}$

    (a) $1 \overset{?}{=} \dfrac{2(-1)^2}{(-1)^4 + 1}$

         $1 \overset{?}{=} \dfrac{2}{2}$

         $1 = 1$

         Yes, $(-1, 1)$ is on the graph of f.

    (b) $f(2) = \dfrac{2(2)^2}{(2)^4 + 1} = \dfrac{8}{17}$

    (c) $1 = \dfrac{2x^2}{x^4 + 1}$

          $x^4 + 1 = 2x^2$

      $x^4 - 2x^2 + 1 = 0$

        $(x^2 - 1)^2 = 0$

              $x = \pm 1$

    (d)   Domain of

            $f = \{x \mid x \in \text{Real Numbers}\}$

25. Not a function since there are vertical lines that intersect the graph in more than one point.

27. Function   (a)   Domain: $\{x \mid -\pi \le x \le \pi\}$; Range: $\{y \mid -1 \le y \le 1\}$

             (b) $\left[-\dfrac{\pi}{2}, 0\right]$, $\left[\dfrac{\pi}{2}, 0\right]$, $(0, 1)$

             (c) y-axis

29. Not a function since there are vertical lines that intersect the graph in more than one point.

31. Function   (a)   Domain: $\{x \mid 0 < x < \infty\}$; Range: all real numbers
             (b) $(1, 0)$ There are no y-intercepts
             (c)   None

33. Function   (a)   Domain: all real numbers; Range: $\{y \mid -\infty < y \le 2\}$
             (b) $(-3, 0)$, $(3, 0)$, $(0, 2)$
             (c) y-axis

35. Function   (a)   Domain: $\{x \mid x \neq 2\}$; Range: $\{y \mid y \neq 1\}$
             (b) $(0, 0)$
             (c)   None

37. $f(x) = 3x + 4$
    all real numbers

39. $f(x) = \dfrac{x}{x^2 + 1}$
    all real numbers

41. $g(x) = \dfrac{x}{x^2 - 1}$ $\{x \mid x \neq -1, x \neq 1\}$

$x^2 - 1 \neq 0$

$x^2 \neq 1$

$x \neq \pm 1$

43. $F(x) = \dfrac{x - 2}{x^3 + x}$ $\{x \mid x \neq 0\}$

$x^3 + x \neq 0$

$x(x^2 + 1) \neq 0$

$x \neq 0 \quad x^2 \neq -1$

45. $h(x) = \sqrt{3x - 12}$ $\{x \mid x \geq 4\}$

$3x - 12 \geq 0$

$3x \geq 12$

$x \geq 4$

47. $f(x) = \sqrt{x^2 - 9}$

$x^2 - 9 \geq 0$

Solve this second-degree inequality.

$(x - 3)(x + 3) = 0$

$x = 3 \qquad x = -3$

	Test Number	$(x + 3)$	$(x - 3)$	$(x + 3)(x - 3)$
$x < -3$	-4	$-$	$-$	$+$
$-3 < x < 3$	0	$+$	$-$	$-$
$x > 3$	4	$+$	$+$	$+$

Hence, $x \leq -3$ or $x \geq 3$ or $(-\infty, -3]$ or $[3, \infty)$.

49. $p(x) = \sqrt{\dfrac{x - 2}{x - 1}}$

$\dfrac{x - 2}{x - 1} \geq 0$

Solve this second-degree inequality.

$x - 2 = 0 \qquad x - 1 = 0$

	Test Number	$(x + 1)$	$(x - 2)$	$\left| \dfrac{x - 2}{x - 1} \right|$
$x < 1$	0	$-$	$-$	$+$
$1 < x < 2$	$1\frac{1}{2}$	$+$	$-$	$-$
$x > 2$	3	$+$	$+$	$+$

Hence, $x < 1$ or $x \geq 2$ or $(-\infty, 1)$ or $[2, \infty)$.

51. $f(x) = 2x^3 + Ax^2 + 4x - 5$ and $f(2) = 5$

$f(2) = 2(2)^3 + A(2)^2 + 4(2) - 5 = 5$

$16 + 4A + 8 - 5 = 5$

$4A + 19 = 5$

$4A = -14$

$A = -\dfrac{7}{2}$

53. $f(x) = \dfrac{3x + 8}{2x - A}$ and $f(0) = 2$

$f(0) = \dfrac{3(0) + 8}{2(0) - A} = 2$

$\dfrac{8}{-A} = \dfrac{2}{1}$ (cross multiply)

$-2A = 8$

$A = -4$

55. $f(x) = \dfrac{2x - A}{x - 3}$ and $f(4) = 0$

$f(4) = \dfrac{2(4) - A}{4 - 3} = 0$

$\dfrac{8 - A}{1} = 0$

$8 - A = 0$

$8 = A$

Since $x - 3 = 0$ when $x = 3$, then f is not defined at 3.

57. (a)

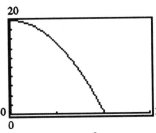

(b) $H(x) = 20 - 4.9x^2$

$x = 1$

$H(1) = 20 - 4.9(1)^2 = 20 - 4.9 = 15.1 \; m$

$x = 1.1$

$H(1.1) = 20 - 4.9(1.1)^2 = 20 - 4.9(1.21) = 20 - 5.929 = 14.071 = 14.07 \; m$

$x = 1.2$

$H(1.2) = 20 - 4.9(1.2)^2 = 20 - 4.9(1.44) = 20 - 7.056 = 12.944 = 12.94 \; m$

$x = 1.3$

$H(1.3) = 20 - 4.9(1.3)^2 = 20 - 4.9(1.69) = 20 - 8.281 = 11.719 = 11.72 \; m$

(c) $H(x) = 15$ $H(x) = 10$ $H(x) = 5$

$15 = 20 - 4.9x^2$ $10 = 20 - 4.9x^2$ $5 = 20 - 4.9x^2$

$-5 = -4.9x^2$ $-10 = -4.9x^2$ $-15 = -4.9x^2$

$x = 1.01$ seconds $x = 1.42$ seconds $x = 1.74$ seconds

(d) The rock strikes the ground when $H = 0$.

$H(x) = 20 - 4.9x^2 = 0$

$20 = 4.9x^2$

$4.0816 = x^2$

$\sqrt{4.0816} = x$

$2.02 \text{ sec} = x$

59. $\ell = x$

$x = 2w \text{ or } \left[\dfrac{x}{2}\right] = w$

$A = \ell \cdot w$

$A(x) = x \left[\dfrac{x}{2}\right] = \dfrac{x^2}{2} = \dfrac{1}{2}x^2$

61. $G(x) = (\text{amt. per hr.})(\text{no. of hrs.})$

$G(x) = 5x$

63. (a) The total cost of installing cable along the road is $10x$. If cable is installed x miles along the road, there are $5 - x$ miles left from the road to the house and where the cable ends. Therefore, using the

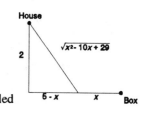

Pythagorean Theorem, there is $\sqrt{(5 - x)^2 + 2^2}$

$= \sqrt{25 - 10x + x^2 + 4} = \sqrt{x^2 - 10x + 29}$ miles of cable installed off the road. The total cost of installation is:

$$C(x) = 10x + 14\sqrt{x^2 - 10x + 29}$$

(b) $C(1) = 10(1) + 14\sqrt{(1)^2 - 10(1) + 29}$
$= 10 + 14\sqrt{20}$
$= 10 + 62.61$
$= \$72.61$

(c) $C(3) = 10(3) + 14\sqrt{(3)^2 - 10(3) + 29}$
$= 30 + 14\sqrt{8}$
$= 30 + 39.60$
$= \$69.60$

(d)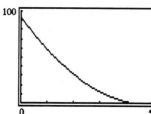

(e) Least cost: $x = 2.95$ miles

65. (a) $A(x) = (8.5 - 2x)(11 - 2x)$

(b) Domain: $0 \le x \le 4.25$
Range: $0 \le A \le 93.5$

(c) $A(1) = (8.5 - 2)(11 - 2)$
$= 58.5$ square inches
$A(1.2) = (8.5 - 2.4)(11 - 2.4)$
$= (6.1)(8.6)$
$= 52.46$ square inches
$A(1.5) = (8.5 - 3)(11 - 3)$
$= 44$ square inches

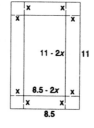

(d)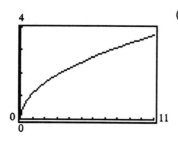

(e) $A(x) = 70$ when
$x = 0.64$ inches
$A(x) = 50$ when
$x = 1.28$ inches

67. (a)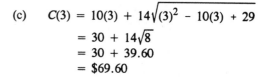

(b) As ℓ changes from 1 to 10, T varies from 1.1 to 3.5.

(c) $T(\ell) = 2\pi\sqrt{\dfrac{\ell}{g}}$, $g = 32.2$

$T(1) = 2\pi\sqrt{\dfrac{1}{32.2}} = 2\pi\sqrt{0.0310559} = 2\pi(0.1762)$

$= 1.1070973 \approx 1.11$ sec.

$T(2) = 2\pi\sqrt{\dfrac{2}{32.2}} = 2\pi\sqrt{0.06211} = 2\pi(0.2492)$

$= 1.5659 \approx 1.57$ sec.
The increase is 1.57 sec. $- 1.11$ sec. $= 0.46$ sec.

69. $y = x^2 + 2x$

Look at some ordered pairs (x, y) in this set:
$(-2, 0), (-1, -1), (0, 0), (1, 3), (2, 8)$
No two pairs have the same *first* element.
Function.

71. $y = \dfrac{2}{x}$

Look at some ordered pairs (x, y) in this set:

$$\left(-3, -\dfrac{2}{3}\right), (-2, -1), (-1, -2),$$

$$(1, 2), (2, 1), \left(3, \dfrac{2}{3}\right)$$

No two pairs have the same *first* element. **Function**.

73. $y^2 = 1 - x^2$

Look at some ordered pairs (x, y) in this set:
$(-1, 0), (0, -1), (0, 1), (1, 0)$

same first element

Not a function.

75. $x^2 + y = 1$

Look at some ordered pairs (x, y) in this set:
$(-1, 0), (0, 1), (1, 0), (2, -3)$
No two pairs have the same *first* element. **Function**.

77. (a)　　　$h(x) = 2x$
$h(a + b) \overset{?}{=} h(a) + h(b)$
$h(a + b) = 2(a + b) = 2a + 2b = h(a) + h(b)$

(b)　　　$g(x) = x^2$
$g(a + b) \overset{?}{=} g(a) + g(b)$
$g(a + b) = (a + b)^2 + a^2 + 2ab + b^2 \overset{?}{=} a^2 + b^2 = g(a) + g(b)$

(c)　　　$F(x) = 5x - 2$
$F(a + b) \overset{?}{=} F(a) + F(b)$
$F(a + b) = 5(a + b) - 2 = 5a + 5b - 2 = 5a - 2 + 5b - 2$
$= F(a) + F(b)$

(d)　　　$G(x) = \dfrac{1}{x}$
$G(a + b) \overset{?}{=} G(a) + G(b)$

$$G(a + b) = \dfrac{1}{a + b} \neq \dfrac{1}{a} + \dfrac{1}{b} = G(a) + G(b)$$

1. C   3. E   5. B   7. F

9. (a) Domain: $\{x \mid -3 \le x \le 4\}$; Range: $\{y \mid 0 \le y \le 3\}$
    (b) In interval notation, increasing on $[-3, 0]$ and on $[2, 4]$; and decreasing on $[0, 2]$. In inequality notation, increasing on $-3 \le x \le 0$ and on $2 \le x \le 4$ and decreasing on $0 \le x \le 2$.
    (c) Since the graph is not symmetric with respect to the y-axis and is not symmetric with respect to the origin, it is NEITHER even nor odd.
    (d) The intercepts are $(-3, 0)$, $(0, 3)$, $(2, 0)$.

11. (a) Domain: all real numbers; Range: $\{y \mid 0 < y < \infty\}$
    (b) In inequality notation, increasing on $-\infty < x < \infty$. In interval notation, increasing on $(-\infty, \infty)$.
    (c) Since the graph is neither symmetric with respect to the y-axis nor the origin, it is NEITHER even nor odd.
    (d) The intercept is $(0, 1)$.

13. (a) Domain: $\{x \mid -\pi \le x \le \pi\}$; Range: $\{y \mid -1 \le y \le 1\}$
    (b) In interval notation, increasing on $\left[-\dfrac{\pi}{2}, \dfrac{\pi}{2}\right]$; and decreasing on $\left(-\pi, -\dfrac{\pi}{2}\right]$ and on $\left[\dfrac{\pi}{2}, \pi\right]$. In inequality notation, increasing on $-\dfrac{\pi}{2} \le x \le \dfrac{\pi}{2}$; and decreasing on $-\pi \le x \le -\dfrac{\pi}{2}$ and $\dfrac{\pi}{2} \le x \le \pi$.
    (c) Since the graph is symmetric with respect to the origin, the graph is ODD.
    (d) The intercepts are $(-\pi, 0)$, $(0, 0)$, $(\pi, 0)$.

15. (a) Domain: $\{x \mid x \ne 2\}$; Range: $\{y \mid y \ne 1\}$
    (b) In interval notation, decreasing on $(-\infty, 2)$ and on $(2, \infty)$. In inequality notation, decreasing on $-\infty < x < 2$ and on $2 < x < \infty$.
    (c) Since the graph is neither symmetric with respect to the y-axis nor symmetric with respect to the origin, it is NEITHER even nor odd.
    (d) The intercept is $(0, 0)$.

17. (a) Domain: $\{x \mid x \ne 0\}$; Range: all real numbers
    (b) In interval notation, increasing on $(-\infty, 0)$ and $(0, \infty)$. In inequality notation, increasing on $-\infty < x < 0$ and on $0 < x < \infty$.
    (c) Since the graph is symmetric with respect to the origin, the graph is ODD.
    (d) The intercepts are $(-1, 0)$ and $(1, 0)$.

19. (a) Domain: $\{x \mid x \ne -2, x \ne 2\}$; Range: $\{y \mid -\infty < y \le 0 \text{ and } 1 < y < \infty\}$.
    (b) In interval notation, increasing on $(-\infty, -2)$ and on $(-2, 0)$, and decreasing on $(0, 2)$ and on $(2, \infty)$. In inequality notation, increasing on $-\infty < x < -2$ and on $-2 < x < 0$ and decreasing on $0 < x < 2$ and on $2 < x < \infty$.
    (c) Since the graph is symmetric with respect to the y-axis, it is EVEN.
    (d) The intercept is $(0, 0)$.

21. (a) Domain: $\{x \mid -4 \le x \le 4\}$; Range: $\{y \mid 0 \le y \le 2\}$
    (b) In interval notation, increasing on $[-2, 0]$ and $[2, 4]$ and decreasing on $[-4, -2]$ and $[0, 2]$.
        In inequality notation, increasing on $-2 \le x \le 0$ and $2 \le x \le 4$ and decreasing on $-4 \le x \le -2$ and $0 \le x \le 2$.
    (c) Since the graph is symmetric with respect to the y-axis, it is even.
    (d) The intercepts are $(-2, 0)$, $(0, 2)$ and $(2, 0)$.

23. (a) Domain: $\{x \mid -4 \le x \le 4\}$; Range: $\{y \mid 0 < y \le 4\}$
    (b) Increasing on $[-4, 0)$; decreasing on $(0, 4]$
    (e) Even
    (d) None

25. (a) $f(1.2) = [\![2(1.2)]\!] = [\![2.4]\!] = 2$
    (b) $f(1.6) = [\![2(1.6)]\!] = [\![3.2]\!] = 3$
    (c) $f(-1.8) = [\![2(-1.8)]\!] = [\![-3.6]\!] = -4$

27. (a) $f(-2) = (-2)^2 = 4$ since $-2 < 0$
    (b) $f(0) = 2$ since $x = 0$
    (c) $f(2) = 2(2) + 1 = 5$ since $2 > 0$

29. $f(x) = 2x + 5$
    (a) $f(-x) = 2(-x) + 5 = -2x + 5$ (b) $-f(x) = -(2x + 5) = -2x - 5$
    (c) $f(2x) = 2(2x) + 5 = 4x + 5$
    (d) $f(x - 3) = 2(x - 3) + 5 = 2x - 6 + 5 = 2x - 1$
    (e) $f\left(\dfrac{1}{x}\right) = 2\left(\dfrac{1}{x}\right) + 5 = \dfrac{2}{x} + 5 = \dfrac{2}{x} + \dfrac{5x}{x} = \dfrac{5x + 2}{x}$
    (f) $\dfrac{1}{f(x)} = \dfrac{1}{2x + 5}$

31. $f(x) = 2x^2 - 4$
    (a) $f(-x) = 2(-x)^2 - 4 = 2x^2 - 4$ (b) $-f(x) = -(2x^2 - 4) = -2x^2 + 4$
    (c) $f(2x) = 2(2x)^2 - 4 = 8x^2 - 4$
    (d) $f(x - 3) = 2(x - 3)^2 - 4 = 2(x^2 - 6x + 9) - 4$
        $= 2x^2 - 12x + 18 - 4 = 2x^2 - 12x + 14$
    (e) $f\left(\dfrac{1}{x}\right) = 2\left(\dfrac{1}{x}\right)^2 - 4 = \dfrac{2}{x^2} - 4 = \dfrac{2}{x^2} - \dfrac{4x^2}{x^2} = \dfrac{2 - 4x^2}{x^2}$
    (f) $\dfrac{1}{f(x)} = \dfrac{1}{2x^2 - 4}$

33. $f(x) = x^3 - 3x$
    (a) $f(-x) = (-x)^3 - 3(-x) = -x^3 + 3x$ (b) $-f(x) = -(x^3 - 3x) = -x^3 + 3x$
    (c) $f(2x) = (2x)^3 - 3(2x) = 8x^3 - 6x$
    (d) $f(x - 3) = (x - 3)^3 - 3(x - 3) = x^3 - 9x^2 + 27x - 27 - 3x + 9$
        $= x^3 - 9x^2 + 24x - 18$
    (e) $f\left(\dfrac{1}{x}\right) = \left(\dfrac{1}{x}\right)^3 - 3\left(\dfrac{1}{x}\right) = \dfrac{1}{x^3} - \dfrac{3}{x} = \dfrac{1}{x^3} - \dfrac{3x^2}{x^3} = \dfrac{1 - 3x^2}{x^3}$
    (f) $\dfrac{1}{f(x)} = \dfrac{1}{x^3 - 3x}$

35. $f(x) = \dfrac{x}{x^2 + 1}$

(a) $f(-x) = \dfrac{-x}{(-x)^2 + 1} = -\dfrac{x}{x^2 + 1}$

(b) $-f(x) = -\left(\dfrac{x}{x^2 + 1}\right) = -\dfrac{x}{x^2 + 1}$

(c) $f(2x) = \dfrac{2x}{(2x)^2 + 1} = \dfrac{2x}{4x^2 + 1}$

(d) $f(x - 3) = \dfrac{x - 3}{(x - 3)^2 + 1} = \dfrac{x - 3}{x^2 - 6x + 9 + 1} = \dfrac{x - 3}{x^2 - 6x + 10}$

(e) $f\left(\dfrac{1}{x}\right) = \dfrac{\dfrac{1}{x}}{\left(\dfrac{1}{x}\right)^2 + 1} = \dfrac{\dfrac{1}{x}}{\dfrac{1}{x^2} + \dfrac{x^2}{x^2}} = \dfrac{\dfrac{1}{x}}{\dfrac{1 + x^2}{x^2}} = \dfrac{1}{x} \cdot \dfrac{x^2}{1 + x^2} = \dfrac{x}{1 + x^2} = \dfrac{x}{x^2 + 1}$

(f) $\dfrac{1}{f(x)} = \dfrac{1}{\dfrac{x}{x^2 + 1}} = \dfrac{x^2 + 1}{x}$

37. $f(x) = |x|$

(a) $f(-x) = |-x| = |x|$

(b) $-f(x) = -|x|$

(c) $f(2x) = |2x| = 2|x|$

(d) $f(x - 3) = |x - 3|$

(e) $f\left(\dfrac{1}{x}\right) = \left|\dfrac{1}{x}\right| = \dfrac{1}{|x|}$

(f) $\dfrac{1}{f(x)} = \dfrac{1}{|x|}$

39. $f(x) = 1 + \dfrac{1}{x}$

(a) $f(-x) = 1 + -\dfrac{1}{x} = 1 - \dfrac{1}{x}$

(b) $-f(x) = -\left(1 + \dfrac{1}{x}\right) = -1 - \dfrac{1}{x}$

(c) $f(2x) = 1 + \dfrac{1}{2x}$

(d) $f(x - 3) = 1 + \dfrac{1}{x - 3}$

(e) $f\left(\dfrac{1}{x}\right) = 1 + \dfrac{1}{\dfrac{1}{x}} = 1 + x$

(f) $\dfrac{1}{f(x)} = \dfrac{1}{1 + \dfrac{1}{x}} = \dfrac{1}{\dfrac{x + 1}{x}} = \dfrac{x}{x + 1}$

41. $\dfrac{f(x) - f(1)}{x - 1} = \dfrac{3x - 3}{x - 1} = \dfrac{3(x - 1)}{x - 1} = 3$

43. $\dfrac{f(x) - f(1)}{x - 1} = \dfrac{(1 - 3x) - (-2)}{x - 1} = \dfrac{3 - 3x}{x - 1} = \dfrac{-3(x - 1)}{x - 1} = -3$

45. $\dfrac{f(x) - f(1)}{x - 1} = \dfrac{(3x^2 - 2x) - (1)}{x - 1} = \dfrac{3x^2 - 2x - 1}{x - 1} = \dfrac{(3x + 1)(x - 1)}{x - 1} = 3x + 1$

47. $\dfrac{f(x) - f(1)}{x - 1} = \dfrac{(x^3 - x) - (0)}{x - 1} = \dfrac{x^3 - x}{x - 1} = \dfrac{x(x - 1)(x + 1)}{x - 1} = x(x + 1)$

49. $\dfrac{f(x) - f(1)}{x - 1} = \dfrac{\dfrac{2}{x + 1} - 1}{x - 1} = \dfrac{\dfrac{2 - (x + 1)}{x + 1}}{x - 1} = \dfrac{1 - x}{(x + 1)(x - 1)} = \dfrac{-1}{x + 1}$

51. $\dfrac{f(x) - f(1)}{x - 1} = \dfrac{\sqrt{x} - 1}{x - 1} = \dfrac{\left(\sqrt{x} - 1\right)}{\left(\sqrt{x} - 1\right)\left(\sqrt{x} + 1\right)} = \dfrac{1}{\sqrt{x} + 1}$

53. $f(x) = 4x^3$
odd: $f(-x) = -f(x)$
$4(-x)^3 = -(4x^3)$
$-4x^3 = -4x^3$

55. $g(x) = 2x^2 - 5$
even: $f(-x) = f(x)$
$2(-x)^2 - 5 = 2x^2 - 5$
$2x^2 - 5 = 2x^2 - 5$

57. $F(x) = \sqrt[3]{x}$
odd: $f(-x) = -f(x)$
$\sqrt[3]{-x} = -\sqrt[3]{x}$
$-\sqrt[3]{x} = -\sqrt[3]{x}$

59. $f(x) = x + |x|$
even: $f(-x) = f(x)$
$-x + |-x| = x + |x|$
$-x + |x| \ne x + |x|$
odd: $f(-x) = -f(x)$
$-x + |-x| = -(x + |x|)$
$-x + |x| \ne -x - |x|$
neither

61. $g(x) = \dfrac{1}{x^2}$
even: $f(-x) = f(x)$
$\dfrac{1}{(-x)^2} = \dfrac{1}{x^2}$
$\dfrac{1}{x^2} = \dfrac{1}{x^2}$

63. $h(x) = \dfrac{x^3}{3x^2 - 9}$
odd: $f(-x) = -f(x)$
$\dfrac{(-x)^3}{3(-x)^2 - 9} = -\dfrac{x^3}{3x^2 - 9}$
$\dfrac{-x^3}{3x^2 - 9} = \dfrac{-x^3}{3x^2 - 9}$

65. One at most because if f is increasing it could only cross the x-axis at most one time. It could not "turn" and cross it again or it would start to decrease.

67. $f(x) = 3x - 3$

(a) The domain is all real numbers.

(b) x-intercept(s): y-intercept:
$0 = 3x - 3$ $y = 3(0) - 3$
$1 = x$ $y = -3$
The intercepts are $(0, -3)$, $(1, 0)$.

(c)

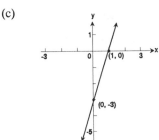

(d) The range is all real numbers.

69. $g(x) = x^2 - 4$

(a) The domain is all real numbers.

(b) x-intercept(s): y-intercept:

$$0 = x^2 - 4 \qquad y = (0)^2 - 4$$
$$4 = x^2 \qquad\qquad y = -4$$
$$x = -2, 2$$

The intercepts are $(-2, 0)$, $(2, 0)$, $(0, -4)$.

(c)

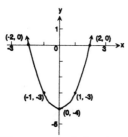

(d) The range is $\{y \mid -4 \le y < \infty\}$.

71. $h(x) = -x^2$

(a) The domain is all real numbers.

(b) x-intercept(s): y-intercept:

$$0 = -x^2 \qquad y = -0^2$$
$$0 = x \qquad\qquad y = 0$$

The intercept is $(0, 0)$.

(c)

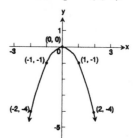

(d) The range is $\{y \mid -\infty < y \le 0\}$.

73. $f(x) = \sqrt{x - 2}$

(a) The domain is $[2, \infty)$.

(b) x-intercept(s): y-intercept:

$$0 = \sqrt{x - 2} \qquad y = \sqrt{0 - 2}$$
$$x = 2 \qquad\qquad y = \sqrt{-2}$$

The intercept is $(2, 0)$.

(c)

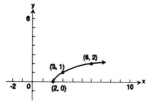

(d) The range is $\{y \mid 0 \le y < \infty\}$.

75. $h(x) = \sqrt{2 - x}$

(a) The domain is $\{x \mid -\infty < x \le 2\}$.

(b) x-intercept(s): y-intercept:

$$0 = \sqrt{2 - x} \qquad y = \sqrt{2 - 0}$$
$$0 = 2 - x \qquad\qquad y = \sqrt{2}$$
$$x = 2$$

The intercepts are $(2, 0)$, $(0, \sqrt{2})$.

(c)

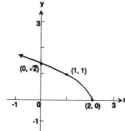

(d) The range is $\{y \mid 0 \le y < \infty\}$.

77. $f(x) = |x| + 3$
    (a) The domain is all real numbers.
    (b) x-intercept(s): y-intercept:
    $$0 = |x| + 3 \qquad y = |0| + 3$$
    $$-3 = |x| \qquad\qquad y = 3$$
    No solution.
    The intercept is $(0, 3)$.
    (c)

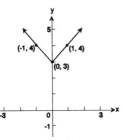

    (d) The range is $\{y \,|\, 3 \le y < \infty\}$.

79. $h(x) = -|x|$
    (a) The domain is all real numbers.
    (b) x-intercept(s): y-intercept:
    $$0 = -|x| \qquad y = -|0|$$
    $$x = 0 \qquad\qquad y = 0$$
    The intercept is $(0, 0)$.
    (c)

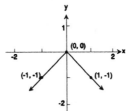

    (d) The range is $\{y \,|\, -\infty < y \le 0\}$

81. $f(x) = \begin{cases} 2x & \text{if } x \ne 0 \\ 0 & \text{if } x = 0 \end{cases}$
    (a) The domain is all real numbers.
    (b) x-intercept(s): y-intercept:
    $$0 = 2x \qquad\qquad y = 0$$
    $$0 = x$$
    The intercept is $(0, 0)$.
    (c)

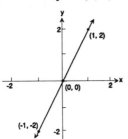

    (d) The range is all real numbers.

83. $f(x) = \begin{cases} 1 + x & \text{if } x < 0 \\ x^2 & \text{if } x \ge 0 \end{cases}$
    (a) The domain is all real numbers.
    (b) x-intercept(s): y-intercept:
    $$0 = 1 + x$$
    $$\text{or } 0 = x^2 \qquad y = 0^2$$
    $$-1 = x$$
    $$\text{or } x = 0 \qquad y = 0$$
    The intercepts are $(-1, 0)$, $(0, 0)$.
    (c)

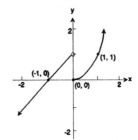

    (d) The range is all real numbers.

85. $f(x) = \begin{cases} |x| & \text{if } -2 \le x < 0 \\ 1 & \text{if } x = 0 \\ x^3 & \text{if } x > 0 \end{cases}$

(a) The domain is $\{x \mid -2 \le x < \infty\}$

(b) x-intercept(s): y-intercept:

       None $y = 1$ if $x = 0$

   The intercept is $(0, 1)$.

(c)

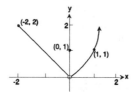

(d) The range is $\{y \mid 0 < y < \infty\}$.

87. $g(x) = \begin{cases} 1 & \text{if } x \text{ is an integer} \\ -1 & \text{if } x \text{ is not an integer} \end{cases}$

(a) The domain is all real numbers.

(b) x-intercept(s): y-intercept:

       None $y = 1$ if $x = 0$

   The intercept is $(0, 1)$.

(c)

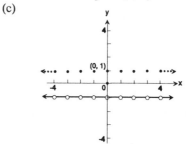

(d) The range is $\{-1, 1\}$.

89. $h(x) = 2[\![x]\!]$

(a) The domain is all real numbers.

(b) x-intercept(s): y-intercept:

     $0 = 2[\![x]\!]$ $y = 2[\![0]\!]$

     $0 = [\![x]\!]$ $y = 0$

     $0 \le x < 1$

   The intercepts are all ordered pairs $(x, 0)$ when $0 \le x < 1$.

(c)

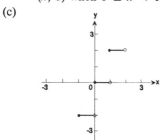

(d) The range is the set of even integers.

91. $F(x) = \begin{cases} 4 - x^2 & \text{if } |x| \le 2 \\ x^2 - 4 & \text{if } |x| > 2 \end{cases}$

(a) The domain is all real numbers.

(b) x-intercept(s): y-intercept(s):

       $0 = 4 - x^2$

or $0 = x^2 - 4$ $y = 4 - 0^2$

     $-4 = -x^2$ $y = 4$

     $\pm 2 = x$

   The intercepts are $(-2, 0)$, $(2, 0)$ and $(0, 4)$.

(c)

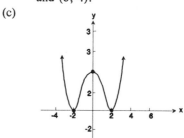

(d) The range is $\{y \mid 0 \le y < \infty\}$

93. $\dfrac{f(x + h) - f(x)}{h} = \dfrac{2(x + h) + 5 - (2x + 5)}{h} = \dfrac{2x + 2h + 5 - 2x - 5}{h} = \dfrac{2h}{h} = 2$

95. $\dfrac{f(x + h) - f(x)}{h} = \dfrac{(x + h)^2 + 2(x + h) - (x^2 + 2x)}{h}$

$= \dfrac{x^2 + 2xh + h^2 + 2x + 2h - x^2 - 2x}{h} = \dfrac{2xh + h^2 + 2h}{h}$

$= 2x + h + 2$

97. $f(x) = \begin{cases} -x & \text{if } -1 \le x \le 0 \\ \dfrac{1}{2}x & \text{if } \;\; 0 < x \le 2 \end{cases}$

Other answers are possible.

99. $f(x) = \begin{cases} -x & \text{if } \;\; x \le 0 \\ 2 - x & \text{if } \;\; 0 < x \le 2 \end{cases}$

Other answers are possible.

101. $f(x) = \begin{cases} x^2 + 4 & \text{if } \;\; x \ne 2 \\ 6 & \text{if } \;\; x = 2 \end{cases}$

To see if f is even, we need to show that $f(x) = f(-x)$ for all possible values of x.

$f(-x) = (-x)^2 + 4 = x^2 + 4 = f(x)$

However, when $x = -2$,

$f(-2) = (-2)^2 + 4 = 8$, but

$f(2) = 6$

Because $f(-2) \ne f(2)$, the function is not even.

103.

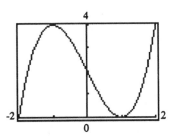

Increasing: $[-2, -1]$, $[1, 2]$
Decreasing: $[-1, 1]$
Local maxima: $(-1, 4)$
Local minima: $(1, 0)$

105.

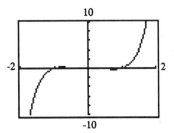

Increasing: $[-2, -0.77]$, $[0.77, 2]$
Decreasing: $[-0.77, 0.77]$
Local maxima: $(-0.77, 0.18)$
Local minima: $(0.77, -0.18)$

107. Each graph is that of $y = x^2$, but shifted vertically. If $y = x^2 + k$, $k > 0$, the shift is up k units; if $y = x^2 + k$, $k < 0$, the shift is down $|k|$ units. The graph of $y = x^2 - 4$ is the same as the graph of $y = x^2$ but shifted down 4. The graph of $y = x^2 + 5$ is the graph of $y = x^2$, but shifted up 5.

109. Each graph is that of $y = |x|$, but either compressed or stretched. If $y = k|x|$ and $k > 1$, the graph is stretched; if $y = k|x|$, $0 < k < 1$, the graph is compressed. The graph of $y = \dfrac{1}{4}|x|$ is the same as the graph of $y = |x|$, but compressed [for example, from $(2, 2)$ to $\left(2, \dfrac{1}{2}\right)$]. The graph of $y = 5|x|$ is the same as the graph of $y = |x|$, but stretched [for example, from $(2, 2)$ to $(2, 10)$].

111. The graph of $y = \sqrt{-x}$ is the reflection about the y-axis of the graph of $y = \sqrt{x}$. The same type of reflection occurs when graphing $y = 2x + 1$ and $y = 2(-x) + 1$. The conclusion is that the graph of $y = f(-x)$ is the reflection about the y-axis of the graph of $y = f(x)$.

113. For the graph $y = x^n$, n an even integer, as n increases, the graph of the function is narrower for $|x| > 1$ and flatter for $|x| < 1$.

115. (a) For 50 therms, the charge $C = 7.00 + 0.21054(50) + 0.26341(50) = \$30.70$

(b) For 500 therms, the charge
$$C = 7.00 + 0.21054(90) + 0.11242(410) + 0.26341(500) = \$203.75$$

(c) If C is the monthly charge, then
$$C = \begin{cases} 7 + 0.21054x + 0.26341x & \text{if } 0 \le x \le 90 \\ 7 + 0.21054(90) + 0.11242(x - 90) + 0.26341x & \text{if } x > 90 \end{cases}$$
$$= \begin{cases} 7 + 0.47395x & \text{if } 0 \le x \le 90 \\ 25.95 + 0.11242(x - 90) + 0.26341x & \text{if } x > 90 \end{cases}$$
$$= \begin{cases} 7 + 0.47395x & \text{if } 0 \le x \le 90 \\ 15.83 + 0.37583x & \text{if } x > 90 \end{cases}$$

(d)

117. (a) $E(x)$ is even if $E(-x) = E(x)$
$$\frac{1}{2}\big[f(-x) + f(-(-x))\big] = \frac{1}{2}\big[f(x) + f(-x)\big]$$
$$\frac{1}{2}\big(f(-x) + f(x)\big) = \frac{1}{2}\big[f(x) + f(-x)\big]$$
$$\frac{1}{2}\big[f(x) + f(-x)\big] = \frac{1}{2}\big[f(x) + f(-x)\big]$$

(b) $O(x)$ is odd if $O(-x) = -O(x)$
$$\frac{1}{2}\big[f(-x) - f(-(-x))\big] = -\frac{1}{2}\big[f(x) - f(-x)\big]$$
$$\frac{1}{2}\big(f(-x) - f(x)\big) = -\frac{1}{2}\big[f(x) - f(-x)\big]$$
$$\frac{1}{2}f(-x) - \frac{1}{2}f(x) = -\frac{1}{2}f(x) + \frac{1}{2}f(-x)$$
$$\frac{1}{2}f(-x) - \frac{1}{2}f(x) = \frac{1}{2}f(-x) - \frac{1}{2}f(x)$$

(c) Show: $\quad f(x) = E(x) + O(x)$
$$f(x) = \frac{1}{2}\big[f(x) + f(-x)\big] + \frac{1}{2}\big[f(x) - f(-x)\big]$$
$$= \frac{1}{2}f(x) + \frac{1}{2}f(-x) + \frac{1}{2}f(x) - \frac{1}{2}f(-x)$$
$$f(x) = f(x)$$

(d) From parts (a), (b), and (c) we have shown that $f(x) = E(x) + O(x)$ and that $E(x)$ is even and $O(x)$ is odd.

2.3 Graphing Techniques

1. B   3. H   5. I   7. L

9. F   11. G   13. C   15. B

17. $y = (x - 4)^3$ 19. $y = x^3 + 4$ 21. $y = -x^3$ 23. $y = 4x^3$

25. $f(x) = x^2 - 1$
Using the graph of $y = x^2$, vertically shift downward 1 unit.

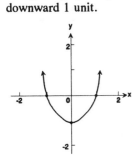

27. $g(x) = x^3 + 1$
Using the graph of $y = x^3$, vertically shift upward 1 unit.

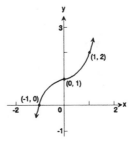

29. $h(x) = \sqrt{x - 2}$

Using the graph of $y = \sqrt{x}$, horizontally shift to the right 2 units.

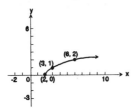

31. $f(x) = (x - 1)^3$
Using the graph of $y = x^3$, horizontally shift to the right 1 unit.

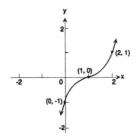

33. $g(x) = 4\sqrt{x}$

Using the graph of $y = \sqrt{x}$, vertically stretch so that $(1, 1)$ becomes $(1, 4)$.

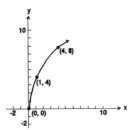

35. $h(x) = \dfrac{1}{2x}$

Using the graph of $y = \dfrac{1}{x}$, vertically

compress so that $(1, 1)$ becomes $\left[1, \dfrac{1}{2}\right]$.

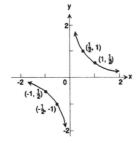

37. $f(x) = -|x|$
Reflect the graph of $y = |x|$ about the x-axis.

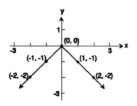

39. $g(x) = \dfrac{-1}{x}$

Reflect the graph of $y = \dfrac{1}{x}$ about the x-axis.

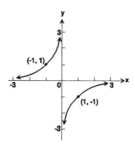

41. $h(x) = [\![\,x\,]\!]$

The greatest integer function $y = [\![\,x\,]\!]$ takes the integer values less than or equal to the given x. Thus, if we have the inequality, $0 \le x < 1$, taking the negative, we have $-(0 \le x < 1) = 0 \ge x > -1$ or $-1 < x \le 0$. In other words, reflect the graph of $y = [\![\,x\,]\!]$ about the axis.

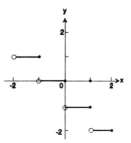

43. $f(x) = (x + 1)^2 - 3$
Using the graph of x^2, horizontally shift to the left 1 unit and vertically shift downward 3 units.

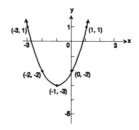

45. $g(x) = \sqrt{x - 2} + 1$

Using the graph of $y = \sqrt{x}$, horizontally shift to the right 2 units, and vertically shift upward 1 unit.

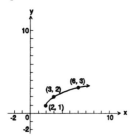

47. $h(x) = \sqrt{-x} - 2$

Reflect the graph $y = \sqrt{x}$ about the y-axis, and vertically shift downward 2 units.

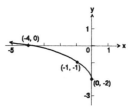

49. $f(x) = (x + 1)^3 - 1$

Using the graph of $y = x^3$, horizontally shift to the left 1 unit, and vertically shift downward 1 unit.

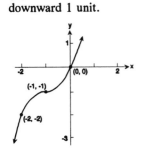

51. $g(x) = 2 \mid 1 - x \mid$

Using the graph of $y = \mid x \mid$, since $\mid 1 - x \mid = \mid x - 1 \mid$, horizontally shift to the right 1 unit and then vertically stretch so that $(0, 1)$ becomes $(0, 2)$.

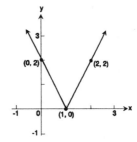

53. $h(x) = 2 [\![x - 1]\!]$

Using the graph of $y = [\![x]\!]$, horizontally shift to the right 1 unit, then vertically stretch so that the range becomes even integers instead of all integers.

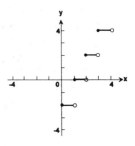

55. (a) $F(x) = f(x) + 3$
Shift up 3 units.

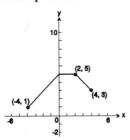

(b) $G(x) = f(x + 2)$
Shift left 2 units.

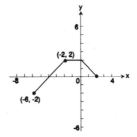

(c) $P(x) = -f(x)$
Reflect about the x-axis.

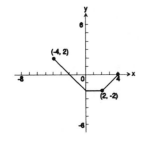

(d) $Q(x) = \frac{1}{2} f(x)$
Vertically compress so $(2, 2)$ becomes $(2, 1)$.

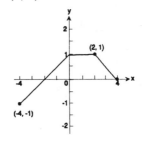

(e) $g(x) = f(-x)$
Reflect about the y-axis.

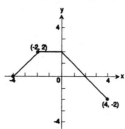

(f) $h(x) = 3f(x)$
Vertically stretch the graph by a factor of 3.

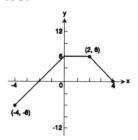

57.   (a) $F(x) = f(x) + 3$
Shift up 3 units.

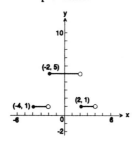

(b) $G(x) = f(x + 2)$
Shift left 2 units.

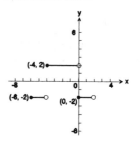

(c) $P(x) = -f(x)$
Reflect about the x-axis.

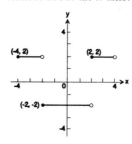

(d) $Q(x) = \dfrac{1}{2}f(x)$

Vertically compress by a factor of $\dfrac{1}{2}$.

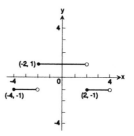

(e) $g(x) = f(-x)$
Reflect about the y-axis.

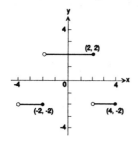

(f) $h(x) = 3f(x)$
Vertically stretch by a factor of 3.

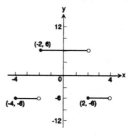

59. (a) $F(x) = f(x) + 3$
Shift up 3 units.

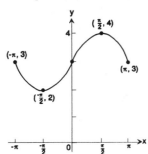

(b) $G(x) = f(x + 2)$
Shift left 2 units.

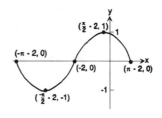

(c) $P(x) = -f(x)$
Reflect about the x-axis.

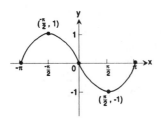

(d) $Q(x) = \frac{1}{2}f(x)$

Vertically compress by a factor of $\frac{1}{2}$.

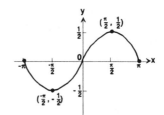

(e) $g(x) = f(-x)$
Reflect about the y-axis.

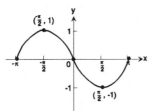

(f) $h(x) = 3f(x)$
Vertically stretch by a factor of 3.

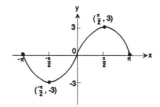

61. (a)

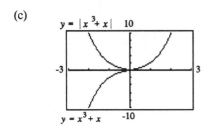

(b)

(c)

(d) Any part of the graph of $y = f(x)$ that lies below the x-axis is reflected about the x-axis to obtain the graph of $y = |f(x)|$

63. (a) Given the graph of $y = f(x)$, if $y = |f(x)|$, then all negative values for y become positive values for y. So the portion of the graph in quadrant III, where y coordinates are negative, become positive y coordinates in quadrant II. In other words, reflect the negative portion about the x-axis.

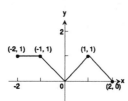

(b) Given the graph $y = f(x)$, if $y = f(|x|)$, then we must reflect about the y-axis, because $f(|+x|) = f(|-x|)$.

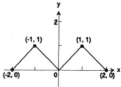

65. $f(x) = x^2 + 2x$
$f(x) = (x^2 + 2x + 1) - 1$
$f(x) = (x + 1)^2 - 1$
Using $f(x) = x^2$, shift left 1 unit and shift down 1 unit.

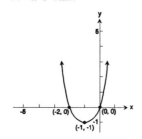

67. $f(x) = x^2 - 8x + 1$
$f(x) = (x^2 - 8x) + 1$
$f(x) = (x^2 - 8x + 16) + 1 - 16$
$f(x) = (x - 4)^2 - 15$
Using $f(x) = x^2$, shift right 4 units and shift down 15 units.

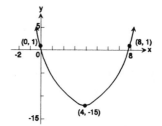

69. $f(x) = x^2 + x + 1$
$f(x) = (x^2 + x) + 1$
$f(x) = \left[x^2 + x + \dfrac{1}{4}\right] + 1 - \dfrac{1}{4}$
$f(x) = \left[x + \dfrac{1}{2}\right]^2 + \dfrac{3}{4}$
Using $f(x) = x^2$, shift left $\dfrac{1}{2}$ unit and shift up $\dfrac{3}{4}$ unit.

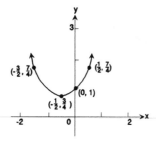

71. $y = (x - c)^2$
If $c = 0$, $y = x^2$.
If $c = 3$, $y = (x - 3)^2$, shift right 3 units.
If $c = -2$, $y = (x + 2)^2$, shift left 2 units.

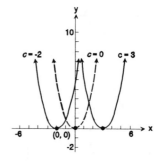

73. $F = \dfrac{9}{5}C + 32$

Graph:

C	0	40	100
F	32	104	212

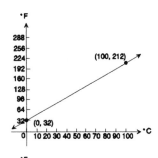

C	0	40	100
$K = C + 273$	273	313	373
F	32	104	212

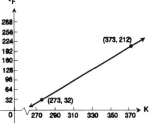

or

$C = K - 273$

$F = \dfrac{9}{5}(K - 273) + 32$

Shift the graph of $F = F = \dfrac{9}{5}C + 32$, 273 units to the right.

2.4 Operations on Functions

1. $f(x) = 3x + 4 \qquad g(x) = 2x - 3$

   (a) $(f + g)(x) = 3x + 4 + 2x - 3$
   $= 5x + 1$
   The domain is all real numbers.

   (b) $(f - g)(x) = (3x + 4) - (2x - 3)$
   $= 3x + 4 - 2x + 3$
   $= x + 7$
   The domain is all real numbers.

   (c) $(f \cdot g)(x) = (3x + 4)(2x - 3)$
   $= 6x^2 - 9x + 8x - 12$
   $= 6x^2 - x - 12$
   The domain is all real numbers.

   (d) $\left[\dfrac{f}{g}\right](x) = \dfrac{3x + 4}{2x - 3}$
   The domain is all real numbers except $\dfrac{3}{2}$.

3. $f(x) = x - 1 \qquad g(x) = 2x^2$

   (a) $(f + g)(x) = (x - 1) + 2x^2$
   $= 2x^2 + x - 1$
   The domain is all real numbers.

   (b) $(f - g)(x) = (x - 1) - (2x^2)$
   $= -2x^2 + x - 1$
   The domain is all real numbers.

   (c) $(f \cdot g)(x) = (x - 1)(2x^2)$
   $= 2x^3 - 2x^2$
   The domain is all real numbers.

   (d) $\left[\dfrac{f}{g}\right](x) = \dfrac{x - 1}{2x^2}$
   The domain is all real numbers except 0, $\{x \mid x \neq 0\}$.

5. $f(x) = \sqrt{x}$, $x \geq 0$ $g(x) = 3x - 5$

   (a) $(f + g)(x) = \sqrt{x} + 3x - 5$
The domain is $\{x \mid 0 \leq x < \infty\}$.

   (b) $(f - g)(x) = \sqrt{x} - (3x - 5)$
$$= \sqrt{x} - 3x + 5$$
The domain is $\{x \mid 0 \leq x < \infty\}$.

   (c) $(f \cdot g)(x) = \sqrt{x}\,(3x - 5)$
$$= 3x\sqrt{x} - 5\sqrt{x},$$
The domain is $\{x \mid 0 \leq x < \infty\}$.

   (d) $\left[\dfrac{f}{g}\right](x) = \dfrac{\sqrt{x}}{3x - 5}$
The domain is $\{x \mid 0 \leq x < \infty$ and $x \neq \dfrac{5}{3}\}$.

7. $f(x) = 1 + \dfrac{1}{x}$, $x \neq 0$ $g(x) = \dfrac{1}{x}$, $x \neq 0$

   (a) $(f + g)(x) = \left[1 + \dfrac{1}{x}\right] + \dfrac{1}{x}$
$$= 1 + \dfrac{2}{x}$$
The domain is $\{x \mid x \neq 0\}$.

   (b) $(f - g)(x) = \left[1 + \dfrac{1}{x}\right] - \dfrac{1}{x} = 1$
The domain is $\{x \mid x \neq 0\}$.

   (c) $(f \cdot g)(x) = \left[1 + \dfrac{1}{x}\right]\left[\dfrac{1}{x}\right]$
$$= \dfrac{1}{x} + \dfrac{1}{x^2}$$
The domain is $\{x \mid x \neq 0\}$.

   (d) $\left[\dfrac{f}{g}\right](x) = \dfrac{1 + \dfrac{1}{x}}{\dfrac{1}{x}} = \dfrac{\dfrac{x + 1}{x}}{\dfrac{1}{x}}$
$$= \dfrac{x + 1}{x} \cdot \dfrac{x}{1} = x + 1$$
The domain is $\{x \mid x \neq 0\}$.

9. $f(x) = \dfrac{2x + 3}{3x - 2}$, $x \neq \dfrac{2}{3}$ $g(x) = \dfrac{4x}{3x - 2}$, $x \neq \dfrac{2}{3}$

   (a) $(f + g)(x) = \dfrac{2x + 3}{3x - 2} + \dfrac{4x}{3x - 2} = \dfrac{6x + 3}{3x - 2}$
The domain is $\left\{x \mid x \neq \dfrac{2}{3}\right\}$.

   (b) $(f - g)(x) = \dfrac{2x + 3}{3x - 2} - \dfrac{4x}{3x - 2} = \dfrac{-2x + 3}{3x - 2}$
The domain is $\left\{x \mid x \neq \dfrac{2}{3}\right\}$.

   (c) $(f \cdot g)(x) = \left[\dfrac{2x + 3}{3x - 2}\right]\left[\dfrac{4x}{3x - 2}\right] = \dfrac{8x^2 + 12x}{(3x - 2)^2}$
The domain is $\left\{x \mid x \neq \dfrac{2}{3}\right\}$.

   (d) $\left[\dfrac{f}{g}\right](x) = \dfrac{\dfrac{2x + 3}{3x - 2}}{\dfrac{4x}{3x - 2}} = \dfrac{2x + 3}{3x - 2} \cdot \dfrac{3x - 2}{4x} = \dfrac{2x + 3}{4x}$
The domain is $\left\{x \mid x \neq \dfrac{2}{3}$ and $x \neq 0\right\}$.

11. $f(x) = 3x + 1$
$$(f + g)(x) = 6 - \dfrac{1}{2}x$$
$$6 - \dfrac{1}{2}x = (3x + 1) + g(x)$$
$$-\dfrac{7}{2}x + 5 = g(x)$$
$$g(x) = 5 - \dfrac{7}{2}x$$

13. $f + g = x + \dfrac{1}{x}$

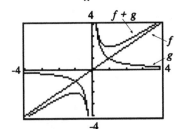

15. $f + g = x^2 + \dfrac{1}{x}$

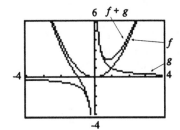

17. $f \cdot g = \dfrac{x}{x^2 + 1}$

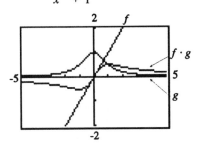

19. $f \cdot g = \dfrac{x^2}{x^2 + 1}$

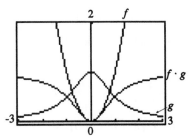

21. Graph: $y = |x|$ on the interval $[0, 2]$
$y = x^2$ on the interval $[0, 2]$
Add the y-coordinates of the plotted points to get $y = |x| + x^2$.

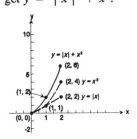

23. Graph: $y = x$ on the interval $[0, 2]$
$y = x^3$ on the interval $[0, 2]$
Add the y-coordinates of the plotted points to get $y = x^3 + x$.

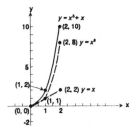

25. $f(x) = 2x \qquad g(x) = 3x^2 + 1$
(a) $(f \circ g)(4) = f(g(4)) = f(49) = 2(49) = 98$
(b) $(g \circ f)(2) = g(f(2)) = g(4) = 3(4)^2 + 1 = 48 + 1 = 49$
(c) $(f \circ f)(1) = f(f(1)) = f(2) = 2(2) = 4$
(d) $(g \circ g)(0) = g(g(0)) = g(1) = 3(1)^2 + 1 = 4$

27. $f(x) = 4x^2 - 3 \qquad g(x) = 3 - \dfrac{1}{2}x^2$
(a) $(f \circ g)(4) = f(g(4)) = f(-5) = 4(-5)^2 - 3 = 97$
(b) $(g \circ f)(2) = g(f(2)) = g(13) = 3 - \dfrac{1}{2}(13)^2 = 3 - \dfrac{169}{2} = \dfrac{6 - 169}{2} = -\dfrac{163}{2}$
(c) $(f \circ f)(1) = f(f(1)) = f(1) = 1$
(d) $(g \circ g)(0) = g(g(0)) = g(3) = 3 - \dfrac{1}{2}(3)^2 = 3 - \dfrac{9}{2} = \dfrac{6 - 9}{2} = -\dfrac{3}{2}$

29. $f(x) = \sqrt{x}$ $g(x) = 2x$

  (a) $(f \circ g)(4) = f(g(4)) = f(8) = \sqrt{8} = 2\sqrt{2}$

  (b) $(g \circ f)(2) = g(f(2)) = g\left(\sqrt{2}\right) = 2\sqrt{2}$

  (c) $(f \circ f)(1) = f(f(1)) = f(1) = \sqrt{1} = 1$
  (d) $(g \circ g)(0) = g(g(0)) = g(0) = 2(0) = 0$

31. $f(x) = |x|$ $g(x) = \dfrac{1}{x^2 + 1}$

  (a) $(f \circ g)(4) = f(g(4)) = f\left(\dfrac{1}{17}\right) = \left|\dfrac{1}{17}\right| = \dfrac{1}{17}$

  (b) $(g \circ f)(2) = g(f(2)) = g(2) = \dfrac{1}{2^2 + 1} = \dfrac{1}{5}$

  (c) $(f \circ f)(1) = f(f(1)) = f(1) = 1$

  (d) $(g \circ g)(0) = g(g(0)) = g(1) = \dfrac{1}{1^2 + 1} = \dfrac{1}{2}$

33. $f(x) = \dfrac{3}{x^2 + 1}$ $g(x) = \sqrt{x}$

  (a) $(f \circ g)(4) = f(g(4)) = f(2) = \dfrac{3}{2^2 + 1} = \dfrac{3}{5}$

  (b) $(g \circ f)(2) = g(f(2)) = g\left[\dfrac{3}{5}\right] = \sqrt{\dfrac{3}{5}} = \dfrac{\sqrt{15}}{5}$

  (c) $(f \circ f)(1) = f(f(1)) = f\left[\dfrac{3}{2}\right] = \dfrac{3}{\left[\dfrac{3}{2}\right]^2 + 1} = \dfrac{3}{\dfrac{9}{4} + \dfrac{4}{4}} = \dfrac{3}{\dfrac{13}{4}} = \dfrac{3}{1} \cdot \dfrac{4}{13} = \dfrac{12}{13}$

  (d) $(g \circ g)(0) = g(g(0)) = g(0) = \sqrt{0} = 0$

35. $f(x) = 2x + 3$ $g(x) = 3x$
  (a) $(f \circ g)(x) = f(g(x)) = f(3x) = 2(3x) + 3 = 6x + 3$
  (b) $(g \circ f)(x) = g(f(x)) = g(2x + 3) = 3(2x + 3) = 6x + 9$
  (c) $(f \circ f)(x) = f(f(x)) = f(2x + 3) = 2(2x + 3) + 3 = 4x + 6 + 3 = 4x + 9$
  (d) $(g \circ g)(x) = g(g(x)) = g(3x) = 3(3x) = 9x$

37. $f(x) = 3x + 1$ $g(x) = x^2$
  (a) $(f \circ g)(x) = f(g(x)) = f(x^2) = 3x^2 + 1$
  (b) $(g \circ f)(x) = g(f(x)) = g(3x + 1) = (3x + 1)^2 = 9x^2 + 6x + 1$
  (c) $(f \circ f)(x) = f(f(x)) = f(3x + 1) = 3(3x + 1) + 1 = 9x + 3 + 1 = 9x + 4$
  (d) $(g \circ g)(x) = g(g(x)) = g(x^2) = (x^2)^2 = x^4$

39. $f(x) = \sqrt{x}$ $g(x) = x^2 - 1$

  (a) $(f \circ g)(x) = f(g(x)) = f(x^2 - 1) = \sqrt{x^2 - 1}$

  (b) $(g \circ f)(x) = g(f(x)) = g\left(\sqrt{x}\right) - 1 = \left(\sqrt{x}\right)^2 - 1 = x - 1$

  (c) $(f \circ f)(x) = f(f(x)) = f\left(\sqrt{x}\right) = \sqrt{\sqrt{x}} = \sqrt[4]{x}$
  (d) $(g \circ g)(x) = g(g(x)) = g(x^2 - 1) = (x^2 - 1)^2 - 1 = x^4 - 2x^2 + 1 - 1 = x^4 - 2x^2$

41. $f(x) = \dfrac{x-1}{x+1}$ $\qquad$ $g(x) = \dfrac{1}{x}$

$\quad$ **(a)** $\quad (f \circ g)(x) = f(g(x)) = f\left(\dfrac{1}{x}\right) = \dfrac{\dfrac{1}{x} - 1}{\dfrac{1}{x} + 1} = \dfrac{\dfrac{1-x}{x}}{\dfrac{1+x}{x}} = \dfrac{1-x}{x} \cdot \dfrac{x}{1+x} = \dfrac{1-x}{1+x}$

$\quad$ **(b)** $\quad (g \circ f)(x) = g(f(x)) = g\left(\dfrac{x-1}{x+1}\right) = \dfrac{1}{\dfrac{x-1}{x+1}} = \dfrac{x+1}{x-1}$

$\quad$ **(c)** $\quad (f \circ f)(x) = f(f(x)) = f\left(\dfrac{x-1}{x+1}\right) = \dfrac{\dfrac{x-1}{x+1} - 1}{\dfrac{x-1}{x+1} + 1} = \dfrac{\dfrac{(x-1)-(x+1)}{x+1}}{\dfrac{(x-1)+(x+1)}{x+1}}$

$\qquad\qquad = \dfrac{\dfrac{-2}{x+1}}{\dfrac{2x}{x+1}} = \dfrac{-2}{(x+1)} \cdot \dfrac{(x+1)}{2x} = -\dfrac{1}{x}$

$\quad$ **(d)** $\quad (g \circ g)(x) = g(g(x)) = g\left(\dfrac{1}{x}\right) = \dfrac{1}{\dfrac{1}{x}} = x$

43. $f(x) = x^2$ $\qquad$ $g(x) = \sqrt{x}$

$\quad$ **(a)** $\quad (f \circ g)(x) = f(g(x)) = f\left(\sqrt{x}\right) = \left(\sqrt{x}\right)^2 = x$

$\quad$ **(b)** $\quad (g \circ f)(x) = g(f(x)) = g(x^2) = \sqrt{x^2} = |x|$
$\quad$ **(c)** $\quad (f \circ f)(x) = f(f(x)) = f(x)^2 = (x^2)^2 = x^4$

$\quad$ **(d)** $\quad (g \circ g)(x) = g(g(x)) = g\left(\sqrt{x}\right) = \sqrt{\sqrt{x}} = \sqrt[4]{x}$

45. $f(x) = \dfrac{1}{2x+3}$ $\qquad$ $g(x) = 2x + 3$

$\quad$ **(a)** $\quad (f \circ g)(x) = f(g(x)) = f(2x+3) = \dfrac{1}{2(2x+3)} = \dfrac{1}{4x+6+3} = \dfrac{1}{4x+9}$

$\quad$ **(b)** $\quad (g \circ f)(x) = g(f(x)) = g\left(\dfrac{1}{2x+3}\right) = 2\left(\dfrac{1}{2x+3}\right) + 3 = \dfrac{2}{2x+3} + \dfrac{6x+9}{2x+3} = \dfrac{6x+11}{2x+3}$

$\quad$ **(c)** $\quad (f \circ f)(x) = f(f(x)) = f\left(\dfrac{1}{2x+3}\right) = \dfrac{1}{2\left(\dfrac{1}{2x+3}\right) + 3} = \dfrac{1}{\dfrac{2}{2x+3} + \dfrac{6x+9}{2x+3}}$

$\qquad\qquad = \dfrac{1}{\dfrac{6x+11}{2x+3}} = \dfrac{2x+3}{6x+11}$

$\quad$ **(d)** $\quad (g \circ g)(x) = g(g(x)) = g(2x+3) = 2(2x+3) + 3 = 4x+6+3 = 4x+9$

47. $f(x) = ax + b$ $\qquad$ $g(x) = cx + d$
$\quad$ **(a)** $\quad (f \circ g)(x) = f(g(x)) = f(cx+d) = a(cx+d) + b = acx + ad + b$
$\quad$ **(b)** $\quad (g \circ f)(x) = g(f(x)) = g(ax+b) = c(ax+b) + d = acx + bc + d$
$\quad$ **(c)** $\quad (f \circ f)(x) = f(f(x)) = f(ax+b) = a(ax+b) + b = a^2x + ab + b$
$\quad$ **(d)** $\quad (g \circ g)(x) = g(g(x)) = g(cx+d) = c(cx+d) + d = c^2x + cd + d$

49. $(f \circ g)(x) = f(g(x)) = f\left[\dfrac{1}{2}x\right] = 2\left[\dfrac{1}{2}x\right]$
$\qquad = x$

$\quad (g \circ f)(x) = g(f(x)) = g(2x) = \dfrac{1}{2}(2x) = x$

51. $(f \circ g)(x) = f(g(x)) = f\left(\sqrt[3]{x}\right) = \left(\sqrt[3]{x}\right)^3$
$\qquad = x$

$\quad (g \circ f)(x) = g(f(x)) = g(x^3) = \sqrt[3]{x^3} = x$

53. $(f \circ g)(x) = f(g(x)) = f\left[\dfrac{1}{2}(x + 6)\right] = 2\left[\dfrac{1}{2}(x + 6)\right] - 6 = x + 6 - 6 = x$

$\quad (g \circ f)(x) = g(f(x)) = g(2x - 6) = \dfrac{1}{2}(2x - 6 + 6) = x$

55. $(f \circ g)(x) = f(g(x)) = f\left[\dfrac{1}{a}(x - b)\right] = a\left[\dfrac{1}{a}(x - b)\right] + b = a\left[\dfrac{x}{a} - \dfrac{b}{a}\right] + b = x - b + b = x$

$\quad (g \circ f)(x) = g(f(x)) = g(ax + b) = \dfrac{1}{a}(ax + b - b) = x$

57. $f(x) = 2x^3 - 3x^2 + 4x - 1 \quad g(x) = 2$
$\quad (f \circ g)(x) = f(g(x)) = f(2) = 2(2)^3 - 3(2)^2 + 4(2) - 1 = 16 - 12 + 8 - 1 = 11$
$\quad (g \circ f)(x) = g(f(x)) = 2$

59. $f(x) = x^2, g(x) = \sqrt{x} + 2, h(x) = 1 - 3x$
$\quad [f \circ (g \circ h)](x) = f \circ \left(g(h(x))\right) = f\left(\sqrt{1 - 3x} + 2\right) = \left(\sqrt{1 - 3x} + 2\right)^2$
$\qquad = (1 - 3x) + 4\sqrt{1 - 3x} + 4 = 5 - 3x + 4\sqrt{1 - 3x}$

61. $[(f + g) \circ h](x) = \left[(f + g)(h(x))\right] = [h(x)]^2 + \sqrt{h(x)} + 2$
$\qquad = (1 - 3x)^2 + \sqrt{1 - 3x} + 2 = 1 - 6x + 9x^2 + \sqrt{1 - 3x} + 2$
$\qquad = 9x^2 - 6x + 3 + \sqrt{1 - 3x}$

63. $f(x) = x^2, g(x) = 3x, h(x) = \sqrt{x} + 1$
$\quad F(x) = 9x^2$
$\quad F(x) = (3x)^2$
$\quad F(x) = (g(x))^2$
$\quad F(x) = f(g(x))$
$\qquad F = f \circ g$

65. $H(x) = |x| + 1$
$\quad H(x) = \sqrt{x^2} + 1$
$\quad H(x) = \sqrt{f(x)} + 1$
$\quad H(x) = h(f(x))$
$\qquad H = h \circ f$

67. $q(x) = x + 2\sqrt{x} + 1$
$\quad q(x) = \left(\sqrt{x} + 1\right)^2$
$\quad q(x) = (h(x))^2$
$\quad q(x) = f(h(x))$
$\qquad q = f \circ h$

69. $P(x) = x^4$
$\quad P(x) = (x^2)^2$
$\quad P(x) = (f(x))^2$
$\quad P(x) = f(f(x))$
$\qquad P = f \circ f$

71. $H(x) = (2x + 3)^4 = f(g(x))$
$\quad f(x) = x^4, g(x) = 2x + 3$

73. $H(x) = \sqrt{x^2 + x + 1} = f(g(x))$
$\quad f(x) = \sqrt{x}, g(x) = x^2 + x + 1$

75. $H(x) = \left[1 - \dfrac{1}{x^2}\right]^2 = f(g(x))$

$f(x) = x^2,\ g(x) = 1 - \dfrac{1}{x^2}$

77. $H(x) = [\![x^2 + 1]\!] = f(g(x))$

$f(x) = [\![x]\!],\ g(x) = x^2 + 1$

79. $(f \circ g)(x) = f(g(x)) = f(3x + a) = 2(3x + a)^2 + 5$

When $x = 0$, $(f \circ g)(x) = f \circ g)(0) = 23$

Then, $2(3 \cdot 0 + a)^2 + 5 = 23$

$$2a^2 + 5 = 23$$
$$2a^2 = 18$$
$$a^2 = 9$$
$$a = -3, 3$$

81. $S(r) = 4\pi r^2$

$r(t) = \dfrac{2}{3}t^3,\ t \geq 0$

$S(r(t)) = 4\pi \left[\dfrac{2}{3}t^3\right]^2$

$= 4\pi \left[\dfrac{4}{9}t^6\right]$

$= \dfrac{16}{9}\pi t^6$

83. $N(t) = 100\,t - 5t^2,\ 0 \leq t \leq 10$

$C(x) = 15000 + 8000x$

$C(N(t)) = 15000 + 8000(100t - 5t^2)$

$C(N(t)) = 15000 + 800{,}000t - 40{,}000t^2$

85. $p = -\dfrac{1}{4}x + 100 \qquad 0 \leq x \leq 400$

$\dfrac{1}{4}x = 100 - p$

$x = 4(100 - p)$

$C = \dfrac{\sqrt{x}}{25} + 600 = \dfrac{\sqrt{4(100 - p)}}{25} + 600$

$= \dfrac{2\sqrt{100 - p}}{25} + 600$

87. Since f and g are odd, $f(-x) = -f(x)$ and $g(-x) = -g(x)$.

$(f \circ g)(-x) = f(g(-x)) = f(-g(x)) = -f(g(x)) = -(f \circ g)(x)$.

Because $(f \circ g)(-x) = -(f \circ g)(x)$, by definition, $f \circ g$ is odd.

2.5 One-to-One Functions; Inverse Functions

1. Yes, any horizontal line intersects the graph of f at most in one point. *One-to-one.*

3. No, there are horizontal lines which intersect the graph of f at more than one point. *Not one-to-one.*

5. Yes, any horizontal line intersects the graph of f at most in one point. *One-to-one.*

7. Reflect about the line $y = x$.

9. Reflect about the line $y = x$.

11. Reflect about the line $y = x$.

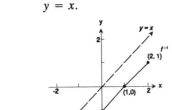

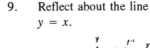
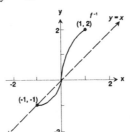

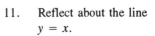
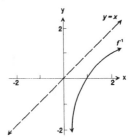

13. $f(x) = 3x + 4, \qquad g(x) = \dfrac{1}{3}(x - 4)$

$f(g(x)) = f\left[\dfrac{1}{3}(x - 4)\right] = 3\left[\dfrac{1}{3}(x - 4)\right] + 4$

$\qquad\qquad = (x - 4) + 4 = x$

$g(f(x)) = g(3x + 4) = \dfrac{1}{3}(3x + 4 - 4) = \dfrac{1}{3}(3x) = x$

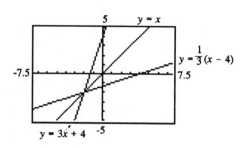

15. $f(x) = 4x - 8, \qquad g(x) = \dfrac{x}{4} + 2$

$f(g(x)) = f\left[\dfrac{x}{4} + 2\right] = 4\left[\dfrac{x}{4} + 2\right] - 8$

$\qquad\qquad = (x + 8) - 8 = x$

$g(f(x)) = g(4x - 8) = \dfrac{4x - 8}{4} + 2 = x - 2 + 2 = x$

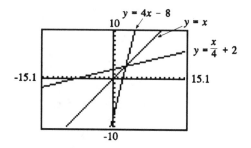

17. $f(x) = x^3 - 8, \qquad\qquad g(x) = \sqrt[3]{x + 8}$

$f(g(x)) = f\left(\sqrt[3]{x + 8}\right) = \left[\sqrt[3]{x + 8}\right]^3 - 8$

$\qquad\qquad = (x + 8) - 8 = x$

$g(f(x)) = g(x^3 - 8) = \sqrt[3]{x^3 - 8 + 8} = \sqrt[3]{x^3} = x$

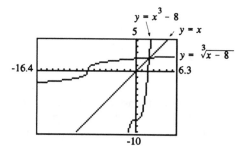

19. $f(x) = \dfrac{1}{x}, \qquad\qquad g(x) = \dfrac{1}{x}$

$f(g(x)) = f\left[\dfrac{1}{x}\right] = \dfrac{1}{\dfrac{1}{x}} = x$

$g(f(x)) = g\left[\dfrac{1}{x}\right] = \dfrac{1}{\dfrac{1}{x}} = x$

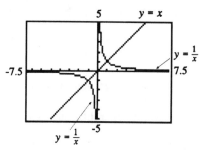

21. $f(x) = \dfrac{2x + 3}{x + 4} \qquad g(x) = \dfrac{4x - 3}{2 - x}$

$f(g(x)) = f\left[\dfrac{4x - 3}{2 - x}\right] = \dfrac{2\left[\dfrac{4x - 3}{2 - x}\right] + 3}{\dfrac{4x - 3}{2 - x} + 4} = \dfrac{\dfrac{2(4x - 3) + 3(2 - x)}{2 - x}}{\dfrac{(4x - 3) + 4(2 - x)}{2 - x}}$

$\qquad = \dfrac{\dfrac{8x - 6 + 6 - 3x}{2 - x}}{\dfrac{4x - 3 + 8 - 4x}{2 - x}} = \dfrac{\dfrac{5x}{2 - x}}{\dfrac{5}{2 - x}} = \dfrac{5x}{2 - x} \cdot \dfrac{(2 - x)}{5} = x$

$$g(f(x)) = g\left[\frac{2x + 3}{x + 4}\right] = \frac{4\left[\dfrac{2x + 3}{x + 4}\right] - 3}{2 - \left[\dfrac{2x + 3}{x + 4}\right]} = \frac{\dfrac{4(2x + 3) - 3(x + 4)}{x + 4}}{\dfrac{2(x + 4) - (2x + 3)}{x + 4}}$$

$$= \frac{\dfrac{8x + 12 - 3x - 12}{x + 4}}{\dfrac{2x + 8 - 2x - 3}{x + 4}} = \frac{\dfrac{5x}{x + 4}}{\dfrac{5}{x + 4}} = \frac{5x}{x + 4} \cdot \frac{(x + 4)}{5} = x$$

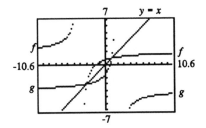

23. $f(x) = 3x$
$y = 3x$
$x = 3y$
$$y = \frac{x}{3}$$
$$f^{-1}(x) = \frac{x}{3}$$

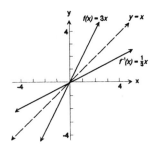

Verify: $f(f^{-1}(x)) = f\left[\dfrac{x}{3}\right] = 3\left[\dfrac{x}{3}\right] = x$

and $f^{-1}(f(x)) = f^{-1}(3x) = \dfrac{3x}{3} = x$

Domain of f = range of $f^{-1} = (-\infty, \infty)$
Range of f = domain of $f^{-1} = (-\infty, \infty)$

25. $f(x) = 4x + 2$
$y = 4x + 2$
$x = 4y + 2$
$4y = x - 2$
$$y = \frac{x}{4} - \frac{1}{2}$$
$$f^{-1}(x) = \frac{x}{4} - \frac{1}{2}$$

Verify:

$$f(f^{-1}(x)) = f\left[\frac{x}{4} - \frac{1}{2}\right] = 4\left[\frac{x}{4} - \frac{1}{2}\right] + 2 = x - 2 + 2 = x$$

$$f^{-1}(f(x)) = f^{-1}(4x + 2) = \frac{4x + 2 - 2}{4} = x$$

Domain of f = range of $f^{-1} = (-\infty, \infty)$
Range of f = domain of $f^{-1} = (-\infty, \infty)$

27. $f(x) = x^3 - 1$
$\quad\;\; y = x^3 - 1$
$\quad\;\; x = y^3 - 1$
$\quad\; y^3 = x + 1$
$\quad\;\; y = \sqrt[3]{x + 1}$
$f^{-1}(x) = \sqrt[3]{x + 1}$

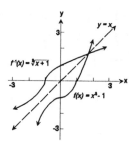

Verify:

$$f(f^{-1}(x)) = f\left(\sqrt[3]{x + 1}\right) = \left(\sqrt[3]{x + 1}\right)^3 - 1 = x + 1 - 1 = x$$

$$f^{-1}(f(x)) = f^{-1}(x^3 - 1) = \sqrt[3]{x^3 - 1 + 1} = \sqrt[3]{x^3} = x$$

Domain of f = range of f^{-1} = $(-\infty, \infty)$
Range of f = domain of f^{-1} = $(-\infty, \infty)$

29. $f(x) = x^2 + 4, x \geq 0$
$\quad\;\; y = x^2 + 4$
$\quad\;\; x = y^2 + 4$
$\quad\; y^2 = x - 4$
$\quad\;\; y = \sqrt{x - 4}$
$f^{-1}(x) = \sqrt{x - 4}$

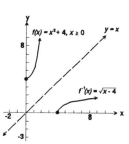

Verify:

$$f(f^{-1}(x)) = f\left(\sqrt{x - 4}\right) = \left(\sqrt{x - 4}\right)^2 + 4 = x - 4 + 4 = x$$

$$f^{-1}(f(x)) = f^{-1}(x^2 + 4) = \sqrt{x^2 + 4 - 4} = \sqrt{x^2} = |x| = x, x \geq 0$$

Domain of f = range of f^{-1} = $[0, \infty)$
Range of f = domain of f^{-1} = $[4, \infty)$

31. $f(x) = \dfrac{4}{x}$

$\quad\;\; y = \dfrac{4}{x}$

$\quad\;\; x = \dfrac{4}{y}$

$\quad yx = 4$

$\quad\;\; y = \dfrac{4}{x}$

$f^{-1}(x) = \dfrac{4}{x}$

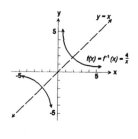

Verify: $f(f^{-1}(x)) = f\left(\dfrac{4}{x}\right) = \dfrac{4}{\dfrac{4}{x}} = 4 \cdot \dfrac{x}{4} = x$

$$f^{-1}(f(x)) = f^{-1}\left(\dfrac{4}{x}\right) = \dfrac{4}{\dfrac{4}{x}} = 4 \cdot \dfrac{x}{4} = x$$

Domain of f = range of f^{-1} = all real numbers except 0
Range of f = domain of f^{-1} = all real numbers except 0

33.
$$f(x) = \frac{1}{x - 2}$$
$$y = \frac{1}{x - 2}$$
$$x = \frac{1}{y - 2}$$
$$xy - 2x = 1$$
$$y = \frac{2x + 1}{x}$$
$$f^{-1}(x) = \frac{2x + 1}{x}$$

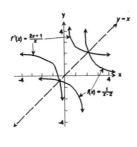

Verify:
$$f(f^{-1}(x)) = f\left(\frac{2x + 1}{x}\right) = \frac{1}{\dfrac{2x + 1}{x} - 2} = \frac{1}{\dfrac{2x + 1 - 2x}{x}} = \frac{x}{1} = x$$

$$f^{-1}(f(x)) = f^{-1}\left[\frac{1}{x - 2}\right] = \frac{2\left[\dfrac{1}{x - 2}\right] + 1}{\dfrac{1}{x - 2}} = \frac{\dfrac{2 + (x - 2)}{x - 2}}{\dfrac{1}{x - 2}} = x$$

Domain of f = range of f^{-1} = all real numbers except 2
Range of f = domain of f^{-1} = all real numbers except 0

35.
$$f(x) = \frac{2}{3 + x}$$
$$y = \frac{2}{3 + x}$$
$$x = \frac{2}{3 + y}$$
$$3x + xy = 2$$
$$y = \frac{2 - 3x}{x}$$
$$f^{-1}(x) = \frac{2 - 3x}{x}$$

Verify: $f(f^{-1}(x)) = f\left(\dfrac{2 - 3x}{x}\right) = \dfrac{2}{3 + \left[\dfrac{2 - 3x}{x}\right]} = \dfrac{2}{\dfrac{3x + 2 - 3x}{x}} = x$

$$f^{-1}(f(x)) = f^{-1}\left[\frac{2}{3 + x}\right] = \frac{2 - 3\left[\dfrac{2}{3 + x}\right]}{\dfrac{2}{3 + x}} = \frac{\dfrac{6 + 2x - 3}{3 + x}}{\dfrac{2}{3 + x}} = x$$

Domain of f = range of f^{-1} = all real numbers except -3
Range of f = domain of f^{-1} = all real numbers except 0

37.
$$f(x) = (x + 2)^2, \ x \geq -2$$
$$y = (x + 2)^2, \ x \geq -2$$
$$x = (y + 2)^2$$
$$\sqrt{x} = y + 2$$
$$y = \sqrt{x} - 2$$
$$f^{-1}(x) = \sqrt{x} - 2$$

Verify: $f(f^{-1}(x)) = f\left(\sqrt{x} - 2\right) = \left(\sqrt{x} - 2 + 2\right)^2 = \sqrt{x^2} = x$

$$f^{-1}(f(x)) = f^{-1}\left[(x + 2)^2\right] = \sqrt{(x + 2)^2} - 2 = x + 3 - 2$$
$$= x$$

Domain of f = range of f^{-1} = $[-2, \infty)$
Range of f = domain of f^{-1} = $[0, \infty)$

39.
$$f(x) = \frac{2x}{x - 1}$$
$$y = \frac{2x}{x - 1}$$
$$x = \frac{2y}{y - 1}$$
$$xy - x = 2y$$
$$xy - 2y = x$$
$$y(x - 2) = x$$
$$y = \frac{x}{x - 2}$$
$$f^{-1}(x) = \frac{x}{x - 2}$$

Verify: $f(f^{-1}(x)) = f\left(\dfrac{x}{x - 2}\right) = \dfrac{2\left[\dfrac{x}{x - 2}\right]}{\left[\dfrac{x}{x - 2}\right] - 1} = \dfrac{\dfrac{2x}{x - 2}}{\dfrac{x - (x - 2)}{x - 2}} = \dfrac{2x}{2} = x$

$$f^{-1}(f(x)) = f^{-1}\left[\frac{2x}{x - 1}\right] = \frac{\dfrac{2x}{x - 1}}{\dfrac{2x}{x - 1} - 2} = \frac{\dfrac{2x}{x - 1}}{\dfrac{2x - 2x + 2}{x - 1}} = \frac{2x}{2} = x$$

Domain of f = range of f^{-1} = all real numbers except 1
Range of f = domain of f^{-1} = all real numbers except 2

41.

$$f(x) = \frac{3x + 4}{2x - 3}$$

$$y = \frac{3x + 4}{2x - 3}$$

$$x = \frac{3y + 4}{2y - 3}$$

$$2xy - 3x = 3y + 4$$

$$2xy - 3y = 3x + 4$$

$$y(2x - 3) = 3x + 4$$

$$y = \frac{3x + 4}{2x - 3}$$

$$f^{-1}(x) = \frac{3x + 4}{2x - 3}$$

Verify:

$$f(f^{-1}(x)) = f\left(\frac{3x + 4}{2x - 3}\right) = \frac{3\left[\frac{3x + 4}{2x - 3}\right] + 4}{2\left[\frac{3x + 4}{2x - 3}\right] - 3} = \frac{\frac{9x + 12 + 8x - 12}{2x - 3}}{\frac{6x + 8 - 6x + 9}{2x - 3}}$$

$$= \frac{17x}{2x - 3} \cdot \frac{2x - 3}{17} = \frac{17x}{17} = x$$

$$f^{-1}(f(x)) = f^{-1}\left(\frac{3x + 4}{2x - 3}\right) = \frac{3\left[\frac{3x + 4}{2x - 3}\right] + 4}{2\left[\frac{3x + 4}{2x - 3}\right] - 3} = \frac{\frac{9x + 12 + 8x - 12}{2x - 3}}{\frac{6x + 8 - 6x + 9}{2x - 3}}$$

$$= \frac{17x}{2x - 3} \cdot \frac{2x - 3}{17} = \frac{17x}{17} = x$$

Domain of f = range of f^{-1} = all real numbers except $\frac{3}{2}$

Range of f = domain of f^{-1} = all real numbers except $\frac{3}{2}$

43.

$$f(x) = \frac{2x + 3}{x + 2}$$

$$y = \frac{2x + 3}{x + 2}$$

$$x = \frac{2y + 3}{y + 2}$$

$$xy + 2x = 2y + 3$$

$$xy - 2y = 3 - 2x$$

$$y(x - 2) = 3 - 2x$$

$$y = \frac{3 - 2x}{x - 2}$$

$$f^{-1}(x) = \frac{3 - 2x}{x - 2} = \frac{-2x + 3}{x - 2}$$

Verify:

$$f(f^{-1}(x)) = f\left(\frac{3 - 2x}{x - 2}\right) = \frac{2\left[\frac{3 - 2x}{x - 2}\right] + 3}{\frac{3 - 2x}{x - 2} + 2} = \frac{\frac{6 - 4x + 3x - 6}{x - 2}}{\frac{3 - 2x + 2x - 4}{x - 2}} = \frac{-x}{-1} = x$$

$$f^{-1}(f(x)) = f^{-1}\left(\frac{2x + 3}{x + 2}\right) = \frac{3 - 2\left[\frac{2x + 3}{x + 2}\right]}{\left[\frac{2x + 3}{x + 2}\right] - 2} = \frac{\frac{3x + 6 - 4x - 6}{x + 2}}{\frac{2x + 3 - 2x - 4}{x + 2}} = \frac{-x}{-1} = x$$

Domain of f = range of f^{-1} = all real numbers except -2

Range of f = domain of f^{-1} = all real numbers except 2

45.

$f(x) = 2\sqrt[3]{x}$

$y = 2\sqrt[3]{x}$

$x = 2\sqrt[3]{y}$

$x^3 = 8y$

$y = \dfrac{x^3}{8}$

$f^{-1}(x) = \dfrac{x^3}{8}$

Verify: $f(f^{-1}(x)) = f\left(\dfrac{x^3}{8}\right) = 2\sqrt[3]{\dfrac{x^3}{8}} = 2\left(\dfrac{x}{2}\right) = x$

$f^{-1}(f(x)) = f^{-1}\left(2\sqrt[3]{x}\right) = \dfrac{\left(2\sqrt[3]{x}\right)^3}{8} = \dfrac{8x}{8} = x$

Domain of f = range of $f^{-1} = (-\infty, \infty)$

Range of f = domain of $f^{-1} = (-\infty, \infty)$

47.

$f(x) = mx + b, \; m \ne 0$

$y = mx + b$

$x = my + b$

$my = x - b$

$y = \dfrac{x - b}{m}$

$f^{-1}(x) = \dfrac{x - b}{m}, \; m \ne 0$

49. No. If a function is even, $f(-x) = f(x)$. Whenever x and $-x$ are in the domain of f, two equal y values, $f(x)$ and $f(-x)$ are present.

51. f^{-1} also lies in quadrant one because whenever (a, b) is on f, then (b, a) is on f^{-1}. In quadrant one (a, b) is $(+, +)$, so (b, a) is $(+, +)$, also in quadrant one.

53. $f(x) = |x|, \; x \ge 0$ is one to one. Thus,

$f(x) = x$

$f^{-1}(x) = x$

55. $f(x) = \dfrac{9}{5}x + 32 \qquad g(x) = \dfrac{5}{9}(x - 32)$

$f(g(x)) = f\left[\dfrac{5}{9}(x - 32)\right] = \dfrac{9}{5}\left[\dfrac{5}{9}(x - 32)\right] + 32 = x - 32 + 32 = x$

$g(f(x)) = g\left[\dfrac{9}{5}x + 32\right] = \dfrac{5}{9}\left[\dfrac{9}{5}x + 32 - 32\right] = x$

57. $T(\ell) = 2\pi\sqrt{\dfrac{\ell}{g}}, \; g \approx 32.2$

$T = 2\pi\sqrt{\dfrac{\ell}{g}}$

$\dfrac{T}{2\pi} = \sqrt{\dfrac{\ell}{g}}$

$\dfrac{T^2}{4\pi^2} = \dfrac{\ell}{g}$

$\dfrac{gT^2}{4\pi^2} = \ell$

$\ell = \dfrac{gT^2}{4\pi^2}$

$\ell(T) = \dfrac{gT^2}{4\pi^2}$

59. $f(x) = \dfrac{ax + b}{cx + d}$

$y = \dfrac{ax + b}{cx + d}$

$x = \dfrac{ay + b}{cy + d}$

$cxy + dx = ay + b$

$cxy - ay = -dx + b$

$y(cx - a) = -dx + b$

$y = \dfrac{-dy + b}{cx - a}$

$f^{-1}(x) = \dfrac{-dx + b}{cx - a}$

$f = f^{-1}$ if $\dfrac{ax + b}{cx + d} = \dfrac{-dx + b}{cx - a}$

This is true if $a = -d$

1. If $V = \pi r^2 h$ and $h = 2r$, then $V(r) = \pi r^2 (2r) = 2\pi r^3$

3. (a) If $p = \dfrac{-1}{6}x + 100$ and $R = xp$, then $R(x) = x\left[\dfrac{-1}{6}x + 100\right] = -\dfrac{1}{6}x^2 + 100x$

   (b)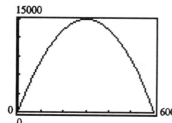

   (c) 300; \$15,000

   (d) $p = -\dfrac{1}{6} \cdot 300 + 100 = \$50$

5. (a) If $x = -5p + 100$ and $R = xp$, then $p = \dfrac{100 - x}{5}$ and $R(x) = x\left[\dfrac{100 - x}{5}\right] = \dfrac{-1}{5}x^2 + 20x$

   (b)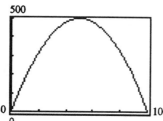

   (c) 50; \$500

   (d) $p = \dfrac{100 - 50}{5} = \dfrac{50}{5} = \$10$

7. (a) We know that width $= x$. In order to enclose a rectangular area, we need the perimeter, P. Let $\ell = $ length. $P = 2\ell + 2x = 400$.

   Then $\ell = \dfrac{400 - 2x}{2} = 200 - x$. The area of the rectangle as a function of the width x, represented by $A(x) = \ell x = (200 - x)x = -x^2 + 200x$.

   (b) $\{x \mid 0 < x < 200\}$

   (c)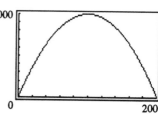

   The value of a is largest when $x = 100$ yards.

9. (a) Let $C = $ circumference, $r = $ radius. We know that $C = 2\pi r$ by definition. If a wire of length x is bent into a circle, then x is the circumference, so $C = x = 2\pi r$. Writing the circumference as a function of x we have $C(x) = x$.

   (b) We know that $C = x = 2\pi r$, so $r = \dfrac{x}{2\pi}$. By definition, the area of a circle is $A = \pi r^2$.

   Expressing the area of the circle as a function of x, we have $A(x) = \pi\left[\dfrac{x}{2\pi}\right]^2 = \dfrac{x^2}{4\pi}$.

11. By definition, a triangle has area $A = \frac{1}{2}bh$, b = base, h = height. Because a vertex of the triangle is at the origin, we know that $b = x$ and $h = y$. Expressing the area of the triangle as a function of x, we have $A(x) = \frac{1}{2}xy = \frac{1}{2}x(x^3) = \frac{1}{2}x^4$.

13. (a) The distance d from P to the origin is $d = \sqrt{x^2 + y^2}$. Since P is a point on the graph of $y = x^2 - 8$, we have
$$d(x) = \sqrt{x^2 + (x^2 - 8)^2} = \sqrt{x^4 - 15x^2 + 64}$$

(b) If $x = 0$, the distance d is $d(0) = \sqrt{64} = 8$

(c) If $x = 1$, the distance d is $d(1) = \sqrt{1 - 15 + 64} = \sqrt{50} = 5\sqrt{2} \approx 7.07$

(d)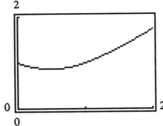

(e) d is smallest when x is 2.73.

15. (a) The distance d from P to the point $(1, 0)$ is $d = \sqrt{(x - 1)^2 + y^2}$

Since P is a point on the graph of $y = \sqrt{x}$, we have
$$d(x) = \sqrt{(x - 1)^2 + \left(\sqrt{x}\right)^2} = \sqrt{x^2 - x + 1}$$

(b)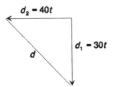

(c) d is smallest when x is 0.50.

17. We know that distance = (velocity)(time), $d = vt$. $d_1 = 25t$ and $d_2 = 40t$. By the Pythagorean Theorem,

$$d^2 = d_1^2 + d_2^2$$
$$d^2 = (30t)^2 + (40t)^2$$
$$d(t) = \sqrt{900t^2 + 1600t^2}$$
$$d(t) = \sqrt{2500t^2} = 50t$$

19. (a) By definition, Volume,
$V = $ (length)(width)(height)
length $= 24 - 2x$, width $= 24 - 2x$,
height $= x$
Therefore, $V(x) = x(24 - 2x)^2$

(b)

The volume is largest when x is about 4 inches.

21. (a) Volume $(V) = $ (length)(width)(height) $= 10$
length $= $ width $= x$, height $= h$,

so $10 = x^2 h$ and $h = \dfrac{10}{x^2}$

Area, $A = 2x^2 + 2xh + 2xh$

$A(x) = 2x^2 + 4x\left[\dfrac{10}{x^2}\right] = 2x^2 + \dfrac{40}{x}$

(b)

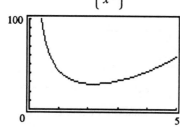

A is smallest when x is about 2.15 feet.

23. (a) $A = $ Area
$A(x) = xy = x(16 - x^2)$

(b) Domain of $A = \{x \mid 0 \le x \le 4\}$ because $x \ge 0$
and $16 - x^2 \ge 0$

(c)

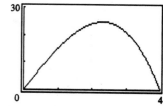

The area is largest for x about 2.31

25. $A = $ Area, $p = $ perimeter
(a) $A(x) = $ (length)(width) $= (2x)(2y) = 4x(4 - x^2)^{1/2}$
(b) $p(x) = 2$ length $+ 2$ width $= 2(2x) = 2(2y) = 4x + 4(4 - x^2)^{1/2}$

(c)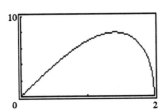

The area is largest for x about 1.41

(d)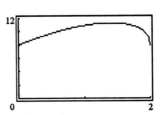

The perimeter is largest for x about 1.41

27. (a) C = Cost of the material, r = radius, h = height

$$500 = \pi r^2 h, \quad h = \frac{500}{\pi r^2}$$

$$C(r) = 6(2\pi r^2) + 4(2\pi rh)$$

$$= 12\pi r^2 + 8\pi r \left[\frac{500}{\pi r^2}\right]$$

$$= 12\pi r^2 + \frac{4000}{r}$$

(b)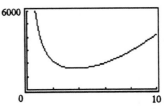

The cost is least for r about 3.75 cm.

29. (a) C = Circumference, A = Area, r = radius, x = side of square

$$C = 2\pi r = 10 - 4x; \quad r = \frac{5 - 2x}{\pi}$$

$$A(x) = x^2 + \pi r^2 = x^2 + \pi\left[\frac{5 - 2x}{\pi}\right]^2 = x^2 + \frac{25 - 20x + 4x^2}{\pi}$$

(b) Since all lengths must be positive, we have $x > 0$ and

$$10 - 4 > 0$$
$$-4x > -10$$
$$x < 2.5$$

Thus, domain
$$A = \{x \mid 0 < x < 2.5\}.$$

(c)

The area is smallest for x about 1.40 meters.

31. (a) A = Area, r = radius; diameter = $2r$

$$A(r) = (2r)r = 2r^2$$

(b) p = perimeter

$$p(r) = 2(2r) + 2r = 6r$$

33. Area of equilateral triangle $= \pi r^2 - \frac{\sqrt{3}}{4}x^2$

Area of equilateral triangle $= \frac{1}{2}x\sqrt{r^2 - \frac{x^2}{4}} = \frac{1}{3} \cdot \frac{\sqrt{3}}{4}x^2$

$$\sqrt{r^2 - \frac{x^2}{4}} = \frac{2\sqrt{3}}{3\,4}x = \frac{x}{2\sqrt{3}}$$

$$r^2 - \frac{x^2}{4} = \frac{x^2}{12}$$

$$r^2 = \frac{4x^2}{12} = \frac{x^2}{3}$$

$$\text{Area} = \frac{\pi x^2}{3} - \frac{\sqrt{3}}{4}x^2 = \left(\frac{\pi}{3} - \frac{\sqrt{3}}{4}\right)x^2$$

35. $$C = \begin{cases} 95 & \text{if} & x = 7 \\ 119 & \text{if} & 7 < x \le 8 \\ 143 & \text{if} & 8 < x \le 9 \\ 167 & \text{if} & 9 < x \le 10 \\ 190 & \text{if} & 10 < x \le 14 \end{cases}$$

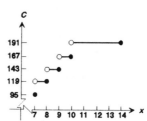

37. $r = $ radius, $h = $ height, $V = $ volume of a cone

$$\frac{r}{h} = \frac{4}{16}$$
$$16r = 4h$$
$$r = \frac{1}{4}h$$
$$V = \frac{1}{3}\pi r^2 h$$
$$V(h) = \frac{1}{3}\pi \left(\frac{h}{4}\right)^2 h$$
$$= \frac{1}{48}\pi h^3$$

39. (a) $Q = -37.6355P + 1571.7196$
    (b) $R = -37.6355P^2 + 1571.7196P$

    (c)

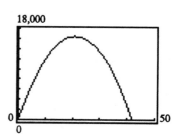

    (d) $20.88

    (e) 786; $16,409.39

2 Chapter Review

1. $f(4) = -5$
   $f(0) = 3$
   $f(x) = mx + b$

   $f(4): \ 4m + b = -5$ } Solve two equations
   $f(0): \ 0m + b = 3$ } in two unknowns.

   $\begin{aligned} 4m + b &= -5 \\ -b &= -3 \\ \hline 4m &= -8 \\ m &= -2 \end{aligned}$

   $-2(0) + b = 3$
   $b = 3$

   Hence, $f(x) = -2x + 3$

3. $f(x) = \dfrac{Ax + 5}{6x - 2}$ and $f(1) = 4$

   $f(1): \ \dfrac{A(1) + 5}{6(1) - 2} = 4$

   $\dfrac{A + 5}{4} = 4$

   $A + 5 = 16$

   $A = 11$

5. (a) *B, C,* and *D* pass the vertical line test and therefore are functions.
   (b) Only *D* passes the horizontal line test and therefore is one-to-one.

7. $f(x) = \dfrac{3x}{x^2 - 4}$

   (a) $f(-x) = \dfrac{-3x}{(-x)^2 - 4} = \dfrac{-3x}{x^2 - 4}$

   (b) $-f(x) = -\left[\dfrac{3x}{x^2 - 4}\right] = -\dfrac{3x}{x^2 - 4}$

   (c) $f(x + 2) = \dfrac{3(x + 2)}{(x + 2)^2 - 4} = \dfrac{3x + 6}{x^2 + 4x + 4 - 4} = \dfrac{3x + 6}{x^2 + 4x}$

   (d) $f(x - 2) = \dfrac{3(x - 2)}{(x - 2)^2 - 4} = \dfrac{3x - 6}{x^2 - 4x + 4 - 4} = \dfrac{3x - 6}{x^2 - 4x}$

9. $f(x) = \sqrt{x^2 - 4}$

   (a) $f(-x) = \sqrt{(-x)^2 - 4} = \sqrt{x^2 - 4}$

   (b) $-f(x) = -\sqrt{x^2 - 4}$

   (c) $f(x + 2) = \sqrt{(x + 2)^2 - 4} = \sqrt{x^2 + 4x + 4 - 4} = \sqrt{x^2 + 4x}$

   (d) $f(x - 2) = \sqrt{(x - 2)^2 - 4} = \sqrt{x^2 - 4x + 4 - 4} = \sqrt{x^2 - 4x}$

11. $f(x) = \dfrac{x^2 - 4}{x^2}$

   (a) $f(-x) = \dfrac{(-x)^2 - 4}{(-x)^2} = \dfrac{x^2 - 4}{x^2}$

   (b) $-f(x) = -\left[\dfrac{x^2 - 4}{x^2}\right] = \dfrac{4 - x^2}{x^2} = -\dfrac{x^2 - 4}{x^2}$

   (c) $f(x + 2) = \dfrac{(x + 2)^2 - 4}{(x + 2)^2} = \dfrac{x^2 + 4x + 4 - 4}{x^2 + 4x + 4} = \dfrac{x^2 + 4x}{x^2 + 4x + 4}$

   (d) $f(x - 2) = \dfrac{(x - 2)^2 - 4}{(x - 2)^2} = \dfrac{x^2 - 4x + 4 - 4}{x^2 - 4x + 4} = \dfrac{x^2 - 4x}{x^2 - 4x + 4}$

13. $f(x) = x^3 - 4x$
   if even: $\qquad f(-x) = f(x)$
   $$(-x)^3 - (-4x) = x^3 - 4x$$
   $$-x^3 + 4x \neq x^3 - 4x$$
   if odd: $\qquad f(-x) = -f(x)$
   $$(-x)^3 - (-4x) = -(x^3 - 4x)$$
   $$-x^3 + 4x = -x^3 + 4x$$
   Hence, function is odd.

15. $h(x) = \dfrac{1}{x^4} + \dfrac{1}{x^2} + 1$
   if even: $\qquad\qquad h(-x) = h(x)$
   $$\dfrac{1}{(-x)^4} + \dfrac{1}{(-x)^2} + 1 = \dfrac{1}{x^4} + \dfrac{1}{x^2} + 1$$
   $$\dfrac{1}{x^4} + \dfrac{1}{x^2} + 1 = \dfrac{1}{x^4} + \dfrac{1}{x^2} + 1$$
   Hence, $h(x)$ is even.

17. $G(x) = 1 - x + x^3$
    if even: $\qquad\qquad G(-x) = G(x)$
    $$1 - (-x) + (-x)^3 = 1 - x + x^3$$
    $$1 + x - x^3 \neq 1 - x + x^3$$
    if odd: $\qquad\qquad G(-x) = -G(x)$
    $$1 - (-x) + (-x)^3 = -(1 - x + x^3)$$
    $$1 + x - x^3 \neq -1 + x - x^3$$
    Hence, $G(x)$ is neither even nor odd since
    $G(-x) \neq G(x)$ and $G(-x) \neq -G(x)$.

19. $f(x) = \dfrac{x}{x^2 - 9}$
    The domain is the set of all values x
    such that
    $$x^2 - 9 \neq 0$$
    $$(x - 3)(x + 3) \neq 0$$
    $$x \neq 3, -3$$
    The domain is $\{x \mid x \neq -3, x \neq 3\}$.

21. $f(x) = \sqrt{2 - x}$
    The domain consists of all values such that
    $2 - x \geq 0 \ x \leq 2$
    The domain is $\{x \mid x \leq 2\}$ or $(-\infty, 2]$.

23. $h(x) = \dfrac{\sqrt{x}}{|x|}$
    The domain is $\{x \mid x > 0\}$ or $(0, \infty)$.

25. $f(x) = \dfrac{x}{x^2 + 2x - 3}$
    The domain consists of all values x such that
    $$x^2 + 2x - 3 \neq 0$$
    $$(x + 3)(x - 1) \neq 0$$
    $$x \neq -3, 1$$
    The domain is $\{x \mid x \neq -3, x \neq 1\}$.

27. $G(x) = \begin{cases} |x| & \text{if } -1 \leq x \leq 1 \\ \dfrac{1}{x} & \text{if } x > 1 \end{cases}$
    The domain is $\{x \mid x \geq -1\}$ or $[-1, \infty)$.

29. $f(x) = \begin{cases} \dfrac{1}{x - 2} & \text{if } x > 2 \\ 0 & \text{if } x = 2 \\ 3x & \text{if } 0 \leq x \leq 2 \end{cases}$
    The domain is $\{x \mid x \geq 0\}$ or $[0, \infty)$.

31. $F(x) = |x| - 4$
    (a) The domain is all real numbers.
    (b) x-intercept(s): $\qquad$ y-intercept:
    $\quad 0 = |x| - 4 \qquad\quad y = |0| - 4$
    $\quad 4 = |x| \qquad\qquad\quad y = -4$
    $\quad x = -4, 4$
    The intercepts are $(-4, 0), (4, 0), (0, -4)$.

    (c)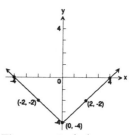

    (d) The range is $\{y \mid -4 \leq y < \infty\}$.

33. $g(x) = -|x|$
    (a) The domain is all real numbers.
    (b) x-intercept(s): $\qquad$ y-intercept:
    $\quad 0 = -|x| \qquad\qquad y = -|0|$
    $\quad x = 0 \qquad\qquad\qquad y = 0$
    The intercept is $(0, 0)$.

    (c)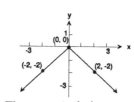

    (d) The range is $\{y \mid -\infty < y \leq 0\}$.

35. $h(x) = \sqrt{x - 1}$

    (a)   The domain is $\{x \mid 1 \leq x < \infty\}$.

    (b) x-intercept(s): y-intercept:

$$0 = \sqrt{x - 1} \qquad y = \sqrt{0 - 1}$$
$$0 = x - 1 \qquad y = \sqrt{-1}$$
$$1 = x \qquad \text{No solution.}$$

    The intercept is $(1, 0)$.

    (c)

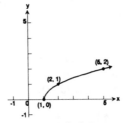

    (d)   The range is $\{y \mid 0 \leq y < \infty\}$.

37. $f(x) = \sqrt{1 - x}$

    (a)   The domain is $\{x \mid -\infty < x \leq 1\}$.

    (b) x-intercept(s): y-intercept:

$$0 = \sqrt{1 - x} \qquad y = \sqrt{1 - 0}$$
$$0 = 1 - x \qquad y = 1$$
$$x = 1$$

    The intercepts are $(1, 0)$, $(0, 1)$.

    (c)

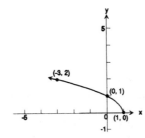

    (d)   The range is $\{y \mid 0 \leq y < \infty\}$.

39. $F(x) = \begin{cases} x^2 + 4 & \text{if } x < 0 \\ 4 - x^2 & \text{if } x \geq 0 \end{cases}$

    (a)   The domain is all real numbers.

    (b) x-intercept(s): y-intercept:

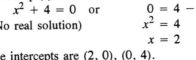

$$x^2 + 4 = 0 \quad \text{or} \qquad 0 = 4 - x^2 \qquad\qquad y = 4 - 0^2$$
$$\text{(No real solution)} \qquad x^2 = 4 \qquad\qquad\qquad y = 4$$
$$x = 2$$

    The intercepts are $(2, 0)$, $(0, 4)$.

    (c)   (d)   The range is all real numbers.

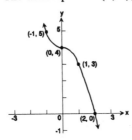

41. $h(x) = (x - 1)^2 + 2$

    (a)   The domain is all real numbers.

    (b) x-intercept(s): y-intercept(s):

$$0 = (x - 1)^2 + 2 \qquad\qquad y = (0 - 1)^2 + 2$$
$$0 = x^2 - 2x + 1 + 2 \qquad\quad y = 1 + 2$$
$$0 = x^2 - 2x + 3 \qquad\qquad\quad y = 3$$

           No real solution.

           The intercept is $(0, 3)$.

    (c)   (d)   The range is $\{y \mid 2 \le y < \infty\}$.

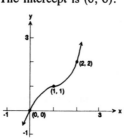

43. $g(x) = (x - 1)^3 + 1$

    (a)   The domain is all real numbers.

    (b) x-intercept(s): y-intercept(s):

$$0 = (x - 1)^3 + 1 \qquad\qquad y = (0 - 1)^3 + 1$$
$$0 = x^3 - x^2 + x - 1 + 1 \qquad y = -1 + 1$$
$$0 = x^3 - x^2 + x \qquad\qquad\quad y = 0$$
$$0 = x(x^2 - x + 1)$$
$$x = 0$$

           The intercept is $(0, 0)$.

    (c)   (d)   The range is all real numbers.

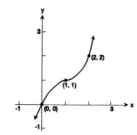

45. $f(x) = \begin{cases} 2\sqrt{x} & \text{if} \quad x \ge 4 \\ x & \text{if} \quad 0 < x < 4 \end{cases}$

    (a)   The domain is $\{x \mid 0 < x < \infty\}$.   (c)

    (b) x-intercept(s): y-intercept:

           None   None

           There are no intercepts.

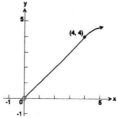

                                          (d)   The range is $\{x \mid 0 < y < \infty\}$.

47. $g(x) = \dfrac{1}{x - 1} + 1$

(a) The domain is $\{x \mid x \neq 1\}$.

(b) x-intercept(s): y-intercept(s):

$$0 = \frac{1}{x - 1} + 1$$
$$-1 = \frac{1}{x - 1} \qquad y = \frac{1}{0 - 1} + 1$$
$$-x + 1 = 1 \qquad y = -1 + 1$$
$$x = 0 \qquad\quad y = 0$$

The intercept is (0, 0).

(c)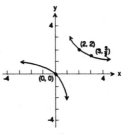

(d) Because $\dfrac{1}{x - 1}$ can never be zero, the range is $\{y \mid y \neq 1\}$.

49. $h(x) = [\![-x]\!]$

(a) The domain is all real numbers.

(b) x-intercept(s): y-intercept:

$$0 = [\![-x]\!] \qquad y = [\![-x]\!]$$
$$-1 < x \le 0 \qquad y = 0$$

(c)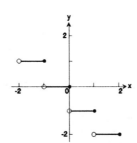

(d) The range is set of all integers.

51.
$$f(x) = \frac{2x + 3}{5x - 2}$$
$$y = \frac{2x + 3}{5x - 2}$$
$$x = \frac{2y + 3}{5y - 2}$$
$$5xy - 2x = 2y + 3$$
$$5xy - 2y = 2x + 3$$
$$y(5x - 2) = 2x + 3$$
$$y = \frac{2x + 3}{5x - 2}$$
$$f^{-1}(x) = \frac{2x + 3}{5x - 2}$$

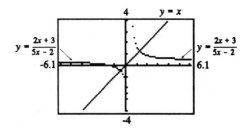

Verify: $f(f^{-1}(x)) = f\left[\dfrac{2x + 3}{5x - 2}\right] = \dfrac{2\left[\dfrac{2x + 3}{5x - 2}\right] + 3}{\left[\dfrac{2x + 3}{5x - 2}\right] - 2} = \dfrac{\dfrac{4x + 6 + 3x - 6}{5x - 2}}{\dfrac{2x + 3 - 2x + 4}{5x - 2}} = \dfrac{7x}{7} = x$

$f^{-1}(f(x)) = f^{-1}\left[\dfrac{x + 3}{5x - 2}\right] = \dfrac{2\left[\dfrac{2x + 3}{5x - 2}\right] + 3}{\left[\dfrac{2x + 3}{5x - 2}\right] - 2} = x$

Domain of f = range of f^{-1} = all real numbers except $\dfrac{2}{5}$

Range of f = domain of f^{-1} = all real numbers except $\dfrac{2}{5}$

53.

$$f(x) = \frac{1}{x - 1}$$

$$y = \frac{1}{x - 2}$$

$$x = \frac{1}{y + 1}$$

$$xy - x = 1$$

$$xy = x + 1$$

$$y = x + 1$$

$$f^{-1} = \frac{x + 1}{x}$$

Verify:

$$f(f^{-1}(x)) = \frac{1}{\dfrac{x + 1}{x} + 1} = \frac{1}{\dfrac{x + 1 - x}{x}} = x$$

$$f^{-1}(f(x)) = \frac{\dfrac{1}{x - 1} + 1}{\dfrac{1}{x - 1}} = \frac{\dfrac{1 + x - 1}{x - 1}}{\dfrac{1}{x - 1}} = x$$

Domain of f = range of f^{-1} = all real numbers except 1
Range of f = domain of f^{-1} = all real numbers except 0

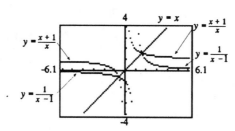

55.

$$f(x) = \frac{3}{x^{1/3}}$$

$$y = \frac{3}{x^{1/3}}$$

$$x = \frac{3}{y^{1/3}}$$

$$y^{1/3}x = 3$$

$$y^{1/3} = \frac{3}{x}$$

$$y = \frac{27}{x^3}$$

$$f^{-1}(x) = \frac{27}{x^3}$$

Verify:

$$f(f^{-1}(x)) = \frac{3}{\left[\dfrac{27}{x^3}\right]^{1/3}} = \frac{3}{\dfrac{3}{x}} = x$$

$$f^{-1}(f(x)) = \frac{27}{\left[\dfrac{3}{x^{1/3}}\right]^3} = \frac{27}{\dfrac{27}{x}} = x$$

Domain of f = range of f^{-1} = all real numbers except 0
Range of f = domain of f^{-1} = all real numbers except 0

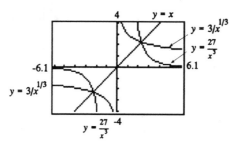

57. $f(x) = 3x - 5,\ g(x) = 1 - 2x^2$

(a) $(f \circ g)(2) = f(g(2)) = f(1 - 2(2)^2) = f(1 - 8) = f(-7) = 3(-7) - 5 = -26$

(b) $(g \circ f)(-2) = g(f(-2)) = g(3(-2) - 5) = g(-11) = 1 - 2(-11)^2 = 1 - 242 = -241$

(c) $(f \circ f)(4) = f(f(4)) = f(3(4) - 5) = f(7) = 3(7) - 5 = 16$

(d) $(g \circ g)(-1) = g(g(-1)) = g(1 - 2(-1)^2) = g(1 - 2) = g(-1) = 1 - 2(-1)^2 = 1 - 2$
$$= -1$$

59. $f(x) = \sqrt{x + 2}$, $g(x) = 2x^2 + 1$

   (a) $(f \circ g)(2) = f(g(2)) = f(9) = \sqrt{11}$

   (b) $(g \circ f)(-2) = g(f(-2)) = g(0) = 1$

   (c) $(f \circ f)(4) = f(f(4)) = f(\sqrt{6}) = \sqrt{\sqrt{6} + 2}$

   (d) $(g \circ g)(-1) = g(g(-1)) = g(3) = 19$

61. $f(x) = \dfrac{1}{x^2 + 4}$, $g(x) = 3x - 2$

   (a) $(f \circ g)(2) = f(g(2)) = f(4) = \dfrac{1}{4^2 + 4} = \dfrac{1}{20}$

   (b) $(g \circ f)(-2) = g(f(-2)) = g\left[\dfrac{1}{8}\right] = \dfrac{3}{8} - 2 = \dfrac{3}{8} - \dfrac{16}{8} = -\dfrac{13}{8}$

   (c) $(f \circ f)(4) = f(f(4)) = f\left[\dfrac{1}{20}\right] = \dfrac{1}{\dfrac{1}{400} + 4} = \dfrac{1}{\dfrac{1601}{400}} = \dfrac{400}{1601}$

   (d) $(g \circ g)(-1) = g(g(-1)) = g(-5) = -17$

63. $f(x) = \dfrac{2 - x}{x}$, $g(x) = 3x + 1$

$(f \circ g)(x) = f(g(x)) = f(3x + 1) = \dfrac{2 - (3x + 1)}{3x + 1} = \dfrac{-3x + 1}{3x + 1}$

$(g \circ f)(x) = g(f(x)) = g\left[\dfrac{2 - x}{x}\right] = 3\left[\dfrac{2 - x}{x}\right] + 1 = \dfrac{6 - 3x}{x} + \dfrac{x}{x} = \dfrac{6 - 2x}{x}$

$(f \circ f)(x) = f(f(x)) = f\left[\dfrac{2 - x}{x}\right] = \dfrac{2 - \left[\dfrac{2 - x}{x}\right]}{\dfrac{2 - x}{x}} = \dfrac{\dfrac{2x - 2 + x}{x}}{\dfrac{2 - x}{x}}$

$= \dfrac{\dfrac{3x - 2}{x}}{\dfrac{2 - x}{x}} = \dfrac{3x - 2}{x} \cdot \dfrac{x}{2 - x} = \dfrac{3x - 2}{2 - x}$

$(g \circ g)(x) = g(g(x)) = g(3x + 1) = 3(3x + 1) + 1 = 9x + 3 + 1 = 9x + 4$

65. $f(x) = 3x^2 + x + 1$, $g(x) = |3x|$

$(f \circ g)(x) = f(g(x)) = f(|3x|) = 3(|3x|)^2 + |3x| + 1 = 27x^2 + 3|x| + 1$

$(g \circ f)(x) = g(f(x)) = g(3x^3 + x + 1) = |3(3x^2 + x + 1)| = |9x^2 + 3x + 3|$

$= 3|3x^2 + x + 1|$

$(f \circ f)(x) = f(f(x)) = f(3x^2 + x + 1) = 3(3x^2 + x + 1)^2 + (3x^2 + x + 1) + 1$

$= 3(3x^2 + x + 1)^2 + 3x^2 + x + 2$

$(g \circ g)(x) = g(g(x)) = g(|3x|) = |3|3x|| = 9|x|$

67. $f(x) = \dfrac{x + 1}{x - 1}$, $g(x) = \dfrac{1}{x}$

$(f \circ g)(x) = f(g(x)) = f\left[\dfrac{1}{x}\right] = \dfrac{\dfrac{1}{x} + 1}{\dfrac{1}{x} - 1} = \dfrac{\dfrac{1 + x}{x}}{\dfrac{1 - x}{x}} = \dfrac{1 + x}{x} \cdot \dfrac{x}{1 - x} = \dfrac{1 + x}{1 - x}$

$$(g \circ f)(x) = g(f(x)) = g\left(\frac{x+1}{x-1}\right) = \frac{1}{\dfrac{x+1}{x-1}} = \frac{x-1}{x+1}$$

$$(f \circ f)(x) = f(f(x)) = f\left(\frac{x+1}{x-1}\right) = \frac{\dfrac{x+1}{x-1}+1}{\dfrac{x+1}{x-1}-1} = \frac{\dfrac{x+1+x-1}{x-1}}{\dfrac{x+1-(x-1)}{x-1}} = \frac{\dfrac{2x}{x-1}}{\dfrac{2}{x-1}}$$

$$= \frac{2x}{x-1} \cdot \frac{x-1}{2} = x$$

$$(g \circ g)(x) = g(g(x)) = g\left(\frac{1}{x}\right) = \frac{1}{\dfrac{1}{x}} = x$$

69. (a) $y = f(-x)$
Reflect about the y-axis.

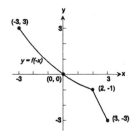

(b) $y = -f(x)$
Reflect about the x-axis.

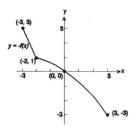

(c) $y = f(x + 2)$
Shift left 2 units.

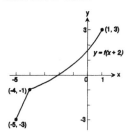

(d) $y = f(x) + 2$
Shift up 2 units.

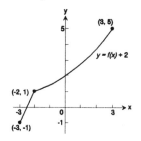

(e) $y = f(2 - x)$
$y = f(-(x - 2))$
Reflect about the y-axis and shift right 2 units.

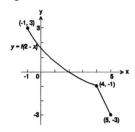

(f) $y = f^{-1}(x)$
Reflect about the line $y = x$.

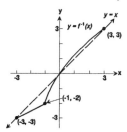

71. $T = T(h)$, $h = 0$ when $T = 30°$ and $h = 10,000$ when $T = 5°$

$T(h) = mh + b$ (linear function)

$$T(0): \qquad\qquad m(0) + b = 30° \quad\Bigg\}$$
$$T(10,000): \quad m(10,000) + b = 5°$$

$$b = \quad 30°$$
$$10,000m + 30 = \quad 5$$
$$10,000m = \quad -25$$
$$m = \quad \dfrac{-25}{10,000}$$
$$m = -0.0025$$

Hence, $T(h) = -0.0025h + 30$

73. Strength $= s = kxd^3$ where $x = $ width, $d = $ depth, and k is the constant of proportionality.

If the radius $= 3$ ft., then the diameter $= 6$ ft. and is a diagonal of the rectangle.

By the Pythagorean Theorem, we have $x^2 + d^2 = 6^2 \Rightarrow d = \sqrt{36 - x^2}$. Hence,

$$s(x) = kxd^3 = kx\left(\sqrt{36 - x^2}\right)^3 = kx(36 - x^2)^{3/2}.$$

Since all lengths must be positive, we have $x > 0$ and

$$36 - x^2 > 0$$
$$(6 - x)(6 + x) > 0$$
$$6 - x > 0 \text{ and } 6 + x > 0$$
$$x < 6 \text{ and } x > -6$$

Thus, the domain of $s = \{x \mid 0 < x < 6\}$.

EQUATIONS AND INEQUALITIES

3.1 Solving Equations Using a Graphing Utility

1. 0.42

3. 2.23

5. 1.25

7. $f(x) = 8x^4 - 2x^2 + 5x - 1;$ [0, 1]

   $f(0) = -1$

   $f(1) = 10$

   Since $f(0) < 0$ and $f(1) > 0$, there must be a zero in (0, 1).

9. $f(x) = 2x^3 + 6x^2 - 8x + 2;$ [-5, -4]

   $f(-5) = -58$

   $f(-4) = 2$

   Since $f(-5) < 0$ and $f(-4) > 0$, there must be a zero in (-5, -4).

11. $f(x) = x^5 - x^4 + 7x^3 - 7x^2 - 18x + 18;$

   [1.4, 1.5]

   $f(1.4) = -0.17536$

   $f(1.5) = 1.40625$

   Since $f(1.4) < 0$ and $f(1.5) > 0$, there must be a zero in (1.4, 1.5).

13.

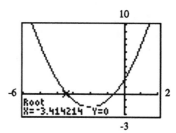

The smaller solution is -3.41 correct to two decimal places.

15.

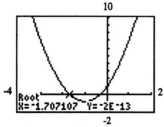

The smaller solution is -1.70 correct to two decimal places.

17.

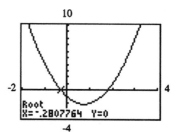

The smaller solution is -0.28 correct to two decimal places.

19.

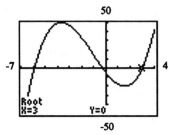

The positive solution is 3.00 correct to two decimal places.

21.

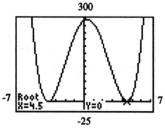

The positive solution is 4.50 correct to two decimal places.

23.

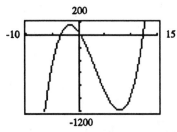

The two positive solutions are 0.31 and 12.30 correct to two decimal places.

25.

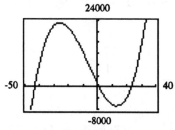

The two positive solutions are 1.00 and 23.00 correct to two decimal places.

3.2 Linear and Quadratic Equations

Solving mentally in Problems 1−7:

1.   For $3x = 21$
    $3(7) = 21$
    so that $x = 7$

3.   For $5x + 15 = 0$
    we want $-15 + 15 = 0$
    or $5(-3) + 15 = 0$
    so that $x = -3$

5.   For $2x - 3 = 5$
    we want $8 - 3 = 5$
    or $2(4) - 3 = 5$
    so that $x = 4$

7.   For $\dfrac{1}{3}x = \dfrac{5}{12}$
    $\dfrac{1}{3} \cdot \dfrac{5}{4} = \dfrac{5}{12}$
    so that $x = \dfrac{5}{4}$

9. $3x + 2 = x + 6$
    $(3x + 2) - 2 = (x + 6) - 2$
    $3x = x + 4$
    $3x - x = (x + 4) - x$
    $2x = 4$
    $\dfrac{2x}{2} = \dfrac{4}{2}$
    $x = 2$

Check:
    $3x + 2 = x - 6$
    $3(-4) + 2 \overset{?}{=} -4 - 6$
    $-12 + 2 \overset{?}{=} -10$
    $-10 = -10$
The solution checks.

11.

$$2t - 6 = 3 - t$$
$$(2t - 6) + 6 = (3 - t) + 6$$
$$2t = 9 - t$$
$$2t + t = (9 - t) + t$$
$$3t = 9$$
$$\frac{3t}{3} = \frac{9}{3}$$
$$t = 3$$

Check:
$$2t - 6 = 3 - t$$
$$2(3) - 6 \overset{?}{=} 3 - 3$$
$$6 - 6 \overset{?}{=} 0$$
$$0 = 0$$
The solution checks.

13.

$$6 - x = 2x + 9$$
$$(6 - x) - 6 = (2x + 9) - 6$$
$$-x = 2x + 3$$
$$-x - 2x = 2x + 3 - 2x$$
$$-3x = 3$$
$$\frac{-3x}{-3} = \frac{3}{-3}$$
$$x = -1$$

Check:
$$6 - x = 2x + 9$$
$$-(-1) \overset{?}{=} 2(-1) + 9$$
$$6 + 1 \overset{?}{=} -2 + 9$$
$$7 = 7$$

15.

$$3 + 2n = 5n + 7$$
$$(3 + 2n) - 3 = (5n + 7) - 3$$
$$2n = 5n + 4$$
$$2n - 5n = 5n + 4 - 5n$$
$$-3n = 4$$
$$\frac{-3n}{-3} = \frac{4}{-3}$$
$$n = -\frac{4}{3}$$

Check:
$$3 + 2n = 5n + 7$$
$$3 + 2\left[\frac{-4}{3}\right] \overset{?}{=} 5\left[\frac{-4}{3}\right] + 7$$
$$3 - \frac{8}{3} \overset{?}{=} \frac{-20}{3} + 7$$
$$\frac{9 - 8}{3} \overset{?}{=} \frac{-20 + 21}{3}$$
$$\frac{1}{3} = \frac{1}{3}$$

17.

$$2(3 + 2x) = 3(x - 4)$$
$$6 + 4x = 3x - 12$$
$$(6 + 4x) - 6 = (3x - 12) - 6$$
$$4x = 3x - 18$$
$$4x - 3x = 3x - 18 - 3x$$
$$x = -18$$

Check:
$$2(3 + 2x) = 3(x - 4)$$
$$2(3 + 2(-18)) \overset{?}{=} 3(-18 - 4)$$
$$2(3 - 36) \overset{?}{=} 3(-22)$$
$$2(-33) \overset{?}{=} -66$$
$$-66 = -66$$

19.

$$8x - (2x + 1) = 3x - 10$$
$$8x - 2x - 1 = 3x - 10$$
$$6x - 1 = 3x - 10$$
$$(6x - 1) + 1 = (3x - 10) + 1$$
$$6x = 3x - 9$$
$$6x - 3x = (3x - 9) - 3x$$
$$3x = -9$$
$$\frac{3x}{3} = \frac{-9}{3}$$
$$x = -3$$

Check:
$$8x - (2x + 1) = 3x - 10$$
$$8(-3) - (2(-3) + 1) \overset{?}{=} 3(-3) - 10$$
$$-24 - (-6 + 1) \overset{?}{=} -9 - 10$$
$$-24 - (-5) \overset{?}{=} -19$$
$$-24 + 5 \overset{?}{=} -19$$
$$-19 = -19$$

21.

$$\frac{3}{2}x + 2 = \frac{1}{2} - \frac{1}{2}x$$

$$2\left[\frac{3}{2}x + 2\right] = 2\left[\frac{1}{2} - \frac{1}{2}x\right]$$

$$3x + 4 = 1 - x$$

$$(3x + 6) - 4 = (1 - x) - 4$$

$$3x = -3 - x$$

$$3x + x = -3 - x + x$$

$$4x = -3$$

$$\frac{4x}{4} = \frac{-3}{4}$$

$$x = -\frac{3}{4}$$

Check:

$$\frac{3}{2}x + 2 = \frac{1}{2} - \frac{1}{2}x$$

$$\frac{3}{2}\left[-\frac{3}{4}\right] + 2 \stackrel{?}{=} \frac{1}{2} - \frac{1}{2}\left[\frac{-3}{4}\right]$$

$$\frac{-9}{8} + 2 \stackrel{?}{=} \frac{1}{2} + \frac{3}{8}$$

$$\frac{-9 + 16}{8} \stackrel{?}{=} \frac{4 + 3}{8}$$

$$\frac{7}{8} = \frac{7}{8}$$

23.

$$\frac{1}{2}x - 5 = \frac{3}{4}x$$

$$4\left[\frac{1}{2}x - 5\right] = 4\left[\frac{3}{4}x\right]$$

$$2x - 20 = 3x$$

$$(2x - 20) + 20 = 3x + 20$$

$$2x = 3x + 20$$

$$2x - 3x = 3x + 20 - 3x$$

$$-x = 20$$

$$\frac{-x}{-1} = \frac{20}{-1}$$

$$x = -20$$

Check:

$$\frac{1}{2}x - 5 = \frac{3}{4}x$$

$$\frac{1}{2}(-20) - 5 = \frac{3}{4}(-20)$$

$$-10 - 5 = -15$$

$$-15 = -15$$

25.

$$\frac{2}{3}p = \frac{1}{2}p + \frac{1}{3}$$

$$6\left[\frac{2}{3}p\right] = 6\left[\frac{1}{2}p + \frac{1}{3}\right]$$

$$4p = 3p + 2$$

$$4p - 3p = (3p + 2) - 3p$$

$$p = 2$$

Check:

$$2p = \frac{1}{2}p + \frac{1}{3}$$

$$\frac{2}{3}(2) \stackrel{?}{=} \frac{1}{2}(2) + \frac{1}{3}$$

$$\frac{4}{3} \stackrel{?}{=} 1 + \frac{1}{3}$$

$$\frac{4}{3} = \frac{4}{3}$$

27.

$$0.9t = 0.4 + 0.1t$$

$$0.9t - 0.1t = (0.4 + 0.1t) - 0.1t$$

$$0.8t = 0.4$$

$$\frac{0.8t}{0.8} = \frac{0.4}{0.8}$$

$$t = 0.5$$

Check:

$$0.9t = 0.4 + 0.1t$$

$$0.9(0.5) \stackrel{?}{=} 0.4 + 0.01(0.5)$$

$$0.45 \stackrel{?}{=} 0.4 + 0.05$$

$$0.45 = 0.45$$

29.

$$\frac{x+1}{3} + \frac{x+2}{7} = 5$$

$$21\left[\frac{x+1}{3} + \frac{x+2}{7}\right] = 21[5]$$

$$7(x+1) + 3(x+2) = 105$$

$$7x + 7 + 3x + 6 = 105$$

$$10x + 13 = 105$$

$$(10x + 13) - 13 = 105 - 13$$

$$10x = 92$$

$$\frac{10x}{10} = \frac{92}{10}$$

$$x = \frac{46}{5}$$

Check:

$$\frac{x+1}{3} + \frac{x+2}{7} = 5$$

$$\frac{\frac{46}{5} + 1}{3} + \frac{\frac{46}{5} + 2}{7} \overset{?}{=} 5$$

$$\frac{\frac{46+5}{5}}{3} + \frac{\frac{46+10}{5}}{7} \overset{?}{=} 5$$

$$\frac{51}{15} + \frac{56}{35} \overset{?}{=} 5$$

$$\frac{357 + 168}{105} \overset{?}{=} 5$$

$$\frac{525}{105} \overset{?}{=} 5$$

$$5 = 5$$

31. The domain is $\{y \mid y \neq 0\}$

$$\frac{2}{y} + \frac{4}{y} = 3$$

$$y\left[\frac{2}{y} + \frac{4}{y}\right] = y[3]$$

$$2 + 4 = 3y$$

$$6 = 3y$$

$$\frac{6}{3} = \frac{3y}{3}$$

$$2 = y$$

$$y = 2$$

Check:

$$\frac{2}{y} + \frac{4}{y} = 3$$

$$\frac{2}{2} + \frac{4}{2} \overset{?}{=} 3$$

$$1 + 2 \overset{?}{=} 3$$

$$3 = 3$$

33.

$$\frac{1}{2} + \frac{2}{x} = \frac{3}{5}$$

$$10x\left[\frac{1}{2} + \frac{2}{x}\right] = \frac{3}{5}(10x)$$

$$5x + 20 = 6x$$

$$5x + 20 - 20 = 6x - 20$$

$$5x = 6x - 20$$

$$5x - 6x = 6x - 20 - 6x$$

$$\frac{-x}{-1} = \frac{-20}{-1}$$

$$x = 20$$

Check:

$$\frac{1}{2} + \frac{2}{x} = \frac{3}{5}$$

$$\frac{1}{2} + \frac{2}{20} = \frac{3}{5}$$

$$\frac{1}{2} + \frac{1}{10} = \frac{3}{5}$$

$$10\left[\frac{1}{2} + \frac{1}{10}\right] = \frac{3}{5}(10)$$

$$5 + 1 = 6$$

$$6 = 6$$

35.

$$x^2 = 9x$$

$$x^2 - 9x = 0$$

$$x(x - 9) = 0$$

$$x = 0 \quad \text{or} \quad x - 9 = 0$$

$$x = 9$$

$$\{0, 9\}$$

37.

$$x^2 - 25 = 0$$

$$(x + 5)(x - 5) = 0$$

$$x + 5 = 0 \quad \text{or} \quad x - 5 = 0$$

$$x = -5 \qquad x = 5$$

$$\{-5, 5\}$$

39.
$$z^2 + z - 12 = 0$$
$$(z + 4)(z - 3) = 0$$
$$z + 4 = 0 \quad \text{or} \quad z - 3 = 0$$
$$z = -4 \qquad\qquad z = 3$$
$$\{-4, 3\}$$

41.
$$2x^2 - 5x - 3 = 0$$
$$(2x + 1)(x - 3) = 0$$
$$2x + 1 = 0 \quad \text{or} \quad x - 3 = 0$$
$$x = -\frac{1}{2} \qquad\qquad x = 3$$
$$\left\{-\frac{1}{2}, 3\right\}$$

43.
$$3t^2 - 48 = 0$$
$$3(t^2 - 16) = 0$$
$$3(t + 4)(t - 4) = 0$$
$$t + 4 = 0 \quad \text{or} \quad t - 4 = 0$$
$$t = -4 \qquad\qquad t = 4$$
$$\{-4, 4\}$$

45.
$$x(x - 7) + 12 = 0$$
$$x^2 - 7x + 12 = 0$$
$$(x - 3)(x - 4) = 0$$
$$x - 3 = 0 \quad \text{or} \quad x - 4 = 0$$
$$x = 3 \qquad\qquad x = 4$$
$$\{3, 4\}$$

47.
$$4x^2 + 9 = 12x$$
$$4x^2 - 12x + 9 = 0$$
$$(2x - 3)(2x - 3) = 0$$
$$2x - 3 = 0$$
$$x = \frac{3}{2}$$
$$\left\{\frac{3}{2}\right\}$$

49.
$$6(p^2 - 1) = 5p$$
$$6p^2 - 6 = 5p$$
$$6p^2 - 5p - 6 = 0$$
$$(2p - 3)(3p + 2) = 0$$
$$2p - 3 = 0 \quad \text{or} \quad 3p + 2 = 0$$
$$p = \frac{3}{2} \qquad\qquad p = -\frac{2}{3}$$
$$\left\{-\frac{2}{3}, \frac{3}{2}\right\}$$

51.
$$6x - 5 = \frac{6}{x}$$
$$6x^2 - 5x = 6$$
$$6x^2 - 5x - 6 = 0$$
$$(2x - 3)(3x + 2) = 0$$
$$2x - 3 = 0 \quad \text{or} \quad 3x + 2 = 0$$
$$x = \frac{3}{2} \qquad\qquad x = -\frac{2}{3}$$
$$\left\{-\frac{2}{3}, \frac{3}{2}\right\}$$

53.
$$\frac{4(x - 2)}{x - 3} + \frac{3}{x} = \frac{-3}{x(x - 3)}$$
$$4x(x - 2) + 3(x - 3) = -3$$
$$4x^2 - 8x + 3x - 9 = -3$$
$$4x^2 - 5x - 6 = 0$$
$$(4x + 3)(x - 2) = 0$$
$$4x + 3 = 0 \quad \text{or} \quad x - 2 = 0$$
$$x = -\frac{3}{4} \qquad\qquad x = 2$$
$$\left\{-\frac{3}{4}, 2\right\}$$

55.
$$(x + 7)(x - 1) = (x + 1)^2$$
$$x^2 + 6x - 7 = x^2 + 2x + 1$$
$$(x^2 + 6x - 7) - x^2 = (x^2 + 2x + 1) - x^2$$
$$6x - 7 = 2x + 1$$
$$(6x - 7) + 7 = (2x + 1) + 7$$
$$6x = 2x + 8$$
$$6x - 2x = (2x + 8) - 2x$$
$$4x = 8$$
$$\frac{4x}{4} = \frac{8}{4}$$
$$x = 2$$

57.
$$x(2x - 3) = (2x + 1)(x - 4)$$
$$2x^2 - 3x = 2x^2 - 7x - 4$$
$$(2x - 3x) - 2x^2 = (2x^2 - 7x - 4) - 2x^2$$
$$-3x = -7x - 4$$
$$-3x + 7x = (-7x - 4) + 7x$$
$$4x = -4$$
$$\frac{4x}{4} = \frac{-4}{4}$$
$$x = -1$$

59.
$$z(z^2 + 1) = 3 + z^3$$
$$z^3 + z = 3 + z^3$$
$$(z^3 + z) - z^3 = (3 + z^3) - z^3$$
$$z = 3$$

61.
$$\frac{x}{x - 3} + 3 = \frac{3}{x - 3}$$
Note that $x - 3$ cannot equal zero so $x = 3$ is **not** in the domain of the variable.
$$(x - 3)\left[\frac{x}{x - 3} + 3\right] = \frac{3}{x - 3}(x - 3)$$
$$x + 3(x - 3) = 3$$
$$x + 3x - 9 = 3$$
$$4x = 12$$
$$x = 3$$
But $x = 3$ is not in the domain of the variable. Hence, the equation has no solution.

63.
$$x^2 = 4x$$
$$x^2 - 4x = 0$$
$$x(x - 4) = 0$$
$$x = 0, x = 4$$
The solution set is $\{0, 4\}$.

65.
$$t^3 - 9t^2 = 0$$
$$t^2(t - 9) = 0$$
$$t^2 = 0, \quad t - 9 = 0$$
$$t = 0, \qquad t = 9$$
The solution set is $\{0, 9\}$.

67.
$$3.2x + \frac{21.3}{65.871} = 19.23$$
$$3.2x + \frac{21.3}{65.871} - \frac{21.3}{65.871} = 19.23 - \frac{21.3}{65.871}$$
$$3.2x = 19.23 - \frac{21.3}{65.871}$$
$$\frac{3.2x}{3.2} = \frac{19.23 - \frac{21.3}{65.871}}{3.2}$$
$$x = 5.9083252$$
$$x \approx 5.90$$

69.
$$14.72 - 21.58x = \frac{18}{2.11}x + 2.4$$
$$14.72 - 21.58x - 2.4 = \frac{18}{2.11}x + 2.4 - 2.4$$
$$14.72 - 21.58x - 2.4 + 21.58x = \frac{18}{2.11}x + 21.58x$$
$$14.72 - 2.4 = x\left[\frac{18}{2.11} + 21.58\right]$$
$$\frac{14.72 - 2.4}{\frac{18}{2.11} + 21.58} = x$$
$$0.4091554 = x$$
$$x \approx 0.40$$

71.

$$ax - b = c \text{ where } a \neq 0$$
$$(ax - b) + b = c + b$$
$$ax = c + b$$
$$\frac{ax}{a} = \frac{c + b}{a} \text{ since } a \neq 0$$
$$x = \frac{c + b}{a} = \frac{b + c}{a}$$

Check:

$$ax - b = c$$
$$a\frac{c + b}{a} - b \overset{?}{=} c$$
$$(c + b) - b \overset{?}{=} c$$
$$c = c$$

73.

$$\frac{x}{a} + \frac{x}{b} = c \text{ where } a \neq 0, \ b \neq 0$$
$$(ax - b) + b = c + b$$
$$ab\left[\frac{x}{a} + \frac{x}{b}\right] = ac[c]$$
$$bx + ax = abc$$
$$(b + a)x = abc$$
$$\frac{(b + a)x}{b + a} = \frac{abc}{b + a}$$
$$x = \frac{abc}{b + a} = \frac{abc}{a + b}$$

Check:

$$\frac{x}{a} + \frac{x}{b} = c$$
$$\frac{\frac{abc}{a + b}}{a} + \frac{\frac{abc}{a + b}}{b} \overset{?}{=} c$$
$$\frac{abc}{a(a + b)} + \frac{abc}{b(a + b)} \overset{?}{=} c$$
$$\frac{ab^2c + a^2bc}{ab(a + b)} \overset{?}{=} c$$
$$\frac{abc[b + a]}{ab(a + b)} \overset{?}{=} c$$
$$c = c$$

75.

$$\frac{1}{x - a} + \frac{1}{x + a} = \frac{2}{x - 1}$$
$$(x - a)(x + a)(x - 1)\left[\frac{1}{x - a} + \frac{1}{x + a}\right] = (x - a)(x + a)(x - 1)\left[\frac{2}{x - 1}\right]$$
$$(x + a)(x - 1) + (x - a)(x - 1) = 2(x - a)(x + a)$$
$$x^2 - x + ax - a + x^2 - x - ax + a = 2(x^2 - a^2)$$
$$2x^2 - 2x = 2x^2 - 2a^2$$
$$(2x^2 - 2x) - 2x^2 = (2x^2 - 2a^2) - 2x^2$$
$$-2x = -2a^2$$
$$\frac{-2x}{-2} = \frac{-2a^2}{-2}$$
$$x = a^2$$

Check:

$$\frac{1}{x - a} + \frac{1}{x + a} = \frac{2}{x - 1}$$
$$\frac{1}{a^2 - a} + \frac{1}{a^2 + a} \overset{?}{=} \frac{2}{a^2 - 1}$$
$$\frac{1}{a(a - 1)} + \frac{1}{a(a + 1)} \overset{?}{=} \frac{2}{a^2 - 1}$$
$$\frac{a + 1 + a - 1}{a(a - 1)(a + 1)} \overset{?}{=} \frac{2}{a^2 - 1}$$
$$\frac{2a}{a(a - 1)(a + 1)} \overset{?}{=} \frac{2}{a^2 - 1}$$
$$\frac{2}{(a - 1)(a + 1)} = \frac{2}{(a - 1)(a + 1)}$$

For Problems 77–95, we get each equation into standard form $ax^2 + bx + c = 0$ and use the quadratic formula,

$$x = \frac{-b \pm \sqrt{b^2 - 4ac}}{2a}$$

77. $x^2 - 4x + 2 = 0$
Here, $a = 1$, $b = -4$, $c = 2$.

$$x = \frac{-(-4) \pm \sqrt{(-4)^2 - 4(1)(2)}}{2(1)}$$

$$= \frac{4 \pm \sqrt{16 - 8}}{2}$$

$$= \frac{4 \pm 2\sqrt{2}}{2} = 2 \pm \sqrt{2}$$

$$\left\{2 - \sqrt{2},\ 2 + \sqrt{2}\right\}$$

79. $x^2 - 4x - 1 = 0$
Here $a = 1$, $b = -4$, $c = -1$.

$$x = \frac{-(-4) \pm \sqrt{(-4)^2 - 4(1)(-1)}}{2(1)}$$

$$= \frac{4 \pm \sqrt{16 + 4}}{2} = \frac{4 \pm \sqrt{20}}{2}$$

$$= \frac{4 \pm 2\sqrt{5}}{2} = 2 \pm \sqrt{5}$$

$$\left(2 - \sqrt{5},\ 2 + \sqrt{5}\right)$$

81. $2x^2 - 5x + 3 = 0$
Here $a = 2$, $b = -5$, $c = 3$.

$$x = \frac{-(-5) \pm \sqrt{(-5)^2 - 4(2)(3)}}{2(2)}$$

$$= \frac{-5 \pm \sqrt{25 - 24}}{4} = \frac{5 \pm 1}{4}$$

$$\left\{1,\ \frac{3}{2}\right\}$$

83. $4y^2 - y + 2 = 0$
Here $a = 4$, $b = -1$, $c = 2$.

$$x = \frac{-(-1) \pm \sqrt{(-1)^2 - 4(4)(2)}}{2(4)}$$

$$= \frac{1 \pm \sqrt{1 - 32}}{8} = \frac{1 \pm \sqrt{-31}}{8}$$

No real solution.

85. $$4x^2 = 1 - 2x$$
$$4x^2 + 2x - 1 = 0$$
Here, $a = 4$, $b = 2$, $c = -1$.

$$x = \frac{-2 \pm \sqrt{2^2 - 4(4)(-1)}}{2(4)}$$

$$= \frac{-2 \pm \sqrt{4 + 16}}{8}$$

$$= \frac{-2 \pm 2\sqrt{5}}{8} = \frac{-1 \pm \sqrt{5}}{4}$$

$$\left\{\frac{-1 - \sqrt{5}}{4},\ \frac{-1 + \sqrt{5}}{4}\right\}$$

87. $$4x^2 = 9x$$
$$4x^2 - 9x = 0$$
Here $a = 4$, $b = -9$, $c = 0$.

$$x = \frac{-9 \pm \sqrt{(-9)^2 - 4(4)(0)}}{2(4)}$$

$$= \frac{9 \pm \sqrt{81}}{8} = \frac{9 \pm 9}{8}$$

$$\left\{0,\ \frac{9}{4}\right\}$$

89. $9t^2 - 6t + 1 = 0$
Here $a = 9$, $b = -6$, $c = 1$.

$$x = \frac{-(-6) \pm \sqrt{(-6)^2 - 4(9)(1)}}{2(9)}$$

$$= \frac{6 \pm \sqrt{36 - 36}}{18} = \frac{1}{3}$$

$$\left\{ \frac{1}{3} \right\}$$

91. $3x^2 - 2x - 2 = 0$
Here $a = 3$, $b = -2$, $c = -2$.

$$x = \frac{-(-2) \pm \sqrt{(-2)^2 - 4(3)(-2)}}{2(3)}$$

$$= \frac{2 \pm \sqrt{4 + 24}}{6} = \frac{2 \pm \sqrt{28}}{6}$$

$$= \frac{2 \pm 2\sqrt{7}}{6} = \frac{1 \pm \sqrt{7}}{3}$$

$$\left\{ \frac{1 - \sqrt{7}}{3}, \frac{1 + \sqrt{7}}{3} \right\}$$

93. $4 - \dfrac{1}{x} - \dfrac{2}{x^2} = 0$
$4x^2 - x - 2 = 0$
Here $a = 4$, $b = -1$, $c = -2$.

$$x = \frac{-(-1) \pm \sqrt{(-1)^2 - 4(4)(-2)}}{2(4)}$$

$$= \frac{1 \pm \sqrt{1 + 32}}{8} = \frac{1 \pm \sqrt{33}}{8}$$

$$\left\{ \frac{1 - \sqrt{33}}{8}, \frac{1 + \sqrt{33}}{8} \right\}$$

95. $3x = 1 - \dfrac{1}{x}$
$3x^2 = x - 1$
$3x^2 - x + 1 = 0$
Here $a = 3$, $b = -1$, $c = 1$.

$$x = \frac{-1(1) \pm \sqrt{(-1)^2 - 4(3)(1)}}{2(3)}$$

$$= \frac{1 \pm \sqrt{1 - 12}}{6} = \frac{1 \pm \sqrt{-11}}{6}$$

No real solution.

97. $x^2 - 4x + 2 = 0$
Using a graphing utility, the real solutions are 0.58 and 3.41 correct to two decimal places.
$x^2 - 4x + 2 = 0$
Here $a = 1$, $b = -4$, $c = 2$.

$$x = \frac{-(-4) \pm \sqrt{(-4)^2 - 4(4)(2)}}{2(1)}$$

$$= \frac{4 \pm \sqrt{16 - 8}}{2} = \frac{4 \pm 2\sqrt{2}}{2}$$

$$= 2 \pm \sqrt{2}$$

99. $x^2 + \sqrt{3}\, x - 3 = 0$
Using a graphing utility, the real solutions are -2.80 and 1.07 correct to two decimal places.
$x^2 + \sqrt{3}x - 3 = 0$
Here, $a = 1$, $b = \sqrt{3}$, $c = -3$.

$$x = \frac{-\sqrt{3} \pm \sqrt{\left(\sqrt{3}\right)^2 - 4(1)(-3)}}{2(1)}$$

$$= \frac{-\sqrt{3} \pm \sqrt{3 + 12}}{2} = \frac{-\sqrt{3} \pm \sqrt{15}}{2}$$

101. $\pi x^2 - x - \pi = 0$
Using a graphing utility, the real solutions are -0.85 and 1.17 correct to two decimal places.
$\pi x^2 - x - \pi = 0$
Here $a = \pi$, $b = -1$, $c = -\pi$.

$$x = \frac{-(-1) \pm \sqrt{(-1)^2 - 4(\pi)(-\pi)}}{2(\pi)}$$

$$= \frac{1 \pm \sqrt{1 + 4\pi^2}}{2\pi}$$

103. $3x^2 + 8\pi x + \sqrt{29} = 0$
Using a graphing utility, the real solutions are -8.15 and -0.22 correct to two decimal places.
$3x^2 + 8\pi x + \sqrt{29} = 0$
Here, $a = 3$, $b = 8\pi$, $c = \sqrt{29}$.

$$x = \frac{-8\pi \pm \sqrt{64\pi^2 - 4(3)\left(\sqrt{29}\right)}}{2(3)}$$

$$= \frac{-8\pi \pm \sqrt{64\pi^2 - 12\sqrt{29}}}{6}$$

105. Graphing utility: $\{-2.64, 2.64\}$

$$x^2 - 7 = 0$$
$$x^2 = 7$$
$$x = \pm\sqrt{7}$$
$$\left\{-\sqrt{7}, \sqrt{7}\right\}$$

107. Graphing utility: $\{0.25\}$

$$16x^2 - 8x + 1 = 0$$
$$(4x - 1)(4x - 1) = 0$$
$$4x - 1 = 0$$
$$x = \frac{1}{4}$$
$$\frac{1}{4}$$

109. Graphing utility $\{-0.60, 2.50\}$

$$10x^2 - 19x - 15 = 0$$
$$(5x + 3)(2x - 5) = 0$$
$$5x + 3 = 0 \quad \text{or} \quad 2x - 5 = 0$$
$$x = \frac{-3}{5} \qquad x = \frac{5}{2}$$
$$\left\{\frac{-3}{5}, \frac{5}{2}\right\}$$

111. Graphing utility: $\{-0.50, 0.66\}$

$$2 + z = 6z^2$$
$$6z^2 - z - 2 = 0$$
$$(2z + 1)(3z - 2) = 0$$
$$2z + 1 = 0 \quad \text{or} \quad 3z - 2 = 0$$
$$z = -\frac{1}{2} \qquad z = \frac{2}{3}$$
$$\left\{-\frac{1}{2}, \frac{2}{3}\right\}$$

113. Graphing utility: $\{-1.70, 0.29\}$

$$x^2 + \sqrt{2}\,x = \frac{1}{2}$$
$$x^2 + \sqrt{2}\,x - \frac{1}{2} = 0$$
$$2x^2 + 2\sqrt{2}\,x - 1 = 0$$

Here $a = 2$, $b = 2\sqrt{2}$, $c = -1$.

$$x = \frac{-2\sqrt{2} \pm \sqrt{\left(2\sqrt{2}\right)^2 - 4(2)(-1)}}{2(2)}$$
$$= \frac{-2\sqrt{2} \pm \sqrt{16}}{4} = \frac{-\sqrt{2} \pm 2}{2}$$
$$\left\{\frac{-\sqrt{2} - 2}{2}, \frac{-\sqrt{2} + 2}{2}\right\}$$

115. Graphing utility: $\{-2.56, 1.56\}$

$$x^2 + x = 4$$
$$x^2 + x - 4 = 0$$

Here $a = 1$, $b = 1$, $c = -4$.

$$x = \frac{-1 \pm \sqrt{1^2 - 4(1)(-4)}}{2(1)}$$
$$= \frac{-1 \pm \sqrt{17}}{2}$$
$$\left\{\frac{-1 - \sqrt{17}}{2}, \frac{-1 + \sqrt{17}}{2}\right\}$$

In Problems 117–121, we use the discriminant $b^2 - 4ac$:

117. $2x^2 - 6x + 7 = 0$

Here $a = 2$, $b = -6$, $c = 7$.
$$b^2 - 4ac = (-6)^2 - 4(2)(7)$$
$$= -20 < 0$$

No real solution.

119. $9x^2 - 30x + 25 = 0$

Here $a = 9$, $b = -30$, $c = 25$.
$$b^2 - 4ac = (-30)^2 - 4(9)(25)$$
$$= 0$$

Repeated real solution.

121. $3x^2 + 5x - 2 = 0$

Here $a = 3$, $b = 5$, $c = -2$.
$$b^2 - 4ac = 5^2 - 4(3)(-2) = 1$$

Two unequal real solutions.

123. Solve for R in $\dfrac{1}{R} = \dfrac{1}{R_1} + \dfrac{1}{R_2}$

$$RR_1R_2\left[\frac{1}{R}\right] = RR_1R_2\left[\frac{1}{R_1} + \frac{1}{R_2}\right]$$
$$R_1R_2 = RR_2 + RR_1$$
$$R_1R_2 = R(R_2 + R_1)$$
$$\frac{R_1R_2}{R_2 + R_1} = \frac{R(R_2 + R_1)}{R_2 + R_1}$$
$$\frac{R_1R_2}{R_1 + R_2} = R$$
$$R = \frac{R_1R_2}{R_1 + R_2}$$

Check:
$$\frac{1}{R} = \frac{1}{R_1} + \frac{1}{R_2}$$
$$\frac{1}{\dfrac{R_1R_2}{R_1 + R_2}} \overset{?}{=} \frac{1}{R_1} + \frac{1}{R_2}$$
$$\frac{R_1 + R_2}{R_1R_2} \overset{?}{=} \frac{R_2 + R_1}{R_1R_2}$$
$$\frac{R_1 + R_2}{R_1R_2} = \frac{R_1 + R_2}{R_1R_2}$$

125. Solve for R in $F = \dfrac{mv^2}{R}$

$$R[F] = R\left[\frac{mv^2}{R}\right]$$
$$RF = mv^2$$
$$\frac{RF}{F} = \frac{mv^2}{F}$$
$$R = \frac{mv^2}{F}$$

Check:
$$F = \frac{mv^2}{R}$$
$$F \overset{?}{=} \frac{mv^2}{\dfrac{mv^2}{F}}$$
$$F \overset{?}{=} \frac{mv^2}{1} \cdot \frac{F}{mv^2}$$
$$F = F$$

127. Solve for r in $S = \dfrac{a}{1 - r}$

$$(1 - r)[S] = 1 - r\left[\frac{a}{1 - r}\right]$$
$$S - rS = a$$
$$(S - rS) - S = a - S$$
$$-rS = a - S$$
$$\frac{-rS}{-S} = \frac{a - S}{-S}$$
$$r = \frac{a - S}{-S}$$
$$r = \frac{-(a - S)}{S}$$
$$r = \frac{S - a}{S}$$

Check:
$$S = \frac{a}{1 - r}$$
$$S \overset{?}{=} \frac{a}{1 - \dfrac{S - a}{S}}$$
$$S \overset{?}{=} \frac{a}{\dfrac{S - (S - a)}{S}}$$
$$S \overset{?}{=} \frac{a}{\dfrac{S - S + a}{S}}$$
$$S \overset{?}{=} \frac{a}{\dfrac{a}{S}}$$
$$S \overset{?}{=} a \cdot \frac{S}{a}$$
$$S = S$$

129. In step (6), we have $(x - 2)(x + 5) = (x - 2)(x + 4)$. In step (7), we cannot divide by $x - 2$ because $x = 2$ from step (1) which means we actually divided by 0. Step (7) should read:
$$(x - 2)(x + 5) - (x - 2)(x + 4) = 0$$
Then, $\quad (x - 2)([(x + 5) - (x + 4)] = 0$
$$(x - 2)(1) = 0$$
$$x - 2 = 0$$
$$x = 2$$

1. Let A represent area of the circle and r the radius:
   Area of a circle is the product of π times the square of the radius.

   $A \qquad = \qquad \pi \quad \cdot \qquad r^2$

   $A = \pi r^2$

3. Let A represent the area of the square and s the length of a side:
   Area of a square is the square of the length of a side.

   $A \qquad = \qquad s^2$

   $A = s^2$

5. Let F represent the force, m the mass, and a the acceleration:
   Force equals the product of mass times acceleration.

   $F \qquad = \qquad m \quad \cdot \qquad a$

   $F = ma$

7. Let W represent the work, F the force, and d the distance:
   Work equals force times distance.

   $W \qquad = \qquad F \quad \cdot \qquad d$

   $W = Fd$

9. $C = $ total variable cost, $x = $ number of dishwashers manufactured.

   $C = 150x$

11.

Amount in Bonds	Amount in CD's	Total
x	$x - 2000$	20,000

$$x + (x - 2000) = 20000$$
$$2x - 2000 = 20000$$
$$2x = 22000$$
$$x = 11000$$

$11,000 will be invested in bonds. $9,000 will be invested in CD's.

13.

Katy	Mike	Dan	Total
x	$\dfrac{3}{4}x$	$\dfrac{1}{2}x$	900,000

$$x + \frac{3}{4}x + \frac{1}{2}x = 900{,}000$$
$$\left(1 + \frac{3}{4} + \frac{1}{2}\right)x = 900{,}000$$
$$\frac{9}{4}x = 900{,}000$$
$$x = \frac{4}{9}(900{,}000)$$
$$x = 400{,}000$$

Katy receives $400,000. Mike receives $300,000. Dan receives $200,000.

15. **Step 1:** We are being asked for an hourly wage in dollars per hour.
    Step 2: Let x represent the hourly wage.
    Step 3: We set up a table:

	Hourly wage	Salary
Regular hours, 40	x	$40x$
Overtime hours, 8	$1.5x$	$8(1.5x) = 12x$

The total of regular salary plus overtime is $442.00, then

$$40x + 12x = 442$$

Step 4:
$$52x = 442$$
$$x = 8.50$$

The hourly wage is $8.50 per hour.

Step 5: Forty hours yields a salary of $40(8.50) = \$340$, and 8 hours of overtime yields a salary of $8(1.5)(8.50) = \$102$, for a total of $442.

17. **Step 1:** We are being asked to find the number of touchdowns scored.
    Step 2: Let x represent the number of touchdowns scored.
    Step 3: We set up a table:

	Point value	Points earned
Safeties, 1	2	$(1)(2) = 2$
Field goals, 2	3	$(2)(3) = 6$
Touchdowns without extra points, 2	6	$(2)(6) = 12$
Touchdowns with extra points, $x - 2$	7	$(x - 2)(7) = 7x - 14$

The total points scores is 41; thus

$$2 + 6 + 12 + 7x - 14 = 41$$

Step 4:
$$7x + 6 = 41$$
$$7x = 35$$
$$x = 5$$

There were two touchdowns without extra points and three touchdowns with extra points for a total of 5 touchdowns.

Step 5: Two safeties (for 2 points) and two field goals (for 6 points) and two touchdowns without extra points (for 12 points) and three touchdowns with extra points (for 21 points) give a total of $2 + 6 + 12 + 21 = 41$ points.

19. $\ell = $ length, $w = $ width

$$2\ell + 2w = 60 \qquad \text{Perimeter} = 2\ell + 2w$$
$$\ell = w + 8 \qquad \text{The length is 8 more than the width.}$$
$$2(w + 8) + 2w = 60$$
$$2w + 16 + 2w = 60$$
$$4w + 16 = 60$$
$$4w = 44$$
$$w = 11 \text{ feet}$$
$$\ell = 19 \text{ feet}$$

21. ℓ = length of garden, w = width of garden

  (a) If the length of garden is to be twice its width, then the width is to be half the length. Thus, $w = \frac{1}{2}\ell$

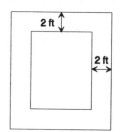

The dimensions of the fence are $\ell + 4$ and $w + 4$, which means the dimensions are $\ell + 4$ and $\frac{1}{2}\ell + 4$.

Its perimeter is 46 feet, so $\quad 2[(\ell + 4) + \frac{1}{2}\ell + 4] = 46$

$$2\ell + 8 + \ell + 8 = 46$$
$$3\ell = 30$$
$$\ell = 10$$

The dimensions of the garden are 10 feet by 5 feet.

  (b) Area $= \ell \cdot w = 10 \cdot 5 = 50$ square feet

  (c) If the dimensions of the fence are the same, then the dimensions are $\ell + 4$ and $\ell + 4$. Its perimeter is 46 feet, so

$$2[(\ell + 4) + (\ell + 4)] = 46$$
$$2\ell + 8 + 2\ell + 8 = 46$$
$$4\ell = 30$$
$$\ell = 7.5$$

The dimensions of the garden are 7.5 feet by 7.5 feet.

  (d) The area of this square garden is $\ell \cdot w$ (7.5)(7.5) = 56.25 square feet.

23. **Step 1:** We want to find two dollar amounts, the principle to invest in B−rated bonds pay 15% per year and the principle to invest in a certificate paying 7% per year.

      Step 2: Let x represent the amount invested in bonds at 15%. Then $50{,}000 - x$ is the amount that will be invested in a certificate at 7%.

      Step 3: We make a table:

	Principle	Rate	Time (yr)	Interest
Bonds at 15%	x	0.15	1	0.15
Certificate at 7%	$50{,}000 - x$	0.07	1	$0.07(50{,}000 - x)$

              Since the total interest is to be $6000, we have:

$$0.15x + 0.07(50{,}000 - x) = 6000$$

      Step 4:
$$15x + 7(50{,}000 - x) = 600{,}000$$
$$15x + 350{,}000 - 7x = 600{,}000$$
$$350{,}000 + 8x = 600{,}000$$
$$8x = 250{,}000$$
$$x = 31{,}250$$

              Thus, $31,250 will be invested in bonds at 15% and $18,750 in a certificate at 7%.

      Step 5: The interest on the bond after one year is $0.15(31{,}250) = \$4687.50$, and the interest on the certificate is $0.07(18{,}750) = \$1312.50$, for a total interest of $\$4687.50 + \$1312.50 = \$6000.00$.

25. **Step 1:** We want to find the dollar amount loaned at 8%.

      Step 2: Let x represent the amount invested at 8%. Then the amount invested at 18% is $12{,}000 - x$.

Step 3: We make a table:

	Principle	Rate	Time (yr.)	Interest
Loan at 8%	x	0.08	1	$0.08x$
Loan at 18%	$12{,}000 - x$	0.18	1	$0.18(12{,}000 - x)$

Since the total interest is to be $1000, we have:

$$0.8x + 0.18(12{,}000 - x) = 1000$$

Step 4:
$$8x + 18(12{,}000 - x) = 100{,}000$$
$$8x + 216{,}000 - 18x = 100{,}000$$
$$-10x = -116{,}000$$
$$x = 11{,}600$$

Thus, $11,600 will be loaned at 8% and $400 at 18%.

Step 5: The interest on the loan at 8% is $0.08(11{,}600) = \$928$, and the interest on the loan at 18% is $0.18(400) = \$72$; thus the total interest is $1000.

---

27. Let w = width of opening and ℓ = length of opening.

$\ell = w + 2$

Area of opening $= w(w + 2) = 143$
$$w^2 + 2w = 143$$
$$w^2 + 2w - 143 = 0$$
$$(w + 13)(w - 11) = 0$$
$$w + 13 = 0 \qquad w - 11 = 0$$
$$w = -13 \qquad w = 11$$

Since measurements must be positive, the dimensions of the opening are 11 feet by 13 feet.

29. Let ℓ = length, w = width,

Perimeter $= 2\ell + 2w$.

(1) $\qquad 2\ell + 2w = 26$
$$\ell w = 40$$

Simplifying (1), we get,
$$\ell + w = 13$$
$$w = 13 - \ell$$
$$\ell(13 - \ell) = 40$$
$$13\ell - \ell^2 = 40$$
$$\ell^2 - 13\ell + 40 = 0$$
$$(\ell - 5)(\ell - 8) = 0$$
$$\ell = 5 \qquad \ell = 8$$
$$w = 8 \qquad w = 5$$

The dimensions of the rectangle are 5 m by 8 m.

---

31. We want the dimensions of a box with square base and volume of 4 cubic feet. Let x be the side of the sheet of metal. Then $v = \ell w$ gives

$$4 = (x - 2)(x - 2)$$
$$4 = x^2 - 4x + 4$$
$$0 = x^2 - 4x$$
$$0 = x(x - 4)$$
$$x = 0 \text{ or } x - 4 = 0$$
$$x = 4$$

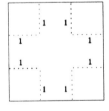

Thus, the sheet should be 4 ft. $\times$ 4 ft. to give a box of $2 \times 2 = 4$ cu. ft.

---

33. (a) For $s = 96 + 80t - 16t^2$, $s = 0$ when ball strikes ground:

$$0 = 96 + 80t - 16t^2$$
$$0 = 16(6 + 5t - t^2)$$
$$0 = 6 + 5t - t^2$$
$$0 = t^2 - 5t - 6$$
$$0 = (t - 6)(t + 1)$$
$$t - 6 = 0 \text{ or } t + 1 = 0$$
$$t = 6 \qquad t = -1$$

Thus, after 6 seconds, the ball strikes the ground.

(b) $s = 96$ when the ball is at the level of the top of the building:

$$96 = 96 + 80t - 16t^2$$
$$0 = 80t - 16t^2$$
$$0 = -16t(t - 5)$$
$$-16t = 0 \text{ or } \qquad t - 5 = 0$$
$$t = 0 \qquad\qquad t = 5$$

Thus, the ball passes the top of the building after 5 seconds.

35. **Step 1:** We are asked to find the original price and *difference* between the original price and the new price.
Step 2: Let x be the *original price* of the house.
Step 3: Note that
$$\text{original price} = x$$
$$\text{amount reduced} = 0.15x$$
$$\text{new price} = \$125{,}000$$
$$\text{Original price} - \text{amount reduced} = \text{new price}$$
$$x \quad - \quad 0.15x = \$125{,}000$$
Step 4:
$$.85x = \$125{,}000$$
$$x = \$147{,}058.82$$
Thus, the original price is \$147,058.82, and the amount of the savings is $0.15(\$147{,}058.82) = \$22{,}058.82$.
Step 5: The original price \$147,058.82 less the 15% reduction $0.15(147{,}058.82)$ gives the new price \$125,000.

37. **Step 1:** We are asked to find the amount the bookstore paid for the book.
Step 2: Let x be the bookstore's price for the book.
Step 3: Note that
$$\text{bookstore price} = x$$
$$\text{percent increase} = 0.25x$$
$$\text{selling price} = \$56.00$$
$$\text{bookstore price} + \text{percent increase} = \text{selling price}$$
$$x \quad + \quad 0.25x = \$56.00$$
Step 4:
$$1.25x = \$56.00$$
$$x = \$44.80$$
Step 5: The amount that the bookstore paid for the book was \$44.80. The 25% markup of the price increases the cost of the book \$11.20, making the selling price of the book \$56.00.

39. **Step 1:** We want to find a time in minutes.
Step 2: Let t be the time in minutes it takes for them to do the job together. Then in one minute Mike can do $\frac{1}{30}$ of the job, Danny can do $\frac{1}{20}$, and together they do $\frac{1}{t}$ of the job.
Step 3: We make a table:

	Minutes to do job	Part of job done in one minute
Mike	30	$\frac{1}{30}$
Danny	20	$\frac{1}{20}$
Together	t	$\frac{1}{t}$

Thus, $\dfrac{1}{30} + \dfrac{1}{20} = \dfrac{1}{t}$
Step 4:
$$\frac{2+3}{60} = \frac{1}{t}$$
$$\frac{5}{60} = \frac{1}{t}$$
$$\frac{1}{12} = \frac{1}{t}$$
$$t = 12$$
Thus, working together, the job can be done in 12 minutes.

Step 5: Mike does $\frac{1}{30}$ of the job in one minute.

Danny does $\frac{1}{20}$ of the job in one minute.

In one minute working together they do $\frac{1}{30} + \frac{1}{20}$ of the job.

In 12 minutes they do $12\left(\frac{1}{30} + \frac{1}{20}\right) = 12\left(\frac{5}{60}\right) = 1$ of the job, or **all** of it.

41. **Step 1:** We want to find a final exam score.
    Step 2: Let x be the final exam score to yield an average of 80 when counted twice and combined with the other scores.
    Step 3: The total points scored with the exam counted twice is:
    $$80 + 83 + 71 + 61 + 95 + 2x = 390 + 2x$$

    Averaging, we get $\quad \dfrac{390 + 2x}{7} = 80$

    Step 4:
    $$390 + 2x = 560$$
    $$2x = 170$$
    $$x = 85$$

    Thus, a score of 85 is needed to get an average of 80 if the final counts as two tests.

    Step 5: $\dfrac{80 + 83 + 71 + 61 + 95 + 2(85)}{7} = 80$

43. **Step 1:** We want to find a position on a football field.
    Step 2: Let s be the distance the tight end runs after catching the ball.
    Step 3: We can make a table (using $v = \dfrac{s}{t}$ and $t = \dfrac{s}{v}$):

	Velocity	Time (seconds)	Distance (yards)
Tight end	$\dfrac{100}{12} = \dfrac{25}{3}$	$\dfrac{s}{\frac{25}{3}} = \dfrac{3s}{25}$	s
Defensive back	$\dfrac{100}{10} = 10$	$\dfrac{s + 5}{10}$	$s + 5$

Since the time is the same for both runners:

$$\frac{3s}{25} = \frac{s + 5}{10}$$

Step 4:
$$30s - 25s = 125$$
$$5s = 125$$
$$s = 25$$

Thus, the tight end goes 25 yards from his 20 yard line, or to his 45 yard line.

Step 5: Checking the time for the appropriate distances:

Tight end: $\quad \dfrac{25 \text{ yds}}{\frac{25}{3} \text{ yds/sec}} = 3 \text{ sec}$

Defensive back: $\quad \dfrac{30 \text{ yds}}{10 \text{ yds/sec}} = 3 \text{ sec}$

45. We are looking for a width. Let x represent the width of the border in feet. It is best to convert all units to feet now:

$$1 \text{ cubic yard} = 27 \text{ cubic feet}$$

$$3 \text{ inches} = \frac{3}{12} = \frac{1}{4} \text{ ft.}$$

From the figure: The total area is $A_T = (6 + 2x)(10 + 2x)$

The area of the garden is $A_G = 6 \times 10 = 60$ sq. ft.

Then, the area of the border, A_B, is $A_B = (6 + 2x)(10 + 2x) - 60$

The volume of the border is 27 cubic feet or $\left[\frac{1}{4} \text{ ft}\right](A_B)$ so that

$$27 = \frac{1}{4}A_B$$

$$108 = A_B$$

Thus, $108 = (6 + 2x)(10 + 2x) - 60$

$$108 = 60 + 12x + 20x + 4x^2 - 60$$

$$0 = 4x^2 + 32x - 108$$

$$0 = x^2 + 8x - 27$$

Here, $a = 1$, $b = 8$, $c = -27$, and $b^2 - 4ac = 8^2 - 4(1)(-27) = 172$

Then, $x = \dfrac{-8 \pm \sqrt{172}}{2(1)} = \dfrac{-8 \pm 2\sqrt{43}}{2} = -4 \pm \sqrt{43}$

so that $x = -4 - \sqrt{43}$ or $x = -4 + \sqrt{43}$
cannot be negative $x = -4 + 6.56$
                                        $x = 2.56$ ft.

The width of the border is 2.56 ft.

Check: Thus, the area of the border is:

$$A_B = (6 + 2(2.56))(10 + 2(2.56)) - (6)(10)$$
$$= 108.13 \text{ which is close to } 108 \text{ sq. ft.}$$

The given volume of cement is then $(108)\left[\dfrac{1}{4}\right] = 27$ cubic ft.

47. We want to find the new dimensions of length and width in centimeters. Let x be the amount of reduction of the length and width measured in centimeters.

The current bar has volume $V_c = (12)(7)(3) = 252$ cubic centimeters.

The new volume is to be $V_N = .90V_c = .9(252) = 226.8$ cubic centimeters.

Then $V_N = 226.8 = (3)(12 - x)(7 - x)$

$$226.8 = 3(84 - 19x + x^2)$$

$$75.6 = 84 - 19x + x^2$$

$$0 = 8.4 - 19x + x^2$$

Here, $a = 1$, $b = -19$, $c = 8.4$ and $b^2 - 4ac = (-19)^2 - 4(1)(8.4) = 327.4$

Then, $x = \dfrac{-(-19) \pm \sqrt{327.4}}{2(1)} = \dfrac{19 \pm 18.09}{2}$

$x = 0.455$ or $x = 18.55$; but since 18.55 exceeds the measurements it would be subtracted from, it is not a practical solution.

The new dimensions are:

$7 - 0.455$, and $12 - 0.455$, and 3 centimeters

or   width $= 6.55$, length $= 11.55$, and thickness $= 3$ centimeters.

Check: $(6.55)(11.55)(3) = 226.9 \cong 226.8$.

49. We want to find the width of a concrete pool border. Let x represent the width in feet of the border. It is best to convert all units to feet now:

$$1 \text{ cubic yard} = 27 \text{ cubic feet}$$

$$3 \text{ inches} = \frac{3}{12} = \frac{1}{4} \text{ foot}$$

We will use $A = \pi r^2$.

We are given that the distance across the pool is 10 feet, which means the radius is $\frac{10}{2} = 5$ feet.

The total area, pool and border, is $A_T = \pi(5 + x)^2$.
The area of the pool alone is $A_p = \pi(5)^2 = 25\pi$.
The area of the border is $A_B = A_T - A_p = \pi(5 + x)^2 - 25\pi$.

The volume of the border, $\frac{1}{4}A_B = \frac{1}{4}(\pi(5 + x)^2 - 25\pi)$.

Then, $\frac{1}{4}(\pi(5 + x)^2 - 25\pi) = 27$

$$\pi(25 + 10x + x^2 - 25) = 108$$

$$x^2 + 10x - \frac{108}{\pi} = 0$$

$$x^2 + 10x - 34.38 = 0$$

Here, $a = 1$, $b = 10$, $c = -34.38$, and $b^2 - 4ac = 10^2 - 4(1)(-34.38) = 237.52$

Then, $x = \dfrac{-10 \pm \sqrt{237.52}}{2(1)} = \dfrac{-10 \pm 15.41}{2}$

Thus, $x = 2.71$ or $x = -12.71$

We ignore -12.71 since a measurement must be positive. The width of the border is 2.71 ft.

Check: Area of border $= \pi[(5 + 2.71)^2 - 25] = 108.21$

Volume of border $= \dfrac{1}{4}(108.2) = 27.1$ cubic ft.

51. **Step 1:** We want to find a speed in miles per hour.
    Step 2: Let v be the speed of the current in miles per hour.
    Step 3: We make a table (using $s = vt$):

	Velocity of boat	Time (hr.)	Distance (mi.)
Upstream	$16 - v$	$\dfrac{20}{60} = \dfrac{1}{3}$	$\dfrac{16 - v}{3}$
Downstream	$16 + v$	$\dfrac{15}{60} = \dfrac{1}{4}$	$\dfrac{16 + v}{4}$

Since the distance is the same in each direction:

$$\frac{16 - v}{3} = \frac{16 + v}{4}$$

Step 4:
$$4(16 - v) = 3(16 + v)$$
$$64 - 4v = 48 + 3v$$
$$16 = 7v$$
$$v = \frac{16}{7} = 2.286$$

Thus, the speed of the current is 2.286 miles per hour.

Step 5: The distance is the same in each direction:

$$\frac{16 - v}{3} = \frac{16 - 2.286}{3} = 4.57 \text{ miles}$$

$$\frac{16 + v}{4} = \frac{16 + 2.286}{4} = 4.57 \text{ miles}$$

53. **Step 1:** We want to find a time in minutes and a distance in miles.
Step 2: Let t be the time in minutes to run the mile.
Step 3: We can make a table:

	Minutes to run the mile	Time	Part of mile run in one minute	Distance
Mike	6	t	$\dfrac{1}{6}$	$\dfrac{1}{6}t$
Dan	9	$t + 1$	$\dfrac{1}{9}$	$\dfrac{1}{9}(t + 1)$

Step 4:

$$\frac{1}{6}t = \frac{1}{9}(t + 1)$$

$$\frac{1}{6}t = \frac{1}{9}t + \frac{1}{9}$$

$$\frac{3t - 2t}{18} = \frac{1}{9}$$

$$\frac{1}{18}t = \frac{1}{9}$$

$$t = 2$$

$$d = \frac{1}{6}t = \frac{1}{9}(t + 1)$$

$$= \frac{1}{6} \cdot 2 = \frac{1}{9}(2 + 1)$$

$$= \frac{1}{3}$$

Thus, after 2 minutes and a distance of $\dfrac{1}{3}$ mile, Mike will pass Dan.

Step 5: If Mike gives Dan a headstart of 1 minute, Mike will pass Dan $\dfrac{1}{3}$ of a mile from the start in 2 minutes.

55. **Step 1**: We want to find a time.
Step 2: Let t be the time it takes the rescue craft to reach the ship.
Step 3: We can make a table.

	Velocity (mph)	Time (hrs) (hrs)	Distance
Ship	10	t	$10t$
Coastguard	20	t	$20t$

The distance to be traveled, in total, by the Coastguard and the ship is 60 miles.
Therefore, $20t + 10t = 60$
Step 4: $30t = 60$
$t = 2$
The rescue craft (Coastguard) will reach the ship after 2 hours.
Step 5: After 2 hours, the Coastguard travels $20(2) = 40$ miles. After 2 hours, the ship travels $10(2)$ or 20 miles. Thus, the total distance traveled is 60 miles.

57. **Step 1**: We want to find a time.
Step 2: Let t be the length of time the auxiliary pump must run.

Step 3:   We can make a table:

	Hours to do job	Part of job done in 1 hour	Part of job done in 3 hours
Main pump	4	$\dfrac{1}{4}$	$\dfrac{3}{4}$
Auxiliary pump	9	$\dfrac{1}{9}$	$--$

$\dfrac{1}{4}$ of the job must be done by the auxiliary pump.

Step 4:

$$\left[\dfrac{1}{9}\right](t) = \dfrac{1}{4}$$

$$\dfrac{t}{9} = \dfrac{1}{4}$$

$$4t = 9$$

$$t = \dfrac{9}{4} = 2\dfrac{1}{4}$$

Thus, the auxiliary pump must run $2\dfrac{1}{4}$ hours, and to finish at noon it must be **started at** 9:45 A.M.

Step 5:   In 3 hours the main pump does $3\left[\dfrac{1}{4}\right] = \dfrac{3}{4}$ of the job.

In $2\dfrac{1}{4}$ hours the auxiliary pump does $2\dfrac{1}{4}\left[\dfrac{1}{9}\right] = \dfrac{1}{4}$ of the job.

59. **Step 1:**   We want to find a time in minutes.
Step 2:   Let t be the time in minutes to fill the tub with both faucets open and the stopper removed.
Step 3:   We can make a table:

	Minutes to do job	Part of job done in one minute
Both faucets open	15	$\dfrac{1}{15}$
Stopper removed	20	$\dfrac{-1}{20}$
Both faucets open and stopper removed	t	$\dfrac{1}{t}$

Step 4:

$$\dfrac{1}{15} - \dfrac{1}{20} = \dfrac{1}{6}$$

$$\dfrac{4-3}{60} = \dfrac{1}{6}$$

$$\dfrac{1}{60} = \dfrac{1}{6}$$

$$t = 60$$

Thus, one hour is required to fill the tub.

Step 5:   In one minute $\dfrac{1}{15} - \dfrac{1}{20}$ of the tub is filled.

In 60 minutes $60\left[\dfrac{1}{15} - \dfrac{1}{20}\right] = 1$ of the tub is filled.

61.

	Amount	Rate	Monthly Interest
CD	x	9% = 0.09	.09x
Bond	100,000 − x	12% = 0.12	.12(100,000) − x
Equity Loan	100,000	10% = 0.10	.10(100,000)

$$.09x + .12(100,000 - x) = .10(100,000)$$
$$.09x + 12,000 - .12x = 10,000$$
$$-.03x = -2000$$
$$x = 66,667$$

You can invest no more than $66,667 in the CD to ensure the monthly home equity loan payment is made.

63. We want to find the average speed from Chicago to Florida.

	Velocity	Time	Distance
Chicago to Atlanta	45	t_1	$45t_1$
Atlanta to Florida	55	t_2	$55t_2$

If Atlanta is halfway between Chicago and Florida, then the distances from Chicago to Atlanta and Atlanta to Florida are equal.

$$45t_1 = 55t_2 \Rightarrow t_1 = \frac{55}{45}t_2$$

$$\text{Average Speed} = \frac{\text{Distance}}{\text{Time}} = \frac{45t_1 + 55t_2}{t_1 + t_2} = \frac{45\left[\frac{55}{45}t_2\right] + 55t_2}{\frac{55}{45}t_2 + t_2} = \frac{55t_2 + 55t_2}{\frac{55t_2 + 45t_2}{45}}$$

$$\frac{100t_2}{\frac{100}{45}t_2} = \frac{110}{\frac{100}{45}}$$

$$\frac{(45)(110)}{100} = 49.5 \text{ miles per hour}$$

The average speed from Chicago to Florida is 49.5 mph.

65. Let x = original price. Then .6x = sale price. Profit of $4.00 at the sale price.
Cost of shirts = $20.00 each

$$.6 - 20 = \$4.00$$
$$.6x = 24$$
$$x = \frac{24}{.6}$$
$$x = 40$$

Set the original price at $40. At 50% off, there will be no profit at all.

3.4 Complex Numbers; Quadratic Equations with a Negative Discriminant

1. $(2 - 3i) + (6 + 8i) = (2 + 6) + (-3 + 8)i = 8 + 5i$

3. $(-3 + 2i) - (4 - 4i) = (-3 - 4) + (2 + 4)i = -7 + 6i$

5. $(2 - 5i) - (8 + 6i) = (2 - 8) + (-5 - 6)i = -6 - 11i$ 7. $3(2 - 6i) = 6 - 18i$

9. $2i(2 - 3i) = 4i - 6i^2 = 4i - 6(-1) = 6 + 4i$

11. $(3 - 4i)(2 + i) = 3(2 + i) - 4i(2 + i) = 6 + 3i - 8i - 4i^2 = 6 - 5i - 4(-1) = 10 - 5i$

13. $(-6 + i)(-6 - i) = -6(-6 - i) + i(-6 - i) = 36 + 6i - 6i - i^2 = 36 - (-1) = 37$

15. $\dfrac{10}{3 - 4i} = \dfrac{10}{3 - 4i} \cdot \dfrac{3 + 4i}{3 + 4i} = \dfrac{30 + 40i}{9 + 12i - 12i - 16i^2} = \dfrac{30 + 40i}{9 - 16(-1)} = \dfrac{30 + 40i}{25} = \dfrac{30}{25} + \dfrac{40}{25}i$
$$= \dfrac{6}{5} + \dfrac{8}{5}i$$

17. $\dfrac{2 + i}{i} = \dfrac{2 + i}{i} \cdot \dfrac{-i}{-i} = \dfrac{-2i - i^2}{-i^2} = -2i + 1 = 1 - 2i$

19. $\dfrac{6 - i}{1 + i} = \dfrac{6 - i}{1 + i} \cdot \dfrac{1 - i}{1 - i} = \dfrac{6 - 6i - i + i^2}{1 - i^2} = \dfrac{6 - 7i + (-1)}{1 - (-1)} = \dfrac{5 - 7i}{2} = \dfrac{5}{2} - \dfrac{7}{2}i$

21. $\left[\dfrac{1}{2} + \dfrac{\sqrt{3}}{2}i\right]^2 = \dfrac{1}{4} + 2\left(\dfrac{1}{2}\right)\left(\dfrac{\sqrt{3}}{2}i\right) + \dfrac{3}{4}i^2 = \dfrac{1}{4} + \dfrac{\sqrt{3}}{2}i + \dfrac{3}{4}(-1) = -\dfrac{1}{2} + \dfrac{\sqrt{3}}{2}i$

23. $(1 + i)^2 = 1 + 2i + i^2 = 1 + 2i - 1 = 2i$

25. $i^{23} = i^{22+1} = i^{22} \cdot i = (i^2)^{11} \cdot i = (-1)^{11}i = -i$

27. $i^{-15} = \dfrac{1}{i^{15}} = \dfrac{1}{i^{14+1}} = \dfrac{1}{i^{14}i} = \dfrac{1}{(i^2)^7 i} = \dfrac{1}{(-1)^7 i} = \dfrac{1}{-i} = \dfrac{1}{-i}\dfrac{i}{i} = \dfrac{i}{-i^2} = \dfrac{i}{-(-1)} = i$

29. $i^6 - 5 = (i^2)^3 - 5 = (-1)^3 - 5 = -1 - 5 = -6$

31. $6i^3 - 4i^5 = i^3(6 - 4i^2) = i^2 \cdot i(6 - 4(-1)) = -1 \cdot i(10) = -10i$

33. $(1 + i)^3 = 1^3 + 3i + 3i^2 + i^3 = 1 + 3i + 3(-1) + i^2 \cdot i = -2 + 3i + (-1)i = -2 + 2i$

35. $i^7(1 + i^2) = i^7(1 + (-1)) = i^7(0) = 0$

37. $i^6 + i^4 + i^2 + 1 = (i^2)^3 + (i^2)^2 + (-1) + 1 = (-1)^3 + (-1)^2 = -1 + 1 = 0$

39. $\sqrt{-4} = \sqrt{4}i = 2i$ 41. $\sqrt{-25} = \sqrt{25}i = 5i$

43. $\sqrt{(3 + 4i)(4i - 3)} = \sqrt{12i - 9 + 16i^2 - 12i} = \sqrt{-9 - 16} = \sqrt{-25} = \sqrt{25}i = 5i$

For Problems 45–57 we use $x = \dfrac{-b \pm \sqrt{b^2 - 4ac}}{2a}$, Equation (9).

45. $x^2 + 4 = 0$
Here $a = 1$, $b = 0$, $c = 4$, and $b^2 - 4ac = 0 - 4(1)(4) = -16$.
Then $x = \dfrac{-0 \pm \sqrt{-16}}{2(1)} = \dfrac{\pm\sqrt{16}\,i}{2} = \dfrac{\pm 4i}{2} = \pm 2i$
The solution set is $\{-2i, 2i\}$.

47. $x^2 - 16 = 0$
Here $a = 1$, $b = 0$, $c = -16$, and $b^2 - 4ac = 0 - 4(1)(-16) = 64$.
Then $x = \dfrac{-0 \pm \sqrt{64}}{2(1)} = \dfrac{\pm 8}{2} = \pm 4$
The solution set is $\{-4, 4\}$.

49. $x^2 - 6x + 13 = 0$
Here $a = 1$, $b = -6$, $c = 13$, and $b^2 - 4ac = (-6)^2 - 4(1)(13) = -16$.
Then $x = \dfrac{-(-6) \pm \sqrt{-16}}{2(1)} = \dfrac{6 \pm \sqrt{16}\,i}{2} = \dfrac{6 \pm 4i}{2} = 3 \pm 2i$
The solution set is $\{3 - 2i, 3 + 2i\}$.

51. $x^2 - 6x + 10 = 0$
Here $a = 1$, $b = -6$, $c = 10$, and $b^2 - 4ac = (-6)^2 - 4(1)(10) = -4$.
Then $x = \dfrac{-(-6) \pm \sqrt{-4}}{2(1)} = \dfrac{6 \pm \sqrt{4}\,i}{2} = \dfrac{6 \pm 2i}{2} = 3 \pm i$
The solution set is $\{3 - i, 3 + i\}$.

53. $8x^2 - 4x + 1 = 0$
Here $a = 8$, $b = -4$, $c = 1$, and $b^2 - 4ac = (-4)^2 - 4(8)(1) = -16$.
Then $x = \dfrac{-(-4) \pm \sqrt{-16}}{2(8)} = \dfrac{4 \pm \sqrt{16}\,i}{16} = \dfrac{4 \pm 4i}{16} = \dfrac{1}{4} \pm \dfrac{1}{4}i$
The solution set is $\left\{\dfrac{1}{4} - \dfrac{1}{4}i, \dfrac{1}{4} + \dfrac{1}{4}i\right\}$.

55. $5x^2 + 2x + 1 = 0$
Here $a = 5$, $b = 2$, $c = 1$, and $b^2 - 4ac = 2^2 - 4(5)(1) = -16$.
Then $x = \dfrac{-2 \pm \sqrt{-16}}{2(5)} = \dfrac{-2 \pm \sqrt{16}\,i}{10} = \dfrac{-2 \pm 4i}{10} = -\dfrac{1}{5} \pm \dfrac{2}{5}i$
The solution set is $\left\{-\dfrac{1}{5} - \dfrac{2}{5}i, -\dfrac{1}{5} + \dfrac{2}{5}i\right\}$.

57. $x^2 + x + 1 = 0$
Here $a = 1$, $b = 1$, $c = 1$, and $b^2 - 4ac = 1^2 - 4(1)(1) = -3$.
Then $x = \dfrac{-1 \pm \sqrt{-3}}{2(1)} = \dfrac{-1 \pm \sqrt{3}\,i}{2} = -\dfrac{1}{2} \pm \dfrac{\sqrt{3}}{2}i$
The solution set is $\left\{-\dfrac{1}{2} - \dfrac{\sqrt{3}}{2}i, -\dfrac{1}{2} + \dfrac{\sqrt{3}}{2}i\right\}$.

59. $x^3 - 8 = 0$

$$x^3 - 8 = (x - 2)(x^2 + 2x + 4) = 0$$

$$x - 2 = 0 \qquad x^2 + 2x + 4 = 0$$

$$x = 2$$

Here $a = 1$, $b = 2$, $c = 4$, and $b^2 - 4ac = 2^2 - 4(1)(4) = -12$

Then, $x = \dfrac{-2 \pm \sqrt{-12}}{2(1)} = \dfrac{-2 \pm \sqrt{12}\,i}{2} = \dfrac{-2 \pm \sqrt{3}\,i}{2} = -1 \pm \sqrt{3}\,i$

The solution set is $\left\{2,\ -1 - \sqrt{3}\,i,\ -1 + \sqrt{3}\,i\right\}$.

61. $x^4 - 16 = 0$

$$x^4 - 16 = (x^2 - 4)(x^2 + 4) = (x - 2)(x + 2)(x^2 + 4) = 0$$

$$x - 2 = 0 \qquad x + 2 = 0 \qquad\qquad x^2 + 4 = 0$$

$$x = 2 \qquad\qquad x = -2$$

Here $a = 1$, $b = 0$, $c = 4$, and $b^2 - 4ac = 0^2 - 4(1)(4) = -16$

Then, $x = \dfrac{0 \pm \sqrt{-16}}{2(1)} = \dfrac{\sqrt{16}\,i}{2} = \dfrac{\pm 4i}{2} = \pm 2i$

The solution set is $\{-2,\ 2,\ -2i,\ 2i\}$.

63. $x^4 + 13x^2 + 36 = 0$

$$(x^2 + 9)(x^2 + 4) = 0$$

$$x^2 = -9 \qquad\quad x^2 = -4$$

$$x^2 = 9i^2 \qquad\quad x^2 = 4i^2$$

$$x = \pm 3i \qquad\quad x = \pm 2i$$

The solution set is $\{-3i,\ -2i,\ 2i,\ 3i\}$.

65. $3x^2 - 3x + 4 = 0$

Here $a = 3$, $b = -3$, $c = 4$,
and $b^2 - 4ac = (-3)^2 - 4(3)(4) = -39$.
Hence, this equation has two complex solutions.

67. $2x^2 + 3x - 4 = 0$

Here $a = 2$, $b = 3$, $c = -4$, and
$b^2 - 4ac = 3^2 - 4(2)(-4) = 41$.
Hence, this equation has two unequal real solutions.

69. $9x^2 - 12x + 4 = 0$

Here $a = 9$, $b = -12$, $c = 4$, and
$b^2 - 4ac = (-12)^2 - 4(9)(4) = 0$.
Hence, this equation has a repeated real solution.

71. The other solution must be the conjugate of $2 + 3i$, or $2 - 3i$.

In Problems 73-76, $z = 3 - 4i$ and $w = 8 + 3i$.

73. $z + \overline{z} = 3 - 4i + \overline{(3 - 4i)} = 3 - 4i + (3 + 4i) = (3 + 3) + (-4 + 4)i = 6$

75. $z\overline{z} = (3 - 4i)\overline{(3 - 4i)} = (3 - 4i)(3 + 4i) = 9 + 12i - 12i - 16i^2 = 9 - 16(-1) = 25$

77. For $z = a + bi$, $z + \overline{z} = a + bi + \overline{a + bi} = (a + bi) + (a - bi) = (a + a) + (b - b)i$
$$= 2a + 0i = 2a$$

and $z - \overline{z} = a + bi - \overline{(a + bi)} = a + bi - (a - bi) = (a - a) + (b - (-b))i$
$$= 0 + 2bi = 2bi$$

79. For $z = a + bi$ and $w = c + di$,
$$\overline{z + w} = \overline{(a + bi) + (c + di)} = \overline{(a + c) + (b + d)i} = (a + c) - (b + d)i$$
$$= (a - bi) + (c - di) = \overline{a + bi} + \overline{c + di} = \overline{z} + \overline{w}$$

1. (a)

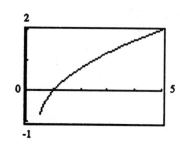

Graph $f(x) = \sqrt{2x - 1} - 1$. Using BOX and TRACE, we conjecture the x-intercept is 1. This is verified since $f(1) = 0$. Thus, the solution is $t = 1$.

(b)
$$\sqrt{2t - 1} = 1$$
$$\left(\sqrt{2t - 1}\right)^2 = 1^2$$
$$2t - 1 = 1$$
$$t = 1$$
Check: $\sqrt{2(1) - 1} = \sqrt{1} = 1$

3. (a)

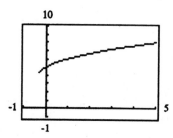

Graph $f(x) = \sqrt{3x + 1} + 4$.
The graph of $f(x)$ does not cross the x-intercept; thus, there is no real solution.

(b)
$$\sqrt{3t + 1} = -4$$
$$3t + 1 = 16$$
$$3t = 15$$
$$t = 5$$
Check: $\sqrt{3(5) + 1} = \sqrt{16}$
$$= 4 \neq -4.$$
No real solution.

5. (a)

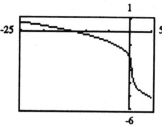

Graph $f(x) = \sqrt[3]{1 - 2x} - 3$. Using BOX and TRACE, we conjecture the x-intercept is -13. Since $f(-13) = 0$, the solution is $x = -13$.

(b)
$$\sqrt[3]{1 - 2x} - 3 = 0$$
$$\sqrt[3]{1 - 2x} = 3$$
$$\left(\sqrt[3]{1 - 2x}\right)^3 = 3^3$$
$$1 - 2x = 27$$
$$-26 = 2x$$
$$-13 = x$$
Check: $\sqrt[3]{1 - 2(-13)} - 3$
$$= \sqrt{27} - 3 = 0$$

7.   (a)
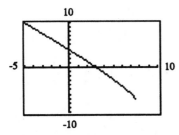

Graph $f(x) = \sqrt{15 - 2x} - x$.
Conjecture $x = 3$ is a solution. Since
$f(3) = 0$, $x = 3$ is the only solution.

(b)
$$\sqrt{15 - 2x} = x$$
$$\left(\sqrt{15 - 2x}\right) = x^2$$
$$15 - 2x = x^2$$
$$-x^2 - 2x + 15 = 0$$
$$x^2 + 2x - 15 = 0$$
$$(x + 5)(x - 3) = 0$$
$$x = -5 \ \text{or} \ x = 3$$
Check 3: $\sqrt{15 - 2(3)} \overset{?}{=} 3$
$$\sqrt{9} \overset{?}{=} 3$$
$$3 = 3$$
Check -5: $\sqrt{15 - 2(-5)} \ne -5$
Does not check.
The solution is $x = 3$.

9.   (a)
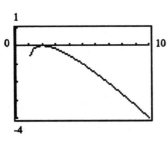

Graph $f(x) = 2\sqrt{x - 1} - x$.
Conjecture $x = 2$ is a solution. Since
$f(2) = 0$, $x = 2$ is the only solution.

(b)
$$x = 2\sqrt{x - 1}$$
$$x^2 = \left(2\sqrt{x - 1}\right)^2$$
$$x^2 = 4(x - 1)$$
$$x = 4x - 4$$
$$x^2 - 4x + 4 = 0$$
$$(x - 2)^2 = 0$$
$$x = 2$$

Check: $2 \overset{?}{=} 2\sqrt{2 - 1}$
$$2 = 2$$

11. (a)
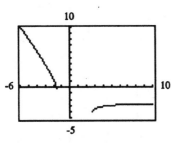

Graph $f(x) = \sqrt{x^2 - x - 4} - x - 2$.
Using BOX and TRACE, the solution is
-1.60 correct to two decimal places.

(b)
$$\sqrt{x^2 - x - 4} = x + 2$$
$$\left(\sqrt{x^2 - x - 4}\right)^2 = (x + 2)^2$$
$$x^2 - x - 4 = x^2 + 4x + 4$$
$$-5x = 8$$
$$x = -\frac{8}{5}$$

Check: $\sqrt{\left(-\dfrac{8}{5}\right)^2 + \dfrac{8}{5} - 4} \overset{?}{=} -\dfrac{8}{5} + 2$
$$\sqrt{\dfrac{64}{25} + \dfrac{40}{25} - \dfrac{100}{25}} \overset{?}{=} \dfrac{2}{5}$$
$$\sqrt{\dfrac{4}{25}} \overset{?}{=} \dfrac{2}{5}$$
$$\dfrac{2}{5} = \dfrac{2}{5}$$

13.   (a)

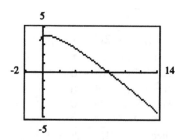

Graph $f(x) = 3 + \sqrt{3x + 1} - x$.
Conjecture $x = 8$ is the only solution.
Since $f(8) = 0$, $x = 8$ is the solution.

(b)

$$3 + \sqrt{3x + 1} = x$$
$$\sqrt{3x + 1} = x - 3$$
$$3x + 1 = (x - 3)^2$$
$$3x + 1 = x^2 - 6x + 9$$
$$0 = x^2 - 9x + 8$$
$$0 = (x - 1)(x - 8)$$
$$x = 1 \qquad x = 8$$

Check 1: $3 + \sqrt{3(1) + 1} \overset{?}{=} 1$
$$3 + 2 \overset{?}{=} 1$$
$$5 \neq 1$$

Check 8: $3 + \sqrt{3(8) + 1} \overset{?}{=} 8$
$$3 + 5 \overset{?}{=} 8$$
$$8 = 8$$

The solution is $x = 8$.

15.   (a)

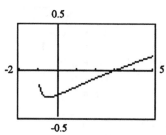

Graph

$$f(x) = \sqrt{2x + 3} - \sqrt{x + 1} - 1.$$
Conjecture $x = 3$ is a solution. Since
$f(3) = 0$, $x = 3$ is a solution. TRACE
reveals $x = -1$ may be a solution.
Since $f(-1) = 0$, $x = -1$ is also a
solution. So, the solution set is
$\{-1, 3\}$.

(b)

$$\sqrt{2x + 3} - \sqrt{x + 1} = 1$$
$$\left(\sqrt{2x + 3}\right)^2 = \left(1 + \sqrt{x + 1}\right)^2$$
$$2x + 3 = 1 + 2\sqrt{x + 1} + x + 1$$
$$x + 1 = 2\sqrt{x + 1}$$
$$(x + 1)^2 = \left(2\sqrt{x + 1}\right)^2$$
$$x^2 + 2x + 1 = 4(x + 1)$$
$$x^2 + 2x + 1 = 4x + 4$$
$$x^2 - 2x - 3 = 0$$
$$(x + 1)(x - 3) = 0$$
$$x = -1 \text{ or } x = 3$$

Check -1:

$$\sqrt{2(-1) + 3} - \sqrt{-1 + 1} \overset{?}{=} 1$$
$$\sqrt{1} - \sqrt{0} \overset{?}{=} 1$$
$$1 = 1$$

Check 3:

$$\sqrt{2(3) + 3} - \sqrt{3 + 1} \overset{?}{=} 1$$
$$\sqrt{9} - \sqrt{4} \overset{?}{=} 1$$
$$3 - 2 \overset{?}{=} 1$$
$$1 = 1$$

The solution set is $\{-1, 3\}$.

17. (a)

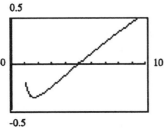

0.5

0

10

-0.5

Graph

$f(x) = \sqrt{3x + 1} - \sqrt{x - 1} - 2.$
Conjecture $x = 5$ is a solution. Since
$f(5) = 0$, $x = 5$ is a solution. TRACE
reveals $x = 1$ may be a solution. Since
$f(1) = 0$, the solution set is $\{1, 5\}$.

(b)
$$\sqrt{3 + 1} - \sqrt{x - 1} = 2$$
$$\sqrt{3x + 1} = 2 + \sqrt{x - 1}$$
$$\left(\sqrt{3x + 1}\right)^2 = \left(2 + \sqrt{x - 1}\right)^2$$
$$3x + 1 = 4 + 4\sqrt{x - 1} + x - 1$$
$$2x - 2 = 4\sqrt{x - 1}$$
$$x - 1 = 2\sqrt{x - 1}$$
$$(x - 1)^2 = \left(2\sqrt{x - 1}\right)^2$$
$$x^2 - 2x + 1 = 4(x - 1)$$
$$x^2 - 6x + 5 = 0$$
$$(x - 1)(x - 5) = 0$$
$$x = 1 \quad \text{or} \quad x = 5$$

Check 1: $\sqrt{3(1) + 1} - \sqrt{x - 1} \overset{?}{=} 2$
$$\sqrt{4} \overset{?}{=} 2$$
$$2 = 2$$

Check 5: $\sqrt{3(5) + 1} - \sqrt{5 - 1} \overset{?}{=} 2$
$$\sqrt{16} - \sqrt{4} \overset{?}{=} 2$$
$$4 - 2 \overset{?}{=} 2$$
$$2 = 2$$

The solution set is $\{1, 5\}$.

19. (a)

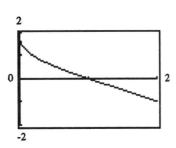

2

0

2

-2

Graph $f(x) = \sqrt{3 - 2\sqrt{x}} - \sqrt{x}$.
Conjecture $x = 1$ is a solution. Since
$f(1) = 0$, $x = 1$ is the solution.

(b)
$$\sqrt{3 - 2\sqrt{x}} = \sqrt{x}$$
$$3 - 2\sqrt{x} = x$$
$$-2\sqrt{x} = x - 3$$
$$4x = x^2 - 6x + 9$$
$$x^2 - 10x + 9 = 0$$
$$(x - 9)(x - 1) = 0$$
$$x = 9, x = 1$$

Check 9: $\sqrt{3 - 2\sqrt{9}} \overset{?}{=} 9$
$$\sqrt{3 - 6} \overset{?}{=} 3$$
$$\sqrt{-3} \neq 3$$
Extraneous Solution.

Check 1: $\sqrt{3 - 2\sqrt{1}} \overset{?}{=} \sqrt{1}$
$$\sqrt{3 - 2} \overset{?}{=} \sqrt{1}$$
$$\sqrt{1} = \sqrt{1}$$
The solution is $x = 1$.

21. (a)

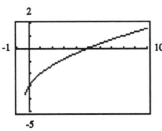

2

-1

10

-5

Graph $f(x) = (3x + 1)^{1/2} - 4$.
Conjecture $x = 5$ is a solution. Since
$f(5) = 0$, $x = 5$ is a solution.

(b)
$$(3x + 1)^{1/2} = 4$$
$$\sqrt{3x + 1} = 4$$
$$3x + 1 = 16$$
$$3x = 15$$
$$x = 5$$

Check: $(3(5) + 1)^{1/2} \overset{?}{=} 4$
$$16^{1/2} \overset{?}{=} 4$$
$$4 = 4$$
The solution is $x = 5$.

23. (a)

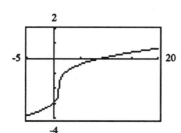

(b)
$$(x - 1)^{1/3} = 2$$
$$\sqrt[3]{x - 1} = 2$$
$$x - 1 = 2^3 = 8$$
$$x = 9$$
Check: $(9 - 1)^{1/3} \overset{?}{=} 2$
$$8^{1/3} \overset{?}{=} 2$$
$$2 = 2$$
The solution is $x = 9$.

Graph $f(x) = (x - 1)^{1/3} - 2$. Conjecture $x = 9$ is the only solution (after using BOX and TRACE). Since $f(9) = 0$, $x = 9$ is the solution.

25. (a)

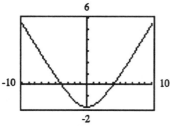

Graph $f(x) = (x^2 + 9)^{1/2} - 5$.
Conjecture $x = 4$ and $x = -4$ are
solutions. Since $f(4) = f(-4) = 0$, the
solution set is $\{-4, 4\}$.

(b)
$$(x^2 + 9)^{1/2} = 5$$
$$\sqrt{x^2 + 9} = 5$$
$$x^2 + 9 = 25$$
$$x^2 = 16$$
$$x = -4, 4$$
Check -4: $((-4)^2 + 9)^{1/2} \overset{?}{=} 5$
$$(25)^{1/2} \overset{?}{=} 5$$
$$5 = 5$$
Check 4: $((4)^2 + 9)^{1/2} \overset{?}{=} 5$
$$25^{1/2} \overset{?}{=} 5$$
$$5 = 5$$
The solution set is $\{-4, 4\}$.

27. (a)

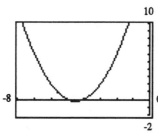

(b)
$$(x + 1)^2 + 7(x + 1) + 12 = 0$$
Let $u = x + 1$
$$u^2 + 7u + 12 = 0$$
$$(u + 4)(u + 3) = 0$$
$$u + 4 = 0 \qquad u + 3 = 0$$
$$u = -4 \qquad u = -3$$
$$x + 1 = -4 \qquad x + 1 = -3$$
$$x = -5 \qquad x = -4$$
$$\{-5, -4\}$$

Graph $f(x) = (x + 1)^2 + 7(x + 1) + 12$. Conjecture $x = -5$ and $x = -4$ are solutions.
Since $f(-5) = f(-4) = 0$, the solution set is $\{-5, -4\}$.

29. (a)

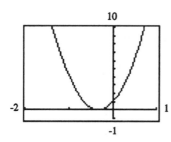

(b)
$$(3x + 4)^2 - 6(3x + 4) + 9 = 0$$
Let $u = 3x + 4$
$$u^2 - 6u + 9 = 0$$
$$(u - 3)^2 = 0$$
$$u = 3$$
$$3x + 4 = 3$$
$$3x = -1$$
$$x = -\frac{1}{3}$$

Graph $f(x) = (3x + 4)^2 - 6(3x + 4) + 9$. Using BOX and TRACE, the solution is
$x = -0.33$ correct to decimal places.

31. **(a)**

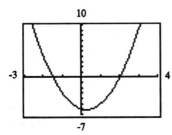

Graph
$$f(x) = 2(x + 1)^2 - 5(x + 1) - 3.$$
Using BOX and TRACE, conjecture
$x = -1.5$ and $x = 2$ are solutions.
Since $f(-1.5) = f(2) = 0$, the solution
set is $\{-1.5, 2\}$.

(b) $2(s + 1)^2 - 5(s + 1) = 3$
Let $u = s + 1$.
Then $u^2 = (s + 1)^2$ and
$$2u^2 - 5u = 3$$
$$2u^2 - 5u - 3 = 0$$
$$(2u + 1)(u - 3) = 0$$

$$u = -\frac{1}{2} \quad \text{or} \quad u = 3$$

$$s + 1 = -\frac{1}{2} \quad \text{or} \quad s + 1 = 3$$

$$s = \frac{-3}{2} \quad \text{or} \quad s = 2$$

Check $\frac{1}{2}$: $2\left[\frac{-3}{2} + 1\right]^2 - 5\left[\frac{-3}{2} + 1\right] \overset{\geq}{=} 3$

$$2\left[\frac{-1}{2}\right](^2 - 5\left[\frac{-1}{2}\right] \overset{\geq}{=} 3$$

$$\frac{1}{2} + \frac{5}{2} \overset{\geq}{=} 3$$

$$3 = 3$$

Check 2: $(2 + 1)^2 - 5(2 + 1) \overset{\geq}{=} 3$

$$18 - 15 \overset{\geq}{=} 3$$

$$3 = 3$$

The solution set is $\left\{-\frac{3}{2}, 2\right\}$.

33. **(a)**

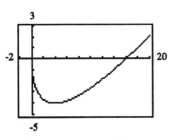

Graph $f(x) = x - 4\sqrt{x}$. It appears the
graph of $f(x)$ crosses the x-axis at
$x = 16$, since $f(16) = 0$, $x = 16$ is a
solution. Using TRACE, it appears
$x = 0$ is also a solution. Since
$f(0) = 0$, the solution set is $\{0, 16\}$.

(b)
$$x - 4\sqrt{x} = 0$$
$$x = 4\sqrt{x}$$
$$x^2 = 16x$$
$$x^2 - 16x = 0$$
$$x(x - 16) = 0$$
$$x = 0 \quad \text{or} \quad x = 16$$
Check 0: $0 - 4\sqrt{0} \overset{\geq}{=} 0$
$$0 = 0$$
16: $16 - 4\sqrt{16} \overset{\geq}{=} 0$
$$0 = 0$$
The solution set is $\{0, 16\}$.

35. (a)

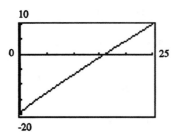

Graph $f(x) = x + \sqrt{x} - 20$. The graph of $f(x)$ appears to cross the x-axis at $x = 16$. Since $f(16) = 0$, $x = 16$ is a solution.

(b)
$$x + \sqrt{x} = 20$$
$$x + \sqrt{x} - 20 = 0$$
Let $u = x^{1/2}$. Then $u^2 = x$, and
$$u^2 + u - 20 = 0$$
$$(u - 4)(u + 5) = 0$$
$$u = 4 \quad \text{or} \quad u = -5$$
$$x^{1/2} = 4 \qquad\qquad x^{1/2} = -5$$
$$\text{(impossible)}$$
$$x = 16$$

Check: $16 + \sqrt{16} \overset{?}{=} 20$
$$16 + 4 \overset{?}{=} 20$$
$$20 = 20$$
The solution is $x = 16$.

37. (a)

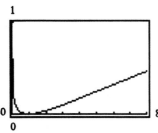

Graph $f(x) = x^{1/2} - 2x^{1/4} + 1$. From the graph, it appears $f(x)$ touches the x-axis at $x = 1$. Since $f(1) = 0$, the solution set is $\{1\}$.

(b) $t^{1/2} - 2t^{1/4} + 1 = 0$
Let $u = t^{1/4}$. Then $u^2 = (t^{1/4})^2 = t^{1/2}$, and
$$u^2 - 2u + 1 = 0$$
$$(u - 1)(u - 1) = 0$$
$$u = 1$$
$$t^{1/4} = 1$$
$$t = 1^4 = 1$$
Check: $1^{1/2} - 2(1)^{1/4} + 1 \overset{?}{=} 0$
$$0 = 0$$

The solution is $t = 1$.

39. (a)

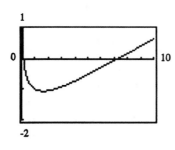

Graph $f(x) = 4x^{1/2} - 9x^{1/4} + 4$. Using BOX and TRACE (or ROOT), the solutions are 0.13 and 7.24 correct to two decimal places.

(b) $4x^{1/2} - 9x^{1/4} + 4 = 0$
Let $u = x^{1/4}$; then $u^2 = x^{1/2}$, and $4u^2 - 9u + 4 = 0$
Here, $a = 4$, $b = -9$, $c = 4$, and $b^2 - 4ac = (-9)^2 - 4(4)(4) = 17$.
Then,

$$u = \frac{-(-9) \pm \sqrt{17}}{2(4)} = \frac{9 \pm \sqrt{17}}{8}$$
$$x^{1/4} = \frac{9 \pm \sqrt{17}}{8}$$
$$x = \left(\frac{9 \pm \sqrt{17}}{8} \right)^4$$

Check $\left[\dfrac{9 - \sqrt{17}}{8}\right]^4$: $4\left(\left[\left(\dfrac{9 - \sqrt{17}}{8}\right)^4\right]^{1/2}\right) - 9\left(\left[\left(\dfrac{9 - \sqrt{17}}{8}\right)^4\right]^{1/4}\right) + 4 \overset{?}{=} 0$

$$4\left[\dfrac{9 - \sqrt{17}}{8}\right]^2 - 9\left[\dfrac{9 - \sqrt{17}}{8}\right] + 4 \overset{?}{=} 0$$

$$\dfrac{81 - 18\sqrt{17} + 17}{16} - \dfrac{81 + 9\sqrt{17}}{8} + 4 \overset{?}{=} 0$$

$$81 - 18\sqrt{17} + 17 - 162 + 18\sqrt{17} + 64 \overset{?}{=} 0$$

$$0 = 0$$

Check $\left[\dfrac{9 + \sqrt{17}}{8}\right]^4$: $4\left(\left[\left(\dfrac{9 + \sqrt{17}}{8}\right)^4\right]^{1/2}\right) - 9\left(\left[\left(\dfrac{9 + \sqrt{17}}{8}\right)^4\right]^{1/4}\right) + 4 \overset{?}{=} 0$

$$4\left[\dfrac{9 + \sqrt{17}}{8}\right]^2 - 9\left[\dfrac{9 + \sqrt{17}}{8}\right] + 4 \overset{?}{=} 0$$

$$\dfrac{81 + 18\sqrt{17} + 17}{16} - \dfrac{81 + 9\sqrt{17}}{8} + 4 \overset{?}{=} 0$$

$$81 + 18\sqrt{17} + 17 - 162 - 18\sqrt{17} + 64 \overset{?}{=} 0$$

$$0 = 0$$

The solution set is $\left\{\left[\dfrac{9 - \sqrt{17}}{8}\right]^4, \left[\dfrac{9 + \sqrt{17}}{8}\right]^4\right\}$.

41. (a)

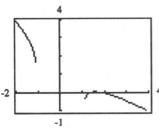

Given $f(x) = \sqrt[4]{5x^2 - 6} - x = (5x^2 - 6)^{1/4} - x$.
Using BOX and TRACE (or ROOT), the solutions are 1.41 and 1.73 correct to two decimal places.

(b) $\sqrt[4]{5x^2 - 6} = x$
$5x^2 - 6 = x^4$
$x^4 - 5x^2 + 6 = 0$
Let $u = x^2$. Then $u^2 = x^4$ and
$u^2 - 5u + 6 = 0$
$(u - 3)(u - 2) = 0$
$u = 2$ or $u = 3$
$x^2 = 2$ or $x^2 = 3$
$x = \pm\sqrt{2}$ or $x = \pm\sqrt{3}$
The solution set is $\left\{\sqrt{2}, \sqrt{3}\right\}$.

Check:

$-\sqrt{3}$: $\sqrt[4]{5\left(-\sqrt{3}\right)^2 - 6} \neq -\sqrt{3}$

$\sqrt{3}$: $\sqrt[4]{5\left(\sqrt{3}\right)^2 - 6} \overset{?}{=} \sqrt{3}$

$\sqrt[4]{9} \overset{?}{=} \sqrt{3}$

$(3^2)^{1/4} \overset{?}{=} \sqrt{3}$

$3^{1/2} = \sqrt{3}$

$-\sqrt{2}$: $\sqrt[4]{5\left(-\sqrt{2}\right)^2 - 6} \neq -\sqrt{2}$

$\sqrt{2}$: $\sqrt[4]{5\left(\sqrt{2}\right)^2 - 6} \overset{?}{=} \sqrt{2}$

$\sqrt[4]{4} \overset{?}{=} \sqrt{2}$

$(2^2)^{1/4} \overset{?}{=} \sqrt{2}$

$2^{1/2} = \sqrt{2}$

43. (a)

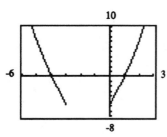

Graph $f(x) = x^2 + 3x + \sqrt{x^2 + 3x} - 6$. It appears the graph crosses the x-axis at $x = -4$ and $x = 1$. Since $f(-4) = f(1) = 0$, the solution set is $\{-4, 1\}$.

(b) $x^2 + 3x + \sqrt{x^2 + 3x} = 6$

Let $u = \sqrt{x^2 + 3x}$. Then $u^2 = x^2 + 3x$ and

$$u^2 + u = 6$$
$$u^2 + u - 6 = 0$$
$$(u + 3)(u - 2) = 0$$
$$u = -3 \quad \text{or} \quad u = 2$$

$\sqrt{x^2 + 3x} = -3$ or $\sqrt{x^2 + 3x} = 2$

No solution. $x^2 + 3x = 4$

$$x^2 + 3x - 4 = 0$$
$$(x + 4)(x - 1) = 0$$
$$x = -4 \text{ or } x = 1$$

Check -4: $(-4)^2 + 3(-4) + \sqrt{(-4)^2 + 3(-4)} \stackrel{?}{=} 6$

$$16 - 12 + \sqrt{4} \stackrel{?}{=} 6$$
$$6 = 6$$

1: $1^2 + 3(1) + \sqrt{1^2 + 3(1)} \stackrel{?}{=} 6$

$$6 = 6$$

The solution set is $\{-4, 1\}$.

45. (a)

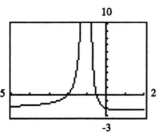

Graph $f(x) = \dfrac{1}{(x + 1)^2} - \dfrac{1}{x + 1} - 2$.

The graph appears to cross the x-axis at $x = -2$ and $x = -\dfrac{1}{2}$. Since $f(-2) = f\left(-\dfrac{1}{2}\right) = 0$, the solution set is $\left\{-2, -\dfrac{1}{2}\right\}$.

(b) $\dfrac{1}{(x + 1)^2} = \dfrac{1}{x + 1} + 2$

Let $u = \dfrac{1}{x + 1}$. Then $u^2 = \dfrac{1}{(x + 1)^2}$ and

$$u^2 = u + 2$$
$$u^2 - u - 2 = 0$$
$$(u + 1)(u - 2) = 0$$
$$u = -1 \text{ or } u = 2$$

$\dfrac{1}{x + 1} = -1$ or $\dfrac{1}{x + 1} = 2$

$x + 1 = -1$ or $x + 1 = \dfrac{1}{2}$

$x = -2$ or $x = -\dfrac{1}{2}$

Check -2: $\dfrac{1}{(-2 + 1)^2} \stackrel{?}{=} \dfrac{1}{-2 + 1} + 2$

$-\dfrac{1}{2}$: $\dfrac{1}{\left(-\dfrac{1}{2} + 1\right)^2} \stackrel{?}{=} \dfrac{1}{-\dfrac{1}{2} + 1} + 2$

$$4 = 4$$

The solution set is $\left\{-2, -\dfrac{1}{2}\right\}$.

47. (a)

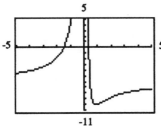

Graph $f(x) = 3x^{-2} - 7x^{-1} - 6$.
Using BOX and TRACE (or ROOT),
the solutions are -1.50 and 0.33
correct to two decimal places.

(b) $3x^{-2} - 7x^{-1} - 6 = 0$
Let $u = x^{-1}$
$$3u^2 - 7u - 6 = 0$$
$$(3u + 2)(u - 3) = 0$$
$$3u = -2 \qquad u = 3$$
$$u = \frac{-2}{3} \qquad \text{or} \quad u = 3$$
$$x = \left(\frac{-2}{3}\right)^{-1} \qquad \text{or} \quad x = (3)^{-1}$$
$$x = \frac{-3}{2} \qquad \text{or} \quad x = \frac{1}{3}$$

The solution set is $\left\{ \frac{-3}{2}, \frac{1}{3} \right\}$.

49. (a)

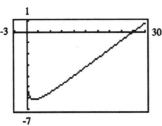

Graph $f(x) = 2x^{2/3} - 5x^{1/3} - 3$. Using
BOX and TRACE (or ROOT), the
graph crosses the x-axis at -0.125 and
27. Since $f(-0.125) = f(27) = 0$, the
solution set is $\{-0.125, 27\}$.

(b) $2x^{2/3} - 5x^{1/3} - 3 = 0$
Let $u = x^{1/3}$
$$2u^2 - 5u - 3 = 0$$
$$(2u + 1)(u - 3) = 0$$
$$2u + 1 = 0 \qquad \text{or} \quad u - 3 = 0$$
$$u = \frac{-1}{2} \qquad \text{or} \qquad u = 3$$
$$x^{1/3} = \frac{-1}{2} \qquad \text{or} \quad x^{1/3} = 3$$
$$x = \left(\frac{-1}{2}\right)^3 \qquad \text{or} \qquad x = (3)^3$$
$$x = \frac{-1}{8} \qquad \text{or} \qquad x = 27$$

The solution set is $\left\{ \frac{-1}{8}, 27 \right\}$.

51. (a)

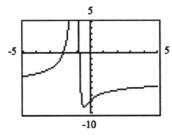

Graph $f(x) = \left(\dfrac{x}{x + 1}\right)^2 + \dfrac{2x}{x + 1} - 8$. The graph
appears to cross the x-axis at $x = -2$ and
$x = -0.80$. Since $f(-2) = f(-0.8) = 0$, the
solution set is $\{-2, -0.8\}$.

(b) $\left[\dfrac{v}{v+1}\right]^2 + \dfrac{2v}{v+1} = 8$

Let $u = \dfrac{v}{v+1}$.

Then $u^2 = \left[\dfrac{v}{v+1}\right]^2$

and
$$u + 2u = 8$$
$$u^2 + 2u - 8 = 0$$
$$(u - 2)(u + 4) = 0$$

$u = 2$ or $u = -4$

$v = 2(v + 1)$ or $v = -4(v + 1)$

$v = 2v + 2$ or $v = -4v - 4$

$-2 = v$ or $5v = -4$

$v = -2$ or $v = -\dfrac{4}{5}$

Check -2: $\left[\dfrac{-2}{-2+1}\right]^2 + \dfrac{2(-2)}{-2+1} \overset{?}{=} 8$

$8 = 8$

Check $-\dfrac{4}{5}$: $\left[\dfrac{-\dfrac{4}{5}}{-\dfrac{4}{5}+1}\right]^2 + \dfrac{2\left[-\dfrac{4}{5}\right]}{-\dfrac{4}{5}+1} \overset{?}{=} 8$

$\left[\dfrac{-\dfrac{4}{5}}{\dfrac{1}{5}}\right]^2 + \dfrac{-\dfrac{8}{5}}{\dfrac{1}{5}} \overset{?}{=} 8$

$16 - 8 \overset{?}{=} 8$

$8 = 8$

The solution set is $\left\{-2, -\dfrac{4}{5}\right\}$.

53.
$$x^3 = x$$
$$x^3 - x = 0$$
$$x(x^2 - 1) = 0$$

$x = 0$ $x^2 - 1 = 0$

$x = 0$ $(x - 1)(x + 1) = 0$

               $x - 1 = 0$ $x + 1 = 0$

$x = 0$ $x = 1$ $x = -1$

$\{-1, 0, 1\}$

55.
$$x^4 - 5x^2 + 4 = 0$$
$$(x^2 - 4)(x^2 - 1) = 0$$

$x^2 - 4 = 0$ $x^2 - 1 = 0$

$(x - 2)(x + 2) = 0$ $(x - 1)(x + 1) = 0$

$x - 2 = 0$ $x + 2 = 0$ $x - 1 = 0$ $x + 1 = 0$

$x = 2$ $x = -2$ $x = 1$ $x = -1$

$\{-2, -1, 1, 2\}$

57.
$$3x^4 - 2x^2 - 1 = 0$$
$$(3x^2 + 1)(x^2 - 1) = 0$$

$x^2 + 1 = 0$ $x^2 - 1 = 0$

$x^2 = -1$ $(x - 1)(x + 1) = 0$

$x^2 = -\dfrac{1}{3}$ $x - 1 = 0$ $x + 1 = 0$

No real solution $x = 1$ $x = -1$

$\{-1, 1\}$

59.
$$x^6 + 7x^3 - 8 = 0$$
$$(x^3 + 8)(x^3 - 1) = 0$$

$$\begin{array}{ll} x^3 + 8 = 0 & \quad x^3 - 1 = 0 \\ x^3 = -8 & \quad x^3 = 1 \\ x = -2 & \quad x = 1 \end{array}$$

$$\{-2, 1\}$$

61.
$$x = 6\sqrt{x}$$
$$x^2 = 36x$$
$$x^2 - 36x = 0$$
$$x(x - 36) = 0$$
$$x = 0, \ x = 36$$

Check 0: $0 = 6\sqrt{0}$
$0 = 0$

Check 36: $36 = 6\sqrt{36}$
$36 = 6 \cdot 6$
$36 = 36$

The solution set is $\{0, 36\}$.

63.
$$x^{3/2} - 2x^{1/2} = 0$$
$$x^{1/2}(x - 2) = 0$$
$$x^{1/2} = 0 \quad \text{or} \quad x = 2$$
$$x = 0$$

Check 0: $0^{3/2} - 2(0)^{1/2} \overset{?}{=} 0$
$0 = 0$

2: $0^{3/2} - 2(2)^{1/2} \overset{?}{=} 0$
$0 = 0$

The solution set is $\{0, 2\}$.

65.
$$x^3 + x^2 - 20x = 0$$
$$x(x^2 + x - 20) = 0$$
$$x(x + 5)(x - 4) = 0$$
$$x = 0 \ \text{ or } \ x = -5 \ \text{ or } \ x = 4$$

Check 0: $0^3 + 0^2 - 20(0) \overset{?}{=} 0$
$0 = 0$

Check -5: $(-5)^3 + (-5)^2 - 20(-5) \overset{?}{=} 0$
$-125 + 25 + 100 \overset{?}{=} 0$
$0 = 0$

Check 4: $(4)^3 + (4)^2 - 20(4) \overset{?}{=} 0$
$64 + 16 - 80 \overset{?}{=} 0$
$0 = 0$

The solution set is $\{-5, 0, 4\}$.

67.
$$x^2 + x^2 + x + 1 = 0$$
$$x^2(x + 1) + (x + 1) = 0$$
$$(x + 1)[x^2 + 1] = 0$$

$$\begin{array}{ll} x + 1 = 0 \ \text{ or } & x^2 + 1 = 0 \\ x = -1 & x^2 = -1 \end{array}$$

No real solution.

Check -1: $(-1)^3 + (-1)^2 + (-1) + 1 \overset{?}{=} 0$
$0 = 0$

The solution set is $\{-1\}$.

69.
$$x^3 - 3x^2 - 4x + 12 = 0$$
$$x^2(x - 3) - 4(x - 3) = 0$$
$$(x^2 - 4)[x - 3] = 0$$
$$(x + 2)(x - 2)(x - 3) = 0$$
$$x = -2 \ \text{ or } \ x = 2 \ \text{ or } x = 3$$

Check -2:
$(-2)^3 - 3(-2)^2 - 4(-2) + 12 \overset{?}{=} 0$
$-8 - 12 + 8 + 12 \overset{?}{=} 0$
$0 = 0$

2: $(2)^3 - 3(2)^2 - 4(2) + 12 \overset{?}{=} 0$
$0 = 0$

3: $(3)^3 - 3(3)^2 - 4(3) + 12 \overset{?}{=} 0$
$0 = 0$

The solution set is $\{-2, 2, 3\}$.

71.
$$t^6 - t^4 - t^2 + 1 = 0$$
$$t^4(t^2 - 1) - (t^2 - 1) = 0$$
$$(t^2 - 1)[t^4 - 1] = 0$$
$$(t + 1)(t - 1)(t^2 + 1)(t^2 - 1) = 0$$
$$(t + 1)^2(t - 1)^2(t^2 + 1)(t + 1)(t - 1) = 0$$
$$(t + 1)^2(t - 1)^2(t^2 + 1) = 0$$
$$t = -1 \quad \text{or} \quad t = 1$$

Check -1: $(-1)^6 - (-1)^4 - (-1)^2 + 1 \overset{?}{=} 0$
$1 - 1 - 1 + 1 \overset{?}{=} 0$
$0 = 0$

1: $1^6 - 1^4 - 1^2 + 1 \overset{?}{=} 0$
$0 = 0$

The solution set is $\{-1, 1\}$.

73. $x - 4x^{1/2} + 2 = 0$

Let $u = x^{1/2}$. Then $u^2 = x$ and $u^2 - 4u + 2 = 0$.

Here $a = 1$, $b = -4$, $c = 2$, and $b^2 - 4ac = (-4)^2 - 4(1)(2) = 8$.

Then, $u = \dfrac{-(-4) \pm \sqrt{8}}{2(1)} = \dfrac{4 \pm 2\sqrt{2}}{2} = 2 \pm \sqrt{2}$

so that

$$u = 2 - \sqrt{2} \qquad \text{or} \qquad u = 2 + \sqrt{2}$$

$$x^{1/2} = 2 - \sqrt{2} \qquad\qquad\qquad x^{1/2} = 2 + \sqrt{2}$$

$$(x^{1/2})^2 = \left(2 - \sqrt{2}\right)^2 \qquad\qquad (x^{1/2})^2 = \left(2 + \sqrt{2}\right)^2$$

$$x = 4 - 4\sqrt{2} + 2 \qquad\qquad x = 4 + 4^2 + 2$$

$$x = 6 - 4\sqrt{2} \qquad\qquad\qquad x = 6 + 4\sqrt{2}$$

$$x = 0.34 \qquad\qquad\qquad\qquad x = 11.66$$

Check 0.34: $0.34 - 4(0.34)^{1/2} + 2 \overset{?}{=} 0$

$$0.34 - 2.33 + 2 \overset{?}{=} 0$$

$$0 = 0$$

11.66: $11.66 - 4(11.66)^{1/2} + 2 \overset{?}{=} 0$

$$11.66 - 13.66 + 2 \overset{?}{=} 0$$

$$0 = 0$$

The solution set is $\{0.34, 11.66\}$.

75. $x^4 + \sqrt{3}\,x^2 - 3 = 0$

Let $u = x^2$. Then $u^2 = x^4$ and $u^2 + \sqrt{3}\,u - 3 = 0$.

Here $a = 1$, $b = \sqrt{3}$, $c = -3$, and $b^2 - 4ac = \left(\sqrt{3}\right)^2 - 4(1)(-3) = 15$.

Then, $u = \dfrac{-\left(\sqrt{3}\right) \pm \sqrt{15}}{2(1)} = \dfrac{-\sqrt{3} \pm \sqrt{15}}{2}$

so that

$$u = \dfrac{-\sqrt{3} - \sqrt{15}}{2} \qquad \text{or} \qquad u = \dfrac{-\sqrt{3} + \sqrt{15}}{2}$$

$$x^2 = \dfrac{-\sqrt{3} - \sqrt{15}}{2} \qquad\qquad x^2 = \dfrac{-\sqrt{3} + \sqrt{15}}{2}$$

$$\text{No real solution.} \qquad\qquad x = \pm \left[\dfrac{-\sqrt{3} + \sqrt{15}}{2}\right]^{1/2}$$

$$x = \pm\, 1.03$$

Check -1.03: $(-1.03)^4 + \sqrt{3}\,(-1.03)^2 - 3 \overset{?}{=} 0$

$$1.13 + \sqrt{3}\,(1.06) - 3 \overset{?}{=} 0$$

$$1.13 + 1.84 - 3 \overset{?}{=} 0$$

$$0 = 0$$

1.03: $(1.03)^4 + \sqrt{3}\,(1.03)^2 - 3 \overset{?}{=} 0$

$$1.13 + \sqrt{3}\,(1.06) - 3 \overset{?}{=} 0$$

$$0 = 0$$

The solution set is $\{-1.03, 1.03\}$.

77. $\pi(1 + t)^2 = \pi + 1 + t$

Let $u = 1 + t$. Then $u^2 = (1 + t)^2$ and

$$\pi u^2 = \pi + u$$

$$\pi u^2 - u - \pi = 0$$

Here $a = \pi$, $b = -1$, $c = -\pi$, and $b^2 - 4ac = (-1)^2 - 4(\pi)(-\pi) = 1 + 4\pi^2$

Then, $u = \dfrac{-(-1) \pm \sqrt{1 + 4\pi^2}}{2(\pi)} = \dfrac{1 \pm \sqrt{1 + 4\pi^2}}{2\pi}$

so that

$$u = \dfrac{1 \pm \sqrt{1 + 4\pi^2}}{2\pi}$$

$$1 + t = \dfrac{1 \pm \sqrt{1 + 4\pi^2}}{2\pi}$$

$$t = \dfrac{1 \pm \sqrt{1 + 4\pi^2}}{2\pi} - 1$$

$$t = \dfrac{1 \pm \sqrt{1 + 4\pi^2}}{2\pi} - 1 \quad \text{or} \quad t = \dfrac{1 + \sqrt{1 + 4\pi^2}}{2\pi} - 1$$

$$= -1.85 \qquad\qquad\qquad = 0.17$$

Check -1.85: $\pi(1 - 1.85)^2 \overset{?}{=} \pi + 1 - 1.85$

$$\pi(.72) \overset{?}{=} 2.29$$

$$2.27 \simeq 2.29$$

0.17: $\pi(1 + 0.17)^2 \overset{?}{=} \pi + 1 + 0.17$

$$\pi(1.37) \overset{?}{=} 4.31$$

$$4.30 \simeq 4.31$$

The solution set is $\{-1.85, 0.17\}$.

79. $k^2 - k = 12$ and $k = \dfrac{x + 3}{x - 3}$

$$k^2 - k - 12 = 0$$

$$(k - 4)(k + 3) = 0$$

$k - 4 = 0$	$k + 3 = 0$
$k = 4$	$k = -3$

$$\dfrac{x + 3}{x - 3} = 4 \qquad\qquad \dfrac{x + 3}{x - 3} = -3$$

$$x + 3 = 4(x - 3) \qquad x + 3 = -3(x - 3)$$

$$x + 3 = 4x - 12 \qquad x + 3 = -3x + 9$$

$$15 = 3x \qquad\qquad 4x = 6$$

$$5 = x \qquad\qquad\quad x = \dfrac{3}{2}$$

$$\left\{\dfrac{3}{2}, 5\right\}$$

81. (a) Total time elapsed $= 4 = \dfrac{\sqrt{s}}{4} + \dfrac{s}{1100}$

$$(1100)(4) = \left[\dfrac{\sqrt{s}}{4} + \dfrac{s}{1100}\right](1100)$$

$$4400 = 275\sqrt{s} + s$$

Let $u = \sqrt{s}$. Then $u^2 = s$ and 4400
$= 275u + u^2$.

$$0 = u^2 + 275u - 4400$$

Here, $a = 1, b = 275, c = -4400$,
and $b^2 - 4ac = (275)^2 - 4(1)(-4400)$
$= 93225$.

Then, $u = \dfrac{-275 \pm \sqrt{93225}}{2(1)}$

$$u = -1160.66 \quad \text{or} \quad u = 15.164$$

$$\sqrt{s} = -1160 \quad \text{or} \quad \sqrt{s} = 15.164$$

No real solution. $s = 229.95$

The depth of the well is 230 ft.

(b)

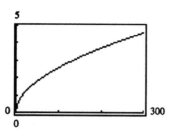

(c)

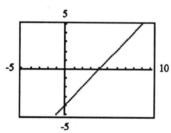

Slope for $x = 0$ to $x = 5$: $m = \dfrac{0.56356 - 0}{5 - 0} = 0.112712$

Slope for $x = 40$ to $x = 45$:

$$m = \dfrac{1.718 - 1.6175}{45 - 40} = 0.0201$$

Slope for $x = 100$ to $x = 105$:

$$m = \dfrac{2.6572 - 2.5909}{105 - 100} = 0.01326$$

As the depth of the well increases, the slope is decreasing.

(d) As the depth of the well increases, the time that elapses before a sound is heard increases at a decreasing rate.

3.6 Inequalities

1. $[0, 2]; 0 \le x \le 2$

3. $(-1, 2); -1 < x < 2$

5. $(-\infty, 0]$ or $(2, \infty); -\infty < x \le 0$ or $2 < x < \infty$

7. $[0, 3); 0 \le x < 3$

9. If $x < 5$, then $x - 5 < 0$

11. If $x > -4$, then $x + 4 > 0$

13. **(a)**

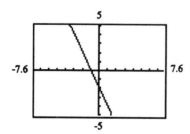

(b)

$$x + 1 < 5$$
$$x + 1 - 1 < 5 - 1$$
$$x < 4$$
$$\{x \mid x < 4\} \text{ or } (-\infty, 4)$$

Graph $f(x) = x - 4$. The x-intercept is $x = 4$. The graph of $f(x)$ is below the x-axis, for $x < 4$. Thus, the solution set is $\{x \mid x < 4\}$.

15. **(a)**

(b)

$$1 - 2x \le 3$$
$$1 - 2x - 1 \le 3 - 1$$
$$-2x \le 2$$
$$\dfrac{-2x}{2} \ge \dfrac{2}{-2}$$
$$x \ge -1$$
$$\{x \mid x \ge -1\} \text{ or } [-1, \infty)$$

Graph $f(x) = -2x - 2$. The x-intercept is $x = -1$. The graph of f is below the x-axis for $x > -1$. Thus, the solution set is $\{x \mid x \ge -1\}$.

17. (a)

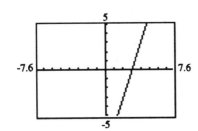

(b)
$$3x - 7 > 2$$
$$3x - 7 + 7 > 2 + 7$$
$$3x > 9$$
$$\frac{3x}{3} > \frac{9}{3}$$
$$x > 3$$
$$\{x \mid x > 3\} \text{ or } (3, \infty)$$

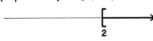

Graph $f(x) = 3x - 9$. The x-intercept is $x = 3$. The graph of f is above the x-axis for $x > 3$. Thus, the solution set is $\{x \mid x > 3\}$.

19. (a)

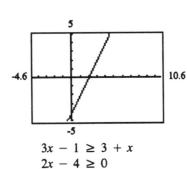

$$3x - 1 \geq 3 + x$$
$$2x - 4 \geq 0$$

(b)
$$3x - 1 \geq 3 + x$$
$$3x - 1 + 1 \geq 3 + x + 1$$
$$3x \geq 4 + x$$
$$3x - x \geq 4 + x - x$$
$$2x \geq 4$$
$$\frac{2x}{2} \geq \frac{4}{2}$$
$$x \geq 2$$
$$\{x \mid x \geq 2\} \text{ or } [2, \infty)$$

Graph $f(x) = 2x - 4$. The x-intercept is $x = 2$. The graph of f is above the x-axis for $x > 2$. Thus, the solution set is $\{x \mid x \geq 2\}$.

21. (a)

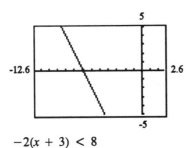

$$-2(x + 3) < 8$$
$$-2x - 6 < 8$$
$$-2x - 14 < 0$$

(b)
$$-2(x + 3) < 8$$
$$\frac{-2(x + 3)}{-2} > \frac{8}{-2}$$
$$x + 3 > -4$$
$$x + 3 - 3 > -4 - 3$$
$$x > -7$$
$$\{x \mid x > -7\} \text{ or } (-7, \infty)$$

Graph $f(x) = -2x - 14$. The x-intercept is $x = -7$. The graph of f is below the x-axis for $x > -7$. Thus, the solution set is $\{x \mid x > -7\}$.

23. (a)

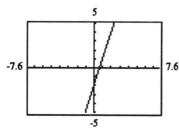

$4 - 3(1 - x) \le 3$
$4 - 3 + 3x \le 3$
$1 + 3x \le 3$
$-2 + 3x \le 0$

Graph $f(x) = 3x - 2$. The x-intercept is $x = 0.66$ correct to two decimal places. The graph of f is below the x-axis for $x < 0.66$. Thus, the solution set is $\{x \mid x \le 0.66\}$.

(b) $4 - 3(1 - x) \le 3$
$4 - 3 + 3x \le 3$
$1 + 3x - 1 \le 3 - 1$
$3x \le 2$
$\dfrac{3x}{3} \le \dfrac{2}{3}$
$x \le \dfrac{2}{3}$

$\left\{x \mid x \le \dfrac{2}{3}\right\}$ or $\left(-\infty, \dfrac{2}{3}\right]$

25. (a)

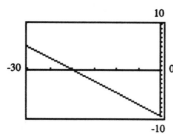

$\dfrac{1}{2}(x - 4) > x + 8$

$\dfrac{1}{2}x - 2 > x + 8$

$-\dfrac{1}{2}x - 10 > 0$

Graph $f(x) = -\dfrac{1}{2}x - 10$. The x-intercept is $x = -20$. The graph is above the x-axis for $x < -20$. Thus, the solution set is $\{x \mid x < -20\}$.

(b) $\dfrac{1}{2}(x - 4) > x + 8$

$\dfrac{1}{2}x - 2 > x + 8$

$\dfrac{1}{2}x - 2 - 8 > x + 8 - 8$

$\dfrac{1}{2}x - 10 > x$

$\dfrac{1}{2}x - 10 - \dfrac{1}{2}x > x - \dfrac{1}{2}x$

$-10 > \dfrac{1}{2}x$

$2(-10) > 2\left[\dfrac{1}{2}x\right]$
$-20 > x$
$x < -20$
$\{x \mid x < -20\}$ or $(-\infty, -20)$

27. (a)

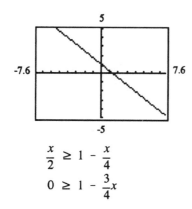

$\dfrac{x}{2} \ge 1 - \dfrac{x}{4}$

$0 \ge 1 - \dfrac{3}{4}x$

(b) $\dfrac{x}{2} \ge 1 - \dfrac{x}{4}$

$\dfrac{x}{2} + \dfrac{x}{4} \ge 1 - \dfrac{x}{4} + \dfrac{x}{4}$

$\dfrac{3x}{4} \ge 1$

$\dfrac{4}{3}\left[\dfrac{3x}{4}\right] \ge \dfrac{4}{3}(1)$

$x \ge \dfrac{4}{3}$

$\left\{x \mid x \ge \dfrac{4}{3}\right\}$ or $\left[\dfrac{4}{3}, \infty\right)$

Graph $f(x) = 1 - \dfrac{3}{4}x$. The x-intercept is 1.33 correct to two decimal places. The graph is below the x-axis for $x > 1.33$. Thus, the solution set is $\{x \mid x \ge 1.33\}$.

29. (a)

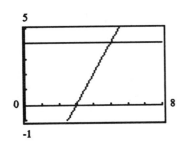

(b)
$$0 \leq 2x - 6 \leq 4$$
$$0 + 6 \leq 2x - 6 + 6 \leq 4 + 6$$
$$6 \leq 2x \leq 10$$
$$\frac{6}{2} \leq \frac{2x}{2} \leq \frac{10}{2}$$
$$3 \leq x \leq 5$$
$$\{x \mid 3 \leq x \leq 5\} \text{ or } [3, 5]$$

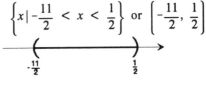

Graph $y_1 = 0$, $y_2 = 2x - 6$, $y_3 = 4$. The intersection points are $x = 3$ and $x = 5$. Since y_2 is between 0 and 4 when $3 \leq x \leq 5$, the solution set is $\{x \mid 3 \leq x \leq 5\}$.

31. (a)

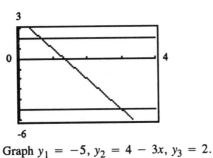

Graph $y_1 = -5$, $y_2 = 4 - 3x$, $y_3 = 2$. The intersection points are $x = 0.66$ and $x = 3$. Since y_2 lies between -5 and 2 when $0.66 \leq x \leq 3$, the solution set is $\{x \mid 0.66 \leq x \leq 3\}$.

(b)
$$-5 \leq 4 - 3x \leq 2$$
$$-5 - 4 \leq -3x \leq 2 - 4$$
$$-9 \leq -3x \leq -2$$
$$\frac{-9}{-3} \leq \frac{-3x}{-3} \leq \frac{-2}{-3}$$
$$3 \geq x \geq \frac{2}{3}$$
$$\frac{2}{3} \leq x \leq 3$$
$$\left\{x \mid \frac{2}{3} \leq x \leq 3\right\} \text{ or } \left[\frac{2}{3}, 3\right]$$

33. (a)

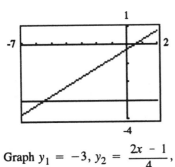

Graph $y_1 = -3$, $y_2 = \dfrac{2x - 1}{4}$, $y_3 = 0$. The intersection points are -5.50 and 0.50. Since y_2 lies between -3 and 0 when $-5.50 < x < 0.50$, the solution set is $\{x \mid -5.50 < x < 0.50\}$.

(b)
$$-3 < \frac{2x - 1}{4} < 0$$
$$(4)(-3) < 4\left[\frac{2x - 1}{4}\right] < 4(0)$$
$$-12 < 2x - 1 < 0$$
$$-12 + 1 < 2x - 1 + 1 < 0 + 1$$
$$-11 < 2x < 1$$
$$\frac{-11}{2} < \frac{2x}{2} < \frac{1}{2}$$
$$\frac{-11}{2} < x < \frac{1}{2}$$
$$\left\{x \mid -\frac{11}{2} < x < \frac{1}{2}\right\} \text{ or } \left(-\frac{11}{2}, \frac{1}{2}\right)$$

35. (a)

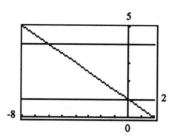

Graph $y_1 = 1$, $y_2 = 1 - \frac{1}{2}x$, $y_3 = 4$.

The intersection points are $x = -6$ and $x = 0$. Since y_2 lies between y_1 and y_3 when $-6 < x < 0$, the solution set is $\{x \mid -6 < x < 0\}$.

(b)

$$1 < 1 - \frac{1}{2}x < 4$$
$$1 - 1 < 1 - \frac{1}{2}x - 1 < 4 - 1$$
$$0 < \frac{-1}{2}x < 3$$
$$(-2)(0) > (-2)\left[-\frac{1}{2}x\right] > (-2)(3)$$
$$0 > x > -6$$
$$-6 < x < 0$$
$$\{x \mid -6 < x < 0\} \text{ or } (-6, 0)$$

37. (a)

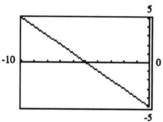

$$(x + 2)(x - 3) > (x - 1)(x + 1)$$
$$x^2 - x - 6 > x^2 - 1$$
$$-x - 5 > 0$$

Graph $f(x) = -x - 5$. The x-intercept is $x = -5$. Since $f(x)$ is above the x-axis for $x < -5$, the solution set is $\{x \mid x < -5\}$.

(b)

$$(x + 2)(x - 3) > (x - 1)(x + 1)$$
$$x^2 - x - 6 > x^2 - 1$$
$$x^2 - x - 6 - x^2 > x^2 - 1 - x^2$$
$$-x - 6 > -1$$
$$-x - 6 + 6 > -1 + 6$$
$$-x > 5$$
$$\frac{-x}{-1} < \frac{5}{-1}$$
$$x < -5$$
$$\{x \mid x < -5\} \text{ or } (-\infty, -5)$$

39. (a)

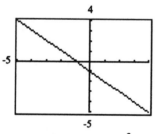

$$x(4x + 3) \leq (2x + 1)^2$$
$$4x^2 + 3x \leq 4x^2 + 4x + 1$$
$$-x - 1 \leq 0$$

Graph $f(x) = -x - 1$. The x-intercept is $x = -1$. Since f is below the x-axis for $x > -1$, the solution set is $\{x \mid x \geq -1\}$.

(b)

$$x(4x + 3) \leq (2x + 1)^2$$
$$4x^2 + 3x \leq 4x^2 + 4x + 1$$
$$4x^2 + 3x - 4x^2 \leq 4x^2 + 4x + 1 - 4x^2$$
$$3x \leq 4x + 1$$
$$3x - 3x \leq 4x + 1 - 3x$$
$$0 \leq x + 1$$
$$0 - 1 \leq x + 1 - 1$$
$$-1 \leq x$$
$$x \geq -1$$
$$\{x \mid x \geq -1\} \text{ or } [-1, \infty)$$

41. **(a)**

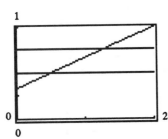

Graph $y_1 = \dfrac{1}{2}$, $y_2 = \dfrac{x+1}{3}$, $y_3 = \dfrac{3}{4}$.
The intersection points are $x = 0.50$
and $x = 1.25$. Since y_2 lies between y_1
and y_3 when $0.50 \leq x < 1.25$, the
solution set is $\{x \mid 0.50 \leq x < 1.25\}$.

(b)

$$\frac{1}{2} \leq \frac{x+1}{3} < \frac{3}{4}$$

$$3\left[\frac{1}{2}\right] \leq 3\left[\frac{x+1}{3}\right] < 3\left[\frac{3}{4}\right]$$

$$\frac{3}{2} \leq x + 1 < \frac{9}{4}$$

$$\frac{3}{2} - 1 \leq x + 1 - 1 < \frac{9}{4} - 1$$

$$\frac{1}{2} \leq x < \frac{5}{4}$$

$$\left\{x \mid \frac{1}{2} \leq x < \frac{5}{4}\right\} \text{ or } \left[\frac{1}{2}, \frac{5}{4}\right]$$

43. **(a)**

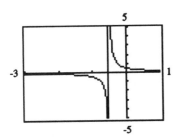

Graph $f(x) = (4x + 2)^{-1}$. The graph of $f(x)$ is below the x-axis when $x < -\dfrac{1}{2}$. Thus, the

solution is $\left\{x \mid x < -\dfrac{1}{2}\right\}$.

(b)

$$(4x + 2)^{-1} < 0$$

$$4x + 2 < 0$$

$$\left\{x \mid x < -\frac{1}{2}\right\} \text{ or } \left(-\infty, -\frac{1}{2}\right]$$

45. **(a)**

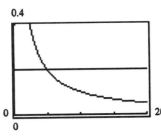

Graph $y_1 = 0$, $y_2 = \dfrac{1}{x}$, $y_3 = \dfrac{1}{5}$. y_2 and y_3 intersect at $x = 5$. y_1 and y_2 do not intersect.
The solution is $\{x \mid x > 5\}$.

(b)

$$0 < \frac{1}{x} < \frac{1}{5}$$

$$x > 5$$

$$\{x \mid x > 5\} \text{ or } (5, \infty)$$

47. **(a)**

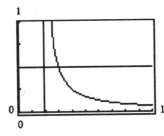

Graph $y_1 = 0$, $y_2 = (2x - 4)^{-1}$, $y_3 = \dfrac{1}{2}$. y_2 and y_3 intersect at $x = 3$, y_1 and y_2 do not

intersect. Thus, the solution is $\{x \mid x > 3\}$.

(b)

$$0 < (2x - 4)^{-1} < \frac{1}{2}$$

$$2x - 4 > 2$$

$$2x > 6$$

$$x > 3$$

$$\{x \mid x > 3\} \text{ or } (3, \infty)$$

49.
$$-1 < x < 1$$
$$-1 + 4 < x + 4 < 1 + 4$$
$$3 < x + 4 < 5$$
$$a = 3, b = 5$$

51.
$$2 < x < 3$$
$$-4(2) > -4x > -4(3)$$
$$-8 > -4x > -12$$
$$-12 < -4x < -8$$
$$a = -12, b = -8$$

53.
$$0 < x < 4$$
$$2(0) < 2x < 2(4)$$
$$0 < 2x < 8$$
$$0 + 3 < 2x + 3 < 8 + 3$$
$$3 < 2x + 3 < 11$$
$$a = 3, b = 11$$

55.
$$-3 < x < 0$$
$$-3 + 4 < x + 4 < 0 + 4$$
$$1 < x + 4 < 4$$
$$\frac{1}{1} > \frac{1}{x + 4} > \frac{1}{4}$$
$$\frac{1}{4} < \frac{1}{x + 4} < 1$$
$$a = \frac{1}{4}, b = 1$$

57.
$$6 < 3x < 12$$
$$\frac{1}{3}(6) < \frac{1}{3}(3x) < \frac{1}{3}(12)$$
$$2 < x < 4$$
$$2^2 < x^2 < 4^2$$
$$4 < x^2 < 16$$
$$a = 4, b = 16$$

59.
$$-10 < x < 5$$

$$-10 + 10 < x + 10 < 5 + 10 \quad \text{and} \quad -10 - 5 < x - 5 < 5 - 5$$
$$0 < x + 10 < 15 \quad \text{and} \quad -15 < x - 5 < 0$$
$$\therefore x + 10 > 0 \quad \text{and} \quad x - 5 < 0$$
$$x > -10 \quad \text{and} \quad x < 5$$

$$2 > \frac{(-1)(x + 10)}{(x - 5)}$$
$$2(x - 5) < (-1)(x + 10)$$
$$2x - 10 < -x - 10$$
$$3x < 0$$
$$x < 0$$
$$\{x \mid -10 < x < 0\}$$

61. $\sqrt{3x + 6}$ is defined when
$$3x + 6 \geq 0$$
$$3x \geq -6$$
$$x \geq -2$$
For real numbers $\{x \mid x \geq -2\}$ or
$[-2, \infty)$, $\sqrt{3x + 6}$ is defined.

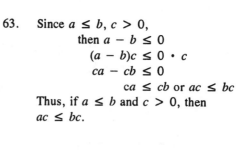

-2

63. Since $a \leq b$, $c > 0$,
$$\text{then } a - b \leq 0$$
$$(a - b)c \leq 0 \cdot c$$
$$ca - cb \leq 0$$
$$ca \leq cb \text{ or } ac \leq bc$$
Thus, if $a \leq b$ and $c > 0$, then
$ac \leq bc$.

Chapter 3 Equations and Inequalities

65. Since $a < b$, $a \cdot \dfrac{1}{2} < b \cdot \dfrac{1}{2}$

$$\frac{a}{2} < \frac{b}{2}$$

$$\frac{a}{2} + \frac{a}{2} < \frac{b}{2} + \frac{a}{2}$$

$$a < \frac{a + b}{2}$$

Also,

$$\frac{a}{2} < \frac{b}{2}$$

$$\frac{a}{2} + \frac{b}{2} < \frac{b}{2} + \frac{b}{2}$$

$$\frac{a + b}{2} < b$$

Thus, $\quad a < \dfrac{a + b}{2} < b$

---

67. If $0 < a < b$, then $\left(\sqrt{ab}\right)^2 - a^2 = ab - a^2 = a(b - a) > 0$.

Therefore, $\left(\sqrt{ab}\right)^2 > a^2$. Hence, $\sqrt{ab} > a$.

$$b^2 - \left(\sqrt{ab}\right)^2 = b^2 - ab = b(b - a) > 0$$

Therefore, $b^2 > \left(\sqrt{ab}\right)^2$. Hence, $b > \sqrt{ab}$.

Combining the inequalities, $a < \sqrt{ab} < b$ and $\sqrt{ab}$ is the geometric mean of a and b.

---

69. For $0 < a < b$, $\dfrac{1}{h} = \dfrac{1}{2}\left[\dfrac{1}{a} + \dfrac{1}{b}\right]$

$$h \cdot \frac{1}{h} = \frac{1}{2}\left[\frac{b + a}{ab}\right] \cdot h$$

$$1 = \frac{1}{2}\left[\frac{b + a}{ab}\right] \cdot h$$

$$2 = \left[\frac{b + a}{ab}\right]h$$

$$\frac{2ab}{a + b} = h$$

$$h - a = \frac{2ab}{a + b} - a$$

$$= \frac{2ab - a(a + b)}{a + b}$$

$$= \frac{2ab - a^2 - ab}{a + b}$$

$$= \frac{ab - a^2}{a + b}$$

$$= \frac{a(b - a)}{a + b} > 0$$

Therefore, $h > a$.

$$b - h = b - \frac{2ab}{a + b}$$

$$= \frac{b(a + b) - 2ab}{a + b}$$

$$= \frac{ab + b^2 - 2ab}{a + b}$$

$$= \frac{b^2 - ab}{a + b}$$

$$= \frac{b(b - a)}{a + b} > 0$$

Therefore, $h < b$.
Combining these inequalities,
$a < h < b$.

---

71. $21 <$ young adult's age < 30

---

73. (a) An average 25-year-old male can expect to live at least 48.4 more years. $25 + 48.4 = 73.4$. Therefore, average male lives ≥ 73.4.

(b) An average 25-year-old female can expect to live at least 54.7 more years. $25 + 54.7 = 79.7$. Therefore, average female lives ≥ 79.7.

(c) By the above information, a female can expect to live longer by 6.3 years.

75. Let P represent the selling price and C the commission. Then

$$C = 45{,}000 + 0.25\,(P - 900{,}000) = 0.25P - 180{,}000 \text{ and we are given that}$$

$$
\begin{array}{rcl}
900{,}000 \le & P & \le 1{,}100{,}000 \\
(0.25)(900{,}000) \le & 0.25P & \le (0.25)(1{,}100{,}000) \\
225{,}000 \le & 0.25P & \le 275{,}000 \\
225{,}000 - 180{,}000 \le & 0.25P - 180{,}000 & \le 275{,}000 - 180{,}000 \\
45{,}000 \le & C & \le 95{,}000
\end{array}
$$

The agent's commission ranges from \$45,000 to \$95,000, inclusive.

$$\frac{45{,}000}{900{,}000} = 0.05 = 5\% \text{ to } \frac{95{,}000}{1{,}100{,}000} = 0.086 = 8.6\%, \text{ inclusive}$$

As a percent of selling price, the commission ranges from 5% to 8.6%.

Check: For a \$900,000 complex, the commission is \$45,000 or 5%.

For a \$1,100,000 complex, the commission is

$$\$45{,}000 + 0.25(1{,}100{,}000 - 900{,}000) = \$95{,}000 \text{ or } 8.6\%.$$

77. Let $W =$ weekly wage and T the withholding tax. Then

$$T = 63.90 + .28(W - 476) = 63.90 + .28W - 133.28 \text{ and since}$$

$$
\begin{array}{rcl}
500 \le & W & \le 550 \\
(0.28)(500) \le & 0.28W & \le 0.28(550) \\
140 \le & 0.28W & \le 154 \\
140 - 69.38 \le & 0.28W - 69.38 & \le 154 - 69.38 \\
70.62 \le & T & \le 84.62
\end{array}
$$

The amount of withholding ranges from \$70.62 to \$84.62, inclusive.

Check: For a wage of \$500, the withholding is

$$63.90 + 0.28(500 - 476) = \$70.62$$

For a wage of \$550, the withholding is

$$63.90 + 0.28(550 - 476) = \$84.62$$

79. Let $K =$ the monthly usage in kilowatt hours and $C =$ the total monthly charges per customer. Then

$$C = 0.12255K + 11.24 \text{ and}$$

$$
\begin{array}{rcl}
94.02 \le & C & \le 317.72 \\
94.02 \le & 0.12255K + 11.24 & \le 317.72 \\
82.78 \le & 0.12255K & \le 306.48 \\
675.48 \le & K & \le 2500.86
\end{array}
$$

The range of usage in kilowatt hours varied from 675.48 to 2500.86.

Check: For 675.48 kilowatt hours, the charge is $0.12255(675.48) + 11.24 = \$94.02$.

For 2500.86 kilowatt hours, the charge is $0.12255(2500.86) + 11.24 = \$317.72$

81. Let $C =$ the dealer cost and $M =$ the markup over dealer's cost. If the price is \$8800, then

$$8800 = C + MC = C(1 + M) \text{ which gives}$$

$$C = \frac{8000}{1 + M}$$

Also

$$
\begin{array}{rcl}
0.12 \le & M & \le 0.18 \\
1.12 \le & 1 + M & \le 1.18 \\
\dfrac{1}{1.12} \ge & \dfrac{1}{1 + M} & \ge \dfrac{1}{1.18} \\
\dfrac{800}{1.12} \ge & \dfrac{8800}{1 + M} & \ge \dfrac{8800}{1.18} \\
7857.14 \ge & C & \ge 7457.63 \\
7457.63 \le & C & \le 7857.14
\end{array}
$$

The cost ranged from \$7457.63 to \$7857.14, inclusive.

Check: For a cost of \$7457.63 and markup of 18%, $7457.63 + 0.18(7457.63) = \$8800$.

For a cost of \$7857.14 and markup of 12%, $7857.14 + 0.12(7857.14) = \$8800$.

83. Let T = the score on the last test and G = the resulting grade. Then
$$G = (68 + 82 + 87 + 89 + T) \div 5 \text{ so that}$$
$$T = 5G - 326 \text{ and}$$
$$80 \le \quad G \quad \le 90$$
$$400 \le \quad 5G \quad \le 450$$
$$74 \le 5G - 326 \quad \le 124$$
The fifth test score must be 74 or greater (≥ 74).

Check: $G = \dfrac{70 + 82 + 85 + 89 + 74}{5} = 80$

85. Let G = the amount (in gallons) of gasoline in the car at the start of the trip, and let D = the distance covered in the trip. Then $D = 25G$ or $G = \dfrac{D}{25}$, and
$$300 \le D$$
$$12 \le \dfrac{D}{25}$$
$$12 \le G$$
The amount of gasoline ranged from 12 to 20 gallons, inclusive, at the beginning of the trip.
Check: For 12 gallons at 25 miles per gallon the car will go $(12)(25) = 300$ miles.

3.7 Other Inequalities

1. (a)

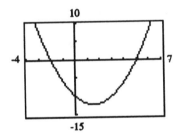

(b) $(x - 5)(x + 2) < 0$

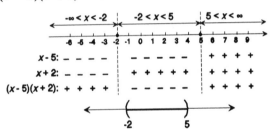

The solution set is $\{x \mid -2 < x < 5\}$.

Graph $f(x) = (x - 5)(x + 2)$. The x-intercepts are $x = 5$ and $x = -2$. The graph of f is below the x-axis for $-2 < x < 5$. Thus, the solution set is $\{x \mid -2 < x < 5\}$.

3. (a)

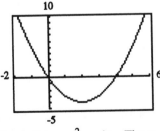

Graph $f(x) = x^2 - 4x$. The x-intercepts are $x = 0$ and $x = 4$. The graph of f is above the x-axis for $x < 0$ and $x > 4$. Thus, the solution set is $\{x \mid x < 0$ or $x > 4\}$.

(b) $x^2 - 4x > 0$
$x(x - 4) > 0$

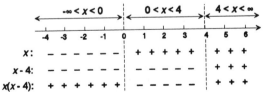

$x(x - 4) > 0$ if $-\infty < x < 0$ or $4 < x < \infty$
The solution set is $\{x \mid -\infty < x < 0$ or $4 < x < \infty\}$.

5. (a)

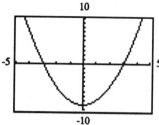

Graph $f(x) = x^2 - 9$. The x-intercepts are $x = -3$ and $x = 3$. The graph of f is below the x-axis for $-3 < x < 3$. The solution set is $\{x \mid -3 < x < 3\}$.

(b)
$$x^2 - 9 < 0$$
$$(x - 3)(x + 3) < 0$$

	$-\infty < x < -3$	$-3 < x < 3$	$3 < x < \infty$
	-7 -6 -5 -4 -3 -2 -1 0 1 2 3 4 5 6 7 8 9		
$x-3$:	– – –	– – – – – –	+ + + + +
$x+3$:	– – –	+ + + + + +	+ + + + +
$(x-3)(x+3)$:	+ + +	– – – – – –	+ + + + +

$(x - 3)(x + 3) < 0$ if $-3 < x < 3$
The solution set is $\{x \mid -3 < x < 3\}$.

7. (a)

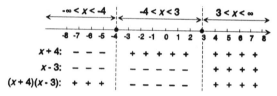

$$x^2 + x > 12$$
$$x^2 + x - 12 > 0$$

Graph $f(x) = x^2 + x - 12$. The x-intercepts are $x = -4$ and $x = 3$. The graph of f is above the x-axis if $x < -4$ or $x > 3$. The solution set is $\{x \mid x < -4$ or $x > 3\}$.

(b)
$$x^2 + x > 12$$
$$x^2 + x - 12 > 0$$
$$(x + 4)(x - 3) > 0$$

	$-\infty < x < -4$	$-4 < x < 3$	$3 < x < \infty$
	-8 -7 -6 -5 -4 -3 -2 -1 0 1 2 3 4 5 6 7 8		
$x+4$:	– – –	+ + + + + +	+ + + +
$x-3$:	– – –	– – – – – –	+ + + +
$(x+4)(x-3)$:	+ + +	– – – – – –	+ + + +

$(x + 4)(x - 3) > 0$ if $-\infty < x < -4$ or $3 < x < \infty$
The solution set is $\{x \mid -\infty < x < -4$ or $3 < x < \infty\}$.

9. (a)

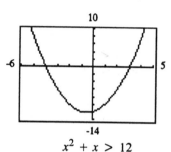

$$2x^2 < 5x + 3$$
$$2x^2 - 5x - 3 < 0$$

Graph $f(x) = 2x^2 - 5x - 3$. The x-intercepts are $x = -0.50$ and $x = 3$. The graph of f is below the x-axis for $-0.50 < x < 3$. The solution set is $\{x \mid -0.50 < x < 3\}$.

(b)
$$2x^2 < 5x + 3$$
$$2x^2 - 5x - 3 < 0$$
$$(x - 3)(2x + 1) < 0$$

	$-\infty < x < -\frac{1}{2}$	$-\frac{1}{2} < x < 3$	$3 < x < \infty$
	-5 -4 -3 -2 -1 0 1 2 3 4 5 6 7 8		
$x-3$:	– – – –	– – – –	+ + + +
$2x+1$:	– – – –	+ + +	+ + + +
$(x-3)(2x+1)$:	+ + + +	– – –	+ + + +

$(x - 3)(2x + 1) < 0$ if $-\dfrac{1}{2} < x < 3$

The solution set is $\left\{x \mid -\dfrac{1}{2} < x < 3\right\}$.

11. **(a)**

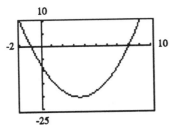

$$x(x - 7) > 8$$
$$x^2 - 7x - 8 > 0$$

Graph $f(x) = x^2 - 7x - 8$. The x-intercepts are $x = -1$ and $x = 8$. The graph of f is above the x-axis for $x < -1$ or $x > 8$. The solution set is $\{x \mid x < -1$ or $x > 8\}$.

(b)

$$x(x - 7) > 8$$
$$x^2 - 7x - 8 > 0$$
$$(x - 8)(x + 1) > 0$$

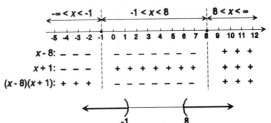

$(x - 8)(x + 1) > 0$ if $-\infty < x < -1$ or $8 < x < \infty$
The solution set is $\{x \mid -\infty < x < -1$ or $8 < x < \infty\}$.

13. **(a)**

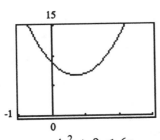

$$4x^2 + 9 < 6x$$
$$4x^2 - 6x + 9 < 0$$

Graph $f(x) = 4x^2 - 6x + 9$. f has no x-intercepts and thus f never is below the x-axis. Therefore, there is no real solution.

(b)

$$4x^2 + 9 < 6x$$
$$4x^2 - 6x + 9 < 0$$

Since for $a = 4, b = -6, c = 9$, $b^2 - 4ac = -108 < 0$, then $4x^2 - 6x + 9$ has no real roots. For $x = 0, 4x^2 - 6 + 9 = 9 > 0$. Thus, there is no real solution.

15. **(a)**

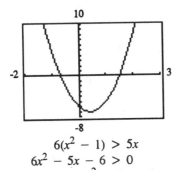

$$6(x^2 - 1) > 5x$$
$$6x^2 - 5x - 6 > 0$$

Graph $f(x) = 6x^2 - 5x - 6$. The graph of f has two x-intercepts at $x = -0.66$ and $x = 1.50$. The graph of f is above the x-axis for $x < -0.66$ or $x > 1.50$. The solution set is $\{x \mid x < -0.66$ or $x > 1.50\}$.

(b)

$$6(x^2 - 1) > 5x$$
$$6x^2 - 5x - 6 > 0$$
$$(2x - 3)(3x + 2) > 0$$

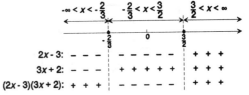

$(2x - 3)(3x + 2) > 0$ if $-\infty < x < -\dfrac{2}{3}$

or $\dfrac{3}{2} < x < \infty$

The solution set is

$$\left\{ x \mid -\infty\ x < -\frac{2}{3} \text{ or } \frac{3}{2} < x < \infty \right\}.$$

17. (a)

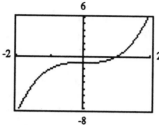

$$(x - 1)(x^2 + x + 1) > 0$$
$$x^3 - 1 > 0$$

Graph $f(x) = x^3 - 1$. The graph of f has one x-intercept at $x = 1$. The graph of f lies above the x-axis for $x > 1$. Thus, the solution set is $\{x \mid x > 1\}$.

(b) $(x - 1)(x^2 + x + 1) > 0$

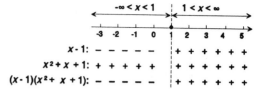

$(x - 1)(x^2 + x + 1) > 0$ if $1 < x < \infty$
The solution set is $\{x \mid 1 < x < \infty\}$.

19. (a)

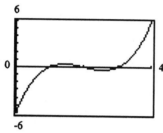

Graph
$$f(x) = (x - 1)(x - 2)(x - 3).$$
The x-intercepts are $x = 1$, $x = 2$, $x = 3$. The graph of f is below the x-axis for $x < 1$ or $2 < x < 3$. The solution set is $\{x \mid x < 1 \text{ or } 2 < x < 3\}$.

(b) $(x - 1)(x - 2)(x - 3) < 0$

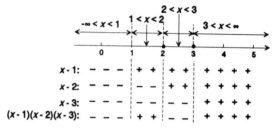

$(x - 1)(x - 2)(x - 3) < 0$ if $-\infty < x < 1$ or $2 < x < 3$
The solution set is $\{x \mid -\infty < x < 1 \text{ or } 2 < x < 3\}$.

21. (a)

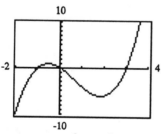

Graph $f(x) = x^3 - 2x^2 - 3x$. The x-intercepts are $x = -1$, $x = 0$, and $x = 3$. The graph of f is above the x-axis for $-1 < x < 0$ or $x > 3$. The solution set is $\{x \mid -1 < x < 0 \text{ or } x > 3\}$.

(b)
$$x^3 - 2x^2 - 3x > 0$$
$$x(x^2 - 2x - 3) > 0$$
$$x(x - 3)(x + 1) > 0$$

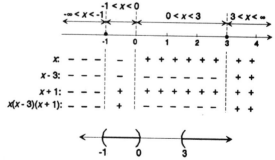

The solution set is $\{x \mid -1 < x < 0 \text{ or } 3 < x < \infty\}$.

23. (a)

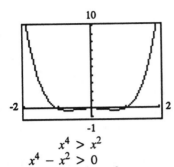

$$x^4 > x^2$$
$$x^4 - x^2 > 0$$

Graph $f(x) = x^4 - x^2$. The
x-intercepts are $x = -1$, $x = 0$,
and $x = 1$. The graph of f is
above the x-axis for $x < -1$ or
$x > 1$. The solution set is
$\{x \mid x < -1 \text{ or } x > 1\}$.

(b)

$$x^4 > x^2$$
$$x^4 - x^2 > 0$$
$$x^2(x^2 - 1) > 0$$
$$x^2(x + 1)(x - 1) > 0$$

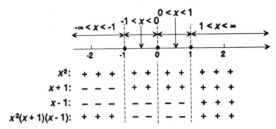

$x^2(x + 1)(x + 1) > 0$ if $-\infty < x < -1$
or $1 < x < \infty$
The solution set is $\{x \mid -\infty < x < -1$
or $1 < x < \infty\}$.

25. (a)

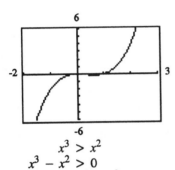

$$x^3 > x^2$$
$$x^3 - x^2 > 0$$

Graph $f(x) = x^3 - x^2$. The
x-intercepts are $x = 0$ and
$x = 1$. The graph of f is above
the x-axis for $x > 1$. The
solution set is $\{x \mid x > 1\}$.

(b)

$$x^3 > x^2$$
$$x^3 - x^2 > 0$$
$$x^2(x - 1) > 0$$

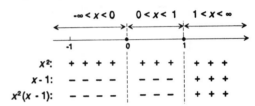

$x^2(x - 1) > 0$ if $1 < x < \infty$
The solution set is $\{x \mid 1 < x < \infty\}$.

27. (a)

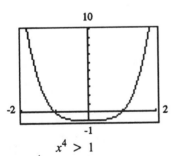

$$x^4 > 1$$
$$x^4 - 1 > 0$$

Graph $f(x) = x^4 - 1$. The
x-intercepts are $x = -1$ and
$x = 1$. The graph of f is above
the x-axis for $x < -1$ or $x > 1$.
The solution set is $\{x \mid x < -1$
or $x > 1\}$.

(b)

$$x^4 > 1$$
$$x^4 - 1 > 0$$
$$(x^2 + 1)(x^2 - 1) > 0$$
$$(x^2 + 1)(x - 1)(x + 1) > 0$$

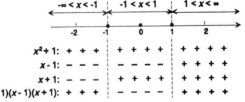

$(x^2 + 1)(x - 1)(x + 1) > 0$ if $-\infty < x < -1$
or $1 < x < \infty$
The solution set is $\{x \mid -\infty < x < -1$
or $1 < x < \infty\}$.

29. (a)

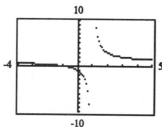

Graph $f(x) = \dfrac{x + 1}{x - 1}$. The x-intercept is $x = -1$. Note, however, $x = 1$ is a critical value. The graph of f is above the x-axis for $x < -1$ or $x > 1$. The solution set is $\{x \mid x < -1$ or $x > 1\}$.

(b) $\dfrac{x + 1}{x - 1} > 0$

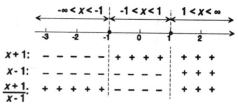

$\dfrac{x + 1}{x - 1} > 0$ if $-\infty < x < -1$ or $1 < x < \infty$

The solution set is $\{x \mid -\infty < x < -1$ or $1 < x < \infty\}$.

31. (a)

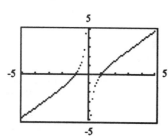

Graph $f(x) = \dfrac{(x - 1)(x + 1)}{x}$.
The x-intercepts are $x = -1$ and $x = 1$. Also, $x = 0$ is a critical value. The graph of f is below the x-axis for $x < -1$ or $0 < x < 1$. The solution set is $\{x \mid x < -1$ or $0 < x < 1\}$.

(b) $\dfrac{(x - 1)(x + 1)}{x} < 0$

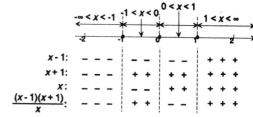

$\dfrac{(x - 1)(x + 1)}{x} < 0$ if $-\infty < x < -1$ or $0 < x < 1$

The solution set is $\{x \mid -\infty < x < -1$ or $0 < x < 1\}$.

33. (a)

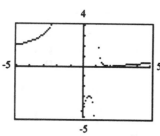

Graph $f(x) = \dfrac{(x - 2)^2}{x^2 - 1}$. The x-intercept is $x = 2$. Also, $x = -1$ and $x = 1$ are critical values. The graph of f is above the x-axis for $x < -1$ and $x > 1$ (the graph touches the x-axis at $x = 2$, but this value of x still satisfies the inequality). The solution set is $\{x \mid x < -1$ or $x > 1\}$.

(b)

$$\dfrac{(x - 2)^2}{x^2 - 1} \geq 0$$

$$\dfrac{(x - 2^2)}{(x - 1)(x + 1)} \geq 0$$

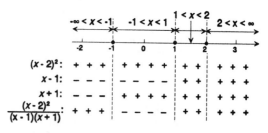

$\dfrac{(x - 2)^2}{(x - 1)(x + 1)} \geq 0$ if $-\infty < x < -1$ or $1 < x < \infty$

The solution set is $\{x \mid -\infty < x < -1$ or $1 < x < \infty\}$.

35. (a)

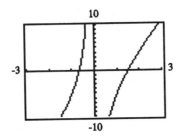

$$6x - 5 < \frac{6}{x}$$

$$6x - 5 - \frac{6}{x} < 0$$

Graph $f(x) = 6x - 5 - \frac{6}{x}$. The x-intercepts are $x = -0.66$ and $x = 1.50$. Also, $x = 0$ is a critical value. The graph of f is below the x-axis for $x < -0.66$ or $0 < x < 1.50$. The solution set is $\{x \mid x < -0.66$ or $0 < x < 1.50\}$.

(b)

$$6x - 5 < \frac{6}{x}$$

$$6x - 5 - \frac{6}{x} < 0$$

$$\frac{6x^2 - 5x - 6}{x} < 0$$

$$\frac{(2x - 3)(3x + 2)}{x} < 0$$

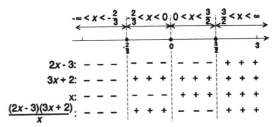

$$\frac{(2x - 3)(3x + 2)}{x} < 0 \text{ if } -\infty < x < \frac{-2}{3}$$

or $0 < x < \frac{3}{2}$

The solution set is

$$\left\{ x \mid -\infty < x < \frac{-2}{3} \text{ or } 0 < x < \frac{3}{2} \right\}.$$

37. (a)

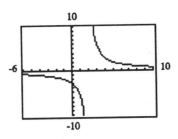

$$\frac{x + 4}{x - 2} \leq 1$$

$$\frac{x + 4}{x - 2} - 1 \leq 0$$

Graph $f(x) = \frac{x + 4}{x - 2} - 1$.

There are no x-intercepts. However, $x = 2$ is a critical value. The graph of f is below the x-axis for $x < 2$. The solution set is $\{x \mid x < 2\}$.

(b)

$$\frac{x + 4}{x - 2} \leq 1$$

$$\frac{x + 4}{x - 2} - 1 \leq 0$$

$$\frac{x + 4 - (x - 2)}{x - 2} \leq 0$$

$$\frac{6}{x - 2} \leq 0$$

$$\frac{6}{x - 2} \leq 0 \text{ if } -\infty < x < 2$$

The solution set is $\{x \mid -\infty < x < 2\}$.

39. (a)

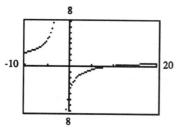

$$\frac{3x - 5}{x + 2} \le 2$$

$$\frac{3x - 5}{x + 2} - 2 \le 0$$

Graph $f(x) = \dfrac{3x - 5}{x + 2} - 2$. The graph of f has an x-intercept at $x = 9$. Also, $x = -2$ is a critical value. The graph of f is below the x-axis for $-2 < x < 9$. The solution set is $\{x \mid -2 < x \le 9\}$.

(b)

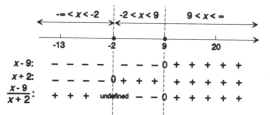

$$\frac{3x - 5}{x + 2} \le 2$$

$$\frac{3x - 5}{x + 2} - 2 \le 0$$

$$\frac{3x + 5 - 2(x + 2)}{x + 2} \le 0$$

$$\frac{3x - 5 - 2x - 4}{x + 2} \le 0$$

$$\frac{x - 9}{x + 2} \le 0$$

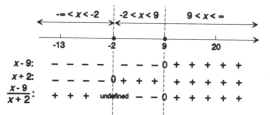

$$\frac{x - 9}{x + 2} \le 0 \text{ if } -2 < x \le 9$$

The solution set is $\{x \mid -2 < x \le 9\}$.

41. (a)

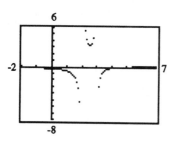

$$\frac{1}{x - 2} < \frac{2}{3x - 9}$$

$$\frac{1}{x - 2} - \frac{2}{3x - 9} < 0$$

Graph $f(x) = \dfrac{1}{x - 2} - \dfrac{2}{3x - 9}$. The graph of f has an x-intercept at $x = 5$. Also, $x = 2$ and $x = 3$ are critical values. The graph of f is below the x-axis for $x < 2$ or $3 < x < 5$. The solution set is $\{x \mid x < 2 \text{ or } 3 < x < 5\}$.

(b)

$$\frac{1}{x - 2} < \frac{2}{3x - 9}$$

$$\frac{1}{x - 2} - \frac{2}{3x - 9} < 0$$

$$\frac{3x - 9 - 2x + 4}{3(x - 2)(x - 3)} < 0$$

$$\frac{x - 5}{3(x - 2)(x - 3)} < 0$$

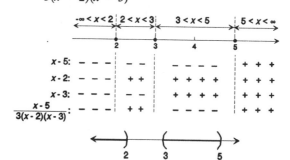

The solution set is $\{x \mid -\infty < x < 2 \text{ or } 3 < x < 5\}$.

43. **(a)**

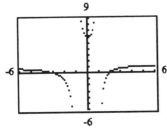

$$\frac{2x + 5}{x + 1} > \frac{x + 1}{x - 1}$$

$$\frac{2x + 5}{x + 1} - \frac{x + 1}{x - 1} > 0$$

Graph $f(x) = \dfrac{2x + 5}{x + 1} - \dfrac{x + 1}{x - 1}$.
The x-intercepts are $x = -3$ and
$x = 2$. Also, $x = -1$ and
$x = 1$ are critical values. The
graph of f is above the x-axis for
$x < -3$ or $-1 < x < 1$ or $x > 2$.
the solution set is $\{x \mid x < -3$
or $-1 < x < 1$ or $x > 2\}$.

(b)

$$\frac{2x + 5}{x + 1} > \frac{x + 1}{x - 1}$$

$$\frac{2x + 5}{x + 1} - \frac{x + 1}{x - 1} > 0$$

$$\frac{(2x + 5)(x - 1) - (x + 1)(x + 1)}{(x + 1)(x - 1)} > 0$$

$$\frac{2x^2 + 3x - 5 - x^2 - 2x - 1}{(x + 1)(x - 1)} > 0$$

$$\frac{x^2 + x - 6}{(x + 1)(x - 1)} > 0$$

$$\frac{(x + 3)(x - 2)}{(x + 1)(x - 1)} > 0$$

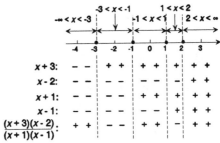

$$\frac{(x + 3)(x - 2)}{(x + 1)(x - 1)} > 0 \text{ if } -\infty < x < -3$$
or $-1 < x < 1$ or $2 < x < \infty$
The solution set is $\{x \mid -\infty < x < -3$
or $-1 < x < 1$ or $2 < x < \infty\}$.

45. **(a)**

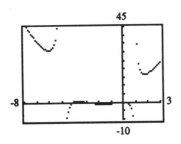

Graph $f(x) = \dfrac{x^2(3 + x)(x + 4)}{(x + 5)(x - 1)}$.
The x-intercepts are $x = -4$,
$x = -3$, $x = 0$. Also, $x = -5$
and $x = 1$ are critical values.
The graph of f is above the x-axis
for $x < -5$ or $-4 < x < -3$
or $x > 1$. The solution set is
$\{x \mid x < -5$ or $-4 < x < -3$
or $x > 1\}$.

(b)

$$\frac{x^2(3 + x)(x + 4)}{(x + 5)(x - 1)} > 0$$

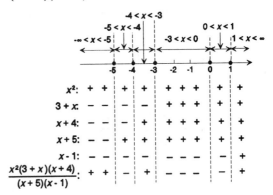

$$\frac{x^2(3 + x)(x + 4)}{(x + 5)(x - 1)} > 0 \text{ if } -\infty < x < -5$$
or $-4 < x < -3$
The solution set is $\{x \mid -\infty < x < -5$
or $-4 < x < -3$ or $1 < x < \infty\}$.

47. Solving algebraically is easier since the expression on the left is easily factored.

$$x^2 - 7x - 8 < 0$$
$$(x - 8)(x + 1) < 0$$

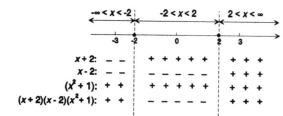

$x^2 - 7x - 8 < 0$ if $-1 < x < 8$
The solution set is $\{x \mid -1 < x < 8\}$.

49. Solving graphically is easier.
Graph $f(x) = x^3 + x - 12$.

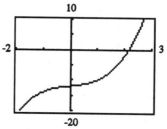

The x-intercept is $x = 2.14$ correct to two decimal places. The graph of f is above the x-axis for $x > 2.14$. The solution set is $\{x \mid 2.14 \le x < \infty\}$.

51. Solving algebraically is easier since the expression on the left is easily factored.

$$x^4 - 3x^2 - 4 > 0$$
$$(x^2 - 4)(x^2 + 1) > 0$$
$$(x + 2)(x - 2)(x^2 + 1) > 0$$

	$-\infty < x < -2$	$-2 < x < 2$	$2 < x < \infty$
	-3 -2	0	2 3
$x + 2$:	$-$ $-$	$+$ $+$ $+$ $+$ $+$	$+$ $+$ $+$
$x - 2$:	$-$ $-$	$-$ $-$ $-$ $-$ $-$	$+$ $+$ $+$
$(x^2 + 1)$:	$+$ $+$	$+$ $+$ $+$ $+$ $+$	$+$ $+$ $+$
$(x+2)(x-2)(x^2+1)$:	$+$ $+$	$-$ $-$ $-$ $-$ $-$	$+$ $+$ $+$

$x^4 - 3x^2 - 4 > 0$ if $-\infty < x < -2$ or $2 < x < \infty$. The solution set is $\{x \mid -\infty < x < -2$ or $2 < x < \infty\}$.

53. Solving algebraically is easier since the numerator is easily factored and the denominator is already prime.

$$\frac{2x^2 - x - 1}{x - 4} \le 0$$
$$\frac{(2x + 1)(x - 1)}{x - 4} \le 0$$

	$-\infty < x < -\frac{1}{2}$	$-\frac{1}{2} < x < 1$	$1 < x < 4$	$4 < x < \infty$
	-1 $-\frac{1}{2}$ 0	1	2	4 5
$(2x + 1)$:	$-$ $-$ $-$	$+$ $+$	$+$ $+$ $+$ $+$ $+$	$+$ $+$ $+$
$(x - 1)$:	$-$ $-$ $-$	$-$ $-$	$+$ $+$ $+$ $+$ $+$	$+$ $+$ $+$
$(x - 4)$:	$-$ $-$ $-$	$-$ $-$	$-$ $-$ $-$ $-$ $-$	$+$ $+$ $+$
$\dfrac{2x^2 - x - 1}{x - 4}$:	$-$ $-$ $-$	$+$ $+$	$-$ $-$ $-$ $-$ $-$	$+$ $+$ $+$

$\dfrac{2x^2 - x - 1}{x - 4} \le 0$ if $-\infty < x < -\dfrac{1}{2}$ or $1 < x < 4$.

The solution set is $\{x \mid -\infty < x < -\dfrac{1}{2}$ or $1 < x < 4\}$.

55. Solving graphically is easier since the numerator of the expression is not factorable. Graph $y = \dfrac{x^2 + 3x - 1}{x + 3}$.

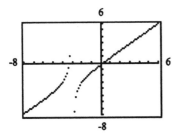

The x-intercepts are $x = -3.30$ and $x = 0.30$ correct to two decimal places. Also, $x = -3$ is a critical value. The graph of f is above the x-axis for $-3.30 < x < -3$ and $x > 0.30$. The solution set is $\{x \mid -3.30 < x < -3$ or $x > 0.30\}$.

57. Solving graphically is easier since the expression is not easily factored.
$$x^3 - 4 \geq 3x^2 + 5x - 3$$
$$x^3 - 3x^2 - 5x - 1 \geq 0$$
Graph $y = x^3 - 3x^2 - 5x - 1$.

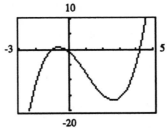

The x-intercepts are $x = -1$, $x = -0.23$ and $x = 4.23$. The graph of f is above the x-axis for $-1 < x < -0.23$ or $x > 4.23$. The solution set is $\{x \mid -1 \leq x \leq -0.23$ or $x \geq 4.23\}$.

59. Let x be the number. Then we want the set so that
$$x^3 > 4x^2$$
$$x^3 - 4x^2 > 0$$
$$x^2(x - 4) > 0$$

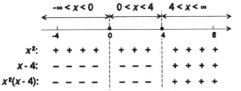

$x^2(x - 4) > 0$ if $4 < x < \infty$
The solution set is $\{x \mid 4 < x < \infty\}$, all numbers larger than 4.

61. The domain of the variable in the expression $\sqrt{x^2 - 16}$ includes all values for which
$$x^2 - 16 \geq 0$$
$$(x - 4)(x + 4) \geq 0$$

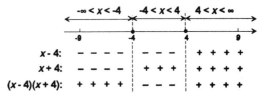

$(x - 4)(x + 4) \geq 0$ if $-\infty < x \leq -4$ or $4 \leq x < \infty$
The domain is $\{x \mid -\infty < x \leq -4$ or $4 \leq x < \infty\}$.

63. The domain of the variable in the expression $\sqrt{\dfrac{x - 2}{x + 4}}$ includes all values for which
$$\frac{x - 2}{x + 4} \geq 0$$

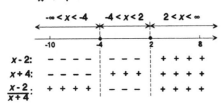

$\dfrac{x - 2}{x + 4} \geq 0$ if $-\infty < x < -4$ or $2 \leq x < \infty$
The domain is $\{x \mid -\infty < x < -4$ or $2 \leq x < \infty\}$.

65. (a) We want to find the set of t values for which
$$80t - 16t^2 > 96$$
$$-16t^2 + 80t - 96 > 0$$
$$+16t^2 - 80t + 96 < 0 \text{ is an equivalent inequality}$$
$$16(t^2 - 5t + 6) < 0$$
$$16(t - 2)(t - 3) < 0 \text{ gives the boundary points 2, 3:}$$

	Test Number	$t - 2$	$t - 3$	$16(t - 2)(t - 3)$
$t < 2$	1	$-$	$-$	$+$
$2 < t < 3$	2.5	$+$	$-$	$-$
$t > 3$	4	$+$	$+$	$+$

The solution set is $2 < t < 3$. The ball is more than 96 feet above the ground for time t between 2 and 3 seconds, $2 < t < 3$.

(b)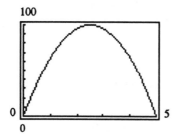
100

0

0   5

(c) The maximum height of the ball is 100 feet.
(d)
$$100 = 80t - 16t^2$$
$$16t^2 - 80t + 100 = 0$$
$$4(4t^2 - 20t + 25) = 0$$
$$4(2t - 5)^2 = 0$$
$$t = 2.5 \text{ seconds}$$

67. (a) Profit = Revenue − Cost
$$= x(40 - 0.2x) - 32x$$
We want the values of x for which
$$x(40 - 0.2x) - 32x \geq 50$$
$$40x - 0.2x^2 - 32x \geq 50$$
$$-0.2x^2 + 8x - 50 \geq 0$$
$$-10(-0.2x^2 + 8x - 50) \leq 0(-10)$$
$$2x^2 - 80x + 500 \leq 0$$
$$x^2 - 40x + 250 \leq 0$$
$$(x - 8)(x - 32) \leq 0 \text{ gives boundary points 7.75, 32.24:}$$

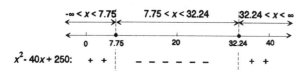

$-\infty < x < 7.75$ $7.75 < x < 32.24$ $32.24 < x < \infty$

0   7.75   20   32.24   40

$x^2 - 40x + 250$: $+$ $+$ $-$ $-$ $-$ $-$ $-$ $-$ $+$ $+$

Thus, for a profit of at least \$50, between 8 and 32 watches must be sold, $8 \leq x \leq 32$.

(b)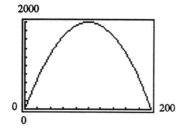
2000

0

0   200

(c) The maximum revenue is \$2000.
(d)
$$\$2000 = x(40 - 0.2x)$$
$$0.2x^2 - 40x + 2000 = 0$$
$$x^2 - 200x + 10,000 = 0$$
$$(x - 100)(x - 100) = 0$$
$$x = 100$$

(e)

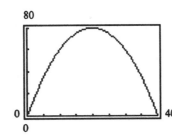

(f) The maximum profit is $80.

(g)
$$80 = x(40 - 0.2x) - 32x$$
$$80 = 40x - 0.2x - 32x$$
$$0.2x^2 - 8x + 80 = 0$$
$$x^2 - 40x + 400 = 0$$
$$(x - 20)(x - 20) = 0$$
$$x = 20$$

69.   Prove that if a, b are real numbers and $a \geq 0$, $b \geq 0$, then $a \leq b$ is equivalent to $\sqrt{a} \leq \sqrt{b}$.

$$a \leq b \Rightarrow b - a \geq 0$$

$$\left(\sqrt{b} - \sqrt{a}\right)\left(\sqrt{b} + \sqrt{a}\right) \geq 0$$

Either   (1) $\sqrt{b} - \sqrt{a} \geq 0$ and $\sqrt{b} + \sqrt{a} \geq 0$

or   (2) $\sqrt{b} - \sqrt{a} \leq 0$ and $\sqrt{b} + \sqrt{a} \leq 0$

In (1) $\sqrt{b} \geq \sqrt{a}$ and $\sqrt{b} \geq -\sqrt{a}$

$\sqrt{a} \leq \sqrt{b}$ and $-\sqrt{a} \leq \sqrt{b}$

In (2) $\sqrt{b} \leq \sqrt{a}$ and $\sqrt{b} \leq -\sqrt{a}$

Case (2) is impossible since $\sqrt{a} \geq 0$ and $\sqrt{b} \geq 0$ which means

$\sqrt{b} \leq -\sqrt{a}$ is impossible.

Case (1) is true and if $a \leq b$, then $\sqrt{a} \leq \sqrt{b}$

So $a \leq b$ is equivalent to $\sqrt{a} \leq \sqrt{b}$.

71.   Given $kx^2 + 2x + 1 = 0$, $a = k$, $b = 2$, $c = 1$

We want a to be such that $b^2 - 4ac > 0$, so
$$2^2 - 4(k)(1) > 0$$
$$4 - 4k > 0$$
$$4 > 4k$$
$$1 > k$$
$$k < 1$$

For $kx^2 + 2x + 1 = 0$ to have two distinct real solutions, $k < 1$.

3.8   Equations and Inequalities Involving Absolute Value

1.   (a)

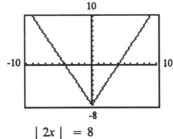

(b) $|2x| = 8$
$$2x = 8 \quad \text{or} \quad 2x = -8$$
$$x = 4 \quad \text{or} \quad x = -4$$
The solution set is $\{-4, 4\}$.

$|2x| = 8$

$|2x| - 8 = 0$

Graph $f(x) = |2x| - 8$. The x-intercepts are $x = -4$ and $x = 4$. Thus, the solution set is $\{-4, 4\}$.

3.　(a)

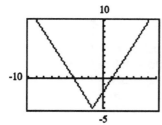

(b)　$|\, 2x + 3 \,| = 5$

$2x + 3 = 5 \quad$ or $\quad 2x + 3 = -5$

$\quad\quad 2x = 2 \quad$ or $\quad\quad 2x = -8$

$\quad\quad\quad x = 1 \quad\quad\quad\quad\quad x = -4$

The solution set is $\{-4, 1\}$.

$|\, 2x + 3 \,| = 5$

$|\, 2x + 3 \,| - 5 = 0$

Graph $f(x) = |\, 2x + 3 \,| - 5$. The x-intercepts are $x = -4$ and $x = 1$. Thus, the solution set is $\{-4, 1\}$.

5.　(a)

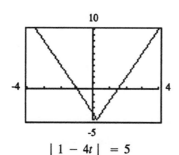

(b)　$|\, 1 - 4t \,| = 5$

$1 - 4t = 5 \quad$ or $\quad 1 - 4t = -5$

$\quad -4t = 4 \quad$ or $\quad\quad -4t = -6$

$\quad\quad t = -1 \,$or $\quad\quad\quad t = \dfrac{3}{2}$

The solution set is $\left\{-1, \dfrac{3}{2}\right\}$

$|\, 1 - 4t \,| = 5$

$|\, 1 - 4t \,| - 5 = 0$

Graph $f(x) = |\, 1 - 4x \,| - 5$. The x-intercepts are $x = -1$ and $x = 1.50$. The solution set is $\{-1, 1.50\}$.

7.　(a)

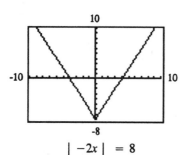

(b)　$|\, -2x \,| = 8$

$-2x = 8 \;$ or$\,-2x = -8$

$\quad x = -4 \quad\quad x = 4$

The solution set is $\{-4, 4\}$.

$|\, -2x \,| = 8$

$|\, -2x \,| - 8 = 0$

Graph $f(x) = |\, -2x \,| - 8$. The x-intercepts are $x = -4$ and $x = 4$. The solution set is $\{-4, 4\}$.

9.　(a)

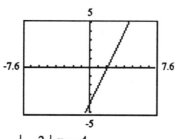

(b)　$|\, -2 \,|\, x = 4$

$\quad 2x = 4$

$\quad\quad x = 2$

The solution is $x = 2$.

$|\, -2 \,|\, x = 4$

$2x - 4 = 0$

Graph $f(x) = 2x - 4$. The x-intercept is $x = 2$. The solution set is $\{2\}$.

11. **(a)**

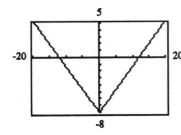

(b)

$$\frac{2}{3} \, |x| \, = 8$$

$$|x| \, = 12$$

$$x = 12 \quad \text{or} \quad x = -12$$

The solution set is $\{-12, 12\}$.

$$\frac{2}{3} \, |x| \, = 8$$

$$\frac{2}{3} \, |x| \, - 8 = 0$$

Graph $f(x) = \frac{2}{3} \, |x| \, - 8$. The x-intercepts are $x = -12$ and $x = 12$. The solution set is $\{-12, 12\}$.

13. **(a)**

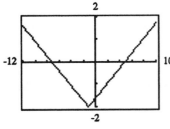

(b)

$$\left| \frac{x}{3} + \frac{2}{5} \right| = 2$$

$$\frac{x}{3} + \frac{2}{5} = 2 \qquad \text{or} \, \frac{x}{3} + \frac{2}{5} = -2$$

$$5x + 6 = 30 \qquad \text{or} \, 5x + 6 = -30$$

$$5x = 24 \qquad \text{or} \qquad 5x = -36$$

$$x = \frac{24}{5} \quad \text{or} \qquad x = \frac{-36}{5}$$

The solution set is $\left\{ \dfrac{-36}{5}, \dfrac{24}{5} \right\}$.

$$\left| \frac{x}{3} + \frac{2}{5} \right| = 2$$

$$\left| \frac{x}{3} + \frac{2}{5} \right| - 2 = 0$$

Graph $f(x) = \left| \dfrac{x}{3} + \dfrac{2}{5} \right| - 2$. The x-intercepts are $x = -7.20$ and $x = 4.80$. Thus, the solution set is $\{-7.20, 4.80\}$.

15. **(a)**

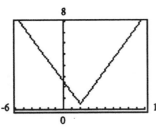

(b)

$$|u - 2| \, = -\frac{1}{2}$$

It is not possible that $|u - 2| < 0$. Therefore, there is no real solution.

$$|u - 2| \, = -\frac{1}{2}$$

$$|u - 2| \, + \frac{1}{2} = 0$$

Graph $f(x) = |x - 2| + \dfrac{1}{2}$. The graph has no x-intercepts; hence, the equation has no real solution.

17. (a)

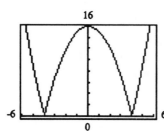

(b)
$$| x^2 - 16 | = 0$$
$$x^2 - 16 = 0$$
$$(x - 4)(x - 4) = 0$$
$$x = 4 \ \text{ or } \ x = -4$$
The solution set is $\{-4, 4\}$.

Graph $f(x) = | x^2 - 16 |$. The x-intercepts are $x = -4$ and $x = 4$. The solution set is $\{-4, 4\}$.

19. (a)

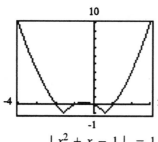

(b)
$$| x^2 - 2x | = 3$$

$x^2 - 2x = 3$ or	$x^2 - 2x = -3$
$x^2 - 2x - 3 = 0$ or	$x^2 - 2x + 3 = 0$
$(x - 3)(x + 1) = 0$ or	No real solution
$x = -1$ or $x = 3$	(Note that $b^2 - 4ac = -8$)

The solution set is $\{-1, 3\}$.

$$| x^2 - 2x | = 3$$
$$| x^2 - 2x | - 3 = 0$$

Graph $f(x) = | x^2 - 2x | - 3$. The x-intercepts are $x = -1$ and $x = 3$. The solution set is $\{-1, 3\}$.

21. (a)

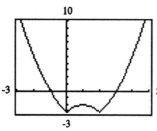

(b)
$$| x^2 + x - 1 | = 1$$

$x^2 + x - 1 = 1$ or	$x^2 + x - 1 = -1$
$x^2 + x - 2 = 0$ or	$x^2 + x = 0$
$(x + 2)(x - 1) = 0$ or	$x(x + 1) = 0$
$x = -2$ or $x = 1$ or	$x = -1$ or $x = 0$

The solution set is $\{-2, -1, 0, 1\}$.

$$| x^2 + x - 1 | = 1$$
$$| x^2 + x - 1 | - 1 = 0$$

Graph $f(x) = | x^2 + x - 1 | - 1$. The x-intercepts are $x = -2, x = -1, x = 0, x = 1$. The solution set is $\{-2, -1, 0, 1\}$.

23. (a)

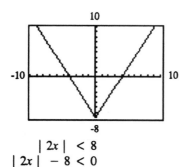

(b)
$$| 2x | < 8$$
$$-8 < 2x < 8$$
$$\frac{-8}{2} < \frac{2x}{2} < \frac{8}{2}$$
$$-4 < x < 4$$

The solutions set consists of all numbers x for which $\{x \mid -4 < x < 4\}$ or $(-4, 4)$.

$$| 2x | < 8$$
$$| 2x | - 8 < 0$$

Graph $f(x) = | 2x | - 8$. The x-intercepts are $x = -4$ and $x = 4$. The graph of f is below the x-axis for $-4 < x < 4$. Thus, the solution set is $\{x \mid -4 < x < 4\}$.

25. (a)

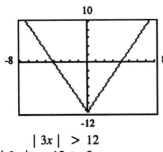

$| 3x | > 12$

$| 3x | - 12 > 0$

Graph $f(x) = | 3x | - 12$. The x-intercepts are $x = -4$ and $x = 4$. The graph of f is above the x-axis for $x < -4$ or $x > 4$. The solution set is $\{x \mid x < -4 \text{ or } x > 4\}$.

(b) $| 3x | > 12$

$3x < -12$ or $3x > 12$

$\dfrac{3x}{3} < \dfrac{-12}{3}$ or $\dfrac{3x}{3} > \dfrac{12}{3}$

$x < -4$ or $x > 4$

The solution set consists of all numbers x for which $\{x \mid x < -4 \text{ or } x > 4\}$ or $(-\infty, -4)$ or $(4, \infty)$.

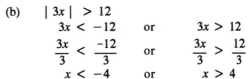

27. (a)

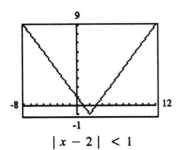

$| x - 2 | < 1$

$| x - 2 | - 1 < 0$

Graph $f(x) = | x - 2 | - 1$. The x-intercepts are $x = 1$ and $x = 3$. The graph of f is below the x-axis for $1 < x < 3$. The solution set is $\{x \mid 1 < x < 3\}$.

(b) $| x - 2 | < 1$

$-1 < x - 2 < 1$

$-1 + 2 < x - 2 + 2 < 1 + 2$

$1 < x < 3$

The solution set consists of all numbers x for which $\{x \mid 1 < x < 3\}$ or $(1, 3)$.

29. (a)

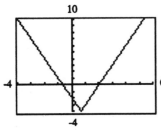

$| 3t - 2 | \leq 4$

$| 3t - 2 | - 4 = 0$

Graph $f(x) = | 3x - 2 | - 4$. The x-intercepts are $x = -0.66$ and $x = 2$. The graph of f is below the x-axis for $-0.66 < x < 2$. The solution set is $\{x \mid -0.66 \leq x \leq 2\}$.

(b) $| 3t - 2 | \leq 4$

$-4 \leq 3t - 2 \leq 4$

$-4 + 2 \leq 3t - 2 + 2 \leq 4 + 2$

$-2 \leq 3t \leq 6$

$\dfrac{-2}{3} \leq \dfrac{3t}{3} \leq \dfrac{6}{3}$

$\dfrac{-2}{3} \leq t \leq 2$

The solution set consists of all numbers t for which

$\left\{ t \mid -\dfrac{2}{3} \leq t \leq 2 \right\}$ or $\left[-\dfrac{2}{3}, 2 \right]$.

31. (a)

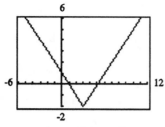

$$|x - 3| \geq 2$$
$$|x - 3| - 2 \geq 0$$

Graph $f(x) = |x - 3| - 2$. The x-intercepts are $x = 1$ and $x = 5$. The graph of f is above the x-axis for $x < 1$ or $x > 5$. The solution set is $\{x \mid x < 1 \text{ or } x > 5\}$.

(b)

$$|x - 3| \geq 2$$
$$x - 3 \leq -2 \text{ or } \quad x - 3 \geq 2$$
$$x \leq 1 \quad \text{or} \qquad x \geq 5$$
$$(-\infty, 1] \text{ or } [5, \infty)$$

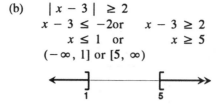

33. (a)

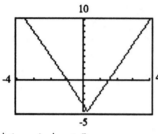

$$|1 - 4x| < 5$$
$$|1 - 4x| - 5 = 0$$

Graph $f(x) = |1 - 4x| - 5$. The x-intercepts are $x = -1$ and $x = 1.50$. The graph of f is below the x-axis for $-1 < x < 1.50$. The solution set is $\{x \mid -1 < x < 1.50\}$.

(b)

$$|1 - 4x| < 5$$
$$-5 < 1 - 4x < 5$$
$$-5 - 1 < 1 - 4x - 1 < 5 - 1$$
$$-6 < -4x < 4$$
$$\frac{-6}{-4} > \frac{-4x}{-4} > \frac{4}{-4}$$
$$\frac{3}{2} > x > -1$$
$$-1 < x < \frac{3}{2}$$

The solution set consists of all numbers x for which

$$\left\{ x \mid -1 < x < \frac{3}{2} \right\} \text{ or } \left(-1, \frac{3}{2} \right).$$

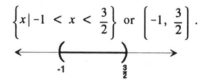

35. (a)

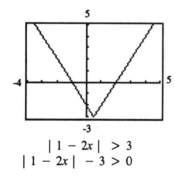

$$|1 - 2x| > 3$$
$$|1 - 2x| - 3 > 0$$

(b)

$$|1 - 2x| > 3$$
$$1 - 2x < -3 \qquad \text{or} \qquad 1 - 2x > 3$$
$$1 - 2x - 1 < -3 - 1 \text{ or } 1 - 2x - 1 > 3 - 1$$
$$-2x < -4 \qquad \text{or} \qquad -2x > 2$$
$$\frac{-2x}{-2} > \frac{4}{2} \qquad \text{or} \qquad \frac{-2x}{-2} < \frac{2}{-2}$$
$$x > 2 \qquad \text{or} \qquad x < -1$$

The solution set consists of all numbers x for which $\{x \mid x < -1 \text{ or } x > 2\}$ or $(-\infty, -1)$ or $(2, \infty)$.

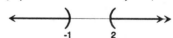

Graph $|1 - 2x| - 3$. The x-intercepts are $x = -1$ and $x = 2$. The graph of f is above the x-axis for $x < -1$ or $x > 2$. The solution set is $\{x \mid x < -1 \text{ or } x > 2\}$.

37. (a)

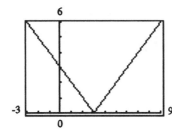

(b) $|x + 3| > 0$
$\qquad x + 3 < 0 \quad$ or $\quad x + 3 > 0$
$\qquad\qquad x < -3$ or $\qquad x > -3$
$\qquad (-\infty, -3) \quad$ or $\quad (-3, \infty)$

Graph $f(x) = |x + 3|$. The x-intercept is $x = 3$. The graph of f is above the x-axis for all x except $x = 3$. The solution set is the set of all real numbers except $x = 3$.

39. (a)

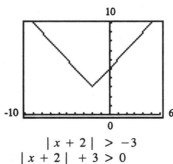

(b) $|x + 2| > -3$
Since $|x + 2| \geq 0$ for all x, we know $|x + 2| > -3$ for all real numbers x or $(-\infty, \infty)$.

$\qquad\qquad |x + 2| > -3$
$\qquad |x + 2| + 3 > 0$

Graph $f(x) = |x + 2| + 3$. The graph has no x-intercepts and is always positive. The solution set is the set of all real numbers.

41. (a)

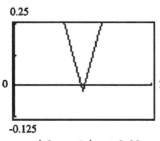

(b) $|2x - 1| < 0.02$
$\qquad -0.02 < 2x - 1 < 0.02$
$-0.02 + 1 < 2x - 1 + 1 < 0.02 + 1$
$\qquad 0.98 < 2x < 1.02$
$\qquad 0.49 < x < 0.51$

The solution set consists of all numbers x for which $\{x \mid 0.49 < x < 0.51\}$ or $(0.49, 0.51)$.

$\qquad\qquad |2x - 1| < 0.02$
$\qquad |2x - 1| - 0.02 < 0$

Graph $f(x) = |2x - 1| - 0.02$. The x-intercepts are $x = 0.49$ and $x = 0.51$. The graph of f is below the x-axis for $0.49 < x < 0.51$. The solution set is $\{x \mid 0.49 < x < 0.51\}$.

43. $\quad |x - 1| < 3$
$\qquad -3 < x - 1 < 3$
$\quad -3 + 1 < x < 3 + 1$
$\qquad -2 < x < 4$
$\qquad 2 < x + 4 < 8$
$\qquad a = 2, b = 8$

45. $\quad |x + 4| \leq 2$
$\qquad -2 \leq x + 4 \leq 2$
$\quad -2 - 4 \leq x \leq 2 - 4$
$\qquad -6 \leq x \leq -2$
$\qquad -12 \leq 2x \leq -4$
$\quad -15 \leq 2x - 3 \leq -7$
$\qquad a = -15, b = -7$

47.
$$|x - 2| \leq 7$$
$$-7 \leq x - 2 \leq 7$$
$$-15 \leq x - 10 \leq -1$$
$$\frac{1}{15} \geq \frac{1}{x - 10} \geq \frac{-1}{1}$$
$$-1 \leq \frac{1}{x - 10} \leq \frac{-1}{15}$$
$$a = -1, b = -\frac{1}{15}$$

49.
$$\left|\frac{a}{b}\right| = \sqrt{\left[\frac{a}{b}\right]^2} = \sqrt{\frac{a^2}{b^2}} = \frac{\sqrt{a^2}}{\sqrt{b^2}} = \frac{|a|}{|b|}$$

51. $(a + b)^2 = a^2 + 2ab + b^2 = |a|^2 + 2|a||b| + |b|^2 = [|a| + |b|]^2;$

therefore, $\sqrt{(a + b)^2} \leq \sqrt{(|a| + |b|)^2}$ or $|a + b| \leq |a| + |b|$

53. x differs from 3 by less than $\frac{1}{2}$

$$|x - 3| \qquad < \qquad \frac{1}{2}$$
$$|x - 3| < \frac{1}{2}$$
$$-\frac{1}{2} < x - 3 < \frac{1}{2}$$
$$-\frac{1}{2} + 3 < x - 3 + 3 < \frac{1}{2} + 3$$
$$\frac{5}{2} < x < \frac{7}{2}$$

The solution set consists of all numbers x

for which $\frac{5}{2} < x < \frac{7}{2}$.

55. x differs from -3 by more than 2
$$|x - (-3)| \qquad > \qquad 2$$
$$|x + 3| > 2$$
$$x + 3 < -2 \text{ or } x + 3 > 2$$
$$x < -5 \text{ or } \qquad x > -1$$
The solution set consists of all numbers
x for which $x < -5$ or $x > -1$.

57. A temperature x that differs from 98.6°F by at least 1.5°
$$|x - 98.6°| \qquad \geq \qquad 1.5°$$
$$x - 98.6° \leq -1.5° \text{ or } x - 98.6° \geq 1.5°$$
$$x \leq 97.1° \text{ or } \qquad x \geq 100.1°$$
The temperatures that are considered unhealthy are those that are less than 97.1°F or greater than 100.1°F inclusive.

59.
$$x^2 < a$$
$$x^2 - a < 0$$

$(x + \sqrt{a})(x - \sqrt{a}) < 0$ gives boundary points $-\sqrt{a}, \sqrt{a}$:

	Test Number	$x + \sqrt{a}$	$x - \sqrt{a}$	$(x + \sqrt{a})(x - \sqrt{a})$
$x < -\sqrt{a}$	$-\sqrt{a} - 1$	$-$	$-$	$+$
$-\sqrt{a} < x < \sqrt{a}$	0	$+$	$-$	$-$
$x > \sqrt{a}$	$\sqrt{a} + 1$	$+$	$+$	$+$

The solution set is $-\sqrt{a} < x < \sqrt{a}$.

61. $x^2 < 1$

Using the results of Problem 59 with $a = 1$, we get the solution set

$-\sqrt{1} < x < \sqrt{1}$ or $-1 < x < 1$.

63. $x^2 \geq 9$

Using the result of Problem 60 with $a = 9$, we get the solution for $x^2 > 9$

to be $x > \sqrt{9}$ or $x < -\sqrt{9}$ which becomes $x > 3$ or $x < -3$. Also $x^2 = 9$ when $x = 3$ or $x = -3$. The solution set is $x \geq 3$ or $x \leq -3$.

65. $x^2 \leq 16$

Using the result of Problem 59 with $a = 16$, we get the solution for $x^2 < 16$ to

be $-\sqrt{16} < x < \sqrt{16}$ or $-4 < x < 4$. Also $x^2 = 16$ when $x = 4$ or $x = -4$. The solution set is $-4 \leq x \leq 4$.

67. $x^2 > 4$

Using the result of Problem 60 with

$a = 4$, we get the solution set $x > \sqrt{4}$

or $x < -\sqrt{4}$; i.e., the solution set is $x > 2$ or $x < -2$.

69. $\left| \, 3x - \left| \, 2x + 1 \, \right| \, \right| = 4$

$3x - \left| \, 2x + 1 \, \right| = -4$		or	$3x - \left| \, 2x + 1 \, \right| = 4$	
$3x + 4 = \left| \, 2x + 1 \, \right|$		or	$3x - 4 = \left| \, 2x + 1 \, \right|$	
$2x + 1 = -(3x + 4)$ or $2x + 1 = 3x + 4$		or	$2x + 1 = (-3x - 4)$ or $2x + 1 = 3x - 4$	
$2x + 1 = -3x - 4$	$-3 = x$		$2x + 1 = -3x + 4$	$5 = x$
$5x = -5$			$5x = 3$	
$x = -1$	$x = -3$		$x = \dfrac{3}{5}$	$x = 5$

Check:

$x = -1$ $\left| \, 3(-1) - \left| \, 2(-1) + 1 \, \right| \, \right| = \left| \, -3 - 1 \, \right| = 4$ Yes

$x = -3$ $\left| \, 3(-3) - \left| \, 2(-3) + 1 \, \right| \, \right| = \left| \, -9 - 5 \, \right| = 14$ No

$x = \dfrac{3}{5}$ $\left| \, 3 \left[\dfrac{3}{5} \right] - \left| \, 2 \left[\dfrac{3}{5} \right] + 1 \, \right| \, \right| = \left| \, \dfrac{9}{5} - \dfrac{11}{5} \, \right| = \dfrac{2}{5}$ No

$x = 5$ $\left| \, 3(5) - \left| \, 2(5) + 1 \, \right| \, \right| = \left| \, 15 - 11 \, \right| = 4$ Yes

The solution set is $\{-1, 5\}$.

---

③ Chapter Review

1. **(a)**

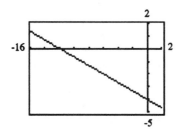

$2 - \dfrac{x}{3} = 6$

$-4 - \dfrac{x}{3} = 0$

Graph $f(x) = -4 - \dfrac{x}{3}$. The x-intercept is $x = -12$. The solution is -12.

(b) $2 - \dfrac{x}{3} = 6$

$2 - \dfrac{x}{3} - 2 = 6 - 2$

$-\dfrac{x}{3} = 4$

$(-3) \left[-\dfrac{x}{3} \right] = 4(-3)$

$x = -12$

Check: $2 - \dfrac{12}{3} \overset{?}{=} 6$

$2 + 4 \overset{?}{=} 6$

$6 = 6$

3. (a)

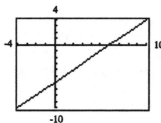

$$-2(5 - 3x) + 8 = 4 + 5x$$
$$-10 + 6x + 8 - 4 - 5x = 0$$
$$x - 6 = 0$$

Graph $f(x) = x - 6$. The x-intercept is $x = 6$. The solution is 6.

(b)
$$-2(5 - 3x) + 8 = 4 + 5x$$
$$-10 + 6x + 8 = 4 + 5x$$
$$-2 + 6x - 4 = 4 + 5x - 4$$
$$6x - 6 = 5x$$
$$6x - 6 - 5x = 5x - 5x$$
$$x - 6 = 0$$
$$x = 6$$

Check:
$$-2(5 - 3 \cdot 6) + 8 \overset{?}{=} 4 + 5 \cdot 6$$
$$-2(-13) + 8 \overset{?}{=} 4 + 30$$
$$26 + 8 \overset{?}{=} 34$$
$$34 = 34$$

5. (a)

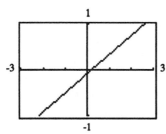

$$\frac{3x}{4} - \frac{x}{3} = \frac{1}{12}$$
$$\frac{3x}{4} - \frac{x}{3} - \frac{1}{12} = 0$$

Graph $f(x) = \frac{3x}{4} - \frac{x}{3} - \frac{1}{12}$. The x-intercept is $x = 0.20$. The solution is 0.20.

(b)
$$\frac{3x}{4} - \frac{x}{3} = \frac{1}{12}$$
$$12\left[\frac{3x}{4} - \frac{x}{3}\right] = \left[\frac{1}{12}\right](12)$$
$$9x - 4x = 1$$
$$5x = 1$$
$$\frac{5x}{5} = \frac{1}{5}$$
$$x = \frac{1}{5}$$

Check:
$$\frac{3\left[\frac{1}{5}\right]}{4} - \frac{\left[\frac{1}{5}\right]}{3} \overset{?}{=} \frac{1}{12}$$
$$\frac{\frac{9}{5}}{4} - \frac{\frac{4}{5}}{3} \overset{?}{=} \frac{1}{12}$$
$$\frac{9}{12} - \frac{4}{12} \overset{?}{=} \frac{1}{12}$$
$$\frac{1}{12} = \frac{1}{12}$$

7. (a)

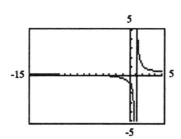

$$\frac{x}{x - 1} = \frac{5}{6}$$
$$\frac{x}{x - 1} - \frac{5}{6} = 0$$

Graph $f(x) = \frac{x}{x - 1} - \frac{5}{6}$. The x-intercept is $x = -5$. The solution is -5.

(b)
$$\frac{x}{x - 1} = \frac{5}{6} \qquad x \neq 1$$
$$6(x - 1)\frac{x}{x - 1} = \frac{5}{6}6(x - 1)$$
$$6x = 5(x - 1)$$
$$6x = 5x - 5$$
$$6x + 5 = 5x - 5 + 5$$
$$6x + 5 - 6x = 5x - 6x$$
$$5 = -x$$
$$x = -5$$

Check:
$$\frac{5}{5 - 1} \overset{?}{=} \frac{5}{6}$$
$$\frac{-5}{-6} \overset{?}{=} \frac{5}{6}$$
$$\frac{5}{6} = \frac{5}{6}$$

9. (a)

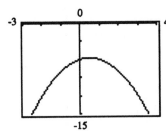

$$x(1 - x) = 6$$
$$-x^2 + x - 6 = 0$$

Graph $f(x) = -x^2 + x - 6$. There are no x-intercepts, so the equation has no real solution.

(b)
$$x(1 - x) = 6$$
$$x - x^2 = 6$$
$$0 = x^2 - x + 6$$

Here $a = 1$, $b = -1$, $c = 6$, and $b^2 - 4ac = -23$. No real solution.

11. (a)

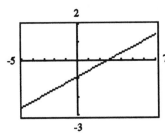

$$\frac{1}{2}\left(x - \frac{1}{3}\right) = \frac{3}{4} - \frac{x}{6}$$
$$\frac{1}{2}\left(x - \frac{1}{3}\right) - \frac{3}{4} + \frac{x}{6} = 0$$

Graph

$$f(x) = \frac{1}{2}\left(x - \frac{1}{3}\right) - \frac{3}{4} + \frac{x}{6}.$$

The x-intercept is $x = 1.375$. The solution is 1.375.

(b)
$$\frac{1}{2}\left(x - \frac{1}{3}\right) = \frac{3}{4} - \frac{x}{6}$$

$$12 \cdot \frac{1}{2}\left(x - \frac{1}{3}\right) = 12\left(\frac{3}{4} - \frac{x}{6}\right)$$

$$6x - 2 = 9 - 2x$$
$$6x - 2 + 2 = 9 - 2x + 2$$
$$6x = 11 - 2x$$
$$6x + 2x = 11 - 2x + 2x$$
$$8x = 11$$
$$\frac{8x}{8} = \frac{11}{8}$$
$$x = \frac{11}{8}$$

Check:
$$\frac{1}{2}\left(\frac{11}{8} - \frac{1}{3}\right) \overset{?}{=} \frac{3}{4} - \frac{\frac{11}{8}}{6}$$

$$\frac{1}{2}\left(\frac{33}{24} - \frac{8}{24}\right) \overset{?}{=} \frac{3}{4} - \frac{11}{48}$$

$$\frac{1}{2}\left(\frac{25}{24}\right) \overset{?}{=} \frac{36}{48} - \frac{11}{48}$$

$$\frac{25}{48} = \frac{25}{48}$$

13. (a)

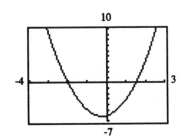

$$(x - 1)(2x + 3) = 3$$
$$(x - 1)(2x + 3) - 3 = 0$$

Graph $f(x) = (x - 1)(2x + 3) - 3$. The x-intercepts are $x = -2$ and $x = 1.50$. The solution set is $\{-2, 1.50\}$.

(b)

$$(x - 1)(2x + 3) = 3$$
$$2x^2 + 3x - 2x - 3 = 3$$
$$2x^2 + x - 3 = 3$$
$$2x^2 + x - 3 - 3 = 3 - 3$$
$$2x^2 + x - 6 = 0$$
$$(2x - 3)(x + 2) = 0$$
$$2x - 3 = 0 \quad \text{or} \quad x + 2 = 0$$
$$x = \frac{3}{2} \quad \text{or} \qquad x = -2$$

$$\left\{ -2, \frac{3}{2} \right\}$$

Check -2:
$$(-2 - 1)(2 \cdot -2 + 3) \overset{?}{=} 3$$
$$(-3)(-1) \overset{?}{=} 3$$
$$3 = 3$$

Check $\dfrac{3}{2}$:

$$\left[\frac{3}{2} - 1 \right]\left[2 \cdot \frac{3}{2} + 3 \right] \overset{?}{=} 3$$
$$\left[\frac{1}{2} \right](6) \overset{?}{=} 3$$
$$3 = 3$$

15. **(a)**

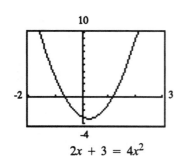

$$2x + 3 = 4x^2$$
$$4x^2 - 2x - 3 = 0$$

Graph $f(x) = 4x^2 - 2x - 3$. The x-intercepts are $x = -0.65$ and $x = 1.15$. The solution set is $\{-0.65, 1.15\}$.

(b)

$$2x + 3 = 4x^2$$
$$2x + 3 - 2x = 4x^2 - 2x$$
$$3 = 4x^2 - 2x$$
$$3 - 3 = 4x^2 - 2x - 3$$
$$0 = 4x^2 - 2x - 3$$

Here $a = 4$, $b = -2$, $c = -3$ and $b^2 - 4ac = 52$.

Then $x = \dfrac{-(-2) \pm \sqrt{52}}{2(4)} = \dfrac{2 \pm 2\sqrt{13}}{8} = \dfrac{1 \pm \sqrt{13}}{4}$

$$\left\{ \frac{1 - \sqrt{13}}{4}, \frac{1 + \sqrt{13}}{4} \right\}$$

Check: $\dfrac{1 - \sqrt{13}}{4}$: $\quad 2\left[\dfrac{1 - \sqrt{13}}{4} \right] + 3 \overset{?}{=} 4\left[\dfrac{1 - \sqrt{13}}{4} \right]^2$

$$\frac{1 - \sqrt{13}}{2} + \frac{6}{2} \overset{?}{=} \frac{\left(1 - \sqrt{13}\right)^2}{4}$$

$$\frac{7 - \sqrt{13}}{2} \overset{?}{=} \frac{1 - 2\sqrt{13} + 13}{4}$$

$$\frac{7 - \sqrt{13}}{2} \overset{?}{=} \frac{14 - 2\sqrt{13}}{4}$$

$$\frac{7 - \sqrt{13}}{2} \overset{?}{=} \frac{2\left(7 - \sqrt{13}\right)}{4}$$

$$\frac{7 - \sqrt{13}}{2} = \frac{7 - \sqrt{13}}{2}$$

Check: $\dfrac{1 + \sqrt{13}}{4}$: $2\left[\dfrac{1 + \sqrt{13}}{4}\right] + 3 \overset{?}{=} 4\left[\dfrac{1 + \sqrt{13}}{4}\right]^2$

$$\dfrac{1 + \sqrt{13} + 6}{2} \overset{?}{=} \dfrac{\left(1 + 2\sqrt{13}\right) + 13}{4}$$

$$\dfrac{7 + \sqrt{13}}{2} \overset{?}{=} \dfrac{14 + 2\sqrt{13}}{4}$$

$$\dfrac{7 + \sqrt{13}}{2} \overset{?}{=} \dfrac{2\left(7 + \sqrt{13}\right)}{4}$$

$$\dfrac{7 + \sqrt{13}}{2} = \dfrac{7 + \sqrt{13}}{2}$$

17. (a)

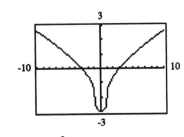

$$\sqrt[3]{x^2 - 1} = 2$$

$$\sqrt[3]{x^2 - 1} - 2 = 0$$

Graph $f(x) = \sqrt[3]{x^2 - 1} - 2$. The x-intercepts are $x = -3$ and $x = 3$. The solution set is $\{-3, 3\}$.

(b) $\sqrt[3]{x^2 - 1} = 2$

$$\left(\sqrt[3]{x^2 - 1}\right)^3 = 2^3$$

$$x^2 - 1 = 8$$

$$x^2 = 9$$

$$x = \pm\sqrt{9} = \pm 3$$

$$\{-3, 3\}$$

Check -3: $\sqrt[3]{(-3)^2 - 1} \overset{?}{=} 2$

$$\sqrt[3]{8} \overset{?}{=} 2$$

$$2 = 2$$

Check 3: $\sqrt[3]{3^2 - 1} \overset{?}{=} 2$

$$\sqrt[3]{8} \overset{?}{=} 2$$

$$2 = 2$$

19. (a)

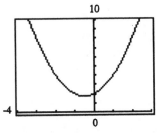

$$x(x + 1) + 2 = 0$$

$$x^2 + x + 2 = 0$$

Graph $f(x) = x^2 + x + 2$. The graph of f has no x-intercepts; hence, the equation has no real solution.

(b) $x(x + 1) + 2 = 0$

$$x^2 + x + 2 = 0$$

Here, $a = 1$, $b = 1$, $c = 2$ and $b^2 - 4ac = -7$.

No real solution.

21. (a)

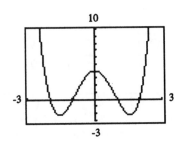

Graph $f(x) = x^4 - 5x^2 + 4$. The x-intercepts are $x = -2$, $x = -1$, $x = 1$, and $x = 2$. The solution set is $\{-2, -1, 1, 2\}$.

(b) $x^4 - 5x^2 + 4 = 0$
Let $u = x^2$. Then $u^2 = x^4$
$$u^2 - 5u + 4 = 0$$
$$(u - 4)(u - 1) = 0$$
$$u - 4 = 0 \quad \text{or} \quad u - 1 = 0$$
$$u = 4 \quad \text{or} \quad u = 1$$
$$x^2 = 4 \quad \text{or} \quad x^2 = 1$$
$$x = \pm 2 \text{ or} \quad x = \pm 1$$
The solution set is $\{-2, -1, 1, 2\}$.

Check -2:
$$(-2)^4 - 5(-2)^2 + 4 \overset{?}{=} 0$$
$$6 - 20 + 4 \overset{?}{=} 0$$
$$0 = 0$$
Check -1:
$$(-1)^4 - 5(-1)^2 + 4 \overset{?}{=} 0$$
$$1 - 5 + 4 \overset{?}{=} 0$$
$$0 = 0$$
Check 1: $(1)^4 - 5(1)^2 + 4 \overset{?}{=} 0$
$$1 - 5 + 4 \overset{?}{=} 0$$
$$0 = 0$$
Check 2: $(2)^4 - 5(2)^2 + 4 \overset{?}{=} 0$
$$16 - 20 + 4 \overset{?}{=} 0$$
$$0 = 0$$

23. (a)

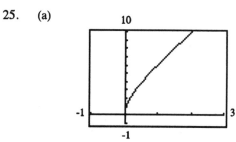

$$\sqrt{2x + 3} + x = 3$$
$$\sqrt{2x + 3} - x - 3 = 0$$

Graph $f(x) = \sqrt{2x + 3} - x - 3$.
The x-intercept is $x = 2$. The solution is 2.

(b)
$$\sqrt{2x - 3} + x = 3$$
$$\sqrt{2x - 3} = 3 - x$$
$$\left(\sqrt{2x - 3}\right)^2 = (3x - x)^2$$
$$2x - 3 = 9 - 6x + x^2$$
$$2x - 3 - 2x + 3 = x^2 - 6x + 9 - 2x + 3$$
$$0 = x^2 - 8x + 12$$
$$0 = (x - 2)(x - 6)$$
$$x - 2 = 0 \quad \text{or} \quad x - 6 = 0$$
$$x = 2 \quad \text{or} \quad x = 6$$

Check 6: $\sqrt{2(6) - 3} + 2 \overset{?}{=} 3$
$$5 \neq 3$$
Check 2: $\sqrt{2(2) - 3} + 2 \overset{?}{=} 3$
$$1 + 2 \overset{?}{=} 3$$
$$3 = 3$$
The solution is $x = 2$.

25. (a)

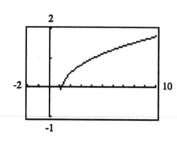

Graph $f(x) = x^{3/2} + 5x^{1/2}$. The x-intercept is $x = 0$. The solution is $x = 0$.

(b)
$$x^{3/2} + 5x^{1/2} = 0$$
$$x^{1/2}(x^{2/2} + 5) = 0$$
$$x^{1/2}(x + 5) = 0$$
$$x^{1/2} = 0 \quad \text{or} \quad x + 5 = 0$$
$$x = 0 \qquad \qquad x = -5$$
Check 0: $\qquad 0^{3/2} + 5 \cdot 0^{1/2} \overset{?}{=} 0$
$$0 = 0$$
Check -5: $-5^{3/2} + 5 \cdot -5^{1/2} \overset{?}{=} 0$
$$5\sqrt{-5} + 5\sqrt{-5} \overset{?}{=} 0$$
$$10\sqrt{-5} \overset{?}{=} 0$$
The solution is $x = 0$.

27. (a)

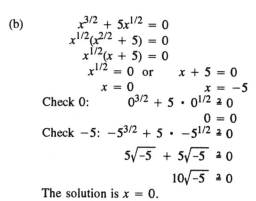

$$\sqrt{x + 1} + \sqrt{x - 1} = \sqrt{2x + 1}$$
$$\sqrt{x + 1} + \sqrt{x - 1} - \sqrt{2x + 1} = 0$$
Graph
$$f(x) = \sqrt{x + 1} + \sqrt{x - 1} - \sqrt{2x + 1}.$$
The x-intercept is $x = 1.11$ correct to two decimal places. The solution is $x = 1.11$.

(b)

$$\sqrt{x + 1} + \sqrt{x - 1} = \sqrt{2x + 1}$$

$$\left(\sqrt{x + 1} + \sqrt{x - 1}\right)^2 = \left(\sqrt{2x + 1}\right)^2$$

$$x + 1 + 2\sqrt{x + 1}\sqrt{x - 1} + x - 1 = 2x + 1$$

$$2\sqrt{x + 1}\sqrt{x - 1} + 2x = 2x + 1$$

$$2\sqrt{x + 1}\sqrt{x - 1} = 1$$

$$\left(2\sqrt{x + 1}\sqrt{x - 1}\right)^2 = 1^2$$

$$4(x + 1)(x - 1) = 1$$

$$4\left(x^2 - 1\right) = 1$$

$$4x^2 - 5 = 0$$

$$x^2 = \frac{5}{4}$$

$$x = \pm\sqrt{\frac{5}{4}} = \frac{\pm\sqrt{5}}{2}$$

Check $\dfrac{-\sqrt{5}}{2}$: $\quad \sqrt{\dfrac{-\sqrt{5}}{2} + 1} + \sqrt{\dfrac{-\sqrt{5}}{2} - 1} = \sqrt{2\left[\dfrac{-\sqrt{5}}{2}\right] + 1}$

$$\sqrt{-0.118} + \sqrt{-2.118} = \sqrt{-0.118}$$

This is not defined.

Check $\dfrac{\sqrt{5}}{2}$: $\quad \sqrt{\dfrac{\sqrt{5}}{2} + 1} + \sqrt{\dfrac{\sqrt{5}}{2} - 1} = \sqrt{2\left[\dfrac{\sqrt{5}}{2}\right] + 1}$

$$\sqrt{\dfrac{\sqrt{5} + 2}{2}} + \sqrt{\dfrac{\sqrt{5} - 2}{2}} \overset{\cdot}{=} \sqrt{\sqrt{5} + 1}$$

$$\dfrac{\sqrt{5} + 2}{2} + 2\sqrt{\left[\dfrac{\sqrt{5} + 2}{2}\right]\left[\dfrac{\sqrt{5} - 2}{2}\right]} + \dfrac{\sqrt{5} - 2}{2} \overset{\cdot}{=} \sqrt{5} + 1$$

$$\dfrac{2\sqrt{5}}{2} + 2\sqrt{\dfrac{5 - 4}{4}} \overset{\cdot}{=} \sqrt{5} + 1$$

$$\sqrt{5} + 2\left[\dfrac{1}{2}\right] \overset{\cdot}{=} \sqrt{5} + 1$$

$$\sqrt{5} + 1 = \sqrt{5} + 1$$

The solution set is $x = \dfrac{\sqrt{5}}{2}$.

29. (a)

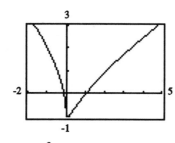

$$2\sqrt[3]{x^2} - \sqrt[3]{x} = 1$$

$$2\sqrt[3]{x^2} - \sqrt[3]{x} - 1 = 0$$

Graph $f(x) = 2\sqrt[3]{x^2} - \sqrt[3]{x} - 1$.
The x-intercepts are $x = -0.125$
and $x = 1$. The solution set is
$\{-0.125, 1\}$.

(b)

$$2\sqrt[3]{x^2} - \sqrt[3]{x} = 1$$
$$2x^{2/3} - x^{1/3} = 1$$

Let $u = x^{1/3}$

$$2u^2 - u = 1$$
$$2u^2 - u - 1 = 0$$
$$(2u + 1)(u - 1) = 0$$
$$2u + 1 = 0 \quad \text{or} \quad u - 1 = 0$$
$$u = -\frac{1}{2} \quad \text{or} \quad u = 1$$
$$x^{1/3} = -\frac{1}{2} \quad \text{or} \quad x^{1/3} = 1$$
$$x = -\frac{1}{8} \quad \text{or} \quad x = 1$$

The solution set is $\left\{\dfrac{-1}{8}, 1\right\}$.

Check $-\frac{1}{8}$: $2\sqrt[3]{\left(\dfrac{-1}{8}\right)^2} - \sqrt[3]{\dfrac{-1}{8}} \overset{?}{=} 1$

$$2\sqrt[3]{\dfrac{1}{64}} - \dfrac{-1}{2} \overset{?}{=} 1$$

$$2\left(\dfrac{1}{4}\right) + \dfrac{1}{2} \overset{?}{=} 1$$

$$1 = 1$$

Check 1: $2\sqrt[3]{1^2} - \sqrt[3]{1} \overset{?}{=} 1$

$$2 - 1 \overset{?}{=} 1$$

$$1 = 1$$

31. (a)

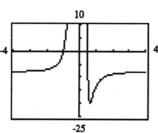

Graph $f(x) = x^{-6} - 7x^{-3} - 8$.
The x-intercepts are $x = -1$ and
$x = 0.50$. The solution set is
$\{-1, 0.50\}$.

(b)

$$x^{-6} - 7x^{-3} - 8 = 0$$
Let $u = x^{-3}$. Then, $u^2 = x^{-6}$
and $u^2 - 7u - 8 = 0$
$$(u - 8)(u + 1) = 0$$
$$u - 8 = 0 \quad \text{or} \quad u + 1 = 0$$
$$u = 8 \quad \text{or} \quad u = -1$$
$$x^{-3} = 8 \quad \text{or} \quad x^{-3} = -1$$
$$(x^{-3})^{-1} = 8^{-1} \quad \text{or} \quad (x^{-3})^{-1} = (-1)^{-1}$$
$$x^3 = \dfrac{1}{8} \quad \text{or} \quad x^3 = -1$$
$$x = \dfrac{1}{2} \quad \text{or} \quad x = -1$$

The solution set is $\left\{-1, \dfrac{1}{2}\right\}$.

Check -1: $(-1)^{-6} - 7(-1)^{-3} - 8 \overset{?}{=} 0$
$$1 + 7 - 8 \overset{?}{=} 0$$
$$0 = 0$$

Check $\dfrac{1}{2}$: $\left(\dfrac{1}{2}\right)^{-6} - 7\left(\dfrac{1}{2}\right)^{-3} - 8 \overset{?}{=} 0$
$$64 - 7(8) - 8 \overset{?}{=} 0$$
$$0 = 0$$

33. (a)

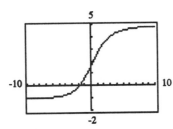

Graph

$$f(x) = \sqrt{x^2 + 3x + 7} - \sqrt{x^2 - 3x + 9} + 2.$$

The x-intercept is $x = -1.80$. The solution is $x = -1.80$.

(b) $\sqrt{x^2 + 3x + 7} - \sqrt{x^2 - 3x + 9} + 2 = 0$

$$\sqrt{x^2 + 3x + 7} = \sqrt{x^2 - 3x + 9} - 2$$

$$\left(\sqrt{x^2 + 3x + 7}\right)^2 = \left(\sqrt{x^2 - 3x + 9} - 2\right)^2$$

$$x^2 + 3x + 7 = x^2 - 3x + 9 - 4\sqrt{x^2 - 3x + 9} + 4$$

$$6x - 6 = -4\sqrt{x^2 - 3x + 9}$$

$$3x - 3 = -2\sqrt{x^2 - 3x + 9}$$

$$(3x - 3)^2 = \left(-2\sqrt{x^2 - 3x + 9}\right)^2$$

$$9x^2 - 18x + 9 = 4(x^2 - 3x + 9)$$

$$9x^2 - 18x + 9 = 4x^2 - 12x + 36$$

$$5x^2 - 6x - 27 = 0$$

$$(5x + 9)(x - 3) = 0$$

$$5x + 9 = 0 \quad \text{or} \quad x - 3 = 0$$

$$x = -\frac{9}{5} \quad \text{or} \quad x = 3$$

Check $-\frac{9}{5}$: $\sqrt{\left[\frac{-9}{5}\right]^2 + 3\left[\frac{-9}{5}\right] + 7} - \sqrt{\left[\frac{-9}{5}\right]^2 - 3\left[\frac{-9}{5}\right] + 9} + 2 = 0$

$$\sqrt{\frac{81}{25} - \frac{135}{25} + \frac{175}{25}} - \sqrt{\frac{81}{25} + \frac{135}{25} + \frac{225}{25}} + 2 = 0$$

$$\sqrt{\frac{121}{25}} - \sqrt{\frac{441}{25}} + 2 = 0$$

$$\frac{11}{5} - \frac{21}{5} + 2 = 0$$

$$0 = 0$$

Check 3: $\sqrt{3^2 + 3(3) + 7} - \sqrt{3^2 - 3(3) + 9} + 2 = 0$

$$5 - 3 + 2 = 0$$

$$4 \neq 0$$

Does not check.

The solution set is $x = \frac{-9}{5}$.

35. $$x^2 + m^2 = 2mx + (nx)^2$$

$$x^2 - n^2x^2 - 2mx + m^2 = 0$$

$$(1 - n^2)x^2 - 2mx + m^2 = 0$$

Here $a = 1 - n^2$, $b = -2m$, $c = m^2$ and $b^2 - 4ac = 4n^2m^2$.

Then, $x = \dfrac{-(-2m) \pm \sqrt{4n^2m^2}}{2(1 - n^2)} = \dfrac{2m \pm 2nm}{2(1 - n^2)} = \dfrac{2m(1 \pm n)}{2(1 - n^2)} = \dfrac{m(1 \pm n)}{1 - n^2}$

Thus, $x = \dfrac{m(1 - n)}{1 - n^2} = \dfrac{m(1 - n)}{(1 + n)(1 - n)} = \dfrac{m}{1 + n}$

and $x = \dfrac{m(1 + n)}{1 - n^2} = \dfrac{m(1 + n)}{(1 + n)(1 - n)} = \dfrac{m}{1 - n}$

The solution set is $\left\{ \dfrac{m}{1 - n}, \dfrac{m}{1 + n} \right\}$.

Check $\dfrac{m}{1 - n}$:

$$\left[\dfrac{m}{1 - n} \right]^2 + m^2 \overset{?}{=} 2m \left[\dfrac{m}{1 - n} \right] + \left[n \left[\dfrac{m}{1 - n} \right] \right]^2$$

$$\dfrac{m^2}{(1 - n)^2} + \dfrac{m^2(1 - n)^2}{(1 - n)^2} \overset{?}{=} \dfrac{2m^2}{1 - n} + \dfrac{n^2 m^2}{(1 - n)^2}$$

$$\dfrac{m^2 + m^2 - 2m^2 n + m^2 n^2}{(1 - n)^2} = \dfrac{2m^2(1 - n) + n^2 m^2}{(1 - n)^2}$$

$$\dfrac{2m^2 - 2m^2 n + m^2 n^2}{(1 - n)^2} = \dfrac{2m^2 - 2m^2 n + m^2 n^2}{(1 - n)^2}$$

Check $\dfrac{m}{1 + n}$:

$$\left[\dfrac{m}{1 + n} \right]^2 + m^2 \overset{?}{=} 2m \left[\dfrac{m}{1 + n} \right] + \left[n \left[\dfrac{m}{1 + n} \right] \right]^2$$

$$\dfrac{m^2 + m^2(1 + n)^2}{(1 + n)^2} + \dfrac{2m^2}{1 + n} = \dfrac{n^2 m^2}{(1 + n)^2}$$

$$\dfrac{2m^2 + 2m^2 n + m^2 n^2}{(1 + n)^2} = \dfrac{2m^2 + 2m^2 n + m^2 n^2}{(1 + n)^2}$$

37. $10a^2 x^2 - 2abx - 36b^2 = 0$
Here $A = 10a^2$, $B = -2ab$, $C = -36b^2$ and
$B^2 - 4AC = (-2ab)^2 - 4(10a^2)(-36b^2)$
$\qquad\qquad = 4a^2 b^2 + 1440a^2 b^2 = 1444a^2 b^2$

Then, $x = \dfrac{-(-2ab) \pm \sqrt{148a^2 b^2}}{2(10a^2)} = \dfrac{2ab \pm 38ab}{20a^2}$

or $x = \dfrac{2ab - 38ab}{20a^2} = \dfrac{-36ab}{20a^2} = \dfrac{-9b}{5a}$

and $x = \dfrac{2ab + 38ab}{20a^2} = \dfrac{40ab}{20a^2} = \dfrac{2b}{a}$

The solution set is $\left\{ \dfrac{-9b}{5a}, \dfrac{2b}{a} \right\}$.

Check $\dfrac{-9b}{5a}$: $10a^2\left[\dfrac{-9b}{5a}\right]^2 - 2ab\left[\dfrac{-9b}{5a}\right] - 36b^2 \stackrel{?}{=} 0$

$$10a^2\left[\dfrac{81b^2}{25a^2}\right] + \dfrac{18b^2}{5} - 36b^2 \stackrel{?}{=} 0$$

$$\dfrac{162b^2}{5} + \dfrac{18b^2}{5} - \dfrac{180b^2}{5} \stackrel{?}{=} 0$$

$$0 \stackrel{?}{=} 0$$

Check $\dfrac{2b}{a}$: $10a^2\left[\dfrac{2b}{a}\right]^2 - 2ab\left[\dfrac{2b}{a}\right] - 36b^2 \stackrel{?}{=} 0$

$$40b^2 - 4b^2 - 36b^2 \stackrel{?}{=} 0$$

$$0 \stackrel{?}{=} 0$$

39. (a)

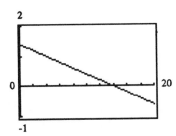

$$\dfrac{2x - 3}{5} + 2 \le \dfrac{x}{2}$$

$$\dfrac{2x - 3}{5} + 2 - \dfrac{x}{2} \le 0$$

Graph $f(x) = \dfrac{2x - 3}{5} + 2 - \dfrac{x}{2}$. The x-intercept is $x = 14$. The graph is below the x-axis for $x > 14$. The solution set is $\{x \mid x \ge 14\}$.

(b)
$$\dfrac{2x - 3}{5} + 2 \le \dfrac{x}{2}$$

$$10\left[\dfrac{2x - 3}{5} + 2\right] \le \left[\dfrac{x}{2}\right]10$$

$$2(2x - 3) + 20 \le 5x$$
$$4x - 6 + 20 \le 5x$$
$$-x \le -14$$
$$x \ge 14$$

The solution set is $\{x \mid 14 \le x < \infty\}$.

41.

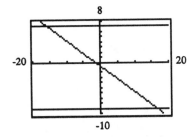

Graph $y_1 = -9$, $y_2 = \dfrac{2x + 3}{-4}$, $y_3 = 7$. The intersection points are $x = -15.5$ and $x = 16.5$. The graph of y_2 is between y_1 and y_3 for $-15.5 < x < 16.5$. The solution set is $\{x \mid -15.5 \le x \le 16.5\}$.

(b)
$$-9 \le \dfrac{2x + 3}{-4} \le 7$$

$$(-4)(-9) \ge (-4)\dfrac{2x + 3}{-4} \ge 7(-4)$$

$$36 \ge 2x + 3 \ge -28$$
$$36 - 3 \ge 2x + 3 - 3 \ge -28 - 3$$
$$33 \ge 2x \ge -31$$
$$\dfrac{33}{2} \ge \dfrac{2x}{2} \ge \dfrac{-31}{2}$$

$$\left\{x \mid \dfrac{-31}{2} \le x \le \dfrac{33}{2}\right\}$$

43. **(a)**

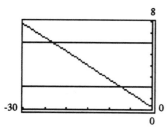

Graph $y_1 = 6$, $y_2 = \dfrac{3 - 3x}{12}$, $y_3 = 2$. The intersection points are at $x = -23$ and $x = -7$. y_2 is between y_1 and y_3 for $-23 < x < 7$. The solution set is $\{x \mid -23 < x < 7\}$.

(b)

$$6 > \frac{3 - 3x}{12} > 2$$

$$12(6) > 12\left[\frac{3 - 3x}{12}\right] > (12)(2)$$

$$72 > 3 - 3x > 24$$
$$72 - 3 > 3 - 3x - 3 > 24 - 3$$
$$69 > -3x > 21$$
$$\frac{69}{-3} < \frac{-3x}{-3} < \frac{21}{-3}$$
$$\{x \mid -23 < x < -7\}$$

45. **(a)**

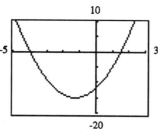

Graph $f(x) = 2x^2 + 5x - 12$. The x-intercepts are $x = -4$ and $x = 1.50$. The graph of f is below the x-axis for $-4 < x < 1.50$. The solution set is $\{x \mid -4 < x < 1.50\}$.

(b)

$$2x^2 + 5x - 12 < 0$$
$$(2x - 3)(x + 4) < 0$$

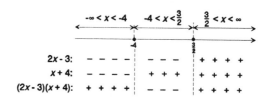

$(2x - 3)(x + 4) < 0$ if $-4 < x < \dfrac{3}{2}$

The solution set is $\left\{x \mid -4 < x < \dfrac{3}{2}\right\}$.

47. **(a)**

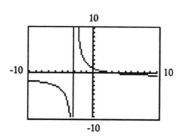

$$\frac{6}{x + 3} \geq 1$$

$$\frac{6}{x + 3} - 1 \geq 0$$

Graph $f(x) = \dfrac{6}{x + 3} - 1$. The x-intercept is $x = 3$; also, $x = -3$ is a critical value. The graph of f is above the x-axis for $-3 < x < 3$. The solution set is $\{x \mid -3 < x \leq 3\}$.

(b)

$$\frac{6}{x + 3} \geq 1, \ x \neq -3$$

$$\frac{6}{x + 3} - 1 \geq 0$$

$$\frac{6 - (x + 3)}{x + 3} \geq 0$$

$$\frac{3 - x}{x + 3} \geq 0$$

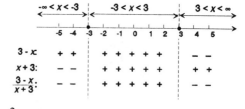

$$\frac{3 - x}{x + 3} \geq 0 \text{ if } -3 < x \leq 3$$

The solution set is $\{x \mid -3 < x \leq 3\}$.

49. (a)

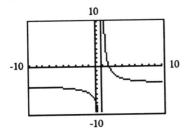

$$\frac{2x - 6}{1 - x} < 2$$

$$\frac{2x - 6}{1 - x} - 2 < 0$$

Graph $f(x) = \dfrac{2x - 6}{1 - x} - 2$. The x-intercept is $x = 2$; also, $x = 1$ is a critical value. The graph of f is below the x-axis for $x < 1$ or $x > 2$. The solution set is $\{x \mid x < 1 \text{ or } x > 2\}$.

(b)

$$\frac{2x - 6}{1 - x} < 2, \quad x \neq 1$$

$$\frac{2x - 6}{1 - x} - 2 < 0$$

$$\frac{2x - 6 - 2(1 - x)}{1 - x} < 0$$

$$\frac{2x - 6 - 2 + 2x}{1 - x} < 0$$

$$\frac{4x - 8}{1 - x} < 0$$

$$\frac{4(x - 2)}{1 - x} < 0$$

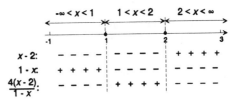

$$\frac{4(x - 2)}{1 - x} < 0 \text{ if } -\infty < x < 1$$
or $2 < x < \infty$.
The solution set is $\{x \mid -\infty < x < 1$ or $2 < x < \infty\}$.

51. (a)

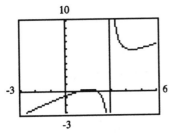

Graph $f(x) = \dfrac{(x - 2)(x - 1)}{x - 3}$.
The x-intercepts are $x = 1$ and $x = 2$; also, $x = 3$ is a critical value. The graph of f is above the x-axis for $1 < x < 2$ and $x > 3$. The solution set is $\{x \mid 1 < x < 2 \text{ or } x > 3\}$.

(b) $\dfrac{(x - 2)(x - 1)}{x - 3} > 0, \quad x \neq 3$

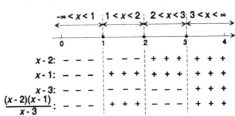

$$\frac{(x - 2)(x - 1)}{x - 3} > 0, \text{ if } 1 < x < 2$$
or $3 < x < \infty$
The solution set is $\{x \mid 1 < x < 2$ or $3 < x < \infty\}$.

53. (a)

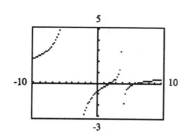

Graph $\dfrac{x^2 - 8x + 12}{x^2 - 16}$. The x-intercepts are $x = 2$ and $x = 6$; also, $x = -4$ and $x = 4$ are critical values. The graph of f is above the x-axis for $x < -4$, $2 < x < 4$, and $x > 6$. The solution set is $\{x \mid x < -4$ or $2 < x < 4$ or $x > 6\}$.

(b) $\dfrac{x^2 - 8x + 12}{x^2 - 16} > 0, \quad x \neq -4, 4$

$\dfrac{(x - 2)(x - 6)}{(x - 4)(x + 4)} > 0$

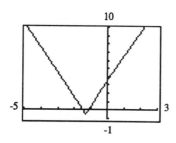

$\dfrac{(x - 2)(x - 6)}{(x - 4)(x + 4)} > 0$ if $-\infty < x < -4$ or $2 < x < 4$ or $6 < x < \infty$. The solution set is $\{x \mid -\infty < x < -4$ or $2 < x < 4$ or $6 < x < \infty\}$.

55. (a)

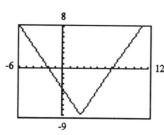

$|3x + 4| < \dfrac{1}{2}$

$|3x + 4| - \dfrac{1}{2} < 0$

Graph $f(x) = |3x + 4| - \dfrac{1}{2}$. The x-intercepts are $x = -1.50$ and $x = -1.16$. The graph of f is below the x-axis for $-1.50 < x < -1.16$. The solution set is $\{x \mid -1.50 < x < -1.16\}$.

(b) $|3x + 4| < \dfrac{1}{2}$

$-\dfrac{1}{2} < 3x + 4 < \dfrac{1}{2}$

$-\dfrac{9}{2} < 3x < -\dfrac{7}{2}$

$-\dfrac{3}{2} < x < -\dfrac{7}{6}$

The solution set is $\left\{x \mid -\dfrac{3}{2} < x < -\dfrac{7}{6}\right\}$.

57. (a)

$|2x - 5| \geq 9$

$|2x - 5| - 9 \geq 0$

Graph $f(x) = |2x - 5| - 9$. The x-intercepts are $x = -2$ and $x = 7$. The graph is above the x-axis for $x < -2$ and $x > 7$. The solution set is $\{x \mid x \leq -2$ or $x \geq 7\}$.

(b) $|2x - 5| \geq 9$

$2x - 5 \leq -9 \quad$ or $2x - 5 \geq 9$

$2x \leq -4 \quad$ or $\quad 2x \geq 14$

$x \leq -2 \qquad\qquad x \geq 7$

The solution set is $\{x \mid -\infty < x \leq -2$ or $7 \leq x < \infty\}$.

59. $x^2 + x + 1 = 0$
Here $a = 1$, $b = 1$, $c = 1$ and
$b^2 - 4ac = -3$.

Then, $x = \dfrac{-1 \pm \sqrt{-3}}{2(1)} = \dfrac{-1 \pm \sqrt{3}\,i}{2}$

The solution set is $\left\{ \dfrac{-1 - \sqrt{3}i}{2}, \dfrac{-1 + \sqrt{3}i}{2} \right\}$.

61. $2x^2 + x - 2 = 0$
Here $a = 2$, $b = 1$, $c = -2$ and
$b^2 - 4ac = 17$.

Then, $x = \dfrac{-1 \pm \sqrt{17}}{2(2)} = \dfrac{-1 \pm \sqrt{17}}{4}$

The solution set is $\left\{ \dfrac{-1 - \sqrt{17}}{4}, \dfrac{-1 + \sqrt{17}}{4} \right\}$.

63. $x^2 + 3 = x$
$x^2 - x + 3 = 0$
Here, $a = 1$, $b = -1$, $c = 3$ and
$b^2 - 4ac = -11$.

Then, $x = \dfrac{-(-1) \pm \sqrt{-11}}{2(1)} = \dfrac{1 \pm \sqrt{11}\,i}{2}$

The solution set is $\left\{ \dfrac{1 - \sqrt{11}i}{2}, \dfrac{1 + \sqrt{11}i}{2} \right\}$.

65. $x(1 - x) = 6$
$x - x^2 = 6$
$-x^2 + x - 6 = 0$
$x^2 - x + 6 = 0$
Here, $a = 1$, $b = -1$, $c = 6$ and
$b^2 - 4ac = -23$.

Then, $x = \dfrac{-(-1) \pm \sqrt{-23}}{2(1)} = \dfrac{1 \pm \sqrt{23}\,i}{2}$

The solution set is $\left\{ \dfrac{1 - \sqrt{23}i}{2}, \dfrac{1 + \sqrt{23}i}{2} \right\}$.

67. $x^4 + 2x^2 - 8 = 0$
Let $u = x^2$ and $u^2 = x^4$.
Then $u^2 + 2u - 8 = 0$
$(u + 4)(u - 2) = 0$

$(u + 4) = 0$	or	$u - 2 = 0$
$u = -4$	or	$u = 2$
$x^2 = -4$	or	$x^2 = 2$
$x = \pm\sqrt{-4}$	or	$x = \pm\sqrt{2}$
$x = \pm\sqrt{4}\,i$		
$= \pm 2i$		

The solution set is $\left\{ -\sqrt{2},\ \sqrt{2},\ -2i,\ 2i \right\}$.

69. Using $s = vt$, we have $t = 3$ and $v = 1100$. We want the distance s in feet. Then,
$s = (1100)(3)$
$s = 3300$
The storm is 3300 feet away.

71. Using $s = vt$, we have a downwind speed of $v_d = 250 + 30 = 280$ and an upwind speed of $v_u = 250 - 30 = 220$ and the time total $t \le 5$. We want to find the distance s the plane can travel.

	Velocity	Time	Distance
Downwind	280	$\dfrac{s/2}{280}$	$\dfrac{s}{2}$
Upwind	220	$\dfrac{s/2}{280}$	$\dfrac{s}{2}$

Then,
$$\frac{s/2}{280} + \frac{s/2}{220} \leq 5$$
$$(6160)\left[\frac{s}{560} + \frac{s}{440}\right] \leq (5)(6160)$$
$$11s + 14s \leq (5)(6160)$$
$$25s \leq 5(6160)$$
$$s \leq \frac{(5)(6160)}{25}$$
$$= 1232$$

The distance out is 616 miles, and the round trip is 1232 miles. Hence, the search plane can go as far as 616 miles.

Check: Downwind: time $= \dfrac{\frac{1232}{2}}{280} = 2.2$ hours

Upwind: time $= \dfrac{\frac{1232}{2}}{220} = 2.8$ hours

Roundtrip time $= 2.2 + 2.8 = 5$ hours maximum.

73. We are asked to find a time in hours. Using $s = vt$, we can make a table as follows:

	Velocity	Time	Distance
Raft	5	t	$5t$
Helicopter	90	t	$90t$

Note that the times are equal. Also
$$5t + 90t = 150$$
$$95t = 150$$
$$t = 1.58 \text{ hours} \approx 1 \text{ hour, } 35 \text{ minutes}$$
The helicopter will reach the lite raft in 1 hour, 35 minutes.
Check: $5(1.58) + 90(1.58) \approx 150$ miles

75. We want a length in feet. The effective speed of the train (i.e, relative to the man) is $30 - 4 = 26$ miles per hour. The time is 5 seconds $= \dfrac{5}{60}$ minutes $= \dfrac{5/60}{60}$ hours $= \dfrac{1}{720}$ hrs. Using $s = vt$, we get $s = (26)\left[\dfrac{1}{720}\right] = \dfrac{26}{720}$ miles $= \dfrac{26/720}{5280} = 190.67$ feet. The freight train is 190.67 feet long.
Check: $v = \dfrac{26}{720} \div \dfrac{5}{3600} = 26$

77. We want to find the number x of passengers over 20 to make a total of \$482.40. We make a table:

Number passengers	Fare Each	Total Cost
$20 + x$	$\$15 - 0.1x$	$(20 + x)(15 - 0.1x)$

Then,
$$(20 + x)(15 - 0.1x) = 482.40$$
$$300 - 2x + 15x - 0.1x^2 = 482.40$$
$$-0.1x^2 + 13x - 182.40 = 0$$
$$x^2 - 130x + 1824 = 0$$
Here $a = 1$, $b = -130$, $c = 1824$, and $b^2 - 4ac = (130)^2 - 4(1)(1824) = 9604$

Thus, $x = \dfrac{-(-130) \pm \sqrt{9604}}{2} = \dfrac{130 \pm 98}{2}$

$= \dfrac{130 + 98}{2}$ or $\dfrac{130 - 98}{2}$

$= 114$ or 16

Since the capacity of the bus is 44, we discard the 114. The total number of passengers is $20 + 16 = 36$, and the ticket price per passenger is $15 - 0.1(16) = \$13.40$.

Check: $(36)(15 - 0.1(16) = 482.40$

36 seniors went on the trip; each one paid $\$13.40$.

79.

Distance = Velocity × Time

Mike	100	v_m	$t \Rightarrow v_m = \dfrac{100}{t}$
Dan	95	v_d	$t \Rightarrow v_d = \dfrac{95}{t}$
Mike	105	$\dfrac{100}{t}$	$\dfrac{105}{100}t$
Dan	100	$\dfrac{95}{t}$	$\dfrac{100}{95}t$

$\dfrac{105t}{100} \overset{?}{\geq} \dfrac{100}{95}t$

$105t = 1.0526316t$

Mike	100	$\dfrac{100}{t}$	t
Dan	95	$\dfrac{95}{t}$	t

(a)   No.

(b)   Mike wins again.

(c)   Mike wins by 0.25 meters.

(d)   Mike should line up 5.26316 meters behind the start line.

(e)   Yes.

POLYNOMIAL AND RATIONAL FUNCTIONS

4.1 Quadratic Functions

1. D   3. A   5. B   7. E   9. D   11. B

13.

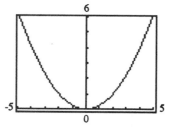

Since $a = \dfrac{1}{4} > 0$, the parabola opens up.

$$\dfrac{-b}{2a} = \dfrac{-0}{2\left[\dfrac{1}{4}\right]} = 0$$

$f(0) = 0$

Thus, vertex: $(0, 0)$

Axis of symmetry: $x = 0$

y-intercept: $f(0) = \dfrac{1}{4}(0)^2 = 0$

x-intercept: $0 = \dfrac{1}{4}x^2$

$\qquad\qquad\qquad x = 0$

15.

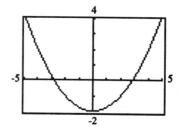

Since $a = \dfrac{1}{4} > 0$, the parabola opens up.

$$\dfrac{-b}{2a} = \dfrac{-0}{2\left[\dfrac{1}{4}\right]} = 0$$

$f(0) = \dfrac{1}{4}(0)^2 - 2 = -2$

Thus, vertex: $(0, -2)$

Axis of symmetry: $x = 0$

y-intercept: $\quad f(0) = \dfrac{1}{4}(0)^2 - 2 = -2$

x-intercepts: $\quad 0 = \dfrac{1}{4}x^2 - 2$

$\qquad\qquad\qquad 2 = \dfrac{1}{4}x^2$

$\qquad\qquad\qquad x^2 = 8$

$\qquad\qquad\qquad x = \pm 2\sqrt{2}$

17.

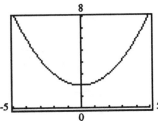

Since $a = \dfrac{1}{4} > 0$, the parabola opens up.

$$\dfrac{-b}{2a} = \dfrac{-0}{2\left(\dfrac{1}{4}\right)} = 0$$

$$f(0) = \dfrac{1}{4}(0)^2 + 2 = 2$$

Thus, vertex: $(0, 2)$

Axis of symmetry: $x = 0$

y-intercept: $\quad f(0) = \dfrac{1}{4}(0)^2 + 2 = 2$

x-intercepts: $\quad 0 = \dfrac{1}{4}x^2 + 2$

$$-2 = \dfrac{1}{4}x^2$$
$$-8 = x^2$$

No real solution; therefore, no x-intercepts.

19.

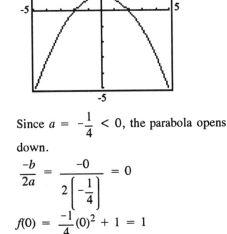

Since $a = -\dfrac{1}{4} < 0$, the parabola opens down.

$$\dfrac{-b}{2a} = \dfrac{-0}{2\left(-\dfrac{1}{4}\right)} = 0$$

$$f(0) = \dfrac{-1}{4}(0)^2 + 1 = 1$$

Thus, vertex: $(0, 1)$

Axis of symmetry: $x = 0$

y-intercept: $\quad f(0) = \dfrac{-1}{4}(0)^2 + 1 = 1$

x-intercepts: $\quad 0 = \dfrac{-1}{4}x^2 + 1$

$$-1 = \dfrac{-1}{4}x^2$$
$$4 = x^2$$
$$x = \pm 2$$

21.

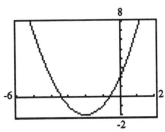

Since $a = 1 > 0$, the parabola opens up.

$$\dfrac{-b}{2a} = \dfrac{-4}{2(1)} = -2$$
$$f(-2) = (-2)^2 + 4(-2) + 2 = -2$$

Thus, vertex: $(-2, -2)$

Axis of symmetry: $x = -2$

y-intercept: $\quad f(0) = (0)^2 + 4(0) + 2 = 2$

x-intercepts: $\quad 0 = x^2 + 4x + 2$

$$x = \dfrac{-4 \pm \sqrt{4^2 - 4(1)(2)}}{2(1)}$$

$$= \dfrac{-4 \pm \sqrt{8}}{2}$$

$$= -2 \pm \sqrt{2}$$

23.

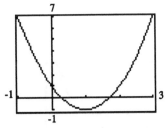

Since $a = 2 > 0$, the parabola opens up.

$$\dfrac{-b}{2a} = \dfrac{-(-4)}{2(2)} = 1$$
$$f(-1) = 2(1)^2 - 4(1) + 1 = -1$$

Thus, vertex: $(1, -1)$

Axis of symmetry: $x = 1$

y-intercept: $\quad f(0) = 2(0)^2 - 4(0) + 1 = 1$

x-intercepts: $\quad 0 = 2x^2 - 4x + 1$

$$x = \dfrac{-(-4) \pm \sqrt{(-4)^2 - 4(2)(1)}}{2(2)}$$

$$= \dfrac{4 \pm \sqrt{8}}{4}$$

$$= 1 \pm \dfrac{\sqrt{2}}{2}$$

25.

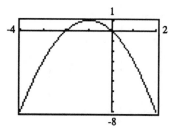

Since $a = -1 < 0$, the parabola opens down.

$$\frac{-b}{2a} = \frac{-(-2)}{2(-1)} = -1$$
$$f(-1) = -(-1)^2 - 2(-1) = 1$$
Thus, vertex: $(-1, 1)$
Axis of symmetry: $x = -1$
y-intercept: $f(0) = -(0)^2 - 2(0) = 0$
x-intercepts: $0 = -x^2 - 2x$
$$= -x(x + 2)$$
$$x = 0, -2$$

27.

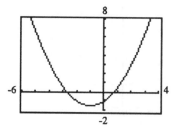

Since $a = \frac{1}{2} > 0$, the parabola opens up.

$$\frac{-b}{2a} = \frac{-1}{2\left(\frac{1}{2}\right)} = -1$$
$$f(-1) = \frac{1}{2}(-1)^2 + (-1) - 1 = -\frac{3}{2}$$

Thus, vertex: $\left(-1, -\frac{3}{2}\right)$
Axis of symmetry: $x = -1$

y-intercept: $f(0) = \frac{1}{2}(0)^2 + 0 - 1 = -1$

x-intercepts: $0 = \frac{1}{2}x^2 + x - 1$

$$x = \frac{-1 \pm \sqrt{1^2 - 4\left(\frac{1}{2}\right)(-1)}}{2\left(\frac{1}{2}\right)}$$

$$= -1 \pm \sqrt{3}$$

29.

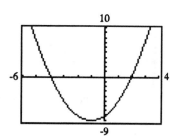

Since $a = 1 > 0$, the parabola opens up.
$$\frac{-b}{2a} = \frac{-2}{2(1)} = -1$$
$$f(-1) = (-1)^2 + 2(-1) - 8 = -9$$
Thus, vertex: $(-1, -9)$
Axis of symmetry: $x = -1$
y-intercept: $f(0) = (0)^2 + 2(0) - 8 = -8$
x-intercepts: $0 = x^2 + 2x - 8$

$$x = \frac{-2 \pm \sqrt{2^2 - 4(1)(-8)}}{2(1)}$$

$$= \frac{-2 \pm \sqrt{36}}{2}$$

$$= -4, 2$$

31.

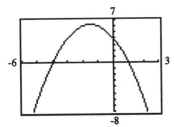

Since $a = -1 < 0$, the parabola opens down.

$$\frac{-b}{2a} = \frac{-(-3)}{2(-1)} = -\frac{3}{2}$$

$$f\left(-\frac{3}{2}\right) = -\left(-\frac{3}{2}\right)^2 - 3\left(-\frac{3}{2}\right) + 4 = \frac{25}{4}$$

Thus, vertex: $\left(-\frac{3}{2}, \frac{25}{4}\right)$

Axis of symmetry: $x = -\frac{3}{2}$

y-intercept: $f(0) = -(0)^2 - 3(0) + 4 = 4$
x-intercepts: $0 = -x^2 - 3x + 4$

$$x = \frac{-(-3) \pm \sqrt{(-3)^2 - 4(-1)(4)}}{2(-1)}$$

$$= \frac{3 \pm \sqrt{25}}{-2}$$

$$x = 1, -4$$

33.

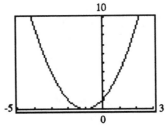

Since $a = 1 > 0$, the parabola opens up.

$$\frac{-b}{2a} = \frac{-2}{2(1)} = -1$$

$$f(-1) = (-1)^2 + 2(-1) + 1 = 0$$

Thus, vertex: $(-1, 0)$

Axis of symmetry: $x = -1$

y-intercept: $\quad f(0) = 0^2 + 2(0) + 1 = 1$
x-intercept: $\quad 0 = x^2 + 2x + 1$

$$x = \frac{-2 \pm \sqrt{2^2 - 4(1)(1}}{2(1)}$$

$$= -\frac{2}{2}$$

$$= -1$$

35.

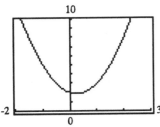

Since $a = 2 > 0$, the parabola opens up.

$$\frac{-b}{2a} = \frac{-(-1)}{2(2)} = \frac{1}{4}$$

$$f\left(\frac{1}{4}\right) = 2\left(\frac{1}{4}\right)^2 - \frac{1}{4} + 2 = \frac{15}{8}$$

Thus, vertex: $\left(\frac{1}{4}, \frac{15}{8}\right)$

Axis of symmetry: $x = \frac{1}{4}$

y-intercept: $\quad f(0) = 2(0)^2 - (0) + 2 = 2$
x-intercepts: $\quad 0 = 2x^2 - x + 2$

$$x = \frac{-(-1) \pm \sqrt{(-1)^2 - 4(2)(2)}}{2(2)}$$

$$= \frac{1 \pm \sqrt{-15}}{4}$$

No real solution; therefore, no x-intercepts.

37.

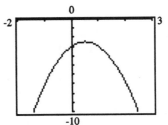

Since $a = -2 < 0$, the parabola opens down.

$$\frac{-b}{2a} = \frac{-2}{2(-2)} = \frac{1}{2}$$

$$f\left(\frac{1}{2}\right) = -2\left(\frac{1}{2}\right)^2 + 2\left(\frac{1}{2}\right) - 3 = -\frac{5}{2}$$

Thus, vertex: $\left(\frac{1}{2}, -\frac{5}{2}\right)$

Axis of symmetry: $x = \frac{1}{2}$

y-intercept: $f(0) = -2(0)^2 + 2(0) - 3 = -3$
x-intercepts: $\quad 0 = -2x^2 + 2x - 3$

$$x = \frac{-2 \pm \sqrt{2^2 - 4(-2)(-3)}}{2(-2)}$$

$$= \frac{-2 \pm \sqrt{20}}{-4}$$

No real solution; therefore, no x-intercepts.

39.

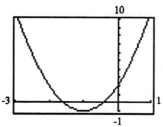

Since $a = 3 > 0$, the parabola opens up.

$\dfrac{-b}{2a} = \dfrac{-6}{2(3)} = -1$

$f(-1) = 3(-1)^2 + 6(-1) + 2 = -1$

Thus, vertex: $(-1, -1)$

Axis of symmetry: $x = -1$

y-intercept: $\quad f(0) = 3(0)^2 + 6(0) + 2 = 2$

x-intercepts: $\quad 0 = 3x^2 + 6x + 2$

$$x = \dfrac{-6 \pm \sqrt{6^2 - 4(3)(2)}}{2(3)}$$

$$= \dfrac{-6 \pm \sqrt{12}}{6}$$

$$= -1 \pm \dfrac{\sqrt{3}}{3}$$

41.

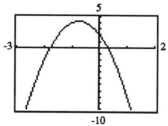

Since $a = -4 < 0$, the parabola opens down.

$\dfrac{-b}{2a} = \dfrac{-(-6)}{2(-4)} = \dfrac{-6}{8} = \dfrac{-3}{4}$

$f\left(-\dfrac{3}{4}\right) = -4\left(-\dfrac{3}{4}\right)^2 - 6\left(-\dfrac{3}{4}\right) + 2 = \dfrac{17}{4}$

Thus, vertex: $\left[-\dfrac{3}{4}, \dfrac{17}{4}\right]$

Axis of symmetry: $x = -\dfrac{3}{4}$

y-intercept: $\quad f(0) = -4(0)^2 - 6(0) + 2 = 2$

x-intercepts: $0 = -4x^2 - 6x + 2$

$$x = \dfrac{-(-6) \pm \sqrt{(-6)^2 - 4(-4)(2)}}{2(-4)}$$

$$= \dfrac{6 \pm \sqrt{68}}{-8}$$

$$= \dfrac{3 \pm \sqrt{17}}{-4}$$

43. $\quad f(x) = 2x^2 + 12x - 3 \ (a = 2, b = 12, c = -3)$

Since $a = 2 > 0$, the parabola opens *upward*, and therefore has a *minimum* at the vertex (h, k).

By (3), $h = \dfrac{-b}{2a} = \dfrac{-12}{4} = -3$, and $k = f(h) = f(-3)$ or $k = -21$. Therefore the minimum value of $f(x)$ is $k = -21$.

45. $\quad f(x) = -x^2 + 10x - 4 \ (a = -1, b = 10, c = -4)$

Here, $a = -1 < 0$, so the parabola opens downward and has a *maximum* at (h, k).

$h = \dfrac{-10}{-2} = 5$ and $k = f(5) = 21$. The maximum value of $f(x)$ is $k = 21$.

47. $\quad f(x) = -3x^2 + 12x + 1 \ (a = -3, b = 12, c = 1)$

Here, $a = -3 < 0$, so the parabola opens downward and $f(x)$ has a maximum at (h, k).

$h = \dfrac{-12}{-6} = 2$ and $k = f(2) = 13$. So the maximum value of $f(x)$ is $k = 13$.

Chapter 4 Polynomial and Rational Functions

49. Graph $f(x) = x^2 + 2x + c$ for $c = -3, 0,$ and 1. In each case, $a = 1$, so the parabola opens upward.

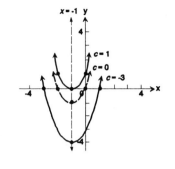

$h = \dfrac{-2}{2} = -1$, and

$k = f(-1) = 1 - 2 + c = c - 1$

Thus, the vertex is at the point $(-1, c - 1)$. To find the y-intercept, set $x = 0$: then $y = c$.

$c = -3$:   Vertex at $(-1, c - 1) = (-1, -4)$
             y-intercept at $(0, -3)$

$c = 0$:   Vertex at $(-1, -1)$
             y-intercept at $(0, 0)$

$c = 1$:   Vertex at $(-1, 0)$
             y-intercept at $(0, 1)$

51. Each parabola opens up and passes through $(0, 1)$. Each one has the same shape.

53. We are given $R = -4p^2 + 4000p$, which represents a parabola that opens downward (since $a = -4 < 0$). Thus, R will be a maximum at the vertex. Now, $h = \dfrac{-b}{2a} = \dfrac{-4000}{-8} = 500$, and $k = f(h) = f(500) = -4(500)^2 + 4000(500) = 1,000,000$. The maximum revenue is $R = k = \$1,000,000$. It occurs when $p = h = 500$ dollars per dryer.

55. Let ℓ be the length and w be the width of the rectangle. Since the perimeter is 400 feet, we have $2\ell + 2w = 400$, so that $\ell = 200 - w$. We want to maximize the area, $A = \ell w = (200 - w)w = -w^2 + 200w$. Now $A = -w^2 + 200w$ is the equation of a parabola that opens downward, and hence has a maximum value w at $w = \dfrac{-b}{2a} = \dfrac{-200}{-2} = 100$, and the value is $f(100) = -(100)^2 + 200(100) = 10000$. The maximum area is 10000 sq. ft., and this occurs when $w = 100$ and $\ell = 200 - w = 200 - 100 = 100$. Thus, the dimensions of the rectangle are 100 ft. by 100 ft.

57. The area of the rectangular plot is given by width times length, or $A = x(4000 - 2x)$. (Refer to the figure.) Thus, $A = -2x^2 + 4000x$; hence, A will be a maximum at the vertex.

$h = \dfrac{-4000}{-4} = 1000$, and $k = f(1000) = -2,000,000 + 4,000,000 = 2,000,000$. Thus, the largest area that can be enclosed is $A = k = 2,000,000$ square meters.

59. Consider the figure: The total length of the fence is $x + x + x + y + y$, and this must equal 10,000 meters: $3x + 2y = 10,000$; or $2y = 10,000 - 3x$ so that $y = \dfrac{10,000 - 3x}{2}$. The total area enclosed is $A = xy = x\left[\dfrac{10,000 - 3x}{2}\right]$,

or $A = \dfrac{-3}{2}x^2 + 5,000x$, which attains a *maximum* at the vertex (since

$a = \dfrac{-3}{2} < 0$). Now, $h = \dfrac{-5,000}{-3} = 1666.67$. Thus, the maximum area is $A = k = f(1666.67) = 4,166,666.67$ square meters.

61. **(a)**

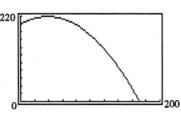

(b) The maximum height of the projectile occurs at $x = \dfrac{-b}{2a}$ where $a = \dfrac{-32}{2500}$ and $b = 1$.

Therefore, the maximum height of the projectile occurs at $x = \dfrac{-1}{2\left(\dfrac{-32}{2500}\right)} = 39.0625$ feet.

The maximum height is $f(39.0625) = \dfrac{-32(39.065)^2}{2500} + 39.06252 + 200 = 219.53$ feet.

(c) The projectile will strike the water when the height is zero. Therefore, solve the quadratic equation $\dfrac{-32}{2500}x^2 + x + 200 = 0$ using the quadratic formula with $a = \dfrac{-32}{2500}$, $b = 1$ and $c = 200$.

$$x = \frac{-1 \pm \sqrt{1^2 - 4\left(\dfrac{-32}{2500}\right)(200)}}{2\left(\dfrac{-32}{2500}\right)}$$

$$x = \frac{-1 \pm \sqrt{11.24}}{\dfrac{-64}{2500}}$$

$$x \approx -91.90 \quad \text{or} \quad x \approx 170.02$$

Since distance is not negative, the projectile will strike the water 170.02 feet from the base of the cliff.

(d) When the height is 100 feet, the projectile is 135.69 feet from the cliff.

63. The situation at 4:00 P.M. is depicted below.

Let x denote the number of hours that have elapsed since 4:00 P.M. In x hours, the aircraft carrier (A) will have travelled $10x$ nautical miles, and the destroyer (D) will have gone $20x$ miles. The situation will then be as indicated:

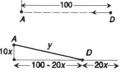

If we let y denote the distance from A to D, then by the Pythagorean Theorem,

$$y^2 = (10x)^2 + (100 - 20x)^2$$
$$= 100x^2 + 10000 - 4000x + 400x^2, \text{ or}$$
$$y^2 = 500x^2 - 4000x + 10000$$

This equation is a parabola that opens upward, since $a = 500 > 0$, so it will have a minimum value at its vertex, (h, k). Now $h = \dfrac{-b}{2a} = \dfrac{4000}{1000} = 4$, and $k = f(h) = f(4) = 2000$. Thus, the smallest possible value of y^2 is 2000, so the minimum value for $y = \sqrt{2000} \approx 44.72$ nautical miles. But the question is to find what time it is. The minimum distance y occurs when $x = h = 4$ hours, i.e., at 8:00 P.M.

65. For simplicity, locate the origin at the point where the cable touches the road:

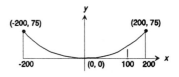

Then the equation of the parabola must be of the form:
$$y = ax^2, \text{ for some } a > 0$$
Since the point (200, 75) is on the parabola, we can determine the constant a:

$$75 = a(200)^2, \text{ or } a = \frac{75}{(200)^2} = 0.001875 \ldots$$

Then when $x = 80$, we have $y = ax^2$, or
$$y = (0.001875)(100)^2 = 18.75 \text{ meters.}$$

67. The area A of the gutter of height x and length y is $A = xy$. Since $2x + y = 12$, we have $y = 12 - 2x$. We want to maximize the area $A = xy = x(12 - 2x)$ $= -2x^2 + 12x$. This is the equation of a parabola that opens downward, and hence

the maximum area occurs when $x = \dfrac{-b}{2a} = \dfrac{-12}{-4} = 3$, and the value is $f(3) = -2(3)^2 + 12(3) = 18$.

Thus, a depth of 3 inches will provide maximum cross-sectional area.

69. Since the diameter ($2r$) equals the width of the rectangle, then $w = 2r$. The perimeter is

$$20 = 2\ell + w = \frac{1}{2}(2\pi r) \text{ or}$$

$$20 = 2\ell + 2r + \pi r \text{ so that}$$
$$2\ell = 20 - 2r - \pi r$$

$$A = \ell w + \frac{1}{2}(\pi r^2) = \ell(2r) + \frac{1}{2}(\pi r^2)$$

$$= r(20 - 2r - \pi r) + \frac{1}{2}\pi r^2$$

$$= \left(-2 - \frac{\pi}{2}\right)r^2 + 20r$$

The area is maximum at $r = \dfrac{-b}{2a} = \dfrac{-20}{-4 - \pi} \approx 2.80$

Thus, $w = \dfrac{40}{\pi + 4} \approx 5.6$ ft., and $\ell = 10 - r - \dfrac{\pi}{2}r \approx 10 - 2.8 - \dfrac{\pi}{2}(2.8) \approx 2.8$ ft.

71. If ℓ is the length of the rectangle, and x is the width, the perimeter of the window is $16 = 2\ell + 3x$ so that $\ell = 8 - \dfrac{3}{2}x$. The area of the window is

$$A = \ell x + \frac{\sqrt{3}}{4}x^2 \text{ or}$$

$$A = \left[8 - \frac{3}{2}x\right]x + \frac{\sqrt{3}}{4}x^2$$

$$A = \left[\frac{-3}{2} + \frac{\sqrt{3}}{4}\right]x^2 + 8x$$

The area is maximum at $x = \dfrac{-b}{2a} - \dfrac{-8}{-3 + \dfrac{\sqrt{3}}{2}} \approx \dfrac{16}{6 - \sqrt{3}} \approx 3.75$ ft. and

$$\ell = 8 - \frac{3}{2}x \approx 8 - \frac{3}{2}(3.75) \approx 2.38 \text{ ft.}$$

73. If the club has exactly 60 members, its revenue would be $60(400) = 24,000$ dollars. Let x be the number in *excess* of 60. Then the total membership is $60 + x$. Now the price for *every* member will be *reduced* by $5 for each member in excess of 60, i.e., by $5x$. Hence, the price per person will be $400 - 5x$. The total revenue is equal to the number of members times the charge per member: $R = (60 + x)(400 - 5x)$, or $R = -5x^2 + 100x + 24,000$. This represents a parabola with a *maximum* at its vertex. $h = \dfrac{-b}{2a} = \dfrac{-100}{-10} = 10$; $k = f(20) = -2000 + 500 + 24,000 = 22,500$. The maximum possible revenue is $R = k = 22,500$, and this occurs when $x = h = 10$, so that the number of members is $60 + 10 = 70$.

75. Since $f(x) = ax^2 + bx + c$ has vertex at $x = 0$, we have $\dfrac{-b}{2a}$ or $b = 0$. At $(0, 2)$, we have $2 = a(0)^2 + c$ or $c = 2$. At $(1, 8)$, we have $8 = a(1)^2 = 2$ or $a = 6$. Thus, $f(x) = 6x^2 + 2$.

77. We are given $V(x) = kx(a - x) = -kx^2 + kax$

This is a maximum when

$$x = \frac{-ka}{-2k} = \frac{a}{2}$$

79. We have
$$ah^2 - bh + c = y_0$$
$$c = y_1$$
$$ah^2 + bh + c = y_2$$
so that
$$y_0 + y_2 = 2ah^2 + 2c$$
$$4y_1 = 4c$$
Hence, Area $= \dfrac{h}{3}(2ah^2 + 6c) = \dfrac{h}{3}(y_0 + 4y_1 + y_2)$

4.2 Polynomial Functions

1. $f(x) = 4x + x^3$ is a polynomial of degree 3.

3. $g(x) = \dfrac{1 - x^2}{2} = \dfrac{1}{2} - \dfrac{1}{2}x^2$ is a polynomial of degree 2.

5. $f(x) = 1 - \dfrac{1}{x} = 1 - x^{-1}$ is *not* a polynomial, since it contains x raised to a negative power.

7. $g(x) = x^{3/2} - x^2 + 2$ is *not* a polynomial since it contains x raised to a fractional power.

9. $F(x) = 5x^4 - \pi x^3 + \dfrac{1}{2}$ is a polynomial, of degree 4.

In Problems 11–17, start with the graph of $y = x^4$ as given below, and perform change of scale, shifting and reflection to obtain the graph of the given function.

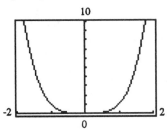

11. For $f(x) = (x + 1)^4$, all that is needed is a horizontal shift, one unit to the left.

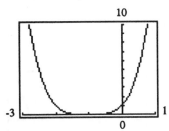

13. For $f(x) = \frac{1}{2}x^4$, vertically compress the graph of $y = x^4$. The graph will pass through $\left(1, \frac{1}{2}\right)$ instead of $(1, 1)$.

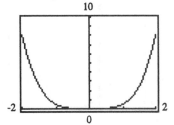

15. To graph $f(x) = 2(x + 1)^4 + 1$, shift left one unit, vertically stretch, and finally shift up one unit.

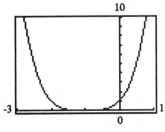
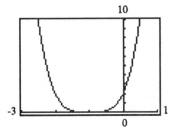
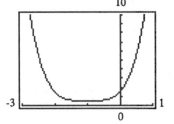

17. Perform the following steps:
   (1) Starting graph: $y = x^4$ (shown above)
   (2) Shift right 2 units: $y = (x - 2)^4$

   (3) Vertically compress and reflect about the x-axis: $y = -\frac{1}{2}(x - 2)^4$

   (4) Shift down one unit: $y = -\frac{1}{2}(x - 2)^4 - 1$

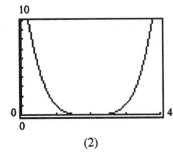

(2)

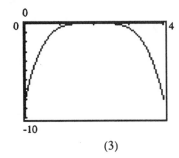

(3)

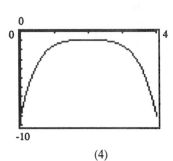

(4)

19. The zeros of $f(x) = 3(x - 7)(x + 3)^2$ are: 7, with multiplicity one; and -3, with multiplicity two. The graph touches the x-axis at -3 and crosses it at 7.

21. The zeros of $f(x) = 4(x^2 + 1)(x - 2)^3$ are: 2, with multiplicity three. Note: $x^2 + 1 = 0$ has no real solution. The graph crosses the x-axis at 2.

23. The zeros of $f(x) = -2\left(x + \dfrac{1}{2}\right)^2(x^2 + 4)^2$ are $\dfrac{-1}{2}$, with multiplicity two and $x^2 + 4 = 0$ has no real solutions. The graph touches the x-axis at $\dfrac{-1}{2}$.

25. The zeros of $f(x) = (x - 5)^3(x + 4)^2$ are: 5, with multiplicity three; and -4, with multiplicity two. The graph touches the x-axis at -4 and crosses it at 5.

27. $f(x) = 3(x^2 + 8)(x^2 + 9)^2$ has no real zeros. The graph neither touches nor crosses the x-axis.

29. (a)
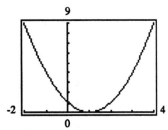
   (b) x-intercept: 1; y-intercept: 1
   (c) 1: Even
   (d) $y = x^2$
   (e) 1
   (f) Local minima: $(1, 0)$

31. (a)
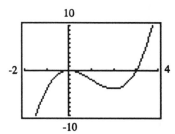
   (b) x-intercepts: 0, 3; y-intercept: 0
   (c) 0: Even; 3: Odd
   (d) $y = x^3$
   (e) 2
   (f) Local maxima: $(0, 0)$;
       Local minima: $(2, -4)$

33. (a)
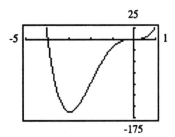
   (b) x-intercepts: -4, 0; y-intercept: 0
   (c) 0: Odd; -4: Odd
   (d) $y = 6x^4$
   (e) 3
   (f) Local minima: $(-3, -162)$

35. (a)
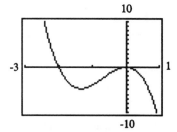
   (b) x-intercepts: -2, 0; y-intercept: 0
   (c) -2: Odd; 0: Even
   (d) $y = -4x^3$
   (e) 2
   (f) Local minima: $(-1.33, -4.74)$;
       Local maxima: $(0, 0)$

37. (a)
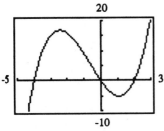

(b) x-intercepts: $-4, 0, 2$; y-intercept: 0
(c) 0: Odd; 2: Odd; -4: Odd
(d) $y = x^3$
(e) 2
(f) Local maxima: $(-2.43, 16.90)$;
Local minima: $(1.09, -5.04)$

39. $f(x) = 4x - x^3 = -x(x^2 - 4)$
$= -x(x + 2)(x - 2)$
(a)
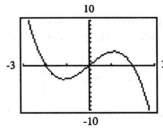

(b) x-intercepts: $-2, 0, 2$; y-intercept: 0
(c) $-2, 0, 2$: Odd
(d) $y = -x^3$
(e) 2
(f) Local minima: $(-1.15, -3.07)$;
Local maxima: $(1.15, 3.07)$

41. (a)
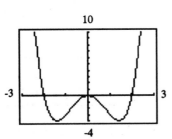

(b) x-intercepts: $-2, 0, 2$; y-intercept: 0
(c) $-2, 2$: Odd; 0: Even
(d) $y = x^4$
(e) 3
(f) Local minima: $(-1.41, -4)$,
$(1.41, -4)$;
Local maxima: $(0, 0)$

43. (a)
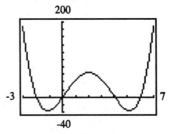

(b) x-intercepts: $0, 2$; y-intercept: 0
(c) $0, 2$: Even
(d) $y = x^4$
(e) 3
(f) Local minima: $(0, 0)$, $(2, 0)$;
Local maxima: $(1, 1)$

45. (a)
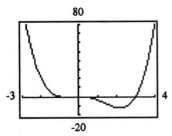

(b) x-intercepts: $-1, 0, 3$; y-intercept: 0
(c) $-1, 3$: Odd, 0: Even
(d) $y = x^4$
(e) 3
(f) Local minima: $(2.18, -12.39)$,
$(-0.68, -0.54)$;
Local maxima: $(0, 0)$

47. (a)
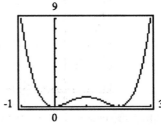

(b) x-intercepts: $-2, 0, 4, 6$;
y-intercept: 0
(c) $-2, 0, 4, 6$: Odd
(d) $y = x^4$
(e) 3
(f) Local minima: $(-1.16, -36)$,
$(5.16, -36)$;
Local maxima: $(2, 64)$

49. (a)

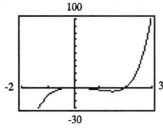

(b) x-intercepts: 0, 2; y-intercept: 0
(c) 0: Even; 2: Odd
(d) $y = x^5$
(e) 2
(f) Local minima: $(1.47, -5.91)$;
Local maxima: $(0, 0)$

51. (a)

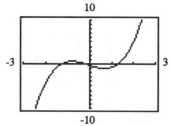

(b) x-intercepts: $-1.26, -0.20, 1.26$;
y-intercept: -0.31752
(c) $-1.26, -0.20, 1.26$: Odd
(d) $y = x^3$
(e) 2
(f) Local minima: $(0.66, -0.99)$;
Local maxima: $(-0.79, 0.56)$

53. (a)

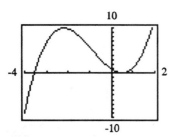

(b) x-intercepts: $-3.56, 0.50$;
y-intercept: 0.89
(c) -3.56: Odd; 0.50: Even
(d) $y = x^3$
(e) 2
(f) Local minima: $(0.50, 0)$;
Local maxima: $(-2.20, 9.91)$

55. (a)

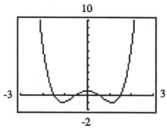

(b) x-intercepts: $-1.50, -0.50, 0.50,$
1.50;
y-intercept: 0.5625
(c) $-1.50, -0.50, 0.50, 1.50$: Odd
(d) $y = x^4$
(e) 3
(f) Local minima: $(-1.11, -1)$,
$(1.11, -1)$;
Local maxima: $(0, 0.5625)$

57. (a)

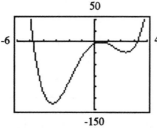

(b) x-intercepts: $-4.78, 0.45, 3.23$;
y-intercept: -3.1264785
(c) $-4.78, 3.23$: Odd; 0.45: Even
(d) $y = x^4$
(e) 3
(f) Local minima: $(-3.31, -135.91)$,
$(2.37, -22.66)$;
Local maxima: $(0.45, 0)$

59. (a)

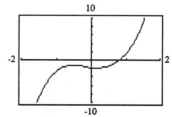

(b) x-intercepts: 0.83; y-intercept: -2
(c) 0.83: Odd
(d) $y = \pi x^3$
(e) 2
(f) Local minima: $(-0.50, -1.53)$
Local maxima: $(0.20, -2.11)$

61. (a)

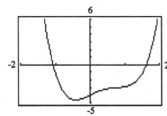

(b) x-intercepts: $-1.06, 1.61$;
    y-intercept: -4
(c) $-1.06, 1.61$: Odd
(d) $y = 2x^4$
(e) 1
(f) Local minima: $(-0.41, -4.64)$

63. (a)

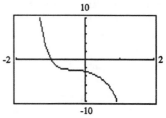

(b) x-intercepts: -0.97
(c) -0.97: Odd
(d) $y = -2x^5$
(e) None
(f) None

65. c, e, f

67. c, c

4.3 Rational Functions

In Problems 1–10, the domain consists of all real numbers except those for which the denominator, q(x), is zero.

1. $R(x) = \dfrac{4x}{x - 3}$. Here the denominator is $q(x) = x - 3$, which has 3 as its only zero. Thus, the domain of $R(x)$ consists of all real numbers except 3.

3. For $H(x) = \dfrac{-4x^2}{(x - 2)(x + 4)}$, the denominator, $q(x) = (x - 2)(x + 4)$, has zeros at 2 and -4. Thus, the domain of $H(x)$ is all real numbers except 2 and -4.

5. In $F(x) = \dfrac{3x(x - 1)}{2x^2 - 5x - 3}$, the denominator is $q(x) = 2x^2 - 5x - 3 = (2x + 1)(x - 3)$, whose zeros are $\dfrac{-1}{2}$ and 3. Thus, the domain of $F(x)$ consists of all real numbers except $\dfrac{-1}{2}$ and 3.

7. For $R(x) = \dfrac{x}{x^3 - 8}$, the denominator is $q(x) = x^3 - 8$, which can be factored as a difference of cubes:
$$q(x) = x^3 - 8 = (x - 2)(x^2 + 2x + 4)$$
Now, $x^2 + 2x + 4$ has no real zeros since its discriminant is $b^2 - 4ac = 4 - 16 = -12 < 0$. Thus, the domain of $R(x)$ is all real numbers except 2.

9. In $H(x) = \dfrac{3x^2 + x}{x^2 + 4}$, the denominator has no real zeros, so the domain is *all* real numbers.

11. (a) Domain: $\{x \mid x \neq 2\}$; Range: $\{y \mid y \neq 1\}$
    (b) $(0, 0)$
    (c) $y = 1$
    (d) $x = 2$
    (e) none

13. (a) Domain: $\{x \mid x \neq 0\}$; Range: all real numbers
    (b) $(-1, 0)$, $(1, 0)$ (c) none
    (d) none   (e) $y = 2x$

15. (a) Domain: $\{x \mid x \neq -2, x \neq 2\}$; Range: $\{y \mid -\infty < y \leq 0 \text{ or } 1 < y < \infty\}$
    (b) $(0, 0)$ (c) $y = 1$
    (d) $x = -2$, $x = 2$ (e) none

17. (a) Domain: $\{x \mid x \neq -1\}$; Range: $\{y \mid -\infty < y < 2 \text{ or } 2 < y < \infty\}$
    (b) $(-1.5, 0)$, $(0, 3)$ (c) $y = 2$
    (d) $x = -1$ (e) none

19. (a) Domain: $\{x \mid x \neq -4, x \neq 3\}$; Range: all real numbers
    (b) $(0, 0)$ (c) $y = 0$
    (d) $x = -4$, $x = 3$ (e) none

Problems 21–29 are all based on the graphs of $R(x) = \dfrac{1}{x}$ and $H(x) = \dfrac{1}{x^2}$, shown below.

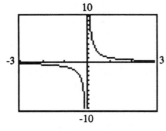

$$R(x) = \frac{1}{x}$$

$$H(x) = \frac{1}{x^2}$$

21. $R(x) = \dfrac{1}{(x - 1)^2}$. This graph can be obtained by shifting the graph

of $H(x) = \dfrac{1}{x^2}$ horizontally, one unit to the right.

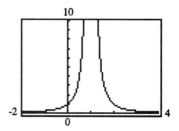

23. $H(x) = \dfrac{-2}{x + 1}$.

(1) Start with $R(x) = \dfrac{1}{x}$.

(2) Vertically stretch by a factor of 2 and reflect about the x-axis.

(3) Shift left one unit.

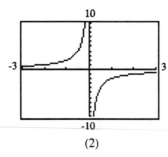

(2)

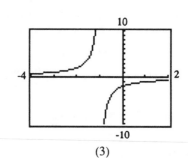

(3)

Chapter 4 Polynomial and Rational Functions

25. $R(x) = \dfrac{1}{x^2 + 4x + 4} = \dfrac{1}{(x + 2)^2}$

Shift the graph of $H(x) = \dfrac{1}{x^2}$ left two units.

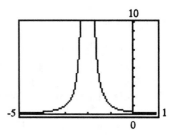

27. $F(x) = 1 - \dfrac{1}{x}$

Start with the graph of $R(x) = \dfrac{1}{x}$, reflect about the x-axis, then shift up 1 unit.

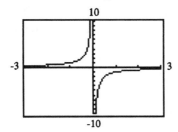
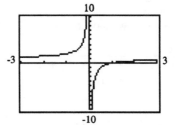

29. $R(x) = \dfrac{x^2 - 4}{x^2} = 1 - \dfrac{4}{x^2} = \dfrac{-4}{x^2} + 1$

(1) Start with the graph of $H(x) = \dfrac{1}{x^2}$

(2) Vertically stretch by a factor of 4 and reflect about the x-axis.

(3) Shift up 1 unit.

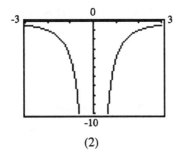

(2)

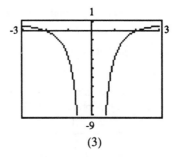

(3)

31. $R(x) = \dfrac{3x}{x + 4}$

The degree of the numerator, $p(x) = 3x$, is $n = 1$. The degree of the denominator, $q(x) = x + 4$ is $m = 1$. Since $n = m$,

the line $y = \dfrac{3}{1} = 3$ is a horizontal

asymptote. The denominator is zero at $x = -4$, so $x = -4$ is a vertical asymptote.

33. $H(x) = \dfrac{x^4 + 2x^2 + 1}{x^2 - x + 1}$

$p(x) = x^4 + 2x^2 + 1$; degree $= n = 4$
$q(x) = x^2 - x + 1$; degree $= m = 2$
Since $n > m + 1$, $H(x)$ has no horizontal nor oblique asymptote. Since $q(x)$ has no real zeros, there is no vertical asymptote.

35. $T(x) = \dfrac{x^3}{x^4 - 1}$

$p(x) = x^3$; degree $= n = 3$

$q(x) = x^4 - 1$; degree $= m = 4$

Since $n < m$, the line $y = 0$ is a horizontal asymptote. We can factor $q(x) = x^4 - 1 = (x^2 - 1)(x^2 + 1) = (x - 1)(x + 1)(x^2 + 1)$, so the vertical asymptotes are the lines $x = -1$ and $x = 1$.

37. $Q(x) = \dfrac{5 - x^2}{3x^4}$

$p(x) = 5 - x^2$; $n = 2$

$q(x) = 3x^4$; $m = 4$

Since $n < m$, the line $y = 0$ is a horizontal asymptote. The vertical asymptote is $x = 0$.

39. $R(x) = \dfrac{3x^4 + 4}{x^3 + 3x}$

Since $n = m + 1$, we have an oblique asymptote, so it is necessary to perform long division:

$$
\begin{array}{r}
3x \\
x^3 + 3x \overline{\big)\,3x^4 + 0x^3 + 0x^2 + 0x + 4} \\
3x^4 + 9x^2 \\
\hline
-9x^2 + 0x + 4 \quad \leftarrow \text{Remainder}
\end{array}
$$

Thus, $R(x) = \dfrac{3x^4 + 4}{x^3 + 3x} = 3x + \dfrac{-9x^2 + 4}{x^3 + 3x}$

Therefore, $y = 3x$ is an oblique asymptote for $R(x)$.

We can factor the denominator: $q(x) = x^3 + 3x = x(x^2 + 3)$.

no real zeros

The vertical asymptote is $x = 0$.

In Problems 41–83, we will use the following terminology: $R(x) = \dfrac{p(x)}{q(x)}$, *where the degree of* $p(x) = n$ *and the degree of* $q(x) = m$. *In every problem, we will follow the steps listed on page 294 of your text:*

41. $R(x) = \dfrac{x + 1}{x(x + 4)}$ $p(x) = x + 1$; $q(x) = x(x + 4) = x^2 + 4x$

$\qquad\qquad\qquad\qquad n = 1, \, m = 2$

Step 1: (a) The x-intercept is the zero of $p(x)$: -1

$\qquad\qquad$ (b) For y-intercept, $R(0)$ is not defined, since $q(0) = 0$, so there is no y-intercept.

Step 2: $R(-x) = \dfrac{-x + 1}{x^2 - 4x}$; this is neither $R(x)$ nor $-R(x)$, so there is no symmetry.

Step 3: The vertical asymptotes are the zeros of $q(x)$: $x = -4$ and $x = 0$.

Step 4: Since $n < m$, the line $y = 0$ is the horizontal asymptote; intersected at $(-1, 0)$.

Step 5:

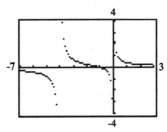

Chapter 4 Polynomial and Rational Functions

43. $R(x) = \dfrac{3x + 3}{2x + 4}$

    Step 1:   (a)   The x-intercept is the zero of $p(x)$: -1

               (b)   For y-intercept is $y = R(0) = \dfrac{3}{4}$

    Step 2:   No symmetry.

    Step 3:   The vertical asymptote is the zero of $q(x)$: $x = -2$.

    Step 4:   Since $n = m$, the horizontal asymptote is the line $y = \dfrac{3}{2}$; not intersected.

    Step 5:

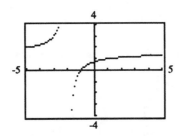

45. $R(x) = \dfrac{3}{x^2 - 4} = \dfrac{3}{(x + 2)(x - 2)}$

    Step 1:   (a)   There are no x-intercepts, since $p(x)$ has no zeros.

               (b)   The y-intercept is $y = R(0) = \dfrac{-3}{4}$

    Step 2: $R(-x) = \dfrac{3}{x^2 - 4} = R(x)$

              Therefore, the graph of $R(x)$ is symmetric with respect to the y-axis.

    Step 3:   The vertical asymptotes are the zeros of $q(x)$: $x = -2$ and $x = 2$.

    Step 4:   Since $n < m$, the line $y = 0$ is a horizontal asymptote; not intersected.

    Step 5:

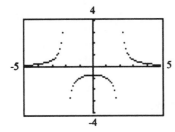

47. $P(x) = \dfrac{x^4 + x^2 + 1}{x^2 - 1} = \dfrac{x^4 + x^2 + 1}{(x + 1)(x - 1)}$

    Step 1:   (a)   There are no x-intercepts, since $p(x)$ has no real zeros.

               (b)   The y-intercept is $y = P(0) = -1$.

    Step 2: $P(-x) = \dfrac{x^4 + x^2 + 1}{x^2 - 1} = P(x)$, so we do have symmetry with respect to the y-axis.

    Step 3:   The vertical asymptotes are the zeros of $q(x)$: $x = -1, x = 1$.

    Step 4:   Since $n > m + 1$, we have no horizontal and no oblique asymptote.

Step 5:

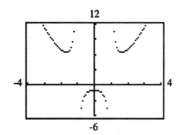

49. $H(x) = \dfrac{x^3 - 1}{x^2 - 9} = \dfrac{(x - 1)(x^2 + x + 1)}{(x - 3)(x + 3)}$

Step 1:   (a)   The x-intercept is $x = 1$, since this is the only solution to $x^3 - 1 = 0$.

   (b)   The y-intercept is $y = H(0) = \dfrac{1}{9}$

Step 2:   No symmetry.

Step 3:   The vertical asymptotes are $x = -3$, $x = 3$, since $x^2 - 9 = 0$ has $x = -3$, 3 as zeros.

Step 4:   Since $n = m + 1$, we have an oblique asymptote. Using long division, we can determine the oblique asymptote.

$$
\begin{array}{r}
x \\
x^2 - 9 \overline{)\,x^3 + 0x^2 + 0x - 1\,} \\
\underline{x^3 -\, 9x } \\
9x - 1
\end{array}
$$

So, $H(x) = x + \dfrac{9x - 1}{x^2 - 9}$, and we have an oblique asymptote, $y = x$.

To find any intersection points, we solve:

$$\dfrac{x^3 - 1}{x^2 - 9} = x$$
$$x^3 - 1 = x^3 - 9x$$
$$-1 = -9x$$
$$x = \dfrac{1}{9}$$

$$H\left(\dfrac{1}{9}\right) = \dfrac{\left(\dfrac{1}{9}\right)^3 - 1}{\left(\dfrac{1}{9}\right)^2 - 9} = \dfrac{1}{9}$$

Thus, the oblique asymptote $y = x$ intersects $H(x)$ at $\left(\dfrac{1}{9}, \dfrac{1}{9}\right)$.

Step 5:

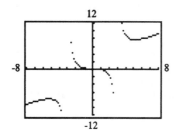

51. $R(x) = \dfrac{x^2}{x^2 + x - 6} = \dfrac{x^2}{(x + 3)(x - 2)}$

Step 1: (a) The x-intercept is $x = 0$

(b) The y-intercept is $y = R(0) = \dfrac{0}{-6} = 0$

The intercept is $(0, 0)$.

Step 2: No symmetry.

Step 3: Vertical asymptotes $x = -3$, $x = 2$, the zeros of $x^2 + x - 6 = (x + 3)(x - 2)$.

Step 4: Since $n = m$, the line $y = \dfrac{1}{1} = 1$ is the horizontal asymptote; intersected at $(6, 1)$, since

$$1 = \dfrac{x^2}{x^2 + x - 6}$$
$$x^2 + x - 6 = x^2$$
$$x - 6 = 0$$
$$x = 6$$

$$R(6) = \dfrac{6^2}{6^2 + 6 - 6} = \dfrac{6^2}{6^2} = 1$$

Step 5:

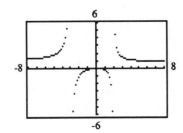

53. $G(x) = \dfrac{x}{x^2 - 4} = \dfrac{x}{(x + 2)(x - 2)}$

Step 1: (a) The x-intercept is $x = 0$

(b) The y-intercept is $y = G(0) = 0$

The intercept is $(0, 0)$.

Step 2: $G(-x) = \dfrac{-x}{x^2 - 4} = -G(x)$

$G(x)$ is symmetric about the origin.

Step 3: The vertical asymptotes are $x = -2$, $x = 2$, the zeros of $x^2 - 4 = (x + 2)(x - 2)$.

Step 4: The horizontal asymptote is $y = 0$; intersected at $(0, 0)$, since

$$0 = \dfrac{x}{x^2 - 4}$$
$$x = 0$$
$$G(0) = 0$$

Step 5:

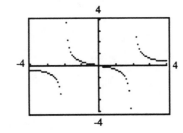

55. $R(x) = \dfrac{3}{(x-1)(x^2-4)} = \dfrac{3}{(x-1)(x-2)(x+2)}$

Step 1: (a) No x-intercept.

            (b) y-intercept is $y = R(0) = \dfrac{3}{4}$

Step 2: No symmetry.

Step 3: Vertical asymptotes: $x = -2$, $x = 1$, $x = 2$, the zeros of $q(x)$.

Step 4: Horizontal asymptote: $y = 0$; not intersected.

Step 5:

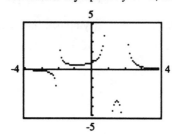

57. $H(x) = \dfrac{4(x^2-1)}{x^4-16} = \dfrac{4(x+1)(x-1)}{(x^2-4)(x^2+4)} = \dfrac{4(x+1)(x-1)}{(x-2)(x+2)(x^2+4)}$

Step 1: (a) The x-intercepts are -1 and 1, the zeros of $p(x)$.

            (b) The y-intercept is $y = H(0) = \dfrac{1}{4}$

Step 2: $H(-x) = \dfrac{4(x^2-1)}{x^4-16} = H(x)$, so the graph is symmetric with respect to the y-axis.

Step 3: The vertical asymptote are: $x = -2$, $x = 2$, the zeros of $q(x)$.

Step 4: The horizontal asymptote is $y = 0$; intersected at $(-1, 0)$ and $(1, 0)$.

Step 5:

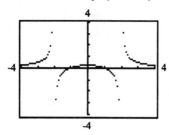

59. $F(x) = \dfrac{x^2-3x-4}{x+2} = \dfrac{(x-4)(x+1)}{x+2}$

Step 1: (a) The x-intercepts are -1, 4.
            (b) The y-intercept is -2

Step 2: No symmetry.

Step 3: The vertical asymptote is $x = -2$.

Step 4: Since $n = m + 1$, $F(x)$ has an oblique asymptote. Using long division:

$$
\begin{array}{r}
x - 5 \\
x+2 \overline{\smash{)}x^2 - 3x - 4} \\
\underline{x^2 + 2x} \\
-5x - 4 \\
\underline{-5x - 10} \\
6
\end{array}
$$

So, $F(x) = x - 5 + \dfrac{6}{x + 2}$, and $y = x - 5$ is an oblique asymptote. Since

$$x - 5 = \frac{x^2 - 3x - 4}{x + 2}$$

has no solution, it is not intersected.

Step 5:

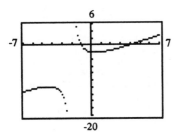

61. $R(x) = \dfrac{x^2 + x - 12}{x - 4} = \dfrac{(x + 4)(x - 3)}{x - 4}$

Step 1:   (a)   The x-intercepts are -4 and 3

           (b)   The y-intercept is $y = R(0) = 3$

Step 2:   No symmetry.

Step 3:   The vertical asymptote is $x = 4$.

Step 4:   Since $n = m + 1$, $R(x)$ has an oblique asymptote. Using long division:

$$
\begin{array}{r}
x + 5 \\
x + 2 \overline{\smash{)}\, x^2 + x - 12} \\
\underline{x^2 - 4x } \\
5x - 12 \\
\underline{5x - 20} \\
8
\end{array}
$$

So $R(x) = x + 5 + \dfrac{8}{x + 2}$ and $y = x + 5$, we have an oblique asymptote.

Since $x + 5 = \dfrac{x^2 + x - 12}{x - 4}$ has no solution, it is not intersected.

Step 5:

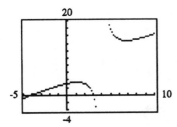

63. $F(x) = \dfrac{x^2 + x - 12}{x + 2} = \dfrac{(x + 4)(x - 3)}{x + 2}$

Step 1:   (a)   The x-intercepts are -4 and 3

           (b)   The y-intercept is $y = F(0) = -6$

Step 2:   No symmetry

Step 3:   The vertical asymptote is $x = -2$

Step 4: Since $n = m + 1$, $F(x)$ has an oblique asymptote. Using long division:

$$
\begin{array}{r}
x - 1 \\
x + 2\overline{\smash{)}x^2 + x - 12} \\
\underline{x^2 + 2x} \\
-x - 12 \\
\underline{-x - 2} \\
-10
\end{array}
$$

So, $F(x) = x - 1 + \dfrac{-10}{x + 2}$ and $y = x - 1$ is an oblique asymptote.

Since $x - 1 = \dfrac{x^2 + x - 12}{x + 2}$ has no solution, it is not intersected.

Step 5:

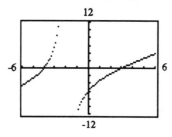

65. $R(x) = \dfrac{x(x - 1)^2}{(x + 3)^3}$

Step 1: (a) The x-intercepts are 0 and 1
        (b) The y-intercept is $y = R(0) = 0$

Step 2: No symmetry

Step 3: The vertical asymptote is $x = -3$

Step 4: Since $n = m$, the horizontal asymptote is the line $y = 1$; not intersected since

$$1 = \dfrac{x(x - 1)^2}{(x + 3)^3} \text{ has no solution.}$$

Step 5:

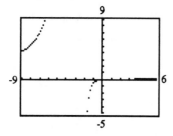

67. $R(x) = \dfrac{4x^3 - 0.5x + 2}{2x^3 + 0.3x^2 - 1}$

Step 1: (a) To find the x-intercepts of $R(x)$, graph $y = p(x)$ and estimate the roots. The x-intercept is -0.84.
        (b) The y-intercept is $y = R(0) = -2$.

Step 2: No symmetry.

Step 3: To find the vertical asymptote of $R(x)$, graph $y = q(x)$ and estimate the roots. The vertical asymptote is $x = 0.74$.

Step 4: Since $n = m$, the horizontal asymptote is the line $y = 2$; not intersected since

$$2 = \frac{4x^3 - 0.5x + 2}{2x^3 + 0.3x^2 - 1}$$ has no solution.

Step 5:

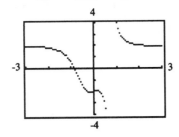

69. $R(x) = \dfrac{3x^3 + 5x^2 - 3}{0.1x^4 - 2x^2 + 1}$

Step 1: (a) To find the x-intercepts of $R(x)$, graph $y = p(x)$ and estimate the roots. The x-intercept is 0.65.

(b) The y-intercept is $y = R(0) = -3$.

Step 2: No symmetry.

Step 3: To find the vertical asymptotes of $R(x)$, graph $y = q(x)$ and estimate the roots. The vertical asymptotes are $x = -4.41$, $x = -0.71$, $x = 0.71$, $x = 4.41$.

Step 4: Since $n < m$, the horizontal asymptote is the line $y = 0$; intersected at $(0.65, 0)$ since

$$0 = \frac{3x^3 + 5x^2 - 3}{0.1x^4 - 2x^2 + 1}$$ has $x = 0.65$ as a solution.

Step 5:

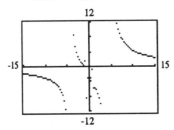

71. $R(x) = \dfrac{0.5x^4 - x^3 + 1}{0.1x^3 - \pi x + 1}$

Step 1: (a) To find the x-intercepts of $R(x)$, graph $y = p(x)$ and estimate the roots. There are no x-intercepts since the graph of $y = p(x)$ does not cross the x-axis.

(b) The y-intercept is $y = R(0) = 1$.

Step 2: No symmetry.

Step 3: To find the vertical asymptote of $R(x)$, graph $y = q(x)$ and estimate the roots. The vertical asymptotes are $x = -5.75$, $x = 0.31$, $x = 5.43$.

Step 4: Since $n = m + 1$, $R(x)$ has an oblique asymptote. Using long division:

$$
\begin{array}{r}
5x - 10 \\
0.1x^3 - \pi x + 1 \overline{)\, 0.5x^4 - x^3 + 1} \\
\underline{-(0.5x^4 - 5\pi x^2 + 5x)} \\
-x^3 + 5\pi x^2 - 5x + 1 \\
\underline{-(-x^3 + 10\pi x - 10)} \\
5\pi x^2 - 5x - 10\pi x + 11
\end{array}
$$

So $R(x) = 5x - 10 + \dfrac{5\pi x^2 - 5x - 10\pi x + 11}{0.1x^3 - \pi x + 1}$ and $y = 5x - 10$ is an oblique

asymptote. Since $5x - 10 = \dfrac{0.5x^4 - x^3 + 1}{0.1x^3 - \pi x + 1}$ has $x = 0.35$ as a solution, the oblique

asymptote is intersected at $(0.35, -8.21)$.

Step 5:

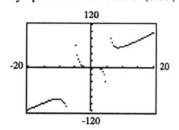

73. $R(x) = \dfrac{x^2 + x - 12}{x^2 - x - 6} = \dfrac{(x + 4)(x - 3)}{(x - 3)(x + 2)} = \dfrac{x + 4}{x + 2}, x \neq 3$

Step 1:   (a)   The x-intercept of $R(x)$ is $x = -4$.

           (b)   The y-intercept of $R(x)$ is $y = R(0) = 2$.

Step 2:   No symmetry.

Step 3:   The vertical asymptote of $R(x)$ is the zero of $q(x)$ when $R(x)$ is in reduced form. Thus, $x = -2$ is a vertical asymptote. Note: There is a hole in the graph of $R(x)$ at $x = 3$.

Step 4:   Since $n = m$, the horizontal asymptote is $y = 0$; not intersected.

Step 5:

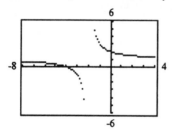

75. $R(x) = \dfrac{6x^2 - 7x - 3}{2x^2 - 7x + 6} = \dfrac{(3x + 1)(2x - 3)}{(2x - 3)(x - 2)} = \dfrac{3x + 1}{x - 2}, x \neq \dfrac{3}{2}$

Step 1:   (a)   The x-intercept of $R(x)$ is $x = \dfrac{-1}{3}$.

           (b)   The y-intercept of $R(x)$ is $y = R(0) = \dfrac{-1}{2}$.

Step 2:   No symmetry.

Step 3:   The vertical asymptote of $R(x)$ is the zero of $q(x)$ when $R(x)$ is in reduced form. Thus, $x = 2$ is a vertical asymptote. Note: There is a hole in the graph of $R(x)$ at $x = \dfrac{3}{2}$.

Step 4:   Since $n = m$, the horizontal asymptote is $y = 0$; not intersected.

Step 5:

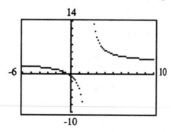

77. $R(x) = \dfrac{x^3 + 2x^2 - 5x - 6}{x^3 + 7x^2 + 7x - 15} = \dfrac{(x - 2)(x + 1)(x + 3)}{(x + 5)(x - 1)(x + 3)} = \dfrac{(x - 2)(x + 1)}{(x + 5)(x - 1)}, x \neq -3$

Step 1:   (a)   The x-intercepts of $R(x)$ is $x = -1, x = 2$.

              (b)   The y-intercept of $R(x)$ is $y = R(0) = \dfrac{6}{15} = \dfrac{2}{5}$.

Step 2:   No symmetry.

Step 3:   The vertical asymptote of $R(x)$ are the zeros of $q(x)$ when $R(x)$ is in reduced form. Thus, $x = -5, x = 1$ are the vertical asymptotes. Note: There is a hole in the graph of $R(x)$ at $x = -3$.

Step 4:   Since $n = m$, the horizontal asymptote is the line $y = 1$; intersected at $\left(\dfrac{3}{5}, 1 \right)$ since

$$1 = \dfrac{(x - 2)(x + 1)}{(x + 5)(x - 1)} \text{ has } x = \dfrac{3}{5} \text{ as a solution.}$$

Step 5:

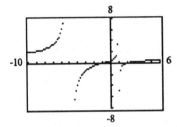

79. $f(x) = x + \dfrac{1}{x} = \dfrac{x^2}{x} + \dfrac{1}{x} = \dfrac{x^2 + 1}{x}$

Step 1:   (a)   No x-intercepts since $0 = x + \dfrac{1}{x}$ has no real solution.

              (b)   No y-intercept since $x = 0$ is not in the domain of $f(x)$.

Step 2:   Symmetry about the origin since $R(-x) = -x + \dfrac{1}{-x} = -\left(x + \dfrac{1}{x} \right) = -R(x)$.

Step 3:   Vertical asymptote at $x = 0$ since $x = 0$ is a root of $q(x)$.

Step 4:   Since $n = m + 1$, $f(x)$ has an oblique asymptote. We use long division:

$$\begin{array}{r} x \\ x{\overline{\smash{\big)}\,x^2 + 1}} \\ \underline{x^2 } \end{array}$$

So $y = x$ is an oblique asymptote; not intersected.

Step 5:

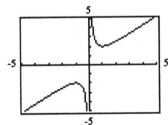

81. $f(x) = x^2 + \dfrac{1}{x} = \dfrac{x^3}{x} + \dfrac{1}{x} = \dfrac{x^3 + 1}{x}$

Step 1:   (a)   The x-intercept is $x = -1$ since $0 = x^2 + \dfrac{1}{x}$ has $x = -1$ as a solution.

              (b)   There is no y-intercept since $x = 0$ is not in the domain of $f(x)$.

Step 2:   No symmetry.

Step 3: The vertical asymptote is $x = 0$ since $x = 0$ is a root of $q(x)$.
Step 4: There are no horizontal or oblique asymptotes.
Step 5:

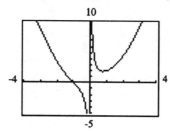

83. $f(x) = x + \dfrac{1}{x^3} = \dfrac{x^4}{x^3} + \dfrac{1}{x^3} = \dfrac{x^4 + 1}{x^3}$

Step 1: (a) There is no x-intercepts since $0 = \dfrac{x^4 + 1}{x^3}$ has no solution.

(b) There is no y-intercept since $x = 0$ is not in the domain of $f(x)$.

Step 2: Symmetric about the origin since $R(-x) = -x + \dfrac{1}{(-x)^3} = -x - \dfrac{1}{x^3} = -\left(x + \dfrac{1}{x^3} \right)$

$= -R(x)$.

Step 3: The vertical asymptote is $x = 0$ since $x = 0$ is a root of $q(x)$.

Step 4: Since $n = m + 1$, there is an oblique asymptote. We use long division:

$$
\begin{array}{r}
x \\
x^3 \overline{\smash{\big)}\, x^4 + 1} \\
\underline{-x^4 } \\
1
\end{array}
$$

Thus, $f(x) = x + \dfrac{1}{x^3}$ and $y = x$ is an oblique asymptote; not intersected.

Step 5:

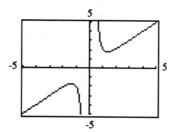

85. 4 must be a zero of the denominator; hence, $x - 4$ must be a factor.

87. c, d

4.4 The Zeros of a Polynomial Function

In Problems 1–9, we use the following facts: the remainder in synthetic division when f(x) is divided by x − c, is f(c); and x − c is a factor of f(x) only if f(c) = 0.

1. We divide by $x - c = x - 2$:

$$
\begin{array}{r|rrrr}
2 & 4 & -3 & -8 & 4 \\
  &   & 8 & 10 & 4 \\
\hline
  & 4 & 5 & 2 & 8
\end{array}
$$

Remainder = 8 ≠ 0; therefore, $x - 2$ is *not* a factor of $f(x)$.

3. $x - c = x - 2$

$$
\begin{array}{r|rrrrr}
2 & 3 & -6 & 0 & -5 & 10 \\
  &   & 6 & 0 & 0 & -10 \\
\hline
  & 3 & 0 & 0 & -5 & \boxed{0}
\end{array}
$$

The remainder = 0; therefore, $x - 2$ *is* a factor of $f(x)$.

5. $x - c = x - (-3) = x + 3$

$$
\begin{array}{r|rrrrrrr}
-3 & 3 & 0 & 0 & 82 & 0 & 0 & 27 \\
   &   & -9 & 27 & -81 & -3 & 9 & -27 \\
\hline
   & 3 & -9 & 27 & 1 & -3 & 9 & \boxed{0}
\end{array}
$$

Remainder = 0; therefore, $x + 3$ is a factor.

7. $x - c = x - (-4) = x + 4$

$$
\begin{array}{r|rrrrrrr}
-4 & 4 & 0 & -64 & 0 & 1 & 0 & -15 \\
   &   & -16 & 64 & 0 & 0 & -4 & 16 \\
\hline
   & 4 & -16 & 0 & 0 & 1 & -4 & \boxed{1}
\end{array}
$$

$x + 4$ is not a factor, since the remainder = 1 ≠ 0.

9. $x - c = x - \dfrac{1}{2}$

$$
\begin{array}{r|rrrrr}
\frac{1}{2} & 2 & -1 & 0 & 2 & -1 \\
            &   & 1 & 0 & 0 & 1 \\
\hline
            & 2 & 0 & 0 & 2 & \boxed{0}
\end{array}
$$

Since the remainder = 0; therefore, $x - \dfrac{1}{2}$ is a factor of $f(x)$.

In Problems 11–21, we must count the number of sign changes in both f(x) and f(−x).

11. $f(x) = -4x^7 + x^3 - x^2 + 2$
The maximum number of zeros is 7.

For $f(x) = -4x^7 + x^3 - x^2 + 2$, there are three variations in sign of

$$- \text{ to } + \quad + \text{ to } - \quad - \text{ to } +$$

the coefficients, so there will be either three positive zeros, or one positive zero.

To find $f(-x)$, replace x in $f(x)$ by $-x$:
$$f(-x) = 4x^7 - x^3 - x^2 + 2$$

$$+ \text{ to } - \qquad - \text{ to } +$$

There are two variations in the sign of $f(-x)$, so there will be either two negative zeros, or none.

13. $f(x) = 2x^6 - 3x^2 - x + 1$
The maximum number of zeros is 6.
Here, $f(x) = 2x^6 - 3x^2 - x + 1$, and

$$+ \text{ to } - \qquad - \text{ to } +$$
$$f(-x) = 2x^6 - 3x^2 + x + 1, \text{ and}$$

$$+ \text{ to } - \quad - \text{ to } +$$

Thus, there are either two positive zeros, or none; and either two negative zeros or none.

15. $f(x) = 3x^3 - 2x^2 + x + 2$
    The maximum number of zeros is 3.
    For $f(x) = 3x^3 - 2x^2 + x + 2$,

    $+$ to $-$ $-$ to $+$

    there are two variations in sign of the coefficients, so there will be either two positive zeros, or none.

    To find $f(-x)$, replace x in $f(x)$ by $-x$:
    $$f(-x) = -3x^3 - 2x^2 - x + 2$$

    $-$ to $+$

    There is only one variation of sign in $f(-x)$, so there will be exactly one negative zero.

17. $f(x) = -x^4 + x^2 - 1$
    The maximum number of zeros is 4.
    Here, $f(x) = -x^4 + x^2 - 1$, and

    $-$ to $+$ $+$ to $-$
    $$f(-x) = -x^4 + x^2 - 1$$
    $-$ to $+$ $+$ to $-$

    Thus, there are either two positive zeros, or none; and either two negative zeros, or none.

19. $f(x) = x^5 + x^4 + x^2 + x + 1$
    The maximum number of zeros is 5.
    We have $f(x) = x^5 + x^4 + x^2 + x + 1$, which has *no* changes of signs; hence, there will be *no* positive zeros.

    Meanwhile, $f(-x) = -x^5 + x^4 + x^2 - x + 1$ has three changes in sign; so $f(x)$ will have either three negative zeros, or one negative zero.

21. $f(x) = x^6 - 1$
    The maximum number of zeros is 6.
    $f(x) = x^6 - 1$; $f(-x) = x^6 - 1$
    We will have exactly one positive zero and one negative zero.

For Problems 23−33, use the Rational Zeros Theorem. The possible rational (whole number or fractional) zeros must be of the form $\dfrac{p}{q}$, where p is a factor of the constant term, and q is a factor of the coefficient of the highest power of x.

23. For $f(x) = 3x^4 - 3x^3 + x^2 - x + 1$,
    p must be a factor of $+1$: $p = +1$ or -1
    q must be a factor of $+3$: $q = \pm 1$ or ± 3

    So the possible zeros $\dfrac{p}{q}$ are $\pm 1, \pm \dfrac{1}{3}$

25. For $f(x) = x^5 - 6x^2 + 9x - 3$,
    p must be a factor of -3: $p = \pm 1$ or ± 3
    q must be a factor of $+1$: $q = \pm 1$
    So the possible rational zeros are $\pm 1, \pm 3$

27. For $f(x) = -4x^3 - x^2 + x + 2$,
    p must be a factor of $+2$: $p = \pm 1, \pm 2$
    q must be a factor of -4: $q = \pm 1, \pm 2, \pm 4$

    Hence, the possible rational zeros, $\dfrac{p}{q}$, are $\pm 1, \pm \dfrac{1}{2}, \pm \dfrac{1}{4}, \pm 2$

29. For $f(x) = 3x^4 - x^2 + 2$,
    p must be a factor of $+2$: $p = \pm 1, \pm 2$
    q must be a factor of $+3$: $q = \pm 1, \pm 3$

    Possible rational zeros: $\pm 1, \pm \dfrac{1}{3}, \pm 2, \pm \dfrac{2}{3}$

Chapter 4 Polynomial and Rational Functions

31. For $f(x) = 2x^5 - x^3 + 2x^2 + 4$, we have the following possibilities:
$$p = \pm1, \pm2, \pm4$$
$$q = \pm1, \pm2$$
Rational zeros, $\dfrac{p}{q}$: $\pm1, \pm\dfrac{1}{2}, \pm2, \pm4$

33. For $f(x) = 6x^4 + 2x^3 - x^2 + 2$, we have the following possibilities:
$$p: \pm1, \pm2; \quad q: \pm1, \pm2, \pm3, \pm6; \quad \frac{p}{q}: \pm1, \pm\frac{1}{2}, \pm\frac{1}{3}, \pm\frac{1}{6}, \pm2, \pm\frac{2}{3}$$

For Problems 35–39, we must find upper and lower bounds to the zeros of each polynomial.
(a) *For upper bounds, we check 1, 2, 3, ... , until we find the bound. Thus, we divide $f(x)$ by $x - 1$,*
   $x - 2$, $x - 3$, ... , until the third row in synthetic division contains only positive numbers or zero.
(b) *For lower bounds, we check $-1, -2, -3, ...$, so we must perform synthetic division using*
   $x - (-1), x - (-2), x - (-3), ...$; in other words, $x + 1, x + 2, x + 3, ...$, until the third row
   alternates between positive (or zero) and negative (or zero).

35. $f(x) = 2x^3 + x^2 - 1$
   (a) Upper bound:

```
1│2   1   0  -1
  │    2   3   3
   ─────────────────
   2   3   3  [2]   ← All positive.
```

   Therefore, 1 is an upper bound.
   (b) Lower bound:

```
-1│2   1   0  -1
   │   -2   1  -1
   ─────────────────
    2  -1   1  [-2]   ← Signs alternate.
```

   Therefore, -1 is a lower bound. Any real zeros of $f(x)$ must lie between -1 and 1.

37. $f(x) = x^3 - 5x^2 - 11x + 11$

Upper Bound:

```
1│1   -5   -11    11
 │      1   -4   -15
  ──────────────────────
  1   -4   -15   -4
```

```
2│1   -5   -11    11
 │      2   -6   -34
  ──────────────────────
  1   -3   -17   -23
```

```
3│1   -5   -11    11
 │      3   -6   -51
  ──────────────────────
  1   -2   -17   -40
```

```
5│1   -5   -11    11
 │      5    0   -55
  ──────────────────────
  1    0   -11   -44
```

```
6│1   -5   -11    11
 │      6    6   -30
  ──────────────────────
  1    1   -5   -19
```

```
7│1   -5   -11    11
 │      7   14    21
  ──────────────────────
  1    2    3    32
```

Lower Bound:

```
-1│1   -5   -11    11
  │     -1    6     5
   ──────────────────────
   1   -6   -5    16
```

```
-2│1   -5   -11    11
  │     -2   14    -6
   ──────────────────────
   1   -7    3     5
```

```
-3│1   -5   -11    11
  │     -3   24   -39
   ──────────────────────
   1   -8   13   -28
```

Upper Bound: 7
Lower Bound: -3

39. $f(x) = x^4 + 3x^3 - 5x^2 + 9$

    (a)   Upper bound:

$$
\begin{array}{r|rrrrr}
1 & 1 & 3 & -5 & 0 & 9 \\
  &   & 1 & 4 &   &   \\
\hline
  & 1 & 4 & -1 & & \quad \leftarrow \text{Stop!}
\end{array}
$$

$$
\begin{array}{r|rrrrr}
2 & 1 & 3 & -5 & 0 & 9 \\
  &   & 2 & 10 & 10 & 20 \\
\hline
  & 1 & 5 & 5 & 10 & 29
\end{array}
$$

Therefore, 2 is an upper bound.

    (b)   Lower bound:

$$
\begin{array}{r|rrrrr}
-1 & 1 & 3 & -5 & 0 & 9 \\
   &   & -1 &   &   &   \\
\hline
   & 1 & 2 &   &   & \quad \leftarrow \text{Stop. \ \ Try 3 next:}
\end{array}
$$

$$
\begin{array}{r|rrrrr}
-3 & 1 & 3 & -5 & 0 & 9 \\
   &   & -3 & -0 &   &   \\
\hline
   & 1 & 0 & -5 &   &
\end{array}
$$

                    +   -   -   Signs do not alternate!

$$
\begin{array}{r|rrrrr}
-4 & 1 & 3 & -5 & 0 & 9 \\
   &   & -4 & 4 &   &   \\
\hline
   & 1 & -1 & -1 & \leftarrow \text{Stop!}
\end{array}
$$

$$
\begin{array}{r|rrrrr}
-5 & 1 & 3 & -5 & 0 & 9 \\
   &   & -5 & 10 & -25 & 125 \\
\hline
   & 1 & -2 & 5 & -25 & 134
\end{array}
$$

Therefore, -5 is a lower bound, and all zeros lie between -5 and 2.

41. $f(x) = x^3 + 2x^2 - 5x - 6; f(-x) = -x^3 + 2x^2 + 5x - 6$

    Step 1: $f(x)$ has at most three zeros.

    Step 2: $f(x)$ has one positive zero, and either two negative zeros, or none.

    Step 3:   Possible rational zeros: p: $\pm 1, \pm 2, \pm 3, \pm 6$; q: ± 1; $\dfrac{p}{q}$: $\pm 1, \pm 2, \pm 3, \pm 6$

    Step 4:   (a)

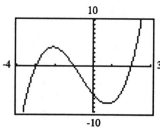

    (b)   From the graph, it appears -3, -1, and 2 are zeros.

    (c)   We check if $x = -1$ is a zero.

    (d)   Synthetic Division:

$$
\begin{array}{r|rrrr}
-1 & 1 & 2 & -5 & -6 \\
   &   & -1 & -1 & 6 \\
\hline
   & 1 & 1 & -6 & \boxed{0}
\end{array}
$$

Since the remainder is 0, $x - (-1) = x + 1$ is a factor. The other factor is the quotient: $x^2 + x - 6$.

Thus, $f(x) = (x + 1)(x^2 + x - 6) = (x + 1)(x + 3)(x - 2)$ and the zeros are -1, -3, and 2.

Step 5: Since we found three zeros and $f(x)$ has at most three zeros, there is no need to perform the Upper and Lower Bound Test.

43. $f(x) = 2x^3 - x^2 + 2x - 1; f(-x) = -2x^3 - x^2 - 2x - 1$

Step 1: $f(x)$ has at most three zeros.

Step 2: $f(x)$ has either three positive zeros, or just one, and no negative zeros.

Step 3: Possible rational zeros: p: ± 1; q: ± 1, ± 2; $\dfrac{p}{q}$: ± 1, $\pm \dfrac{1}{2}$

Step 4: (a)

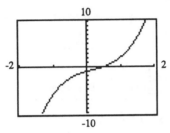

(b) From the graph, it appears $\dfrac{1}{2}$ is a zero.

(c) We check if $x = \dfrac{1}{2}$ is a zero:

$$\dfrac{1}{2} \overline{\smash{\big)}\ 2 \quad -1 \quad 2 \quad -1}$$
$$\phantom{\dfrac{1}{2}}\quad\quad 1 \quad\ 0 \quad\ 1$$
$$\overline{\phantom{\dfrac{1}{2}}\quad 2 \quad\ 0 \quad\ 2 \quad \boxed{0}} \sim \left(x - \dfrac{1}{2}\right) \text{ is a factor.}$$

Quotient $= 2x^2 + 2$

Thus, $f(x) = \left(x - \dfrac{1}{2}\right)(2x^2 + 2) = 2\left(x - \dfrac{1}{2}\right)(x^2 + 1)$

The only real zero is $\dfrac{1}{2}$, since $x^2 + 1$ has no real zeros.

Step 5: From Step 2, we know there are no negative zeros, thus we need only determine the Upper Bound. We will check if 2 is an Upper Bound.

$$2 \overline{\smash{\big)}\ 2 \quad -1 \quad 2 \quad -1}$$
$$\quad\quad\quad\ 4 \quad\ 6 \quad 16$$
$$\overline{\quad 2 \quad\ \ 3 \quad\ 8 \quad 15}$$

Since the last row contains all positive numbers, we know there are no zeros greater than 2.

45. $f(x) = x^4 + x^2 - 2; f(-x) = x^4 + x^2 - 2$

Step 1: $f(x)$ has at most four zeros.

Step 2: $f(x)$ has one positive zero and one negative zero. (The other two zeros must be complex numbers.)

Step 3: Possible rational zeros: p: ± 1, ± 2; q: ± 1; $\dfrac{p}{q}$: ± 1, ± 2

Step 4: (a)

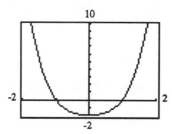

(b) From the graph, it appears -1 and 1 are zeros.

(c) We check if -1 is a zero.

$$
\begin{array}{r|rrrrr}
-1 & 1 & 0 & 1 & 0 & -2 \\
   &   & -1 & 1 & -2 & 2 \\
\hline
   & 1 & -1 & 2 & -2 & \boxed{0}
\end{array}
$$

$(x + 1)$ *is* a factor!

Quotient: $x^3 - x^2 + 2x - 2$

We have $f(x) = (x + 1)(x^3 - x^2 + 2x - 2)$

We can factor $x^3 - x^2 + 2x - 2$ by grouping terms:

$$x^3 - x^2 + 2x - 2 = x^2(x - 1) + 2(x - 1) = (x^2 + 2)(x - 1)$$

Thus, $f(x) = (x + 1)(x - 1)(x^2 + 2)$

no real zeros

We have two real zeros: $-1, 1$, which agrees with what we discovered in Step 2.

Note: $f(x) = x^4 + x^2 - 2$ could have been factored at the beginning:

$$x^4 + x^2 - 2 = (x^2 + 2)(x^2 - 1) = (x^2 + 2)(x + 1)(x - 1)$$

Step 5: Since we know the nature of the zeros from Step 2, it is not necessary to perform the Upper and Lower Bounds Test.

47. $f(x) = 4x^4 + 7x^2 - 2; f(-x) = 4x^4 + 7x^2 - 2$

Step 1: There are at most four zeros.

Step 2: There is one positive zero and one negative zero.

Step 3: Possible rational zeros: p: $\pm 1, \pm 2$; q: $\pm 1, \pm 2, \pm 4$; $\dfrac{p}{q}$: $\pm 1, \pm \dfrac{1}{2}, \pm \dfrac{1}{4}, \pm 2$

Step 4: We can start by factoring:

$$f(x) = 4x^4 + 7x^2 - 2 = (4x^2 - 1)(x^2 + 2)$$

$$= 4 \left[x^2 - \dfrac{1}{4} \right] (x^2 + 2)$$

$$= 4 \left[x + \dfrac{1}{2} \right] \left[x - \dfrac{1}{2} \right] (x^2 + 2)$$

no real zeros

Thus, $f(x)$ has the two real zeros $-\dfrac{1}{2}, \dfrac{1}{2}$.

Step 5: Since we know the nature of the zeros from Step 2, it is not necessary to perform the Upper and Lower Bounds Test.

49. $f(x) = x^4 + x^3 - 3x^2 - x + 2; f(-x) = x^4 - x^3 - 3x^2 + x + 2$

Step 1: There are at most four zeros.

Step 2: We will have either two positive zeros, or none, and either two negative zeros, or none.

Step 3: Possible rational zeros: p: $\pm 1, \pm 2$; q: ± 1; $\dfrac{p}{q}$: $\pm 1, \pm 2$.

Step 4:   (a)

(b)   From the graph, it appears -2, -1 and 1 are zeros.

(c)   We check if $x = -1$ is a zero:

$$
\begin{array}{r|rrrr}
-1 & 1 & 1 & -3 & -1 & 2 \\
   &   & -1 & 0 & 3 & -2 \\
\hline
   & 1 & 0 & -3 & 2 & \boxed{0}
\end{array}
$$

so $x + 1$ *is* a factor!

$$x^3 - 3x + 2$$

We now work on the depressed equation, $x^3 - 3x + 2 = 0$. This is cubic, and not easily factored.

We now check if $x = -2$ is a zero. So we divide by $x - (-2) = x + 2$:

$$
\begin{array}{r|rrrr}
-2 & 1 & 0 & -3 & 2 \\
   &   & -2 & 4 & -2 \\
\hline
   & 1 & -2 & 1 & \boxed{0}
\end{array}
$$

Thus, $(x + 2)$ *is* a factor!

$$x^2 - 2x + 1$$

We now have:

$$
\begin{aligned}
f(x) &= (x + 1)(x^3 - 3x + 2) \\
     &= (x + 1)(x + 2)(x^2 - 2x + 1) \\
     &= (x + 1)(x + 2)(x - 1)(x - 1) \\
     &= (x + 1)(x + 2)(x - 1)^2
\end{aligned}
$$

and the zeros are -1, -2, 1, 1.

Step 5:   Since we found all four zeros, there is no need to perform the Upper and Lower Bounds Test.

51. $f(x) = 4x^5 - 8x^4 - x + 2;\ f(-x) = -4x^5 - 8x^4 + x + 2$

    Step 1:   There are at most five zeros.

    Step 2:   There are either two or no positive zeros, and there is one negative zero.

    Step 3:   Possible rational zeros: p: ± 1, ± 2; q: ± 1, ± 2, ± 4; $\dfrac{p}{q}$: ± 1, $\pm\dfrac{1}{2}$, $\pm\dfrac{1}{4}$, ± 2

Step 4: Here we can factor by grouping:

$$f(x) = 4x^5 - 8x^4 - x + 2 = 4x^4(x-2) - 1(x-2)$$
$$= (4x^4 - 1)(x-2)$$
$$= 4\left[x^4 - \frac{1}{4}\right](x-2)$$
$$= 4\left[x^2 - \frac{1}{2}\right]\left[x^2 + \frac{1}{2}\right](x-2)$$
$$= 4\left[x - \sqrt{\frac{1}{2}}\right]\left[x + \sqrt{\frac{1}{2}}\right](x-2)\left[x^2 + \frac{1}{2}\right]$$

We have $f(x) = 4\left[x - \dfrac{1}{\sqrt{2}}\right]\left[x + \dfrac{1}{\sqrt{2}}\right](x-2)\underbrace{\left[x^2 + \dfrac{1}{2}\right]}_{\text{no real zeros}}$

and the real zeros of $f(x)$ are:

$$\frac{1}{\sqrt{2}} = \frac{\sqrt{2}}{2}, \frac{-\sqrt{2}}{2}, \text{ and } 2$$

Step 5: Since we know the nature of the solutions from Step 2, it is not necessary to perform the Upper and Lower Bounds Test.

53. $f(x) = x^3 + 3.2x^2 - 16.83x - 5.31; f(-x) = -x^3 + 3.2x^2 + 16.83x - 5.31$

Step 1: There are at most three zeros.

Step 2: There is one positive zero, and there are two or zero negative zeros.

Step 3: Rational Zeros Theorem does not apply.

Step 4:

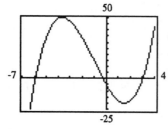

The zeros are estimated to be -5.90, -0.30, and 3.00.

Step 5: Since we found all three zeros, there is no need to perform the Upper and Lower Bound Test.

55. $f(x) = x^4 - 1.4x^3 - 33.71x^2 + 23.94x + 292.41;$
$f(-x) = x^4 + 1.4x^3 - 33.71x^2 - 23.94x + 292.41$

Step 1: There are at most four zeros.

Step 2: There are two or zero positive zeros, and two or zero negative zeros.

Step 3: Rational Zeros Theorem does not apply.

Step 4:

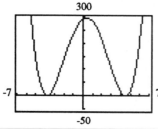

The zeros are -3.80 and 4.50.

Step 5: We will test if -7 is a Lower Bound.

$$
\begin{array}{r|rrrr}
-7) & 1 & -1.4 & -33.71 & 23.94 & 292.41 \\
 & & -7 & 58.8 & -175.63 & 1061.83 \\
\hline
 & 1 & -8.4 & 25.09 & -151.68 & 1354.24
\end{array}
$$

Since the third row alternates in sign, -7 is a Lower Bound. Now we'll determine if 7 is an Upper Bound:

$$
\begin{array}{r|rrrr}
-7) & 1 & -1.4 & -33.71 & 23.94 & 292.41 \\
 & & 7 & 39.2 & 38.43 & 436.59 \\
\hline
 & 1 & 5.6 & 5.49 & 62.37 & 729
\end{array}
$$

Since the third row is all positive numbers, 7 is an Upper Bound.

Thus, there are no zeros less than -7 and no zeros larger than 7, so -3.80 and 4.50 must be zeros of multiplicity two.

57. $f(x) = \pi x^3 - (8.88\pi + 1)x^2 - (42.066\pi - 8.88)x + 42.066;$
$f(-x) = -\pi x^3 - (8.88\pi + 1)x^2 + (42.066\pi - 8.88)x + 42.066$

Step 1: There are at most three zeros.

Step 2: There are two or zero positive real zeros, and one negative real zero.

Step 3: Rational Zeros Theorem does not apply.

Step 4:

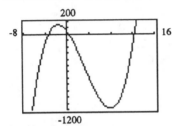

The zeros are estimated to be $-3.42, 0.31$ and 12.30.

Step 5: Since all the zeros were found, there is no need to perform the Upper and Lower Bounds Test.

59. $f(x) = x^3 + 19.5x^2 - 1021x + 1000.5; f(-x) = -x^3 + 19.5x^2 + 1021x + 1000.5$

Step 1: There are at most three zeros.

Step 2: There are two or zero positive real zeros, and one negative real zero.

Step 3: Rational Zeros Theorem does not apply.

Step 4:

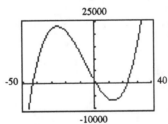

The zeros are approximated to be $-43.50, 1.00,$ and 23.00.

Step 5: Since all the zeros have been found, there is no need to perform the Upper and Lower Bounds Test.

61. $x^4 - x^3 + 2x^2 - 4x - 8 = 0$

The solutions of this equation are the zeros of the polynomial function $f(x)$.

$$f(x) = x^4 - x^3 + 2x^2 - 4x - 8$$
$$f(-x) = x^4 + x^3 + 2x^2 + 4x - 8$$

Step 1: There are at most four zeros.

Step 2: There are either three or one positive zero, and there is one negative zero.

Step 3: p: ± 1, ± 2, ± 4, ± 8; q: ± 1; $\dfrac{p}{q}$: ± 1, ± 2, ± 4, ± 8

Step 4:   (a)

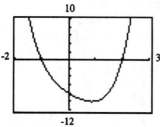

(b)   From the graph, it appears -1 and 2 are zeros.

(c)   We check if $x = -1$ is a zero:

$$
\begin{array}{r|rrrr}
-1 & 1 & -1 & 2 & -4 & -8 \\
   &   & -1 & 2 & -4 & 8 \\
\hline
   & 1 & -2 & 4 & -8 & \boxed{0}
\end{array}
$$

Thus, $x + 1$ is a factor.

$$x^3 - 2x^2 + 4x - 8 = 0$$

We now have:

$$(x + 1)(x^3 - 2x^2 + 4x - 8) = 0$$
$$(x + 1)[x^2(x - 2) + 4(x - 2)] = 0$$
$$(x + 1)(x^2 + 4)(x - 2) = 0$$

no real zeros

The real zeros are -1, 2.

63. $3x^3 + 4x^2 - 7x + 2 = 0$

The solutions of this equation are the zeros of the polynomial function $f(x)$.

$$f(x) = 3x^3 + 4x^2 - 7x + 2$$
$$f(-x) = -3x^3 + 4x^2 + 7x + 2$$

Step 1:   There are at most three zeros.

Step 2:   There are two or no positive zeros, and there is one negative zero.

Step 3: p: ± 1, ± 2; q: ± 1, ± 3; $\dfrac{p}{q}$: ± 1, $\pm \dfrac{1}{3}$, ± 2, $\pm \dfrac{2}{3}$

Step 4:   (a)

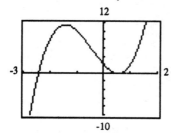

(b)   From the graph, it appears $\dfrac{2}{3}$ may be a factor. The negative zero and the other positive zero must be irrational.

(c)   We will check if $x = \dfrac{2}{3}$ is a factor.

$$\dfrac{2}{3} \overline{\smash{\big)}\ 3 \quad 4 \quad -7 \quad 2}$$

$$\phantom{\dfrac{2}{3}}\quad\ \ \ 2 \quad\ \ 4 \quad -2$$

$$\overline{\phantom{\dfrac{2}{3})}\ 3 \quad 6 \quad -3 \quad \boxed{0}} \qquad \left(x - \dfrac{2}{3}\right) \text{ is a factor.}$$

$$3x^2 + 6x - 3 = 0$$

$$f(x) = \left[x - \dfrac{2}{3}\right](3x^2 + 6x - 3) = 3\left[x - \dfrac{2}{3}\right](x^2 + 2x - 1)$$

Now, $x^2 + 2x - 1$ cannot be factored over the integers, so we use the quadratic formula:

$$x = \dfrac{-2 \pm \sqrt{4 - 4(-1)}}{2} = \dfrac{-2 \pm \sqrt{8}}{2} = \dfrac{-2 \pm 2\sqrt{2}}{2}$$

The three roots are $\dfrac{2}{3}$, $-1 + \sqrt{2}$, $-1 - \sqrt{2}$.

Recall that once a zero, c, is known, then $x - c$ is a factor; so in factored form:

$$3\left[x - \dfrac{2}{3}\right]\left(x - (-1 + \sqrt{2})\right)\left(x - (-1 - \sqrt{2})\right) = 0$$

$$3\left[x - \dfrac{2}{3}\right]\left(x + 1 - \sqrt{2}\right)\left(x + 1 + \sqrt{2}\right) = 0$$

65. $3x^3 - x^2 - 15x + 5 = 0$

The solutions of this equation are the zeros of the polynomial function $f(x)$.

$$f(x) = 3x^3 - x^2 - 15x + 5$$
$$f(-x) = -3x^3 - x^2 + 15x + 5$$

Let's just start factoring:

$$3x^3 - x^2 - 15x + 5 = 0$$
$$x^2(3x - 1) = 5(3x - 1) = 0$$
$$(x^2 - 5)(3x - 1) = 0$$
$$\left(x + \sqrt{5}\right)\left(x - \sqrt{5}\right)(3x - 1) = 0$$
$$3\left[x - \dfrac{1}{3}\right]\left(x + \sqrt{5}\right)\left(x - \sqrt{5}\right) = 0$$

The zeros of $f(x)$ are $\dfrac{1}{3}$, $-\sqrt{5}$, $\sqrt{5}$.

67. $x^4 + 4x^3 + 2x^2 - x + 6 = 0$

The solutions of this equation are the zeros of the polynomial function $f(x)$.

$$f(x) = x^4 + 4x^3 + 2x^2 - x + 6$$
$$f(-x) = x^4 - 4x^3 + 2x^2 + x + 6$$

Step 1:   There are at most four zeros.

Step 2:   There are either two positive zeros or none, and either two negative zeros or none.

Step 3: p: $\pm 1, \pm 2, \pm 3, \pm 6$; q: ± 1; $\dfrac{p}{q}$: $\pm 1, \pm 2, \pm 3, \pm 6$

Step 4:   (a)

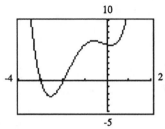

(b)   From the graph, it appears -3 and -2 are zeros.

(c)   We will check if $x = -2$ is a zero.

$$
\begin{array}{r|rrrr}
-2 & 1 & 4 & 2 & -1 & 6 \\
   &   & -2 & -4 & 4 & -6 \\
\hline
   & 1 & 2 & -2 & 3 & \boxed{0}
\end{array}
\quad x + 2 \text{ is a factor!}
$$

$$x^3 + 2x^2 - 2x + 3 = 0 \quad \text{← Depressed equation}$$

We continue with synthetic division and check if $x = -3$ is a zero.

$$
\begin{array}{r|rrrr}
-3 & 1 & 2 & -2 & 3 \\
   &   & -3 & 3 & -3 \\
\hline
   & 1 & -1 & 1 & \boxed{0}
\end{array}
\quad \text{So, } x + 3 \text{ is a factor.}
$$

We now have: $(x + 2)(x^3 + 2x^2 - 2x + 3) = 0$

$(x + 2)(x + 3)(x^2 - x + 1) = 0$

Now $x^2 - x + 1$ has no real zeros, since its discriminant is negative: $b^2 - 4ac = 1 - 4 = -3$. The real roots are -2 and -3.

Step 5:   Notice that -3 is a Lower Bound since the third row alternates in sign. We will check if $x = 1$ is an Upper Bound so we can be sure there are no real zeros off the viewing window:

$$
\begin{array}{r|rrrr}
1 & 1 & 4 & 2 & -1 & 6 \\
  &   & 1 & 5 & 7 & 6 \\
\hline
  & 1 & 5 & 7 & 6 & 12
\end{array}
$$

Since the third row contains all positive numbers, we can be certain we have found all the real zeros.

69. $x^3 - \dfrac{2}{3}x^2 + \dfrac{8}{3}x + 1 = 0$

The solutions of this equation are the zeros of the polynomial function $f(x)$.

$$f(x) = x^3 - \frac{2}{3}x^2 + \frac{8}{3}x + 1$$

$$f(-x) = -x^3 - \frac{2}{3}x^2 - \frac{8}{3}x + 1$$

Step 1:   There are at most three zeros.

Step 2:   There are either two positive zeros, or none, and there is one negative zero.

Step 3:   There is a trick to finding the possible rational zeros: All coefficients must be integers in order to use the Rational Zeros Theorem. But

$$x^3 - \frac{2}{3}x^2 + \frac{8}{3}x + 1 = 0$$

is equivalent to $3x^3 - 2x^2 + 8x + 3 = 0$

and we can determine the possible rational zeros:

$$p: \pm 1, \pm 3; \quad q: \pm 1, \pm 3; \quad \frac{p}{q}: \pm 1, \pm \frac{1}{3}, \pm 3$$

Step 4:   (a)

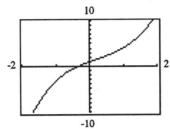

(b) From the graph, it appears $-\frac{1}{3}$ may be a zero.

(c) We check if $x = -\frac{1}{3}$ is a zero:

$$-\frac{1}{3} \overline{\left) \begin{array}{cccc} 1 & -\frac{2}{3} & \frac{8}{3} & 1 \\ & -\frac{1}{3} & \frac{1}{3} & -1 \\ \hline 1 & -1 & 3 & \boxed{0} \end{array} \right.} \quad \text{So} \left(x + \frac{1}{3}\right) \text{ is a factor.}$$

$$x^2 - x + 3$$

$$\left(x + \frac{1}{3}\right)(x^2 - x + 3) = 0$$

For $x^2 - x + 3$, we have: $x = \dfrac{1 \pm \sqrt{1 - 12}}{2} = \dfrac{1 \pm \sqrt{-11}}{2}$,

so we only have one real zero, $x = -\dfrac{1}{3}$

71. $(x - 2)$ will be a factor of $f(x) = x^3 - kx^2 + kx + 2$ only if the remainder that results when $f(x)$ is divided by $x - 2$ is 0:

$$2 \overline{\left) \begin{array}{cccc} 1 & -k & k & 2 \\ & 2 & -2k + 4 & -2k + 8 \\ \hline 1 & -k + 2 & -k + 4 & \boxed{-2k + 10} \end{array} \right.} \leftarrow \text{Remainder}$$

Therefore, we want $-2k + 10 = 0$ or $-2k = -10$, or $k = 5$

73.   Either long division *or* synthetic division would be a tedious way to find the remainder in this problem, but we know by the Remainder Theorem that if $f(x)$ is divided by $x - c$, then the remainder must be $f(c)$. Here, $x - c = x - 1$, so that $c = 1$, so we want $f(1)$. This can be found by substitution:
$$f(1) = 2 - 8 + 1 - 2 = -7$$
Thus, if we divided $f(x)$ by $x - 1$, the remainder would be -7.

75.   We want to prove that $x - c$ is a factor of $x^n - c^n$, for any positive integer n. By the Factor Theorem, $x - c$ will be a factor of $f(x)$ provided $f(c) = 0$. Here, $f(x) = x^n - c^n$ so that $f(c) = c^n - c^n = 0$, and therefore, $x - c$ is indeed a factor of $x^n - c^n$.

77. $f(x) = 2x^3 + 3x^2 - 6x + 7$
By the Rational Zero Theorem, the only
possible rational zeros of $f(x)$ are:

$$\frac{p}{q}: \pm 1, \pm \frac{1}{2}, \pm 7, \pm \frac{7}{2}$$

Therefore, $\frac{1}{3}$ is *not* a zero of $f(x)$.

79. By the Rational Zero Theorem, the *possible*
rational zeros of
$f(x) = 2x^6 - 5x^4 + x^3 - x + 1$ are:

$$\frac{p}{q}: \pm 1, \pm \frac{1}{2},$$

so $\frac{3}{5}$ is *not* a zero.

81. To start with, a cube has all three dimensions of equal length.
Let x = length of each side of the original cube. After removing the slice, we
will have this situation:

Now the volume is width times length times height, so we have
$$(x - 1)x \cdot x = 294$$
$$(x - 1)x^2 = 294$$
$$x^3 - x^2 = 294$$
$$x^3 - x^2 - 294 = 0$$

We see that there is exactly one positive zero and we can list the possibilities if we take the time to
factor 294:
$$294 = 2 \cdot 147 = 2 \cdot 3 \cdot 7 \cdot 7$$
Therefore, the possibilities for p are ± 1, ± 2, ± 3, ± 6, ± 7, ± 14, ± 21, ± 42, ± 49, ± 98, ± 147,
± 294. These are also the possible rational zeros, since $q = \pm 1$. Let's try 6 (divide by $x - 6$).

$$
\begin{array}{r|rrrr}
6 & 1 & -1 & 0 & -294 \\
  &   & 6 & 30 & 180 \\
\hline
  & 1 & 5 & 30 & -114 \\
\end{array}
$$

How about 7 (divide by $x - 7$)?

$$
\begin{array}{r|rrrr}
7 & 1 & -1 & 0 & -294 \\
  &   & 7 & 42 & 294 \\
\hline
  & 1 & 6 & 42 & \boxed{0} \\
\end{array}
$$ Therefore, $x - 7$ is a factor.

Since $x - 7$ is a factor, 7 is a zero, and we know there is only one positive solution, by Descartes'
Rule of signs.

Thus, the original length of each edge was $x = 7$ inches.

83. Since all we know about $f(x)$ is that its leading coefficient is 1, we can write:
$$f(x) = x^n = a_{n-1}x^{n-1} + \cdots + a_1x + a_0$$
where each coefficient is an integer. (See the Rational Zeros Theorem.)

Now let r be a real zero of $f(x)$. We need to show that r is either an integer or an irrational number.

Since r is a real number, it is either rational or irrational. But if r is rational (i.e., of the form $\frac{p}{q}$,
where p and q are integers), then q must be a factor of the leading coefficient, which is 1. Therefore,
$q = \pm 1$, and $r = \frac{p}{q} = \pm p$, where p is a factor of a_0. Thus, if r is rational, then r is an *integer*!

Therefore, r is either an integer, or r is irrational.

85. $y^3 + by^2 cy + d = 0$

Using the substitution, $y = x - \dfrac{b}{3}$, we have:

$$\left[x - \frac{b}{3}\right]^3 + b\left[x - \frac{b}{3}\right]^2 + c\left[x - \frac{b}{3}\right] + d = 0$$

$$x^3 - 3\frac{b}{3}x^2 + 3\left[\frac{b^2}{9}\right]x - \frac{b^3}{27} + b\left[x^2 - 2\frac{b}{3}x + \frac{b^2}{9}\right] + cx - \frac{bc}{3} + d = 0$$

$$x^3 - \frac{b^2}{3}x + cx - \frac{b^3}{27} + \frac{b^3}{9} - \frac{bc}{3} + d = 0$$

$$x^3 + \left[c - \frac{b^2}{3}\right]x + \left[\frac{2b^3}{27} - \frac{bc}{3} + d\right] = 0$$

87. Based on Problem 86, we have two equations:

$\qquad 3HK = -p$ and $H^3 + K^3 = -q$

Thus, $K = \dfrac{-p}{3H}$

so that

$$H^3 + K^3 = -q$$

$$H^3 + \left[\frac{-p}{3H}\right]^3 = -q$$

$$H^3 - \frac{p^3}{qH^3} = -q$$

$$H^6 + qH^3 - \frac{p^3}{27} = 0$$

$$H^3 = \frac{-q \pm \sqrt{q^2 + \dfrac{4p^3}{27}}}{2}; \quad \text{(choose + sign)}$$

$$H = \sqrt[3]{\frac{-q}{2} + \sqrt{\frac{q^2}{4} + \frac{p^3}{27}}}$$

89. From Problem 86, we know $x = H + K$

From Problem 87, $H = \sqrt[3]{\dfrac{-q}{2} + \sqrt{\dfrac{q^2}{4} + \dfrac{p^3}{27}}}$

From Problem 88, $K = \sqrt[3]{\dfrac{-q}{2} - \sqrt{\dfrac{q^2}{4} + \dfrac{p^3}{27}}}$

Therefore, $x = \sqrt[3]{\dfrac{-q}{2} + \sqrt{\dfrac{q^2}{4} + \dfrac{p^3}{27}}} + \sqrt[3]{\dfrac{-q}{2} - \sqrt{\dfrac{q^2}{4} + \dfrac{p^3}{27}}}$

91. $x^3 + 3x - 14 = 0$, so $p = 3$ and $q = -14$.

$$x = H + K = \sqrt[3]{\frac{-(-14)}{2} + \sqrt{\frac{(-14)^2}{4} + \frac{3^3}{27}}} + \sqrt[3]{\frac{-(-14)}{2} - \sqrt{\frac{(-14)^2}{4} + \frac{3^3}{27}}}$$

$$= \sqrt[3]{7 + \sqrt{50}} + \sqrt[3]{7 - \sqrt{50}} = \sqrt[3]{7 + 5\sqrt{2}} + \sqrt[3]{7 - 5\sqrt{2}}$$

93. $x^3 - 6x + 4 = 0$. Using the formula in Problem 89 with $p = -6$ and $q = 4$, we obtain

$$x = \sqrt[3]{\frac{-4}{2} + \sqrt{\frac{4^2}{4} + \frac{(-6)^3}{27}}} + \sqrt[3]{\frac{-4}{2} - \sqrt{\frac{4^2}{4} + \frac{(-6)^3}{27}}}$$

$$= \sqrt[3]{-2 + \sqrt{4 + (-8)}} + \sqrt[3]{-2 - \sqrt{4 + (-8)}}$$

$$= \sqrt[3]{-2 + 2i} + \sqrt[3]{-2 - 2i}$$

The solutions of $x^3 - 6x + 4 = 0$ are the zeros of $f(x) = x^3 - 6x + 4$
$f(-x) = -x^3 + 6x + 4$.

Step 1:   There are at most three zeros.

Step 2:   There are two or zero positive zeros and one negative zero.

Step 3: p: $\pm 1, \pm 2, \pm 4$; q: ± 1; $\dfrac{p}{2}$: $\pm 1, \pm 2, \pm 4$

Step 4:   (a)

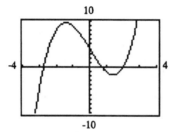

(b)   From the graph $x = 2$ appears to be the only rational zero.

(c)   We will check if $x = 2$ is a zero:

$$\begin{array}{r|rrrr}
2 & 1 & 0 & -6 & 4 \\
  &   & 2 & 4 & -4 \\
\hline
  & 1 & 2 & -2 & \boxed{0}
\end{array}$$
So, $x - 2$ is a factor.

The depressed equation is $x^2 + 2x - 2$.

We find the remaining zeros using the quadratic formula:

$$x = \frac{-2 \pm \sqrt{2^2 - 4(1)(-2)}}{2(1)} = \frac{-2 \pm \sqrt{12}}{2} = -1 \pm \sqrt{3}$$

Thus, the solution to $x^3 - 6x + 4 = 0$ are:

$$\{-2, -1 - \sqrt{3}, -1 + \sqrt{3}\}$$

4.5 Complex Polynomials; Fundamental Theorem of Algebra

1. Since complex zeros appear as conjugate pairs, it follows that $4 + i$, the conjugate of $4 - i$, is the remaining zero of f.

3. Since complex zeros appear as conjugate pairs, it follows that $-i$, the conjugate of i, and $1 - i$, the conjugate of $1 + i$, are the remaining zeros of f.

5. Since complex zeros appear as conjugate pairs, it follows that $-i$, the conjugate of i, and $-2i$, the conjugate of $2i$, are the remaining zeros of f.

7. Since complex zeros appear as conjugate pairs, it follows that $-i$, the conjugate of i, is the remaining zero of f.

9. Since complex zeros appear as conjugate pairs, it follows that $2 - i$, the conjugate of $2 + i$, and $-3 + i$, the conjugate of $-3 - i$, are the remaining zeros of f.

11. If the coefficients are real and $2 + i$ is a zero, then $2 - i$ would also be a zero.

13. If the coefficients are real, complex zeros must come in *pairs*; i.e., there will always be an *even* number of complex zeros. If the remaining zero was complex, $f(z)$ would have three complex zeros, which is impossible. Thus, the remaining zero must be real.

15. $f(z) = z^3 - 1$ is a difference of cubes: $f(z) = z^3 - 1 = (z - 1)(z^2 + z + 1)$
    The zeros of $z^2 + z + 1$ are:
    $$z = \frac{-1 \pm \sqrt{1 - 4}}{2} = \frac{-1 \pm \sqrt{-3}}{2} = -\frac{1}{2} + \frac{\sqrt{3}}{2}i \quad or \quad -\frac{1}{2}\frac{\sqrt{3}}{2}i$$
    Thus, the zeros of $z^3 - 1$ are 1, $-\frac{1}{2} + \frac{\sqrt{3}}{2}i$, and $-\frac{1}{2} - \frac{\sqrt{3}}{2}i$.

17. $f(z) = iz - 3$. Let $z = 1 + i$:
    $$\begin{aligned} f(1 + i) &= i(1 + i) - 3 \\ &= i + i^2 - 3 \\ &= i - 1 - 3, \text{ since } i^2 = -1 \\ &= i - 4 \\ &= -4 + i \end{aligned}$$

19. $f(z) = 3z^2 - z$. Let $z = 1 + i$:
    $$\begin{aligned} f(1 + i) &= 3(1 + i^2) - (1 + i) \\ &= 3(1 + 2i + i^2) - 1 - i \\ &= 3(1 + 2i - 1) - 1 - i \\ &= 6i - 1 - i \\ &= 5i - 1 \\ &= -1 + 5i \end{aligned}$$

21. $f(z) = z^3 + iz - 1 + i$
    We can evaluate $f(z)$ at $z = 1 + i$ by using synthetic division to divide $f(z)$ by $z - (1 + i)$
    $= z - 1 - i$:

    $$\begin{array}{r|rrrr} 1 + i & 1 & 0 & i & -1 + i \\ & & 1 + i & 2i & -3 + 3i \\ \hline & 1 & 1 + i & 3i & -4 + 4i \end{array}$$

    Note: $(-1 - i)(1 + i) = -1 - i - i - i^2 = -2i$
    and $\quad (-1 - i)(3i) = -3i - 3i^2 = 3 - 3i$
    Since the Remainder is $-4 + 4i$, we have $f(1 + i) = -4 + 4i$.

23. Here, $f(z) = 5z^5 - iz^4 + 2$. To find $f(r) = f(1 + i)$, divide $f(z)$ by $z - r = z - (1 + i)$
$= z - 1 - i$:

$$
\begin{array}{r|cccccc}
1 + i & 5 & -i & 0 & 0 & 0 & 2 \\
& & 5 + 5i & 1 + 9i & -8 + 10i & -18 + 2i & -20 - 16i \\
\hline
& 5 & 5 + 4i & 1 + 9i & -8 + 10i & -18 + 2i & -18 - 16i
\end{array}
$$

We have computed the following products:
$$(1 + i)(5 + 4i) = 5 + 4i + 5i + 4i^2 = 1 + 9i$$
$$(1 + i)(1 + 9i) = 1 + 9i + i + 9i^2 = -8 + 10i$$
$$(1 + i)(-8 + 10i) = -8 + 10i - 8i + 10i^2 = -18 + 2i$$
$$(1 + i)(-18 + 2i) = -18 + 2i - 18i + 2i^2 = -20 - 16i$$
Then $f(1 + i)$ is the Remainder, $-18 - 16i$.

25. Divide $f(z) = (1 + i)z^4 - z^3 + iz$ by $z - r = z - (2 - i) = z - 2 + i$:

$$
\begin{array}{r|ccccc}
2 - i & 1 + i & -1 & 0 & i & 0 \\
& & 3 + i & 5 & 10 - 5i & 16 - 18i \\
\hline
& 1 + i & 2 + i & 5 & 10 - 4i & 16 - 18i
\end{array}
$$

We have computed the following products:
$$(2 - i)(1 + i) = 2 + 2i - i - i^2 = 3 + i$$
$$(2 - i)(2 + i) = 4 + 2i - 2i - i^2 = 5$$
$$(2 - i)(10 - 4i) = 20 - 8i - 10i + 4i^2 = 16 - 18i$$

27. Divide $f(z) = iz^5 + iz^3 + iz$ by $z - r = z - (1 + 2i) = z - 1 - 2i$:

$$
\begin{array}{r|cccccc}
1 + 2i & i & 0 & i & 0 & i & 0 \\
& & -2 + i & -4 - 3i & -10i & 20 - 10i & 38 + 31i \\
\hline
& i & -2 + i & -4 - 2i & -10i & 20 - 9i & 38 + 31i
\end{array}
$$

Therefore, $f(r) = f(1 + 2i) = 38 + 31i$.

29. We know all three zeros: $3 + 2i$, 4, and 4.
Therefore, $f(z) = (z - (3 + 2i))(z - 4)(z - 4)$
$$= (z - 3 - 2i)(z^2 - 8z + 16)$$
$$= z^3 - 8z^2 + 16z - 3z^2 + 24z - 48 - 2iz^2 + 16iz - 32i$$
$$= z^3 + (-11 - 2i)z^2 + (40 + 16i)z - 48 - 32i$$

31. We have: $f(z) = (z - 2)(z - (-i))(z - (1 + i))$
$$= (z - 2)(z + i)(z - 1 - i)$$
$$= (z^2 + iz - 2z - 2i)(z - 1 - i)$$
$$= z^3 - z^2 - iz^2 + iz^2 - iz + z - 2z^2 + 2z + 2iz - 2iz + 2i - 2$$
$$= z^3 - 3z^2 + (3 - i)z - 2 + 2i$$

33. We have: $f(z) = (z - 3)(z - 3)(z - (-i))(z - (-i))$
$$= (z^2 - 6z + 9)(z^2 + 2iz - 1)$$
$$= z^4 + 2iz^3 - z^2 - 6z^3 - 12iz^2 + 6z + 9z^2 + 18iz - 9$$
$$= z^4 + (2i - 6)z^3 + (8 - 12i)z^2 + (6 + 18i)z - 9$$

1.

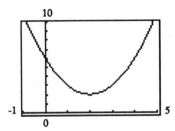

Since $a = 1 > 0$, the parabola opens up.
Vertex: $(2, 2) = (h, k)$
Axis of Symmetry: $x = 2$
y-intercept: $f(0) = (0 - 2)^2 + 2 = 6$
x-intercepts: $\quad 0 = (x - 2)^2 + 2$
$$-2 = (x - 2)^2$$
$$\sqrt{-2} = (x - 2)$$
Thus, the parabola has no x-intercepts.

3.

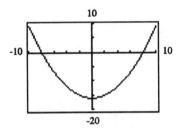

Since $a = \dfrac{1}{4} > 0$, the parabola opens up.

$$\frac{-b}{2a} = \frac{-0}{2\left[\dfrac{1}{4}\right]} = 0$$

$$f(0) = \frac{1}{4}(0)^2 - 16 = -16$$

Thus, vertex: $(0, -16)$
Axis of symmetry: $x = 0$

y-intercept: $f(0) = \dfrac{1}{4}(0)^2 - 16 = -16$

x-intercepts: $\quad 0 = \dfrac{1}{4}x^2 - 16$

$$16 = \frac{1}{4}x^2$$
$$x^2 = 64$$
$$x = \pm 8$$

5.

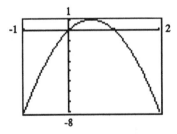

Since $a = -4 < 0$, the parabola opens down.

$$\frac{-b}{2a} = \frac{4}{2(-4)} = \frac{1}{2} = 0$$

$$f\left(-\frac{1}{2}\right) = -4\left[\frac{1}{2}\right]^2 + 4\left[\frac{1}{2}\right] = \frac{-4}{4} + 2$$
$$= 1$$

Thus, vertex: $\left[\dfrac{1}{2}, 1\right]$

Axis of Symmetry: $x = \dfrac{1}{2}$

y-intercept: $f(0) = -4(0)^2 + 4(0) = 0$
x-intercepts: $\quad 0 = 4x^2 + 4x$
$$= -4x(x - 1)$$
$$x = 0, 1$$

7.

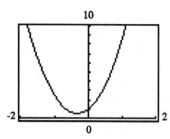

Since $a = \dfrac{9}{2} > 0$, the parabola opens up.

$$\dfrac{-b}{2a} = \dfrac{-3}{2\left(\dfrac{9}{2}\right)} = -\dfrac{1}{3}$$

$$f\left(-\dfrac{1}{3}\right) = \dfrac{9}{2}\left(-\dfrac{1}{3}\right)^2 + 3\left(-\dfrac{1}{3}\right) + 1$$

$$= \dfrac{1}{2} - 1 + 1$$

$$= \dfrac{1}{2}$$

Thus, vertex: $\left(-\dfrac{1}{3}, \dfrac{1}{2}\right)$

y-intercept: $f(0) = \dfrac{9}{2}(0)^2 + 3(0) + 1 = 1$

x-intercepts: $0 = \dfrac{9}{2}x^2 + 3x + 1$

$$x = \dfrac{-3 \pm \sqrt{3^2 - 4\left(\dfrac{9}{2}\right)(1)}}{2\left(\dfrac{9}{2}\right)}$$

$$= \dfrac{-3 \pm \sqrt{-9}}{9}$$

Thus, no x-intercepts.

9.

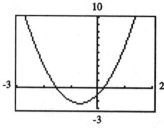

Since $a = 3 > 0$, the parabola opens up.

$$\dfrac{-b}{2a} = \dfrac{-4}{2(3)} = \dfrac{-2}{3}$$

$$f\left(\dfrac{-2}{3}\right) = 3\left(\dfrac{-2}{3}\right)^2 + 4\left(\dfrac{-2}{3}\right) - 1 = \dfrac{-7}{3}$$

Thus, vertex: $\left(\dfrac{-2}{3}, \dfrac{-7}{3}\right)$

y-intercept: $f(0) = 3(0)^2 + 4(0) - 1 = -1$

x-intercepts:

$$0 = 3x^2 + 4x - 1$$

$$x = \dfrac{-4 \pm \sqrt{4^2 - 4(3)(-1)}}{2(3)}$$

$$= \dfrac{-4 \pm \sqrt{28}}{6}$$

$$= \dfrac{-2 \pm \sqrt{7}}{3}$$

11. $f(x) = (x + 2)^3$ is the graph of $f(x) = x^3$ shifted 2 units left.

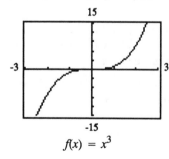

$f(x) = x^3$

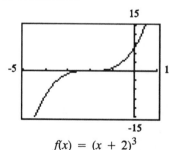

$f(x) = (x + 2)^3$

13. $f(x) = -(x - 1)^4$. Graph $y = x^4$. The graph of $y = -x^4$ is the graph of $y = x^4$ reflected about the x-axis. Finally, shift $y = -x^4$ one unit right to obtain $y = -(x - 1)^4$.

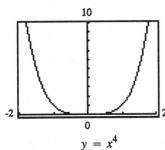

$y = x^4$

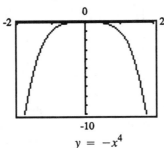

$y = -x^4$

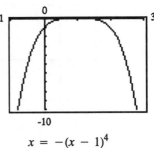

$x = -(x - 1)^4$

15. $f(x) = (x - 1)^4 + 2$ is the graph of $f(x) = x^4$ shifted right one unit and up two units.

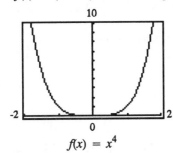

$f(x) = x^4$

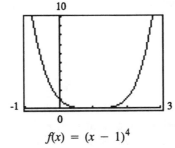

$f(x) = (x - 1)^4$

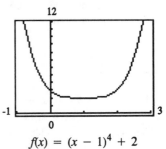

$f(x) = (x - 1)^4 + 2$

17. Minimum value; 1

19. Maximum value; 12

21. Maximum value; 16

23. (a)

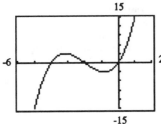

(b) x-intercepts: $-4, -2, 0$;
    y-intercept: 0
(c) $-4, -2, 0$: odd
(d) $y = x^3$
(e) 2
(f) Local minima: $(-0.84, -3.07)$
    Local maxima: $(-3.15, 3.07)$

25. (a)

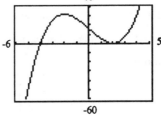

(b) x-intercepts: $-4, 2$; y-intercept: 16
(c) -4: odd; 2 even
(d) $y = x^3$
(e) 2
(f) Local minima: $(2, 0)$
    Local maxima: $(-2, 32)$

27. $f(x) = x^3 - 4x^2 = x^2(x - 4)$

(a)

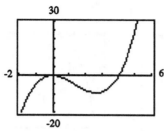

(b) x-intercepts: 0, 4; y-intercept: 0

(c) 0: even; 4 odd

(d) $y = x^3$

(e) 2

(f) Local minima: $(2.66, -9.48)$
Local maxima: $(0.00, 0.00)$

29. (a)

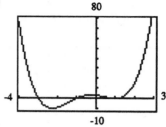

(b) x-intercepts: -3, -1, 1;
y-intercept: 3

(c) Crosses at -3, -1; touches at 1

(d) $y = x^4$

(e) 3

(f) Local minima: $(-2.28, -9.91)$,
$(1.00, 0)$
Local maxima: $(-0.21, 3.22)$

31. $R(x) = \dfrac{2x - 6}{x}$

Here, $p(x) = 2x - 6$, $n = 1$
$q(x) = x$, $m = 1$

Step 1: (a) The x-intercept is the zero of $p(x)$, $x = 3$.

(b) To find the y-intercept, we would let $x = 0$, but $R(0)$ is undefined. Thus, there is no y-intercept.

Step 2: $R(-x) = \dfrac{-2x - 6}{-x} = \dfrac{2x + 6}{x}$

This is neither $R(x)$ nor $-R(x)$, so there is no symmetry present.

Step 3: The vertical asymptotes are the zeros of $q(x)$: $x = 0$.

Step 4: Since $n = m$, the horizontal asymptote is the line $y = \dfrac{2}{1} = 2$; not intersected.

Step 5:

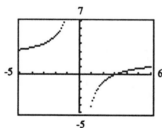

33. $H(x) = \dfrac{x + 2}{x(x - 2)}$

Here, $p(x) = x + 2$, $(n = 1)$
$q(x) = x(x - 2)$ $(m = 2)$

Step 1: (a) The x-intercept is the zero of $p(x)$, $x = -2$.

(b) $H(0)$ is undefined, so there is no y-intercept.

Step 2: $H(-x) = \dfrac{-x + 2}{-x(-x - 2)} = \dfrac{-x + 2}{x(x + 2)}$ There is no symmetry.

Step 3: The vertical asymptotes are the zeros of $q(x)$: $x = 0$ and $x = 2$.

Step 4: Since $m > n$, the horizontal asymptote is the line $y = 0$ (the x-axis); intersected at $(-2, 0)$.

Step 5:

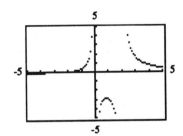

35. $R(x) = \dfrac{x^2 + x - 6}{x^2 - x - 6}$

$p(x) = x^2 + x - 6 \qquad\qquad (n = 2)$

$q(x) = x^2 - x - 6 \qquad\qquad (m = 2)$

Step 1: (a) The x-intercepts are the zeros of $p(x)$, $(x + 3)(x - 2) = 0$, $x = -3, 2$

            (b) The y-intercept is $y = R(0) = 1$.

Step 2: $R(-x) = \dfrac{x^2 - x - 6}{x^2 + x - 6}$ No symmetry present.

Step 3: The vertical asymptote is the zero of $q(x)$: $(x - 3)(x + 2) = 0$; $x = 3$, $x = -2$

Step 4: Since $m = n$, the line $y = \dfrac{1}{1} = 1$ is the horizontal asymptote; intersected at $(0, 1)$.

Step 5:

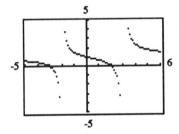

37. $F(x) = \dfrac{x^3}{x^2 - 4}$

$p(x) = x^3 \quad (n = 3)$

$q(x) = x^2 - 4 = (x + 2)(x - 2) \ (m = 2)$

Step 1: (a) The x-intercept is $x = 0$.

            (b) The y-intercept is $y = F(0) = 0$.

Step 2: $F(-x) = \dfrac{-x^3}{x^2 - 4} = -F(x)$, so the graph is symmetric about the origin.

Step 3: The vertical asymptotes are $x = -2$ and $x = 2$.

Step 4: Since $n = m + 1$, perform long division:

$$
\begin{array}{r}
x \\
x^2 - 4 \overline{)\, x^3 \quad 0x^2 \quad 0x \quad 0} \\
\underline{x^3 -4x } \\
4x
\end{array}
$$

$$F(x) = \dfrac{x^3}{x^2 - 4} = x + \dfrac{4x}{x^2 - 4}$$

Thus, we have an oblique asymptote, $y = x$; intersected at $(0, 0)$.

Step 5:

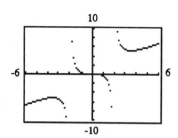

39. $R(x) = \dfrac{2x^4}{(x - 1)^2}$

    $p(x) = 2x^4 \quad (n = 4)$

    $q(x) = (x - 1)^2 \quad (m = 2)$

    Step 1:   (a)   The x-intercept is $x = 0$.

                (b)   The y-intercept is $y = 0$.

    Step 2: $R(-x) = \dfrac{2x^4}{(-x - 1)^2} = \dfrac{2x^4}{(x + 1)^2}$ No symmetry.

    Step 3:   The vertical asymptote is the zero of $q(x)$: $x = 1$.

    Step 4:   Since $n > m + 1$, there is no horizontal nor oblique asymptote.

    Step 5:

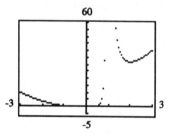

41. $f(x) = 12x^8 - x^7 + 8x^4 - 2x^3 + x + 3$

                + to −   − to +   + to −   − to +

    $f(-x) = 12x^8 + x^7 + 8x^4 + 2x^3 - x + 3$

                        + to −   − to +

    We have two or no negative zeros, and either four, two, or no positive zero(s).

43. $f(x) = 12x^8 - x^7 + 6x^4 - x^3 + x - 3$

            p: $\pm 1, \pm 3$

            q: $\pm 1, \pm 2, \pm 3, \pm 4, \pm 6, \pm 12$

    Possible rational zeros:

    $$\dfrac{p}{q}: \; \pm 1, \; \pm\dfrac{1}{2}, \; \pm\dfrac{1}{3}, \; \pm\dfrac{1}{4}, \; \pm\dfrac{1}{6}, \; \pm\dfrac{1}{12}, \; \pm 3, \; \pm\dfrac{3}{2}, \; \pm\dfrac{3}{4}$$

45. First apply Descartes' Rule of Signs:

    $f(x) = x^3 - 3x^2 - 6x + 8$

             + to −   − to +

    $f(-x) = -x^3 - 3x^2 + 6x + 8$

              − to +

    We see that there are either two positive zeros, or none; and exactly one negative zero.

To find the possible *rational* zeros, p must be a factor of 8: p: $\pm1, \pm2, \pm4, \pm8$
and q must be a factor of 1: q: ±1

Therefore, the *possible rational* zeros are: $\dfrac{p}{q}$: $\pm1, \pm2, \pm4, \pm8$

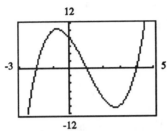

From the graph, it appears -2, 1, and 4 are zeros.
We will check if $x = 1$ is a factor.

$$
\begin{array}{r|rrrr}
1 & 1 & -3 & -6 & 8 \\
  &   & 1 & -2 & -8 \\
\hline
  & 1 & -2 & -8 & \boxed{0}
\end{array}
$$
$\leftarrow$ $(x - 1)$ is a factor.

We now have: $f(x) = x^3 - 3x^2 - 6x + 8$
$$= (x - 1)(x^2 - 2x - 8)$$
$$= (x - 1)(x - 4)(x + 2)$$
Therefore, the real zeros of $f(x)$ are 1, 4, and -2.

47.   We have: $f(x) = 4x^3 + 4x^2 - 7x + 2$ (2 sign changes)
and $f(-x) = -4x^3 + 4x^2 + 7x + 2$ (1 sign change)
By Descartes' Rule of Signs, $f(x)$ must have one negative zero, and either two positive zeros or none.

Check for possible *rational* zeros:

p: $\pm1, \pm2$; q: $\pm1, \pm2, \pm4$; $\dfrac{p}{q}$: $\pm1, \pm\dfrac{1}{2}, \pm\dfrac{1}{4}, \pm2$

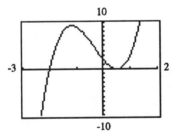

From the graph it appears -2 and $\dfrac{1}{2}$ are zeros.

We will check if $x = -2$ is a zero.

$$
\begin{array}{r|rrrr}
-2 & 4 & 4 & -7 & 2 \\
   &   & -8 & 8 & -2 \\
\hline
   & 4 & -4 & 1 & \boxed{0}
\end{array}
$$
$\leftarrow$ $(x + 2)$ is a factor.

Quotient: $q(x) = 4x^2 - 4x + 1$
We now have: $f(x) = 4x^3 + 4x^2 - 7x + 2 = (x + 2)(4x^2 - 4x + 1)$
Let's find the zeros of the depressed equation, $4x^2 - 4x + 1 = 0$, by the quadratic formula:

$$x = \frac{-b \pm \sqrt{b^2 - 4ac}}{2a} = \frac{4 \pm \sqrt{16 - 4(4)(1)}}{8} \text{ or } x = \frac{4 \pm 0}{8} = \frac{1}{2} \text{ is a double root.}$$

Thus, $\left(x - \dfrac{1}{2}\right)$ is a repeated factor.

Since the leading coefficient is 4, we have: $4x^2 - 4x + 1 = 4\left(x - \dfrac{1}{2}\right)^2$

and $f(x) = (x + 2)(4x^2 - 4x + 1) = 4(x + 2)\left(x - \dfrac{1}{2}\right)^2$

The real zeros of $f(x)$ are -2 and $\dfrac{1}{2}$ (with multiplicity 2).

49. $f(x) = x^4 - 4x^3 + 9x^2 - 20x + 20$
$f(-x) = x^4 + 4x^3 + 9x^2 + 20x + 20$
We see that we either have four positive zeros, or two positive zeros, or none, and we have no negative zeros (since there are no changes of sign in $f(-x)$).
Possible rational zeros:

$$p: \pm1, \pm2, \pm4, \pm5, \pm10, \pm20; \quad q: \pm1; \quad \frac{p}{q}: \pm1, \pm2, \pm4, \pm5, \pm10, \pm20$$

(But we can exclude the negative possibilities, since $f(x)$ has no negative zeros).

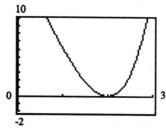

From the graph it appears $x = 2$ is a zero.
We will check if $x = 2$ is a zero.

$$\begin{array}{r|rrrrr}
2 & 1 & -4 & 9 & -20 & 20 \\
  &   & 2 & -4 & 10 & -20 \\
\hline
  & 1 & -2 & 5 & -10 & \boxed{0}
\end{array} \leftarrow (x - 2) \ \text{is a factor.}$$

Quotient: $q(x) = x^3 - 2x^2 + 5x - 10$
We now have: $f(x) = x^4 - 4x^3 + 9x^2 - 20x + 20$
$$= (x - 2)(x^3 - 2x^2 + 5x - 10)$$

factor by grouping
$$= (x - 2)[x^2(x - 2) + 5(x - 2)]$$
$$= (x - 2)(x - 2)(x^2 + 5)$$

no real zeros
$$= (x - 2)^2(x^2 + 5)$$
The real zero of $f(x)$ is 2 (multiplicity two).

51. $f(x) = 2x^3 - 11.84x^2 - 9.116x + 82.46$
$f(-x) = -2x^3 - 11.84x^2 + 9.116x + 82.46$
The function has at most three zeros.
The function will have two or zero positive zeros, and one negative zero.
The Rational Zeros Theorem does not apply.

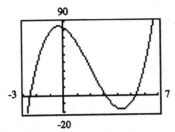

The zeros are approximated to be -2.50, 3.10, and 5.32.

Since we found the maximum number of zeros, we do not need to use the Upper and Lower Bounds Test.

53. $g(x) = 15x^4 - 21.5x^3 - 1718.3x^2 + 5308x + 3796.8$
     $g(-x) = 15x^4 + 21.5x^3 - 1718.3x^2 - 5308x + 3796.8$
     The function has at most four zeros.
     The function will have two or zero positive zeros, and two or zero negative zeros.
     The Rational Zeros Theorem does not apply.

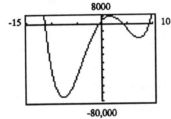

We approximate the zeros to be -11.30, -0.60, 4.00, and 9.33.

55. $f(x) = 3x^3 + 18.02x^2 + 11.0467x - 53.8756$
     $f(-x) = -3x^3 + 18.02x^2 - 11.0467x - 53.8756$
     The function will have at most three zeros.
     The function will have one positive zero and two or zero negative zeros.
     The Rational Zero Theorem does not apply.

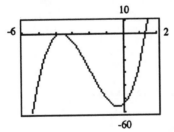

The zeros are approximated to be -3.67 and 1.33.
We will check if -6 is a Lower Bound:

$$
\begin{array}{r|rrrr}
-6) & 3 & 18.02 & 11.0467 & -53.8756 \\
    &   & -18 & & \\
\hline
    & 3 & 0.02 & & \\
\end{array}
$$

-6 fails the Lower Bound Test. We will check -7.

$$
\begin{array}{r|rrrr}
-7) & 3 & 18.02 & 11.0467 & -53.8756 \\
    &   & -21 & 20.86 & -223.3469 \\
\hline
    & 3 & -2.98 & 31.9067 & -277.2225 \\
\end{array}
$$

Thus, -7 is a Lower Bound, so -3.67 must be a zero of multiplicity 2.

57. $2x^4 + 2x^3 - 11x^2 + x - 6 = 0$

The solutions of this equation are the zeros of the polynomial function $f(x)$.
$$f(x) = 2x^4 + 2x^3 - 11x^2 + x - 6$$
$$f(-x) = 2x^4 - 2x^3 - 11x^2 - x - 6$$
We have either three positive zeros, or one; and exactly one negative zero.
Possible *rational* zeros:

$$p: \pm 1, \pm 2, \pm 3, \pm 6; \quad q: \pm 1, \pm 2; \quad \frac{p}{q}: \pm 1, \pm \frac{1}{2}, \pm 2, \pm 3, \pm \frac{3}{2}, \pm 6$$

Graph $f(x) = 2x^4 + 2x^3 - 11x^2 + x - 6$

From the graph, it appears -3 and 2 are solutions. We will check if $x = 2$ is a solution.

$$
\begin{array}{r|rrrrr}
2 & 2 & 2 & -11 & 1 & -6 \\
  &   & 4 & 12 & 2 & 6 \\
\hline
  & 2 & 6 & 1 & 3 & \boxed{0}
\end{array}
$$
← $(x - 2)$ *is* a factor.

We now have: $2x^4 + 2x^3 - 11x^2 + x - 6 = 0$
$$(x - 2)(2x^3 + 6x^2 + x + 3) = 0$$

factor by grouping
$$(x - 2)[2x^2(x + 3) + 1(x = 3)] = 0$$
$$(x - 2)(x + 3)(2x^2 + 1) = 0$$
The real zeros are -3 and 2.

59. $2x^4 + 7x^3 + x^2 - 7x - 3 = 0$

The solutions of this equation are the zeros of the polynomial function $f(x)$.
$$f(x) = 2x^4 + 7x^3 + x^2 - 7x - 3$$
$$f(-x) = 2x^4 - 7x^3 + x^2 + 7x - 3$$
We have exactly one positive zero, and either three or one negative zero(s).

For possible *rational* zeros: $p: \pm 1, \pm 3; \quad q: \pm 1, \pm 2; \quad \frac{p}{q}: \pm 1, \pm \frac{1}{2}, \pm 3, \pm \frac{3}{2}$

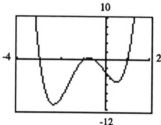

From the graph, it appears -3, -1, $-\frac{1}{2}$, and 1 are solutions to the equation.

We will check if $x = 1$ is a solution.

$$
\begin{array}{r|rrrrr}
1 & 2 & 7 & 1 & -7 & -3 \\
  &   & 2 & 9 & 10 & 3 \\
\hline
  & 2 & 9 & 10 & 3 & \boxed{0}
\end{array}
$$

Therefore, $(x - 1)$ is a factor, so $x = 1$ is a zero. All other real zeros *must* be negative.
We have: $f(x) = (x - 1)(2x^3 + 9x^2 + 10x + 3)$
We now concentrate on the depressed equation, $q(x) = (2x^3 + 9x^2 + 10x + 3)$. The *possible* rational zeros are the same as before, but since we have found the only positive zero, we only need to consider the negative ones: $-1, -3, -\frac{1}{2}, -\frac{3}{2}$

To see if $x = -1$ is a zero, divide by $x - (-1) = x + 1$:

$$\begin{array}{r|rrrr}
-1 & 2 & 9 & 10 & 3 \\
 & & -2 & -7 & -3 \\
\hline
 & 2 & 7 & 3 & \boxed{0}
\end{array}$$

So $(x + 1)$ *is* a factor, and we have:
$$f(x) = (x - 1)(2x^3 + 9x^2 + 10x + 3) = (x - 1)(x + 1)(2x^2 + 7x + 3)$$
Now $2x^2 + 7x + 3$ can be factored by trial and error:
$$2x^2 + 7x + 3 = (2x + 1)(x + 3)$$

We have:
$$2\left[x + \frac{1}{2}\right]$$

$$(x - 1)(x + 1)\overbrace{(2x + 1)}(x + 3) = 0$$

$$2(x - 1)(x + 1)\left[x + \frac{1}{2}\right](x + 3) = 0$$

The real zeros are: -3, -1, $-\dfrac{1}{2}$, and 1.

61. $f(x) = 2x^3 - x^2 - 4x + 2$

To find an upper bound to the zeros of $f(x)$, we divide $f(x)$ by $x - 1$, $x - 2$, ... , until the third row in the synthetic division process contains no negative numbers:

$$\begin{array}{r|rrrr}
1 & 2 & -1 & -4 & 2 \\
 & & 2 & 1 & \\
\hline
 & 2 & 1 & -3 & \leftarrow \text{Stop!}
\end{array}$$

$$\begin{array}{r|rrrr}
2 & 2 & -1 & -4 & 2 \\
 & & 4 & 6 & 4 \\
\hline
 & 2 & 3 & 2 & 6
\end{array}$$

The entries in the last row are all positive, so there is no zero of $f(x)$ greater than 2.

To find a lower bound, divide $f(x)$ by $x + 1$, $x + 2$, ... , until the third row contains numbers that are alternately positive (or zero) and negative (or zero).

$$\begin{array}{r|rrrr}
-1 & 2 & -1 & -4 & 2 \\
 & & -2 & 3 & \\
\hline
 & 2 & -3 & -1 & \leftarrow \text{Stop!}
\end{array}$$

$$\begin{array}{r|rrrr}
-2 & 2 & -1 & -4 & 2 \\
 & & -4 & 10 & -12 \\
\hline
 & 2 & -5 & 6 & -10
\end{array}$$

There are no zeros less than -2. Therefore, all real zeros must lie between -2 and 2.

63. $f(x) = 2x^3 - 7x^2 - 10x + 35$

(a) Upper bound:

$$\begin{array}{r|rrrr}
1 & 2 & -7 & -10 & 35 \\
 & & 2 & & \\
\hline
 & 2 & -5 & & \leftarrow \text{Stop}
\end{array}$$

Now 2 and 3 will not work either (do you see why?), so we try 4:

$$\begin{array}{r|rrrr}
4 & 2 & -7 & -10 & 35 \\
 & & 8 & 4 & \\
\hline
 & 2 & 1 & -6 & \leftarrow \text{Stop}
\end{array}$$

$$\begin{array}{r|rrrr}
5 & 2 & -7 & -10 & 35 \\
 & & 10 & 15 & 25 \\
\hline
 & 2 & 3 & 5 & 60
\end{array}$$

So, $x = 5$ is an upper bound.

(b) Lower bound:

$$\begin{array}{r|rrrr}
-1 & 2 & -7 & -10 & 35 \\
 & & -2 & 9 & \\
\hline
 & 2 & -9 & -1 & \leftarrow \text{Stop}
\end{array}$$

$$\begin{array}{r|rrrr}
-2 & 2 & -7 & -10 & 35 \\
 & & -4 & 22 & -24 \\
\hline
 & 2 & -11 & 12 & 11 \quad \text{(Not good)}
\end{array}$$

$$\begin{array}{r|rrrr}
-3 & 2 & -7 & -10 & 35 \\
 & & -6 & 39 & -87 \\
\hline
 & 2 & -13 & 29 & -52
\end{array}$$

So $x = -3$ is a lower bound. All real zeros must lie between -3 and 5.

65. We are trying to locate the positive zero of $f(x) = x^3 - x - 2$. We start by finding the two consecutive whole numbers on either side of the zero:
$$f(0) = -2$$
$$f(1) = 1 - 1 - 2 = -2 < 0$$
$$f(2) = 8 - 2 - 2 = 4 > 0$$
Therefore, the zero lies between 1 and 2. We now check $x = 1.1, 1.2, 1.3, \ldots, 1.9$.

First, write $f(x)$ in nested form:
$$f(x) = x^3 - x - 2$$
$$= (x^2 - 1)x - 2$$
$$= (x \cdot x - 1)x - 2$$
Then, $f(1.1) = -1.769 < 0$
$\quad\quad f(1.2) = -1.472 < 0$
Let's skip to: $f(1.5) = -.125 < 0$
(See how that saved some time?)
Now, $f(1.6) = .496 > 0$.

The zero must lie between 1.5 and 1.6. We continue once more, to isolate the zero to one of the intervals:
$$[1.5, 1.51], [1.51, 1.52], \ldots, [1.59, 1.6].$$
We have $f(1.5) = -.125 < 0$
Then $f(1.51) = -.06705 < 0$
$\quad\quad f(1.52) = -.00819 < 0$
$\quad\quad f(1.53) = .05158 > 0$
The zero is 1.52 and correct to two decimal places.

67. We start with whole numbers:
$$f(0) = -1 < 0$$
$$f(1) = 1 > 0$$
The zero lies in the interval [0, 1].

Now proceed by tenths:
$$f(0) = -1 < 0$$
$$f(0.1) = -1.2032 < 0$$
Skip to: $f(0.5) = -2 < 0$
How about: $f(0.7) = -1.8512 < 0$
$\quad\quad\quad\quad\quad f(0.8) = -1.3712 < 0$
$\quad\quad\quad\quad\quad f(0.9) = -0.4672 < 0$
The zero lies between 0.9 and 1.
We go again:
$$f(0.9) = -.4672 < 0$$
$$f(0.91) = -.34829 < 0$$
$$\vdots$$
$$f(0.95) = 0.18655 > 0$$
This time I went too far, so I back up:
$$f(0.94) = .04366 > 0$$
$$f(0.93) = -.09301 < 0$$
So the zero is 0.93 correct to two decimal places.

69. Since complex zeros appear as conjugate pairs, it follows that $4 - i$, the conjugate of $4 + i$, is the remaining zero of f.

71. $-i$, the conjugate of i, and $1 - i$, the conjugate of $1 + i$, are the remaining zeros of f.

73. $f(z) = (z - 1)(z - 1)(z - i)(z - 3)$
$\quad = (z^2 - 2z + 1)(z^2 - 3z - iz + 3i)$
$\quad = z^4 - 3z^3 - iz^3 + 3iz^2 - 2z^3 + 6z^2$
$\quad\quad + 2iz^2 - 6iz + z^2 - 3z - iz + 3i$
$\quad = z^4 - (5 + i)z^3 + (7 + 5i)z^2$
$\quad\quad - (3 + 7i)z + 3i$

75. $f(z) = (z - 2)(z - 3)(z - (1 + i))$
$\quad = (z^2 - 5z + 6)(z - 1 - i)$
$\quad = z^3 - z^2 - iz^2 - 5z^2 + 5z + 5iz$
$\quad\quad + 6z - 6 - 6i$
$\quad = z^3 - (6 + i)z^2 + (11 + 5i)z - 6$
$\quad\quad - 6i$

77. We can first divide by $(x - 1)$, and then divide by $(x - 2)$:

$$\begin{array}{r|rrrr} 1 & 1 & 2 & -7 & -8 & 12 \\ & & 1 & 3 & -4 & -12 \\ \hline & 1 & 3 & -4 & -12 & \boxed{0} \end{array}$$

Thus, $\dfrac{x^4 + 2x^3 - 7x^2 - 8x + 12}{x - 1} = x^3 + 3x^2 - 4x - 12$

Now divide by $x - 2$:

$$\begin{array}{r|rrrr} 2 & 1 & 3 & -4 & -12 \\ & & 2 & 10 & 12 \\ \hline & 1 & 5 & 6 & \boxed{0} \end{array}$$

Thus, $\dfrac{x^4 + 2x^3 - 7x^2 - 8x + 12}{(x - 2)(x - 1)} = \dfrac{x^3 + 3x^2 - 4x - 12}{x - 2} = x^2 + 5x + 6$

79. $x^3 - x^2 - 8x + 12 = 0$

The solutions of this equation are the zeros of the polynomial function $f(x)$.

$$f(x) = x^3 - x^2 - 8x + 12$$

Determine the *possible rational* zeros:

$$p: \ \pm1, \pm2, \pm3, \pm4, \pm6, \pm12; \ q: \ \pm1; \ \frac{p}{q}: \ \pm1, \pm2, \pm3, \pm4, \pm6, \pm12$$

Graph $y = x^3 - x^2 - 8x + 12$.

From the graph, it appears -3 and 2 are solutions to the equation. We will check if $x = 2$ is a solution.

```
2)1  -1  -8   12
       2   2  -12
   ─────────────────
   1   1  -6   [0]   ← (x - 2) is a factor.
```

So

$$x^3 - x^2 - 8x + 12 = 0$$
$$(x - 2)(x^2 + x - 6) = 0$$
$$(x - 2)(x + 3)(x - 2) = 0$$

The zeros are $x = 2$ (multiplicity two), and $x = -3$.

81. $3x^4 - 4x^3 + 4x^2 + 1 = 0$

The solutions of this equation are the zeros of the polynomial function $f(x)$.

$$f(x) = 3x^4 - 4x^3 + 4x^2 - 4x + 1$$

For possible rational zeros: $p: \ \pm1; \ q: \ \pm1, \pm3; \ \frac{p}{q}: \ \pm1, \pm\frac{1}{3}$

Graph $y = 3x^4 - 4x^3 + 4x^2 - 4x + 1$. From the graph, it appears $\frac{1}{3}$ and 1 are solutions to the equation. We will check if $x = 1$ is a solution.

```
1)3  -4   4  -4   1
      3  -1   3  -1
   ──────────────────
   3  -1   3  -1  [0]   ← (x - 1) is a factor.
```

We have:
$$(x - 1)(3x^3 - x^2 + 3x - 1) = 0$$
$$(x - 1)[x^2(3x - 1) + 1(3x - 1)] = 0$$
$$(x - 1)(3x - 1)(x^2 + 1) = 0$$

$$3(x - 1)\left[x - \frac{1}{3}\right](x + i)(x - i) = 0$$

The zeros are $1, \frac{1}{3}, -i, i$.

83. (a) To find an upper bound on the zeros of $f(x)$, divide $f(x)$ by $x - 1, x - 2, \ldots$:

```
1)4  -3   0   8   1    2
      4   1   1   9   10
   ────────────────────────
   4   1   1   9  10   12
```

Since the last row contains only positive entries, there is no zero greater than 1.

(b) Lower bound:

```
-1)4  -3   0   8   1    2
      -4   7  -7
   ─────────────────────────
    4  -7   7   1   ←  Stop
```

```
-2)4  -3   0    8    1     2
      -8  22  -44   72  -146
   ──────────────────────────────
    4 -11  22  -36   73  -144
```

There is no zero less than -2.

Therefore, all zeros lie between -2 and 1.

85. Let (x, y) be any point on the line $y = x$. Then the distance from (x, y) to $(3, 1)$ is given by the distance formula.

$$d = \sqrt{(x - 3)^2 + (y - 1)^2} = \sqrt{x^2 - 6x + 9 + y^2 - 2y + 1}$$

But since (x, y) in on the line $y = x$, we can replace all y's in the above formula by x:

$$d = \sqrt{x^2 - 6x + 9 + x^2 - 2x + 1}$$

or $\quad d = \sqrt{2x^2 - 8x + 10}$

Now d is a fairly complicated function, but notice: $d = 2x^2 - 8x + 10$ so d^2 is a parabola that opens upward, so we can find the minimum value of d^2, and then it will be easy to find the minimum value of d by taking a square root.

Consider the parabola $y = 2x^2 - 8x + 10$. We have $a = 2$, $b = -8$, $c = 10$, so that

$$h = \frac{-b}{2a} = \frac{8}{4} = 2, \text{ and } k = f(2) = 2.$$

Thus, the vertex is at $(2, 2)$. That means that the minimum value of d^2 is 2, and that value occurs when $x = 2$. (So the minimum value of d is $\sqrt{2}$.) Since $y = x$, and $x = 2$, the point on the line is the point $(2, 2)$.

87. Let the origin be the highest point on the parabolic arch (i.e., the vertex). Then the equation of the parabola is $y = ax^2$, where $a < 0$ (since the parabola opens downward). From the illustration, we see that when $x = 10$, $y = -10$. Therefore:

$$-10 = a(10)^2$$
$$-10 = 100a$$
$$a = -\frac{1}{10}$$

Now find y when $x = -8$: $y = -\frac{1}{10}(-8)^2 = -\frac{64}{10} = -6.4$

Since the water is 10 feet below the x-axis, $h = 10 - 6.4 = 3.6$ ft.

91. (a)   Even
   (b)   Positive
   (c)   Even
   (d)   Since 0 is a zero of even multiplicity (it touches the x-axis)
   (e)   8 since there are 8 zeros.

EXPONENTIAL AND LOGARITHMIC FUNCTIONS

5.1 Exponential Functions

1. (a) $3^{2.2} \approx 11.212$ (b) $3^{2.23} \approx 11.587$ (c) $3^{2.236} \approx 11.664$ (d) $3^{\sqrt{5}} \approx 11.665$

3. (a) $2^{3.14} \approx 8.815$ (b) $2^{3.141} \approx 8.821$ (c) $2^{3.1415} \approx 8.824$ (d) $2^{\pi} \approx 8.825$

5. (a) $3.1^{2.7} \approx 21.217$ (b) $3.14^{2.71} \approx 22.217$
   (c) $3.141^{2.718} \approx 22.440$ (d) $\pi^{e} \approx 22.459$

7. 3.320   9. 0.427   11. B   13. D   15. A

17. E   19. A   21. E   23. B

Problems 25–31 are transformations of the graph $y = e^x$, given below.

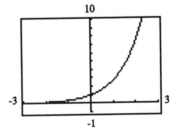

25. $y = e^{-x}$
    Using the graph of $y = e^x$, reflect about the y-axis.

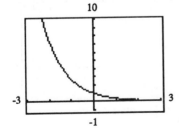

27. $y = e^{x+2}$
    Using the graph of $y = e^x$, shift 2 units to the left.

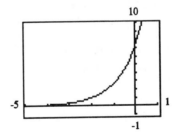

29. $y = 5 - e^{-x}$

Using the graph of $y = e^x$, reflect about the x- and y-axis, then shift up 5 units.

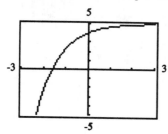

31. $y = 2 - e^{-x/2}$

Using the graph of $y = e^x$, horizontally stretch by a factor of 2, reflect about the x-axis and y-axis, then shift up 2 units.

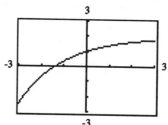

33.
$$4^x = 7$$
$$(4^x)^{-2} = 7^{-2}$$
$$4^{-2x} = \frac{1}{7^2}$$
$$4^{-2x} = \frac{1}{49}$$

35.
$$3^{-x} = 2$$
$$(3^{-x})^{-2} = 2^{-2}$$
$$3^{2x} = \frac{1}{2^2}$$
$$3^{2x} = \frac{1}{4}$$

37. $p = 100e^{-0.03n}$

    (a) $p = 100e^{-0.03(10)}$
          $p = 100e^{-0.3}$
          $p = 100(.741)$
          $p = 74\%$ of light

    (b) $p = 100e^{-0.03(25)}$
          $p = 100e^{-0.75}$
          $p = 100(.472)$
          $p = 47\%$ of light

39. $w = 50e^{-0.004d}$

    (a) $w = 50e^{-0.004(30)}$
          $w = 50e^{-0.12}$
          $w = 50(.887)$
          $w = 44$ watts

    (b) $w = 50e^{-0.004(365)}$
          $w = 50e^{-1.46}$
          $w = 50(.232)$
          $w = 11.6$ watts

41. $0 = 5e^{-0.4h}$

After 1 hour: $D = 5e^{-0.4(1)}$
                    $D = 5e^{-0.4}$
                    $D = 5(.670)$
                    $D = 3.35$ milligrams

After 6 hours: $D = 5e^{-0.4(6)}$
                      $D = 5e^{-2.4}$
                      $D = 5(.091)$
                      $D = 0.45$ milligrams

43. $R = 70 - 100e^{-0.2t}$

    (a) $R = 70 - 100e^{-0.2(10)}$
          $R = 70 - 100e^{-2}$
          $R = 70 - 100(.135)$
          $R = 56\%$ of viewers

    (b) $R = 70 - 100e^{-0.2(20)}$
          $R = 70 - 100e^{-4}$
          $R = 70 - 100e(.018)$
          $R = 68\%$ of viewers

    (c) As t increases, $e^{-0.2t}$ decreases to 0. Therefore, the highest percent of viewers expected to respond is 70%.

    (d)

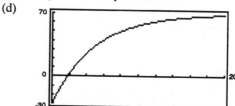

$R = 40\%$ just after 6 days.

45. $I = \dfrac{E}{R}\big[1 - e^{-(R/L)t}\big]$

(a) $I_1(t) = \dfrac{120}{10}\big[1 - e^{-(10/5)t}\big]$

After 0.3 second: $I = 5.414$ amperes
After 0.5 second: $I = 7.5854$ amperes
After 1 second: $I = 10.38$ amperes

(b) As t increases, $e^{-(10/5)t}$ goes to zero. Therefore, the maximum current is 12 amperes.

(c) See part (f)

(d) $I_2(t) = \dfrac{120}{5}\big[1 - e^{-(5/10)t}\big]$

After 0.3 second: $I = 3.343$ amperes
After 0.5 second: $I = 5.309$ amperes
After 1 second: $I = 9.443$ amperes

(e) The maximum current is 24 amperes.

(f)

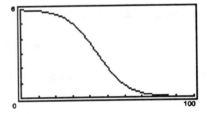

47. $y = 6\big[1 + e^{-(5.085 - 0.1156x)}\big]$

(a) $y = 6\big[1 + e^{-(5.085 - 0.1156(100))}\big]^{-1}$
$y = 9.23 \times 10^{-3}$

(b) $y = 6\big[1 + e^{-(5.085 - 0.1156(60))}\big]^{-1}$
$y = 0.81$

(c) $y = 6\big[1 + e^{-(5.085 - 0.1156(30))}\big]^{-1}$
$y = 5$

(d) For $y = 1$, $x = 57.91°F$
For $y = 3$, $x = 43.98°F$
For $y = 5$, $x = 30.06°F$

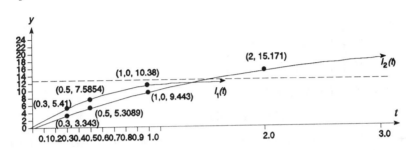

49. $2 + \dfrac{1}{2!} + \dfrac{1}{3!} + \ldots + \dfrac{1}{n!}$

$n = 4,\ 2 + \dfrac{1}{2!} + \dfrac{1}{3!} + \dfrac{1}{4!} = 2.7083$

$n = 6,\ 2 + \dfrac{1}{2!} + \dfrac{1}{3!} + \dfrac{1}{4!} + \dfrac{1}{5!} + \dfrac{1}{6!} = 2.7181$

$n = 8,\ 2 + \dfrac{1}{2!} + \dfrac{1}{3!} + \dfrac{1}{4!} + \dfrac{1}{5!} + \dfrac{1}{6!} + \dfrac{1}{7!} + \dfrac{1}{8!} = 2.7182788$

$n = 10,\ 2 + \dfrac{1}{2!} + \dfrac{1}{3!} + \dfrac{1}{4!} + \dfrac{1}{5!} + \dfrac{1}{6!} + \dfrac{1}{7!} + \dfrac{1}{8!} + \dfrac{1}{9!} + \dfrac{1}{10!} = 2.7182818$

$e = 2.718281828$

51. For $f(x) = a^x$, $\dfrac{f(x + h) - f(x)}{h} = \dfrac{a^{x+h} - a^x}{h} = \dfrac{a^x a^h - a^x}{h} = a^x\left[\dfrac{a^h - 1}{h}\right]$

53. For $f(x) = a^x$, $f(-x) = a^{-x} = \dfrac{1}{a^x} = \dfrac{1}{f(x)}$

55. (a) $\sinh(-x) = \dfrac{1}{2}\left(e^{-x} - e^{-(-x)}\right)$

$= \dfrac{1}{2}\left(e^{-x} - e^{x}\right)$

$= -\dfrac{1}{2}\left(e^{x} - e^{-x}\right)$

$= -\sinh x$

so that $\sinh x$ is an odd function.

(b)

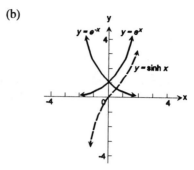

57. $f(x) = 2^{(2^x)} + 1$

$f(1) = 2^{(2^1)} + 1 = 2^2 + 1 = 4 + 1 = 5$

$f(2) = 2^{(2^2)} + 1 = 2^4 + 1 = 16 + 1 = 17$

$f(3) = 2^{(2^3)} + 1 = 2^8 + 1 = 256 + 1 = 257$

$f(4) = 2^{(2^4)} + 1 = 2^{16} + 1 = 65{,}536 + 1 = 65{,}537$

$f(5) = 2^{(2^5)} + 1 = 2^{32} + 1 = 4{,}294{,}967{,}296 + 1 = 4{,}294{,}967{,}297 = 641 \times 6{,}700{,}417$

5.2 Logarithmic Functions

In Problems 1–23, we use the equivalence of $a^x = M$ and $x = \log_a M$.

1. $9 = 3^2$ is equivalent to $2 = \log_3 9$

3. $a^2 = 1.6$ is equivalent to $2 = \log_a 1.6$

5. $1.1^2 = M$ is equivalent to $2 = \log_{1.1} M$

7. $2^x = 7.2$ is equivalent to $x = \log_2 7.2$

9. $x^{\sqrt{2}} = \pi$ is equivalent to $\sqrt{2} = \log_x \pi$

11. $e^x = 8$ is equivalent to $x = \ln 8$

13. $\log_2 8 = 3$ is equivalent to $2^3 = 8$

15. $\log_a 3 = 6$ is equivalent to $a^6 = 3$

17. $\log_3 2 = x$ is equivalent to $3^x = 2$

19. $\log_2 M = 1.3$ is equivalent to $2^{1.3} = M$

21. $\log_{\sqrt{2}} \pi = x$ is equivalent to $\left(\sqrt{2}\right)^x = \pi$

23. $\ln 4 = x$ is equivalent to $e^x = 4$

In Problems 25–35, we use (3)–(6):

25. $\log_2 1 = 0$ by (3)

27. $\log_5 25 = \log_5 5^2 = 2$ by (6)

29. $\log_{1/2} 16 = \log_{1/2} 2^4 = \log_{1/2}\left[\dfrac{1}{2}\right]^{-4}$

$= -4$ by (6)

31. $\log_{10}\sqrt{10} = \log_{10} 10^{1/2} = \dfrac{1}{2}$ by (6)

33. $\log_{\sqrt{2}} 4 = \log_{\sqrt{2}} \left(\sqrt{2}\right)^4$ since $4 = \left(\sqrt{2}\right)^4$
$= 4 \qquad$ by (6)

35. $\ln\sqrt{e} = \ln e^{1/2} = \dfrac{1}{2} \qquad$ by (6)

37. For $f(x) = \ln(3 - x)$, the domain is all x such that
$$3 - x > 0$$
$$-x > -3$$
$$\{x \mid x < 3\}$$

39. For $F(x) = \log_2 x^2$, the domain is all x such that $x^2 > 0$, or all real numbers except zero.

41. For $h(x) = \log_{1/2}(x^2 - x - 6)$, the domain is all x such that
$$x^2 - x - 6 > 0$$
$$(x - 3)(x + 2) > 0$$
$$\{x \mid x < -2 \text{ or } x > 3\}$$

43. For $f(x) = \dfrac{1}{\ln x}$ the domain is all x such that $x > 0$ except that $\ln x$ cannot be zero. Recall that $\ln 1 = 0$ so that $x \neq 1$. The domain is $\{x \mid x > 0, x \neq 1\}$.

45. For $g(x) = \log_5\left[\dfrac{x + 1}{x}\right]$, the domain is all x such that $\dfrac{x + 1}{x} > 0$.

	$x + 1$	x	$\dfrac{x + 1}{x}$
$x < -1$	$-$	$-$	$+$
$-1 < x < 0$	$+$	$-$	$-$
$x > 0$	$+$	$+$	$+$

$\dfrac{x + 1}{x} > 0$ when $x < -1$ or $x > 0$. The domain is $\{x \mid x < -1 \text{ or } x > 0\}$

47. 0.511

49. 30.099

51. For $f(x) = \log_a x$, we want to find a so that $f(2) = \log_a 2 = 2$ or $a^2 = 2$ or $a = \sqrt{2}$. Recall that $a > 0$ by definition.

53. B   55. D   57. A   59. E   61. C   63. A   65. D

Problems 67–75 are the transformations of the graph $y = \ln x$, given below.

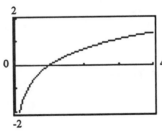

67. $y = \ln(x + 4)$ is the graph of $y = \ln x$ shifted 4 units left.

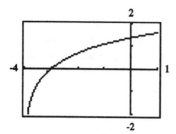

69. $y = \ln(-x)$ is the graph of $y = \ln x$ reflected about the y-axis.

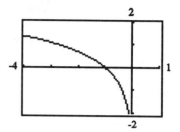

71. $y = \ln(2x)$ is the graph of $y = \ln x$ horizontally compressed by a factor of $\frac{1}{2}$.

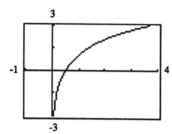

73. $y = 3 \ln x$ is the graph of $y = \ln x$ vertically stretched by a factor of 3.

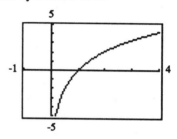

75. $y = \ln(3 - x) = \ln(-(x - 3))$ is obtained by reflecting $y = \ln x$ about the y-axis and then shifting the graph 3 units right.

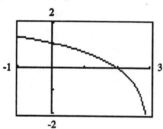

77. $p = 100e^{-0.1n}$

(a) $50 = 100e^{-0.1n}$

$0.5 = e^{-0.1n}$

$\ln 0.5 = -0.1 \, n$

$n = \dfrac{\ln 0.5}{-0.1}$

$n = 6.93$, so 7 panes of glass are necessary.

(b) $25 = 100e^{-0.1n}$ ($p = 25$ since 25% of the light passes through.)

$0.25 = e^{-0.1n}$

$\ln 0.25 = -0.1 \, n$

$n = \dfrac{\ln 0.25}{-0.1}$

$n = 13.86$, so 14 panes of glass are necessary.

79. $w = 50e^{-0.004d}$

(a)
$$30 = 50e^{-0.004d}$$
$$0.6 = e^{-0.004d}$$
$$\ln 0.6 = -0.004\,d$$
$$d = \frac{\ln 0.6}{-0.004}$$
$$d = 127.7, \text{ so it takes about}$$
$$128 \text{ days.}$$

(b)
$$5 = 50e^{-0.004d}$$
$$0.1 = e^{-0.004d}$$
$$\ln 0.1 = -0.004\,d$$
$$d = \frac{\ln 0.1}{-0.004}$$
$$d = 575.6, \text{ so it takes about } 576 \text{ days.}$$

81.
$$D = 5e^{-0.4h}$$
$$2 = 5e^{-0.4h}$$
$$0.4 = e^{-0.4h}$$
$$\ln 0.4 = -0.4\,h$$
$$h = \frac{\ln 0.4}{-0.4}$$
$$d = 2.29 \text{ hours or 2 hours,}$$
$$17 \text{ minutes}$$

83. 0.5 ampere:
$$0.5 = \frac{12}{10}\left[1 - e^{-(10/5)t}\right]$$
$$0.4167 = 1 - e^{-2t}$$
$$e^{-2t} = 0.583$$
$$-2t = \ln 0.5833$$
$$t = \frac{\ln 0.5833}{-2}$$
$$t = 0.2695 \text{ second}$$

1.0 ampere:
$$1.0 = \frac{12}{10}\left[1 - e^{-(10/5)t}\right]$$
$$0.8333 = 1 - e^{-2t}$$
$$e^{-2t} = 0.1\overline{666}$$
$$-2t = \ln 0.1\overline{666}$$
$$t = \frac{\ln 0.1\overline{666}}{-2}$$
$$t = 0.8959 \text{ second}$$

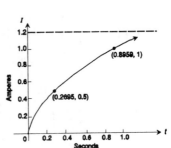

85. $r = 3e^{kx}$

(a)
$$10 = 3e^{k(0.06)}$$
$$3.\overline{333} = e^{k(0.06)}$$
$$\ln 3.\overline{3333} = k(0.06)$$
$$k = 20.07$$

(c)
$$100 = 3e^{20.07x}$$
$$33.\overline{3333} = e^{20.07x}$$
$$\ln 33.\overline{3333} = 20.07x$$
$$x = 0.175$$

(b)
$$R = 3e^{(20.07)(0.17)}$$
$$R = 3e^{3.4119}$$
$$R = 91\%$$

(d)
$$15 = 3e^{20.07x}$$
$$5 = e^{20.07x}$$
$$\ln 5 = 20.07x$$
$$x = 0.08$$

87. $y = 20e^{0.023t}$; $y = 89.2$ is predicted.

5.3 Properties of Logarithms

For Problems 1–11, we use $\ln 2 = a$ and $\ln 3 = b$.

1. $\ln 6 = \ln(3 \cdot 2) = \ln 3 + \ln 2 = b + a$

3. $\ln 1.5 = \ln\dfrac{3}{2} = \ln 3 - \ln 2 = b - a$

5. $\ln 2e = \ln 2 + \ln e = a + 1$

7. $\ln 12 = \ln(3 \cdot 4) = \ln 3 + \ln 4$
$$= \ln 3 + \ln 2^2 = \ln 3 + 2 \ln 2$$
$$= b + 2a$$

9. $\ln\sqrt[5]{18} = \ln(18)^{1/5} = \dfrac{1}{5}\ln(2 \cdot 3^2) = \dfrac{1}{5}(\ln 2 + \ln 3^2) = \dfrac{1}{5}(\ln 2 + 2 \ln 3) = \dfrac{1}{5}[a + 2b]$

11. $\log_2 3 = \dfrac{\log_e 3}{\log_e 2} = \dfrac{\ln 3}{\ln 2} = \dfrac{b}{a}$

13. $\ln\left[x^2\sqrt{1 - x}\right] = \ln x^2 + \ln\sqrt{1 - x} \qquad$ by (7)

$\qquad\qquad = \ln x^2 + \ln(1 - x)^{1/2}$

$\qquad\qquad = 2 \ln x + \dfrac{1}{2}\ln(1 - x) \qquad$ by (10)

15. $\log_2\left(\dfrac{x^3}{x - 3}\right) = \log_2 x^3 - \log_2(x - 3) \qquad$ by (8)

$\qquad\qquad\qquad = 3 \log_2 x - \log_2(x - 3) \qquad$ by (10)

17. $\log\left[\dfrac{x(x + 2)}{(x + 3)^2}\right] = \log x(x + 2) - \log(x + 3)^2 \qquad$ by (8)

$\qquad\qquad\qquad = \log x + \log(x + 2) - 2 \log(x + 3) \qquad$ by (7) and (10)

19. $\ln\left[\dfrac{x^2 - x - 2}{(x + 4)^2}\right]^{1/3} = \dfrac{1}{3}\ln\left[\dfrac{(x + 1)(x - 2)}{(x + 4)^2}\right] \qquad$ by (10)

$\qquad\qquad\qquad = \dfrac{1}{3}\left(\ln(x + 1)(x - 2) - \ln(x + 4)^2\right) \qquad$ by (8)

$\qquad\qquad\qquad = \dfrac{1}{3}\left(\ln(x + 1) + \ln(x - 2) - 2 \ln(x + 4)\right) \qquad$ by (7) and (10)

$\qquad\qquad\qquad = \dfrac{1}{3}\ln(x + 1) + \dfrac{1}{3}\ln(x - 2) - \dfrac{2}{3}\ln(x + 4)$

21. $\ln\dfrac{5x\sqrt{1 - 3x}}{(x - 4)^3} = \ln\left(5x\sqrt{1 - 3x}\right) - \ln(x - 4)^3 \qquad$ by (8)

$\qquad\qquad = \ln 5 + \ln x + \ln\sqrt{1 - 3x} - \ln(x - 4)^3 \qquad$ by (7)

$\qquad\qquad = \ln 5 + \ln x + \ln(1 - 3x)^{1/2} - \ln(x - 4)^3$

$\qquad\qquad = \ln 5 + \ln x + \dfrac{1}{2}\ln(1 - 3x) - 3 \ln(x - 4) \qquad$ by (10)

23. $3 \log_5 u + 4 \log_5 v = \log_5 u^3 + \log_5 v^4 \qquad$ by (10)

$\qquad\qquad\qquad = \log_5 u^3 v^4 \qquad$ by (7)

25. $\log_{1/2}\sqrt{x} - \log_{1/2} x^3 = \log_{1/2}\dfrac{\sqrt{x}}{x^3} \qquad$ by (8)

$\qquad\qquad\qquad = \log_{1/2} x^{(1/2)-3}$

$\qquad\qquad\qquad = \log_{1/2} x^{-5/2}$

$\qquad\qquad\qquad = -\dfrac{5}{2}\log_{1/2} x \qquad$ by (10)

27. $\ln\left[\dfrac{x}{x-1}\right] + \ln\left[\dfrac{x+1}{x}\right] - \ln(x^2-1) = \ln\dfrac{x}{x-1} \cdot \dfrac{x+1}{x} - \ln(x^2-1)$ by (7)

$$= \ln\dfrac{x+1}{x-1} \div (x^2-1) \qquad\qquad \text{by (8)}$$

$$= \ln\dfrac{1}{(x-1)^2}$$

$$= \ln(x-1)^2$$

$$= -2\ln(x-1) \qquad\qquad\qquad \text{by (10)}$$

29. $8\log_2\sqrt{3x-2} - \log_2\left[\dfrac{4}{x}\right] + \log_2 4 = \log_2\left(\sqrt{3x-2}\right)^8 - \log_2\dfrac{4}{x} + \log_2 4$ by (10)

$$= \log_2\dfrac{\left(\sqrt{3x-2}\right)^8 \cdot 4}{\dfrac{4}{x}} \qquad\qquad \text{by (7) and (8)}$$

$$= \log_2\left[x\left(\sqrt{3x-2}\right)^8\right]$$

$$= \log_2\left[x\left((3x-2)^{1/2}\right)^8\right]$$

$$= \log_2\left[x(3x-2)^4\right]$$

31. $2\log_a 5x^3 - \dfrac{1}{2}\log_a(2x+3) = \log_a(5x^3)^2 - \log_a(2x+3)^{1/2}$ by (10)

$$= \log_a 25x^6 - \log_a\sqrt{2x+3}$$

$$= \log_a\dfrac{25x^6}{\sqrt{2x+3}} \qquad\qquad \text{by (8)}$$

$$= \log_a\left[\dfrac{25x^6}{(2x+3)^{1/2}}\right]$$

33. $\log_3 21 = \dfrac{\log 21}{\log 3} = \dfrac{1.32222}{0.47712} = 2.771$

35. $\log_{1/3} 71 = \dfrac{\log 71}{\log\dfrac{1}{3}} = \dfrac{\log 7}{-\log 3} = \dfrac{1.85126}{-0.47712} = -3.880$

37. $\log_{\sqrt{2}} 7 = \dfrac{\log 7}{\log\sqrt{2}} = \dfrac{\log 7}{\log 2^{1/2}} = \dfrac{\log 7}{\dfrac{1}{2}\log 2} = \dfrac{0.84510}{0.5(0.30103)} = 5.615$

39. $\log_\pi e = \dfrac{\ln e}{\ln \pi} = \dfrac{1}{1.14473} = 0.874$

41. $y = \log_4 x = \dfrac{\ln x}{\ln 4} = \dfrac{\log x}{\log 4}$

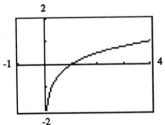

43. $y = \log_2(x + 2) = \dfrac{\ln(x + 2)}{\ln 2} = \dfrac{\log(x + 2)}{\log 2}$

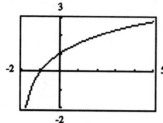

45. $y = \log_{x-1}(x + 1) = \dfrac{\ln(x + 1)}{\ln(x - 1)}$

$= \dfrac{\log(x + 1)}{\log(x - 1)}$

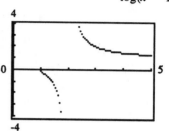

47. $\log_a\!\left(x + \sqrt{x^2 - 1}\right) + \log_a\!\left(x - \sqrt{x^2 - 1}\right)$

$= \log_a\!\left(x + \sqrt{x^2 - 1}\right)\!\left(x - \sqrt{x^2 - 1}\right) = \log_a\!\left[x^2 - \left(\sqrt{x^2 - 1}\right)^2\right]$

$= \log_a\!\left(x^2 - \left(x^2 - 1\right)\right) = \log_a\!\left(x^2 - x^2 + 1\right) = \log_a 1 = 0$

49. $\ln(1 + e^{2x}) = \ln[e^{2x}(e^{-2x} + 1)]$

$= \ln e^{2x} + \ln(e^{-2x} + 1)$

$= 2x + \ln(1 + e^{-2x})$

51. If $y = f(x) = \log_a x,$

$a^y = x$

$\left(\dfrac{1}{a}\right)^{-y} = x$

$-y = \log_{1/a} x$

Thus, $-f(x) = \log_{1/a} x$

53. If $f(x) = \log_a x,$

$f(AB) = \log_a AB$

$= \log_a A + \log_a B$

$= f(A) + f(B)$

55. $\ln y = \ln x + \ln C$

$\ln y = \ln xC$

$y = xC = Cx$

57. $\ln y = \ln x + \ln(x + 1) + \ln C$

$\ln y = \ln x(x + 1)C$

$y = x(x + 1)C = Cx(x + 1)$

59. $\ln y = 3x + \ln C$

$\ln y = \ln e^{3x} + \ln C$

$\ln y = \ln Ce^{3x}$

$y = Ce^{3x}$

61. $\ln(y - 3) = -4x + \ln C$

$\ln(y - 3) = \ln e^{-4x} + \ln C$

$\ln(y - 3) = \ln Ce^{-4x}$

$y - 3 = Ce^{-4x}$

$y = Ce^{-4x} + 3$

63. $3 \ln y = \dfrac{1}{2} \ln(2x + 1) - \dfrac{1}{3} \ln(x + 4) + \ln C$

$\ln y^3 = \ln(2x + 1)^{1/2} - \ln(x + 4)^{1/3} + \ln C$

$\ln y^3 = \ln\left[\dfrac{C(2x + 1)^{1/2}}{(x + 4)^{1/3}}\right]$

$y^3 = \left[\dfrac{C(2x + 1)^{1/2}}{(x + 4)^{1/3}}\right]$

$y = \left\{\dfrac{C(2x + 1)^{1/2}}{(x + 4)^{1/3}}\right\}^{1/3}$

$= \dfrac{\sqrt[3]{C}(2x + 1)^{1/6}}{(x + 4)^{1/9}}$

65. $\log_2 3 \cdot \log_3 4 \cdot \log_4 5 \cdot \log_5 6 \cdot \log_6 7 \cdot \log_7 8$

$$= \frac{\log 3}{\log 2} \cdot \frac{\log 4}{\log 3} \cdot \frac{\log 5}{\log 4} \cdot \frac{\log 6}{\log 5} \cdot \frac{\log 7}{\log 6} \cdot \frac{\log 8}{\log 7} = \frac{\log 8}{\log 2} = \frac{\log 2^3}{\log 2} = \frac{3 \log 2}{\log 2} = 3$$

67. $\log_2 3 \cdot \log_3 4 \cdot \cdots \cdot \log_n(n + 1) \cdot \log_{n+1} 2$

$$= \frac{\log 3}{\log 2} \cdot \frac{\log 4}{\log 3} \cdot \cdots \cdot \frac{\log n + 1}{\log n} \cdot \frac{\log 2}{\log n + 1} = \frac{\log 2}{\log 2} = 1$$

69. If $A = \log_a M$ and $B = \log_a N$, then $a^A = M$ and $a^B = N$.

Then, $\log_a\left[\dfrac{M}{N}\right] = \log_a\left[\dfrac{a^A}{a_B}\right] = \log_a a^{A-B} = A - B = \log_a M - \log_a N$

5.4 Logarithmic and Exponential Equations

1. $\log_2(2x + 1) = 3$
$$2x + 1 = 2^3$$
$$2x + 1 = 8$$
$$2x = 7$$
$$x = \frac{7}{2}$$

3. $\log_3(x^2 + 1) = 2$
$$x^2 + 1 = 3^2$$
$$x^2 + 1 = 9$$
$$x^2 = 8$$
$$x = \pm 2\sqrt{2}$$
$$\left\{-2\sqrt{2}, \, 2\sqrt{2}\right\}$$

5. $\dfrac{1}{2} \log_3 x = 2 \log_3 2$
$$\log_3 x^{1/2} = \log_3 2^2$$
$$x^{1/2} = 4$$
$$x = 16$$

7. $2 \log_5 x = 3 \log_5 4$
$$\log_5 x^2 = \log_5 4^3$$
$$x^2 = 64$$
$$x = 8 \text{ since } x > 0$$

9. $3 \log_2(x - 1) + \log_2 4 = 5$
$$\log_2(x - 1)^3 + \log_2 4 = 5$$
$$\log_2(x - 1)^3 \cdot 4 = 5$$
$$4(x - 1)^3 = 2^5$$
$$(x - 1)^3 = \frac{2^5}{4}$$
$$(x - 1)^3 = 8$$
$$x - 1 = 2$$
$$x = 3$$

11. $\log_{10} x + \log_{10}(x + 15) = 2$
$$\log_{10} x(x + 15) = 2$$
$$x(x + 15) = 10^2$$
$$x^2 + 15x - 100 = 0$$
$$(x + 20)(x - 5) = 0$$
$$x = -20 \text{ or } x = 5$$
Since $\log_{10}(-20)$ is undefined, we choose only $x = 5$.

13. $\log_x 4 = 2$
$$4 = x^2$$
$$\pm 2 = x$$
Since $\log_x(-2)$ is undefined, we have only $x = 2$.

15. $\log_3(x - 1)^2 = 2$
$$(x - 1)^2 = 3^2$$
$$x - 1 = 3 \text{ or } \qquad x = 1 - 3$$
$$x = 3 \text{ or } \qquad x = -2$$
$$\{-2, 4\}$$

17. $\log_{1/2}(3x + 1)^{1/3} = -2$

$$(3x + 1)^{1/3} = \left(\frac{1}{2}\right)^{-2}$$

$$3x + 1 = \left(\frac{1}{2}\right)^{-6}$$

$$3x = 2^6 - 1$$

$$\frac{3x}{6} = \frac{63}{3}$$

$$x = 21$$

19. $2^{2x+1} = 4$

$$2^{2x+1} = 2^2$$

$$2x + 1 = 2$$

$$2x = 1$$

$$x = \frac{1}{2}$$

21. $3^{x^3} = 9x$

$$3^{x^3} = 3^{2x}$$

$$x^3 = 2x$$

$$x^3 - 2x = 0$$

$$x(x^2 - 2) = 0$$

$$x = 0 \qquad x^2 - 2 = 0$$

$$x^2 = 2$$

$$x = \pm\sqrt{2}$$

$$\left\{-\sqrt{2},\, 0,\, \sqrt{2}\right\}$$

23. $8^{x^2-2x} = \frac{1}{2}$

$$2^{3(x^2-2x)} = 2^{-1}$$

$$3x^2 - 6x = -1$$

$$3x^2 - 6x + 1 = 0$$

$$x = \frac{6 \pm \sqrt{36 - 4(3)(1)}}{2(3)}$$

$$= \frac{6 \pm \sqrt{24}}{6} = 1 \pm \frac{\sqrt{6}}{3}$$

$$\left\{1 - \frac{\sqrt{6}}{3},\, 1 + \frac{\sqrt{6}}{3}\right\}$$

25. $2^x \cdot 8^{-x} = 4^x$

$$2^x \cdot 2^{-3x} = 2^{2x}$$

$$2^{-2x} = 2^{2x}$$

$$-2x = 2x$$

$$x = 0$$

27. $2^{2x} - 2^x - 12 = 0$

$$(2^x)^2 - 2^x - 12 = 0$$

$$(2^x - 4)(2^x + 3) = 0$$

$$2^x - 4 = 0 \quad \text{or} \qquad 2^x + 3 = 0$$

$$2^x = 4 \quad \text{or} \qquad\qquad 2^x = -3$$

$$2^x = 2^2 \qquad\qquad\qquad \text{No solution}$$

$$x = 2$$

29. $3^{2x} + 3^{x+1} - 4 = 0$

$$3^{2x} + 3 \cdot 3^x - 4 = 0$$

$$(3^x + 4)(3^x - 1) = 0$$

$$3^x + 4 = 0 \text{ or } \quad 3^x - 1 = 0$$

$$\text{No solution} \qquad\quad 3^x = 1$$

$$x = 0$$

31. $4^x = 8$

$$2^{2x} = 2^3$$

$$2x = 3$$

$$x = \frac{3}{2}$$

33. $2^x = 10$

$$\log 2^x = \log 10$$

$$x \log 2 = 1$$

$$x = \frac{1}{\log 2}$$

$$= 3.322$$

35. $8^{-x} = 1.2$

$$\log 8^{-x} = \log 1.2$$

$$-x \log 8 = \log 1.2$$

$$-x = \frac{\log 1.2}{\log 8}$$

$$-x = 0.088$$

$$x = -0.088$$

37.
$$3^{1-2x} = 4^x$$
$$\log 3^{1-2x} = \log 4^x$$
$$(1 - 2x)\log 3 = x \log 4$$
$$\log 3 - 2x \log 3 = x \log 4$$
$$\log 3 = x \log 4 + 2x \log 3$$
$$\log 3 = x \log (4 + 2 \log 3)$$
$$\frac{\log 3}{\log 4 + 2 \log 3} = x$$
$$x = \frac{0.47712}{0.60206 + 2(0.47712)} = 0.307$$

39.
$$\left(\frac{3}{5}\right)^x = 7^{1-x}$$
$$\log \left(\frac{3}{5}\right)^x = \log 7^{1-x}$$
$$x \log\frac{3}{5} = (1 - x)\log 7$$
$$x \log\frac{3}{5} = \log 7 - x \log 7$$
$$x \log\frac{3}{5} + x \log 7 = \log 7$$
$$x\left[\log \frac{3}{5} + \log 7\right] = \log 7$$
$$x = \frac{\log 7}{\log \frac{3}{5} + \log 7} = \frac{\log 7}{\log 3 - \log 5 + \log 7} = 1.356$$

41.
$$1.2^x = (0.5)^{-x}$$
$$\log 1.2^x = \log (0.5)^{-x}$$
$$x \log 1.2 = -x \log 0.5$$
$$x \log 1.2 + x \log 0.5 = 0$$
$$x(\log 1.2 + \log 0.5) = 0$$
$$x = 0$$

43.
$$\pi^{1-x} = e^x$$
$$\ln \pi^{1-x} = \ln e^x$$
$$(1 - x)\ln \pi = x \ln e$$
$$\ln \pi - x \ln \pi = x(1)$$
$$\ln \pi = x + x \ln \pi$$
$$\ln \pi = x(1 + \ln \pi)$$
$$\frac{\ln \pi}{1 + \ln \pi} = x$$
$$0.534 = x$$

45.
$$5(2^{3x}) = 8$$
$$2^{3x} = \frac{8}{5}$$
$$\ln 2^{3x} = \ln \frac{8}{5}$$
$$3x \ln 2 = \ln \frac{8}{5}$$
$$x = \frac{\ln \frac{8}{5}}{3 \ln 2}$$
$$x = 0.226$$

47.
$$400e^{0.2x} = 600$$
$$e^{0.2x} = \frac{600}{400}$$
$$e^{0.2x} = \frac{3}{2}$$
$$0.2x = \ln \frac{3}{2}$$
$$x = \frac{\ln \frac{3}{2}}{0.2}$$
$$= 2.027$$

49.
$$\log_a(x - 1) - \log_a(x + 6) = \log_a(x - 2) - \log_a(x + 3)$$
$$\log_a\frac{x - 1}{x + 6} = \log_a\frac{x - 2}{x + 3}$$
$$\frac{x - 1}{x + 6} = \frac{x - 2}{x + 3}$$
$$(x - 1)(x + 3) = (x - 2)(x + 6)$$
$$x^2 + 2x - 3 = x^2 + 4x - 12$$
$$9 = 2x$$
$$\frac{9}{2} = x$$

51.
$$\log_{1/3}(x^2 + x) - \log_{1/3}(x^2 - x) = -1$$
$$\log_{1/3}\frac{x^2 + x}{x^2 - x} = -1$$
$$\frac{x^2 + x}{x^2 - x} = \left(\frac{1}{3}\right)^{-1}$$
$$\frac{x(x + 1)}{x(x - 1)} = 3$$
$$x + 1 = 3(x - 1)$$
$$x + 1 = 3x - 3$$
$$4 = 2x$$
$$2 = x$$

53.
$$\log_2 8^x = -3$$
$$x \log_2 8 = -3$$
$$x = \frac{-3}{\log_2 8}$$
$$= \frac{-3}{\log_2 2^3}$$
$$= \frac{-3}{3}$$
$$= -1$$

55.
$$\log_2(x^2 + 1) - \log_4 x^2 = 1$$
$$\log_2(x^2 + 1) - \frac{\log_2 x^2}{\log_2 4} = 1$$
$$\log_2(x^2 + 1) - \frac{\log_2 x^2}{2} = 1$$
$$\log_2(x^2 + 1) - \frac{1}{2}\left(\log_2 x^2\right) = 1$$
$$\log_2(x^2 + 1) - \log_2 x = 1$$
$$\log_2\frac{x^2 + 1}{x} = 1$$
$$\frac{x^2 + 1}{x} = 2$$
$$x^2 - 2x + 1 = 0$$
$$(x - 1)^2 = 0$$
$$x = 1$$

57.
$$\log_{16} x + \log_4 x + \log_2 x = 7$$
$$\frac{\log_2 x}{\log_2 16} + \frac{\log_2 x}{\log_2 4} + \log_2 x = 7$$
$$\frac{\log_2 x}{4} + \frac{\log_2 x}{2} + \log_2 x = 7$$
$$\left[\frac{1}{4} + \frac{1}{2} + 1\right]\log_2 x = 7$$
$$\frac{7}{4}\log_2 x = 7$$
$$\log_2 x^{\frac{7}{4}} = 7$$
$$x^{\frac{7}{4}} = 2^7$$
$$x^{\frac{1}{4}} = 2$$
$$x = 2^4 = 16$$

59. 1.92   **61.** 2.78   **63.** −0.56   **65.** −0.70

67. 0.56   **69.** 0.39, 1.00   **71.** 1.31   **73.** 1.30

5.5 Compound Interest

1. Here, $P = \$100$, $r = 0.04$, $n = 4$ and $t = 2$ in the formula
$$A = P\left(1 + \frac{r}{n}\right)^{nt} = 100\left(1 + \frac{0.04}{4}\right)^{(4)(2)} = \$108.29$$

3. Here, $P = \$500$, $r = 0.08$, $n = 4$ and $t = 2.5$ in
$$A = P\left(1 + \frac{r}{n}\right)^{nt} = 500\left(1 + \frac{0.08}{4}\right)^{4(2.5)} = \$609.50$$

5. Here, $P = \$600$, $r = 0.05$, $n = 365$ and $t = 3$ in
$$A = P\left(1 + \frac{r}{n}\right)^{nt} = 600\left(1 + \frac{0.05}{365}\right)^{(365)(3)} = \$697.09$$

7. Here $P = \$10$, $r = 0.11$, and $t = 2$ in
$$A = Pe^{rt} = 10e^{(0.11)(2)} = \$12.46$$

9. Here $P = \$100$, $r = 0.10$, and $t = 2.25$ in
$$A = Pe^{rt} = 100e^{(0.10)(2.25)} = \$125.23$$

11. Here, $A = \$100$, $t = 2$, $r = 0.06$, and $n = 12$ in
$$V = A\left[1 + \frac{r}{n}\right]^{-nt} = 100\left[1 + \frac{0.06}{12}\right]^{-12(2)} = \$88.72$$

13. Here, $A = \$1000$, $t = 2.5$, $r = 0.06$, and $n = 365$ in
$$V = A\left[1 + \frac{r}{n}\right]^{-nt} = 1000\left[1 + \frac{0.06}{365}\right]^{-365(2.5)} = \$860.72$$

15. Here, $A = \$600$, $t = 2$, $r = 0.04$, and $n = 4$ in
$$V = A\left[1 + \frac{r}{n}\right]^{-nt} = 600\left[1 + \frac{0.04}{4}\right]^{-4(2)} = \$554.09$$

17. Here $A = \$80$, $t = 3.25$, and $r = 0.09$ in
$$V = Ae^{-rt} = 80e^{-0.09(3.25)} = \$59.71$$

19. Here $A = \$400$, $t = 1$, and $r = 0.10$ in
$$V = Ae^{-rt} = 400e^{-0.10(1)} = \$361.93$$

21. $r_e = \left[1 + \frac{.0525}{4}\right]^4 - 1$
$= 1.0535 - 1$
$= .0535$
$= 5.35\%$

23. $2P = P(1 + r)^3$
$2 = (1 + r)^3$
$\sqrt[3]{2} = 1 + r$
$1.26 = 1 + r$
$r = 26\%$

25. 6% compounded quarterly:
$$A = \$10,000\left[1 + \frac{.06}{4}\right]^4$$
$$= \$10,000(1.0614)$$
$$= \$10,614$$

$6\frac{1}{4}\%$ compounded annually:
$$A = \$10,000(1 + .0625)$$
$$= \$10,000(1.0625)$$
$$= \$10,625$$

$6\frac{1}{4}\%$ compounded annually yields a larger amount.

27. 9% compounded monthly:
$$A = \$10,000\left[1 + \frac{.09}{12}\right]^{12}$$
$$= \$10,000(1.0938)$$
$$= \$10,938$$

8.8% compounded daily:
$$A = \$10,000\left[1 + \frac{.088}{365}\right]^{365}$$
$$= \$10,000(1.0920)$$
$$= \$10,920$$

9% compounded monthly yields a larger amount.

29. Compounded monthly:

$$2P = P\left[1 + \frac{.08}{12}\right]^{12t}$$

$$2 = (1.00\overline{66})^{12t}$$

$$\ln 2 = 12t \ln (1.00\overline{66})$$

$$12t = \frac{\ln 2}{\ln(1.00\overline{66})}$$

$$t = 104.32 \text{ months}$$

Compounded continuously:

$$2P = Pe^{.08t}$$

$$2 = e^{.08t}$$

$$\ln 2 = .08t$$

$$t = 8.66 \text{ years or } 103.97 \text{ months}$$

31. Compounded monthly:

$$\$150 = \$100\left[1 + \frac{.08}{12}\right]^{.12t}$$

$$1.5 = (1.00\overline{66})^{12t}$$

$$\ln 1.5 = 12t \ln (1.00\overline{66})$$

$$t = \frac{\ln 1.5}{12 \ln(1.00\overline{66})}$$

$$t = 5.0852 \text{ years or } 61.02 \text{ months}$$

Compounded continuously:

$$\$150 = \$100e^{.08t}$$

$$1.5 = e^{.08t}$$

$$\ln 1.5 = .08t$$

$$t = 5.0683 \text{ years or } 60.82 \text{ months}$$

33.
$$\$25,000 = \$10,000e^{.06t}$$
$$2.5 = e^{.06t}$$
$$\ln 2.5 = .06t$$
$$t = 15.27 \text{ years or } 15 \text{ years,}$$
$$4 \text{ months}$$

35.
$$A = \$90,000(1 + .03)^5$$
$$A = \$90,000(1.15927)$$
$$A \approx \$104,335$$

37.
$$P = \$15,000e^{-.05(3)}$$
$$= \$15,000(.86071)$$
$$= \$12,910.62$$

39.
$$A = \$1500(1 + .15)^5$$
$$= \$1500(2.01136)$$
$$\approx \$3017$$

41.
$$\$850,000 = \$650,000(1 + r)^3$$
$$1.3077 = (1 + r)^3$$
$$\sqrt[3]{1.3077} = 1 + r$$
$$r = .0935$$
$$r = 9.35\%$$

43. Compounded continuously:

$$A = \$1000e^{.056}$$
$$= \$1057.60$$

You do not quite have enough money.

Compounded monthly:

$$A = \$1000\left[1 + \frac{.059}{12}\right]^{12}$$
$$= \$1000(1.06062)$$
$$= \$1060.62$$

So, the second bank offers a better deal.

45. For you: $P = \$2000$, $r = 0.09$, $n = 2$, and $t = 20$ in

$$A = P\left[1 + \frac{r}{n}\right]^{nt} = 2000\left[1 + \frac{0.09}{2}\right]^{2(20)} = \$11,632.73$$

For your friend: $P = \$2000$, $r = 0.085$, and $t = 20$ in
$$A = Pe^{rt} = 2000e^{0.085(20)} = \$10,947.89$$
You have more money after 20 years.

47. Here $P = \$50,000$ and $t = 5$ in each of the following:

(a) Now $r = 0.12$ and $n = \dfrac{1}{5}$ (no compounding) in

$$A = P\left[1 + \frac{r}{n}\right]^{nt} = 50,000\left[1 + \frac{0.12}{\dfrac{1}{5}}\right]^{1/5(5)} = \$80,000.00$$

(b) Now $r = 0.115$ and $n = 12$ in

$$A = P\left[1 + \frac{r}{n}\right]^{nt} = 50,000\left[1 + \frac{0.115}{12}\right]^{12(5)} = \$88,613.59$$

(c) Now, $r = 0.1125$ in

$A = 50,000e^{0.1125(5)} = \$87,752.73$
Subtracting the original 50,000 from each, we get the interest of
   (a) $30,000.00$
   (b) $38,613.59$
   (c) $37,752.73$
Option (a) results in the least interest.

49. (a) $\$10,000 = P\left[1 + \dfrac{.10}{12}\right]^{12(20)}$

$\$10,000 = P(7.328074)$

$P = \$1364.62$

(b) $P = \$10,000e^{-.10(20)}$

$\quad = \$1353.35$

51. $\$10,000 = P(1 + .08)^{10}$

$P = \$10,000(1 + .08)^{-10}$

$\quad = \$10,000(.4631935)$

$\quad = \$4631.93$

59. (a) $y = \dfrac{\ln 2}{1 \cdot \ln\left[1 + \dfrac{.12}{1}\right]}$

$y = \dfrac{\ln 2}{\ln(1.12)}$

$y = 6.1$ years

(b) $y = \dfrac{\ln 3}{4 \cdot \ln\left[1 + \dfrac{.06}{4}\right]}$

$y = \dfrac{\ln 3}{4 \cdot \ln 1.015}$

$y = 18.45$ years

(c) $mP = P\left[1 + \dfrac{r}{n}\right]^{nt}$

$m = \left[1 + \dfrac{r}{n}\right]^{nt}$

$\ln m = nt \ln\left[1 + \dfrac{r}{n}\right]$

$t = \dfrac{\ln m}{n \cdot \ln\left[1 + \dfrac{r}{n}\right]}$

5.6 Growth and Decay

1. (a) For $P = 500e^{0.02t}$ we want to find t (in days)
   when $P = 1000$:
   $$1000 = 500e^{0.02t}$$
   $$2 = e^{0.02t}$$
   $$0.02t = \ln 2$$
   $$t = \frac{\ln 2}{0.02}$$
   $$= 34.7 \text{ days}$$

   when $P = 2000$:
   $$2000 = 500e^{0.02t}$$
   $$4 = e^{0.02t}$$
   $$0.02t = \ln 4$$
   $$t = \frac{\ln 4}{0.02} = 69.3 \text{ days}$$

   (b)
   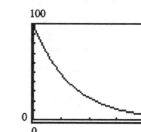

3. For $A = A_0 = e^{-0.0244t}$, the half-life is the time until $\frac{1}{2}A_0$ remains, so that $A = \frac{1}{2}A_0 = A_0e^{-0.0244t}$

   $$\frac{1}{2} = e^{-0.0244t}$$
   $$-0.0244t = \ln \frac{1}{2}$$
   $$t = \frac{-\ln 2}{-0.0244} = 28.4 \text{ years}$$

5. (a) We have $A = A_0e^{-0.0244t}$, with $A_0 = 100$ and $A = 10$:
   $$10 = 100e^{-0.0244t}$$
   $$\frac{1}{10} = e^{-0.0244t}$$
   $$-0.0244t = \ln\frac{1}{10}$$
   $$t = \frac{-\ln 10}{-0.0244} = 94.4 \text{ years}$$

   (b)

7. Using $N(t) = N_0e^{kt}$ where $N_0 = 1000$, $N(t) = 1800$, and $t = 1$, we get
   $$1800 = 1000e^{k1}$$
   $$1.8 = e^k$$
   $$k = \ln 1.8$$
   $$= 0.5878$$
   Thus, $N(t) = N_0e^{0.5878t}$ and when $t = 3$
   $$N(3) = 1000e^{0.5878(3)} = 1000e^{1.7634} = 5832 \text{ mosquitoes}$$

   When $N(t) = 10{,}000$, we have
   $$10{,}000 = 1000e^{0.5878t}$$
   $$10 = e^{0.5878t}$$
   $$0.5878t = \ln 10$$
   $$t = \frac{\ln 10}{0.5878} = 3.9 \text{ days}$$

Chapter 5 Exponential and Logarithmic Functions

9. Using $A = A_0 e^{kt}$ if after 18 months (= 1.5 years) we have $A = 2A_0$, then

$$2A_0 = A_0 e^{k(1.5)}$$
$$2 = e^{1.5k}$$
$$1.5k = \ln 2$$
$$k = \frac{\ln 2}{1.5} = 0.4621$$

Here $A_0 = 10,000$ and when $t = 2$,

$$A = 10,000e^{0.4621(2)} = 10,000e^{0.9242} = 25,198 \text{ is the population 2 years from now.}$$

11. Using $A = A_0 e^{kt}$ where the half-life, when $\frac{1}{2}A_0$ is present, is $t = 1690$ years, gives

$$\frac{1}{2}A_0 = A_0 e^{k(1690)}$$

$$\frac{1}{2} = e^{1690k}$$

$$1690k = \ln \frac{1}{2}$$

$$k = \frac{-\ln 2}{1690} = -0.00041$$

If $A_0 = 10$, the amount present after 50 years is

$$A = 10e^{-0.00041(50)} = 10e^{-0.0205} = 9.797 \text{ grams}$$

13. (a) Using $A_0 = A_0 e^{kt}$ the half-life $t = 5600$ years

(b)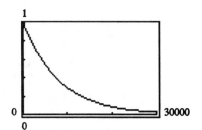

when $\frac{1}{2}A_0$ will be present:

$$\frac{1}{2}A_0 = A_0 e^{k(5600)}$$

$$\frac{1}{2} = e^{5600k}$$

$$5600k = \ln \frac{1}{2}$$

$$k = \frac{-\ln 2}{5600} = -0.000124$$

Thus, $A = A_0 e^{-0.000124t}$ and let $A_0 = 100$ so that $A = 30$:

$$30 = 100e^{-0.000124t}$$
$$0.3 = e^{-0.000124t}$$
$$-0.000124t = \ln 0.3$$
$$t = \frac{\ln 0.3}{-0.000124} = 9727 \text{ years ago}$$

15. (a) Using $u = T + (u_0 - T)e^{kt}$ where $t = 5$,   (b)
     $T = 70$, $u_0 = 450$, and $u = 300$:

$$300 = 70 + (450 - 70)e^{k(5)}$$
$$230 = 380e^{5k}$$
$$0.6053 = e^{5k}$$
$$5k = \ln 0.6053$$
$$k = -0.1004$$

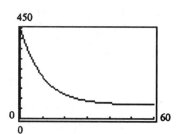

Thus, $u = T + (u_0 - T)e^{-0.1004t} = 70 + (450 - 70)e^{-0.1004t} = 70 + 380e^{-0.1004t}$

And, when $u = 135$ (with $T = 70$, $u_0 = 450$ still)

$$135 = 70 + (450 - 70)e^{-0.1004t}$$
$$65 = 380e^{-0.1004t}$$
$$0.17105 = e^{-0.1004t}$$
$$-0.1004t = \ln 0.17105$$
$$t = 17.59 \text{ minutes past 5:00 p.m.,}$$
$$\approx 5:18 \text{ p.m.}$$

(c) After 14.3 minutes, the pizza will be 160°F.   (d) As time passes, the temperature of
                                                               the pizza gets closer to 70°F.

17. (a) Using $u = T + (u_0 - T)e^{kt}$ where $T = 35$,   (b)
     $u_0 = 8$, and $u = 15$, when $t = 3$:

$$15 = 35 + (8 - 35)e^{k(3)}$$
$$-20 = -27e^{3k}$$
$$0.7407 = e^{3k}$$
$$3k = \ln 0.7407$$
$$k = -0.100035$$

Thus, $u = T + (u_0 - T)e^{kt}$

$$= 35 + (8 - 35)e^{-0.100035t}$$

When $t = 5$, $u = 35 - 27e^{-0.100035(5)}$

$$= 18.63°C$$

When $t = 10$, $u = 35 - 27e^{-0.100035(10)}$

$$= 25.1°C$$

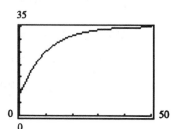

19. Using $A = A_0e^{kt}$ where $A_0 = 25$ and $A = 15$ when $t = 10$:

$$15 = 25e^{k(10)}$$
$$0.6 = e^{10k}$$
$$10k = \ln 0.6$$
$$k = -0.0511$$

Thus, $A = A_0e^{kt} = 25e^{-0.0511t}$

When $t = 24$, $A = 25e^{-0.0511(24)} = 7.34$ kg. remain.

When $A = \frac{1}{2}$kg, $\frac{1}{2} = 25e^{-0.0511t}$

$$.02 = e^{-0.0511t}$$
$$-0.0511t = \ln .02$$
$$t = 76.6 \text{ hours have passed.}$$

21. Using
$$A = A_0 e^{-0.087t}$$
$$.10A_0 = A_0 e^{-0.087t}$$
$$.10 = e^{-0.087t}$$
$$\ln .10 = -0.087t$$
$$t = \frac{\ln(.10)}{-0.087}$$
$$t \approx 26.5 \text{ days}$$

Farmers need to wait 26.5 days to use the hay.

5.7 Nonlinear Curve Fitting

1. (a)

(b) $P = 1000.187781(1.41414215)^t$

(c) $1000.187781(1.41414215)^t = A_0 e^{kt}$

So, $A_0 = 1000.187781$ and
$$e^k = 1.41414215$$
$$k = \ln 1.41414215$$
$$= 0.3465231$$
$$P = 1000.187781 e^{0.3465231t}$$

(d) $P = 1000.187781 e^{0.3465231(7)}$
$$= 11312$$

3. (a)

(b) $P = 96.0100692(0.981490683)^Q$

(c) $60 = 96.0100692(0.981490683)^Q$
$$0.624934452188 = 0.981490683^Q$$
$$\ln 0.624934452188 = Q \ln 0.981490683$$
$$Q = 25$$

5. (a)

(b) $y = 100.3262508(0.8768651017)^t$

(c) $100.3262508(0.8768651017)^t = A_0 e^{kt}$

Thus, $A_0 = 100.3262508$ and
$$0.8768651017 = e^k$$
$$k = \ln 0.8768651017$$
$$= -0.1314021163$$

So, $A = 100.3262508 e^{-0.1314021163t}$

(d) $\frac{1}{2}(100.3262508)$
$$= 100.3262508 e^{-0.1314021163t}$$
$$\frac{1}{2} = e^{-0.1314021163t}$$
$$\ln \frac{1}{2} = -0.1314021163t$$
$$t = 5.3 \text{ weeks}$$

(e) $A = 100.3262508 e^{-0.1314021163(50)}$
$$= 0.14 \text{ grams}$$

7. (a)

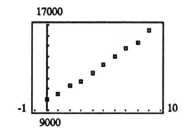

(b) Value $= 10014.4963(1.056554737)^t$

(c) From Section 5.5, we know
$$A = P(1 + r)^t$$
where r is the annual rate of return.
Thus, $r = 0.05655$ or 5.655%.

(d) Value $= 10014.4963(1.056554737)^{35}$
$$= \$68,682.99$$

9. (a)

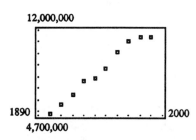

(b) $T = 173.2867183(0.9889878268)^t$

(c) $110° = 173.2867183(0.9889878268)^t$
$0.6347860995 = 0.9889878268^t$
$\ln 0.6347860995 = t \ln 0.9889878268$
$t = 41.04$

11. (a)

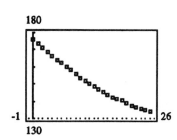

(b) $P = -1163825970 + 154808355.6 \ln t$

(c) $P = -1163825970 + 154808355.6 \ln 1995$
$= 12,469,735$ people

13. (a)

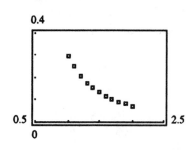

(b) $I = 0.2995056364 x^{-2.01200304}$

(c) Close; -2.01 versus -2

(d) $I = 0.2995056364(2.3)^{-2.01200304}$
$= 0.056054$

15. (a)

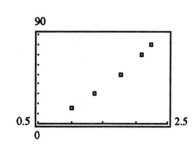

(b) $s = 15.97274236 t^{2.001820627}$

(c) $s = \dfrac{1}{2}(31.94548472)t^{2.001820627}$
$g \approx 31.9455$ ft/sec^2

(d) $100 = \dfrac{1}{2}(31.9548472)t^{2.001820627}$
$t^{2.001820627} = 6.2606656857$
$t = (6.2606656857)^{1/2.001820627}$
$= 2.5$ seconds

17. (a)

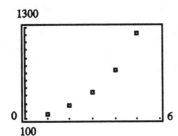

(b) The "best" model is the exponential model since it had the largest value for r. So,
$y = 99.06645375(1.652917358)^x$ is the best model.

19. (a)

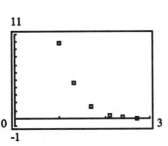

(b) The "best" model is the logarithmic model since it has the largest value for r. So,
$y = -16.55688464 + 10.31473128 \ln x$ is the best model.

21. (a)

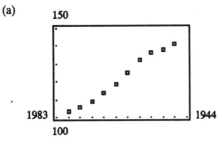

(b) The "best" model is the exponential model since it has the largest value for r. So,
$y = 346.113717(0.0352268374)^x$ is the "best" model.

23. (a)

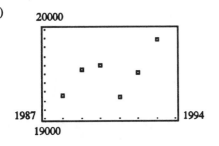

(b) The "best" model is the logarithmic model since it has the largest value for r. So,
$CPI = -69946.44499 + 9225.465596 \ln t.$

(c) $CPI = -69946.44499 + 9225.465596 \ln(1994)$
$= 147.7$

25. (a)

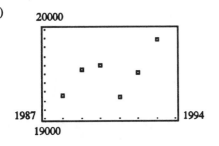

(b) The "best" model is the linear model since it has the largest value for r. So,
$GDP = 77.45714286t - 134670.9429.$

(c) $GDP = 77.45714286(1994) - 134670.9429$
$= 19779$

27. (a)

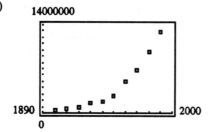

(b) The "best" model is the exponential model since it has the largest value for r. So,
$$P = 2.829551 \times 10^{-25}(1.037329901)^t$$
(c) $P = 2.829551 \times 10^{-25}(1.037329901)^{1995}$
$$= 16,066,684$$

5.8 Logarithmic Scales

1. $L(10^{-5}) = 10 \log \dfrac{10^{-5}}{10^{-12}}$
$= 10 \log 10^7$
$= 10(7)$
$= 70$ decibels

3. $L(0.15) = 10 \log \dfrac{0.15}{10^{-12}}$
$= 10 \log (0.15(10^{12}))$
$= 10(\log 0.15 + \log 10^{12})$
$= 10(-0.8239 + 12)$
$= 111.76$ decibels

5.
$$L(x) = 10 \log \frac{x}{10^{-12}} = 130$$
$$\log (x(10^{12})) = 13$$
$$\log x + \log 10^{12} = 13$$
$$\log x + 12 = 13$$
$$\log x = 1$$
$$x = 10^1$$
$$x = 10 \text{ Watts per}$$
$$\text{square meter}$$

7. For $M(x) = \log \dfrac{x}{x_0}$ we have $x = 10.0$ and $x_0 = 10^{-3}$:
$$M(10.0) = \log \frac{10.0}{10^{-3}} = \log 10^4 = 4.0$$
on the Richter scale.

9. Using $M(x) = \log \dfrac{x}{x_0}$ we have $M(x) = 7.85$ and $x_0 = 10^{-3}$:

$$7.85 = \log \frac{x}{10^{-3}}$$

$$\frac{x}{10^{-3}} = 10^{7.85}$$

$$x = 10^{4.85} = 70,794.58 \text{ mm}$$

For $M(x) = 8.9 = \log \dfrac{x}{10^{-3}}$

$$\frac{x}{10^{-3}} = 10^{8.9}$$

$$x = 10^{5.9}$$

The San Francisco earthquake was $\dfrac{10^{5.9}}{10^{4.85}} = 10^{1.05} = 11.22$ times as intense as the one in Mexico City.

5 Chapter Review

1. $\log_2 \dfrac{1}{8} = \log_2(2)^{-3} = -3$

3. $\ln e^{\sqrt{2}} = \sqrt{2}$

5. $2^{\log_2 0.4} = 0.4$

7. $3\log_4 x^2 + \dfrac{1}{2}\log_4\sqrt{x} = \log_4(x^2)^3 + \log_4\left(\sqrt{x}\right)^{1/2} = \log_4 x^6 + \log_4 x^{1/4} = \log_4 x^6 \cdot x^{1/4}$

$$= \log_4 x^{25/4} = \dfrac{25}{4}\log_4 x$$

9. $\ln\left[\dfrac{x-1}{x}\right] + \ln\left[\dfrac{x}{x+1}\right] - \ln(x^2-1) = \ln\dfrac{\dfrac{x-1}{x} \cdot \dfrac{x}{x+1}}{x^2-1} = \ln\dfrac{x-1}{(x^2-1)(x+1)}$

$$= \ln\dfrac{1}{(x+1)^2} = \ln(x+1)^{-2} = -2\ln(x+1)$$

11. $2\log 2 + 3\log x - \dfrac{1}{2}[\log(x+3) + \log(x-2)]$

$$= \log 2^2 + \log x^3 - \dfrac{1}{2}[\log(x+3)(x-2)]$$

$$= \log 2^2 x^3 - \log\left((x+3)(x-2)\right)^{1/2}$$

$$= \log \dfrac{4x^3}{\left((x+3)(x-2)\right)^{1/2}}$$

13. $\ln y = 2x^2 + \ln C$

$\ln y = \ln e^{2x^2} + \ln C$

$\ln y = \ln Ce^{2x^2}$

$y = Ce^{2x^2}$

15. $\dfrac{1}{2}\ln y = 3x^2 + \ln C$

$\ln y^{1/2} = \ln e^{3x^2} + \ln C$

$\ln y^{1/2} = \ln Ce^{3x^2}$

$y^{1/2} = Ce^{3x^2}$

$y = \left(Ce^{3x^2}\right)^2$

17. $\ln(y-3) + \ln(y+3) = x + C$

$\ln(y-3)(y+3) = x + C$

$(y-3)(y+3) = e^{x+C}$

$y^2 - 9 = e^{x+C}$

$y^2 = e^{x+C} + 9$

$y = \sqrt{e^{x+C} + 9}$

19. $e^{y+C} = x^2 + 4$

$\ln e^{y+C} = \ln(x^2 + 4)$

$y + C = \ln(x^2 + 4)$

$y = \ln(x^2 + 4) - C$

Problems 21–29 are transformations of $f(x) = e^x$ shown below.

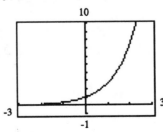

21. The graph of $f(x) = e^{-x}$ is the graph of $f(x) = e^x$ reflected about the y-axis.

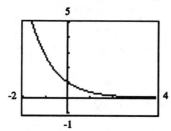

23. The graph of $f(x) = 1 - e^x$ is the graph of $f(x) = e^x$ reflected about the x-axis and shifted up 1 unit.

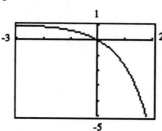

25. The graph of $f(x) = 3e^x$ is the graph of $f(x) = e^x$ vertically stretched 3 units.

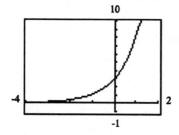

27. The graph of $f(x) = e^{|x|} = \begin{cases} e^x & x \geq 0 \\ e^{-x} & x < 0 \end{cases}$ is that of $y = e^x$ in either direction from $x = 0$.

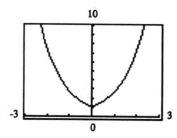

29. The graph of $f(x) = 3 - e^{-x}$ is the graph of $f(x) = e^x$ reflected about both coordinate axes, then shifted up 3 units.

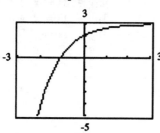

31.
$$4^{1-2x} = 2$$
$$2^{2(1-2x)} = 2^1$$
$$2(1 - 2x) = 1$$
$$2 - 4x = 1$$
$$-4x = -1$$
$$x = \frac{1}{4}$$

33.
$$3^{x^2+x} = \sqrt{3}$$
$$\log_3 3^{x^2+x} = \log_3 \sqrt{3}$$
$$(x^2 + x)\log_3 3 = \log_3 3^{1/2}$$
$$x^2 + x = \frac{1}{2}$$
$$x^2 + x - \frac{1}{2} = 0$$
$$2x^2 + 2x - 1 = 0$$

Here $a = 2$, $b = 2$, $c = -1$ and $b^3 - 4ac = 12$, so that
$$x = \frac{-2 \pm \sqrt{12}}{2(2)} = \frac{-2 \pm 2\sqrt{3}}{4} = \frac{-1 \pm \sqrt{3}}{2}$$
$$x = \frac{-1 - \sqrt{3}}{2} \quad \text{or} \quad x = \frac{-1 + \sqrt{3}}{2}$$
$$\left\{ \frac{-1 - \sqrt{3}}{2}, \frac{-1 + \sqrt{3}}{2} \right\}$$

35.
$$\log_x 64 = -3$$
$$64 = x^{-3}$$
$$64^{-1/3} = (x^{-3})^{-1/3}$$
$$\frac{1}{\sqrt[3]{64}} = x$$
$$\frac{1}{4} = x$$

37.
$$5^x = 3^{x+2}$$
$$\log 5^x = \log 3^{x+2}$$
$$x \log 5 = (x + 2)\log 3$$
$$x \log 5 = x \log 3 + 2 \log 3$$
$$x \log 5 - x \log 3 = \log 3^2$$
$$x(\log 5 - \log 3) = \log 9$$
$$x = \frac{\log 9}{\log 5 - \log 3} = 4.301$$

39.
$$9^{2x} = 27^{3x-4}$$
$$3^{2(2x)} = 3^{3(3x-4)}$$
$$2(2x) = 3(3x - 4)$$
$$4x = 9x - 12$$
$$-5x = -12$$
$$x = \frac{12}{5}$$

41.
$$\log_3 \sqrt{x - 2} = 2$$
$$\sqrt{x - 2} = 3^2$$
$$x - 2 = 9^2$$
$$x = 83$$

43.
$$8 = 4^{x^2} \cdot 2^{5x}$$
$$8 = 2^{2x^2} \cdot 2^{5x}$$
$$2^3 = 2^{2x^2+5x}$$
$$3 = 2x^2 + 5x$$
$$0 = 2x^2 + 5x - 3$$
$$0 = (2x - 1)(x + 3)$$
$$2x - 1 = 0 \quad \text{or} \quad x + 3 = 0$$
$$2x = 1$$
$$x = \frac{1}{2} \quad \text{or} \quad x = -3$$
$$\left\{-3, \frac{1}{2}\right\}$$

45.
$$\log_6(x + 3) + \log_6(x + 4) = 1$$
$$\log_6(x + 3)(x + 4) = 1$$
$$(x + 3)(x + 4) = 6^1$$
$$x^2 + 7x + 12 = 6$$
$$x^2 + 7x + 6 = 0$$
$$(x + 1)(x + 6) = 0$$
$$x = -1 \quad \text{or} \quad x = -6$$
But, -6 does not check, so $x = -1$ is the only solution.

47.
$$e^{1-x} = 5$$
$$\ln e^{1-x} = \ln 5$$
$$(1 - x)\ln e = \ln 5$$
$$1 - x = \ln 5$$
$$1 - \ln 5 = x$$
$$-0.609 = x$$

49.
$$2^{3x} = 3^{2x+1}$$
$$\log 2^{3x} = \log 3^{2x+1}$$
$$3x \log 2 = (2x + 1)\log 3$$
$$3x \log 2 = 2x \log 3 + \log 3$$
$$3x \log 2 - 2x \log 3 = \log 3$$
$$x(3 \log 2) - 2 \log 3 = \log 3$$
$$x = \frac{\log 3}{3 \log 2 - 2 \log 3}$$
$$x = -9.327$$

51.
$$h = (30 \cdot 0 + 8000)\log\left[\frac{760}{300}\right]$$
$$= 8000(.403692)$$
$$\approx 3229.5 \text{ meters}$$

53.
$$10{,}000 = (30 \cdot (-100) + 8000)\log\left[\frac{760}{x}\right]$$
$$10{,}000 = 5000 \log\left[\frac{760}{x}\right]$$
$$2 = \log\left[\frac{760}{x}\right]$$
$$10^2 = \frac{760}{x}$$
$$x = 7.6 \text{ mm Hg}$$

55. $P = 25e^{0.1d}$

    (a) $P = 25e^{0.1(4)}$

          $= 37.3$ watts

    (b) $50 = 25e^{0.1d}$

          $2 = e^{0.1d}$

        $\ln 2 = 0.1\,d$

           $d = 6.9$ decibels

57. $P = 90 - 80\left(\dfrac{3}{4}\right)^{t}$

    (a) $P = 90 - 80\left(\dfrac{3}{4}\right)^{5}$

          $= 71\%$

    (b) $P = 90 - 80\left(\dfrac{3}{4}\right)^{10}$

          $= 85.5\%$

    (c)   As t increases, $\left(\dfrac{3}{4}\right)^{t}$ approaches 0. Therefore, the maximum percent of purchases is 90%.

    (d) $40 = 90 - 80\left(\dfrac{3}{4}\right)^{t}$

        $-50 = -80\left(\dfrac{3}{4}\right)^{t}$

     $0.625 = 0.75t$

    $\ln 0.625 = t \ln 0.75$

           $t = \dfrac{\ln 0.625}{\ln 0.75}$

           $t \approx 1.6$ months

    (e) $70 = 90 - 80\left(\dfrac{3}{4}\right)^{t}$

        $-20 = -80(0.75)^{t}$

     $0.25 = 0.75t$

    $\ln 0.25 = t \ln 0.75$

           $t = \dfrac{\ln 0.25}{\ln 0.75}$

           $t \approx 4.8$ months

59. $n = \dfrac{\log_{10} s - \log_{10} i}{\log_{10}(1 - d)}$

    (a) $n = \dfrac{\log_{10} 10{,}000 - \log_{10} 90{,}000}{\log_{10}(1 - 0.2)}$

          $n = 9.85$ years

    (b) $n = \dfrac{\log_{10}(.5i) - \log_{10}}{\log_{10}(1 - 0.15)}$

          $n = \dfrac{\log_{10}\left[\dfrac{.5i}{i}\right]}{\log_{10}(.85)}$

          $n = \dfrac{\log_{10}(0.5)}{\log_{10}(0.85)}$

          $n = 4.27$ years

61. $\$85{,}000 = P\left[1 + \dfrac{.04}{2}\right]^{2(18)}$

        $P = \$85{,}000\left[1 + \dfrac{.04}{2}\right]^{-36}$

        $P = \$85{,}000(1.02)^{-36}$

        $P = \$41{,}668.97$

The grandparents should purchase about $41,669.

63.   Using $L(x) = 10 \log\dfrac{x}{I^{0}}$, where $I_0 = 10^{-12}$

     $L(10^{-4}) = 10 \log\dfrac{10^{-4}}{10^{-12}}$

            $= 10 \log 10^{8}$

            $= 10(8)$

            $= 80$ decibels

65. Using $A = A_0e^{kt}$ where $A = \frac{1}{2}A_0$ when $t = 5600$:

$$\frac{1}{2}A_0 = A_0e^{k(5600)}$$

$$\frac{1}{2} = e^{5600k}$$

$$\ln \frac{1}{2} = \ln e^{5600k}$$

$$-\ln 2 = 5600k$$

$$k = \frac{-\ln 2}{5600} = -0.000124$$

Thus, $A = A_0e^{-0.000124t}$
In this case, $A = 0.05A_0$:

$$0.05A_0 = A_0e^{-0.000124t}$$
$$\ln 0.05 = \ln e^{-0.000124t}$$
$$\ln 0.05 = -0.000124t$$

$$t = \frac{\ln 0.05}{-0.000124t} = 24{,}203 \text{ years ago}$$

67. (a)

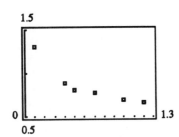

(b) The "best" model is the power model since it has the largest value for r. So,
$$y = 1.298985429x^{-0.2721702954}$$

69. (a)

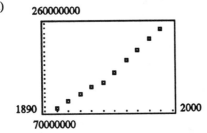

(b) The "best" model is the power model since it has the largest value for r. So,
$$P = 9.545211 \times 10^{-76} \cdot t^{25.29069649}$$

(c) $$P = 9.545211 \times 10^{-76} \cdot (1995)^{25.29069649}$$
$$= 273{,}930{,}160 \text{ people}$$

TRIGONOMETRIC FUNCTIONS

1.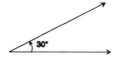

3.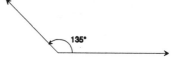

5.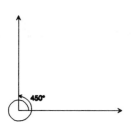

7.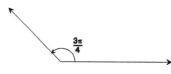

9.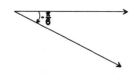

11.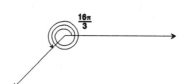

13. $30° = 30 \cdot 1 \text{ degree} = 30 \cdot \dfrac{\pi}{180} \text{ radian} = \dfrac{\pi}{6} \text{ radians}$

15. $240° = 240 \cdot 1 \text{ degree} = 240 \cdot \dfrac{\pi}{180} \text{ radian} = \dfrac{4\pi}{3} \text{ radians}$

17. $-60° = -60 \cdot 1 \text{ degree} = -60 \cdot \dfrac{\pi}{180} \text{ radian} = -\dfrac{\pi}{3} \text{ radians}$

19. $180° = 180 \cdot 1 \text{ degree} = 180 \cdot \dfrac{\pi}{180} \text{ radian} = \pi \text{ radians}$

21. $135° = 135 \cdot 1 \text{ degree} = 135 \cdot \dfrac{\pi}{180} \text{ radian} = \dfrac{3\pi}{4} \text{ radians}$

23. $\dfrac{\pi}{3} \text{ radian} = \dfrac{\pi}{3} \cdot 1 \text{ radian} = \dfrac{\pi}{3} \cdot \dfrac{180}{\pi} \text{ degrees} = 60°$

25. $\dfrac{-5\pi}{4}$ radian $= \dfrac{-5\pi}{4} \cdot 1$ radian $= \dfrac{-5\pi}{4} \cdot \dfrac{180}{\pi}$ degrees $= -225°$

27. $\dfrac{\pi}{2}$ radians $= \dfrac{\pi}{2} \cdot 1$ radian $= \dfrac{\pi}{2} \cdot \dfrac{180}{\pi}$ degrees $= 90°$

29. $\dfrac{\pi}{12}$ radian $= \dfrac{\pi}{12} \cdot 1$ radian $= \dfrac{\pi}{12} \cdot \dfrac{180}{\pi}$ degrees $= 15°$

31. $\dfrac{2\pi}{3}$ radian $= \dfrac{2\pi}{3} \cdot 1$ radian $= \dfrac{2\pi}{3} \cdot \dfrac{180}{\pi}$ degrees $= 120°$

33. $r = 10$ meters

$\theta = \dfrac{1}{2}$ radian

$s = ?$

Use $s = r\theta$, $s = 10\left[\dfrac{1}{2}\right]$

$\qquad = 5$ meters

35. $\theta = \dfrac{1}{3}$ radian

$s = 2$ feet

$\theta = ?$

Use $s = r\theta$, $2 = r\left[\dfrac{1}{3}\right]$

$\qquad 6 = r$

$\qquad r = 6$ feet

37. $r = 5$ miles

$s = 3$ miles

$\theta = ?$

Use $s = r\theta$ or $\theta = \dfrac{s}{r}$

$\theta = \dfrac{3}{5} = 0.6$ radian

39. $r = 2$ inches

$\theta = 30°$

$s = ?$

Convert $30°$ to $\dfrac{\pi}{6}$ radian.

Use $s = r\theta$

$s = 2\left[\dfrac{\pi}{6}\right] = \dfrac{\pi}{3}$ inches ≈ 1.047 inches

41. $17° = 17 \cdot 1$ degree $= 17 \cdot \dfrac{\pi}{180}$ radians $= \dfrac{17\pi}{180}$ radians $= 0.30$

43. $-40° = -40 \cdot 1$ degree $= -40 \cdot \dfrac{\pi}{180}$ radians $= \dfrac{-2\pi}{9}$ radians $= -0.70$

45. $125° = 125 \cdot 1$ degree $= 125 \cdot \dfrac{\pi}{180}$ radians $= \dfrac{25\pi}{36}$ radians $= 2.18$

47. $340° = 340 \cdot 1$ degree $= 340 \cdot \dfrac{\pi}{180}$ radians $= \dfrac{17\pi}{9}$ radians $= 5.93$

49. 3.14 radians $= 3.14 \cdot 1$ radian $= 3.14 \cdot \dfrac{180}{\pi}$ degrees $= 179.91°$

51. 10.25 radians $= 10.25 \cdot 1$ radian $= 10.25 \cdot \dfrac{180}{\pi}$ degrees $= 587.28°$

53. 2 radians $= 2 \cdot 1$ radian $= 2 \cdot \dfrac{180}{\pi}$ degrees $= 114.59°$

55. 6.32 radians = 6.32 · 1 radian = 6.32 · $\dfrac{180}{\pi}$ degrees = 362.11°

57. $40°10'25'' = \left[40 + 10 \cdot \dfrac{1}{60} + 25 \cdot \dfrac{1}{60} \cdot \dfrac{1}{60}\right]° = (40 + 0.16667 + 0.00694)° = 40.1736$

$$= 40.17°$$

59. $1°2'3'' = \left[1 + 2 \cdot \dfrac{1}{60} + 3 \cdot \dfrac{1}{60} \cdot \dfrac{1}{60}\right]° = (1 + 0.03333 + 0.00083)° = 1.03416 = 1.03°$

61. $9°9'9'' = \left[9 + 9 \cdot \dfrac{1}{60} + 9 \cdot \dfrac{1}{60} \cdot \dfrac{1}{60}\right] = (9 + 0.15 + 0.0025) = 9.15°$

63. $40.32° = ?$
$0.32° = (0.32)(1°) = (0.32)(60') = 19.2'$
$0.2' = (0.2)(1') = (0.2)(60'') = 12''$
$40.32° = 40° + 0.32° = 40° + 19.2' = 40° + 19' + 0.2' = 40° + 19' + 12'' = 40°19'12''$

65. $18.255° = ?$
$0.255° = (0.255)(1 \text{ degree}) = (0.255)(60') = 15.3'$
$0.3' = (0.3)(1') = (0.3)(60'') = 18''$
$18.255° = 18° + 0.255° = 18° + 15.3' = 18° + 15' + 0.3' = 18° + 15' + 18'' = 18°15'18''$

67. $19.99° = ?$
$0.99° = (0.99)(1 \text{ degree}) = (0.99)(60') = 59.4'$
$0.4' = (0.4)(1') = (0.4)(60') = 24''$
$19.99° = 19° + 0.99° = 19° + 59.4' = 19° + 59' + 0.4' = 19° + 59' + 24'' = 19°59'24''$

69. $s = r\theta \qquad \theta = 90° = \dfrac{\pi}{2}$ in 15 minutes

$s = (6)\left[\dfrac{\pi}{2}\right]$; $s = 3\pi$ inches ≈ 9.4248 in.

$\theta = \dfrac{25}{60} = \dfrac{5}{12} \cdot \dfrac{360}{1} = 150° = \dfrac{5}{6}\pi$ in 25 minutes

$s = 6\left[\dfrac{5\pi}{6}\right]$; $s = 5\pi$ inches ≈ 15.7080 in.

71. $r = 5$ cm $\qquad t = 20$ sec $\qquad \theta = \dfrac{1}{3}$ rad $\qquad \omega = \dfrac{\theta}{t}$

$\omega = \dfrac{\frac{1}{3}}{20} = \dfrac{1}{3} \cdot \dfrac{1}{20} = \dfrac{1}{60}$ rad/sec

$\nu = \dfrac{s}{t} \qquad$ where $s = r\theta$; $s = 5\left[\dfrac{1}{3}\right]$

$\nu = \dfrac{\frac{5}{3}}{20} = \dfrac{5}{3} \cdot \dfrac{1}{20} = \dfrac{1}{12}$ cm/sec

73. $d = 26$ inches; $\nu = 35$ mi/hr

$$\dfrac{35 \text{ mi}}{\text{hr}} \cdot \dfrac{5280 \text{ ft}}{\text{mi}} \cdot \dfrac{12 \text{ in}}{\text{ft}} \cdot \dfrac{\text{rev}}{\pi(26 \text{ in})} \cdot \dfrac{1 \text{ hr}}{60 \text{ min}} = 452.5 \text{ rpm}$$

75. $s = r\theta;\ r = 18$ inches

$$\theta = \frac{1}{3} \cdot 2 = \frac{2\pi}{3}$$

$$s = 18 \cdot \frac{2\pi}{3} = 12\pi \approx 37.7 \text{ inches}$$

77. $v = r\omega = 2.39 \times 10^5 \cdot \dfrac{1 \text{ rev}}{27.3 \text{ days}} \cdot \dfrac{1 \text{ day}}{24 \text{ hrs}} \cdot \dfrac{2\pi \text{ rad}}{\text{rev}} = 2292$ mph

79. Find distance (d) traveled by 2″ pulley in 3 revolutions.
$$d = \pi D \cdot N = \pi \cdot 2 \cdot 3 = 6\pi''$$

This distance is the same traveled by 8″ pulley. Solve for N.
$$6\pi = \pi \cdot 8 \cdot N$$
$$N = \frac{6}{8} = \frac{3}{4} \text{ rpm}$$

81. Find linear speed v using $v = r\omega$
$$v = r\omega = 4' \left[\frac{2\pi \text{ rad}}{\text{rev}} \right] \left[\frac{10 \text{ rev}}{\text{min}} \right] \left[\frac{1 \text{ mile}}{5280 \text{ ft}} \right] \left[\frac{60 \text{ min}}{\text{hr}} \right] = 2.86 \text{ mph}$$

83. The linear speed is 9.55 mi/hr. Since the diameter is 8.5 feet, the radius of the wheel is
$$\frac{8.5 \text{ feet}}{2} = 4.25 \text{ feet. Thus,}$$

$$v = r \cdot \omega$$
$$\omega = \frac{v}{r} = \frac{9.55 \text{ mi/hr}}{4.25 \text{ feet}} = \frac{50424 \text{ feet/hr}}{4.25 \text{ feet}} = 11864.47 \text{ radians/hr}$$
One revolution is 2π radians. So,
$$\omega = 11864.47 \frac{\text{radians}}{\text{hr}} \cdot \frac{1 \text{ revolution}}{2\pi} \cdot \frac{1 \text{ hr}}{60 \text{ minutes}} = 31.47 \text{ rev/min}$$

85. The earth makes one full rotation in 24 hours. The distance, s, traveled in 24 hours is the circumference of the earth. The circumference of the earth is 2π (3960 miles). Therefore, the linear velocity a person must travel to keep up with the sun is:
$$v = \frac{s}{t} = \frac{2\pi(3960 \text{ miles})}{24 \text{ hours}} \approx 1037 \text{ mph}$$

6.2 Right Triangle Trigonometry

1. opposite = 5; adjacent = 12
By the Pythagorean Theorem:
$$5^2 + 12^2 = (\text{hypotenuse})^2$$
$$25 + 144 = (\text{hypotenuse})^2$$
$$169 = (\text{hypotenuse})^2$$
$$13 = \text{hypotenuse}$$

$$\sin \theta = \frac{\text{opp}}{\text{hyp}} = \frac{5}{13} \qquad \cos \theta = \frac{\text{adj}}{\text{hyp}} = \frac{12}{13} \qquad \tan \theta = \frac{\text{opp}}{\text{adj}} = \frac{5}{12}$$

$$\csc \theta = \frac{\text{hyp}}{\text{opp}} = \frac{13}{5} \qquad \sec \theta = \frac{\text{hyp}}{\text{adj}} = \frac{13}{12} \qquad \cot \theta = \frac{\text{adj}}{\text{opp}} = \frac{12}{5}$$

3.   opposite = 2; adjacent = 3
By the Pythagorean Theorem:
$$2^2 + 3^2 = (\text{hypotenuse})^2$$
$$4 + 9 = (\text{hypotenuse})^2$$
$$13 = \text{hypotenuse}$$
$$\sqrt{13} = \text{hypotenuse}$$

$\sin \theta = \dfrac{\text{opp}}{\text{hyp}} = \dfrac{2}{\sqrt{13}} \cdot \dfrac{\sqrt{13}}{\sqrt{13}} = \dfrac{2\sqrt{13}}{13}$ $\csc \theta = \dfrac{\text{hyp}}{\text{opp}} = \dfrac{\sqrt{13}}{2}$

$\cos \theta = \dfrac{\text{adj}}{\text{hyp}} = \dfrac{3}{\sqrt{13}} \cdot \dfrac{\sqrt{13}}{\sqrt{13}} = \dfrac{3\sqrt{13}}{13}$ $\sec \theta = \dfrac{\text{hyp}}{\text{adj}} = \dfrac{\sqrt{13}}{3}$

$\tan \theta = \dfrac{\text{opp}}{\text{adj}} = \dfrac{2}{3}$ $\cot \theta = \dfrac{\text{adj}}{\text{opp}} = \dfrac{3}{2}$

5.   adjacent = 2; hypotenuse = 4
By the Pythagorean Theorem:
$$2^2 + (\text{opp})^2 = 4^2$$
$$4 + (\text{opp})^2 = 16$$
$$(\text{opp})^2 = 12$$
$$\text{opp} = \sqrt{12} = 2\sqrt{3}$$

$\sin \theta = \dfrac{\text{opp}}{\text{hyp}} = \dfrac{2\sqrt{3}}{4} = \dfrac{\sqrt{3}}{2}$ $\csc \theta = \dfrac{\text{hyp}}{\text{opp}} = \dfrac{4}{2\sqrt{3}} \cdot \dfrac{\sqrt{3}}{\sqrt{3}} = \dfrac{4\sqrt{3}}{6} = \dfrac{2\sqrt{3}}{3}$

$\cos \theta = \dfrac{\text{adj}}{\text{hyp}} = \dfrac{2}{4} = \dfrac{1}{2}$ $\sec \theta = \dfrac{\text{hyp}}{\text{adj}} = \dfrac{4}{2} = 2$

$\tan \theta = \dfrac{\text{opp}}{\text{adj}} = \dfrac{2\sqrt{3}}{2} = \sqrt{3}$ $\cot \theta = \dfrac{\text{adj}}{\text{opp}} = \dfrac{2}{2\sqrt{3}} \cdot \dfrac{\sqrt{3}}{\sqrt{3}} = \dfrac{\sqrt{3}}{3}$

7.   opposite = $\sqrt{2}$; adjacent = 1
By the Pythagorean Theorem:
$$\left(\sqrt{2}\right)^2 + 1^2 = (\text{hypotenuse})^2$$
$$2 + 1 = (\text{hypotenuse})^2$$
$$3 = (\text{hypotenuse})^2$$
$$\sqrt{3} = (\text{hypotenuse})$$

$\sin \theta = \dfrac{\text{opp}}{\text{hyp}} = \dfrac{\sqrt{2}}{\sqrt{3}} \cdot \dfrac{\sqrt{3}}{\sqrt{3}} = \dfrac{\sqrt{6}}{3}$ $\csc \theta = \dfrac{\text{hyp}}{\text{opp}} = \dfrac{\sqrt{3}}{\sqrt{2}} \cdot \dfrac{\sqrt{2}}{\sqrt{2}} = \dfrac{\sqrt{6}}{2}$

$\cos \theta = \dfrac{\text{adj}}{\text{hyp}} = \dfrac{1}{\sqrt{3}} \cdot \dfrac{\sqrt{3}}{\sqrt{3}} = \dfrac{\sqrt{3}}{3}$ $\sec \theta = \dfrac{\text{hyp}}{\text{adj}} = \dfrac{\sqrt{3}}{1} = \sqrt{3}$

$\tan \theta = \dfrac{\text{opp}}{\text{adj}} = \dfrac{\sqrt{2}}{1} = \sqrt{2}$ $\cot \theta = \dfrac{\text{adj}}{\text{opp}} = \dfrac{1}{\sqrt{2}} \cdot \dfrac{\sqrt{2}}{\sqrt{2}} = \dfrac{\sqrt{2}}{2}$

9.   opposite = 1; hypotenuse = $\sqrt{5}$
By the Pythagorean Theorem:
$$1^2 + (\text{adjacent})^2 = \left(\sqrt{5}\right)^2$$
$$1 + (\text{adjacent})^2 = 5$$
$$(\text{adjacent})^2 = 4$$
$$\text{adjacent} = 2$$

     Chapter 6   Trigonometric Functions

$$\sin \theta = \frac{\text{opp}}{\text{hyp}} = \frac{1}{\sqrt{5}} \cdot \frac{\sqrt{5}}{\sqrt{5}} = \frac{\sqrt{5}}{5} \qquad \csc \theta = \frac{\text{hyp}}{\text{opp}} = \frac{\sqrt{5}}{1} = \sqrt{5}$$

$$\cos \theta = \frac{\text{adj}}{\text{hyp}} = \frac{2}{\sqrt{5}} \cdot \frac{\sqrt{5}}{\sqrt{5}} = \frac{2\sqrt{5}}{5} \qquad \sec \theta = \frac{\text{hyp}}{\text{adj}} = \frac{\sqrt{5}}{2}$$

$$\tan \theta = \frac{\text{opp}}{\text{adj}} = \frac{1}{2} \qquad \cot \theta = \frac{\text{adj}}{\text{opp}} = \frac{2}{1} = 2$$

11. $\sin \theta = \dfrac{1}{2}, \cos \theta = \dfrac{\sqrt{3}}{2}$

$$\tan \theta = \frac{\sin \theta}{\cos \theta} = \frac{\frac{1}{2}}{\frac{\sqrt{3}}{2}} = \frac{1}{\sqrt{3}} = \frac{\sqrt{3}}{3} \qquad \csc \theta = \frac{1}{\sin \theta} = \frac{1}{\frac{1}{2}} = 2$$

$$\sec \theta = \frac{1}{\cos \theta} = \frac{1}{\frac{\sqrt{3}}{2}} = \frac{2}{\sqrt{3}} = \frac{2\sqrt{3}}{3} \qquad \cot \theta = \frac{1}{\tan \theta} = \frac{1}{\frac{\sqrt{3}}{3}} = \frac{3}{\sqrt{3}} = \sqrt{3}$$

13. $\sin \theta = \dfrac{2}{3}, \cos \theta = \dfrac{\sqrt{5}}{3}$

$$\tan \theta = \frac{\sin \theta}{\cos \theta} = \frac{\frac{2}{3}}{\frac{\sqrt{5}}{3}} = \frac{2}{\sqrt{5}} \cdot \frac{\sqrt{5}}{\sqrt{5}} = \frac{2\sqrt{5}}{5} \qquad \csc \theta = \frac{1}{\sin \theta} = \frac{1}{\frac{2}{3}} = \frac{3}{2}$$

$$\sec \theta = \frac{1}{\cos \theta} = \frac{1}{\frac{\sqrt{5}}{3}} = \frac{3}{\sqrt{5}} \cdot \frac{\sqrt{5}}{\sqrt{5}} = \frac{3\sqrt{5}}{5} \qquad \cot \theta = \frac{1}{\tan \theta} = \frac{1}{\frac{2}{\sqrt{5}}} = \frac{\sqrt{5}}{2}$$

15. $\sin \theta = \dfrac{\text{opposite}}{\text{hypotenuse}} = \dfrac{\sqrt{2}}{2}$

$(\text{adjacent})^2 = (\text{hypotenuse})^2 - (\text{opposite})^2$

$a^2 = 4 - 2$

$a^2 = 2$

$a = \sqrt{2}$

$$\cos \theta = \frac{\text{adjacent}}{\text{hypotenuse}} = \frac{\sqrt{2}}{2} \qquad \sec \theta = \frac{\text{hypotenuse}}{\text{adjacent}} = \frac{2}{\sqrt{2}} = \sqrt{2}$$

$$\tan \theta = \frac{\text{opposite}}{\text{adjacent}} = \frac{\sqrt{2}}{\sqrt{2}} = 1 \qquad \cot \theta = \frac{\text{adjacent}}{\text{opposite}} = \frac{\sqrt{2}}{\sqrt{2}} = 1$$

$$\csc \theta = \frac{\text{hypotenuse}}{\text{opposite}} = \frac{2}{\sqrt{2}} = \sqrt{2}$$

17. $\cos \theta = \dfrac{\text{adjacent}}{\text{hypotenuse}} = \dfrac{1}{3}$

$(\text{opposite})^2 = (\text{hypotenuse})^2 - (\text{adjacent})^2$

$b^2 = 9 - 1$

$b^2 = 8$

$b = 2\sqrt{2}$

$$\sec \theta = \frac{\text{hypotenuse}}{\text{adjacent}} = 3$$

$$\sin \theta = \frac{\text{opposite}}{\text{hypotenuse}} = \frac{2\sqrt{2}}{3} \qquad\qquad \csc \theta = \frac{\text{hypotenuse}}{\text{opposite}} = \frac{3}{2\sqrt{2}} = \frac{3\sqrt{2}}{4}$$

$$\tan \theta = \frac{\text{opposite}}{\text{adjacent}} = \frac{2\sqrt{2}}{1} = 2\sqrt{2} \qquad\qquad \cot \theta = \frac{\text{adjacent}}{\text{opposite}} = \frac{1}{2\sqrt{2}} = \frac{\sqrt{2}}{4}$$

19. $\tan \theta = \dfrac{\text{opposite}}{\text{adjacent}} = \dfrac{1}{2}$

$$(\text{hypotenuse})^2 = (\text{adjacent})^2 + (\text{opposite})^2$$
$$c^2 = 2^2 + 1^2$$
$$c^2 = 5$$
$$c = \sqrt{5}$$

$$\cot \theta = \frac{\text{adjacent}}{\text{opposite}} = \frac{2}{1} = 2$$

$$\sin \theta = \frac{\text{opposite}}{\text{hypotenuse}} = \frac{1}{\sqrt{5}} = \frac{\sqrt{5}}{5} \qquad\qquad \csc \theta = \frac{\text{hypotenuse}}{\text{opposite}} = \frac{\sqrt{5}}{1} = \sqrt{5}$$

$$\cos \theta = \frac{\text{adjacent}}{\text{hypotenuse}} = \frac{2}{\sqrt{5}} = \frac{2\sqrt{5}}{5} \qquad\qquad \sec \theta = \frac{\text{hypotenuse}}{\text{adjacent}} = \frac{\sqrt{5}}{2}$$

21. $\sec \theta = \dfrac{\text{hypotenuse}}{\text{adjacent}} = \dfrac{3}{1}$

$$(\text{opposite})^2 = (\text{hypotenuse})^2 - (\text{adjacent})^2$$
$$b^2 = 3^2 - 1^2$$
$$b^2 = 8$$
$$b = 2\sqrt{2}$$

$$\cos \theta = \frac{\text{adjacent}}{\text{hypotenuse}} = \frac{1}{3}$$

$$\sin \theta = \frac{\text{opposite}}{\text{hypotenuse}} = \frac{2\sqrt{2}}{3} \qquad\qquad \csc \theta = \frac{\text{hypotenuse}}{\text{opposite}} = \frac{3}{2\sqrt{2}} = \frac{3\sqrt{2}}{4}$$

$$\tan \theta = \frac{\text{opposite}}{\text{adjacent}} = \frac{2\sqrt{2}}{1} = 2\sqrt{2} \qquad\qquad \cot \theta = \frac{\text{adjacent}}{\text{opposite}} = \frac{1}{2\sqrt{2}} = \frac{\sqrt{2}}{4}$$

23. $\tan \theta = \dfrac{\text{opposite}}{\text{adjacent}} = \dfrac{\sqrt{2}}{1}$

$$(\text{hypotenuse})^2 = (\text{adjacent})^2 + (\text{opposite})^2$$
$$c^2 = 1^2 + \left(\sqrt{2}\right)^2$$
$$c^2 = 3$$
$$c = \sqrt{3}$$

$$\cot \theta = \frac{\text{adjacent}}{\text{opposite}} = \frac{1}{\sqrt{2}} = \frac{\sqrt{2}}{2}$$

$$\sin \theta = \frac{\text{opposite}}{\text{hypotenuse}} = \frac{\sqrt{2}}{\sqrt{3}} = \frac{\sqrt{6}}{3} \qquad\qquad \csc \theta = \frac{\text{hypotenuse}}{\text{opposite}} = \frac{\sqrt{3}}{\sqrt{2}} = \frac{\sqrt{6}}{2}$$

$$\cos \theta = \frac{\text{adjacent}}{\text{hypotenuse}} = \frac{1}{\sqrt{3}} = \frac{\sqrt{3}}{3} \qquad \sec \theta = \frac{\text{hypotenuse}}{\text{adjacent}} = \frac{\sqrt{3}}{1} = \sqrt{3}$$

25. $\sin^2 20° + \cos^2 20° = 1$

27. $\sin 80° \csc 80° = \sin 80° \cdot \dfrac{1}{\sin 80°} = 1$

29. $\tan 50° - \dfrac{\sin 50°}{\cos 50°} = \tan 50° - \tan 50° = 0$

31. $\sin 38° - \cos 52° = \sin 38° - \sin 38° = 0$

    $\underbrace{\qquad\qquad\qquad\qquad}_{\text{cofunctions}}$

33. $\dfrac{\cos 10°}{\sin 80°} = \dfrac{\cos 10°}{\cos 10°} = 1$

    $\underbrace{\qquad\qquad\qquad}_{\text{cofunctions}}$

35. $1 - \cos^2 20° - \cos^2 70° = (1 - \cos^2 20°) - \cos^2 70° = \sin^2 20° - \cos^2 70°$
    $= \cos^2 70° - \cos^2 70° = 0$

37. $\tan 20° - \dfrac{\cos 70°}{\cos 20°} = \tan 20° - \dfrac{\sin 20°}{\cos 20°} = \tan 20° - \tan 20° = 0$

39. $\cos 35° \sin 55° + \cos 55° \sin 35° = (\cos 35°) \cdot (\cos 35°) + (\sin 35°) \cdot (\sin 35°)$
    $= \cos^2 35° + \sin^2 35° = 1$

41. (a) $\cos 60° = \sin 30° = \dfrac{1}{2}$

    (b) $\cos^2 30° = 1 - \sin^2 30° = 1 - \left[\dfrac{1}{2}\right]^2 = 1 - \dfrac{1}{4} = \dfrac{3}{4}$

    (c) $\csc \dfrac{\pi}{6} = \csc\left[\dfrac{\pi}{6} \cdot \dfrac{180}{\pi}\right] = \csc 30° = \dfrac{1}{\sin 30°} = \dfrac{1}{\dfrac{1}{2}} = 2$

    (d) $\sec \dfrac{\pi}{3} = \sec\left[\dfrac{\pi}{3} \cdot \dfrac{180}{\pi}\right] = \sec 60° = \csc 30° = 2$

43. (a) $\sec^2 \theta = 1 + \tan^2 \theta = 1 + (4)^2 = 17$

    (b) $\cot \theta = \dfrac{1}{\tan \theta} = \dfrac{1}{4}$

    (c) $\cot\left[\dfrac{\pi}{2} - \theta\right] = \tan \theta = 4$

    (d) $\csc^2 \theta = 1 + \cot^2 \theta = 1 + \left[\dfrac{1}{4}\right]^2 = \dfrac{17}{16}$

45. $\csc \theta = 4$

    (a) $\sin \theta = \dfrac{1}{\csc \theta} = \dfrac{1}{4}$

    (b) $\cot^2 \theta = \csc^2 \theta - 1 = (4)^2 - 1 = 15$

    (c) $\sec(90° - \theta) = \csc \theta = 4$

    (d) $\sec^2 \theta = \tan^2 \theta + 1 = \dfrac{1}{\cot^2 \theta} + 1 = \dfrac{1}{15} + 1 = \dfrac{16}{15}$

47. $\sin 38° \approx 0.62$

(a) $\cos 38° = \sqrt{1 - \sin^2 38°} = \sqrt{1 - (0.62)^2} \approx \sqrt{0.6156} \approx 0.78$

(b) $\tan 38° = \dfrac{\sin 38°}{\cos 38°} \approx \dfrac{0.62}{0.79} \approx 0.78$

(c) $\cot 38° = \dfrac{\cos 38°}{\sin 38°} \approx \dfrac{0.79}{0.62} \approx 1.27$

(d) $\sec 38° = \dfrac{1}{\cos 38°} \approx \dfrac{1}{0.79} \approx 1.27$

(e) $\sec 38° = \dfrac{1}{\sin 38°} \approx \dfrac{1}{0.62} \approx 1.61$

(f) $\sin 52° = \cos 38° \approx 0.78$

(g) $\cos 52° = \sin 38° \approx 0.62$

(h) $\tan 52° = \dfrac{\sin 52°}{\cos 52°} \approx \dfrac{0.79}{0.62} \approx 1.27$

49. $\sin \theta = 0.3$

$\sin \theta + \cos\left[\dfrac{\pi}{2} - \theta\right] = \sin \theta + \sin \theta$

$= \sin \theta = 2(0.3) = 0.6$

51. $\sin \theta = \cos(2\theta + 30°)$

$\theta + 2\theta + 30° = 90°$

$3\theta = 60°$

$\theta = 20°$

53. (a) $T = \dfrac{1500 \text{ feet}}{300 \text{ ft/min}} + \dfrac{500 \text{ feet}}{100 \text{ ft/min}} = 5 \text{ min} + 5 \text{ min} = 10 \text{ minutes}$

(b) $T = \dfrac{500 \text{ feet}}{100 \text{ ft/min}} + \dfrac{1500 \text{ feet}}{100 \text{ ft/min}} = 5 \text{ min} + 15 \text{ min} = 20 \text{ minutes}$

(c) Since $\tan \theta = \dfrac{\text{opposite}}{\text{adjacent}}$, we have $\tan \theta = \dfrac{500}{x}$; therefore, $x = \dfrac{500}{\tan \theta}$.

Since $\sin \theta = \dfrac{\text{opposite}}{\text{hypotenuse}}$, we have $\sin \theta = \dfrac{500}{\text{length of trip in the sand}}$; therefore, length of

trip in the sand $= \dfrac{500}{\sin \theta}$. Thus, the time of the trip is:

$$T = \dfrac{1500 - x}{300} + \dfrac{500/\sin \theta}{100} = \dfrac{1500 - \dfrac{500}{\tan \theta}}{300} + \dfrac{500}{100 \sin \theta}$$

$$T(\theta) = 5 - \dfrac{5}{3 \tan \theta} + \dfrac{5}{\sin \theta} = 5\left[1 - \dfrac{1}{3 \tan \theta} + \dfrac{1}{\sin \theta}\right]$$

(d) Since $\tan \theta = \dfrac{500}{1500} = \dfrac{1}{3}$, then $\sin \theta = \dfrac{1}{\sqrt{1 + 9}} = \dfrac{1}{\sqrt{10}}$. Thus,

$$T = 5\left[1 - \dfrac{1}{3\left[\dfrac{1}{3}\right]} + \dfrac{1}{\dfrac{1}{\sqrt{10}}}\right] = 5\left(\sqrt{10}\right) \approx 15.8 \text{ min.}$$

(e) From the figure, we see $\tan \theta = \dfrac{500}{500} = 1$ and $\sin \theta = \dfrac{1}{\sqrt{2}}$.

Thus, $T = 5\left[1 - \dfrac{1}{3 \cdot 1} + \dfrac{1}{\dfrac{1}{\sqrt{2}}}\right] = 5\left[1 - \dfrac{1}{3} + \sqrt{2}\right]$

$\approx 10.4 \text{ minutes}$

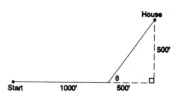

(f)

The time, T, is least when $\theta \approx 70.52°$ correct to two decimal places. To find x, we solve the following equation: $\tan 70.52 = \dfrac{500}{x}$

Thus, $x = \dfrac{500}{\tan 70.52} \approx 176.9$ feet.

55.   (a) $|OA| = |OC| = 1$; Angle $OAC = $ Angle OCA.

Thus, Angle $OAC + $ Angle $OAC + 180° - \theta = 180°$

$2($Angle $OAC) = \theta$

Angle $OAC = \dfrac{\theta}{2}$

(b) $\sin \theta = \dfrac{|CD|}{|OC|} = |CD|$, since $|OC| = 1$

$\cos \theta = \dfrac{|OD|}{|OC|} = |OD|$

(c) $\tan \dfrac{\theta}{2} = \dfrac{|CD|}{|AD|}$, since angle $OAC = \dfrac{\theta}{2}$ by part (a)

$= \dfrac{\sin \theta}{1 + |OD|}$, since $|CD| = \sin \theta$ by part (b), and

$|AD| = |AO| + |OD| = 1 + |OD|$, since $|AO| = 1$

$= \dfrac{\sin \theta}{1 + \cos \theta}$, since $|OD| = \cos \theta$ by part (b)

57. $h = x \cdot \dfrac{h}{x} = x \tan \theta$ and $h = (1 - x)\dfrac{h}{1 - x} = (1 - x) \tan n\theta$

Thus, $x \tan \theta = (1 - x) \tan n\theta$

$x \tan \theta = \tan n\theta - x \tan n\theta$

$x(\tan \theta + \tan n\theta) = \tan n\theta$

$x = \dfrac{\tan n\theta}{\tan \theta + \tan n\theta}$

59.   (a)   Area $\triangle OAC = \dfrac{1}{2}|OC|\ |AC| = \dfrac{1}{2}\dfrac{|OC|}{1} \cdot \dfrac{|AC|}{1} = \dfrac{1}{2}\sin \alpha \cos \alpha$

(b)   Area $\triangle OCB = \dfrac{1}{2}|OC|\ |BC| = \dfrac{1}{2}|OB|^2\dfrac{|BC|}{|OB|} \cdot \dfrac{|OC|}{|OB|} = \dfrac{1}{2}|OB|^2 \sin \beta \cos \beta$

(c)   Area $\triangle OAB = \dfrac{1}{2}|BD|\ |OA| = \dfrac{1}{2}|BD| \cdot 1 = \dfrac{1}{2}|OB|\dfrac{|BD|}{|OB|} = \dfrac{1}{2}|OB| \sin(\alpha + \beta)$

(d) $\dfrac{\cos \alpha}{\cos \beta} = \dfrac{\dfrac{|OC|}{|OA|}}{\dfrac{|OC|}{|OB|}} = \dfrac{|OC|}{1} \cdot \dfrac{|OB|}{|OC|} = |OB|$

(e)   Area $\triangle OAB$ = Area $\triangle OAC$ + Area $\triangle OCB$

$$\frac{1}{2} \mid OB \mid \sin(\alpha + \beta) = \frac{1}{2} \sin \alpha (\cos \alpha + \frac{1}{2} \mid OB \mid^2 \sin \beta \cos \beta$$

$$\frac{\cos \alpha}{\cos \beta} \sin(\alpha + \beta) = \sin \alpha \cos \alpha + \frac{\cos^2 \alpha}{\cos^2 \beta} \sin \beta \cos \beta$$

$$\sin(\alpha + \beta) = \frac{\cos \beta}{\cos \alpha} (\sin \alpha \cos \alpha) + \frac{\cos \alpha}{\cos \beta} (\sin \beta \cos \beta)$$

$$\sin(\alpha + \beta) = \cos \beta \sin \alpha + \cos \alpha \sin \beta$$

61. $\sin \alpha = \dfrac{\sin \alpha}{\cos \alpha} \cos \alpha = \tan \alpha \cos \alpha = \cos \beta \cos \alpha = \cos \beta \tan \beta = \cos \beta \cdot \dfrac{\sin \beta}{\cos \beta} = \sin \beta$

$\sin^2 \alpha + \cos^2 \alpha = 1$, thus

$\sin^2 \alpha + \tan^2 \beta = 1$

$$\sin^2 \alpha + \frac{\sin^2 \beta}{\cos^2 \beta} = 1$$

$$\sin^2 \alpha + \frac{\sin^2 \alpha}{1 - \sin^2 \alpha} = 1$$

$\sin^2 \alpha - \sin^4 \alpha + \sin^2 \alpha = 1 - \sin^2 \alpha$

$\sin^4 \alpha - 3 \sin^2 \alpha + 1 = 0$

$$\sin^2 \alpha = \frac{3 \pm \sqrt{5}}{2}$$

$$\sin^2 \alpha = \frac{3 - \sqrt{5}}{2}$$

$$\sin \alpha = \sqrt{\frac{3 - \sqrt{5}}{2}}$$

63.   We are given that $0 < \theta < 90°$.

Since $a^2 + b^2 = c^2$, $a > 0$, $b > 0$, then $0 < a^2 < c^2$ or $0 < a < c$.

Thus, $0 < \dfrac{a}{c} < 1$ and $0 < \sin \theta < 1$.

6.3   Computing the Values of Trigonometric Functions of Given Angles

1. $\sin 45° = \dfrac{\sqrt{2}}{2}$ $\cos 45° = \dfrac{\sqrt{2}}{2}$ $\tan 45° = 1$

$\csc 45° = \sqrt{2}$ $\sec 45° = \sqrt{2}$ $\cot 45° = 1$

3. $\sin 60° = \dfrac{\sqrt{3}}{2}$ $\cos 60° = \dfrac{1}{2}$ $\tan 60° = \sqrt{3}$

$\csc 60° = \dfrac{2\sqrt{3}}{3}$ $\sec 60° = 2$ $\cot 60° = \dfrac{\sqrt{3}}{3}$

5. $6 \tan 45° - 8 \cos 60° = 6(1) - 8\left[\dfrac{1}{2}\right] = 6 - 4 = 2$

7. $\sec \dfrac{\pi}{4} + 2 \csc \dfrac{\pi}{3} = \sqrt{2} + 2\left[\dfrac{2}{\sqrt{3}}\right] = \sqrt{2} + \dfrac{4}{\sqrt{3}} \cdot \dfrac{\sqrt{3}}{\sqrt{3}} = \sqrt{2} + \dfrac{4\sqrt{3}}{3}$

9. $\sec^2 \dfrac{\pi}{6} - 4 = \left[\dfrac{2\sqrt{3}}{3}\right]^2 - 4$

$= \dfrac{12}{9} - 4 = \dfrac{4 - 12}{3} = \dfrac{-8}{3}$

11. $\sin^2 40° + \cos^2 40° = 1$

13. $1 - \cos^2 20° - \cos^2 70°$
$= 1 - (\cos^2 20° + \sin^2 20°)$
$= 1 - 1 = 0$

15. $\sin \theta = \sin 60° = \dfrac{\sqrt{3}}{2}$

17. $\sin \dfrac{\theta}{2} = \sin \dfrac{60°}{2} = \sin 30° = \dfrac{1}{2}$

19. $(\sin \theta)^2 = (\sin 60°)^2 = \left[\dfrac{\sqrt{3}}{2}\right]^2 = \dfrac{3}{4}$

21. $2 \sin \theta = 2 \sin 60° = 2\left[\dfrac{\sqrt{3}}{2}\right] = \sqrt{3}$

23. $\dfrac{\sin \theta}{2} = \dfrac{\sin 60°}{2} = \dfrac{\frac{\sqrt{3}}{2}}{2} = \dfrac{\sqrt{3}}{4}$

25. $\sin 28° \approx 0.4694716 \approx 0.47$

27. $\tan 21° \approx 0.383864 \approx 0.38$

29. $\sec 41° = \dfrac{1}{\cos 41°} \approx 1.33$

31. $\cot 70° = \dfrac{1}{\tan 70°} \approx 0.3639702 \approx 0.36$

33. Set the mode to receive radians.
$\sin \dfrac{\pi}{10} \approx 0.309017 \approx 0.31$

35. Set the mode to receive radians.
$\tan \dfrac{5\pi}{12} \approx 3.7320508 \approx 3.73$

37. Set the mode to receive radians.
$\sec \dfrac{\pi}{12} = \dfrac{1}{\cos(\pi/12)} \approx 1.0352762 \approx 1.04$

39. Set the mode to receive radians.
$\cot \dfrac{\pi}{18} = \dfrac{1}{\tan(\pi/18)} \approx 5.6712819 \approx 5.67$

41. Set the mode to receive radians.
$\sin 1 \approx 0.841471 \approx 0.84$

43. Use the degree mode.
$\sin 1° \approx 0.0174524 \approx 0.02$

45. Use the degree mode.
$\cos 21.5° \approx 0.9304176 \approx 0.93$

47. Set the mode to receive radians.
$\tan 0.3 \approx 0.3093362 \approx 0.31$

49. Using the formula $R = \dfrac{v_0^2 \sin 2\theta}{g}$ and $g \approx 32.2$ ft/sec^2, and given $\theta = 45°$ and $v_0 = 100$ ft/sec, we

get: $R = \dfrac{(100)^2 \sin 2(45°)}{32.2}$

$R = \dfrac{(10,000)(\sin 90°)}{32.2}$, using calculator

$R = \dfrac{(10,000)(1)}{32.2} \approx 310.559$

$R \approx 310.56$ feet

Using the formula $H = \dfrac{v_0^2 \sin^2 \theta}{2g}$ and $g \approx 32.2$ ft/sec^2, and given $\theta = 45°$ and $v_0 = 100$ ft/sec, we

get: $\quad H = \dfrac{(100)^2(\sin 45°)^2}{2(32.2)}$, using calculator or table sin 45° = 0.7071

$\quad H = \dfrac{(10{,}000)(0.7071)^2}{(64.4)}$, using calculator

$\quad H \approx 77.638262$

$\quad H \approx 77.64$ feet

51. Using the formula $R = \dfrac{v_0^2 \sin 2\theta}{g}$ and $g \approx 9.8$ m/sec^2, and given $\theta = 25°$ and $v_0 = 500$ m/sec, we

get: $R = \dfrac{(500)^2 \sin 2(25°)}{9.8}$

$\quad R = \dfrac{(250{,}000)(\sin 50°)}{9.8}$

$\quad R = \dfrac{(250{,}000)(0.7660444)}{9.8}$

$\quad R \approx 19{,}541.95$

$\quad R \approx 19{,}542$ meters

Using the formula $H = \dfrac{v_0^2 \sin^2 \theta}{2g}$ and $g \approx 9.8$ m/sec^2, and given $\theta = 25°$ and $v_0 = 500$ m/sec, we

get: $H = \dfrac{(500)^2 \sin^2(25°)}{2(9.8)}$

$\quad H \approx 2278.1403$

$\quad H \approx 2278$ meters

53. We use the formula, $t = \sqrt{\dfrac{2a}{g \sin \theta \cos \theta}}$, where $g \approx 32$ ft/sec/sec and $a = 10$. Then:

(a) $t = \sqrt{\dfrac{20}{32 \sin 30° \cos 30°}} = \sqrt{\dfrac{20}{32\left(\dfrac{1}{2}\right)\left(\dfrac{\sqrt{3}}{2}\right)}} = \sqrt{\dfrac{20}{8\sqrt{3}}} = \sqrt{\dfrac{5}{2\sqrt{3}}} \approx 1.2$ sec

(b) $t = \sqrt{\dfrac{20}{32 \sin 45° \cos 45°}} = \sqrt{\dfrac{20}{32\left(\dfrac{\sqrt{2}}{2}\right)\left(\dfrac{\sqrt{2}}{2}\right)}} = \sqrt{\dfrac{20}{16}} = \sqrt{\dfrac{5}{4}} \approx 1.12$ sec

(c) $t = \sqrt{\dfrac{20}{32 \sin 60° \cos 60°}} = \sqrt{\dfrac{20}{32\left(\dfrac{\sqrt{3}}{2}\right)\left(\dfrac{1}{2}\right)}} = \sqrt{\dfrac{5}{2\sqrt{3}}} \approx 1.2$ sec

55. (a) $T(\theta) = \dfrac{\dfrac{1}{\sin \theta}}{3} + \dfrac{8 - \dfrac{2}{\tan \theta}}{8} + \dfrac{\dfrac{1}{\sin \theta}}{3} = \dfrac{1}{3 \sin \theta} + 1 - \dfrac{1}{4 \tan \theta} + \dfrac{1}{3 \sin \theta}$

$\quad = \dfrac{2}{3 \sin \theta} + 1 - \dfrac{1}{4 \tan \theta} = 1 + \dfrac{2}{3 \sin \theta} - \dfrac{1}{4 \tan \theta}$

(b) $T(30°) = 1 + \dfrac{2}{3 \sin 30°} - \dfrac{1}{4 \tan 30°} = 1 + \dfrac{2}{3\left(\dfrac{1}{2}\right)} - \dfrac{1}{4\left(\dfrac{1}{\sqrt{3}}\right)}$

$\qquad = 1 + \dfrac{4}{3} - \dfrac{\sqrt{3}}{4} \approx 1.9$ hours

The time Sally is on the paved road is given by $P(\theta) = \dfrac{8 - \dfrac{2}{\tan \theta}}{8}$.

Thus, $P(30°) = \dfrac{8 - \dfrac{2}{\tan 30°}}{8} \approx 0.57$ hours.

(c) $T(45°) = 1 + \dfrac{2}{3 \sin 45°} - \dfrac{1}{4 \tan 45°} = 1 + \dfrac{2}{3\left(\dfrac{1}{\sqrt{2}}\right)} - \dfrac{1}{4(1)}$

$\qquad = 1 + \dfrac{2\sqrt{2}}{3} - \dfrac{1}{4} \approx 1.69$ hours

$P(45°) = \dfrac{8 - \dfrac{2}{\tan 45°}}{8} \approx 0.75$ hours

(d) $T(60°) = 1 + \dfrac{2}{3 \sin 60°} - \dfrac{1}{4 \tan 60°} = 1 + \dfrac{2}{3\left(\dfrac{\sqrt{3}}{2}\right)} - \dfrac{1}{4\left(\dfrac{\sqrt{3}}{1}\right)}$

$\qquad = 1 + \dfrac{4}{3\sqrt{3}} - \dfrac{1}{4\sqrt{3}} \approx 1.63$ hours

$P(60°) = \dfrac{8 - \dfrac{2}{\tan 60°}}{8} \approx 0.86$ hours

(e) $T(90°) = 1 + \dfrac{2}{3 \sin 90°} - \dfrac{1}{4 \tan 90°} = 1 + \dfrac{2}{3(1)} - \dfrac{\cos \dfrac{\pi}{2}}{4 \sin \dfrac{\pi}{2}} = 1 + \dfrac{2}{3} \approx 1.67$ hours

Sally walks directly to the path from the house, takes the path, then walks directly to the house from the path.

(f) If $\tan \theta = \dfrac{1}{4}$, then $\sin \theta = \dfrac{1}{\sqrt{17}}$.

$T(\theta) = 1 + \dfrac{2}{3\left(\dfrac{1}{\sqrt{17}}\right)} - \dfrac{1}{4\left(\dfrac{1}{4}\right)} = 1 + \dfrac{2\sqrt{17}}{3} - 1 = \dfrac{2\sqrt{17}}{3} \approx 2.75$ hours

(g)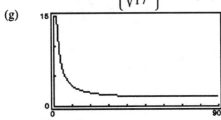

The angle, θ, which results in the least time is 67.97° correct to two decimal places. The least time is 1.61 hours correct to two decimal places. Sally is on the road for 0.9 hours.

57.

θ	0.5	0.4	0.2	0.1	0.01	0.001	0.0001	0.00001
$\sin \theta$	0.4794	0.3894	0.1987	0.0998	0.0100	0.0010	0.0001	0.00001
$\dfrac{\sin \theta}{\theta}$	0.9589	0.9735	0.9933	0.9983	1.0000	1.0000	1.0000	1.0000

$\dfrac{\sin \theta}{\theta}$ approaches 1 as θ approaches 0

59. $\tan 1° \cdot \tan 2° \cdot \tan 3° \cdot \cdots \cdot \tan 89°$
$$= (\tan 1° \cdot \tan 2° \cdot \cdots \cdot \tan 45°) \cdot \left(\cot(90° - 46°) \cdot \cot(90° - 47°) \cdot \cdots\right.$$
$$\left. \cdot \cot(90° - 89°)\right)(\tan 1° \cdot \cot 1°) \cdot (\tan 2° \cot 2°) \cdot \cdots \cdot (\tan 44° \cdot \cot 44°) \cdot \tan 45°$$
$$= 1 \cdot 1 \cdot \cdots \cdot 1 \cdot 1 = 1$$

61. $\cos 1° \cdot \cos 2° \cdot \cdots \cdot \cos 45° \cdot \csc 46° \cdot \cdots \cdot \csc 89°$
$$= (\sin 90° - 1°) \cdot \sin(90° - 2°) \cdot \cdots \cdot \sin(90° - 45°) \cdot \csc 46° \cdot \cdots \cdot \csc 89°$$
$$= (\sin 89° \cdot \csc 89°) \cdot (\sin 88° \cdot \csc 88°) \cdot \cdots \cdot (\sin 46° \cdot \csc 46°) \cdot (\sin 45°)$$
$$= 1 \cdot 1 \cdot \cdots \cdot 1 \cdot \frac{\sqrt{2}}{2} = \frac{\sqrt{2}}{2}$$

6.4 Trigonometric Functions of General Angles

1. $(-3, 4)$

For $(a, b) = (-3, 4)$, we find $a = -3$, $b = 4$, and $r = \sqrt{a^2 + b^2} = \sqrt{9 + 16} = \sqrt{25} = 5$.

Thus,
$$\sin \theta = \frac{b}{r} = \frac{4}{5} \qquad\qquad \csc \theta = \frac{r}{b} = \frac{5}{4}$$
$$\cos \theta = \frac{a}{r} = -\frac{3}{5} \qquad\qquad \sec \theta = \frac{r}{a} = -\frac{5}{3}$$
$$\tan \theta = \frac{b}{a} = -\frac{4}{3} \qquad\qquad \cot \theta = \frac{a}{b} = -\frac{3}{4}$$

3. $(2, -3)$

For $(a, b) = (2, -3)$, we find $a = 2$, $b = -3$, and $r = \sqrt{a^2 + b^2} = \sqrt{4 + 9} = \sqrt{13}$.

Thus,
$$\sin \theta = \frac{b}{r} = \frac{-3}{\sqrt{13}} \cdot \frac{\sqrt{13}}{\sqrt{13}} = -\frac{3\sqrt{13}}{13} \qquad\qquad \csc \theta = \frac{r}{b} = -\frac{\sqrt{13}}{3}$$
$$\cos \theta = \frac{a}{r} = \frac{2}{\sqrt{13}} \cdot \frac{\sqrt{13}}{\sqrt{13}} = \frac{2\sqrt{13}}{13} \qquad\qquad \sec \theta = \frac{r}{a} = \frac{\sqrt{13}}{2}$$
$$\tan \theta = \frac{b}{a} = -\frac{3}{2} \qquad\qquad\qquad\qquad \cot \theta = \frac{a}{b} = -\frac{2}{3}$$

5. $(-3, -3)$

For $(a, b) = (-3, -3)$, we find $a = -3$, $b = -3$, and $r = \sqrt{a^2 + b^2} = \sqrt{9 + 9} = \sqrt{18} = 3\sqrt{2}$.

Thus,
$$\sin\theta = \frac{b}{r} = \frac{-3}{3\sqrt{2}} = -\frac{1}{\sqrt{2}} \cdot \frac{\sqrt{2}}{\sqrt{2}} = -\frac{\sqrt{2}}{2} \qquad\qquad \csc\theta = \frac{r}{b} = \frac{3\sqrt{2}}{-3} = -\sqrt{2}$$

$$\cos\theta = \frac{a}{r} = \frac{-3}{3\sqrt{2}} = -\frac{1}{\sqrt{2}} \cdot \frac{\sqrt{2}}{\sqrt{2}} = -\frac{\sqrt{2}}{2} \qquad\qquad \sec\theta = \frac{r}{a} = \frac{3\sqrt{2}}{-3} = -\sqrt{2}$$

$$\tan\theta = \frac{b}{a} = \frac{-3}{-3} = 1 \qquad\qquad\qquad\qquad\qquad \cot\theta = \frac{a}{b} = \frac{-3}{-3} = 1$$

7. $(-3, -2)$

For $(a, b) = (-3, -2)$, we find $a = -3$, $b = -2$, and $r = \sqrt{a^2 + b^2} = \sqrt{9 + 4} = \sqrt{13}$.

Thus, $\sin\theta = \dfrac{b}{r} = \dfrac{-2}{\sqrt{13}} = -\dfrac{2}{\sqrt{13}} \cdot \dfrac{\sqrt{13}}{\sqrt{13}} = -\dfrac{2\sqrt{13}}{13} \qquad \csc\theta = \dfrac{r}{b} = \dfrac{\sqrt{13}}{-2} = -\dfrac{\sqrt{13}}{2}$

$\cos\theta = \dfrac{a}{r} = \dfrac{-3}{\sqrt{13}} = -\dfrac{3}{\sqrt{13}} \cdot \dfrac{\sqrt{13}}{\sqrt{13}} = -\dfrac{3\sqrt{13}}{13} \qquad \sec\theta = \dfrac{r}{a} = \dfrac{\sqrt{13}}{-3} = -\dfrac{\sqrt{13}}{3}$

$\tan\theta = \dfrac{b}{a} = \dfrac{-2}{-3} = \dfrac{2}{3} \qquad\qquad\qquad\qquad\qquad \cot\theta = \dfrac{a}{b} = \dfrac{-3}{-2} = \dfrac{3}{2}$

9. Using Table 5, we find that $\sin\theta > 0$ in quadrants I and II, and $\cos\theta < 0$ in quadrants II and III. Both conditions are satisfied only if θ lies in quadrant II.

11. Since $\sin\theta < 0$ in quadrants III and IV, and $\tan\theta < 0$ in quadrants II and IV, θ lies in quadrant IV.

13. Since $\cos\theta > 0$ in quadrants I and IV, and $\cot\theta < 0$ in quadrants II and IV, θ lies in quadrant IV.

15. Since $\sec\theta < 0$ in quadrants II and III, and $\tan\theta > 0$ in quadrants I and III, θ lies in quadrant III.

17. The reference angle for $-30°$ is $0° - (-30°) = 30°$.

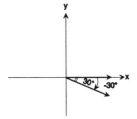

19. Let α represent the reference angle.
$$120° + \alpha = 180°$$
$$\alpha = 60°$$

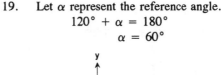

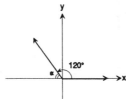

21. $180° + \alpha = 210°$
$\alpha = 30°$

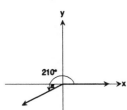

23. Remember that the reference angle is the acute angle formed by the terminal side of θ and the positive or negative x-axis.

Hence, $\pi + \alpha = \dfrac{5\pi}{4}$

$\alpha = \dfrac{5\pi}{4} - \pi = \dfrac{\pi}{4}$

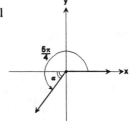

25. $\dfrac{8\pi}{3} + \alpha = 3\pi$

$$\alpha = 3\pi - \dfrac{8\pi}{3}$$
$$= \dfrac{9\pi}{3} - \dfrac{8\pi}{3}$$
$$\alpha = \dfrac{\pi}{3}$$

27. $180° - 135° = \alpha$
$\qquad 45° = \alpha$

29. $\pi - \dfrac{2\pi}{3} = \alpha$

$$\dfrac{\pi}{3} = \alpha$$

31. $440° - 360° = \alpha$
$\qquad 80° = \alpha$

33. $\sin 405° = \sin(360° + 45°) = \sin 45° = \dfrac{\sqrt{2}}{2}$

35. $\tan 405° = \tan(180° + 180° + 45°) = \tan 45° = 1$

37. $\csc 450° = \csc(360° + 90°) = \csc 90° = 1$

39. $\cot 390° = \cot(180° + 180° + 30°) = \cot 30° = \sqrt{3}$

41. $\cos\dfrac{33\pi}{4} = \cos\left(\dfrac{\pi}{4} + \dfrac{32\pi}{4}\right) = \cos\left(\dfrac{\pi}{4} + 8\pi\right) = \cos\left(\dfrac{\pi}{4} + 2\pi \cdot 4\right) = \cos\dfrac{\pi}{4} = \dfrac{\sqrt{2}}{2}$

43. $\tan 21\pi = \tan(\pi + 20\pi) = \tan(\pi + \pi \cdot 20) = \tan \pi = 0$

45. $\sec \dfrac{17\pi}{4} = \left[\dfrac{\pi}{4} + \dfrac{16\pi}{4}\right] = \sec\left(\dfrac{\pi}{4} + 4\pi\right) = \sec\left(\dfrac{\pi}{4} + 2\pi \cdot 2\right) = \sec\dfrac{\pi}{4} = \sqrt{2}$

47. $\tan \dfrac{19\pi}{6} = \tan\left(\dfrac{\pi}{6} + \dfrac{18\pi}{6}\right) = \tan\left(\dfrac{\pi}{6} + 3\pi\right) = \tan\left(\dfrac{\pi}{6} + \pi \cdot 3\right) = \tan\dfrac{\pi}{6} = \dfrac{\sqrt{3}}{3}$

49. $\sin 150°$
The angle 150° is in quadrant II, where the $\sin \theta$ is positive. The reference angle for 150° is 30°.

Thus, $\sin 150° = \sin 30° = \dfrac{1}{2}$.

51. $\cos 315°$
The angle 315° is in quadrant IV where the $\cos \theta$ is positive. The reference angle for 315° is 45°.

Thus, $\cos 315° = \cos 45° = \dfrac{\sqrt{2}}{2}$.

53. sec 240°

The angle 240° is in quadrant III where sec θ is negative. The reference angle for 240° is 60°. Thus, sec 240° = $-$sec 60° = -2.

55. cot 330°

The angle 330° is in quadrant IV where cot θ is negative. The reference angle for 330° is 30°. Thus, cot 330° = $-$cot 30°

$$= -\frac{\sqrt{3}}{1} = -\sqrt{3}.$$

57. $\sin \frac{3\pi}{4}$

The angle $\frac{3\pi}{4}$ is in quadrant II where sin is positive. The reference angle for $\frac{3\pi}{4}$ is $\frac{\pi}{4}$.

Thus, $\sin \frac{3\pi}{4} = \sin \frac{\pi}{4} = \frac{\sqrt{2}}{2}$.

59. $\cot \frac{7\pi}{6}$

The angle $\frac{7\pi}{6}$ is in quadrant III where cot θ is positive. The reference angle for $\frac{7\pi}{6}$ is $\frac{\pi}{6}$. Thus, $\cot \frac{7\pi}{6} = \cot \frac{\pi}{6} = \sqrt{3}$.

61. $\cos(-60°)$

The angle $(-60°)$ is in quadrant IV where $\cos \theta$ is positive. The reference angle for $-60°$ is 60°.

Thus, $\cos(-60°) = \cos 60° = \frac{1}{2}$.

63. $\sin\left[-\frac{2\pi}{3}\right]$

The angle $-\frac{2\pi}{3}$ is in quadrant III where $\sin \theta$ is negative. The reference angle for

$\left[-\frac{2\pi}{3}\right]$ is $\frac{\pi}{3}$.

Thus, $\sin\left[-\frac{2\pi}{3}\right] = -\sin \frac{\pi}{3} = -\frac{\sqrt{3}}{2}$.

65. $\tan \frac{14\pi}{3}$

$\tan \frac{14\pi}{3} = \tan\left[\frac{12\pi}{3} + \frac{2\pi}{3}\right]$

$= \tan\left[4\pi + \frac{2\pi}{3}\right]$

The angle $\frac{14\pi}{3}$ is in quadrant II where tan θ is negative. The reference angle for $\frac{14\pi}{3}$ is

$\frac{\pi}{3}$. Thus, $\tan \frac{14\pi}{3} = -\tan \frac{\pi}{3} = -\sqrt{3}$.

67. $\csc(-315°)$

The angle $-315°$ is in quadrant I where csc θ is positive. The reference angle for $-315°$ is 45°.

Thus, $\csc(-315°) = \csc 45° = \sqrt{2}$.

69. $\sin \theta = \frac{12}{13}$, $90° < \theta < 180°$

First, we solve for $\cos \theta$:

$\sin^2 \theta + \cos^2 \theta = 1$

$\cos^2 \theta = 1 - \sin^2 \theta$

$\cos \theta = \pm\sqrt{1 - \sin^2 \theta}$

Because, $90° < \theta < 180°$, $\cos \theta < 0$.

$\cos \theta = -\sqrt{1 - \sin^2 \theta} = -\sqrt{1 - \frac{144}{169}} = -\sqrt{\frac{25}{169}} = \frac{-5}{13}$

$$\tan \theta = \frac{\sin \theta}{\cos \theta} = \frac{\frac{12}{13}}{\frac{-5}{13}} = \frac{-12}{5} \qquad \csc \theta = \frac{1}{\sin \theta} = \frac{1}{\frac{12}{13}} = \frac{13}{12}$$

$$\sec \theta = \frac{1}{\cos \theta} = \frac{1}{\frac{-5}{13}} = \frac{-13}{5} \qquad \cot \theta = \frac{1}{\tan \theta} = \frac{1}{\frac{-12}{5}} = \frac{-5}{12}$$

71. $\cos \theta = \dfrac{-4}{5}$, $\pi < \theta < \dfrac{3\pi}{2}$

$\sin^2 \theta = 1 - \cos^2 \theta$

$\sin \theta = \pm\sqrt{1 - \cos^2 \theta}$

Because $\pi < \theta < \dfrac{3\pi}{2}$, $\sin \theta < 0$.

$$\sin \theta = -\sqrt{1 - \cos^2 \theta} = -\sqrt{1 - \frac{16}{25}} = -\sqrt{\frac{9}{25}} = \frac{-3}{5}$$

$$\tan \theta = \frac{\sin \theta}{\cos \theta} = \frac{\frac{-3}{5}}{\frac{-4}{5}} = \frac{3}{4} \qquad \cot \theta = \frac{1}{\tan \theta} = \frac{1}{\frac{3}{4}} = \frac{4}{3}$$

$$\csc \theta = \frac{1}{\sin \theta} = \frac{1}{\frac{-3}{5}} = \frac{-5}{3} \qquad \sec \theta = \frac{1}{\cos \theta} = \frac{1}{\frac{-4}{5}} = \frac{-5}{4}$$

73. $\sin \theta = \dfrac{5}{13}$, $\cos \theta < 0$

First, we solve for $\cos \theta$: $\quad \sin^2 \theta + \cos^2 \theta = 1$

$$\cos^2 \theta = 1 - \sin^2 \theta$$

$$\cos \theta = \pm\sqrt{1 - \sin^2 \theta}$$

Because $\cos \theta < 0$, we use the minus sign:

$$\cos \theta = -\sqrt{1 - \sin^2 \theta} = -\sqrt{1 - \left(\frac{5}{13}\right)^2} = -\sqrt{1 - \frac{25}{169}} = -\sqrt{\frac{144}{169}} = \frac{-12}{13}$$

$$\tan \theta = \frac{\sin \theta}{\cos \theta} = \frac{\frac{5}{13}}{\frac{-12}{13}} = \frac{-5}{12} \qquad \csc \theta = \frac{1}{\sin \theta} = \frac{1}{\frac{5}{13}} = \frac{13}{5}$$

$$\sec \theta = \frac{1}{\cos \theta} = \frac{1}{\frac{-12}{13}} = \frac{-13}{12} \qquad \cot \theta = \frac{1}{\tan \theta} = \frac{1}{\frac{-5}{12}} = \frac{-12}{5}$$

75. $\cos \theta = \dfrac{-1}{3}$, $\csc \theta > 0$

$\sin^2 \theta = 1 - \cos^2 \theta$

$\sin \theta = \pm\sqrt{1 - \cos^2 \theta}$

Because $\csc \theta > 0$, and $\sin \theta = \dfrac{1}{\csc \theta}$, it follows that $\sin \theta > 0$.

$$\sin\theta = \sqrt{1-\cos^2\theta} = \sqrt{1-\left[\frac{-1}{3}\right]^2} = \sqrt{1-\frac{1}{9}} = \sqrt{\frac{8}{9}} = \frac{2\sqrt{2}}{3}$$

$$\tan\theta = \frac{\sin\theta}{\cos\theta} = \frac{\frac{2\sqrt{2}}{3}}{\frac{-1}{3}} = -2\sqrt{2} \qquad \cot\theta = \frac{1}{\tan\theta} = \frac{1}{-2\sqrt{2}} = \frac{-\sqrt{2}}{4}$$

$$\csc\theta = \frac{1}{\sin\theta} = \frac{1}{\frac{2\sqrt{2}}{3}} = \frac{3\sqrt{2}}{4} \qquad \sec\theta = \frac{1}{\cos\theta} = \frac{1}{\frac{-1}{3}} = -3$$

77. $\sin\theta = \dfrac{2}{3}$, $\tan\theta < 0$

$\cos^2\theta = 1 - \sin^2\theta$

Because $\tan\theta = \dfrac{\sin\theta}{\cos\theta} < 0$ and $\sin\theta > 0$, it follows that $\cos\theta < 0$. Therefore, we use the minus sign:

$$\cos\theta = -\sqrt{1-\sin^2\theta} = -\sqrt{1-\left[\frac{2}{3}\right]^2} = -\sqrt{1-\frac{4}{9}} = -\sqrt{\frac{5}{9}} = \frac{-\sqrt{5}}{3}$$

$$\csc\theta = \frac{1}{\sin\theta} = \frac{1}{\frac{2}{3}} = \frac{3}{2} \qquad \sec\theta = \frac{1}{\cos\theta} = \frac{1}{\frac{-\sqrt{5}}{3}} = \frac{-3\sqrt{5}}{5}$$

$$\tan\theta = \frac{\sin\theta}{\cos\theta} = \frac{\frac{2}{3}}{\frac{-\sqrt{5}}{3}} = \frac{-2\sqrt{5}}{5} \qquad \cot\theta = \frac{1}{\tan\theta} = \frac{1}{\frac{-2}{\sqrt{5}}} = \frac{-\sqrt{5}}{2}$$

79. $\sec\theta = 2$, $\sin\theta < 0$

Because $\sec\theta = \dfrac{1}{\cos\theta}$ and $\sec\theta = 2$, then $\cos\theta = \dfrac{1}{2}$.

$\cos\theta = \dfrac{1}{2}$

$\sin^2\theta = 1 - \cos^2\theta$

$\sin\theta = \pm\sqrt{1-\cos^2\theta}$

Because $\sin\theta < 0$, we use the minus sign:

$$\sin\theta = -\sqrt{1-\cos^2}\,\theta = -\sqrt{1-\left[\frac{1}{2}\right]^2} = -\sqrt{1-\frac{1}{4}} = -\sqrt{\frac{3}{4}} = \frac{-\sqrt{3}}{2}$$

$$\csc\theta = \frac{1}{\sin\theta} = \frac{1}{\frac{-\sqrt{3}}{2}} = \frac{-2\sqrt{3}}{3} \qquad \tan\theta = \frac{\sin\theta}{\cos\theta} = \frac{\frac{-\sqrt{3}}{2}}{\frac{1}{2}} = -\sqrt{3}$$

$$\cot\theta = \frac{1}{\tan\theta} = \frac{1}{-\sqrt{3}} = \frac{-\sqrt{3}}{3}$$

81. $\tan \theta = \dfrac{3}{4}$, $\sin \theta < 0$

$\cot \theta = \dfrac{1}{\tan \theta} = \dfrac{1}{\dfrac{3}{4}} = \dfrac{4}{3}$

Because $\tan \theta = \dfrac{\sin \theta}{\cos \theta} = \dfrac{3}{4} > 0$ and $\sin \theta < 0$, it follows that $\cos \theta < 0$. We know that $\tan^2 \theta + 1 = \sec^2 \theta$.

$\sec \theta = \pm\sqrt{\tan^2 \theta + 1}$

Because $\cos \theta = \dfrac{1}{\sec \theta} < 0$, it follows that $\sec \theta < 0$. Therefore, we use the minus sign:

$\sec \theta = -\sqrt{\tan^2 \theta + 1} = -\sqrt{\left(\dfrac{3}{4}\right)^2 + 1} = -\sqrt{\dfrac{9}{16} + 1} = -\sqrt{\dfrac{25}{16}} = \dfrac{-5}{4}$

$\cos \theta = \dfrac{1}{\sec \theta} = \dfrac{1}{\dfrac{-5}{4}} = \dfrac{-4}{5}$
$\qquad \sin \theta = -\sqrt{1 - \left(\dfrac{-4}{5}\right)^2} = -\sqrt{\dfrac{9}{25}} = \dfrac{-3}{5}$

$\csc \theta = \dfrac{1}{\sin \theta} = \dfrac{1}{\dfrac{-3}{5}} = \dfrac{-5}{3}$

83. $\tan \theta = \dfrac{-1}{3}$, $\sin \theta > 0$

Because $\tan \theta = \dfrac{\sin \theta}{\cos \theta} < 0$ and $\sin \theta > 0$, it follows that $\cos \theta < 0$.

$\sec^2 \theta = \tan^2 \theta + 1$

$\sec \theta = \pm\sqrt{\tan^2 \theta + 1}$

Because $\cos \theta = \dfrac{1}{\sec \theta}$, it follows that $\sec \theta < 0$. Therefore, we use the minus sign.

$\sec \theta = \pm\sqrt{-\tan^2 \theta + 1} = -\sqrt{\left(\dfrac{-1}{3}\right)^2 + 1} = -\sqrt{\dfrac{1}{9} + 1} = \dfrac{-\sqrt{10}}{3}$

$\cos \theta = \dfrac{1}{\sec \theta} = \dfrac{1}{\dfrac{-\sqrt{10}}{3}} = \dfrac{-3\sqrt{10}}{10}$

$\sin \theta = \sqrt{1 - \left(\dfrac{-3}{\sqrt{10}}\right)^2} = \sqrt{1 - \dfrac{9}{10}} = \dfrac{1}{\sqrt{10}} = \dfrac{\sqrt{10}}{10}$

$\csc \theta = \dfrac{1}{\sin \theta} = \dfrac{1}{\dfrac{\sqrt{10}}{10}} = \sqrt{10}$
$\qquad \cot \theta = \dfrac{1}{\tan \theta} = \dfrac{1}{\dfrac{-1}{3}} = -3$

85. $\sin 45° + \sin 135° + \sin 225° + \sin 315°$

$= \dfrac{\sqrt{2}}{2} + \dfrac{\sqrt{2}}{2} - \dfrac{\sqrt{2}}{2} - \dfrac{\sqrt{2}}{2} = 0$

87. If $\sin \theta = 0.2$, then $\sin (\theta + \pi) = -0.2$

89. If $\tan \theta = 3$, then $\tan (\theta + \pi) = 3$

91. If $\sin \theta = \dfrac{1}{5}$, then $\csc \theta = \dfrac{1}{\frac{1}{5}} = 5$

93. $\sin 1° + \sin 2° + \sin 3° + \cdots + \sin 358° + \sin 359°$
$= (\sin 1° + \sin 359°) + (\sin 2° + \sin 358°) + \cdots$
$= [\sin 1° + \sin (-1°)] + [\sin 2° + (\sin -2°)] + \cdots + [\sin 179° + \sin (-179°)]$
$\qquad + \sin 90° + \sin 270° + \sin 180°$
$= (\sin 1° - \sin 1°) + (\sin 2° - \sin 2°) + \cdots + (\sin 179° - \sin 179°) + 1 + (-1) + 0$
$= 0 + 0 + \cdots + 0 + 0 = 0$

95. (a) $R = \dfrac{(32)^2 \sqrt{2}}{32} (\sin(2 \cdot 60) - \cos(2 \cdot 60) - 1) = 32\sqrt{2} (0.866 - (-0.5) - 1)$

$\qquad\qquad\qquad = 32\sqrt{2}\, (0.366) \approx 16.6 \text{ ft}$

(b)

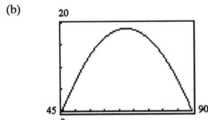

(c) R is largest when $\theta = 67.50°$.

6.5 Properties of the Trigonometric Functions

1. $P = \left[\dfrac{\sqrt{3}}{2}, \dfrac{1}{2} \right]$, which means $a = \dfrac{\sqrt{3}}{2}$ and $b = \dfrac{1}{2}$.

$\sin t = b = \dfrac{1}{2} \qquad\qquad\qquad \csc t = \dfrac{1}{b} = 2$

$\cos t = a = \dfrac{\sqrt{3}}{2} \qquad\qquad\quad \sec t = \dfrac{1}{a} = \dfrac{2}{\sqrt{3}} = \dfrac{2\sqrt{3}}{3}$

$\tan t = \dfrac{b}{a} = \dfrac{\frac{1}{2}}{\frac{\sqrt{3}}{2}} = \dfrac{1}{2} \cdot \dfrac{2}{\sqrt{3}} = \dfrac{1}{\sqrt{3}} \cdot \dfrac{\sqrt{3}}{\sqrt{3}} = \dfrac{\sqrt{3}}{3} \qquad \cot t = \dfrac{a}{b} = \dfrac{\frac{\sqrt{3}}{2}}{\frac{1}{2}} = \dfrac{\sqrt{3}}{2} \cdot \dfrac{2}{1} = \sqrt{3}$

3. $P = \left[\dfrac{-1}{2}, \dfrac{-\sqrt{3}}{2}\right]$, which means $a = \dfrac{-1}{2}$ and $b = \dfrac{-\sqrt{3}}{2}$.

$\sin t = b = \dfrac{-\sqrt{3}}{2}$ $\qquad\qquad$ $\csc t = \dfrac{1}{b} = \dfrac{1}{\dfrac{-\sqrt{3}}{2}} = \dfrac{-2}{\sqrt{3}} = \dfrac{-2\sqrt{3}}{3}$

$\cos t = a = \dfrac{-1}{2}$ $\qquad\qquad$ $\sec t = \dfrac{1}{a} = \dfrac{1}{\dfrac{-1}{2}} = -2$

$\tan t = \dfrac{b}{a} = \dfrac{\dfrac{-\sqrt{3}}{2}}{\dfrac{-1}{2}} = \dfrac{-\sqrt{3}}{2} \cdot \dfrac{-2}{1} = \sqrt{3}$ $\qquad$ $\cot t = \dfrac{a}{b} = \dfrac{\dfrac{-1}{2}}{\dfrac{-\sqrt{3}}{2}} = \dfrac{-1}{2} \cdot \dfrac{-2}{\sqrt{3}} = \dfrac{1}{\sqrt{3}} = \dfrac{\sqrt{3}}{3}$

5. $P = \left[\dfrac{-3}{5}, \dfrac{4}{5}\right]$, which means $a = \dfrac{-3}{5}$ and $b = \dfrac{4}{5}$.

$\sin t = b = \dfrac{4}{5}$ $\qquad\qquad$ $\csc t = \dfrac{1}{\dfrac{4}{5}} = \dfrac{5}{4}$

$\cos t = a = \dfrac{-3}{5}$ $\qquad\qquad$ $\sec t = \dfrac{1}{a} = \dfrac{1}{\dfrac{-3}{5}} = \dfrac{-5}{3}$

$\tan t = \dfrac{b}{a} = \dfrac{\dfrac{4}{5}}{\dfrac{-3}{5}} = \dfrac{4}{5} \cdot \dfrac{-5}{3} = \dfrac{-4}{3}$ $\qquad$ $\cot t = \dfrac{a}{b} = \dfrac{\dfrac{-3}{5}}{\dfrac{4}{5}} = \dfrac{-3}{5} \cdot \dfrac{5}{4} = \dfrac{-3}{4}$

7. $P = \left[\dfrac{\sqrt{5}}{5}, \dfrac{-2\sqrt{5}}{5}\right]$, which means $a = \dfrac{\sqrt{5}}{5}$ and $b = \dfrac{-2\sqrt{5}}{5}$.

$\sin t = b = \dfrac{-2\sqrt{5}}{5}$ $\qquad\qquad$ $\csc t = \dfrac{1}{b} = \dfrac{1}{\dfrac{-2\sqrt{5}}{5}} = \dfrac{-5}{2\sqrt{5}} = \dfrac{-5\sqrt{5}}{10} = \dfrac{-\sqrt{5}}{2}$

$\cos t = a = \dfrac{\sqrt{5}}{5}$ $\qquad\qquad$ $\sec t = \dfrac{1}{a} = \dfrac{1}{\dfrac{\sqrt{5}}{5}} = \dfrac{5}{\sqrt{5}} = \dfrac{5\sqrt{5}}{5} = \sqrt{5}$

$\tan t = \dfrac{b}{a} = \dfrac{\dfrac{-2\sqrt{5}}{5}}{\dfrac{\sqrt{5}}{5}} = -2$ $\qquad$ $\cot t = \dfrac{a}{b} = \dfrac{\dfrac{\sqrt{5}}{5}}{\dfrac{-2\sqrt{5}}{5}} = \dfrac{1}{-2} = \dfrac{-1}{2}$

9. $\sin \theta = \dfrac{2}{3}$, $\dfrac{\pi}{2} < \theta < \pi$

Suppose $P = (a, b)$ is a point on the terminal side of θ that lies a distance of $r = 3$ units from the origin. We know that $\sin \theta = \dfrac{2}{3} = \dfrac{b}{r}$, so $b = 2$ and $r = 3$. Since $\dfrac{\pi}{2} < \theta < \pi$, θ is in quadrant II and $\cos \theta = \dfrac{a}{r} < 0$. It follows that $a < 0$. Thus,

$$a^2 + b^2 = r^2$$
$$a^2 + 2^2 = 3^2$$
$$a^2 = 5$$
$$a = -\sqrt{5}$$

Therefore,

$$\cos \theta = \frac{a}{r} = \frac{-\sqrt{5}}{3}$$

$$\sec \theta = \frac{r}{a} = \frac{-3}{\sqrt{5}} = \frac{-3\sqrt{5}}{5}$$

$$\tan \theta = \frac{b}{a} = \frac{-2}{\sqrt{5}} = \frac{-2\sqrt{5}}{5}$$

$$\cot \theta = \frac{a}{b} = \frac{-\sqrt{5}}{2}$$

$$\csc \theta = \frac{r}{b} = \frac{3}{2}$$

11. $\tan \theta = \dfrac{1}{2}, \ \pi < \theta < \dfrac{3\pi}{2}$

Because $\tan \theta = \dfrac{b}{a} = \dfrac{1}{2}$, we know that $|b| = 1$ and $|a| = 2$.

Since $\pi < \theta < \dfrac{3\pi}{2}$, the point $P = (a, b)$ has x and y coordinates of negative value. Therefore, $b = -1$, and $a = -2$. To find the distance of P from the origin, we use:

$$a^2 + b^2 = r^2$$
$$(-2)^2 + (-1)^2 = r^2$$
$$4 + 1 = r^2$$
$$\sqrt{5} = r$$

Thus,

$$\cot \theta = \frac{a}{b} = \frac{-2}{-1} = 2$$

$$\sin \theta = \frac{b}{r} = \frac{-1}{\sqrt{5}} = \frac{-\sqrt{5}}{5}$$

$$\csc \theta = \frac{r}{b} = \frac{\sqrt{5}}{-1} = -\sqrt{5}$$

$$\cos \theta = \frac{a}{r} = \frac{-2}{\sqrt{5}} = \frac{-2\sqrt{5}}{5}$$

$$\sec \theta = \frac{r}{a} = \frac{\sqrt{5}}{-2} = \frac{-\sqrt{5}}{2}$$

13. $\sin \theta = \dfrac{-1}{4}, \ \cos \theta > 0$

Because $\sin \theta = \dfrac{b}{r}$, we know that $b = -1$ and $r = 4$.

Since $\cos \theta = \dfrac{a}{r} > 0, \ a > 0$. Thus,

$$a^2 + b^2 = r^2$$
$$(a)^2 + (-1)^2 = (4)^2$$
$$a^2 = 15$$
$$a = \sqrt{15}$$

$$\csc \theta = \frac{r}{b} = \frac{4}{-1} = -4$$

$$\cos \theta = \frac{a}{r} = \frac{\sqrt{15}}{4}$$

$$\sec \theta = \frac{r}{a} = \frac{4}{\sqrt{15}} = \frac{4\sqrt{15}}{15}$$

$$\tan \theta = \frac{b}{a} = \frac{-1}{\sqrt{15}} = \frac{-1\sqrt{15}}{15}$$

$$\cot \theta = \frac{a}{b} = \frac{\sqrt{15}}{-1} = -\sqrt{15}$$

15. $\sec \theta = 3$, $\tan \theta < 0$

We know that $\sec \theta = \dfrac{r}{a} = \dfrac{3}{1}$, so $r = 3$ and $a = 1$.

Since $\tan \theta = \dfrac{b}{a} < 0$, $b < 0$.

Thus,
$$a^2 + b^2 = r^2$$
$$1^2 + b^2 = 3^2$$
$$b^2 = 8$$
$$b = -2\sqrt{2}$$

Therefore, $\cos \theta = \dfrac{a}{r} = \dfrac{1}{3}$

$$\sin \theta = \dfrac{b}{r} = \dfrac{-2\sqrt{2}}{3} \qquad\qquad \csc \theta = \dfrac{r}{b} = \dfrac{3}{-2\sqrt{2}} = \dfrac{-3\sqrt{2}}{4}$$

$$\tan \theta = \dfrac{b}{a} = \dfrac{-2\sqrt{2}}{1} = -2\sqrt{2} \qquad \cot \theta = \dfrac{a}{b} = \dfrac{1}{-2\sqrt{2}} = \dfrac{-\sqrt{2}}{4}$$

17. The domain of the sine function is the set of all real numbers.

19. $f(\theta) = \tan \theta$ is not defined for numbers that are odd multiples of $\dfrac{\pi}{2}$.

21. $f(\theta) = \sec \theta$ is not defined for numbers that are odd multiples of $\dfrac{\pi}{2}$.

23. The range of the sine function includes values between -1 and 1 and including -1 and 1. We write $-1 \le \sin \theta \le 1$ or $[-1, 1]$.

25. The range of the tangent function is the set of all real numbers. We write $(-\infty, \infty)$.

27. The range of the secant function are values less than or equal to -1 and values greater than or equal to 1. We write $(-\infty, -1] \cup [1, \infty)$.

29. Because $\sin(-\theta) = -\sin \theta$, the sine function is odd. Its graph is symmetric with respect to the origin because for each point (a, b) $(-a, -b)$ is also included on the sine curve.

31. The tangent function is odd because $\tan(-\theta) = -\tan \theta$. Its graph is symmetric with respect to the origin.

33. The secant function is even because $\sec(-\theta) = \sec \theta$. Its graph is symmetric with respect to the y-axis because for each point (a, b), $(-a, b)$ is also on the secant curve.

35. Let $P = (x, y)$ be the point on the unit circle that corresponds to t.

Consider the equation $\tan t = \dfrac{y}{x} = a$. Then $y = ax$. But $x^2 + y^2 = 1$ so that $x^2 + a^2 x^2 = 1$.

Thus, $x = \pm \dfrac{1}{\sqrt{1 + a^2}}$ and $y = \pm \dfrac{a}{\sqrt{1 + a^2}}$; that is, for any real number a, there is a point $P = (x, y)$ on the unit circle for which $\tan t = a$. In other words, $-\infty < \tan t < +\infty$, and the range of the tangent function is the set of all real numbers.

37. Suppose there is a number p, $0 < p < 2\pi$, for which $\sin(\theta + p) = \sin\theta$ for all θ. If $\theta = 0$, then $\sin(0 + p) = \sin p = \sin 0 = 0$; so that $p = \pi$. If $\theta = \dfrac{\pi}{2}$, then $\sin\left(\dfrac{\pi}{2} + p\right) = \sin\left(\dfrac{\pi}{2}\right)$. But $p = \pi$. Thus, $\sin\left(\dfrac{3\pi}{2}\right) = -1 = \sin\left(\dfrac{\pi}{2}\right) = 1$. This is impossible. The smallest positive number p for which $\sin(\theta + p) = \sin\theta$ for all θ is therefore $p = 2\pi$.

39. $\sec\theta = \dfrac{1}{\cos\theta}$; since $\cos\theta$ has period 2π, so does $\sec\theta$.

41. If $P = (a, b)$ is the point on the unit circle corresponding to θ, then $Q = (-a, -b)$ is the point on the unit circle corresponding to $\theta + \pi$. Thus, $\tan(\theta + \pi) = \dfrac{-b}{-a} = \dfrac{b}{a} = \tan\theta$; that is, the period of the tangent function is π.

43. Slope of $L^* = \dfrac{\sin\theta - 0}{\cos\theta - 0} = \dfrac{\sin\theta}{\cos\theta} = \dfrac{b}{a} = \tan\theta$

Since L is parallel to L^*, then slope of $L = \tan\theta$.

6.6 Graphs of the Trigonometric Functions

1. 0

3. The graph of $y = \sin x$ is increasing for $\dfrac{-\pi}{2} \le x \le \dfrac{\pi}{2}$.

5. The largest value of $y = \sin x$ is 1.

7. $\sin x = 0$ when $x = 0,\ \pi,\ 2\pi$

9. $\sin x = 1$ for $x = \dfrac{-3\pi}{2},\ \dfrac{\pi}{2}$ if $-2\pi \le x \le 2\pi$; $\sin x = -1$ for $x = \dfrac{-\pi}{2},\ \dfrac{3\pi}{2}$ if $-2\pi \le x \le 2\pi$

11. 0　　**13.** 1

15. $\sec x = 1$ for $x = -2\pi,\ 0,\ 2\pi$ if $-2\pi \le x \le 2\pi$
$\sec x = -1$ for $x = -\pi,\ \pi$ if $-2\pi \le x \le 2\pi$

17. $y = \sec x$ has vertical asymptotes for $x = \dfrac{-3\pi}{2},\ \dfrac{-\pi}{2},\ \dfrac{\pi}{2},\ \dfrac{3\pi}{2}$ if $-2\pi \le x \le 2\pi$.

19. $y = \tan x$ has vertical asymptotes for $x = \dfrac{-3\pi}{2},\ \dfrac{-\pi}{2},\ \dfrac{\pi}{2},\ \dfrac{3\pi}{2}$ if $-2\pi \le x \le 2\pi$.

21. B, C, F　　**23.** C　　**25.** D　　**27.** B　　**29.** A

31.

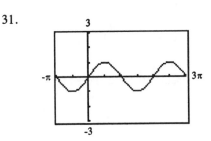

$y = \sin x$

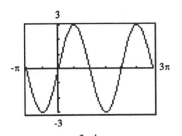

$y = 3 \sin x$

$y = 3 \sin x$ is a vertical stretch by a factor of 3 of the graph of $y = \sin x$.

33.

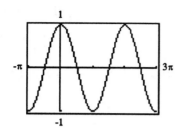

$$y = \cos x$$

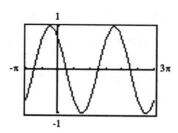

$$y = \cos\left[x + \frac{\pi}{4}\right]$$

$y = \cos\left[x + \dfrac{\pi}{4}\right]$ is a horizontal shift $\dfrac{\pi}{4}$ units left of the graph of $y = \cos x$.

35.

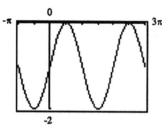

$$y = \sin x - 1$$

$y = \sin x - 1$ is the graph of $y = \sin x$ shifted down one unit.

37.

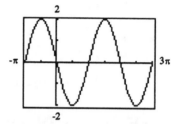

$y = 2 \sin x$ is a vertical stretch by a factor of 2 of the graph of $y = \sin x$. $y = -2 \sin x$ is a reflection of the graph of $y = 2 \sin x$ about the x-axis.

39.

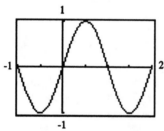

$y = \sin(\pi x)$ is a horizontal compression of $y = \sin x$.

41.

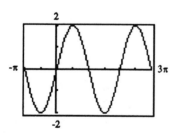

$$y = 2 \sin x$$

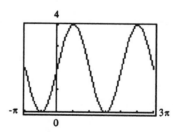

$$y = 2 \sin x + 2$$

$y = 2 \sin x + 2$ is the graph of $y = \sin x$ vertically stretched by a factor of 2 and shifted up two units.

43.

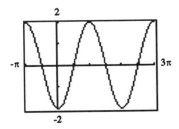

$$y = -2 \cos x$$

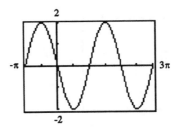

$$y = -2 \cos \left[x - \frac{\pi}{2} \right]$$

$y = -2 \cos \left[x - \dfrac{\pi}{2} \right]$ is the graph of $y = \cos x$ reflected about the x-axis, vertically stretched by a factor of 2, and shifted horizontally $\dfrac{\pi}{2}$ units right.

45.

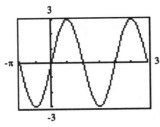

$$y = 3 \sin x$$

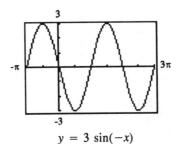

$$y = 3 \sin(-x)$$

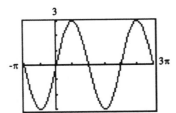

$$y = 3 \sin(\pi - x) = 3 \sin(-(x - \pi))$$

$y = 3 \sin(\pi - x)$ is obtained by vertically stretching $y = \sin x$ by a factor of 3. Then reflect the graph of $y = 3 \sin x$ about the y-axis. Finally, horizontally shift the graph of $y = 3 \sin (-x)$ π units to the right.

47.

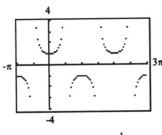

$$y = \sec x = \frac{1}{\cos x}$$

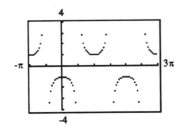

$$y = -\sec x = \frac{-1}{\cos x}$$

The graph of $y = -\sec x$ is the graph of $y = \sec x$ reflected about the x-axis.

49.

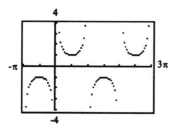

$$y = \sec\left(x - \frac{\pi}{2}\right)$$

The graph of $y = \sec\left(x - \frac{\pi}{2}\right)$ is the graph of $y = \sec x$ shifted horizontally $\frac{\pi}{2}$ units to the right.

51.

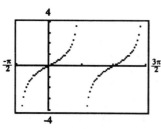
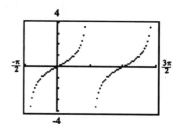

$y = \tan x$ $y = \tan(x - \pi)$

The graph of $y = \tan x$ is the same as the graph of $y = \tan(x - \pi)$.

53.

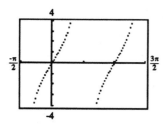
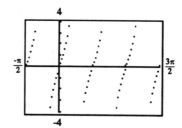

$y = 3 \tan x$ $y = 3 \tan(2x)$

The graph of $y = 3 \tan(2x)$ is the graph of $y = \tan x$ vertically stretched by a factor of 3 and horizontally compressed by a factor of $\frac{1}{2}$ (i.e., each x-value is multiplied by $\frac{1}{2}$).

55.

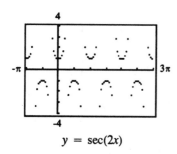

$y = \sec(2x)$

The graph of $y = \sec(2x)$ is the graph of $y = \sec x$ horizontally compressed by a factor of 2.

57.

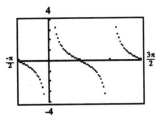
$$y = \cot x$$

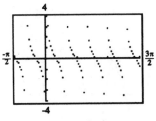
$$y = \cot(\pi x)$$

The graph of $y = \cot(\pi x)$ is the graph of $y = \cot x$ horizontally compressed by a factor of π.

59.

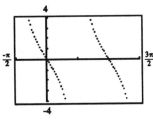
$$y = -3 \tan x$$

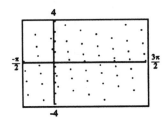
$$y = -3 \tan(4x)$$

The graph of $y = -3 \tan(4x)$ is obtained by vertically stretching the graph of $y = \tan x$ by a factor of 3, reflecting $y = 3 \tan x$ about the x-axis to obtain $y = -3 \tan x$ and horizontally compressing $y = -3 \tan x$ by a factor of 4.

61.

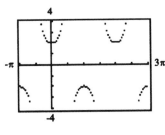
$$y = 2 \sec x$$

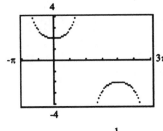
$$y = 2 \sec \frac{1}{2}x$$

The graph of $y = 2 \sec \frac{1}{2}x$ is the graph of $y = \sec x$ vertically stretched by a factor of 2 and horizontally compressed by a factor of 2.

63. (a) $L = \dfrac{3}{\cos \theta} + \dfrac{4}{\sin \theta}$

$L = 3 \sec \theta + 4 \csc \theta$

(c) L is least when $\theta = 0.83$.

(b)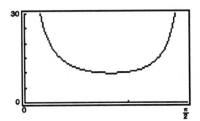

65. The graph of $y = \sin \omega x$ has period $\dfrac{2\pi}{\omega}$.

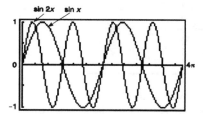 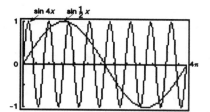

67. It would appear that $\sin x = \cos\left(x - \dfrac{\pi}{2}\right)$ because if the cosine function was shifted to the right $\dfrac{\pi}{2}$ units, it would become the sine function.

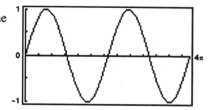

6.7 The Inverse Trigonometric Functions

1. $\sin^{-1} 0$

We seek the angle θ, $-\dfrac{\pi}{2} \le \theta \le \dfrac{\pi}{2}$, whose sine equals 0.

$$\sin \theta = 0 \qquad -\dfrac{\pi}{2} \le \theta \le \dfrac{\pi}{2}$$
$$\theta = 0$$
$$\sin^{-1} 0 = 0$$

3. $\sin^{-1}(-1)$

We seek the angle θ, $-\dfrac{\pi}{2} \le \theta \le \dfrac{\pi}{2}$, whose sine equals (-1).

$$\sin \theta = -1 \qquad -\dfrac{\pi}{2} \le \theta \le \dfrac{\pi}{2}$$
$$\theta = -\dfrac{\pi}{2}$$
$$\sin^{-1}(-1) = -\dfrac{\pi}{2}$$

5. $\tan^{-1} 0$

We seek the angle θ, $-\dfrac{\pi}{2} < \theta < \dfrac{\pi}{2}$, whose tangent equals 0.

$$\tan \theta = 0 \qquad -\dfrac{\pi}{2} < \theta < \dfrac{\pi}{2}$$
$$\theta = 0$$
$$\tan^{-1} 0 = 0$$

7. $\sin^{-1} \dfrac{\sqrt{2}}{2}$

We seek the angle θ, $-\dfrac{\pi}{2} \le \theta \le \dfrac{\pi}{2}$, whose sine equals $\dfrac{\sqrt{2}}{2}$.

$$\sin \theta = \dfrac{\sqrt{2}}{2} \qquad -\dfrac{\pi}{2} \le \theta \le \dfrac{\pi}{2}$$

$$\theta = \dfrac{\pi}{4}$$

$$\sin^{-1} \dfrac{\sqrt{2}}{2} = \dfrac{\pi}{4}$$

9. $\tan^{-1} \sqrt{3}$

We seek the angle θ, $-\dfrac{\pi}{2} < \theta < \dfrac{\pi}{2}$, whose tangent equals $\sqrt{3}$.

$$\tan \theta = \sqrt{3} \qquad -\dfrac{\pi}{2} < \theta < \dfrac{\pi}{2}$$

$$\theta = \dfrac{\pi}{3}$$

$$\tan^{-1} \sqrt{3} = \dfrac{\pi}{3}$$

11. $\cos^{-1} \left[-\dfrac{\sqrt{3}}{2} \right]$

We seek the angle θ, $0 \le \theta \le \pi$, whose tangent equals $-\dfrac{\sqrt{3}}{2}$.

$$\cos \theta = \left[\dfrac{-\sqrt{3}}{2} \right] \qquad 0 \le \theta \le \pi$$

$$\theta = \dfrac{5\pi}{6}$$

$$\cos^{-1} \left[-\dfrac{\sqrt{3}}{2} \right] = \dfrac{5\pi}{6}$$

13. Set the mode of the calculator to radians.
$\sin^{-1} 0.1 = 0.10$

15. Set the mode of the calculator to radians.
$\tan^{-1} 5 = 1.37$

17. Set the mode of the calculator to radians.
$\cos^{-1} \dfrac{7}{8} = 0.51$

19. Set the mode of the calculator to radians.
$\tan^{-1} (-0.4) = -0.38$

21. Set the mode of the calculator to radians.
$\sin^{-1} (-0.12) = -0.12$

23. Set the mode of the calculator to radians.
$\cos^{-1} \dfrac{\sqrt{2}}{3} = 1.08$

25. $\cos\left[\sin^{-1}\dfrac{\sqrt{2}}{2}\right]$

First find the angle θ, $-\dfrac{\pi}{2} \le \theta \le \dfrac{\pi}{2}$, whose sine equals $\dfrac{\sqrt{2}}{2}$.

$$\sin\theta = \dfrac{\sqrt{2}}{2} \qquad -\dfrac{\pi}{2} \le \theta \le \dfrac{\pi}{2}$$

$$\theta = \dfrac{\pi}{4}$$

Now, $\cos\left[\sin^{-1}\dfrac{\sqrt{2}}{2}\right] = \cos\theta = \cos\dfrac{\pi}{4} = \dfrac{\sqrt{2}}{2}$.

27. $\tan\left[\cos^{-1}\left(-\dfrac{\sqrt{3}}{2}\right)\right]$

First find the angle θ, $0 \le \theta \le \pi$, whose cosine equals $-\dfrac{\sqrt{3}}{2}$.

$$\cos\theta = -\dfrac{\sqrt{3}}{2} \qquad 0 \le \theta \le \pi$$

$$\theta = \dfrac{5\pi}{4}$$

Now, $\tan\left[\cos^{-1}\left(-\dfrac{\sqrt{3}}{2}\right)\right] = \tan\theta = \tan\dfrac{5\pi}{6} = -\dfrac{\sqrt{3}}{3}$.

29. $\sec\left[\cos^{-1}\dfrac{1}{2}\right]$

First find the angle θ, $0 \le \theta \le \pi$, whose cosine equals $\dfrac{1}{2}$.

$$\cos\theta = \dfrac{1}{2} \qquad 0 \le \theta \le \pi$$

$$\theta = \dfrac{\pi}{3}$$

Now, $\sec\left[\cos^{-1}\dfrac{1}{2}\right] = \sec\theta = \sec\dfrac{\pi}{3} = 2$.

31. $\csc\left(\tan^{-1}1\right)$

First find the angle θ, $-\dfrac{\pi}{2} < \theta < \dfrac{\pi}{2}$, whose tangent equals 1.

$$\tan\theta = 1 \qquad -\dfrac{\pi}{2} < \theta < \dfrac{\pi}{2}$$

$$\theta = \dfrac{\pi}{4}$$

Now, $\csc\left(\tan^{-1}1\right) = \csc\theta = \cos\dfrac{\pi}{4} = \sqrt{2}$.

33. $\sin\left(\tan^{-1}(-1)\right)$

First find the angle θ, $-\dfrac{\pi}{2} < \theta < \dfrac{\pi}{2}$, whose tangent equals -1.

$$\tan \theta = -1 \qquad\qquad -\dfrac{\pi}{2} < \theta < \dfrac{\pi}{2}$$

$$\theta = -\dfrac{\pi}{4}$$

Now, $\sin\left[\tan^{-1}(-1)\right] = \sin\theta = \sin\left(-\dfrac{\pi}{4}\right) = -\dfrac{\sqrt{2}}{2}$.

35. $\sec\left[\sin^{-1}\left(-\dfrac{1}{2}\right)\right]$

First find the angle θ, $-\dfrac{\pi}{2} \le \theta \le \dfrac{\pi}{2}$, whose sine equals $-\dfrac{1}{2}$.

$$\sin \theta = -\dfrac{1}{2}$$

$$\theta = -\dfrac{\pi}{6}$$

Now, $\sec\left[\sin^{-1}\left(-\dfrac{1}{2}\right)\right] = \sec\theta = \sec\left(-\dfrac{\pi}{6}\right) = \dfrac{2\sqrt{3}}{3}$.

37. $\tan\left[\sin^{-1}\dfrac{1}{3}\right]$

First we know that $\sin\theta = \dfrac{1}{3}$, $-\dfrac{\pi}{2} \le \theta \le \dfrac{\pi}{2}$, so we have:

By the Pythagorean Theorem, the missing side of the triangle is:
$$x^2 + 1 = 9$$
$$x^2 = 8$$
$$x = \pm\sqrt{8}$$
but x is positive in quadrant I, so
$$x = \sqrt{8} = 2\sqrt{2}$$

Now, $\tan\left[\sin^{-1}\dfrac{1}{3}\right] = \tan\theta = \dfrac{1}{2\sqrt{2}}$ (using $\dfrac{\text{opp}}{\text{adj}}$ in triangle)

$$\tan\theta = \dfrac{1}{2\sqrt{2}} \cdot \dfrac{\sqrt{2}}{\sqrt{2}} = \dfrac{\sqrt{2}}{4}$$

39. $\sec\left[\tan^{-1}\dfrac{1}{2}\right]$

First we know that $\tan\theta = \dfrac{1}{2}$, $-\dfrac{\pi}{2} < \theta < \dfrac{\pi}{2}$, so we have:

By the Pythagorean Theorem, the hypotenuse is:
$$1^2 + 2^2 = r^2$$
$$1 + 4 = r^2$$
$$5 = r^2$$
$$\sqrt{5} = r$$

Now, $\sec\left[\tan^{-1}\dfrac{1}{2}\right] = \sec\theta = \dfrac{\text{hyp}}{\text{adj}} = \sec\theta = \dfrac{\sqrt{5}}{2}$

41. $\cot\left[\sin^{-1}\left(-\dfrac{\sqrt{2}}{3}\right)\right]$

First draw the angle θ, $-\dfrac{\pi}{2} \le \theta \le \dfrac{\pi}{2}$, whose sine equals $-\dfrac{\sqrt{2}}{3}$

$$\sin\theta = -\frac{\sqrt{2}}{3}$$

By the Pythagorean Theorem, the missing side is:

$$x^2 + \left(-\sqrt{2}\right)^2 = 3^2$$
$$x^2 + 2 = 9$$
$$x^2 = 7$$

$x = \pm\sqrt{7}$, but x is positive in quadrant IV

$$x = \sqrt{7}$$

Now, $\cot\left[\sin^{-1}\left(-\dfrac{\sqrt{2}}{3}\right)\right] = \cot\theta = \dfrac{\text{adj}}{\text{opp}}$

$$\cot\theta = \frac{\sqrt{7}}{-\sqrt{2}} = -\frac{\sqrt{7}}{\sqrt{2}} \cdot \frac{\sqrt{2}}{\sqrt{2}} = -\frac{\sqrt{14}}{2}$$

43. $\sin\left[\tan^{-1}(-3)\right]$

First draw the angle θ, $-\dfrac{\pi}{2} < \theta < \dfrac{\pi}{2}$, whose tangent is -3.

$$\tan\theta = -3$$

By the Pythagorean Theorem, the hypotenuse is:
$$r^2 = 1^2 + (-3)^2$$
$$r^2 = 1 + 9$$
$$r^2 = 10$$
$$r = \sqrt{10}$$

Now, $\sin\left[\tan^{-1}(-3)\right] = \sin\theta = \dfrac{\text{opp}}{\text{hyp}}$

$$\sin\theta = \frac{-3}{\sqrt{10}} = \frac{-3}{\sqrt{10}} \cdot \frac{\sqrt{10}}{\sqrt{10}} = \frac{-3\sqrt{10}}{10}$$

45. $\sec\left[\sin^{-1}\dfrac{2\sqrt{5}}{5}\right]$

First draw the angle θ, $-\dfrac{\pi}{2} \le \theta \le \dfrac{\pi}{2}$, whose sine is $\dfrac{2\sqrt{5}}{5}$

$$\sin\theta = \frac{2\sqrt{5}}{5}$$

By the Pythagorean Theorem, we find the missing side,

$$\left(2\sqrt{5}\right)^2 + x^2 = (5)^2$$
$$20 + x^2 = 25$$
$$x^2 = 5$$

$x = \pm\sqrt{5}$ but $x > 0$ in quadrant I

$$x = \sqrt{5}$$

Now, $\sec\left[\sin^{-1}\dfrac{2\sqrt{5}}{5}\right] = \sec\theta = \dfrac{\text{hyp}}{\text{adj}}$

$$\sec\theta = \frac{5}{\sqrt{5}} = \frac{5}{\sqrt{5}}\cdot\frac{\sqrt{5}}{\sqrt{5}} = \frac{5\sqrt{5}}{5} = \sqrt{5}$$

47. Use radian mode on calculator.
$\sin^{-1}(\tan 0.5) = 0.58$

49. Use radian mode on calculator.
$\tan^{-1}(\sin 0.1) = 0.10$

51. Use radian mode on calculator.
$\cos^{-1}(\sin 1) = 0.57$

53. Use radian mode on calculator.
$\sin^{-1}\left[\tan\dfrac{\pi}{8}\right] = 0.43$

55. Use radian mode on calculator.
$\tan^{-1}\left[\sin\dfrac{\pi}{8}\right] = 0.37$

57. $\sec(\tan^{-1} v) = \sqrt{1 + v^2}$

Let $\theta = \tan^{-1} v$ Then, $\tan\theta = v$, $-\dfrac{\pi}{2} < \theta < \dfrac{\pi}{2}$.

Hence, $\sec\theta > 0$ and $\tan^2\theta + 1 = \sec^2\theta$
$$v^2 + 1 = \sec^2\theta$$
$$\sqrt{v^2 + 1} = \sec\theta$$

Thus, $\sec(\tan^{-1} v) = \sec\theta = \sqrt{v^2 + 1} = \sqrt{1 + v^2}$

59. Let $\theta = \cos^{-1} v$. Then $\cos\theta = v$, $0 \le \theta \le \pi$

$$\tan(\cos^{-1} v) = \tan\theta = \frac{\sin\theta}{\cos\theta} = \frac{\sqrt{1 - \cos^2\theta}}{\cos\theta} = \frac{\sqrt{1 - v^2}}{v}$$

61. Let $\theta = \sin^{-1} v$. Then $\sin\theta = v$, $\dfrac{-\pi}{2} \le \theta \le \dfrac{\pi}{2}$

$$\cos(\sin^{-1} v) = \cos\theta = \sqrt{1 - \sin^2\theta} = \sqrt{1 - v^2}$$

63. Let $\alpha = \sin^{-1} v$ and $\beta = \cos^{-1} v$

Then $\sin\alpha = v = \cos\beta$, so α, β are complementary. Thus, $\alpha + \beta = \dfrac{\pi}{2}$.

65. Let $\alpha = \tan^{-1}\dfrac{1}{v}$. Then, $\dfrac{1}{v} = \tan\alpha$, $\dfrac{-\pi}{2} < \alpha < \dfrac{\pi}{2}$, $\alpha \ne 0$.

Let $\beta = \tan^{-1} v$. Then $v = \tan\beta$, $\dfrac{-\pi}{2} < \beta < \dfrac{\pi}{2}$.

Thus, $\tan\alpha\tan\beta = 1$, so that $\tan\alpha = \cot\beta$. Thus, $\alpha + \beta = \dfrac{\pi}{2}$

67. $\sec^{-1} 4$

Let $\nu = \sec^{-1} 4$. Then $\sec \theta = 3, 0 \leq \nu \leq \pi, \theta \neq \dfrac{\pi}{2}$.

Thus, $\cos \theta = \dfrac{1}{4}$ and $\sec^{-1} 4 = \theta = \cos^{-1} \dfrac{1}{4} \approx 1.32$.

69. $\cot^{-1} 2$

Let $\theta = \cot^{-1} 2$. Then $\cot \theta = 2, 0 < \nu < \pi$.

Thus, $\tan \theta = \dfrac{1}{2}$ and $\cot^{-1} 2 = \theta = \tan^{-1} \dfrac{1}{2} \approx 0.46$.

71. $\csc^{-1} (-3)$

Let $\theta = \csc^{-1} (-3)$. Then $\csc \theta = -3, \dfrac{-\pi}{2} \leq \theta \leq \dfrac{\pi}{2}, \theta \neq 0$.

Thus, $\sin \theta = \dfrac{-1}{3}$ and $\csc^{-1} (-3) = \theta = \sin^{-1} \left[\dfrac{-1}{3} \right] \approx -0.34$.

73. $\cot^{-1} \left(-\sqrt{5} \right)$

Let $\theta = \cot^{-1} \left(-\sqrt{5} \right)$. Then $\cot \theta = -\sqrt{5}, 0 < \theta < \pi$.

Thus, $\tan \theta = \dfrac{-1}{\sqrt{5}}$ and $\cot^{-1} \left(-\sqrt{5} \right) = \theta = \tan^{-1} \left[\dfrac{-1}{\sqrt{5}} \right] \approx 2.72$

75. Let $\theta = \csc^{-1} \left[-\dfrac{3}{2} \right]$. Then $\csc \theta = -\dfrac{3}{2}$, so $\sin \theta = -\dfrac{2}{3}$ and $\theta = \sin^{-1} \left(-\dfrac{2}{3} \right) \approx -0.73$.

77. Let $\theta = \cot^{-1} \left[-\dfrac{3}{2} \right]$. Then $\cot \theta = -\dfrac{3}{2}, 0 < \theta < \pi$. Thus, $\cos \theta = \dfrac{-3}{\sqrt{13}}, \dfrac{\pi}{2} < \theta < \pi$, and

$\theta = \cos^{-1} \left[\dfrac{-3}{\sqrt{13}} \right] \approx 2.55$.

79.
$$\dfrac{6.5}{2.5} = \dfrac{26.5 + x}{2.5 + x}$$
$$(2.5)(26.5) + 2.5x = 6.5x + (6.5)(2.5)$$
$$4x = 2.5(20)$$
$$x = 12.5$$

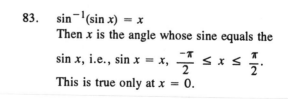

$\cos \theta = \dfrac{2.5}{15}$

$\theta = 1.4$ radians $= 80.4°$, $\alpha = 180 - 80.4° = 99.6° = 1.73$

$s_1 = r_1 \alpha = 6.5(1.73) = 11.3$

$s_2 = r_2 \alpha = 2.5(1.4) = 3.5$

Length of belt $= 2(11.3 + 24 + 3.5) = 77.6$ inches

81. $\sin(\sin^{-1} x) = x$

Let $\theta = \sin^{-1} x$

$\sin \theta = x$ where $-\dfrac{\pi}{2} \leq \theta \leq \dfrac{\pi}{2}$ and

$-1 \leq x \leq 1$.

Hence, $-1 \leq x \leq 1$.

83. $\sin^{-1}(\sin x) = x$

Then x is the angle whose sine equals the

$\sin x$, i.e., $\sin x = x, \dfrac{-\pi}{2} \leq x \leq \dfrac{\pi}{2}$.

This is true only at $x = 0$.

85.
$$y = \sec^{-1} x$$
$$\sec y = x$$
$$\frac{1}{\cos y} = x$$
$$\cos y = \frac{1}{x}$$
$$y = \cos^{-1}\left[\frac{1}{x}\right] = \sec^{-1} x$$

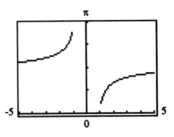

6 Chapter Review

1. $135° = 135 \cdot 1 \text{ degree} = 135 \cdot \dfrac{\pi}{180} \text{ radian} = \dfrac{3\pi}{4} \text{ radians}.$

3. $18° = 18 \cdot 1 \text{ degree} = 18 \cdot \dfrac{\pi}{180} \text{ radian} = \dfrac{\pi}{10} \text{ radians}.$

5. $\dfrac{3\pi}{4} \text{ radian} = \dfrac{3\pi}{4} \cdot 1 \text{ radian} = \dfrac{3\pi}{4} \cdot \dfrac{180}{\pi} \text{ degrees} = 135°$

7. $-\dfrac{5\pi}{2} \text{ radian} = -\dfrac{5\pi}{2} \cdot 1 \text{ radian} = -\dfrac{5\pi}{2} \cdot \dfrac{180}{\pi} \text{ degrees} = -450°$

9. $\tan \dfrac{\pi}{4} - \sin \dfrac{\pi}{6} = 1 - \dfrac{1}{2} = \dfrac{1}{2}$

11. $3 \sin 45° - 4 \tan \dfrac{\pi}{6} = 3\left[\dfrac{\sqrt{2}}{2}\right] - 4\left[\dfrac{\sqrt{3}}{3}\right] = \dfrac{3\sqrt{2}}{2} - \dfrac{4\sqrt{3}}{3}$

13. $6 \cos \dfrac{3\pi}{4} + 2 \tan\left[-\dfrac{\pi}{3}\right]$

   Using reference angles: $\cos \dfrac{3\pi}{4} = -\cos \dfrac{\pi}{4} = -\dfrac{\sqrt{2}}{2}$

   $$\tan\left[-\dfrac{\pi}{3}\right] = -\tan \dfrac{\pi}{3} = -\sqrt{3}$$

   Hence, $6\left[\dfrac{-\sqrt{2}}{2}\right] + 2\left(-\sqrt{3}\right) = -3\sqrt{2} - 2\sqrt{3}$

15. $\sec\left[-\dfrac{\pi}{3}\right] - \cot\left[-\dfrac{5\pi}{4}\right]$

   Using reference angles: $\sec\left[-\dfrac{\pi}{3}\right] = \sec\dfrac{\pi}{3} = 2$

   $$\cot\left[-\dfrac{5\pi}{4}\right] = -\cot\left[\dfrac{\pi}{4}\right] = -1$$

   Hence, $\sec\left[-\dfrac{\pi}{3}\right] - \cot\left[-\dfrac{5\pi}{4}\right] = 2 - (-1) = 3$

17. $\tan \pi + \sin \pi = 0 + 0 = 0$ 19. $\cos 180° - \tan(-45°) = -1 - (-1) = -1 + 1 = 0$

21. $\sin^2 20° + \dfrac{1}{\sec^2 20°} = \sin2\,20° + \cos^2 20° = 1$

23. $\sec 50° \cdot \cos 50° = \dfrac{1}{\cos 50°} \cdot \cos 50° = 1$

25. $\dfrac{\sin 50°}{\cos 40°} = \dfrac{\sin 50°}{\sin(90° - 40°)} = \dfrac{\sin 50°}{\sin 50°} = 1$ 27. $\dfrac{\sin(-40°)}{\cos 50°} = \dfrac{-\sin 40°}{\sin(90° - 50°)} = \dfrac{-\sin 40°}{\sin 40°} = -1$

29. $\sin 400° \sec(-50°) = \sin(400° - 360°)\sec 50° = \sin 40°\,\csc(90° - 50°) = \sin 40°\,\csc 40°$

$$= \sin 40° \cdot \dfrac{1}{\sin 40°} = 1$$

31. $\sin \theta = \dfrac{-4}{5},\ \cos \theta > 0$

First we solve for $\cos \theta$:
$$\cos^2 \theta = 1 - \sin^2 \theta$$

$$\cos \theta = \sqrt{1 - \sin^2 \theta} = \sqrt{1 - \left[\dfrac{-4}{5}\right]^2} = \sqrt{1 - \dfrac{16}{25}} = \dfrac{3}{5}$$

$\tan \theta = \dfrac{\sin \theta}{\cos \theta} = \dfrac{\frac{-4}{5}}{\frac{3}{5}} = \dfrac{-4}{3}$ $\csc \theta = \dfrac{1}{\sin \theta} = \dfrac{1}{\frac{-4}{5}} = \dfrac{-5}{4}$

$\sec \theta = \dfrac{1}{\cos \theta} = \dfrac{1}{\frac{3}{5}} = \dfrac{5}{3}$ $\cot \theta = \dfrac{1}{\tan \theta} = \dfrac{1}{\frac{-4}{3}} = \dfrac{-3}{4}$

33. $\tan \theta = \dfrac{12}{5},\ \sin \theta < 0$

Because $\tan \theta = \dfrac{\sin \theta}{\cos \theta} > 0$ and $\sin \theta < 0$, $\cos \theta < 0$. Since $\cos \theta = \dfrac{1}{\sec \theta} < 0$, $\sec \theta < 0$.
$$\sec^2 \theta = \tan^2 \theta + 1$$

$\sec \theta = -\sqrt{\tan^2\theta + 1} = -\sqrt{\left[\dfrac{12}{5}\right]^2 + 1} = -\sqrt{\dfrac{144}{25} + 1} = -\sqrt{\dfrac{169}{25}} = \dfrac{-13}{5}$

$\cos \theta = \dfrac{1}{\sec \theta} = \dfrac{1}{\frac{-13}{5}} = \dfrac{-5}{13}$

$\sin \theta = -\sqrt{1 - \cos^2 \theta} = -\sqrt{1 - \left[\dfrac{-5}{13}\right]^2} = -\sqrt{1 - \dfrac{25}{169}} = -\sqrt{\dfrac{144}{169}} = \dfrac{-12}{13}$

$\csc \theta = \dfrac{1}{\sin \theta} = \dfrac{1}{\frac{-12}{13}} = \dfrac{-13}{12}$ $\cot \theta = \dfrac{1}{\tan \theta} = \dfrac{1}{\frac{12}{5}} = \dfrac{5}{12}$

35. $\sec \theta = \dfrac{-5}{4},\ \tan \theta < 0$
$$\tan^2 \theta + 1 = \sec^2 \theta$$
$$\tan^2 \theta = \sec^2 \theta - 1$$
$$\tan \theta = \pm\sqrt{\sec^2 \theta - 1}$$

Because tan $t < 0$, we use the minus sign:

$$\tan \theta = -\sqrt{\left(\frac{-5}{4}\right)^2 - 1} = -\sqrt{\frac{25}{16} - 1} = -\sqrt{\frac{9}{16}} = \frac{-3}{4}$$

$$\cot \theta = \frac{1}{\tan \theta} = \frac{1}{\dfrac{-3}{4}} = \frac{-4}{3} \qquad\qquad \cos \theta = \frac{1}{\sec \theta} = \frac{1}{\dfrac{-5}{4}} = \frac{-4}{5}$$

$$\sin \theta = \pm\sqrt{1 - \cos^2 \theta}$$

Since $\tan \theta = \dfrac{\sin \theta}{\cos \theta} < 0$ and $\cos \theta < 0$, $\sin \theta > 0$. Therefore, we use the plus sign:

$$\sin \theta = \sqrt{1 - \left(\frac{-4}{5}\right)^2} = \sqrt{1 - \frac{16}{25}} = \sqrt{\frac{9}{25}} = \frac{3}{5} \qquad\qquad \csc \theta = \frac{1}{\sin \theta} = \frac{1}{\dfrac{3}{5}} = \frac{5}{3} \quad .$$

37.　$\sin \theta = \dfrac{12}{13}$, θ in quadrant II

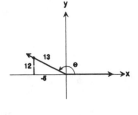

$\sin \theta = \dfrac{b}{r} = \dfrac{12}{13}$ so $b = 12$ and $r = 13$

Since $a < 0$ in quadrant II and using $r = \sqrt{a^2 + b^2}$, we get:

$$13 = \sqrt{a^2 + 12^2}$$
$$169 = a^2 + 144$$
$$25 = a^2$$
$$\pm 5 = a$$
$$-5 = a$$

$$\csc \theta = \frac{r}{b} = \frac{13}{12}$$

$$\cos \theta = \frac{a}{r} = \frac{-5}{13} = -\frac{5}{13} \qquad \sec \theta = \frac{r}{a} = \frac{13}{-5} = -\frac{13}{5}$$

$$\tan \theta = \frac{b}{a} = \frac{12}{-5} = -\frac{12}{5} \qquad \cot \theta = \frac{a}{b} = \frac{-5}{12} = -\frac{5}{12}$$

39.　$\sin \theta = -\dfrac{5}{13}$, $\dfrac{3\pi}{2} < \theta < 2\pi$

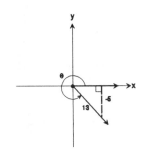

$\sin \theta = \dfrac{b}{r} = \dfrac{-5}{13}$, so $b = -5$ and $r = 13$

Since $a > 0$ in quadrant IV and using $r = \sqrt{a^2 + b^2}$, we get:

$$\sqrt{a^2 + (-5)^2} = 13$$
$$a^2 + 25 = 169$$
$$a^2 = 144$$
$$a = \pm 12$$
$$a = 12$$

$$\csc \theta = \frac{r}{b} = \frac{13}{-5} = -\frac{13}{5}$$

$$\cos \theta = \frac{a}{r} = \frac{12}{13} \qquad\qquad \sec \theta = \frac{r}{a} = \frac{13}{12}$$

$$\tan \theta = \frac{b}{a} = \frac{-5}{12} = -\frac{5}{12} \qquad \cot \theta = \frac{a}{b} = \frac{12}{-5} = -\frac{12}{5}$$

41. $\tan \theta = \dfrac{1}{3}$, $180° < \theta < 270°$

$\tan \theta = \dfrac{b}{a}$; but $a < 0$ and $b < 0$ in quadrant III, so

$\tan \theta = \dfrac{b}{a} = \dfrac{1}{3}$, so $a = -3$ and $b = -1$.

Using $r = \sqrt{a^2 + b^2}$, we find:

$$\sqrt{(-3)^2 + (-1)^2} = r$$
$$\sqrt{9 + 1} = r$$
$$\sqrt{10} = r$$

$\sin \theta = \dfrac{b}{r} = -\dfrac{1}{\sqrt{10}} = -\dfrac{1}{\sqrt{10}} \cdot \dfrac{\sqrt{10}}{\sqrt{10}} = -\dfrac{\sqrt{10}}{10}$ $\csc \theta = \dfrac{r}{b} = \dfrac{\sqrt{10}}{-1} = -\sqrt{10}$

$\cos \theta = \dfrac{a}{r} = \dfrac{-3}{\sqrt{10}} = -\dfrac{3}{\sqrt{10}} \cdot \dfrac{\sqrt{10}}{\sqrt{10}} = -\dfrac{3\sqrt{10}}{10}$ $\sec \theta = \dfrac{r}{a} = \dfrac{\sqrt{10}}{-3} = -\dfrac{\sqrt{10}}{3}$

$\cot \theta = \dfrac{a}{b} = \dfrac{-3}{-1} = 3$

43. $\sec \theta = 3$, $\dfrac{3\pi}{2} < \theta < 2\pi$

$\sec \theta = \dfrac{r}{a} = \dfrac{3}{1}$ so $a = 1$ and $r = 3$

Sin $b < 0$ in quadrant IV and using $\sqrt{a^2 + b^2}$ $= r$, we get:

$$\sqrt{1^2 + b^2} = 3$$
$$1 + b^2 = 9$$
$$b^2 = 8$$
$$b = \pm\sqrt{8}$$
$$b = -2\sqrt{2}$$

$\sin \theta = \dfrac{b}{r} = \dfrac{-2\sqrt{2}}{3}$ $\csc \theta = \dfrac{r}{b} = \dfrac{3}{-2\sqrt{2}} = -\dfrac{3}{2\sqrt{2}} \cdot \dfrac{\sqrt{2}}{\sqrt{2}} = -\dfrac{3\sqrt{2}}{4}$

$\cos \theta = \dfrac{a}{r} = \dfrac{1}{3}$

$\tan \theta = \dfrac{b}{a} = \dfrac{-2\sqrt{2}}{1} = -2\sqrt{2}$ $\cot \theta = \dfrac{a}{b} = \dfrac{1}{-2\sqrt{2}} = -\dfrac{1}{2\sqrt{2}} \cdot \dfrac{\sqrt{2}}{\sqrt{2}} = -\dfrac{\sqrt{2}}{4}$

45. $\cot \theta = -2$, $\dfrac{\pi}{2} < \theta < \pi$

$\cot \theta = \dfrac{a}{b}$, so $a < 0$ and $b > 0$ in quadrant II

$\cot \theta = \dfrac{a}{b} = \dfrac{-2}{1}$, so $a = -2$ and $b = 1$

Using $\sqrt{a^2 + b^2}$ $= r$, we get:

$$\sqrt{(-2)^2 + 1^2} = r$$
$$\sqrt{5} = r$$

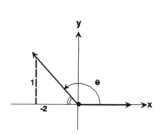

Chapter 6 Trigonometric Functions

$$\sin \theta = \frac{b}{r} = \frac{1}{\sqrt{5}} = \frac{\sqrt{5}}{5} \qquad\qquad \csc \theta = \frac{r}{b} = \sqrt{5}$$

$$\cos \theta = \frac{a}{r} = \frac{-2}{\sqrt{5}} = \frac{-2\sqrt{5}}{5} \qquad\qquad \sec \theta = \frac{r}{a} = \frac{\sqrt{5}}{-2} = \frac{-\sqrt{5}}{2}$$

$$\tan \theta = \frac{b}{a} = \frac{1}{-2} = \frac{-1}{2}$$

47.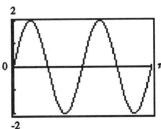

The graph of $y = 2 \sin 4x$ is the graph of $y = \sin x$ vertically stretched by a factor of 2 and horizontally compressed by a factor of 4 (i.e., each x-value is multiplied by $\frac{1}{4}$).

49.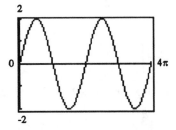

The graph of $y = -2 \cos \left(x + \dfrac{\pi}{2} \right)$ is the graph of $y = \cos x$ vertically stretched by a factor of 2, reflected about the x-axis and shifted horizontally $\dfrac{\pi}{2}$ units left.

51.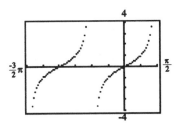

The graph of $y = \tan(x + \pi)$ is the graph of $y = \tan x$ shifted horizontally π units to the left.

53.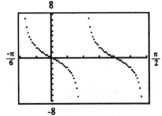

The graph $y = -2 \tan 3x$ is the graph of $y = \tan x$ vertically stretched by a factor of 2, reflected about the x-axis and horizontally compressed by a factor of 3.

55. $\sin^{-1} 1$

We are looking for the angle θ, $-\dfrac{\pi}{2} \le \theta \le \dfrac{\pi}{2}$, whose sine is 1.

$$\sin \theta = 1 \qquad\qquad -\frac{\pi}{2} \le \theta \le \frac{\pi}{2}$$

$$\theta = \frac{\pi}{2}$$

Hence, $\sin^{-1} 1 = \dfrac{\pi}{2}$

57. $\tan^{-1} 1$

We are looking for the angle θ, $-\dfrac{\pi}{2} < \theta < \dfrac{\pi}{2}$, whose tangent is 1.

$\tan \theta = 1 \qquad\qquad -\dfrac{\pi}{2} < \theta < \dfrac{\pi}{2}$

$\theta = \dfrac{\pi}{4}$

Hence, $\tan^{-1} 1 = \dfrac{\pi}{4}$

59. $\cos^{-1}\left(-\dfrac{\sqrt{3}}{2}\right)$

We are looking for the angle θ, $0 \le \theta \le \pi$, whose cosine is $-\dfrac{\sqrt{3}}{2}$.

$\cos \theta = -\dfrac{\sqrt{3}}{2} \qquad\qquad 0 \le \theta \le \pi$

$\theta = \dfrac{5\pi}{6}$

Hence, $\cos^{-1}\left(-\dfrac{\sqrt{3}}{2}\right) = \dfrac{5\pi}{6}$

61. $\sin\left(\cos^{-1}\dfrac{\sqrt{2}}{2}\right)$

First find the angle θ, $0 \le \theta \le \pi$, whose cosine equals $\dfrac{\sqrt{2}}{2}$.

$\cos \theta = \dfrac{\sqrt{2}}{2}, \; 0 \le \theta \le \pi$

$\theta = \dfrac{\pi}{4}$

Now, $\sin\left(\cos^{-1}\dfrac{\sqrt{2}}{2}\right) = \sin \theta = \sin \dfrac{\pi}{4} = \dfrac{\sqrt{2}}{2}$

63. $\tan\left[\sin^{-1}\left(-\dfrac{\sqrt{3}}{2}\right)\right]$

First find the angle θ, $-\dfrac{\pi}{2} \le \theta \le \dfrac{\pi}{2}$, whose sine equals $-\dfrac{\sqrt{3}}{2}$.

$\sin \theta = -\dfrac{\sqrt{3}}{2}, \; -\dfrac{\pi}{2} \le \theta \le \dfrac{\pi}{2}$

$\theta = -\dfrac{\pi}{3}$

Now, $\tan\left[\sin^{-1}\left(-\dfrac{\sqrt{3}}{2}\right)\right] = \tan \theta = \tan\left(-\dfrac{\pi}{3}\right) = -\sqrt{3}$

65. $\sec\left[\tan^{-1}\dfrac{\sqrt{3}}{3}\right]$

First find the angle θ, $-\dfrac{\pi}{2} < \theta < \dfrac{\pi}{2}$, whose tangent equals $\dfrac{\sqrt{3}}{3}$.

$$\tan\theta = \dfrac{\sqrt{3}}{3} \quad -\dfrac{\pi}{2} < \theta < \dfrac{\pi}{2}$$

$$\theta = \dfrac{\pi}{6}$$

Now, $\sec\left[\tan^{-1}\dfrac{\sqrt{3}}{3}\right] = \sec\theta = \sec\dfrac{\pi}{6} = \dfrac{2\sqrt{3}}{3}$

67. $\sin\left[\tan^{-1}\dfrac{3}{4}\right]$

Let $\theta = \tan^{-1}\dfrac{3}{4}$

Then, $\tan\theta = \dfrac{3}{4} = \dfrac{\text{opp}}{\text{adj}},\ -\dfrac{\pi}{2} < \theta < \dfrac{\pi}{2}$

The hypotenuse is $r = \sqrt{x^2 + y^2}$

$$r = \sqrt{16 + 9}$$
$$r = \sqrt{25}$$
$$r = 5$$

Thus, $\sin\left[\tan^{-1}\dfrac{3}{4}\right] = \sin\theta = \dfrac{3}{5}$

69. $\tan\left[\sin^{-1}\left(-\dfrac{4}{5}\right)\right]$

Let $\theta = \sin^{-1}\left(-\dfrac{4}{5}\right)$

Then, $\sin\theta = -\dfrac{4}{5} = \dfrac{\text{opp}}{\text{hyp}},\ -\dfrac{\pi}{2} \le \theta \le \dfrac{\pi}{2}.$
Using the Pythagorean Theorem, we find the missing side:

$$x^2 + (-4)^2 = 5^2$$
$$x^2 + 16 = 25$$
$$x^2 = 9$$
$$x = \pm 3$$
$$x = 3,\ x > 0 \text{ in quadrant IV}$$

Hence, $\tan\left[\sin^{-1}\left(-\dfrac{4}{5}\right)\right] = \tan\theta = -\dfrac{4}{3}$

71. $\theta = 30°$ or $\theta = \dfrac{\pi}{6}$

radius = 2 feet
Using $s = r\theta$, we get:

$$s = 2\left[\dfrac{\pi}{6}\right]$$
$$s = \dfrac{\pi}{3} \text{ feet}$$

73. $v = 180$ mi/hr

diameter $= \dfrac{1}{2}$ mi, so $r = \dfrac{1}{4}$ mi
Find angular speed: ω
Using $v = r\omega$, we get:

$$180 = \dfrac{1}{4}\omega$$

720 rad/hr $= \omega$ (Remember that ω is expressed in radians per unit time.)

$$\dfrac{720\text{ rad}}{\text{hr}} \cdot \dfrac{1\text{ rev}}{2\pi\text{ rad}} = \omega$$

$$\dfrac{720}{2\pi} \text{ rev/hr} = \omega$$

$$114.59 \text{ rev/hr} = \omega$$

ANALYTIC TRIGONOMETRY

7.1 Trigonometric Identities

1. To prove that $\csc \theta \cos \theta = \cot \theta$, we start with the left side and apply a reciprocal identity:

$$\csc \theta \cdot \cos \theta = \frac{1}{\sin \theta} \cdot \cos \theta = \frac{\cos \theta}{\sin \theta} = \cot \theta$$

3. To prove that $1 + \tan^2(-\theta) = \sec^2 \theta$, we begin with the left side and apply an even-odd identity:
$$1 + \tan^2(-\theta) = 1 + (-\tan \theta)^2 = 1 + \tan^2 \theta = \sec^2 \theta$$

5. To prove that $\cos \theta(\tan \theta + \cot \theta) = \csc \theta$, we begin with the left side because it contains the more complicated expression, and write the expression so it contains only sines and cosines:

$$\cos \theta(\tan \theta + \cot \theta) = \cos \theta \left[\frac{\sin \theta}{\cos \theta} + \frac{\cos \theta}{\sin \theta} \right] = \cos \theta \left[\frac{\sin^2 \theta + \cos^2 \theta}{\cos \theta \sin \theta} \right] = \frac{1}{\sin \theta} = \csc \theta$$

7. To prove that $\tan \theta \cot \theta - \cos^2 \theta = \sin^2 \theta$, we write the left side expression in terms of sines and cosines:

$$\tan \theta \cot \theta - \cos^2 \theta = \frac{\sin \theta}{\cos \theta} \cdot \frac{\cos \theta}{\sin \theta} - \cos^2 \theta = 1 - \cos^2 \theta = \sin^2 \theta$$

9. To prove that $(\sec \theta - 1)(\sec \theta + 1) = \tan^2 \theta$, we begin with the left side, multiply, and then apply a form of a Pythagorean identity:
$$(\sec \theta - 1)(\sec \theta + 1) = \sec^2 \theta - 1 = \tan^2 \theta$$

11. To prove that $(\sec \theta + \tan \theta)(\sec \theta - \tan \theta) = 1$, we begin with the left side, multiply, and then apply a form of a Pythagorean identity:
$$(\sec \theta + \tan \theta)(\sec \theta - \tan \theta) = \sec^2 \theta - \tan^2 \theta = 1$$

13. To prove that $\sin^2 \theta(1 + \cot^2 \theta) = 1$, we apply a form of a Pythagorean identity:

$$\sin^2 \theta(1 + \cot^2 \theta) = \sin^2 \theta(\csc^2 \theta) = \sin^2 \theta \left[\frac{1}{\sin^2 \theta} \right] = 1$$

15. To prove that $(\sin \theta + \cos \theta)^2 + (\sin \theta - \cos \theta)^2 = 2$, we begin with the complicated left side and proceed by carrying out the exponents and simplifying:
$$(\sin \theta + \cos \theta)^2 + (\sin \theta - \cos \theta)^2$$
$$= \sin^2 \theta + 2 \sin \theta \cos \theta + \cos^2 \theta + \sin^2 \theta - 2 \sin \theta \cos \theta + \cos^2 \theta$$
$$= \sin^2 \theta + \cos^2 \theta + \sin^2 \theta + \cos^2 \theta = 1 + 1 = 2$$

17. To prove that $\sec^4 \theta - \sec^2 \theta = \tan^4 \theta + \tan^2 \theta$, we begin with the left side and factor:
$$\sec^4 \theta - \sec^2 \theta = \sec^2 \theta (\sec^2 \theta - 1) = (1 + \tan^2 \theta) \tan^2 \theta = \tan^2 \theta + \tan^4 \theta$$
$$= \tan^4 \theta + \tan^2 \theta$$

19. To prove that $\sec \theta - \tan \theta = \dfrac{\cos \theta}{1 + \sin \theta}$, we put the left side's expression in terms of sines and cosines:
$$\sec \theta - \tan \theta = \frac{1}{\cos \theta} - \frac{\sin \theta}{\cos \theta} = \frac{1 - \sin \theta}{\cos \theta} \cdot \frac{1 + \sin \theta}{1 + \sin \theta} = \frac{1 - \sin^2 \theta}{\cos \theta (1 + \sin \theta)}$$
$$= \frac{\cos^2 \theta}{\cos \theta (1 + \sin \theta)} = \frac{\cos \theta}{1 + \sin \theta}$$

21. To prove that $3 \sin^2 \theta + 4 \cos^2 \theta = 3 + \cos^2 \theta$, we begin with the left side and let $4 \cos^2 \theta = 3 \cos^2 \theta + 1 \cos^2 \theta$:
$$3 \sin^2 \theta + 4 \cos^2 \theta = 3 \sin^2 \theta + 3 \cos^2 \theta + \cos^2 \theta = 3(\sin^2 \theta + \cos^2 \theta) + \cos^2 \theta$$
$$= 3 + \cos^2 \theta$$

23. To prove that $1 - \dfrac{\cos^2 \theta}{1 + \sin \theta} = \sin \theta$, we begin with the complicated left side and use a form of a Phythagorean identity:
$$1 - \frac{\cos^2 \theta}{1 + \sin \theta} = 1 - \frac{1 - \sin^2 \theta}{1 + \sin \theta} = 1 - \frac{(1 - \sin \theta)(1 + \sin \theta)}{1 + \sin \theta} = 1 - 1 + \sin \theta = \sin \theta$$

25. To prove that $\dfrac{1 + \tan \theta}{1 - \tan \theta} = \dfrac{\cot \theta + 1}{\cot \theta - 1}$,
$$\frac{1 + \tan \theta}{1 - \tan \theta} = \frac{1 + \dfrac{1}{\cot \theta}}{1 - \dfrac{1}{\cot \theta}} = \frac{\dfrac{\cot \theta + 1}{\cot \theta}}{\dfrac{\cot \theta - 1}{\cot \theta}} = \frac{\cot \theta + 1}{\cot \theta - 1}$$

27. To prove that $\dfrac{\sec \theta}{\csc \theta} + \dfrac{\sin \theta}{\cos \theta} = 2 \tan \theta$,
$$\frac{\sec \theta}{\csc \theta} + \frac{\sin \theta}{\cos \theta} = \frac{\dfrac{1}{\cos \theta}}{\dfrac{1}{\sin \theta}} + \tan \theta = \frac{\sin \theta}{\cos \theta} + \tan \theta = \tan \theta + \tan \theta = 2 \tan \theta$$

29. To prove that $\dfrac{1 + \sin \theta}{1 - \sin \theta} = \dfrac{\csc \theta + 1}{\csc \theta - 1}$,
$$\frac{1 + \sin \theta}{1 - \sin \theta} = \frac{1 + \dfrac{1}{\csc \theta}}{1 - \dfrac{1}{\csc \theta}} = \frac{\dfrac{\csc \theta + 1}{\csc \theta}}{\dfrac{\csc \theta - 1}{\csc \theta}} = \frac{\csc \theta + 1}{\csc \theta - 1}$$

31. To prove that $\dfrac{1 - \sin \theta}{\cos \theta} + \dfrac{\cos \theta}{1 - \sin \theta} = 2 \sec \theta$,
$$\frac{1 - \sin \theta}{\cos \theta} + \frac{\cos \theta}{1 - \sin \theta} = \frac{(1 - \sin \theta)^2 + \cos^2 \theta}{\cos \theta (1 - \sin \theta)} = \frac{1 - 2 \sin \theta + \sin^2 \theta + \cos^2 \theta}{\cos \theta (1 - \sin \theta)}$$
$$= \frac{2 - 2 \sin \theta}{\cos \theta (1 - \sin \theta)} = \frac{2(1 - \sin \theta)}{\cos \theta (1 - \sin \theta)} = \frac{2}{\cos \theta} = 2 \sec \theta$$

33. To prove that $\dfrac{\sin \theta}{\sin \theta - \cos \theta} = \dfrac{1}{1 - \cot \theta}$,

$$\frac{\sin \theta}{\sin \theta - \cos \theta} = \frac{1}{\dfrac{\sin \theta - \cos \theta}{\sin \theta}} = \frac{1}{1 - \dfrac{\cos \theta}{\sin \theta}} = \frac{1}{1 - \cot \theta}$$

35. To prove that $\dfrac{1 - \sin \theta}{1 + \sin \theta} = (\sec \theta - \tan \theta)^2$, we begin with the right side and square the expression:

$$(\sec \theta - \tan \theta)^2 = \sec^2 \theta - 2 \sec \theta \tan \theta + \tan^2 \theta = \frac{1}{\cos^2 \theta} - \frac{2 \sin \theta}{\cos^2 \theta} + \frac{\sin^2 \theta}{\cos^2 \theta}$$

$$= \frac{1 - 2 \sin \theta + \sin^2 \theta}{\cos^2 \theta} = \frac{(1 - \sin \theta)^2}{1 - \sin^2 \theta} = \frac{(1 - \sin \theta)^2}{(1 - \sin \theta)(1 + \sin \theta)}$$

$$= \frac{1 - \sin \theta}{1 + \sin \theta}$$

37. To prove that $\dfrac{\cos \theta}{1 - \tan \theta} + \dfrac{\sin \theta}{1 - \cot \theta} = \sin \theta + \cos \theta$,

$$\frac{\cos \theta}{1 - \tan \theta} + \frac{\sin \theta}{1 - \cot \theta} = \frac{\cos \theta}{1 - \dfrac{\sin \theta}{\cos \theta}} + \frac{\sin \theta}{1 - \dfrac{\cos \theta}{\sin \theta}} = \frac{\cos \theta}{\dfrac{\cos \theta - \sin \theta}{\cos \theta}} + \frac{\sin \theta}{\dfrac{\sin \theta - \cos \theta}{\sin \theta}}$$

$$= \frac{\cos^2 \theta}{\cos \theta - \sin \theta} + \frac{\sin^2 \theta}{\sin \theta - \cos \theta} = \frac{\cos^2 \theta - \sin^2 \theta}{\cos \theta - \sin \theta}$$

$$= \frac{(\cos \theta - \sin \theta)(\cos \theta + \sin \theta)}{\cos - \sin \theta} = \cos \theta + \sin \theta = \sin \theta + \cos \theta$$

39. To prove that $\tan \theta + \dfrac{\cos \theta}{1 + \sin \theta} = \sec \theta$,

$$\tan \theta + \frac{\cos \theta}{1 + \sin \theta} = \frac{\sin \theta}{\cos \theta} + \frac{\cos \theta}{(1 + \sin \theta)} = \frac{\sin \theta(1 + \sin \theta) + \cos^2 \theta}{\cos \theta(1 + \sin \theta)}$$

$$= \frac{\sin \theta + \sin^2 \theta + \cos^2 \theta}{\cos \theta(1 + \sin \theta)} = \frac{\sin \theta + 1}{\cos \theta(1 + \sin \theta)} = \frac{1}{\cos \theta} = \sec \theta$$

41. To prove that $\dfrac{\tan \theta + \sec \theta - 1}{\tan \theta - \sec \theta + 1} = \tan \theta + \sec \theta$,

$$\frac{\tan \theta + \sec \theta - 1}{\tan \theta - \sec \theta + 1} = \frac{\tan \theta + (\sec \theta - 1)}{\tan \theta - (\sec \theta - 1)} \cdot \frac{\tan \theta + (\sec \theta - 1)}{\tan \theta + (\sec \theta - 1)}$$

$$= \frac{\tan^2 \theta + 2 \tan \theta(\sec \theta - 1) + \sec^2 \theta - 2 \sec \theta + 1}{\tan^2 \theta - (\sec^2 \theta - 2 \sec \theta + 1)}$$

$$= \frac{\sec^2 \theta - 1 + 2 \tan \theta(\sec \theta - 1) + \sec^2 \theta - 2 \sec \theta + 1}{\sec^2 \theta - 1 - \sec^2 \theta + 2 \sec \theta - 1}$$

$$= \frac{2 \sec^2 \theta - 2 \sec \theta + 2 \tan \theta(\sec \theta - 1)}{-2 + 2 \sec \theta}$$

$$= \frac{2 \sec \theta(\sec \theta - 1) + 2 \tan \theta(\sec \theta - 1)}{2(\sec \theta - 1)}$$

$$= \frac{2 (\sec \theta - 1)(\sec \theta + \tan \theta)}{2(\sec \theta - 1)} = \sec \theta + \tan \theta = \tan \theta + \sec \theta$$

Chapter 7 Analytic Trigonometry

43. To prove that $\dfrac{\tan \theta - \cot \theta}{\tan \theta + \cot \theta} = \sin^2 \theta - \cos^2 \theta,$

$$\dfrac{\tan \theta - \cot \theta}{\tan \theta + \cot \theta} = \dfrac{\dfrac{\sin \theta}{\cos \theta} - \dfrac{\cos \theta}{\sin \theta}}{\dfrac{\sin \theta}{\cos \theta} + \dfrac{\cos \theta}{\sin \theta}} = \dfrac{\dfrac{\sin^2 \theta - \cos^2 \theta}{\cos \theta \sin \theta}}{\dfrac{\sin^2 \theta + \cos^2 \theta}{\cos \theta \sin \theta}} = \dfrac{\sin^2 \theta - \cos^2 \theta}{1}$$

$$= \sin^2 \theta - \cos^2 \theta$$

45. To prove that $\dfrac{\tan \theta - \cot \theta}{\tan \theta + \cot \theta} = 2 \sin^2 \theta - 1,$

$$\dfrac{\tan \theta - \cot \theta}{\tan \theta + \cot \theta} = \dfrac{\dfrac{\sin \theta}{\cos \theta} - \dfrac{\cos \theta}{\sin \theta}}{\dfrac{\sin \theta}{\cos \theta} + \dfrac{\cos \theta}{\sin \theta}} = \dfrac{\dfrac{\sin^2 \theta - \cos^2 \theta}{\cos \theta \sin \theta}}{\dfrac{\sin^2 \theta + \cos^2 \theta}{\cos \theta \sin \theta}} = \sin^2 \theta - \cos^2 \theta$$

$$= \sin^2 \theta - (1 - \sin^2 \theta) = 2 \sin^2 \theta - 1$$

47. To prove that $\dfrac{\sec \theta + \tan \theta}{\cot \theta + \cos \theta} = \tan \theta \sec \theta,$

$$\dfrac{\sec \theta + \tan \theta}{\cot \theta + \cos \theta} = \dfrac{\dfrac{1}{\cos \theta} + \dfrac{\sin \theta}{\cos \theta}}{\dfrac{\cos \theta}{\sin \theta} + \cos \theta} = \dfrac{\dfrac{1 + \sin \theta}{\cos \theta}}{\dfrac{\cos \theta + \cos \theta \sin \theta}{\sin \theta}}$$

$$= \dfrac{1 + \sin \theta}{\cos \theta} \cdot \dfrac{\sin \theta}{\cos \theta(1 + \sin \theta)} = \dfrac{\sin \theta}{\cos \theta} \cdot \dfrac{1}{\cos \theta} = \tan \theta \sec \theta$$

49. To prove that $\dfrac{1 - \tan^2 \theta}{1 + \tan^2 \theta} = 2 \cos^2 \theta - 1,$

$$\dfrac{1 - \tan^2 \theta}{1 + \tan^2 \theta} = \dfrac{1 - \tan^2 \theta}{\sec^2 \theta} = \dfrac{1}{\sec^2 \theta} - \dfrac{\tan^2 \theta}{\sec^2 \theta} = \cos^2 \theta - \dfrac{\dfrac{\sin^2 \theta}{\cos^2 \theta}}{\dfrac{1}{\cos^2 \theta}} = \cos^2 \theta - \sin^2 \theta$$

$$= \cos^2 \theta - (1 - \cos^2 \theta) = 2 \cos^2 \theta - 1$$

51. To prove that $\dfrac{\sec \theta - \csc \theta}{\sec \theta \csc \theta} = \sin \theta - \cos \theta,$

$$\dfrac{\sec \theta - \csc \theta}{\sec \theta \csc \theta} = \dfrac{\dfrac{1}{\cos \theta} - \dfrac{1}{\sin \theta}}{\dfrac{1}{\cos \theta} \cdot \dfrac{1}{\sin \theta}} = \dfrac{\dfrac{\sin \theta - \cos \theta}{\cos \theta \cdot \sin \theta}}{\dfrac{1}{\cos \theta \cdot \sin \theta}} = \sin \theta - \cos \theta$$

53. To prove that $\sec \theta - \cos \theta = \sin \theta \tan \theta,$

$$\sec \theta - \cos \theta = \dfrac{1}{\cos \theta} - \cos \theta = \dfrac{1 - \cos^2 \theta}{\cos \theta} = \dfrac{\sin^2 \theta}{\cos \theta} = \sin \theta \cdot \dfrac{\sin \theta}{\cos \theta} = \sin \theta \tan \theta$$

55. To prove that $\dfrac{1}{1 - \sin \theta} + \dfrac{1}{1 + \sin \theta} = 2 \sec^2 \theta,$

$$\dfrac{1}{1 - \sin \theta} + \dfrac{1}{1 + \sin \theta} = \dfrac{1 + \sin \theta + 1 - \sin \theta}{(1 + \sin \theta)(1 - \sin \theta)} = \dfrac{2}{1 - \sin^2 \theta} = \dfrac{2}{\cos^2 \theta} = 2 \sec^2 \theta$$

57. To prove that $\dfrac{\sec \theta}{1 - \sin \theta} = \dfrac{1 + \sin \theta}{\cos^3 \theta}$,

$$\dfrac{\sec \theta}{1 - \sin \theta} = \dfrac{\sec \theta}{1 - \sin \theta} \cdot \dfrac{1 + \sin \theta}{1 + \sin \theta} = \dfrac{\sec \theta(1 + \sin \theta)}{1 - \sin^2 \theta} = \dfrac{\sec \theta(1 + \sin \theta)}{\cos^2 \theta}$$

$$= \dfrac{1 + \sin \theta}{\cos^3 \theta}$$

59. To prove that $\dfrac{(\sec \theta - \tan \theta)^2 + 1}{\csc \theta(\sec \theta - \tan \theta)} = 2 \tan \theta$,

$$\dfrac{(\sec \theta - \tan \theta)^2 + 1}{\csc \theta(\sec \theta - \tan \theta)} = \dfrac{\sec^2 \theta - 2 \sec \theta \tan \theta + \tan^2 \theta + 1}{\csc \theta \sec \theta - \csc \theta \tan \theta} = \dfrac{2 \sec^2 \theta - 2 \sec \theta \tan \theta}{\csc \theta \sec \theta - \csc \theta \tan \theta}$$

$$= \dfrac{\dfrac{2}{\cos^2 \theta} - \dfrac{2 \sin \theta}{\cos^2 \theta}}{\dfrac{1}{\sin \theta \cos \theta} - \dfrac{\sin \theta}{\sin \theta \cos \theta}} = \dfrac{\dfrac{2 - 2 \sin \theta}{\cos^2 \theta}}{\dfrac{1 - \sin \theta}{\sin \theta \cos \theta}}$$

$$= \dfrac{2(1 - \sin \theta)}{\cos^2 \theta} \cdot \dfrac{\sin \theta \cos \theta}{1 - \sin \theta} = \dfrac{2 \sin \theta}{\cos \theta} = 2 \tan \theta$$

61. To prove that $\dfrac{\sin \theta + \cos \theta}{\cos \theta} - \dfrac{\sin \theta - \cos \theta}{\sin \theta} = \sec \theta \csc \theta$,

$$\dfrac{\sin \theta + \cos \theta}{\cos \theta} - \dfrac{\sin \theta - \cos \theta}{\sin \theta} = \dfrac{\sin \theta(\sin \theta + \cos \theta) - \cos \theta(\sin \theta - \cos \theta)}{\cos \theta \sin \theta}$$

$$= \dfrac{\sin^2 \theta + \sin \theta \cos \theta - \sin \theta \cos \theta + \cos^2 \theta}{\cos \theta \sin \theta}$$

$$= \dfrac{1}{\cos \theta \sin \theta} = \sec \theta \csc \theta$$

63. To prove that $\dfrac{\sin^3 \theta + \cos^3 \theta}{\sin \theta + \cos \theta} = 1 - \sin \theta \cos \theta$,

$$\dfrac{\sin^3 \theta + \cos^3 \theta}{\sin \theta + \cos \theta} = \dfrac{(\sin \theta + \cos \theta)(\sin^2 \theta - \sin \theta \cos \theta + \cos^2 \theta)}{\sin \theta + \cos \theta}$$

$$= \sin^2 \theta + \cos^2 \theta - \sin \theta \cos \theta = 1 - \sin \theta \cos \theta$$

65. To prove that $\dfrac{\cos^2 \theta - \sin^2 \theta}{1 - \tan^2 \theta} = \cos^2 \theta$,

$$\dfrac{\cos^2 \theta - \sin^2 \theta}{1 - \tan^2 \theta} = \dfrac{\cos^2 \theta - \sin^2 \theta}{1 - \dfrac{\sin^2 \theta}{\cos^2 \theta}} = \dfrac{\cos^2 \theta - \sin^2 \theta}{\dfrac{\cos^2 \theta - \sin^2 \theta}{\cos^2 \theta}} = \cos^2 \theta$$

67. To prove that $\dfrac{(2 \cos^2 \theta - 1)^2}{\cos^4 \theta - \sin^4 \theta} = 1 - 2 \sin^2 \theta$,

$$\dfrac{(2 \cos^2 \theta - 1)^2}{\cos^4 \theta - \sin^4 \theta} = \dfrac{[2 \cos^2 \theta - (\sin^2 \theta + \cos^2 \theta)]^2}{(\cos^2 \theta - \sin^2 \theta)(\cos^2 \theta + \sin^2 \theta)}$$

$$= \dfrac{(\cos^2 \theta - \sin^2 \theta)^2}{(\cos^2 \theta - \sin^2 \theta)(\cos^2 \theta + \sin^2 \theta)} = \dfrac{\cos^2 \theta - \sin^2 \theta}{\cos^2 \theta + \sin^2 \theta}$$

$$= \cos^2 \theta - \sin^2 \theta = (1 - \sin^2 \theta) - \sin^2 \theta = 1 - 2 \sin^2 \theta$$

69. To prove that $\dfrac{1 + \sin\theta + \cos\theta}{1 + \sin\theta - \cos\theta} = \dfrac{1 + \cos\theta}{\sin\theta}$,

$$\dfrac{1 + \sin\theta + \cos\theta}{1 + \sin\theta - \cos\theta} = \dfrac{(1 + \sin\theta) + \cos\theta}{(1 + \sin\theta) - \cos\theta} \cdot \dfrac{(1 + \sin\theta) + \cos\theta}{(1 + \sin\theta) + \cos\theta}$$

$$= \dfrac{1 + 2\sin\theta + \sin^2\theta + 2(1 + \sin\theta)(\cos\theta) + \cos^2\theta}{1 + 2\sin\theta + \sin^2\theta - \cos^2\theta}$$

$$= \dfrac{1 + 2\sin\theta + \sin^2\theta + 2(1 + \sin\theta)(\cos\theta) + (1 - \sin^2\theta)}{1 + 2\sin\theta + \sin^2\theta - (1 - \sin^2\theta)}$$

$$= \dfrac{2 + 2\sin\theta + 2(1 + \sin\theta)(\cos\theta)}{2\sin\theta + 2\sin^2\theta} = \dfrac{2(1 + \sin\theta) + 2(1 + \sin\theta)(\cos\theta)}{2\sin\theta(1 + \sin\theta)}$$

$$= \dfrac{2(1 + \sin\theta)(1 + \cos\theta)}{2\sin\theta(1 + \sin\theta)} = \dfrac{1 + \cos\theta}{\sin\theta}$$

71. To prove that $(a\sin\theta + b\cos\theta)^2 + (a\cos\theta - b\sin\theta)^2 = a^2 + b^2$,

$$(a\sin\theta + b\cos\theta)^2 + (a\cos\theta - b\sin\theta)^2$$
$$= a^2\sin^2\theta + 2ab\sin\theta\cos\theta + b^2\cos^2\theta + a^2\cos^2\theta - 2ab\sin\theta\cos\theta + b^2\sin^2\theta$$
$$= a^2(\sin^2\theta + \cos^2\theta) + b^2(\cos^2\theta + \sin^2\theta) = a^2 + b^2$$

73. To prove that $\dfrac{\tan\alpha + \tan\beta}{\cot\alpha + \cot\beta} = \tan\alpha\tan\beta$,

$$\dfrac{\tan\alpha + \tan\beta}{\cot\alpha + \cot\beta} = \dfrac{\tan\alpha + \tan\beta}{\dfrac{1}{\tan\alpha} + \dfrac{1}{\tan\beta}} = \dfrac{\tan\alpha + \tan\beta}{\dfrac{\tan\beta + \tan\alpha}{\tan\alpha\tan\beta}}$$

$$= (\tan\alpha + \tan\beta) \cdot \dfrac{\tan\alpha\tan\beta}{\tan\alpha + \tan\beta} = \tan\alpha\tan\beta$$

75. To prove that $(\sin\alpha + \cos\beta)^2 + (\cos\beta + \sin\alpha)(\cos\beta - \sin\alpha) = 2\cos\beta(\sin\alpha + \cos\beta)$,

$$(\sin\alpha + \cos\beta)^2 + (\cos\beta + \sin\alpha)(\cos\beta - \sin\alpha)$$
$$= (\sin^2\alpha + 2\sin\alpha\cos\beta + \cos^2\beta) + (\cos^2\beta - \sin^2\alpha)$$
$$= 2\cos^2\beta + 2\sin\alpha\cos\beta = 2\cos\beta(\cos\beta + \sin\alpha)$$

77. To prove that $\ln|\sec\theta| = -\ln|\cos\theta|$,

$$\ln|\sec\theta| = \ln|\cos\theta|^{-1} = -\ln|\cos\theta|$$

79. To prove that $\ln|1 + \cos\theta| + \ln|1 - \cos\theta| = 2\ln|\sin\theta|$,

$$\ln|1 + \cos\theta| + \ln|1 - \cos\theta| = \ln(|1 + \cos\theta|\,|1 - \cos\theta|)$$
$$= \ln|1 - \cos^2\theta| = \ln|\sin^2\theta| = 2\ln|\sin\theta|$$

7.2 Sum and Difference Formulas

1. $\sin\dfrac{5\pi}{12} = \sin\left(\dfrac{3\pi}{12} + \dfrac{2\pi}{12}\right) = \sin\dfrac{\pi}{4}\cos\dfrac{\pi}{6} + \cos\dfrac{\pi}{4}\sin\dfrac{\pi}{6} = \dfrac{\sqrt{2}}{2} \cdot \dfrac{\sqrt{3}}{2} + \dfrac{\sqrt{2}}{2} \cdot \dfrac{1}{2}$

$\qquad = \dfrac{1}{4}\left(\sqrt{6} + \sqrt{2}\right)$

3. $\cos\dfrac{7\pi}{12} = \cos\left(\dfrac{4\pi}{12} + \dfrac{3\pi}{12}\right) = \cos\dfrac{\pi}{3}\cos\dfrac{\pi}{4} - \sin\dfrac{\pi}{3}\sin\dfrac{\pi}{4}$

$\qquad = \dfrac{1}{2} \cdot \dfrac{\sqrt{2}}{2} - \dfrac{\sqrt{3}}{2} \cdot \dfrac{\sqrt{2}}{2} = \dfrac{1}{4}\left(\sqrt{2} - \sqrt{6}\right)$

5. $\cos 165° = \cos(120° + 45°) = \cos 120° \cos 45° - \sin 120° \sin 45°$

$$= -\frac{1}{2} \cdot \frac{\sqrt{2}}{2} - \frac{\sqrt{3}}{2} \cdot \frac{\sqrt{2}}{2} = \frac{-1}{4}\left(\sqrt{2} + \sqrt{6}\right)$$

7. $\tan 15° = \tan(45° - 30°) = \dfrac{\tan 45° - \tan 30°}{1 + \tan 45° \tan 30°} = \dfrac{1 - \dfrac{1}{\sqrt{3}}}{1 + 1 \cdot \dfrac{1}{\sqrt{3}}} = \dfrac{\dfrac{\sqrt{3} - 1}{\sqrt{3}}}{\dfrac{\sqrt{3} + 1}{\sqrt{3}}}$

$$= \frac{\sqrt{3} - 1}{1 + \sqrt{3}} \cdot \frac{1 - \sqrt{3}}{1 - \sqrt{3}} = \frac{2\sqrt{3} - 4}{-2} = \frac{-2\left(2 - \sqrt{3}\right)}{-2} = 2 - \sqrt{3}$$

9. $\sin \dfrac{17\pi}{12} = \sin\left[\dfrac{15\pi}{12} + \dfrac{2\pi}{12}\right] = \sin \dfrac{5\pi}{4} \cos \dfrac{\pi}{6} + \cos \dfrac{5\pi}{4} \sin \dfrac{\pi}{6}$

$$= \frac{-\sqrt{2}}{2} \cdot \frac{\sqrt{3}}{2} + \frac{-\sqrt{2}}{2} \cdot \frac{1}{2} = \frac{-1}{4}\left(\sqrt{6} + \sqrt{2}\right)$$

11. $\sec\left[-\dfrac{\pi}{12}\right] = \dfrac{1}{\cos\left[\dfrac{-\pi}{12}\right]} = \dfrac{1}{\cos\left[\dfrac{3\pi}{12} - \dfrac{4\pi}{12}\right]} = \dfrac{1}{\cos \dfrac{\pi}{4} \cos \dfrac{\pi}{3} + \sin \dfrac{\pi}{4} \sin \dfrac{\pi}{3}}$

$$= \frac{1}{\dfrac{\sqrt{2}}{2} \cdot \dfrac{1}{2} + \dfrac{\sqrt{2}}{2} \cdot \dfrac{\sqrt{3}}{2}} = \frac{1}{\dfrac{\sqrt{2} + \sqrt{2}\sqrt{3}}{4}} = \frac{4}{\sqrt{2} + \sqrt{6}}$$

$$= \frac{4}{\sqrt{2} + \sqrt{6}} \cdot \frac{\sqrt{2} - \sqrt{6}}{\sqrt{2} - \sqrt{6}} = \frac{4\left(\sqrt{2} - \sqrt{6}\right)}{-4} = \sqrt{6} - \sqrt{2}$$

13. $\sin 20° \cos 10° + \cos 20° \sin 10° = \sin(20° + 10°) = \sin 30° = \dfrac{1}{2}$

15. $\cos 70° \cos 20° - \sin 70° \sin 20° = \cos(70° + 20°) = \cos 90° = 0$

17. $\dfrac{\tan 20° + \tan 25°}{1 - \tan 20° \tan 25°} = \tan(20° + 25°) = \tan 45° = 1$

19. $\sin \dfrac{\pi}{12} \cos \dfrac{7\pi}{12} - \cos \dfrac{\pi}{12} \sin \dfrac{7\pi}{12} = \sin\left[\dfrac{\pi}{12} - \dfrac{7\pi}{12}\right] = \sin\left[\dfrac{-\pi}{2}\right] = -1$

21. $\sin\dfrac{\pi}{12} \cos\dfrac{5\pi}{12} - \sin\dfrac{5\pi}{12} \cos \dfrac{\pi}{12} = \sin\left[\dfrac{\pi}{12} - \dfrac{5\pi}{12}\right] = \sin\left[\dfrac{-\pi}{3}\right] = -\dfrac{\sqrt{3}}{2}$

23. $\sin \alpha = \dfrac{3}{5}, 0 < \alpha < \dfrac{\pi}{2}; \cos \beta = \dfrac{2}{\sqrt{5}}, \dfrac{-\pi}{2} < \beta < 0.$

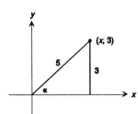

Notice that $y = 3$ and $r = 5$ so that
$$x^2 + 3^2 = 5^2, x > 0$$
$$x^2 = 25 - 9 = 16, x > 0$$
$$x = 4$$

Thus, $\cos \alpha = \dfrac{4}{5}$

$\tan \alpha = \dfrac{3}{4}$

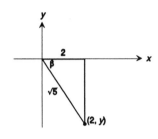

Notice that $x = 2$ and $r = \sqrt{5}$ so that
$$2^2 + y^2 = \left(\sqrt{5}\right)^2, y < 0$$
$$y^2 = 5 - 4 = 1, y < 0$$
$$y = -1$$

Thus, $\sin \beta = \dfrac{-1}{\sqrt{5}}$

$\tan \beta = \dfrac{-1}{2}$

(a) $\sin(\alpha + \beta) = \sin \alpha \cos \beta + \cos \alpha \sin \beta = \dfrac{3}{5} \cdot \dfrac{2}{\sqrt{5}} + \dfrac{4}{5} \cdot \dfrac{-1}{\sqrt{5}} = \dfrac{6 - 4}{5\sqrt{5}} = \dfrac{2}{5\sqrt{5}} = \dfrac{2\sqrt{5}}{25}$

(b) $\cos(\alpha + \beta) = \cos \alpha \cos \beta - \sin \alpha \sin \beta = \dfrac{4}{5} \cdot \dfrac{2}{\sqrt{5}} - \dfrac{3}{5} \cdot \dfrac{-1}{\sqrt{5}} = \dfrac{8 + 3}{5\sqrt{5}} = \dfrac{11}{5\sqrt{5}} = \dfrac{11\sqrt{5}}{25}$

(c) $\sin(\alpha - \beta) = \sin \alpha \cos \beta - \cos \alpha \sin \beta = \dfrac{3}{5} \cdot \dfrac{2}{\sqrt{5}} - \dfrac{4}{5} \cdot \dfrac{-1}{\sqrt{5}} = \dfrac{6 + 4}{5\sqrt{5}} = \dfrac{10}{5\sqrt{5}} = \dfrac{2\sqrt{5}}{5}$

(d) $\tan(\alpha - \beta) = \dfrac{\tan \alpha - \tan \beta}{1 + \tan \alpha \tan \beta} = \dfrac{\dfrac{3}{4} - \dfrac{-1}{2}}{1 + \dfrac{3}{4} \cdot \dfrac{-1}{2}} = \dfrac{\dfrac{5}{4}}{\dfrac{5}{8}} = 2$

25. $\tan \alpha = \dfrac{-4}{3}, \dfrac{\pi}{2} < \alpha < \pi; \cos \beta = \dfrac{1}{2}, 0 < \beta < \dfrac{\pi}{2}$

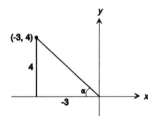

$\tan \alpha = \dfrac{-4}{3}$
$$-3)^2 + (4)^2 = r^2$$
$$25 = r^2$$
$$5 = r$$

$\sin \alpha = \dfrac{4}{5}, \cos \alpha = \dfrac{-3}{5}$

$\cos \beta = \dfrac{1}{2}$
$$1^2 + y^2 = 2^2, y > 0$$
$$y^2 = 4 - 1 = 3, y > 0$$
$$y = \sqrt{3}$$

$\sin \beta = \dfrac{\sqrt{3}}{2}, \tan \beta = \sqrt{3}$

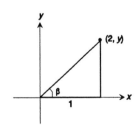

(a) $\sin(\alpha + \beta) = \sin \alpha \cos \beta + \cos \alpha \sin \beta = \dfrac{4}{5} \cdot \dfrac{1}{2} + \dfrac{-3}{5} \cdot \dfrac{\sqrt{3}}{2} = \dfrac{4 - 3\sqrt{3}}{10}$

(b) $\cos(\alpha + \beta) = \cos \alpha \cos \beta - \sin \alpha \sin \beta = \dfrac{-3}{5} \cdot \dfrac{1}{2} - \dfrac{4}{5} \cdot \dfrac{\sqrt{3}}{2} = \dfrac{-3 - 4\sqrt{3}}{10}$

(c) $\sin(\alpha - \beta) = \sin \alpha \cos \beta - \cos \alpha \sin \beta = \dfrac{4}{5} \cdot \dfrac{1}{2} - \dfrac{-3}{5} \cdot \dfrac{\sqrt{3}}{2} = \dfrac{4 + 3\sqrt{3}}{10}$

(d) $\tan(\alpha - \beta) = \dfrac{\tan \alpha - \tan \beta}{1 + \tan \alpha \tan \beta} = \dfrac{\dfrac{-4}{3} - \sqrt{3}}{1 + \dfrac{-4}{3}\sqrt{3}} = \dfrac{\dfrac{-4 - 3\sqrt{3}}{3}}{\dfrac{3 - 4\sqrt{3}}{3}}$

$= \dfrac{-4 - 3\sqrt{3}}{3 - 4\sqrt{3}} = \dfrac{4 + 3\sqrt{3}}{4\sqrt{3} - 3} \cdot \dfrac{4\sqrt{3} + 3}{4\sqrt{3} + 3} = \dfrac{25\sqrt{3} + 48}{39}$

27. $\sin \alpha = \dfrac{5}{13}, \dfrac{-3\pi}{2} < \alpha < -\pi;\ \tan \beta = -\sqrt{3}, \dfrac{\pi}{2} < \beta < \pi$

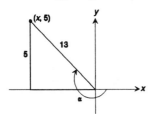

$\sin \alpha = \dfrac{5}{13}$

$x^2 + 5^2 = 13^2, x < 0$

$x^2 = 169 - 25 = 144, x < 0$

$x = -12$

$\cos \alpha = \dfrac{-12}{13}, \tan \alpha = \dfrac{-5}{12}$

$\left(\sqrt{3}\right)^2 + (-1)^2 = r^2$

$4 = r^2$

$2 = r$

$\sin \beta = \dfrac{\sqrt{3}}{2}, \cos \beta = \dfrac{-1}{2}$

(a) $\sin(\alpha + \beta) = \sin \alpha \cos \beta + \cos \alpha \sin \beta = \dfrac{5}{13} \cdot \dfrac{-1}{2} + \dfrac{-12}{13} \cdot \dfrac{\sqrt{3}}{2}$

$= \dfrac{-5 - 12\sqrt{3}}{26} = \dfrac{-1}{26}\left(5 + 12\sqrt{3}\right)$

(b) $\cos(\alpha + \beta) = \cos \alpha \cos \beta - \sin \alpha \sin \beta = \dfrac{-12}{13} \cdot \dfrac{-1}{2} - \dfrac{5}{13} \cdot \dfrac{\sqrt{3}}{2}$

$= \dfrac{12 - 5\sqrt{3}}{26} = \dfrac{1}{26}\left(12 - 5\sqrt{3}\right)$

(c) $\sin(\alpha - \beta) = \sin \alpha \cos \beta - \cos \alpha \sin \beta = \dfrac{5}{13} \cdot \dfrac{-1}{2} - \dfrac{-12}{13} \cdot \dfrac{\sqrt{3}}{2}$

$= \dfrac{-5 + 12\sqrt{3}}{26} = \dfrac{-1}{26}\left(5 - 12\sqrt{3}\right)$

(d) $\tan(\alpha - \beta) = \dfrac{\tan \alpha - \tan \beta}{1 + \tan \alpha \tan \beta} = \dfrac{\dfrac{-5}{12} - \left(-\sqrt{3}\right)}{1 + \dfrac{-5}{12} \cdot -\sqrt{3}} = \dfrac{\dfrac{-5 + 12\sqrt{3}}{12}}{\dfrac{12 + 5\sqrt{3}}{12}}$

$$= \frac{-5 + 12\sqrt{3}}{12 + 5\sqrt{3}} \cdot \frac{12 - 5\sqrt{3}}{12 - 5\sqrt{3}} = \frac{-240 + 169\sqrt{3}}{69}$$

29. $\sin\left[\dfrac{\pi}{2} + \theta\right] = \sin \dfrac{\pi}{2} \cos \theta + \cos \dfrac{\pi}{2} \sin \theta = 1 \cdot \cos \theta + 0 \cdot \sin \theta = \cos \theta$

31. $\sin(\pi - \theta) = \sin \pi \cos \theta - \cos \pi \sin \theta = 0 \cdot \cos \theta - (-1) \sin \theta = \sin \theta$

33. $\sin(\pi + \theta) = \sin \pi \cos \theta + \cos \pi \sin \theta = 0 \cdot \cos \theta + (-1) \sin \theta = -\sin \theta$

35. $\tan(\pi - \theta) = \dfrac{\tan \pi - \tan \theta}{1 + \tan \pi \tan \theta} = \dfrac{0 - \tan \theta}{1 + 0} = -\tan \theta$

37. $\sin\left[\dfrac{3\pi}{2} + \theta\right] = \sin \dfrac{3\pi}{2} \cos \theta + \cos \dfrac{3\pi}{2} \sin \theta = -1 \cdot \cos \theta + 0 \cdot \sin \theta = -\cos \theta$

39. $\sin(\alpha + \beta) + \sin(\alpha - \beta) = \sin \alpha \cos \beta + \cos \alpha \sin \beta + \sin \alpha \cos \beta - \cos \alpha \sin \beta$
$$= 2 \sin \alpha \cos \beta$$

41. $\dfrac{\sin(\alpha + \beta)}{\sin \alpha \cos \beta} = \dfrac{\sin \alpha \cos \beta + \cos \alpha \sin \beta}{\sin \alpha \cos \beta} = \dfrac{\sin \alpha \cos \beta}{\sin \alpha \cos \beta} + \dfrac{\cos \alpha \sin \beta}{\sin \alpha \cos \beta}$
$ = 1 + \cot \alpha \tan \beta$

43. $\dfrac{\cos(\alpha + \beta)}{\cos \alpha \cos \beta} = \dfrac{\cos \alpha \cos \beta - \sin \alpha \sin \beta}{\cos \alpha \cos \beta} = \dfrac{\cos \alpha \cos \beta}{\cos \alpha \cos \beta} - \dfrac{\sin \alpha \sin \beta}{\cos \alpha \cos \beta}$
$ = 1 - \tan \alpha \tan \beta$

45. $\dfrac{\sin(\alpha + \beta)}{\sin(\alpha - \beta)} = \dfrac{\sin \alpha \cos \beta + \cos \alpha \sin \beta}{\sin \alpha \cos \beta - \cos \alpha \sin \beta} = \dfrac{\dfrac{\sin \alpha \cos \beta + \cos \alpha \sin \beta}{\cos \alpha \cos \beta}}{\dfrac{\sin \alpha \cos \beta - \cos \alpha \sin \beta}{\cos \alpha \cos \beta}}$

$ = \dfrac{\dfrac{\sin \alpha \cos \beta}{\cos \alpha \cos \beta} + \dfrac{\cos \alpha \sin \beta}{\cos \alpha \cos \beta}}{\dfrac{\sin \alpha \cos \beta}{\cos \alpha \cos \beta} - \dfrac{\cos \alpha \sin \beta}{\cos \alpha \cos \beta}} = \dfrac{\tan \alpha + \tan \beta}{\tan \alpha - \tan \beta}$

47. $\cot(\alpha + \beta) = \dfrac{\cos(\alpha + \beta)}{\sin(\alpha + \beta)} = \dfrac{\cos \alpha \cos \beta - \sin \alpha \sin \beta}{\sin \alpha \cos \beta + \cos \alpha \sin \beta} = \dfrac{\dfrac{\cos \alpha \cos \beta - \sin \alpha \sin \beta}{\sin \alpha \sin \beta}}{\dfrac{\sin \alpha \cos \beta + \cos \alpha \sin \beta}{\sin \alpha \sin \beta}}$

$ = \dfrac{\dfrac{\cos \alpha \cos \beta}{\sin \alpha \sin \beta} - \dfrac{\sin \alpha \sin \beta}{\sin \alpha \sin \beta}}{\dfrac{\sin \alpha \cos \beta}{\sin \alpha \sin \beta} + \dfrac{\cos \alpha \sin \beta}{\sin \alpha \sin \beta}} = \dfrac{\cot \alpha \cot \beta - 1}{\cot \beta + \cot \alpha}$

49. $\sec(\alpha + \beta) = \dfrac{1}{\cos(\alpha + \beta)} = \dfrac{1}{\cos \alpha \cos \beta - \sin \alpha \sin \beta} = \dfrac{\dfrac{1}{\sin \alpha \sin \beta}}{\dfrac{\cos \alpha \cos \beta - \sin \alpha \sin \beta}{\sin \alpha \sin \beta}}$

$$= \dfrac{\dfrac{1}{\sin \alpha} \cdot \dfrac{1}{\sin \beta}}{\dfrac{\cos \alpha \cos \beta}{\sin \alpha \sin \beta} - \dfrac{\sin \alpha \sin \beta}{\sin \alpha \sin \beta}} = \dfrac{\csc \alpha \csc \beta}{\cot \alpha \cot \beta - 1}$$

51. $\sin(\alpha - \beta)\sin(\alpha + \beta)$

$= (\sin \alpha \cos \beta - \cos \alpha \sin \beta)(\sin \alpha \cos \beta + \cos \alpha \sin \beta)$

$= \sin^2 \alpha \cos^2 \beta - \cos^2 \alpha \sin^2 \beta = (\sin^2 \alpha)(1 - \sin^2 \beta) - (1 - \sin^2 \alpha)(\sin^2 \beta)$

$= \sin^2 \alpha - \sin^2 \alpha \sin^2 \beta - \sin^2 \beta + \sin^2 \alpha \sin^2 \beta = \sin^2 \alpha - \sin^2 \beta$

53. $\sin(\theta + k\pi) = \sin \theta \cos k\pi + \cos \theta \sin k\pi = (\sin \theta)(-1)^k + (\cos \theta)(0)$

$= (-1)^k \cdot \sin \theta, \ k$ any integer

55. $\sin(\sin^{-1} \frac{1}{2} + \cos^{-1} 0) = \sin\left[\dfrac{\pi}{6} + \dfrac{\pi}{2}\right] = \sin \dfrac{\pi}{6} \cos \dfrac{\pi}{2} + \cos \dfrac{\pi}{6} \sin \dfrac{\pi}{2}$

$$= \dfrac{1}{2} \cdot 0 + \dfrac{\sqrt{3}}{2} \cdot 1 = \dfrac{\sqrt{3}}{2}$$

57. $\sin\left[\sin^{-1} \dfrac{3}{5} - \cos^{-1}\left(-\dfrac{4}{5}\right)\right]$

Let $\theta = \sin^{-1} \dfrac{3}{5}$ and $\alpha = \cos^{-1}\left(-\dfrac{4}{5}\right)$

Then, $\sin\left[\sin^{-1} \dfrac{3}{5} - \cos^{-1}\left(-\dfrac{4}{5}\right)\right]$

$= \sin(\theta - \alpha) = \sin \theta \cos \alpha - \cos \theta \sin \alpha$

$= \left(\dfrac{3}{5}\right)\left(\dfrac{-4}{5}\right) - \left(\dfrac{4}{5}\right)\left(\dfrac{3}{5}\right) = \dfrac{-12}{25} - \dfrac{12}{25} = \dfrac{-24}{25}$

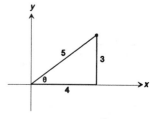

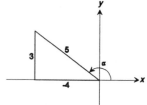

59. $\cos\left[\tan^{-1} \dfrac{4}{3} + \cos^{-1} \dfrac{5}{13}\right]$

Let $\theta = \tan^{-1} \dfrac{4}{3}$ and $\alpha = \cos^{-1} \dfrac{5}{13}$

Then, $\cos\left[\tan^{-1} \dfrac{4}{3} + \cos^{-1} \dfrac{5}{13}\right]$

$= \cos(\theta + \alpha) = \cos \theta \cos \alpha - \sin \theta \sin \alpha$

$= \left(\dfrac{3}{5}\right)\left(\dfrac{5}{13}\right) - \left(\dfrac{4}{5}\right)\left(\dfrac{12}{13}\right)$

$= \dfrac{15}{65} - \dfrac{48}{65} = \dfrac{-33}{65}$

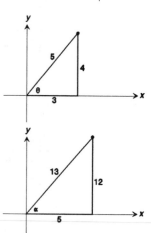

61. $\sec\left[\sin^{-1}\dfrac{5}{13} - \tan^{-1}\dfrac{3}{4}\right]$

Let $\theta = \sin^{-1}\dfrac{5}{13}$ and $\alpha = \cos^{-1}\dfrac{3}{4}$

Then, $\sec\left[\sin^{-1}\dfrac{5}{13} - \tan^{-1}\dfrac{3}{4}\right]$

$$= \sec(\theta - \alpha) = \frac{1}{\cos(\theta - \alpha)}$$

$$= \frac{1}{\cos\theta\cos\alpha + \sin\theta\sin\alpha}$$

$$= \frac{1}{\left[\dfrac{12}{13}\right]\left[\dfrac{4}{5}\right] + \left[\dfrac{5}{13}\right]\left[\dfrac{3}{5}\right]}$$

$$= \frac{1}{\dfrac{48}{65} + \dfrac{15}{65}} = \frac{1}{\dfrac{63}{65}} = \frac{65}{63}$$

63. Let $\theta = \sec^{-1}\dfrac{5}{3}$

Then, $\cot\left[\sec^{-1}\dfrac{5}{3} + \dfrac{\pi}{6}\right]$

$$= \cot\left[\theta + \dfrac{\pi}{6}\right] = \frac{1}{\tan\left[\theta + \dfrac{\pi}{6}\right]} = \frac{1}{\dfrac{\tan\theta + \tan\dfrac{\pi}{6}}{1 - \tan\theta\tan\dfrac{\pi}{6}}}$$

$$= \frac{1 - \tan\theta\tan\dfrac{\pi}{6}}{\tan\theta + \tan\dfrac{\pi}{6}} = \frac{1 - \left[\dfrac{4}{3}\right]\left[\dfrac{\sqrt{3}}{3}\right]}{\dfrac{4}{3} + \dfrac{\sqrt{3}}{3}} = \frac{1 - \dfrac{4\sqrt{3}}{9}}{\dfrac{4 + \sqrt{3}}{3}}$$

$$= \frac{9 - 4\sqrt{3}}{9} \cdot \frac{3}{4 + \sqrt{3}} = \frac{9 - 4\sqrt{3}}{3\left(4 + \sqrt{3}\right)} = \frac{1}{3}\left[\frac{9 - 4\sqrt{3}}{4 + \sqrt{3}} \cdot \frac{4 - \sqrt{3}}{4 - \sqrt{3}}\right]$$

$$= \frac{1}{3}\left[\frac{48 - 25\sqrt{3}}{13}\right] = \frac{1}{39}\left[48 - 25\sqrt{3}\right]$$

65. Let $\alpha = \cos^{-1} u$ and $\beta = \sin^{-1} v$. Then, $\cos\alpha = u$, $0 \le \alpha \le \pi$, and $\sin\beta = v$,

$\dfrac{-\pi}{2} \le \beta \le \dfrac{\pi}{2}$. Since $\sin\alpha \ge 0$ and $\cos\beta \ge 0$, we have:

$$\sin\alpha = \sqrt{1 - \cos^2\alpha} = \sqrt{1 - u^2} \quad \text{and} \quad \cos\beta = \sqrt{1 - \sin^2\beta} = \sqrt{1 - v^2}$$

Thus, $\cos(\cos^{-1} u + \sin^{-1} v) = \cos(\alpha + \beta) = \cos\alpha\cos\beta - \sin\alpha\sin\beta$

$$= u\sqrt{1 - v^2} - v\sqrt{1 - u^2}$$

67. Let $\alpha = \tan^{-1} u$ and $\beta = \sin^{-1} v$. Then, $\tan \alpha = u$, $\dfrac{-\pi}{2} < \alpha < \dfrac{\pi}{2}$ and $\sin \beta = v$,

$\dfrac{-\pi}{2} \le \beta \le \dfrac{\pi}{2}$. Now, $\sin \alpha = \dfrac{u}{\sqrt{1 + u^2}}$, $\cos \alpha = \dfrac{1}{\sqrt{1 + u^2}}$, and $\cos \beta = \sqrt{1 - v^2}$.

Thus, $\sin(\tan^{-1} u - \sin^{-1} v) = \sin(\alpha - \beta) = \sin \alpha \cos \beta - \cos \alpha \sin \beta$

$$= \dfrac{u\sqrt{1 - v^2}}{\sqrt{1 + u^2}} - \dfrac{v}{\sqrt{1 + u^2}} = \dfrac{u\sqrt{1 - v^2} - v}{\sqrt{1 + u^2}}$$

69. Let $\alpha = \sin^{-1} u$ and $\beta = \cos^{-1} v$. Then $\sin \alpha = u$, $\dfrac{-\pi}{2} \le \alpha \le \dfrac{\pi}{2}$ and $\cos \beta = v$, $0 \le \beta \le \pi$.

Then, $\cos \alpha = \sqrt{1 - u^2}$ and $\sin \beta = \sqrt{1 - v^2}$.

Thus, $\tan(\sin^{-1} u - \cos^{-1} v) = \tan(\alpha - \beta) = \dfrac{\tan \alpha - \tan \beta}{1 + \tan \alpha \tan \beta}$

$$= \dfrac{\dfrac{u}{\sqrt{1 - u^2}} - \dfrac{\sqrt{1 - v^2}}{v}}{1 + \dfrac{u\sqrt{1 - v^2}}{v\sqrt{1 - u^2}}} = \dfrac{uv - \sqrt{1 - u^2}\sqrt{1 - v^2}}{v\sqrt{1 - u^2} + u\sqrt{1 - v^2}}$$

71. $\dfrac{\sin(x + h) - \sin x}{h} = \dfrac{\sin x \cos h + \cos x \sin h - \sin x}{h} = \dfrac{\cos x \sin h - (\sin x)(1 - \cos h)}{h}$

$$= \cos x \cdot \dfrac{\sin h}{h} - \sin x \cdot \dfrac{1 - \cos h}{h}$$

73. $\sin(\sin^{-1} u + \cos^{-1} u) = \sin(\sin^{-1} u) \cos(\cos^{-1} u) + \cos(\sin^{-1} u) \sin(\cos^{-1} u)$

$$= (u)(u) + \sqrt{1 - u^2}\sqrt{1 - u^2} = u^2 + 1 - u^2 = 1$$

75. $\tan\left[\dfrac{\pi}{2} - \theta\right] = \dfrac{\tan \dfrac{\pi}{2} - \tan \theta}{1 + \tan \dfrac{\pi}{2} \tan \theta}$ Impossible because $\tan \dfrac{\pi}{2}$ is not defined.

Therefore, $\tan\left[\dfrac{\pi}{2} - \theta\right] = \dfrac{\sin\left[\dfrac{\pi}{2} - \theta\right]}{\cos\left[\dfrac{\pi}{2} - \theta\right]} = \dfrac{\cos \theta}{\sin \theta} = \cot \theta$

77. $\tan \theta = \tan(\theta_2 - \theta_1) = \dfrac{\tan \theta_2 - \tan \theta_1}{1 + \tan \theta_2 \tan \theta_1} = \dfrac{m_2 - m_1}{1 + m_1 m_2}$

1. $\sin \theta = \dfrac{3}{5}$, $0 < \theta < \dfrac{\pi}{2}$. Thus, $0 < \dfrac{\theta}{2} < \dfrac{\pi}{4}$, or $\dfrac{\theta}{2}$ lies in quadrant I.

   (a) Because $\sin 2\theta = 2 \sin \theta \cos \theta$ and because we know $\sin \theta = \dfrac{3}{5}$, we only need to find $\cos \theta$.

   It is given that $0 < \theta < \dfrac{\pi}{2}$, which means $\cos \theta > 0$. Using a form of Pythagorean identity, we find that:

   $$\cos \theta = \sqrt{1 - \sin^2 \theta} = \sqrt{1 - \frac{9}{25}} = \frac{4}{5}$$

   $$\sin 2\theta = 2 \sin \theta \cos \theta = 2\left(\frac{3}{5}\right)\left(\frac{4}{5}\right) = \frac{24}{25}$$

   (b) $\cos 2\theta = 1 - 2\sin^2 \theta = 1 - 2\left(\dfrac{9}{25}\right) = \dfrac{7}{25}$

   (c) $\sin \dfrac{1}{2}\theta = \sqrt{\dfrac{1 - \cos \theta}{2}} = \sqrt{\dfrac{\frac{1}{5}}{10}} = \dfrac{\sqrt{10}}{10}$

   (d) $\cos \dfrac{1}{2}\theta = \sqrt{\dfrac{1 + \cos \theta}{2}} = \sqrt{\dfrac{9}{10}} = \dfrac{3\sqrt{10}}{10}$

3. $\tan \theta = \dfrac{4}{3}$, $\pi < \theta < \dfrac{3\pi}{2}$. Thus, $\dfrac{\pi}{2} < \dfrac{\theta}{2} < \dfrac{3\pi}{4}$, or $\dfrac{\theta}{2}$ lies in quadrant II.

   $$-(3)^2 + (-4)^2 = r^2$$
   $$5 = r$$

   $$\sin 2\theta = \frac{-4}{5} \qquad \cos \theta = \frac{-3}{5}$$

   (a) $\sin 2\theta = 2 \sin \theta \cos \theta = 2\left(\dfrac{-4}{5}\right)\left(\dfrac{-3}{5}\right) = \dfrac{24}{25}$

   (b) $\cos 2\theta = 1 - 2\sin^2 \theta = 1 - 2\left(\dfrac{16}{25}\right) = \dfrac{-7}{25}$

   (c) $\sin \dfrac{1}{2}\theta = +\sqrt{\dfrac{1 - \cos \theta}{2}} = \sqrt{\dfrac{\frac{8}{5}}{2}} = \sqrt{\dfrac{4}{5}} = \dfrac{2\sqrt{5}}{5}$

   (d) $\cos \dfrac{1}{2}\theta = -\sqrt{\dfrac{1 + \cos \theta}{2}} = -\sqrt{\dfrac{\frac{2}{5}}{2}} = -\sqrt{\dfrac{1}{5}} = -\dfrac{\sqrt{5}}{5}$

5. $\cos \theta = \dfrac{-\sqrt{2}}{\sqrt{3}}$, $\dfrac{\pi}{2} < \theta < \pi$. Thus, $\dfrac{\pi}{4} < \dfrac{\theta}{2} < \dfrac{\pi}{2}$, or $\dfrac{\theta}{2}$ lies in quadrant I.

   $$\sin \theta = \sqrt{1 - \cos^2 \theta} = \sqrt{1 - \frac{2}{3}} = \frac{1}{\sqrt{3}} = \frac{\sqrt{3}}{3}$$

   (a) $\sin 2\theta = 2 \sin \theta \cos \theta = 2\left(\dfrac{\sqrt{3}}{3}\right)\left(\dfrac{-\sqrt{2}}{\sqrt{3}}\right) = \dfrac{-2\sqrt{2}}{3}$

   (b) $\cos 2\theta = 2\cos^2 \theta - 1 = 2\left(\dfrac{2}{3}\right) - 1 = \dfrac{1}{3}$

(c) $\quad \sin \frac{1}{2}\theta = \sqrt{\dfrac{1-\cos\theta}{2}} = \sqrt{\dfrac{1-\dfrac{-\sqrt{2}}{\sqrt{3}}}{2}} = \sqrt{\dfrac{\dfrac{3+\sqrt{6}}{3}}{2}} = \sqrt{\dfrac{3+\sqrt{6}}{6}}$

(d) $\quad \cos \frac{1}{2}\theta = \sqrt{\dfrac{1+\cos\theta}{2}} = \sqrt{\dfrac{1+\dfrac{-\sqrt{6}}{3}}{2}} = \sqrt{\dfrac{3-\sqrt{6}}{6}}$

7. $\quad \sec\theta = 3,\ \sin\theta > 0.$ Thus, θ lies in quadrant I or $0 < \theta < \dfrac{\pi}{2}$, and $0 < \dfrac{\theta}{2} < \dfrac{\pi}{4}$, or $\dfrac{\theta}{2}$ lies in quadrant I.

Therefore, $\cos\theta = \dfrac{1}{3}$, $\sin\theta = \sqrt{1-\cos^2\theta} = \sqrt{\dfrac{9-1}{9}} = \dfrac{\sqrt{8}}{3} = \dfrac{2\sqrt{2}}{3}$

(a) $\quad \sin 2\theta = 2\sin\theta\cos\theta = 2\left[\dfrac{\sqrt{8}}{3}\right]\left[\dfrac{1}{3}\right] = \dfrac{4\sqrt{2}}{9}$

(b) $\quad \cos 2\theta = 2\cos^2 - 1 = 2\left[\dfrac{1}{9}\right] - 1 = -\dfrac{7}{9}$

(c) $\quad \sin \frac{1}{2}\theta = \sqrt{\dfrac{1-\cos\theta}{2}} = \sqrt{\dfrac{1-\dfrac{1}{3}}{2}} = \dfrac{\sqrt{3}}{3}$

(d) $\quad \cos\frac{1}{2}\theta = \sqrt{\dfrac{1+\cos\theta}{2}} = \sqrt{\dfrac{1+\dfrac{1}{3}}{2}} = \sqrt{\dfrac{3+1}{6}} = \sqrt{\dfrac{4}{6}} = \dfrac{2}{\sqrt{6}} \cdot \dfrac{\sqrt{6}}{\sqrt{6}} = \dfrac{2\sqrt{6}}{6} = \dfrac{\sqrt{6}}{3}$

9. $\quad \cot\theta = -2,\ \sec\theta < 0.$ Thus, $\cos\theta < 0$ and $\sin\theta > 0$ and θ lies in quadrant II, or $\dfrac{\pi}{2} < \theta < \pi$.

Hence, $\dfrac{\pi}{4} < \dfrac{\theta}{2} < \dfrac{\pi}{2}$, or $\dfrac{\theta}{2}$ lies in quadrant I.

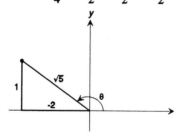

$1^2 + (-2)^2 = \left(\sqrt{5}\right)^2$

$\cos\theta = \dfrac{-2}{\sqrt{5}} = \dfrac{-2\sqrt{5}}{5}$

$\sin\theta = \dfrac{1}{\sqrt{5}} = \dfrac{\sqrt{5}}{5}$

(a) $\quad \sin 2\theta = 2\sin\theta\cos\theta = 2\left[\dfrac{\sqrt{5}}{5}\right]\left[\dfrac{-2\sqrt{5}}{5}\right] = \dfrac{-20}{25} = \dfrac{-4}{5}$

(b) $\quad \cos 2\theta = \cos^2\theta - \sin^2\theta = \dfrac{4}{5} - \dfrac{1}{5} = \dfrac{3}{5}$

(c) $\quad \sin \frac{1}{2}\theta = \sqrt{\dfrac{1-\cos\theta}{2}} = \sqrt{\dfrac{1+\dfrac{2\sqrt{5}}{5}}{2}} = \sqrt{\dfrac{5+2\sqrt{5}}{10}}$

(d) $\cos\frac{1}{2}\theta = \sqrt{\dfrac{1 + \cos\theta}{2}} = \sqrt{\dfrac{1 - \dfrac{2\sqrt{5}}{5}}{2}} = \sqrt{\dfrac{5 - 2\sqrt{5}}{10}}$

11. $\tan\theta = -3$, $\sin\theta < 0$. Thus, θ lies in quadrant IV, or $\dfrac{3\pi}{2} < \theta < 2\pi$. Hence, $\dfrac{3\pi}{4} < \dfrac{\theta}{2} < \pi$, or $\dfrac{\theta}{2}$ lies in quadrant II.

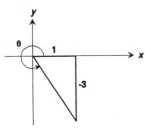

$(1)^2 + (-3)^2 = \left(\sqrt{10}\right)^2$

$\cos\theta = \dfrac{1}{\sqrt{10}} = \dfrac{\sqrt{10}}{10}$

$\sin\theta = \dfrac{-3}{\sqrt{10}} = \dfrac{-3\sqrt{10}}{10}$

(a) $\sin 2\theta = 2\sin\theta\cos\theta = 2\left[\dfrac{-3\sqrt{10}}{10}\right]\left[\dfrac{\sqrt{10}}{10}\right] = \dfrac{-6}{10} = \dfrac{-3}{5}$

(b) $\cos 2\theta = \cos^2\theta - \sin^2\theta = \dfrac{1}{10} - \dfrac{9}{10} = \dfrac{-8}{10} = \dfrac{-4}{5}$

(c) $\sin\dfrac{1}{2}\theta = \sqrt{\dfrac{1 - \cos\theta}{2}} = \sqrt{\dfrac{1 - \dfrac{\sqrt{10}}{10}}{2}} = \sqrt{\dfrac{10 - \sqrt{10}}{20}} = \dfrac{\sqrt{10 - \sqrt{10}}}{2\sqrt{5}} = \dfrac{1}{2}\sqrt{\dfrac{10 - \sqrt{10}}{5}}$

(d) $\cos\dfrac{1}{2}\theta = -\sqrt{\dfrac{1 + \cos\theta}{2}} = -\sqrt{\dfrac{1 + \dfrac{\sqrt{10}}{10}}{2}} = -\sqrt{\dfrac{10 + \sqrt{10}}{20}} = \dfrac{-\sqrt{10 + \sqrt{10}}}{2\sqrt{5}}$

$= -\dfrac{1}{2}\sqrt{\dfrac{10 + \sqrt{10}}{5}}$

13. Because $22.5° = \dfrac{45°}{2}$, we can use the half-angle formula for $\sin\left[\dfrac{\alpha}{2}\right]$ with $\alpha = 45°$. Also, because $22.5°$ is in quadrant I, $\sin 22.5° > 0$, and we choose the $+$ sign in using the half-angle formula:

$\sin 22.5° = \sin\dfrac{45°}{2} = \sqrt{\dfrac{1 - \cos 45°}{2}} = \sqrt{\dfrac{1 - \dfrac{\sqrt{2}}{2}}{2}} = \sqrt{\dfrac{2 - \sqrt{2}}{4}} = \dfrac{\sqrt{2 - \sqrt{2}}}{2}$

15. Because $\dfrac{7\pi}{8}$ is in quadrant II, $\tan \dfrac{7\pi}{8} < 0$, and we choose the $-$ sign in using the half-angle formula:

$$\tan \frac{7\pi}{8} = \tan\frac{\frac{7\pi}{4}}{2} = -\sqrt{\frac{1 - \cos \alpha}{1 + \cos \alpha}} = -\sqrt{\frac{1 - \dfrac{1}{\sqrt{2}}}{1 + \dfrac{1}{\sqrt{2}}}} = -\sqrt{\frac{\sqrt{2} - 1}{1 + \sqrt{2}}\left(\frac{\sqrt{2} - 1}{\sqrt{2} - 1}\right)}$$

$$= -\left[\frac{\sqrt{2} - 1}{1}\right] = 1 - \sqrt{2}$$

17. Because $165°$ is in quadrant II, $\cos 165° < 0$, and we choose the $-$ sign in using the half-angle formula:

$$\cos 165° = \cos \frac{330°}{2} = -\sqrt{\frac{1 + \cos 330°}{2}} = -\sqrt{\frac{1 + \dfrac{\sqrt{3}}{2}}{2}} = -\frac{\sqrt{2 + \sqrt{3}}}{2}$$

19. Because $\dfrac{15\pi}{8}$ is in quadrant IV, $\sec \dfrac{15\pi}{8} > 0$, and we choose the $+$ sign in using the formula:

$$\sec \frac{15\pi}{8} = \frac{1}{\cos \dfrac{15\pi}{8}} = \frac{1}{\cos\dfrac{\frac{15\pi}{4}}{2}} = \frac{1}{\sqrt{\dfrac{1 + \dfrac{\cos\frac{15\pi}{4}}{2}}{}}} = \frac{1}{\sqrt{\dfrac{1 + \dfrac{\sqrt{2}}{2}}{2}}}$$

$$= \frac{1}{\sqrt{\dfrac{2 + \sqrt{2}}{4}}} = \frac{2}{\sqrt{2 + \sqrt{2}}} \cdot \frac{\sqrt{2 + \sqrt{2}}}{\sqrt{2 + \sqrt{2}}} = \frac{2\left(\sqrt{2 + \sqrt{2}}\right)}{2 + \sqrt{2}} \cdot \frac{2 - \sqrt{2}}{2 - \sqrt{2}}$$

$$= \frac{2\left(2 - \sqrt{2}\right)\left(\sqrt{2 + \sqrt{2}}\right)}{4 - 2} = \left(2 - \sqrt{2}\right)\left(\sqrt{2 + \sqrt{2}}\right)$$

21. Because $\dfrac{-\pi}{8}$ is in quadrant IV, $\sin\left(\dfrac{-\pi}{8}\right) < 0$, and we choose the $-$ sign in using the half-angle formula:

$$\sin\left(\frac{-\pi}{8}\right) = \sin\frac{\dfrac{-\pi}{4}}{2} = -\sqrt{\frac{1 - \cos\left[\dfrac{-\pi}{4}\right]}{2}} = -\sqrt{\frac{1 - \dfrac{\sqrt{2}}{2}}{2}} = -\sqrt{\frac{2 - \sqrt{2}}{4}} = -\frac{\sqrt{2 - \sqrt{2}}}{2}$$

23. To show that $\sin^4 \theta = \dfrac{3}{8} - \dfrac{1}{2} \cos 2\theta + \dfrac{1}{8} \cos 4\theta,$

$$\sin^4 \theta = (\sin^2 \theta)^2 = \left[\frac{1 - \cos 2\theta}{2}\right]^2 = \frac{1}{4}\left(1 - 2 \cos 2\theta + \cos^2 2\theta\right)$$

$$= \frac{1}{4} - \frac{1}{2} \cos 2\theta + \frac{1}{4} \cos^2 2\theta = \frac{1}{4} - \frac{1}{2} \cos 2\theta + \frac{1}{4}\left[\frac{1 + \cos 4\theta}{2}\right]$$

$$= \frac{1}{4} - \frac{1}{2} \cos 2\theta + \frac{1}{8} + \frac{1}{8} \cos 4\theta = \frac{3}{8} - \frac{1}{2} \cos 2\theta + \frac{1}{8} \cos 4\theta$$

25. $\sin 4\theta = \sin 2(2\theta) = 2 \sin 2\theta \cos 2\theta = (4 \sin \theta \cos \theta)(1 - 2 \sin^2 \theta)$
    $= 4 \sin \theta \cos \theta - 8 \sin^3 \theta \cos \theta = (\cos \theta)(4 \sin \theta - 8 \sin^3 \theta)$

27. $\sin(5\theta) = \sin(2\theta + 3\theta) = \sin 2\theta \cos 3\theta + \cos 2\theta \sin 3\theta$

$= (2 \sin \theta \cos \theta)\cos(2\theta + \theta) + (1 - 2 \sin^2 \theta) \sin(2\theta + \theta)$

$= (2 \sin \theta \cos \theta)(\cos 2\theta \cos \theta - \sin 2\theta \sin \theta) + (1 - 2 \sin^2 \theta)(\sin 2\theta \cos \theta + \cos 2\theta \sin \theta)$

$= 2 \cos 2\theta \sin \theta \cos^2 \theta - 2 \sin 2\theta \sin^2 \theta \cos \theta + \sin 2\theta \cos \theta + \cos 2\theta \sin \theta$
$\qquad - 2 \sin 2\theta \sin^2 \theta \cos \theta - 2 \cos 2\theta \sin^3 \theta$

$= 2(1 - 2 \sin^2 \theta) \sin \theta(1 - \sin^2 \theta) - 4(2 \sin \theta \cos \theta) \sin^2 \theta \cos \theta + (2 \sin \theta \cos \theta) \cos \theta$
$\qquad + (1 - 2 \sin^2 \theta) \sin \theta - 2(1 - 2 \sin^2 \theta) \sin^3 \theta$

$= (2 - 4 \sin^2 \theta)(\sin \theta - \sin^3 \theta) - 8 \sin^3 \theta \cos^2 \theta + 2 \sin \theta \cos^2 \theta + \sin \theta - 2 \sin^3 \theta$
$\qquad - 2 \sin^3 \theta + 4 \sin^5 \theta$

$= 2 \sin \theta = 6 \sin^3 \theta + 4 \sin^5 \theta - 8 \sin^3 \theta (1 - \sin^2 \theta) + 2 \sin \theta(1 - \sin^2 \theta) + \sin \theta$
$\qquad - 4 \sin^3 \theta + 4 \sin^5 \theta$

$= 3 \sin \theta - 10 \sin^3 \theta + 8 \sin^5 \theta - 8 \sin^3 \theta + 8 \sin^5 \theta + 2 \sin \theta - 2 \sin^3 \theta$

$= 16 \sin^5 \theta - 20 \sin^3 \theta + 5 \sin \theta$

29. Using the difference of two squares,
$\cos^4 \theta - \sin^4 \theta = (\cos^2 \theta + \sin^2 \theta)(\cos^2 \theta - \sin^2 \theta) = \cos^2 \theta - \sin^2 \theta = \cos 2\theta$

31. $\cot 2\theta = \dfrac{1}{\tan 2\theta} = \dfrac{1}{\dfrac{2 \tan \theta}{1 - \tan^2 \theta}} = \dfrac{1 - \tan^2 \theta}{2 \tan \theta} = \dfrac{1 - \dfrac{1}{\cot^2 \theta}}{\dfrac{2}{\cot \theta}} = \dfrac{\dfrac{\cot^2 \theta - 1}{\cot^2 \theta}}{\dfrac{2}{\cot \theta}}$

$= \dfrac{\cot^2 \theta - 1}{\cot^2 \theta} \cdot \dfrac{\cot \theta}{2} = \dfrac{\cot^2 \theta - 1}{2 \cot \theta}$

33. $\sec 2\theta = \dfrac{1}{\cos 2\theta} = \dfrac{1}{2 \cos^2 \theta - 1} = \dfrac{1}{\dfrac{2}{\sec^2 \theta} - 1} = \dfrac{1}{\dfrac{2 - \sec^2 \theta}{\sec^2 \theta}} = \dfrac{\sec^2 \theta}{2 - \sec^2 \theta}$

35. $\cos^2 2\theta - \sin^2 2\theta = \cos 2(2\theta) = \cos 4\theta$

37. $\dfrac{\cos 2\theta}{1 + \sin 2\theta} = \dfrac{\cos^2 \theta - \sin^2 \theta}{1 + 2 \sin \theta \cos \theta} = \dfrac{(\cos \theta - \sin \theta)(\cos \theta + \sin \theta)}{\sin^2 \theta + \cos^2 \theta + 2 \sin \theta \cos \theta}$

$= \dfrac{(\cos \theta - \sin \theta)(\cos \theta + \sin \theta)}{(\sin \theta + \cos \theta)^2} = \dfrac{\cos \theta - \sin \theta}{\cos \theta + \sin \theta}$

$= \dfrac{\dfrac{\cos \theta - \sin \theta}{\sin \theta}}{\dfrac{\cos \theta + \sin \theta}{\sin \theta}} = \dfrac{\dfrac{\cos \theta}{\sin \theta} - \dfrac{\sin \theta}{\sin \theta}}{\dfrac{\cos \theta}{\sin \theta} + \dfrac{\sin \theta}{\sin \theta}} = \dfrac{\cot \theta - 1}{\cot \theta + 1}$

39. $\sec^2 \dfrac{\theta}{2} = \dfrac{1}{\cos^2 \left(\dfrac{\theta}{2} \right)} = \dfrac{1}{\dfrac{1 + \cos \theta}{2}} = \dfrac{2}{1 + \cos \theta}$

41. $\cot^2 \dfrac{\theta}{2} = \dfrac{1}{\tan^2 \left(\dfrac{\theta}{2} \right)} = \dfrac{1}{\dfrac{1 - \cos \theta}{1 + \cos \theta}} = \dfrac{1 + \cos \theta}{1 - \cos \theta} = \dfrac{1 + \dfrac{1}{\sec \theta}}{1 - \dfrac{1}{\sec \theta}} = \dfrac{\dfrac{\sec \theta + 1}{\sec \theta}}{\dfrac{\sec \theta - 1}{\sec \theta}} = \dfrac{\sec \theta + 1}{\sec \theta - 1}$

43. Let $\dfrac{1 - \tan^2\left[\dfrac{\theta}{2}\right]}{1 + \tan^2\left[\dfrac{\theta}{2}\right]} = \dfrac{1 - \dfrac{1 - \cos\theta}{1 + \cos\theta}}{1 + \dfrac{1 - \cos\theta}{1 + \cos\theta}} = \dfrac{\dfrac{1 + \cos\theta - (1 - \cos\theta)}{1 + \cos\theta}}{\dfrac{1 + \cos\theta + 1 - \cos\theta}{1 + \cos\theta}} = \dfrac{2\cos\theta}{1 + \cos\theta} \cdot \dfrac{1 + \cos\theta}{2}$

$\qquad = \cos\theta$

Therefore, $\cos\theta = \dfrac{1 - \tan^2\left[\dfrac{\theta}{2}\right]}{1 + \tan^2\left[\dfrac{\theta}{2}\right]}$

45. $\dfrac{\sin 3\theta}{\sin\theta} - \dfrac{\cos 3\theta}{\cos\theta} = \dfrac{\sin 3\theta \cos\theta - \cos 3\theta \sin\theta}{\sin\theta \cos\theta} = \dfrac{\sin(3\theta - \theta)}{\dfrac{1}{2}(2\sin\theta\cos\theta)} = \dfrac{2\sin 2\theta}{\sin 2\theta} = 2$

47. $\tan 3\theta = \tan(\theta + 2\theta) = \dfrac{\tan 2\theta + \tan\theta}{1 - \tan 2\theta \tan\theta} = \dfrac{\dfrac{2\tan\theta}{1 - \tan^2\theta} + \tan\theta}{1 - \dfrac{(2\tan\theta)\tan\theta}{1 - \tan^2\theta}}$

$\qquad = \dfrac{2\tan\theta + \tan\theta(1 - \tan^2\theta)}{1 - \tan^2\theta - 2\tan^2\theta} = \dfrac{3\tan\theta - \tan^3\theta}{1 - 3\tan^2\theta}$

49. $\sin\left[2\sin^{-1}\dfrac{1}{2}\right] = \sin\left[2 \cdot \dfrac{\pi}{6}\right] = \sin\left[\dfrac{\pi}{3}\right] = \dfrac{\sqrt{3}}{2}$

51. $\cos\left[2\sin^{-1}\dfrac{3}{5}\right] = 1 - 2\sin^2\left[\sin^{-1}\dfrac{3}{5}\right] = 1 - 2\left[\dfrac{3}{5}\right]^2 = 1 - 2\left[\dfrac{9}{25}\right] = \dfrac{25 - 18}{25} = \dfrac{7}{25}$

53. $\tan\left[2\cos^{-1}\left(-\dfrac{3}{5}\right)\right] = \dfrac{2\tan\left[\cos^{-1}\left(-\dfrac{3}{5}\right)\right]}{1 - \tan^2\left[\cos^{-1}\left(-\dfrac{3}{5}\right)\right]} = \dfrac{2\left[\dfrac{-4}{3}\right]}{1 - \left[\dfrac{-4}{3}\right]^2} = \dfrac{\dfrac{-8}{3}}{\dfrac{9 - 16}{9}} = \dfrac{24}{7}$

55. $\sin\left[2\cos^{-1}\dfrac{4}{5}\right] = 2\sin\left[\cos^{-1}\dfrac{4}{5}\right]\cos\left[\cos^{-1}\dfrac{4}{5}\right] = 2\left[\dfrac{3}{5}\right]\left[\dfrac{4}{5}\right] = \dfrac{24}{25}$

57. $\sin^2\left[\dfrac{1}{2}\cos^{-1}\dfrac{3}{5}\right] = \dfrac{1 - \cos\left[\cos^{-1}\dfrac{3}{5}\right]}{2} = \dfrac{1 - \dfrac{3}{5}}{2} = \dfrac{2}{10} = \dfrac{1}{5}$

59. $\sec\left[2\tan^{-1}\dfrac{3}{4}\right] = \dfrac{1}{\cos\left[2\tan^{-1}\dfrac{3}{4}\right]} = \dfrac{1}{1 - 2\sin^2\left[\tan^{-1}\dfrac{3}{4}\right]} = \dfrac{1}{1 - 2\left[\dfrac{3}{5}\right]^2} = \dfrac{1}{1 - 2\left[\dfrac{9}{25}\right]}$

$\qquad = \dfrac{1}{\dfrac{25 - 18}{25}} = \dfrac{1}{\dfrac{7}{25}} = \dfrac{25}{7}$

61. $\cot^2\left[\dfrac{1}{2}\tan^{-1}\dfrac{4}{3}\right] = \dfrac{1}{\tan^2\left[\dfrac{1}{2}\tan^{-1}\dfrac{4}{3}\right]} = \dfrac{1}{\dfrac{1 - \cos 2\left[\dfrac{1}{2}\tan^{-1}\dfrac{4}{3}\right]}{1 + \cos 2\left[\dfrac{1}{2}\tan^{-1}\dfrac{4}{3}\right]}} = \dfrac{1 + \cos\left[\tan^{-1}\dfrac{4}{3}\right]}{1 - \cos\left[\tan^{-1}\dfrac{4}{3}\right]}$

$$= \dfrac{1 + \dfrac{3}{5}}{1 - \dfrac{3}{5}} = \dfrac{\dfrac{8}{5}}{\dfrac{2}{5}} = 4$$

63. $\dfrac{1}{2}\sin^2 x + c = -\dfrac{1}{4}\cos 2x$

$$c = -\dfrac{1}{4}\cos 2x - \dfrac{1}{2}\sin^2 x$$

$$c = -\dfrac{1}{4}(\cos 2x + 2\sin^2 x)$$

$$c = -\dfrac{1}{4}(\cos 2x + 1 - \cos 2x)$$

$$c = -\dfrac{1}{4}$$

65. The area of a triangle is $A = \dfrac{1}{2}bh$.

We need to find the base and height of the triangle in terms of θ and s.

Since $\cos\alpha = \dfrac{x}{r}$, we have $\cos\dfrac{\theta}{2} = \dfrac{h}{s}$, so $h = s \cdot \cos\dfrac{\theta}{2}$.

The side opposite θ is x; thus, the side opposite $\dfrac{\theta}{2}$ is $\dfrac{x}{2}$.

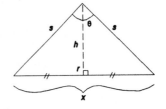

So, $\sin\dfrac{\theta}{2} = \dfrac{\dfrac{x}{2}}{s} = \dfrac{x}{2s}$. Thus, $x = 2s \cdot \sin\dfrac{\theta}{2}$.

Substituting these values into the area formula, we obtain:

$$A = \dfrac{1}{2}\left[2s \cdot \sin\dfrac{\theta}{2}\right]\left[s \cdot \cos\dfrac{\theta}{2}\right] = \dfrac{1}{2}s^2\, 2\sin\dfrac{\theta}{2}\cos\dfrac{\theta}{2}$$

$$= \dfrac{1}{2}s^2 \cdot \sin\theta \text{ since } \sin 2\theta = 2\sin\theta\cos\theta$$

67. To graph $f(x) = \sin^2(x) = \dfrac{(1 - \cos 2x)}{2}$ for $0 \le x \le 2\pi$,

begin with the graph of $y = \cos x$. Apply a horizontal compression to obtain the graph of $y = \cos 2x$. Reflect about the x-axis to obtain the graph of $y = -\cos 2x$. Shift up 1 unit to obtain the graph of $y = 1 - \cos 2x$. Apply a vertical compression to obtain the final graph:

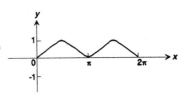

$$y = \dfrac{1}{2}(1 - \cos 2x)$$

69. $\sin\left(\dfrac{\pi}{24}\right) = \sin\left(\dfrac{\dfrac{\pi}{12}}{2}\right) = \sqrt{\dfrac{1 - \cos\dfrac{\pi}{12}}{2}} = \sqrt{\dfrac{1 - \dfrac{\sqrt{2}}{4}\left(\sqrt{3} + 1\right)}{2}} = \sqrt{\dfrac{4 - \sqrt{2}\left(\sqrt{3} + 1\right)}{8}}$

$\qquad = \sqrt{\dfrac{4 - \sqrt{6} - \sqrt{2}}{8}} = \dfrac{1}{2\sqrt{2}}\sqrt{4 - \sqrt{6} - \sqrt{2}} = \dfrac{\sqrt{2}}{4}\sqrt{4 - \sqrt{6} - \sqrt{2}}$

$\quad\cos\left(\dfrac{\pi}{24}\right) = \cos\left(\dfrac{\dfrac{\pi}{12}}{2}\right) = \sqrt{\dfrac{1 + \cos\dfrac{\pi}{12}}{2}} = \sqrt{\dfrac{1 + \dfrac{\sqrt{2}}{4}\left(\sqrt{3} + 1\right)}{2}} = \sqrt{\dfrac{4 + \sqrt{6} + \sqrt{2}}{4} \cdot \dfrac{1}{2}}$

$\qquad = \dfrac{1}{2\sqrt{2}}\sqrt{4 + \sqrt{6} + \sqrt{2}} = \dfrac{\sqrt{2}}{4}\sqrt{4 + \sqrt{6} + \sqrt{2}}$

71. $\sin^3\theta + \sin^3(\theta + 120°) + \sin^3(\theta + 240°)$

$\qquad = \sin^3\theta + [\sin\theta\cos 120° + \cos\theta\sin 120°]^3 + [\sin\theta\cos 240° + \cos\theta\sin 240°]^3$

$\qquad = \sin^3\theta + \left[-\dfrac{1}{2}\sin\theta + \dfrac{\sqrt{3}}{2}\cos\theta\right]^3 + \left[-\dfrac{1}{2}\sin\theta - \dfrac{\sqrt{3}}{2}\cos\theta\right]^3$

$\qquad = \sin^3\theta + \dfrac{1}{8}\left[3\sqrt{3}\cos^3\theta - 9\cos^2\theta\sin\theta + 3\sqrt{3}\cos\theta\sin^2\theta - \sin^3\theta\right]$

$\qquad\quad -\dfrac{1}{8}\left[\sin^3\theta + 3\sqrt{3}\sin^2\theta\cos\theta + 9\sin\theta\cos^2\theta + 3\sqrt{3}\cos^3\theta\right]$

$\qquad = \dfrac{3}{4}\sin^3\theta - \dfrac{9}{4}\cos^2\theta\sin\theta = \dfrac{3}{4}\left[\sin^3\theta - 3\sin\theta(1 - \sin^2\theta)\right]$

$\qquad = \dfrac{3}{4}\left[4\sin^3\theta - 3\sin\theta\right] = \dfrac{-3}{4}\sin 3\theta$

73. Let $\dfrac{1}{2}\left(\ln|1 - \cos 2\theta| - \ln 2\right) = \ln\left[\dfrac{|1 - \cos 2\theta|}{2}\right]^{1/2} = \ln|\sin^2\theta|^{1/2} = \ln|\sin\theta|$

75. **(a)** $R = \dfrac{v_0^2\sqrt{2}}{16}\cos\theta(\sin\theta - \cos\theta)$

$\qquad R = \dfrac{v_0^2\sqrt{2}}{16}\cos\theta(\sin\theta - \cos^2\theta)$

$\qquad R = \dfrac{v_0^2\sqrt{2}}{16}\left[\dfrac{1}{2}\sin 2\theta - \left(\dfrac{\cos 2\theta + 1}{2}\right)\right]$

$\qquad R = \dfrac{v_0^2\sqrt{2}}{32}(\sin 2\theta - \cos 2\theta - 1)$

(b)

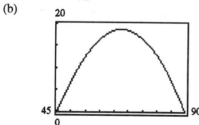

$\quad R$ is largest when $\theta = 67.5°$.

For Problems 1–10, we are using the formulas:

$$\sin \alpha \sin \beta = \frac{1}{2}[\cos(\alpha - \beta) - \cos(\alpha + \beta)]$$

$$\cos \alpha \cos \beta = \frac{1}{2}[\cos(\alpha - \beta) + \cos(\alpha + \beta)]$$

$$\sin \alpha \cos \beta = \frac{1}{2}[\sin(\alpha + \beta) + \sin(\alpha - \beta)]$$

1. $\sin 4\theta \sin 2\theta = \frac{1}{2}[\cos(4\theta - 2\theta) - \cos(4\theta + 2\theta)] = \frac{1}{2}(\cos 2\theta - \cos 6\theta)$

3. $\sin 4\theta \cos 2\theta = \frac{1}{2}[\sin(4\theta + 2\theta) + \sin(4\theta - 2\theta)] = \frac{1}{2}(\sin 6\theta + \sin 2\theta)$

5. $\cos 3\theta \cos 5\theta = \frac{1}{2}[\cos(3\theta - 5\theta) + \cos(3\theta + 5\theta)] = \frac{1}{2}[\cos(-2\theta) + \cos(8\theta)] = \frac{1}{2}(\cos 2\theta + \cos 8\theta)$

7. $\sin \theta \sin 2\theta = \frac{1}{2}[\cos(\theta - 2\theta) - \cos(\theta + 2\theta)] = \frac{1}{2}[\cos(-\theta) - \cos(3\theta)] = \frac{1}{2}(\cos \theta - \cos 3\theta)$

9. $\sin\dfrac{3\theta}{2} \cos\dfrac{\theta}{2} = \dfrac{1}{2}\left[\sin\left(\dfrac{3\theta}{2} + \dfrac{\theta}{2}\right) + \sin\left(\dfrac{3\theta}{2} - \dfrac{\theta}{2}\right)\right] = \dfrac{1}{2}(\sin 2\theta + \sin \theta)$

In Problems 11–18, we are using the formulas:

$$\sin \alpha + \sin \beta = 2 \sin \frac{\alpha + \beta}{2} \cos \frac{\alpha - \beta}{2}$$

$$\sin \alpha - \sin \beta = 2 \sin \frac{\alpha - \beta}{2} \cos \frac{\alpha + \beta}{2}$$

$$\cos \alpha + \cos \beta = 2 \cos \frac{\alpha + \beta}{2} \cos \frac{\alpha - \beta}{2}$$

$$\cos \alpha - \cos \beta = -2 \sin \frac{\alpha + \beta}{2} \cos \frac{\alpha - \beta}{2}$$

11. $\sin 4\theta - \sin 2\theta = 2 \sin \dfrac{4\theta - 2\theta}{2} \cos \dfrac{4\theta + 2\theta}{2} = 2 \sin \theta \cos 3\theta$

13. $\cos 2\theta + \cos 4\theta = 2 \cos \dfrac{2\theta + 4\theta}{2} \cos \dfrac{2\theta - 4\theta}{2} = 2 \cos 3\theta \cos \theta$

15. $\sin \theta + \sin 3\theta = 2 \sin \dfrac{\theta + 3\theta}{2} \cos \dfrac{\theta - 3\theta}{2} = 2 \sin 2\theta \cos \theta$

17. $\cos \dfrac{\theta}{2} - \cos \dfrac{3\theta}{2} = -2 \sin\dfrac{\dfrac{\theta}{2} + \dfrac{3\theta}{2}}{2} \sin \dfrac{\dfrac{\theta}{2} - \dfrac{3\theta}{2}}{2} = -2 \sin \theta\left(-\sin \dfrac{\theta}{2}\right) = 2 \sin \theta \sin \dfrac{\theta}{2}$

19. $\dfrac{\sin \theta + \sin 3\theta}{2 \sin 2\theta} = \dfrac{2 \sin 2\theta \cos(-\theta)}{2 \sin 2\theta} = \cos (-\theta) = \cos \theta$

19. $\dfrac{\sin\theta + \sin 3\theta}{2\sin 2\theta} = \dfrac{2\sin 2\theta\,\cos(-\theta)}{2\sin 2\theta} = \cos(-\theta) = \cos\theta$

21. $\dfrac{\sin 4\theta + \sin 2\theta}{\cos 4\theta + \cos 2\theta} = \dfrac{2\sin 3\theta\,\cos\theta}{2\cos 3\theta\,\cos\theta} = \dfrac{\sin 3\theta}{\cos 3\theta} = \tan 3\theta$

23. $\dfrac{\cos\theta - \cos 3\theta}{\sin\theta + \sin 3\theta} = \dfrac{-2\sin 2\theta\,\sin(-\theta)}{2\sin 2\theta\,\cos(-\theta)} = \dfrac{2\sin 2\theta\,\sin\theta}{2\sin 2\theta\,\cos\theta} = \dfrac{\sin\theta}{\cos\theta} = \tan\theta$

25. $\sin\theta(\sin\theta + \sin 3\theta) = \sin\theta[2\sin 2\theta\,\cos(-\theta)] = 2\sin 2\theta\,\sin\theta\,\cos\theta$

$\qquad = \cos\theta(2\sin 2\theta\,\sin\theta) = \cos\theta\left[2\cdot\dfrac{1}{2}(\cos\theta - \cos 3\theta)\right]$

$\qquad = \cos\theta(\cos\theta - \cos 3\theta)$

27. $\dfrac{\sin 4\theta + \sin 8\theta}{\cos 4\theta + \cos 8\theta} = \dfrac{2\sin 6\theta\,\cos(-2\theta)}{2\cos 6\theta\,\cos(-2\theta)} = \dfrac{\sin 6\theta}{\cos 6\theta} = \tan 6\theta$

29. $\dfrac{\sin 4\theta + \sin 8\theta}{\sin 4\theta - \sin 8\theta} = \dfrac{2\sin 6\theta\,\cos(-2\theta)}{2\sin(-2\theta)\,\cos 6\theta} = \dfrac{\sin 6\theta}{\cos 6\theta}\cdot\dfrac{\cos 2\theta}{-\sin(2\theta)} = (\tan 6\theta)(-\cot 2\theta) = -\dfrac{\tan 6\theta}{\tan 2\theta}$

31. $\dfrac{\sin\alpha + \sin\beta}{\sin\alpha - \sin\beta} = \dfrac{2\sin\dfrac{\alpha+\beta}{2}\cos\dfrac{\alpha-\beta}{2}}{2\sin\dfrac{\alpha-\beta}{2}\cos\dfrac{\alpha+\beta}{2}} = \dfrac{\sin\dfrac{\alpha+\beta}{2}}{\cos\dfrac{\alpha+\beta}{2}}\cdot\dfrac{\cos\dfrac{\alpha-\beta}{2}}{\sin\dfrac{\alpha-\beta}{2}} = \tan\dfrac{\alpha+\beta}{2}\cot\dfrac{\alpha-\beta}{2}$

33. $\dfrac{\sin\alpha + \sin\beta}{\cos\alpha + \cos\beta} = \dfrac{2\sin\dfrac{\alpha+\beta}{2}\cos\dfrac{\alpha-\beta}{2}}{2\cos\dfrac{\alpha+\beta}{2}\cos\dfrac{\alpha-\beta}{2}} = \dfrac{\sin\dfrac{\alpha+\beta}{2}}{\cos\dfrac{\alpha+\beta}{2}} = \tan\dfrac{\alpha+\beta}{2}$

35. $1 + \cos 2\theta + \cos 4\theta + \cos 6\theta$

$\qquad = (1\cos 0\theta + \cos 6\theta) + (\cos 2\theta + \cos 4\theta) = 2\cos 3\theta\,\cos(-3\theta) + 2\cos 3\theta\,\cos(-\theta)$

$\qquad = 2\cos^2 3\theta + 2\cos 3\theta\,\cos\theta = 2\cos 3\theta(\cos 3\theta + \cos\theta)$

$\qquad = 2\cos 3\theta(2\cos 2\theta\,\cos\theta) = 4\cos\theta\,\cos 2\theta\,\cos 3\theta$

37. (a) $\quad y = \sin 2\pi(852)t + \sin(2\pi(1209)t)$

$\qquad = 2\sin\left[\dfrac{2\pi(852)t + 2\pi(1209)t}{2}\right]\cos\left[\dfrac{2\pi(852)t - 2\pi(1209)t}{2}\right]$

$\qquad = 2\sin 2061\pi t\,\cos 357\pi t \quad\text{since }\cos(-\theta) = \cos\theta$

(b)

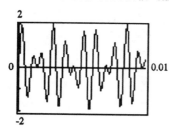

(c) $\quad y_{\max} = 2$

39. $\sin 2\alpha + \sin 2\beta + \sin 2\gamma$

$$= 2 \sin 2 \cdot \left[\frac{\alpha + \beta}{2}\right] \cos 2 \cdot \left[\frac{\alpha - \beta}{2}\right] + \sin 2\gamma$$

$$= 2 \sin(\alpha + \beta) \cos(\alpha - \beta) + 2 \sin \gamma \cos \gamma$$

(Because $\alpha + \beta + \gamma = \pi$)

$$= 2 \sin(\pi - \gamma)\cos(\alpha - \beta) + 2 \sin \gamma \cos \gamma$$

$$= 2 \sin \gamma \cos(\alpha - \beta) + 2 \sin \gamma \cos \gamma = 2 \sin \gamma[\cos(\alpha - \beta) + \cos \gamma]$$

$$= 2 \sin \gamma \left[2 \cos \frac{\alpha - \beta + \gamma}{2} \cos \frac{\alpha - \beta - \gamma}{2}\right] = 4 \sin \gamma \cos \frac{\pi - 2\beta}{2} \cos \frac{2\alpha - \pi}{2}$$

$$= 4 \sin \gamma \cos \left[\frac{\pi}{2} - \beta\right] \cos \left[\alpha - \frac{\pi}{2}\right] = 4 \sin \gamma \sin \beta \sin \alpha$$

$$= 4 \sin \alpha \sin \beta \sin \gamma$$

41.
$$\sin(\alpha - \beta) = \sin \alpha \cos \beta - \cos \alpha \sin \beta$$
$$\sin(\alpha + \beta) = \sin \alpha \cos \beta + \cos \alpha \sin \beta$$
$$\sin(\alpha - \beta) + \sin(\alpha + \beta) = 2 \sin \alpha \cos \beta$$

$$2 \sin \alpha \cos \beta = \frac{1}{2}[\sin(\alpha + \beta) + \sin(\alpha - \beta)]$$

43. $2 \cos \dfrac{\alpha + \beta}{2} \cos \dfrac{\alpha - \beta}{2} = 2 \cdot \dfrac{1}{2}\left[\cos\left[\dfrac{\alpha + \beta}{2} - \dfrac{\alpha - \beta}{2}\right] + \cos\left[\dfrac{\alpha + \beta}{2} + \dfrac{\alpha - \beta}{2}\right]\right]$

$$= \cos \frac{2\beta}{2} + \cos \frac{2\alpha}{2} = \cos \beta + \cos \alpha$$

Therefore, $\cos \alpha + \cos \beta = 2 \cos \dfrac{\alpha + \beta}{2} \cos \dfrac{\alpha - \beta}{2}$

7.5 Trigonometric Equations

1. The period of the sine function is 2π; and, on the interval $[0, 2\pi)$, the sine function has the value $\dfrac{1}{2}$ at $\dfrac{\pi}{6}$ and $\dfrac{5\pi}{6}$. Because the sine function has period 2π, all the solutions of $\sin \theta = \dfrac{1}{2}$ may be given by:

$$\theta = \frac{\pi}{6} + 2k\pi \text{ or } \theta = \frac{5\pi}{6} + 2k\pi, \ k \text{ any integer}$$

On the interval $[0, 2\pi)$, the solutions of $\sin \theta = \dfrac{1}{2}$ are $\theta = \dfrac{\pi}{6}, \dfrac{5\pi}{6}$.

3. The period of the tangent function is π; and, in the interval $[0, \pi]$, the tangent function has the value $-\dfrac{1}{\sqrt{3}}$ at $\dfrac{5\pi}{6}$.

$$\theta = \frac{5\pi}{6} + k\pi, \ k \text{ any integer}$$

On the interval $[0, 2\pi)$, the solutions of $\tan \theta = \dfrac{-1}{\sqrt{3}}$ are $\theta = \dfrac{5\pi}{6}, \dfrac{11\pi}{6}$.

5. $\cos \theta = 0$

The solutions on the interval $[0, 2\pi)$ are $\theta = \dfrac{\pi}{2}$ or $\theta = \dfrac{3\pi}{2}$.

7. $\sin 3\theta = -1$

$$3\theta = \frac{3\pi}{2} + 2k\pi, \ k \text{ any integer}$$

$$\theta = \frac{\pi}{2} + \frac{2}{3}k\pi, \ k \text{ any integer}$$

The solutions on the interval $[0, 2\pi)$ are $\theta = \frac{\pi}{2}, \frac{7\pi}{6}, \frac{11\pi}{6}$.

9. $\cos\left(2\theta - \frac{\pi}{2}\right) = -1$

$$2\theta - \frac{\pi}{2} = \pi + 2k\pi, \ k \text{ any integer}$$

$$2\theta = \frac{3\pi}{2} + 2k\pi, \ k \text{ any integer}$$

$$\theta = \frac{3\pi}{4} + k\pi, \ k \text{ any integer}$$

The solutions on the interval $[0, 2\pi)$ are $\theta = \frac{3\pi}{4}, \frac{7\pi}{4}$.

11. $\sec \frac{3\theta}{2} = -2$, since $\sec\theta = \frac{1}{\cos\theta}$, $\cos \frac{3\theta}{2} = -\frac{1}{2}$

$$\frac{3\theta}{2} = \frac{2\pi}{3} + 2k\pi \text{ or } \frac{3\theta}{2} = \frac{4\pi}{3} + 2k\pi, \ k \text{ any integer}$$

$$\theta = \frac{4\pi}{9} + \frac{4}{3}k\pi \text{ or } \theta = \frac{8\pi}{9} + \frac{4}{3}k\pi, \ k \text{ any integer}$$

The solutions on the interval $[0, 2\pi)$ are $\theta = \frac{4\pi}{9}, \frac{16\pi}{9}$ or $\theta = \frac{8\pi}{9}$.

13. $\sin\theta = 0.4$
$\theta = 0.41$ or $\pi - 0.41$

15. $\tan\theta = 5$
$\theta = 1.37$ or $\pi + 1.37$

17. $\cos\theta = -0.9$
$\theta = 2.69$ or $2\pi - 2.69$

19. $\sec\theta = -4$. Therefore, $\cos\theta = -\frac{1}{4}$
$\theta = 1.82$ or $2\pi - 1.82$

21. $\qquad 2\cos^2\theta + \cos\theta = 0$
$\qquad \cos\theta(2\cos\theta + 1) = 0$
$\cos\theta = 0 \qquad$ or $2\cos\theta + 1 = 0$
$$\cos\theta = -\frac{1}{2}$$
$\theta = \frac{\pi}{2}, \frac{3\pi}{2} \quad$ or $\quad \theta = \frac{2\pi}{3}, \frac{4\pi}{3}$

23. $\qquad 2\sin^2\theta - \sin\theta - 1 = 0$
$\qquad (2\sin\theta + 1)(\sin\theta - 1) = 0$
$2\sin\theta + 1 = 0 \qquad$ or $\quad \sin\theta - 1 = 0$
$$\sin\theta = -\frac{1}{2} \qquad \text{or} \qquad \sin\theta = 1$$
$\theta = \frac{7\pi}{6}, \frac{11\pi}{6} \quad$ or $\quad \theta = \frac{\pi}{2}$

25. $(\tan\theta - 1)(\sec\theta - 1) = 0$
$\tan\theta - 1 = 0 \qquad$ or $\quad \sec\theta - 1 = 0$
$\tan\theta = 1 \qquad$ or $\qquad \sec\theta = 1$
$\theta = \frac{\pi}{4}, \frac{5\pi}{4} \qquad$ or $\qquad \theta = 0$

27. $\cos\theta = \sin\theta$
$\tan\theta = 1$
$$\theta = \frac{\pi}{4} \text{ or } \theta = \frac{5\pi}{4}$$

29.
$$\tan \theta = 2 \sin \theta$$
$$\frac{\sin \theta}{\cos \theta} = 2 \sin \theta$$
$$\sin \theta = 2 \sin \theta \cos \theta$$
$$0 = 2 \sin \theta \cos \theta - \sin \theta$$
$$0 = \sin \theta (2 \cos \theta - 1)$$
$$\sin \theta = 0 \quad \text{or} \quad 2 \cos \theta - 1 = 0$$
$$\cos \theta = \frac{1}{2}$$
$$\theta = 0, \pi, \text{ or } \theta = \frac{\pi}{3}, \frac{5\pi}{3}$$

31.
$$\sin \theta = \csc \theta$$
$$\sin \theta - \csc \theta = 0$$
$$\sin \theta - \frac{1}{\sin \theta} = 0$$
$$\frac{\sin^2 \theta - 1}{\sin \theta} = 0$$
$$\sin^2 \theta - 1 = 0$$
$$\cos^2 \theta = 0$$
$$\cos \theta = 0$$
$$\theta = \frac{\pi}{2} \quad \text{or} \quad \theta = \frac{3\pi}{2}$$

33.
$$\cos 2\theta = \cos \theta$$
$$2 \cos^2 \theta - 1 = \cos \theta$$
$$2 \cos^2 \theta - \cos \theta - 1 = 0$$
$$(\cos \theta - 1)(2 \cos \theta + 1) = 0$$
$$\cos \theta - 1 = 0 \quad \text{or} \quad 2 \cos \theta + 1 = 0$$
$$\cos \theta = 1 \quad \text{or} \quad \cos \theta = \frac{-1}{2}$$
$$\theta = 0, \frac{2\pi}{3}, \frac{4\pi}{3}$$

35.
$$\sin 2\theta + \sin 4\theta = 0$$
$$2 \sin \theta \cos \theta + 2 \sin 2\theta \cos 2\theta = 0$$
$$\sin 2\theta (1 + 2 \cos 2\theta) = 0$$
$$\sin 2\theta = 0 \quad \text{or} \quad 1 + 2 \cos 2\theta = 0$$

$$2\theta = 0 + 2k\pi, \ 2\theta = \pi + 2k\pi, \text{ or } \cos 2\theta = -\frac{1}{2}$$

$$2\theta = 0 + 2k\pi, \ 2\theta = \pi + 2k\pi, \text{ or } 2\theta = \frac{2\pi}{3} + 2k\pi, \ 2\theta = \frac{4\pi}{3} + 2k\pi$$

$$\theta = k\pi, \ \theta = \frac{\pi}{2} + k\pi \text{ or } \theta = \frac{\pi}{3} + k\pi, \ \theta = \frac{2\pi}{3} + k\pi, \ k \text{ any integer}$$

The solutions on the interval $[0, 2\pi)$ are $\theta = 0, \ \frac{\pi}{2}, \frac{\pi}{3}, \frac{2\pi}{3}, \frac{4\pi}{3}, \pi, \frac{5\pi}{3}, \frac{3\pi}{2}$.

37.
$$\cos 4\theta - \cos 6\theta = 0$$
$$-2 \sin \frac{10\theta}{2} \sin \frac{-2\theta}{2} = 0$$
$$\sin 5\theta = 0 \quad \text{or} \quad \sin \theta = 0$$
$$5\theta = k\pi \qquad \qquad \theta = k\pi$$
$$\theta = k\frac{\pi}{5}, \ k \text{ any integer}$$
The solutions on the interval $[0, 2\pi)$ are:
$$\theta = 0, \ \frac{\pi}{5}, \frac{2\pi}{5}, \frac{3\pi}{5}, \frac{4\pi}{5},$$
$$\pi, \frac{6\pi}{5}, \frac{7\pi}{5}, \frac{8\pi}{5}, \frac{9\pi}{5}$$

39.
$$1 + \sin \theta = 2 \cos^2 \theta$$
$$1 + \sin \theta = 2(1 - \sin^2 \theta)$$
$$1 + \sin \theta = 2 - 2 \sin^2 \theta$$
$$2 \sin^2 \theta + \sin \theta - 1 = 0$$
$$(2 \sin \theta - 1)(\sin \theta + 1) = 0$$
$$2 \sin \theta - 1 = 0 \quad \text{or} \quad \sin \theta + 1 = 0$$
$$\sin \theta = \frac{1}{2} \quad \text{or} \quad \sin \theta = -1$$
$$\theta = \frac{\pi}{6}, \frac{5\pi}{6}, \frac{3\pi}{2}$$

41.
$$\tan^2 \theta = \frac{3}{2} \sec \theta$$
$$\sec^2 \theta - 1 = \frac{3}{2} \sec \theta$$
$$\sec^2 \theta - \frac{3}{2} \sec \theta - 1 = 0$$
$$\left[\sec \theta + \frac{1}{2}\right](\sec \theta - 2) = 0$$
$$\sec \theta = \frac{-1}{2} \quad \text{or} \quad \sec \theta = 2$$
$$\cos \theta = -2 \quad \text{or} \quad \cos \theta = \frac{1}{2}$$

For any angle θ, $-1 \le \cos \theta \le 1$; thus, $\cos \theta = -2$ has no solution.

The solutions of $\cos \theta = \frac{1}{2}$ are $\theta = \frac{\pi}{3}, \frac{5\pi}{3}$

43.
$$3 - \sin \theta = \cos 2\theta$$
$$3 - \sin \theta = 1 - 2 \sin^2 \theta$$
$$2 \sin^2 \theta - \sin \theta + 2 = 0$$
This is a quadratic equation in $\sin \theta$. The discriminant is $b^2 - 4ac = 1 - 16 = -15 < 0$. The equation, therefore, has no real solution.

45.
$$\sec^2 \theta + \tan \theta = 0$$
$$(\tan^2 \theta + 1) + \tan \theta = 0$$
$$\tan^2 + \tan \theta + 1 = 0$$
The discriminate is $b^2 - 4ac = 1 - 4 = -3 < 0$. The equation, therefore, has no real solution.

47. $\sin \theta - \sqrt{3} \cos \theta = 1$

We divide each side of the equation by 2. Then, $\frac{1}{2} \sin \theta - \frac{\sqrt{3}}{2} \cos \theta = \frac{1}{2}$.

There is a unique angle ϕ, $0 \le \phi < 2\pi$, for which $\cos \phi = \frac{1}{2}$ and $\sin \phi = \frac{\sqrt{3}}{2}$

$$\phi = \frac{\pi}{3}$$

Thus, the equation may be written as:
$$\sin \theta \cos \phi - \cos \theta \sin \phi = \frac{1}{2}$$

or $\quad \sin(\theta - \phi) = \frac{1}{2}$

$$\theta - \phi = \frac{\pi}{6} \quad \text{or} \qquad \theta - \phi = \frac{5\pi}{6}$$
$$\theta - \frac{\pi}{3} = \frac{\pi}{6} \quad \text{or} \qquad \theta - \frac{\pi}{3} = \frac{5\pi}{6}$$
$$\theta = \frac{\pi}{2} \quad \text{or} \qquad \theta = \frac{7\pi}{6}$$

49.

$$\tan 2\theta + 2 \sin \theta = 0$$

$$\frac{\sin 2\theta}{\cos 2\theta} + 2 \sin \theta = 0$$

$$\frac{2 \sin \theta \cos \theta + 2 \sin \theta \cos 2\theta}{\cos 2\theta} = 0$$

$$2 \sin \theta(\cos \theta + \cos 2\theta) = 0$$

$$2 \sin \theta[\cos \theta + (2 \cos^2 \theta - 1)] = 0$$

$$2 \sin \theta(2 \cos^2 \theta + \cos \theta - 1) = 0$$

$$2 \sin \theta(2 \cos \theta - 1)(\cos \theta + 1) = 0$$

$$2 \sin \theta = 0 \quad \text{or} \quad 2 \cos \theta - 1 = 0 \quad \text{or} \quad \cos \theta + 1 = 0$$

$$\sin \theta = 0 \quad \text{or} \quad \cos \theta = \frac{1}{2} \quad \text{or} \quad \cos \theta = -1$$

$$\theta = 0, \ \frac{\pi}{3}, \ \pi, \ \frac{5\pi}{3}$$

51.

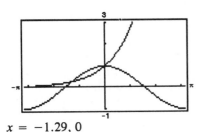

$x = -1.29, 0$

53.

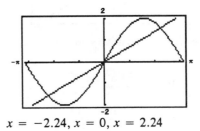

$x = -2.24, x = 0, x = 2.24$

55.

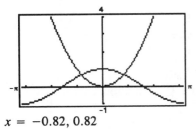

$x = -0.82, 0.82$

57. $-1.30, 1.97, 3.83$

59.   0.59   **61.**   1.25   **63.** $-1.02, 1.02$ **65.**   0, 2.14   **67.**   1.34

69.   (a) $\cos 2\theta + \cos \theta = 0$

$2 \cos^2 \theta - 1 + \cos \theta$

$2 \cos^2 \theta + \cos \theta - 1 = 0$

$(2 \cos \theta - 1)(\cos \theta + 1) = 0$

$2 \cos \theta - 1 = 0 \quad \text{or} \quad \cos \theta + 1 = 0$

$\cos \theta = \frac{1}{2} \qquad\qquad \cos \theta = -1$

$\theta = 60°$

(b) $\cos 2\theta + \cos \theta = 0$

$2 \cos \dfrac{3\theta}{2} \cos \dfrac{\theta}{2} = 0$

$\cos \dfrac{3\theta}{2} = 0 \quad \text{or} \quad \cos \dfrac{\theta}{2} = 0$

$\dfrac{3\theta}{2} = 90° \qquad\qquad \dfrac{\theta}{2} = 90°$

$\theta = 60° \qquad\qquad \theta = 180°$

(c) $A(60°) = 16 \sin 60°(\cos 60° + 1) = 12\sqrt{3}$ square inches

(d)

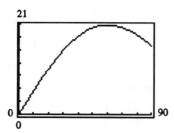

The angle θ that maximizes area is $60°$. The maximum area is 20.78 square inches.

71.

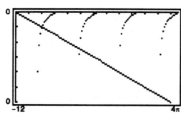

The first two positive solutions are 2.02 and 4.91.

73. (a) $107 = \dfrac{34.8^2 \sin 2\theta}{9.8}$

$\sin 2\theta = \dfrac{(107)(9.8)}{34.8^2}$

$2\theta = \sin^{-1}\left[\dfrac{107 \cdot 9.8}{34.8^2}\right]$

$2\theta = 59.98189°$

$\theta \approx 30°$

(b)

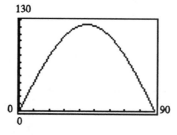

(c) Using TRACE, we find that when $\theta = 30°$, $R = 107$ meters.

(d) R is maximized when $\theta = 45°$. The maximum distance you can throw the ball is 123.6 meters.

75. The ratio $\dfrac{v_1}{v_2}$ is the index of refraction. The index of refraction of light in passing from a vacuum into water is 1.33.

Therefore, $\dfrac{v_1}{v_2} = 1.33$.

The angle of incidence, θ_1, is $40°$.

Because $\dfrac{\sin \theta_1}{\sin \theta_2} = \dfrac{v_1}{v_2}$, $\dfrac{\sin 40°}{\sin \theta_2} = 1.33$.

The angle of refraction, θ_2, can be found as follows:

$\dfrac{\sin 40°}{\sin \theta_2} = 1.33$

$\sin 40° = 1.33 \sin \theta_2$

$\dfrac{\sin 40°}{1.33} = \sin \theta_2$

$0.4833 = \sin \theta_2$

$\sin^{-1} 0.4833 = \sin_2$

$28.9° = \theta_2$

77. To agree with Snell's Law, the measured values in the table for the angle of incidence θ and the angle of refraction θ_2 for a light beam passing from air into water must have the proportion of $\dfrac{\sin \theta_1}{\sin \theta_2}$

which is $\dfrac{v_1}{v_2}$, the index of refraction.

For $\theta = 10°$ and $\theta_2 = 7° 45'$, $\dfrac{\sin 10°}{\sin 7°45'} = \dfrac{\sin 10°}{\sin 7.75°}$.

Now, using a calculator, $\dfrac{\sin 10°}{\sin 7.75°} \approx 1.2877$.

For $\theta_1 = 20°$ and $\theta_2 = 15° 30'$, $\dfrac{\sin 20°}{\sin 15° 30'} = \dfrac{\sin 20°}{\sin 15.5°} \approx 1.2798$.

For $\theta_1 = 30°$ and $\theta_2 = 22° 30'$, $\dfrac{\sin 30°}{\sin 22° 30'} = \dfrac{\sin 30°}{\sin 22.5°} \approx 1.3066$.

For $\theta_1 = 40°$ and $\theta_2 = 29.0'$, $\dfrac{\sin 40°}{\sin 29°} \approx 1.3259$.

For $\theta_1 = 50°$ and $\theta_2 = 35° 0'$, $\dfrac{\sin 50°}{\sin 35°} \approx 1.3356$.

For $\theta_1 = 60°$ and $\theta_2 = 40° 30'$, $\dfrac{\sin 60°}{\sin 40° 30'} = \dfrac{\sin 60°}{\sin 40.5°} \approx 1.3335$.

For $\theta_1 = 70°$ and $\theta_2 = 45° 30'$, $\dfrac{\sin 70°}{\sin 45° 30'} = \dfrac{\sin 70°}{\sin 45.5°} \approx 1.3175$.

For $\theta_1 = 80°$ and $\theta_2 = 50° 0'$, $\dfrac{\sin 80°}{\sin 50°} \approx 1.2856$.

These values do agree with Snell's Law, and the index of refraction varies from 1.27 to 1.34.

79. $\dfrac{\sin \theta_1}{\sin \theta_2} = \dfrac{\sin 40°}{\sin 26°} = \dfrac{v_1}{v_2}$, the index of refraction

$\dfrac{\sin 40°}{\sin 26°} = \dfrac{v_1}{v_2} = 1.47$

The index of refraction of a beam of light of wave length 589 nanometers traveling in air is 1.47.

81. If θ is the original angle of incidence and ϕ is the angle of refraction, then $\dfrac{\sin \theta}{\sin \phi} = n_2$. The angle of incidence of the emerging beam is also ϕ, and the index of refraction is $\dfrac{1}{n_2}$. Thus, θ is the angle of refraction of the emerging beam.

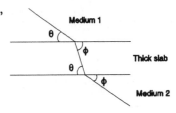

7 Chapter Review

1. $\tan \theta \cot \theta - \sin^2 \theta = 1 - \sin^2 \theta = \cos^2 \theta$

3. $\cos^2 \theta (1 + \tan^2 \theta) = (\cos^2 \theta)(\sec^2 \theta) = 1$

5. $4 \cos^2 \theta + 3 \sin^2 \theta = \cos^2 \theta + 3(\cos^2 \theta + \sin^2 \theta) = 3 + \cos^2 \theta$

7. $\dfrac{1 - \cos\theta}{\sin\theta} + \dfrac{\sin\theta}{1 - \cos\theta} = \dfrac{(1 - \cos\theta)^2 + \sin^2\theta}{\sin\theta(1 - \cos\theta)} = \dfrac{1 - 2\cos\theta + \cos^2\theta + \sin^2\theta}{\sin\theta(1 - \cos\theta)}$

$\qquad = \dfrac{2 - 2\cos\theta}{\sin\theta(1 - \cos\theta)} = \dfrac{2(1 - \cos\theta)}{\sin\theta(1 - \cos\theta)} = \dfrac{2}{\sin\theta} = 2\csc\theta$

9. $\dfrac{\cos\theta}{\cos\theta - \sin\theta} = \dfrac{\dfrac{\cos\theta}{\cos\theta}}{\dfrac{\cos\theta - \sin\theta}{\cos\theta}} = \dfrac{1}{1 - \dfrac{\sin\theta}{\cos\theta}} = \dfrac{1}{1 - \tan\theta}$

11. $\dfrac{\csc\theta}{1 + \csc\theta} = \dfrac{\dfrac{1}{\sin\theta}}{1 + \dfrac{1}{\sin\theta}} = \dfrac{\dfrac{1}{\sin\theta}}{\dfrac{\sin\theta + 1}{\sin\theta}} = \dfrac{1}{1 + \sin\theta} = \dfrac{1}{1 + \sin\theta} \cdot \dfrac{1 - \sin\theta}{1 - \sin\theta}$

$\qquad = \dfrac{1 - \sin\theta}{1 - \sin^2\theta} = \dfrac{1 - \sin\theta}{\cos^2\theta}$

13. $\csc\theta - \sin\theta = \dfrac{1}{\sin\theta} - \sin\theta = \dfrac{1 - \sin^2\theta}{\sin\theta} = \dfrac{\cos^2\theta}{\sin\theta} = \cos\theta \cdot \dfrac{\cos\theta}{\sin\theta} = \cos\theta\cot\theta$

15. $\dfrac{1 - \sin\theta}{\sec\theta} = \cos\theta(1 - \sin\theta) = (\cos\theta)(1 - \sin\theta) \cdot \dfrac{1 + \sin\theta}{1 + \sin\theta} = \dfrac{\cos\theta(1 - \sin^2\theta)}{1 + \sin\theta}$

$\qquad = \dfrac{(\cos\theta)(\cos^2\theta)}{1 + \sin\theta} = \dfrac{\cos^3\theta}{1 + \sin\theta}$

17. $\cot\theta - \tan\theta = \dfrac{\cos\theta}{\sin\theta} - \dfrac{\sin\theta}{\cos\theta} = \dfrac{\cos^2\theta - \sin^2\theta}{\sin\theta\cos\theta} = \dfrac{1 - \sin^2\theta - \sin^2\theta}{\sin\theta\cos\theta} = \dfrac{1 - 2\sin^2\theta}{\sin\theta\cos\theta}$

19. $\dfrac{\cos(\alpha + \beta)}{\cos\alpha\sin\beta} = \dfrac{\cos\alpha\cos\beta - \sin\alpha\sin\beta}{\cos\alpha\sin\beta} = \dfrac{\cos\alpha\cos\beta}{\cos\alpha\sin\beta} - \dfrac{\sin\alpha\sin\beta}{\cos\alpha\sin\beta}$

$\qquad = \dfrac{\cos\beta}{\sin\beta} - \dfrac{\sin\alpha}{\cos\alpha} = \cot\beta - \tan\alpha$

21. $\dfrac{\cos(\alpha - \beta)}{\cos\alpha\cos\beta} = \dfrac{\cos\alpha\cos\beta + \sin\alpha\sin\beta}{\cos\alpha\cos\beta} = \dfrac{\cos\alpha\cos\beta}{\cos\alpha\cos\beta} + \dfrac{\sin\alpha\sin\beta}{\cos\alpha\cos\beta} = 1 + \tan\alpha\tan\beta$

23. $(1 + \cos\theta)\left[\tan\dfrac{\theta}{2}\right] = \left[2\cos^2\dfrac{\theta}{2}\right]\dfrac{\sin\dfrac{\theta}{2}}{\cos\dfrac{\theta}{2}} = 2\sin\dfrac{\theta}{2}\cos\dfrac{\theta}{2} = \sin\theta$

25. $2\cot\theta\cot 2\theta = 2\left[\dfrac{\cos\theta}{\sin\theta}\right]\left[\dfrac{\cos 2\theta}{\sin 2\theta}\right] = \dfrac{2\cos\theta(\cos^2\theta - \sin^2\theta)}{\sin\theta(2\sin\theta\cos\theta)}$

$\qquad = \dfrac{2\cos\theta(\cos^2\theta - \sin^2\theta)}{(2\cos\theta)\sin^2\theta} = \dfrac{\cos^2\theta - \sin^2\theta}{\sin^2\theta} = \cot^2\theta - 1$

27. $1 - 8\sin^2\theta\cos^2\theta = 1 - 2(2\sin\theta\cos\theta)^2 = 1 - 2\sin^2 2\theta = \cos 2(2\theta) = \cos 4\theta$

29. $\dfrac{\sin 2\theta + \sin 4\theta}{\cos 2\theta + \cos 4\theta} = \dfrac{2\sin 3\theta\cos(-\theta)}{2\cos 3\theta\cos(-\theta)} = \dfrac{\sin 3\theta}{\cos 3\theta} = \tan 3\theta$

31. $\dfrac{\cos 2\theta - \cos 4\theta}{\cos 2\theta + \cos 4\theta} - \tan \theta \tan 3\theta$

$\qquad = \dfrac{-2 \sin 3\theta \sin(-\theta)}{2 \cos 3\theta \cos(-\theta)} - \tan \theta \tan 3\theta = -\tan 3\theta \tan(-\theta) - \tan \theta \tan 3\theta$

$\qquad = \tan 3\theta \tan \theta - \tan \theta \tan 3\theta = 0$

33. $\sin 165° = \sin(30° + 135°) = \sin 30° \cos 135° + \cos 30° \sin 135°$

$\qquad = \dfrac{1}{2} \cdot \dfrac{-\sqrt{2}}{2} + \dfrac{\sqrt{3}}{2} \cdot \dfrac{\sqrt{2}}{2} = \dfrac{-\sqrt{2}\left(1 - \sqrt{3}\right)}{4} = \dfrac{\sqrt{2}}{4}\left(\sqrt{3} - 1\right) = \dfrac{1}{4}\left(\sqrt{6} - \sqrt{2}\right)$

35. $\cos \dfrac{5\pi}{12} = \cos\left[\dfrac{\pi}{6} + \dfrac{\pi}{4}\right] = \cos \dfrac{\pi}{6} \cos \dfrac{\pi}{4} - \sin \dfrac{\pi}{6} \sin \dfrac{\pi}{4}$

$\qquad = \dfrac{\sqrt{3}}{2} \cdot \dfrac{\sqrt{2}}{2} - \dfrac{1}{2} \cdot \dfrac{\sqrt{2}}{2} = \dfrac{\sqrt{2}}{4}\left(\sqrt{3} - 1\right) = \dfrac{1}{4}\left(\sqrt{6} - \sqrt{2}\right)$

37. $\cos 80° \cos 20° + \sin 80° \sin 20° = \cos(80° - 20°) = \cos 60° = \dfrac{1}{2}$

39. $\tan \dfrac{\pi}{8} = \tan \dfrac{\frac{\pi}{4}}{2} = \sqrt{\dfrac{1 - \cos \frac{\pi}{4}}{1 + \cos \frac{\pi}{4}}} = \sqrt{\dfrac{1 - \frac{\sqrt{2}}{2}}{1 + \frac{\sqrt{2}}{2}}} = \sqrt{\dfrac{\frac{2 - \sqrt{2}}{2}}{\frac{2 + \sqrt{2}}{2}}} = \sqrt{\dfrac{2 - \sqrt{2}}{2 + \sqrt{2}}}$

$\qquad = \sqrt{\dfrac{2 - \sqrt{2}}{2 + \sqrt{2}} \cdot \left[\dfrac{2 - \sqrt{2}}{2 - \sqrt{2}}\right]} = \sqrt{\dfrac{\left(2 - \sqrt{2}\right)^2}{2}} = \dfrac{2 - \sqrt{2}}{\sqrt{2}}\left[\dfrac{\sqrt{2}}{\sqrt{2}}\right]$

$\qquad = \dfrac{2\sqrt{2} - 2}{2} = \sqrt{2} - 1$

41. $\sin \alpha = \dfrac{4}{5}, 0 < \alpha < \dfrac{\pi}{2}; \sin \beta = \dfrac{5}{13}, \dfrac{\pi}{2} < \beta < \pi$

Therefore, $\cos \alpha = \dfrac{3}{5}, \cos \beta = \dfrac{-12}{13}, 0 < \dfrac{\alpha}{2} < \dfrac{\pi}{4}, \dfrac{\pi}{4} < \dfrac{\beta}{2} < \dfrac{\pi}{2}$

(a) $\sin(\alpha + \beta) = \sin \alpha \cos \beta + \cos \alpha \sin \beta = \dfrac{4}{5} \cdot \dfrac{-12}{13} + \dfrac{3}{5} \cdot \dfrac{5}{13} = \dfrac{-48 + 15}{65} = \dfrac{-33}{65}$

(b) $\cos(\alpha + \beta) = \cos \alpha \cos \beta + \sin \alpha \sin \beta = \dfrac{3}{5} \cdot \dfrac{-12}{13} - \dfrac{4}{5} \cdot \dfrac{5}{13} = \dfrac{-36 - 20}{65} = \dfrac{-56}{65}$

(c) $\sin(\alpha - \beta) = \sin \alpha \cos \beta - \cos \alpha \sin \beta = \dfrac{4}{5} \cdot \dfrac{-12}{13} - \dfrac{3}{5} \cdot \dfrac{5}{13} = \dfrac{-48 - 15}{65} = \dfrac{-63}{65}$

(d) $\tan(\alpha + \beta) = \dfrac{\tan \alpha + \tan \beta}{1 - \tan \alpha \tan \beta} = \dfrac{\frac{4}{3} + \frac{-5}{12}}{1 + \frac{5}{9}} = \dfrac{\frac{11}{12}}{\frac{14}{9}} = \dfrac{33}{56}$

(e) $\sin 2\alpha = 2 \sin \alpha \cos \alpha = 2 \cdot \dfrac{4}{5} \cdot \dfrac{3}{5} = \dfrac{24}{25}$

(f) $\cos 2\beta = 1 - 2 \sin^2 \beta = 1 - 2\left[\dfrac{5}{13}\right]^2 = 1 - \dfrac{50}{169} = \dfrac{119}{169}$

(g) $\sin \dfrac{\beta}{2} = \sqrt{\dfrac{1 - \cos \beta}{2}} = \sqrt{\dfrac{1 - \dfrac{-12}{13}}{2}} = \sqrt{\dfrac{\dfrac{25}{13}}{2}} = \sqrt{\dfrac{25}{26}} = \dfrac{5\sqrt{26}}{26}$

(h) $\cos \dfrac{\alpha}{2} = \sqrt{\dfrac{1 + \cos \alpha}{2}} = \sqrt{\dfrac{1 + \dfrac{3}{5}}{2}} = \sqrt{\dfrac{\dfrac{8}{5}}{2}} = \sqrt{\dfrac{4}{5}} = \dfrac{2\sqrt{5}}{5}$

43. $\sin \alpha = -\dfrac{3}{5}, \pi < \alpha < \dfrac{3\pi}{2}; \cos \beta = \dfrac{12}{13}, \dfrac{3\pi}{2} < \beta < 2\pi$

Therefore, $\cos \alpha = \dfrac{-4}{5}, \tan \alpha = \dfrac{3}{4}, \dfrac{\pi}{2} < \dfrac{\alpha}{2} < \dfrac{3\pi}{4}, \dfrac{3\pi}{4} < \dfrac{\beta}{2} < \pi$

$\sin \beta = \dfrac{-5}{13}, \tan \beta = \dfrac{-5}{12}$

(a) $\sin(\alpha + \beta) = \sin \alpha \cos \beta + \cos \alpha \sin \beta = \dfrac{-3}{5} \cdot \dfrac{12}{13} + \dfrac{-4}{5} \cdot \dfrac{-5}{13} = \dfrac{-36 + 20}{65} = \dfrac{-16}{65}$

(b) $\cos(\alpha + \beta) = \cos \alpha \cos \beta - \sin \alpha \sin \beta = \dfrac{-4}{5} \cdot \dfrac{12}{13} - \dfrac{3}{5} \cdot \dfrac{-5}{13} = \dfrac{-48 - 15}{65} = \dfrac{-63}{65}$

(c) $\sin(\alpha - \beta) = \sin \alpha \cos \beta - \cos \alpha \sin \beta = \dfrac{-3}{5} \cdot \dfrac{12}{13} - \dfrac{-4}{5} \cdot \dfrac{-5}{13} = \dfrac{-36 - 20}{65} = \dfrac{-56}{65}$

(d) $\tan(\alpha + \beta) = \dfrac{\tan \alpha + \tan \beta}{1 - \tan \alpha \tan \beta} = \dfrac{\dfrac{3}{4} - \dfrac{5}{12}}{1 + \dfrac{5}{16}} = \dfrac{\dfrac{1}{3}}{\dfrac{21}{16}} = \dfrac{16}{63}$

(e) $\sin 2\alpha = 2 \sin \alpha \cos \alpha = 2 \cdot \dfrac{-3}{5} \cdot \dfrac{-4}{5} = \dfrac{24}{25}$

(f) $\cos 2\beta = 2 \cos^2 \beta - 1 = 2 \left[\dfrac{12}{13}\right]^2 - 1 = \dfrac{288}{169} - 1 = \dfrac{119}{169}$

(g) $\sin \dfrac{\beta}{2} = \sqrt{\dfrac{1 - \cos \beta}{2}} = \sqrt{\dfrac{1 - \dfrac{12}{13}}{2}} = \sqrt{\dfrac{\dfrac{1}{13}}{2}} = \sqrt{\dfrac{1}{26}} = \dfrac{\sqrt{26}}{26}$

(h) $\cos \dfrac{\alpha}{2} = -\sqrt{\dfrac{1 + \cos \alpha}{2}} = -\sqrt{\dfrac{1 - \dfrac{4}{5}}{2}} = -\sqrt{\dfrac{\dfrac{1}{5}}{2}} = -\sqrt{\dfrac{1}{10}} = \dfrac{-\sqrt{10}}{10}$

45. $\tan \alpha = \dfrac{3}{4}, \pi < \alpha < \dfrac{3\pi}{2}; \tan \beta = \dfrac{12}{5}, 0 < \beta < \dfrac{\pi}{2}$

Therefore, $\sin \alpha = \dfrac{-3}{5}, \cos \alpha = \dfrac{-4}{5}$

$\sin \beta = \dfrac{12}{13}, \cos \beta = \dfrac{5}{13}, \dfrac{\pi}{2} < \dfrac{\alpha}{2} < \dfrac{3\pi}{4}, 0 < \dfrac{\beta}{2} < \dfrac{\pi}{4}$

(a) $\sin(\alpha + \beta) = \sin \alpha \cos \beta + \cos \alpha \sin \beta = \dfrac{-3}{5} \cdot \dfrac{5}{13} + \dfrac{-4}{5} \cdot \dfrac{12}{13} = \dfrac{-15 - 48}{65} = \dfrac{-63}{65}$

(b) $\cos(\alpha + \beta) = \cos \alpha \cos \beta - \sin \alpha \sin \beta = \dfrac{-4}{5} \cdot \dfrac{5}{13} - \dfrac{-3}{5} \cdot \dfrac{12}{13} = \dfrac{-20 + 36}{65} = \dfrac{16}{65}$

(c) $\sin(\alpha - \beta) = \sin \alpha \cos \beta - \cos \alpha \sin \beta = \dfrac{-3}{5} \cdot \dfrac{5}{13} - \dfrac{-4}{5} \cdot \dfrac{12}{13} = \dfrac{-15 + 48}{65} = \dfrac{33}{65}$

(d) $\tan(\alpha + \beta) = \dfrac{\tan \alpha + \tan \beta}{1 - \tan \alpha \tan \beta} = \dfrac{\dfrac{3}{4} + \dfrac{12}{5}}{1 - \dfrac{9}{5}} = \dfrac{\dfrac{15 + 48}{20}}{\dfrac{-4}{5}} = \dfrac{63}{20}\left(-\dfrac{5}{4}\right) = \dfrac{-63}{16}$

(e) $\sin 2\alpha = 2 \sin \alpha \cos \alpha = 2\left(\dfrac{-3}{5}\right)\left(\dfrac{-4}{5}\right) = \dfrac{24}{25}$

(f) $\cos 2\beta = \cos^2 \beta - \sin^2 \beta = \left(\dfrac{5}{13}\right)^2 - \left(\dfrac{12}{13}\right)^2 = \dfrac{25 - 144}{169} = \dfrac{-119}{169}$

(g) $\sin \dfrac{\beta}{2} = \sqrt{\dfrac{1 - \cos \beta}{2}} = \sqrt{\dfrac{1 - \dfrac{5}{13}}{2}} = \sqrt{\dfrac{8}{26}} = \dfrac{4\sqrt{13}}{26} = \dfrac{2\sqrt{13}}{13}$

(h) $\cos \dfrac{\alpha}{2} = -\sqrt{\dfrac{1 + \cos \alpha}{2}} = -\sqrt{\dfrac{1 + \dfrac{-4}{5}}{2}} = -\sqrt{\dfrac{1}{10}} = \dfrac{-\sqrt{10}}{10}$

47. $\sec \alpha = 2,\ \dfrac{-\pi}{2} < \alpha < 0;\ \sec \beta = 3,\ \dfrac{3\pi}{2} < \beta < 2\pi$

Therefore, $\sin \alpha = \dfrac{-\sqrt{3}}{2},\ \cos \alpha = \dfrac{1}{2},\ \dfrac{-\pi}{4} < \dfrac{\alpha}{2} < 0,\ \dfrac{3\pi}{4}, < \dfrac{\beta}{2} < \pi$

$\sin \beta = \dfrac{-2\sqrt{2}}{3},\ \cos \beta = \dfrac{1}{3}$

(a) $\sin(\alpha + \beta) = \sin \alpha \cos \beta + \cos \alpha \sin \beta = \left(\dfrac{-\sqrt{3}}{2}\right)\left(\dfrac{1}{3}\right) + \left(\dfrac{1}{2}\right)\left(\dfrac{-2\sqrt{2}}{3}\right)$

$= \dfrac{-\sqrt{3}}{6} + \dfrac{-2\sqrt{2}}{6} = \dfrac{-\left(\sqrt{3} + 2\sqrt{2}\right)}{6} = \dfrac{-\sqrt{3} - 2\sqrt{2}}{6}$

(b) $\cos(\alpha + \beta) = \cos \alpha \cos \beta + \sin \alpha \sin \beta = \left(\dfrac{1}{2}\right)\left(\dfrac{1}{3}\right) - \left(\dfrac{-\sqrt{3}}{2}\right)\left(\dfrac{-2\sqrt{2}}{3}\right) = \dfrac{1}{6} - \dfrac{2\sqrt{6}}{6}$

$= \dfrac{1 - 2\sqrt{6}}{6}$

(c) $\sin(\alpha - \beta) = \sin \alpha \cos \beta - \cos \alpha \sin \beta = \left(\dfrac{-\sqrt{3}}{2}\right)\left(\dfrac{1}{3}\right) - \left(\dfrac{1}{2}\right)\left(\dfrac{-2\sqrt{2}}{3}\right)$

$= \dfrac{-\sqrt{3}}{6} + \dfrac{2\sqrt{2}}{6} = \dfrac{2\sqrt{2} - \sqrt{3}}{6}$

(d) $\tan(\alpha + \beta) = \dfrac{\tan \alpha + \tan \beta}{1 - \tan \alpha \tan \beta} = \dfrac{-\sqrt{3} + -2\sqrt{2}}{1 - \left(-\sqrt{3}\right)\left(-2\sqrt{2}\right)} = \dfrac{-\sqrt{3} - 2\sqrt{2}}{1 - 2\sqrt{6}}$

$= \dfrac{-\sqrt{3} - 2\sqrt{2}}{1 - 2\sqrt{6}} \cdot \dfrac{1 + 2\sqrt{6}}{1 + 2\sqrt{6}} = \dfrac{8\sqrt{2} + 9\sqrt{3}}{23}$

(e) $\sin 2\alpha = 2 \sin \alpha \cos \alpha = 2\left(\dfrac{-\sqrt{3}}{2}\right)\left(\dfrac{1}{2}\right) = \dfrac{-2\sqrt{3}}{4} = \dfrac{-\sqrt{3}}{2}$

(f) $\cos 2\beta = \cos^2 \beta - \sin^2 \beta = \left(\dfrac{1}{3}\right)^2 - \left(\dfrac{-2\sqrt{2}}{3}\right)^2 = \dfrac{1}{9} - \dfrac{8}{9} = \dfrac{-7}{9}$

(g) $\quad \sin \dfrac{\beta}{2} = \sqrt{\dfrac{1 - \cos \beta}{2}} = \sqrt{\dfrac{1 - \dfrac{1}{3}}{2}} = \sqrt{\dfrac{2}{6}} = \dfrac{\sqrt{3}}{3}$

(h) $\quad \cos \dfrac{\alpha}{2} = \sqrt{\dfrac{1 + \cos \alpha}{2}} = \sqrt{\dfrac{1 + \dfrac{1}{2}}{2}} = \sqrt{\dfrac{3}{4}} = \dfrac{\sqrt{3}}{2}$

49. $\quad \sin \alpha = -\dfrac{2}{3}, \ \pi < \alpha < \dfrac{3\pi}{2}; \ \cos \beta = \dfrac{-2}{3}, \ \pi < \beta < \dfrac{3\pi}{2}$

Therefore, $\cos \alpha = \dfrac{-\sqrt{5}}{3}, \ \sin \beta = \dfrac{-\sqrt{5}}{3}, \ \dfrac{\pi}{2} < \dfrac{\alpha}{2} < \dfrac{3\pi}{4}, \ \dfrac{\pi}{2} < \dfrac{\beta}{2} < \dfrac{3\pi}{4}$

(a) $\quad \sin(\alpha + \beta) = \sin \alpha \cos \beta + \cos \alpha \sin \beta = \left[\dfrac{-2}{3}\right]\left[\dfrac{-2}{3}\right] + \left[\dfrac{-\sqrt{5}}{3}\right]\left[\dfrac{-\sqrt{5}}{3}\right] = \dfrac{4 + 5}{9} = 1$

(b) $\quad \cos(\alpha + \beta) = \cos \alpha \cos \beta + \sin \alpha \sin \beta = \left[\dfrac{-\sqrt{5}}{3}\right]\left[\dfrac{-2}{3}\right] + \left[\dfrac{-2}{3}\right]\left[\dfrac{-\sqrt{5}}{3}\right]$

$$= \dfrac{2\sqrt{5} - 2\sqrt{5}}{9} = 0$$

(c) $\quad \sin(\alpha - \beta) = \sin \alpha \cos \beta - \cos \alpha \sin \beta = \left[\dfrac{-2}{3}\right]\left[\dfrac{-2}{3}\right] - \left[\dfrac{-\sqrt{5}}{3}\right]\left[\dfrac{-\sqrt{5}}{3}\right]$

$$= \dfrac{4 - 5}{9} = \dfrac{-1}{9}$$

(d) $\quad \tan(\alpha + \beta) = \dfrac{\tan \alpha + \tan \beta}{1 - \tan \alpha \tan \beta} = \dfrac{\dfrac{2}{\sqrt{5}} + \dfrac{\sqrt{5}}{2}}{1 - \left[\dfrac{2}{\sqrt{5}}\right]\left[\dfrac{-\sqrt{5}}{3}\right]}$ Not defined.

(e) $\quad \sin 2\alpha = 2 \sin \alpha \cos \alpha = 2\left[\dfrac{-2}{3}\right]\left[\dfrac{-\sqrt{5}}{3}\right] = \dfrac{4\sqrt{5}}{9}$

(f) $\quad \cos 2\beta = 2 \cos^2 \beta - 1 = 2\left[\dfrac{-2}{3}\right]^2 - 1 = -\dfrac{1}{9}$

(g) $\quad \sin \dfrac{\beta}{2} = \sqrt{\dfrac{1 - \cos \beta}{2}} = \sqrt{\dfrac{1 + \dfrac{2}{3}}{2}} = \sqrt{\dfrac{5}{6}} = \dfrac{\sqrt{30}}{6}$

(h) $\quad \cos \dfrac{\alpha}{2} = -\sqrt{\dfrac{1 + \cos \alpha}{2}} = -\sqrt{\dfrac{1 - \dfrac{\sqrt{5}}{3}}{2}} = -\sqrt{\dfrac{3 - \sqrt{5}}{6}} = \dfrac{-\sqrt{6}\sqrt{3 - \sqrt{5}}}{6}$

51. $\quad \cos \theta = \dfrac{1}{2}$

On the interval $[0, 2\pi)$, the solutions are

$\theta = \dfrac{\pi}{3}, \dfrac{5\pi}{3}.$

53. $\quad \cos \theta = \dfrac{-\sqrt{2}}{2}$

On the interval $[0, 2\pi)$, the solutions are

$\theta = \dfrac{3\pi}{4}, \text{ or } \theta = \dfrac{5\pi}{4}.$

55. $\sin 2\theta = -1$

$\qquad 2\theta = \dfrac{3\pi}{2} + 2k\pi,\ k$ any integer

$\qquad \theta = \dfrac{3\pi}{4} + k\pi,\ k$ any integer

On the interval $[0, 2\pi)$, the solutions are

$\theta = \dfrac{3\pi}{4},\ \dfrac{7\pi}{4}.$

57. $\tan 2\theta = 0$

$\qquad 2\theta = 0 + k\pi,\ k$ any integer

$\qquad \theta = \dfrac{k\pi}{2},\ k$ any integer

On the interval $[0, 2\pi)$, the solutions are

$\theta = 0,\ \dfrac{\pi}{2},\ \pi,$ and $\dfrac{3\pi}{2}.$

59. $\sin \theta = 0.9$

$\qquad \theta = 1.1197695$ or $\theta = \pi - 1.1197695$

61.

$$\sin \theta = \tan \theta$$
$$\sin \theta = \frac{\sin \theta}{\cos \theta}$$
$$\sin \theta - \frac{\sin \theta}{\cos \theta} = 0$$
$$\sin \theta \left[1 - \frac{1}{\cos \theta}\right] = 0$$
$$\sin \theta \left[\frac{\cos \theta - 1}{\cos \theta}\right] = 0$$

$\sin \theta = 0 \quad$ or $\qquad \cos \theta - 1 = 0$

$\sin \theta = 0 \quad$ or $\qquad \cos \theta = 1$

$\qquad\qquad \theta = 0,\ \pi$

63.

$$\sin \theta + \sin 2\theta = 0$$
$$\sin \theta + 2 \sin \theta \cos \theta = 0$$
$$\sin \theta (1 + 2 \cos \theta) = 0$$

$\sin \theta = 0 \quad$ or $\qquad 1 + 2 \cos \theta = 0$

$\qquad\qquad\qquad\qquad \cos \theta = \dfrac{-1}{2}$

$\theta = 0,\ \dfrac{2\pi}{3},\ \dfrac{4\pi}{3},\ \pi$

65.

$$\sin 2\theta - \cos \theta - 2 \sin \theta + 1 = 0$$
$$2 \sin \theta \cos \theta - \cos \theta - 2 \sin \theta + 1 = 0$$
$$\cos \theta (2 \sin \theta - 1) - 1(2 \sin \theta - 1) = 0$$
$$(\cos \theta - 1)(2 \sin \theta - 1) = 0$$

$\cos \theta - 1 = 0 \quad$ or $\qquad 2 \sin \theta - 1 = 0$

$\cos \theta = 1 \quad$ or $\qquad\qquad \sin \theta = \dfrac{1}{2}$

$\theta = 0 \quad$ or $\quad \theta = \dfrac{\pi}{6},\ \dfrac{5\pi}{6}$

67.

$$2 \sin^2 - 3 \sin \theta + 1 = 0$$
$$(2 \sin \theta - 1)(\sin \theta - 1) = 0$$

$2 \sin \theta - 1 = 0 \quad$ or $\qquad \sin \theta - 1 = 0$

$\sin \theta = \dfrac{1}{2}$ or $\qquad\qquad \sin \theta = 1$

$\theta = \dfrac{\pi}{6},\ \dfrac{5\pi}{6},\ \dfrac{\pi}{2}$

69.

$$\sin \theta - \cos \theta = 1$$
$$\frac{1}{\sqrt{2}} \sin \theta - \frac{1}{\sqrt{2}} \cos \theta = \frac{1}{\sqrt{2}}$$

(Divide by $\sqrt{2}$)

Let $\cos \phi = \dfrac{1}{\sqrt{2}},\ \sin \phi = \dfrac{1}{\sqrt{2}},\ 0 \le \phi \le 2\pi$

Then, $\phi = \dfrac{\pi}{4}$

and $\cos \phi \sin \theta - \sin \phi \cos \theta = \dfrac{1}{\sqrt{2}}$

$$\sin(\theta - \phi) = \sin\left[\theta - \frac{\pi}{4}\right] = \frac{1}{\sqrt{2}}$$

$\theta - \dfrac{\pi}{4} = \dfrac{\pi}{4} \quad$ or $\quad \theta - \dfrac{\pi}{4} = \dfrac{3\pi}{4}$

$\theta = \dfrac{\pi}{2} \quad$ or $\qquad \theta = \pi$

71. 1.11 $\qquad\qquad$ **73.** 0.86 $\qquad\qquad$ **75.** 2.21

APPLICATIONS OF TRIGONOMETRIC FUNCTIONS

8.1 Solving Right Triangles

1. $b = 5; \beta = 20°$

$$\cot \beta = \frac{a}{b}$$

$$\cot 20° = \frac{a}{5}$$

$$5(\cot 20°) = a$$

$$5(2.7475) = a$$

$$13.7374 = a$$

$$13.74 \approx a$$

$$\alpha = 90° - \beta = 90° - 20° = 70°$$

$$\csc \beta = \frac{c}{b}$$

$$\csc 20° = \frac{c}{5}$$

$$5(\csc 20°) = c$$

$$14.619022 = c$$

$$14.62 \approx c$$

3. $a = 6; \beta = 40°$

$$\tan \beta = \frac{b}{a}$$

$$\tan 40° = \frac{b}{6}$$

$$6(\tan 40°) = b$$

$$5.0345978 \approx b$$

$$5.03 \approx b$$

$$\alpha = 90° - \beta = 90° - 40° = 50°$$

$$\sec \beta = \frac{c}{a}$$

$$\sec 40° = \frac{c}{6}$$

$$6(\sec 40°) = c$$

$$7.8324437 \approx c$$

$$7.83 \approx c$$

5. $b = 4; \alpha = 10°$

$$\tan \alpha = \frac{a}{b}$$

$$\tan 10° = \frac{a}{4}$$

$$4(\tan 10°) = a$$

$$0.7053079 \approx a$$

$$0.705 \approx a$$

$$\beta = 90° - \alpha = 90° - 10° = 80°$$

$$\sec \alpha = \frac{c}{b}$$

$$\sec 10° = \frac{c}{4}$$

$$4(\sec 10°) = c$$

$$4.0617064 \approx c$$

$$4.06 \approx c$$

7.　$a = 5; \alpha = 25°$

$$\cot \alpha = \frac{b}{a} \qquad\qquad \csc \alpha = \frac{c}{a}$$

$$\cot 25° = \frac{b}{5} \qquad\qquad \csc 25° = \frac{c}{5}$$

$$5(\cot 25°) = b \qquad\qquad 5(\csc 25°) = c$$

$$10.722535 \approx b \qquad\qquad 11.831008 \approx c$$

$$10.72 \approx b \qquad\qquad 11.83 \approx c$$

$$\beta = 90° - 25° = 65°$$

9.　$c = 9; \beta = 20°$

$$\sin \beta = \frac{b}{c} \qquad\qquad \cos \beta = \frac{a}{c}$$

$$\sin 20° = \frac{b}{9} \qquad\qquad \cos 20° = \frac{a}{9}$$

$$9(\sin 20°) = b \qquad\qquad 9(\cos 20°) = a$$

$$3.0781813 \approx b \qquad\qquad 8.4572336 \approx a$$

$$3.08 \approx b \qquad\qquad 8.46 \approx a$$

$$\alpha = 90° - 20° = 70°$$

11.　$c^2 = a^2 + b^2 = 5^2 + 3^2 = 25 + 9 = 34$　　13.　$b^2 = c^2 - a^2 = 5^2 - 2^2 = 21$

$$c = \sqrt{34} \approx 5.83 \qquad\qquad\qquad\qquad\qquad b = \sqrt{21} \approx 4.58$$

$$\sin \alpha = \frac{a}{c} = \frac{5}{5.83} \qquad\qquad\qquad\qquad \sin \alpha = \frac{a}{c} = \frac{2}{5}$$

$$\alpha \approx 59.0° \qquad\qquad\qquad\qquad\qquad\qquad \alpha = 23.6°$$

$$\beta = 90° - \alpha \approx 90 - 59° \approx 31.0° \qquad\quad \beta = 90° - \alpha \approx 66.4°$$

15.

$$\sin 35° = \frac{a}{3} \qquad \cos 35° = \frac{b}{3}$$

$$3(\sin 35°) = a \qquad 3(\cos 35°) = b$$

$$1.7207293 \approx a \qquad 2.4574561 \approx b$$

$$1.72 \text{ in.} \approx a \qquad 2.46 \text{ in.} \approx b$$

17.

or

$$\csc 35° = \frac{c}{5} \qquad \text{or} \qquad \sec 35° = \frac{c}{5}$$

$$5(\csc 35°) = c \qquad \text{or} \qquad 5(\sec 35°) = c$$

$$8.71723398 \approx c \qquad \text{or} \qquad 6.1038729 \approx c$$

$$8.72 \text{ in.} \approx c \qquad \text{or} \qquad 6.10 \text{ in.} \approx c$$

19.　$c = 5$. Suppose $a = 2$. Then, $\sin \alpha = \frac{a}{c} = \frac{2}{5}$ so $\alpha = 23.6°$ and $\beta = 90° - \alpha \approx 66.4°$

21.
$$\tan 35° = \frac{a}{100}$$

$$100(\tan 35°) = a$$

$$70.020753 \approx a$$

$$70 \text{ ft.} \approx a$$

23.
$$\tan 85.361° = \frac{x}{80}$$

$$80(\tan 85.361°) = x$$

$$985.91117 \approx x$$

$$985.9 \text{ feet} \approx x$$

25.

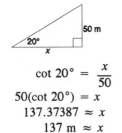

$$\cot 20° = \frac{x}{50}$$
$$50(\cot 20°) = x$$
$$137.37387 \approx x$$
$$137 \text{ m} \approx x$$

27.

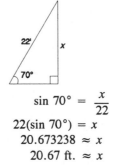

$$\sin 70° = \frac{x}{22}$$
$$22(\sin 70°) = x$$
$$20.673238 \approx x$$
$$20.67 \text{ ft.} \approx x$$

29. We want to solve for x:

$$\cot 46.27° = \frac{100 + y}{x} \qquad \text{so} \qquad x \cot 46.27 = 100 + y$$
$$x \cot 46.27 - 100 = y$$

$$\cot 40.3° = \frac{200 + y}{x} \qquad \text{so} \qquad x \cot 40.3 = 200 + y$$
$$x \cot 40.3 - 200 = y$$

Since $y = y$, we have:
$$x \cot 46.27 - 100 = x \cot 40.3 - 200$$
$$100 = x \cot 40.3 - x \cot 46.27$$
$$100 = x(\cot 40.3 - \cot 46.27)$$
$$\frac{100}{\cot 40.3 - \cot 46.27} = x$$
$$\frac{100}{1.1791595 - 0.956623} = x$$
$$\frac{100}{0.2225365} = x$$
$$449.36449 \approx x$$
$$449.36 \text{ ft.} \approx x$$

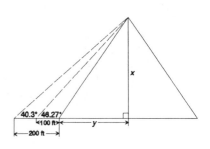

31. If α is the angle of elevation, then $\tan \alpha = \dfrac{300}{50} = 6$, so $\alpha \approx 80.5°$.

33.

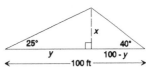

We want to solve for x. Therefore, we relabel the base of the triangle in two parts, y and $100 - y$.

$$\cot 25° = \frac{y}{x} \qquad\qquad \cot 40° = \frac{100 - y}{x}$$
$$x \cot 25° = y \qquad\qquad x \cot 40° = 100 - y$$
$$y = 100 - x \cot 40°$$

Since $y = y$, we have:
$$x \cot 25° = 100 - x \cot 40°$$
$$x \cot 25° + x \cot 40° = 100$$
$$x(\cot 25° + \cot 40°) = 100$$
$$x = \frac{100}{\cot 25° + \cot 40°}$$
$$x = \frac{100}{2.1445069 + 1.1917536}$$
$$x = \frac{100}{3.3362605}$$
$$x \approx 29.973679$$
$$x \approx 30 \text{ ft.}$$

35.
$$\sin 21° = \frac{190}{x}$$
$$x \sin 21° = 190$$
$$x = \frac{190}{\sin 21°}$$
$$x \approx 530 \text{ feet}$$

37.
$$\tan 35.1° = \frac{x}{789}$$
$$x = 789 \tan 35.1°$$
$$x = 555 \text{ feet}$$

39. (a)
$$\cot 15° = \frac{x}{30}$$
$$30(\cot 15°) = x$$
$$111.96153 \approx x$$

Hence, he is traveling at 111.96 ft/sec. Convert to mi/hr by:

$$\frac{1 \text{ mile}}{5280 \text{ ft}} \cdot \frac{111.96 \text{ ft}}{1 \text{ sec}} \cdot \frac{3600 \text{ sec}}{\text{hr}} \approx 76.336 \text{ mi/hr} \approx 76.34 \text{ mph}$$

(b)
$$\cot 20° = \frac{x}{30}$$
$$30(\cot 20°) = x$$
$$82.42432 \approx x$$

Hence, he is traveling at 82.4 ft/sec. Convert to mi/hr by:

$$\frac{1 \text{ mile}}{5280 \text{ ft}} \cdot \frac{82.4 \text{ ft}}{1 \text{ sec}} \cdot \frac{3600 \text{ sec}}{1 \text{ hr}} \approx 56.2 \text{ mph}$$

(c) A ticket is issued for traveling at a speed of 60 mph or faster.

$$60 \text{ mph} = \frac{60 \text{ mi}}{1 \text{ hr}} \cdot \frac{1 \text{ hr}}{3600 \text{ sec}} \cdot \frac{5280 \text{ ft}}{1 \text{ mi}} = 88 \text{ ft/sec}$$

If $\tan \theta < \dfrac{30}{88}$, the trooper should issue a ticket.

Thus, for $\theta < 18.8°$, a ticket will be issued.

41. (a) See the figure. After flying $\dfrac{1}{2}$ mile from the airport O, the aircraft is at P and we see that
$$\angle QOP = 40°$$
After turning toward the Southeast, we see that
$$\angle OQP = 90° - 40° = 50°$$
Therefore, the bearing of the aircraft is
$$180° - 50° = 130°$$

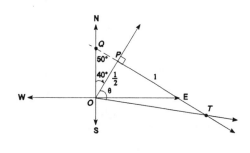

(b) After flying 1 mile at a bearing of 130°, the aircraft is at T. If $\theta = \angle POT$, then

$$\tan \theta = \frac{1}{\frac{1}{2}} = 2, \text{ so } \theta = 63.4°$$

Therefore, the bearing of the plane is $40° + 63.4° = 103.4°$.

43. If α is the angle of elevation, then $\tan \alpha = \dfrac{10 - 6}{15} = \dfrac{4}{15}$, so $\alpha \approx 14.9°$.

45. (a) From the figure, we see $\tan 15° = \dfrac{x}{y}$ and $\tan 35° = \dfrac{3-x}{y}$.

Thus, $x = y \cdot \tan 15°$. Substituting this into the other equation, we have:

$$\tan 35° = \frac{3 - y \tan 15°}{y}$$
$$y \tan 35° = 3 - y \tan 15°$$
$$y(\tan 35° + \tan 15°) = 3$$
$$y = \frac{3}{\tan 35° + \tan 15°} \approx 3.1 \text{ miles}$$

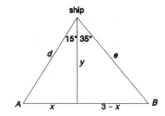

(b) Let d be the distance from the ship to lighthouse A. Then

$$\sin 15° = \frac{x}{d}$$
$$\sin 15° = \frac{3.1 \tan 15°}{d} \text{ since } x = y \tan 15°$$
$$d = \frac{3.1 \tan 15°}{\sin 15°}$$
$$d \approx 3.2 \text{ miles}$$

(c) Let e be the distance from the ship to lighthouse B. Then

$$\sin 35° = \frac{3 - x}{e}$$
$$\sin 35° = \frac{3 - 3.1 \tan 15°}{e}$$
$$e = \frac{3 - 3.1 \tan 15°}{\sin 35°}$$
$$e \approx 3.8 \text{ miles}$$

47. (a) See Figure. Let h be the height from the surface of the earth to the satellite. The two right triangles in the figure are congruent. Therefore, θ is bisected by the line from the satellite to the center of the earth. Thus,

$$\cos \frac{\theta}{2} = \frac{3960}{3960 + h} \text{ since the radius of the earth is 3960 mi.}$$

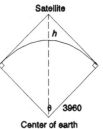

(b) Recall the formula for arc length, $s = \theta r$. In this problem, d is the arc length, s. Thus, $d = \theta r$. Since $r = 3960$, we have:
$$d = 3960\,\theta$$

(c) From (b) we have $\theta = \dfrac{d}{3960}$. Thus, $\dfrac{\theta}{2} = \dfrac{\frac{d}{3960}}{2} = \dfrac{d}{7920}$

So, $\cos \dfrac{d}{7920} = \dfrac{3960}{3960 + h}$

(d) We want to solve the equation in (c) for h:

$$3960 + h = \frac{3960}{\cos \dfrac{d}{7920}}$$
$$h = \frac{3960}{\cos \dfrac{d}{7920}} - 3960$$
$$h = \frac{3960}{\cos \dfrac{2500}{7920}} - 3960 \approx 206 \text{ miles}$$

(e)   Again, using the equation in (c), let $h = 300$,

$$\cos \frac{d}{7920} = \frac{3960}{3960 + 300}$$

$$\cos \frac{d}{7920} = \frac{3960}{4260}$$

$$\frac{d}{7920} = 0.3775$$

$$d = 2990 \text{ miles}$$

8.2   The Law of Sines

1.   The third angle α is easily found because the sum of the angles of a triangle equals $180°$.

$$\alpha + \beta + \gamma = 180°$$
$$\alpha + 45° + 95° = 180°$$
$$\alpha = 40°$$

Now we use the Law of Sines (twice) to find the unknown sides a and b:

$$\frac{\sin \alpha}{a} = \frac{\sin \gamma}{c} \qquad\qquad \frac{\sin \beta}{b} = \frac{\sin \gamma}{c}$$

$$\frac{\sin 40°}{a} = \frac{\sin 95°}{5} \qquad\qquad \frac{\sin 45°}{b} = \frac{\sin 95°}{5}$$

Thus, $a = \dfrac{5 \sin 40°}{\sin 95°} \approx 3.23$ (from calculator),

and $b = \dfrac{5 \sin 45°}{\sin 95°} \approx 3.55$ (from calculator)

3.   Because we know two angles ($\alpha = 50°$ and $\gamma = 85°$), it is easy to find the third angle using the equation:

$$\alpha + \beta + \gamma = 180°$$
$$50° + \beta + 85° = 180°$$
$$\beta = 45°$$

Now we know the three angles and one side ($b = 3$) of the triangle. To find the remaining two sides a and c, we use the Law of Sines (twice):

$$\frac{\sin \alpha}{a} = \frac{\sin \beta}{b} \qquad\qquad \frac{\sin \beta}{b} = \frac{\sin \gamma}{c}$$

$$\frac{\sin 50°}{a} = \frac{\sin 45°}{3} \qquad\qquad \frac{\sin 45°}{3} = \frac{\sin 85°}{c}$$

$$a = \frac{3 \sin 50°}{\sin 45°} \qquad\qquad c = \frac{3 \sin 85°}{\sin 45°}$$

$$a \approx 3.25 \qquad\qquad c \approx 4.23$$

5.

$$\alpha + \beta + \gamma = 180°$$
$$40° + 45° + \gamma = 180°$$
$$\gamma = 95°$$

$$\frac{\sin \alpha}{a} = \frac{\sin \beta}{b} \qquad\qquad \frac{\sin \beta}{b} = \frac{\sin \gamma}{c}$$

$$\frac{\sin 40°}{a} = \frac{\sin 45°}{7} \qquad\qquad \frac{\sin 45°}{7} = \frac{\sin 95°}{c}$$

$$a = \frac{7 \sin 40°}{\sin 45°} \qquad\qquad c = \frac{7 \sin 95°}{\sin 45°}$$

$$a \approx 6.36 \qquad\qquad c \approx 9.86$$

7.
$$\alpha + \beta + \gamma = 180°$$
$$\alpha + 40° + 100° = 180°$$
$$\alpha = 40°$$

$$\frac{\sin \alpha}{a} = \frac{\sin \beta}{b} \qquad \frac{\sin \beta}{b} = \frac{\sin \gamma}{c}$$
$$\frac{\sin 40°}{a} = \frac{\sin 40°}{2} \qquad \frac{\sin 40°}{2} = \frac{\sin 100°}{c}$$
$$a = \frac{2 \sin 40°}{\sin 40°} \qquad c = \frac{2 \sin 100°}{\sin 40°}$$
$$a = 2 \qquad c \approx 3.06$$

9.
$$\alpha + \beta + \gamma = 180°$$
$$40° + 20° + \gamma = 180°$$
$$\gamma = 120°$$

$$\frac{\sin \alpha}{a} = \frac{\sin \beta}{b} \qquad \frac{\sin \alpha}{a} = \frac{\sin \gamma}{c}$$
$$\frac{\sin 40°}{2} = \frac{\sin 20°}{b} \qquad \frac{\sin 40°}{2} = \frac{\sin 120°}{c}$$
$$b = \frac{2 \sin 20°}{\sin 40°} \qquad c = \frac{2 \sin 120°}{\sin 40°}$$
$$b \approx 1.06 \qquad c \approx 2.69$$

11.
$$\alpha + \beta + \gamma = 180°$$
$$\alpha + 70° + 10° = 180°$$
$$\alpha = 100°$$

$$\frac{\sin \alpha}{a} = \frac{\sin \beta}{b} \qquad \frac{\sin \beta}{b} = \frac{\sin \gamma}{c}$$
$$\frac{\sin 100°}{a} = \frac{\sin 70°}{5} \qquad \frac{\sin 70°}{5} = \frac{\sin 10°}{c}$$
$$a \approx 5.24 \qquad c \approx 0.92$$

13.
$$\alpha + \beta + \gamma = 180°$$
$$110° + \beta + 30° = 180°$$
$$\beta = 40°$$

$$\frac{\sin \alpha}{a} = \frac{\sin \gamma}{c} \qquad \frac{\sin \beta}{b} = \frac{\sin \gamma}{c}$$
$$\frac{\sin 110°}{a} = \frac{\sin 30°}{3} \qquad \frac{\sin 40°}{b} = \frac{\sin 30°}{3}$$
$$a = \frac{3 \sin 110°}{\sin 30°} \qquad b = \frac{3 \sin 40°}{\sin 30°}$$
$$a \approx 5.64 \qquad b \approx 3.86$$

15.
$$\alpha + \beta + \gamma = 180°$$
$$40° + 40° + \gamma = 180°$$
$$\gamma = 100°$$

$$\frac{\sin \alpha}{a} = \frac{\sin \gamma}{c} \qquad \frac{\sin \beta}{b} = \frac{\sin \gamma}{c}$$
$$\frac{\sin 40°}{a} = \frac{\sin 100°}{2} \qquad \frac{\sin 40°}{b} = \frac{\sin 100°}{2}$$
$$a = \frac{2 \sin 40°}{\sin 100°} \qquad b = \frac{2 \sin 40°}{\sin 100°}$$
$$a \approx 1.31 \qquad b \approx 1.31$$

17. Because $a = 3$, $b = 2$, and $\alpha = 50°$ are known, we use the Law of Sines to find the angle β:

$$\frac{\sin \alpha}{a} = \frac{\sin \beta}{b}$$

Then, $\quad \dfrac{\sin 50°}{3} = \dfrac{\sin \beta}{2}$

$$\sin \beta = \frac{2 \sin 50°}{3} \approx 0.5107$$

There are two angles β, $0° < \beta < 180°$, for which
$\sin \beta \approx 0.5107$, namely
$\beta \approx 30.7°$ or $\beta \approx 149.3$
The second possibility is ruled out, because $\alpha = 50°$, making $\alpha + \beta \approx 199.3° > 180°$. Now, using $\beta \approx 30.7°$, we find $\gamma = 180° - \alpha - \beta \approx 99.3°$. The third side c is determined using the Law of Sines:

$$\frac{\sin \alpha}{a} = \frac{\sin \gamma}{c}$$

$$\frac{\sin 50°}{3} = \frac{\sin 99.3°}{c}$$

$$c = \frac{3 \sin 99.3°}{\sin 50°} \approx 3.86$$

One triangle; $\beta \approx 30.7°$, $\gamma \approx 99.3°$, $c \approx 3.86$

19. Because $b = 5$, $c = 3$, and $\beta = 100°$ are known, we use the Law of Sines to find the angle γ:

$$\frac{\sin \beta}{b} = \frac{\sin \gamma}{c}$$

Then, $\quad \dfrac{\sin 100°}{5} = \dfrac{\sin \gamma}{3}$

$$\sin \gamma = \frac{3 \sin 100°}{5} \approx 0.5909$$

$$\gamma \approx 36.2° \text{ or } \gamma \approx 143.8°$$

The second possibility is ruled out, because $\beta = 100°$, making $\beta + \gamma \approx 243.8° > 180°$. Now, using $\gamma \approx 36.2°$, we find

$$\alpha = 180° - \beta - \gamma \approx 43.8°$$

The third side is determined using the Law of Sines.

$$\frac{\sin \alpha}{a} = \frac{\sin \beta}{b}$$

$$\frac{\sin 43.8°}{a} = \frac{\sin 100°}{5}$$

$$a = \frac{5 \sin 43.8°}{\sin 100°} \approx 3.51$$

One triangle; $\gamma \approx 36.2°$, $\alpha \approx 43.8°$, $a \approx 3.51$

21. Because $a = 4$, $b = 5$, and $\alpha = 60°$ are known, we use the Law of Sines to find the angle β:

$$\frac{\sin \alpha}{a} = \frac{\sin \beta}{b}$$

Then, $\quad \dfrac{\sin 60°}{4} = \dfrac{\sin \beta}{5}$

$$\sin \beta = \frac{5 \sin 60°}{4} \approx 1.0825$$

There is no angle β for which $\sin \beta > 1$. Hence, there can be no triangle having the given measurements.

23. Because $b = 4$, $c = 6$, and $\beta = 20°$ are known, we use the Law of Sines to find the angle γ:

$$\frac{\sin \beta}{b} = \frac{\sin \gamma}{c}$$

$$\frac{\sin 20°}{4} = \frac{\sin \gamma}{6}$$

$$\sin \gamma = \frac{6 \sin 20°}{4} \approx 0.5130$$

$$\gamma_1 \approx 30.9° \text{ or } \gamma_2 \approx 149.1°$$

For both possibilities we have $\beta + \gamma < 180°$. Hence, there are two triangles—one containing the angle $\gamma = \gamma_1 \approx 30.9°$, the other containing the angle $\gamma = \gamma_2 \approx 149.1°$. The third angle α is either:

$$\alpha_1 = 180° - \beta - \gamma_1 \approx 129.1° \text{ or } \alpha_2 = 180° - \beta - \gamma_2 \approx 10.9°$$

The third side a obeys the Law of Sines, so we have:

$$\frac{\sin \alpha}{a} = \frac{\sin \beta}{b}$$

$$\frac{\sin 129.1°}{a_1} = \frac{\sin 20°}{4} \qquad \text{or} \qquad \frac{\sin 10.9°}{a_2} = \frac{\sin 20°}{4}$$

$$a_1 = \frac{4 \sin 129.1°}{\sin 20°} \qquad \text{or} \qquad a_2 = \frac{4 \sin 10.9°}{\sin 20°}$$

$$a_1 \approx 9.08 \qquad\qquad\qquad a_2 \approx 2.21$$

Two triangles; $\qquad \gamma_1 \approx 30.9°$, $\alpha_1 \approx 129.1°$, $a_1 \approx 9.08$

or $\qquad \gamma_2 \approx 149.1°$, $\alpha_2 \approx 10.9°$, $a_2 \approx 2.21$

25. Because $a = 2$, $c = 1$, and $\gamma = 100°$, are known, we use the Law of Sines to find the angle α:

$$\frac{\sin \alpha}{a} = \frac{\sin \gamma}{c}$$

$$\frac{\sin \alpha}{2} = \frac{\sin 100°}{1}$$

$$\sin \alpha = \frac{2 \sin 100°}{1}$$

$$\sin \alpha \approx 1.9696$$

There is no angle α for which $\sin \alpha > 1$. Hence, there can be no triangle having the given measurements.

27. Because $a = 2$, $c = 1$, and $\gamma = 25°$, we use the Law of Sines to find the angle α:

$$\frac{\sin \alpha}{a} = \frac{\sin \gamma}{c}$$

Then, $\qquad \dfrac{\sin \alpha}{2} = \dfrac{\sin 25°}{1}$

$$\sin \alpha = 2 \sin 25° \approx 0.8452$$

$$\alpha_1 \approx 57.7° \text{ or } \alpha_2 = 122.3°$$

The third angle β is either: $\qquad \beta_1 = 180° - 25° - 57.7° \approx 97.3°$ or

$$\beta_2 = 180° - 25° - 122.3° \approx 32.7°$$

The third side b obeys the Law of Sines, so we have:

$$\frac{\sin \beta}{b} = \frac{\sin \gamma}{c}$$

$$\frac{\sin 97.3°}{b} = \frac{\sin 25°}{1} \qquad \text{or} \qquad \frac{\sin 32.7°}{b_2} = \frac{\sin 25°}{1}$$

$$b_1 = \frac{\sin 97.3°}{\sin 25°} \qquad \text{or} \qquad b_2 = \frac{\sin 32.7°}{\sin 25°}$$

$$b_1 \approx 2.35 \qquad\qquad\qquad b_2 \approx 1.28$$

Two triangles; $\alpha_1 \approx 57.7°$, $\beta_1 \approx 97.3°$, $b_1 \approx 2.35$ or $\alpha_2 \approx 122.3°$, $\beta_2 \approx 32.7°$, $b_2 \approx 1.28$

29.

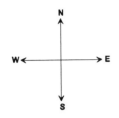
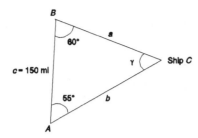

(a)   The angle γ is found to be
$$\gamma = 180° - 60° - 55° = 65°$$
The Law of Sines can now be used to find the two distances a and b. We seek:

$$\frac{\sin 55°}{a} = \frac{\sin 65°}{150} \qquad\qquad \frac{\sin 60°}{b} = \frac{\sin 65°}{150}$$

$$a = \frac{150 \sin 55°}{\sin 65°} \qquad\qquad b = \frac{150 \sin 60°}{\sin 65°}$$

$$a \approx 135.6 \text{ miles} \qquad\qquad b \approx 143.3 \text{ miles}$$

Thus, station Able is 143.3 miles from the ship and station B is 135.6 miles from the ship.

(b)   The time t needed for the helicopter to reach the ship from Station Baker is found by using the formula:
$$(\text{Velocity, } v)(\text{Time, } t) = \text{Distance}$$

Then, $t = \dfrac{a}{v} = \dfrac{135.6}{200} \approx 0.68 \text{ hour} \approx 41 \text{ minutes}$

31.   Because angle DAB is supplementary to angle CAB, angle $CAB = 180° - 25° = 155°$. Angle $ABC = 180° - 155° - 15° = 10°$.

Let c denote the distance from A to B. Using the Law of Sines,

$$\frac{\sin 10°}{1000} = \frac{\sin 15°}{c}$$

$$c = \frac{1000 \sin 15°}{\sin 10°} \approx 1490.5 \text{ feet}$$

The length of the span of a ski lift from A to B is 149 feet.

33.

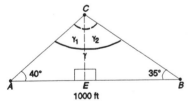

$$\gamma_1 = 180° - 40° - 90° = 50°$$
$$\gamma_2 = 180° - 35° - 90° = 55°$$
$$\gamma = 180° - 40° - 35° = 50° + 55° = 105°$$

Before we find the height of the airplane, we need to know either the distance from A to C or the distance from B to C. Let's find the distance from A to C, denote it by b, using the Law of Sines.

$$\frac{\sin 35°}{b} = \frac{\sin 105°}{1000}$$

$$b = \frac{1000 \sin 35°}{\sin 105°} \approx 593.8$$

Now we can find the height of the airplane. Let a denote the distance from C to E.

$$\frac{\sin 40°}{a} = \frac{\sin 90°}{593.8}$$

$$a = \frac{593.8 \sin 40°}{\sin 90°} \approx 381.7$$

The airplane is 381.7 feet high.

35. (a) Angle $CBA = 180° - 40° = 140°$
Let the distance from city A to city B be denoted by c and let γ denote angle ACB. We can use the Law of Sines to find γ.

$$\frac{\sin 140°}{300} = \frac{\sin \gamma}{150}$$

$$\sin \gamma = \frac{150 \sin 140°}{300} \approx 0.3214$$

$$\gamma = 18.7°$$

Let α denote angle BAC.

$$\alpha = 180° - 140° - 18.7° = 21.3°$$

Using the Law of Sines to find the distance from city B to city C, denoted by a,

$$\frac{\sin 21.3°}{a} = \frac{\sin 140°}{300}$$

$$a = \frac{300 \sin 21.3°}{\sin 140°} \approx 169$$

It is 169.0 miles from city B to city C.

(b)

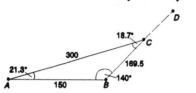

Angle $ACD = 180° - 18.7° = 161.3°$
The pilot should turn through an angle $161.3°$ at city C to return to city A.

37.

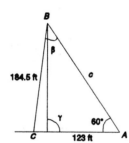

Let β = Angle ACB

$$\frac{\sin 60°}{184.5} = \frac{\sin \beta}{123}$$

$$\sin \beta = \frac{123 \sin 60°}{184.5} \approx 0.5774$$

$$\beta = 35.3°$$

Thus, Angle $CAB = 180° - 60° - 35.3° \approx 84.7°$

Let x be the perpendicular distance from C to AB. Then

$$\sin 84.7 = \frac{x}{184.5}$$

$$\sin 84.7(184.5) = x$$

$$183.7 \text{ ft} = x$$

39. Using supplementary angles, $\alpha = 180° - 140° = 40°$

$$\beta = 180° - 135° = 45°$$

Thus, $\gamma = 180° - 40° - 45° = 95°$
Using the Law of Sines, we can find a and b.

$$\frac{\sin 40°}{a} = \frac{\sin 95°}{2}$$

$$a = \frac{2 \sin 40°}{\sin 95°} \approx 1.29 \text{ miles}$$

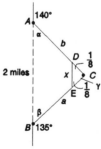

Since the distance $\overline{BC} \approx 1.29$ and $\overline{CE} = 0.125$, we have $\overline{BE} = 1.29 - 0.125 = 1.165$ miles.

$$\frac{\sin 45°}{b} = \frac{\sin 95°}{2}$$

$$b = \frac{2 \sin 45°}{\sin 95°} \approx 1.42 \text{ miles}$$

Since the distance $\overline{AC} \approx 1.42$ and $\overline{CD} = 0.125$, we have $\overline{AD} = 1.42 - 0.125 = 1.295$ miles.

For the equilateral triangle CDE, we have $\angle CDE = \angle CED = \dfrac{180° - 95°}{2} = 42.5°$. We use the Law of Sines, to find $\overline{DE}$.

$$\frac{\sin 95°}{\overline{DE}} = \frac{\sin 42.5°}{0.125}$$

$$\overline{DE} = \frac{0.125 \sin 95°}{\sin 42.5°} \approx 0.18 \text{ miles}$$

Thus, the length of the highway is $1.165 + 1.295 + 0.18 = 2.64$ miles.

41. See the figure. Using supplementary angles, we have $\angle ABD = 180° - 30° = 150°$. Thus, $\gamma = 180° - 150° - 20° = 10°$. Using the Law of Sines, we can now find y.

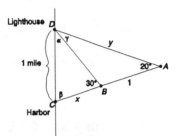

$$\frac{\sin 10°}{1} = \frac{\sin 150°}{y}$$

$$y = \frac{\sin 150°}{\sin 10°} \approx 2.8794 \text{ miles}$$

Now, using the Law of Sines, we can find β.

$$\frac{\sin \beta}{2.8794} = \frac{\sin 20}{1}$$

$$\sin \beta = 2.8794 \sin 20 = 0.9848$$

$$\beta = 80°$$

Thus, $\alpha = 180° - 80° - 30° = 70°$

Finally, using the Law of Sines, we can find x.

$$\frac{\sin 70°}{x} = \frac{\sin 30°}{1}$$

$$x = \frac{\sin 70°}{\sin 30°} \approx 1.8794 \text{ miles.}$$

Thus, the ship is 1.88 miles from the harbor.

43. We want to prove that

$$\frac{a + b}{c} = \frac{\cos \frac{1}{2}(\alpha - \beta)}{\sin \frac{1}{2}\gamma}$$

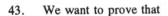

$$\frac{a + b}{c} = \frac{a}{c} + \frac{b}{c} = \frac{\sin \alpha}{\sin \gamma} + \frac{\sin \beta}{\sin \gamma} = \frac{\sin \alpha + \sin \beta}{\sin \gamma}$$

$$= \frac{2 \sin \frac{\alpha + \beta}{2} \cos \frac{\alpha - \beta}{2}}{\sin \left[\frac{\gamma}{2} + \frac{\gamma}{2} \right]} = \frac{2 \sin \frac{\alpha + \beta}{2} \cos \frac{\alpha - \beta}{2}}{\sin \frac{\gamma}{2} \cos \frac{\gamma}{2} + \cos \frac{\gamma}{2} \sin \frac{\gamma}{2}}$$

$$= \frac{2 \sin \frac{\alpha + \beta}{2} \cos \frac{\alpha - \beta}{2}}{2 \sin \frac{\gamma}{2} \cos \frac{\gamma}{2}} = \frac{\sin \left[\frac{\pi}{2} - \frac{\gamma}{2} \right] \cos \frac{\alpha - \beta}{2}}{\sin \frac{\gamma}{2} \cos \frac{\gamma}{2}}$$

$$= \frac{\cos \frac{\gamma}{2} \cos \frac{\alpha - \beta}{2}}{\sin \frac{\gamma}{2} \cos \frac{\gamma}{2}} = \frac{\cos \frac{\alpha - \beta}{2}}{\sin \frac{\gamma}{2}} = \frac{\cos \frac{1}{2}(\alpha - \beta)}{\sin \frac{1}{2}\gamma}$$

45. We want to derive the formula:

$$a = \frac{b \sin \alpha}{\sin \beta} = \frac{b \sin[180° - (\beta + \gamma)]}{\sin \beta}$$

$$= \frac{b}{\sin \beta}\left[\sin 180° \cos(\beta + \gamma) - \cos 180° \sin(\beta + \gamma)\right]$$

$$= \frac{b}{\sin \beta}\left[0 + \sin(\beta + \gamma)\right] = \frac{b}{\sin \beta}(\sin \beta \cos \gamma + \cos \beta \sin \gamma)$$

$$= b \cos \gamma + \frac{b \sin \gamma}{\sin \beta} \cos \beta = b \cos \gamma + c \cos \beta$$

47. $\sin \beta = \sin(\text{Angle } AB'C) = \dfrac{b}{2r}$;

Thus, $\dfrac{\sin \beta}{b} = \dfrac{1}{2r}$. By the Law of Sines,

$$\frac{\sin \alpha}{a} = \frac{\sin \beta}{b} = \frac{\sin \alpha}{c} = \frac{1}{2r}$$

8.3 The Law of Cosines

1. $b^2 = a^2 + c^2 - 2ac \cos \beta$
$b^2 = 4 + 16 - 2 \cdot 2 \cdot 4 \cos 45°$

$$b^2 = 20 - \left[16 \cdot \frac{\sqrt{2}}{2}\right]$$

$b^2 = 20 - 8\sqrt{2} \approx 8.6863$
$b \approx 2.95$

To find α:
$$a^2 = b^2 + c^2 = 2bc \cos \alpha$$
$$2bc \cos \alpha = b^2 + c^2 - a^2$$
$$\cos \alpha = \frac{b^2 + c^2 - a^2}{2bc}$$
$$\cos \alpha = \frac{8.6866 + 16 - 4}{2 \cdot 2.9473 \cdot 4}$$
$$= \frac{20.6866}{23.5784} \approx 0.8774$$
$$\alpha \approx 28.7°$$

To find γ:
$$c^2 = a^2 + b^2 - 2ab \cos \gamma$$
$$2ab \cos \gamma = a^2 + b^2 - c^2$$
$$\cos \gamma = \frac{a^2 + b^2 - c^2}{2ab}$$
$$\cos \gamma = \frac{4 + 8.6863 - 16}{2 \cdot 2 \cdot 2 \cdot 9473} = \frac{-3.3137}{11.7892} \approx -0.2811$$
$$\gamma \approx 106.3°$$

3. $c^2 = a^2 + b^2 - 2ab \cos \gamma$
$c^2 = 4 + 9 - 12 \cos 95°$
$c^2 = 13 - 12(-0.0872) = 14.0464$
$c \approx 3.75$

$$\cos \alpha = \frac{b^2 + c^2 - a^2}{2bc}$$

$$\cos \alpha = \frac{9 + 14.0464 - 4}{2 \cdot 3 \cdot 3.7479} = \frac{19.0464}{22.4874} \approx 0.8470$$

$$\alpha = 32.1°$$

$$\cos \beta = \frac{a^2 + c^2 - b^2}{2ac}$$

$$\cos \beta = \frac{4 + 14.0464 - 9}{2 \cdot 2 \cdot 3.7479} = \frac{9.0464}{14.9916} \approx 0.6034$$

$$\beta \approx 52.9°$$

5. $$\cos \alpha = \frac{b^2 + c^2 - a^2}{2bc}$$

$$\cos \alpha = \frac{25 + 64 - 36}{2 \cdot 5 \cdot 8} = \frac{53}{80} \approx 0.6625$$

$$\alpha = 48.5°$$

$$\cos \beta = \frac{a^2 + c^2 - b^2}{2ac}$$

$$\cos \beta = \frac{36 + 64 - 25}{2 \cdot 6 \cdot 8} = \frac{75}{96} \approx 0.7813$$

$$\beta = 38.6°$$

$$\cos \gamma = \frac{a^2 + b^2 - c^2}{2ab}$$

$$\cos \gamma = \frac{36 + 25 - 64}{2 \cdot 6 \cdot 5} = \frac{-3}{60} = -0.05$$

$$\gamma = 92.9°$$

7. $$\cos \alpha = \frac{b^2 + c^2 - a^2}{2bc}$$

$$\cos \alpha = \frac{36 + 16 - 81}{2 \cdot 6 \cdot 4} = \frac{-29}{48} \approx -0.6042$$

$$\alpha = 127.2°$$

$$\cos \beta = \frac{a^2 + c^2 - b^2}{2ac}$$

$$\cos \beta = \frac{81 + 16 - 36}{2 \cdot 9 \cdot 4} = \frac{61}{72} \approx 0.8472$$

$$\beta = 32.1°$$

$$\cos \gamma = \frac{a^2 + b^2 - c^2}{2ab}$$

$$\cos \gamma = \frac{81 + 36 - 16}{2 \cdot 9 \cdot 6} = \frac{101}{108} \approx 0.9352$$

$$\gamma = 20.7°$$

9. For c: $c^2 = a^2 + b^2 - 2ab \cos \gamma$
$c^2 = 9 + 16 - 2 \cdot 3 \cdot 4 \cdot \cos 40°$
$c^2 = 25 - 24(0.76604) = 6.615$
$c \approx 2.57$

For α: $$\cos \alpha = \frac{b^2 + c^2 - a^2}{2bc}$$

$$\cos \alpha = \frac{16 + 6.616 - 9}{2 \cdot 4 \cdot 2.5722} = \frac{13.616}{20.5776} \approx 0.6617$$

$$\alpha = 48.6°$$

For β: $$\cos \beta = \frac{a^2 + c^2 - b^2}{2ac}$$

$$\cos \beta = \frac{9 + 6.616 - 16}{2 \cdot 3 \cdot 2.5722} = \frac{-0.384}{15.4332} \approx -0.0249$$

$$\beta \approx 91.4°$$

11. $a^2 = b^2 + c^2 - 2bc \cos \alpha$
$a^2 = 1 + 9 - 2 \cdot 1 \cdot 3 \cdot \cos 80°$
$a^2 = 10 - 6(-0.17365) = 8.9581$
$a \approx 2.99$

$$\cos \beta = \frac{a^2 + c^2 - b^2}{2ac}$$

$$\cos \beta = \frac{8.9584 + 9 - 1}{2 \cdot 2.9931 \cdot 3} = \frac{16.9584}{17.9586} \approx 0.9443$$

$$\beta = 19.2°$$

$$\cos \gamma = \frac{a^2 + b^2 - c^2}{2ab}$$

$$\cos \gamma = \frac{8.9584 + 1 - 9}{2 \cdot 2.9931 \cdot 1} = \frac{0.9584}{5.9862} \approx 0.1601$$

$$\gamma = 80.8°$$

13. $b^2 = a^2 + c^2 - 2ac \cos \beta$
$b^2 = 9 + 4 - 2 \cdot 3 \cdot 2 \cos 110°$
$b^2 = 13 - 12(-0.3420) = 17.104$
$b \approx 4.14$

$$\cos \alpha = \frac{b^2 + c^2 - a^2}{2bc}$$

$$\cos \alpha = \frac{17.104 + 4 - 9}{2 \cdot 4.1357 \cdot 2} = \frac{12.104}{16.5428} \approx 0.7317$$

$$\alpha = 43.0°$$

$$\cos \gamma = \frac{a^2 + b^2 - c^2}{2ab}$$

$$\cos \gamma = \frac{9 + 17.104 - 4}{2 \cdot 3 \cdot 4.1357} = \frac{22.104}{24.8142} \approx 0.8908$$

$$\gamma \approx 27.0°$$

15. $c^2 = a^2 + b^2 - 2ab \cos \gamma$
$c^2 = 4 + 4 - 2 \cdot 2 \cdot 2 \cdot \cos 50°$
$c^2 = 8 - 8(0.64279) = 2.85768$
$c \approx 1.69$

$$\cos \alpha = \frac{b^2 + c^2 - a^2}{2bc}$$

$$\cos \alpha = \frac{4 + 2.8576 - 4}{2 \cdot 2 \cdot 1.6904} = \frac{2.8576}{6.7616} \approx 0.4226$$

$$\alpha = 65.0°$$

$$\cos \beta = \frac{a^2 + c^2 - b^2}{2ac}$$

$$\cos \beta = \frac{4 + 2.8576 - 4}{2 \cdot 2 \cdot 1.6904} = \frac{2.8576}{6.7616} \approx 0.4226$$

$$\beta \approx 65.0°$$

17. For α: $\cos \alpha = \dfrac{b^2 + c^2 - a^2}{2bc}$

$$\cos \alpha = \frac{169 + 25 - 144}{2 \cdot 13 \cdot 5} = \frac{50}{130} \approx 0.3846$$

$$\alpha \approx 67.4°$$

For β: $\cos \beta = \dfrac{a^2 + c^2 - b^2}{2ac}$

$\cos \beta = \dfrac{144 + 25 - 169}{2 \cdot 12 \cdot 5} = \dfrac{0}{120} \approx 0$

$\beta = 90°$

For γ: $\cos \gamma = \dfrac{a^2 + b^2 - c^2}{2ab}$

$\cos \gamma = \dfrac{144 + 169 - 25}{2 \cdot 12 \cdot 13} = \dfrac{288}{312} \approx 0.9231$

$\gamma = 22.6°$

19. $\cos \alpha = \dfrac{b^2 + c^2 - a^2}{2bc}$

$\cos \alpha = \dfrac{4 + 4 - 4}{2 \cdot 2 \cdot 2} = \dfrac{4}{8} \approx 0.5$

$\alpha \approx 60°$

$\cos \beta = \dfrac{a^2 + c^2 - b^2}{2ac}$

$\cos \beta = \dfrac{4 + 4 - 4}{2 \cdot 2 \cdot 2} = \dfrac{1}{2} = 0.5$

$\beta = 60°$

$\cos \gamma = \dfrac{a^2 + b^2 - c^2}{2ab}$

$\cos \gamma = \dfrac{4 + 4 - 4}{2 \cdot 2 \cdot 2} = \dfrac{1}{2} = 0.5$

$\gamma \approx 60°$

21. $\cos \alpha = \dfrac{b^2 + c^2 - a^2}{2bc}$

$\cos \alpha = \dfrac{64 + 81 - 25}{2 \cdot 8 \cdot 9} = \dfrac{120}{144} \approx 0.8333$

$\alpha = 33.6°$

$\cos \beta = \dfrac{a^2 + c^2 - b^2}{2ac}$

$\cos \beta = \dfrac{25 + 81 - 64}{2 \cdot 5 \cdot 9} = \dfrac{42}{90} \approx 0.4667$

$\beta \approx 62.2°$

$\cos \gamma = \dfrac{a^2 + b^2 - c^2}{2ab}$

$\cos \gamma = \dfrac{25 + 64 - 81}{2 \cdot 5 \cdot 8} = \dfrac{8}{80} = 0.1$

$\gamma \approx 84.3°$

23. $\cos \alpha = \dfrac{b^2 + c^2 - a^2}{2bc}$

$\cos \alpha = \dfrac{64 + 25 - 100}{2 \cdot 8 \cdot 5} = \dfrac{-11}{80} \approx 0.1375$

$\alpha = 97.9°$

$\cos \beta = \dfrac{a^2 + c^2 - b^2}{2ac}$

$\cos \beta = \dfrac{100 + 25 - 64}{2 \cdot 10 \cdot 5} = \dfrac{61}{100} = 0.61$

$\beta \approx 52.4°$

$\cos \gamma = \dfrac{a^2 + b^2 - c^2}{2ab}$

$\cos \gamma = \dfrac{100 + 64 - 25}{2 \cdot 10 \cdot 8} = \dfrac{139}{160} \approx 0.8688$

$\gamma \approx 29.7°$

25. The distance c we seek is the third side of a triangle in which the other two sides and their included angle are known.

$c^2 = a^2 + b^2 - 2ab \cos \gamma$

$c^2 = 4900 + 2500 - 2 \cdot 70 \cdot 50 \cdot \cos 70°$

$c^2 = 7400 - 7000(0.34202) = 5005.86$

$c \approx 70.7521$ feet

The houses are about 70.75 feet apart.

27. (a) Using the formula,

Velocity × Time = Distance

220 mi/hr × .25 hr. = 55 miles

We are looking for angle $180° - ?$, the angle at which the pilot should turn to head toward city B. But first we need to find the measurement of the third side a. We know the other two sides and their included angle so we use the Law of Cosines:

$a^2 = b^2 + c^2 - 2bc \cos \alpha$
$a^2 = 3,025 + 108,900 - 2 \cdot 55 \cdot 330 \cdot \cos 10°$
$a^2 = 111,925 - 36(0.9848) = 76,176.76$
$a \approx 276$

Now we can find angle γ:

$$\cos \gamma = \frac{a^2 + b^2 - c^2}{2ab}$$

$$\cos \gamma = \frac{76,176.76 + 3,025 - 108,900}{2 \cdot 276 \cdot 55}$$

$$\cos \gamma = \frac{-29,698.24}{30,360} \approx -0.9782$$

$$\gamma = 168°$$

$180° - \gamma = 12°$

The pilot should turn through an angle of $12°$.

(b)
$$\text{Velocity} = \frac{\text{Distance}}{\text{Time}}$$

$$\text{Velocity} = \frac{331 \text{ miles}}{1.5 \text{ m.}} \approx 220.8 \text{ mph}$$

The pilot should maintain an average speed of 220.8 miles per hour so that the total time of the trip is 90 minutes or 1.5 hours.

29.

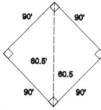

(a)

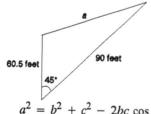

We know that $\alpha = 45°$ because the diagonal of the square diamond bisects the right angle. Because we know two sides and the included angle, we use the Law of Cosines to find the third side a.

$a^2 = b^2 + c^2 - 2bc \cos \alpha$
$a^2 = 3660.25 + 8100 - 2 \cdot 60.5 \cdot 90 \cdot \cos 45°$
$a^2 = 11,760.5 - 10,890(0.7071) = 4060.181$
$a \approx 63.7 \text{ feet}$

It is about 63.7 feet from the pitching rubber to first base.

(b)

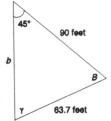

Using the Law of Sines, we find angle γ:

$$\frac{\sin 45°}{63.7} = \frac{\sin \gamma}{90}$$

$$\sin \gamma = \frac{90 \sin 45°}{63.7} \approx 0.9991$$

$$\gamma = 87.5°$$

$$\beta = 180° - \alpha - \gamma = 47.5°$$

Now we can use the Law of Cosines to find side b:

$$b^2 = b^2 + c^2 - 2bc \cos \beta$$
$$b^2 = 4060.181 + 8100 - 2 \cdot 63.7 \cdot 90 \cdot \cos 47.5°$$
$$b^2 = 12,160.181 - 11,466(0.67559) = 4413.86606$$
$$b \approx 66.8 \text{ feet}$$

(c)

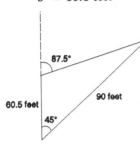

From (b), we know that $\gamma = 87.5°$.
$180° - \gamma = 92.5°$. The pitcher needs to turn through an angle of $92.5°$ to face first base.

31. (a)

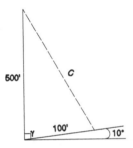

$\gamma = 90° - 10° = 80°$
Now we know two sides of the triangle and the included angle. Therefore, we use the Law of Cosines to find the third side:

$$c^2 = a^2 + b^2 - 2ab \cos \gamma$$
$$c^2 = 10,000 + 250,000 - 2 \cdot 100 \cdot 500 \cdot \cos 80°$$
$$c^2 = 260,000 - 100,000(0.1736) = 242,640$$
$$c \approx 492.6 \text{ feet}$$

The guy wire should be 492.6 feet.

(b)

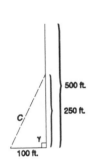

We know two sides of the triangle and the included angle. We use the Law of Cosines to find the third side.

$$c^2 = a^2 + b^2 - 2ab \cos \gamma$$
$$c^2 = 10,000 + 62,500 - 2 \cdot 100 \cdot 250 \cdot \cos 90°$$
$$c^2 = 72,500 - 50,000(0) = 72,500$$
$$c \approx 269.3 \text{ feet}$$

33.

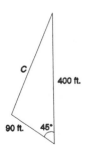

We know two sides and the included angle so we can use the Law of Cosines to find the third side.

$$c^2 = a^2 + b^2 - 2ab \cos \gamma$$
$$c^2 = 8100 + 160,000 - 2 \cdot 90 \cdot 400 \cdot \cos 45°$$
$$c^2 = 168,100 - 72,000(0.7071)$$
$$c^2 \approx 117,188.8$$
$$c \approx 342.3 \text{ feet}$$

In Wrigley Field, it is 342.3 feet from dead center to third base.

35. Using the Law of Cosines, we find angle θ:

$$L^2 = x^2 + r^2 - 2xr \cos \theta$$

$$x^2 - 2rx \cos \theta + r^2 - L^2 = 0$$

$$x = \frac{2r \cos \theta + \sqrt{4r^2 \cos^2 \theta - 4(r^2 - L^2)}}{2}$$

$$x = r \cos \theta + \sqrt{r^2 \cos^2 \theta + L^2 - r^2}$$

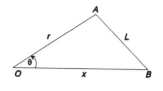

37.
$$\cos \frac{\gamma}{2} = \sqrt{\frac{1 + \cos \gamma}{2}} = \sqrt{\frac{1 + \dfrac{a^2 + b^2 - c^2}{2ab}}{2}} = \sqrt{\frac{2ab + a^2 + b^2 - c^2}{4ab}} = \sqrt{\frac{(a+b)^2 - c^2}{4ab}}$$

$$= \sqrt{\frac{(a + b + c)(a + b - c)}{4ab}} = \sqrt{\frac{2s(2x - c - c)}{4ab}} = \sqrt{\frac{4s(s - c)}{4ab}} = \sqrt{\frac{s(s - c)}{ab}}$$

39. In order to prove that $\dfrac{\cos \alpha}{a} + \dfrac{\cos \beta}{b} + \dfrac{\cos \gamma}{c} = \dfrac{a^2 + b^2 + c^2}{2abc}$, we let

$$\frac{\cos \alpha}{a} + \frac{\cos \beta}{b} + \frac{\cos \gamma}{c} = \frac{b^2 + c^2 - a^2}{2bca} + \frac{a^2 + c^2 - b^2}{2acb} + \frac{a^2 + b^2 - c^2}{2abc}$$

$$= \frac{b^2 + c^2 - a^2 + a^2 + c^2 - b^2 + a^2 + b^2 - c^2}{2abc} = \frac{a^2 + b^2 + c^2}{2abc}$$

8.4 The Area of a Triangle

1. $A = \dfrac{1}{2}ac \sin \beta$

$A = \dfrac{1}{2} \cdot 2 \cdot 4 \sin 45° = 2.83$

3. $A = \dfrac{1}{2}ac \sin \gamma$

$A = \dfrac{1}{2} \cdot 2 \cdot 3 \sin 95° = 2.99$

5. $a = 6, b = 5, c = 8$

$s = \dfrac{1}{2}(a + b + c) = \dfrac{1}{2}(6 + 5 + 8) = \dfrac{19}{2}$

Heron's Formula then gives the area A as:

$$A = \sqrt{s(s - a)(s - b)(s - c)} = \sqrt{\frac{19}{2} \cdot \frac{7}{2} \cdot \frac{9}{2} \cdot \frac{3}{2}} = \sqrt{\frac{3591}{16}} = \sqrt{224.4375} \approx 14.98$$

7. $s = \dfrac{1}{2}(a + b + c) = \dfrac{1}{2}(9 + 6 + 4) = \dfrac{19}{2}$

$$A = \sqrt{s(s - a)(s - b)(s - c)} = \sqrt{\frac{19}{2} \cdot \frac{1}{2} \cdot \frac{7}{2} \cdot \frac{11}{2}} = \sqrt{\frac{1463}{16}} = \sqrt{91.4375} \approx 9.56$$

9. $A = \dfrac{1}{2}ab \sin \gamma$

$A = \dfrac{1}{2} \cdot 3 \cdot 4 \sin 40° = 3.86$

11. $A = \dfrac{1}{2}ab \sin \alpha$

$A = \dfrac{1}{2} \cdot 1 \cdot 3 \sin 80° = 1.48$

13. $A = \frac{1}{2}ac \sin \beta$

$A = \frac{1}{2} \cdot 3 \cdot 2 \sin 110° = 2.82$

15. $A = \frac{1}{2}ab \sin \gamma$

$A = \frac{1}{2} \cdot 2 \cdot 2 \sin 50° = 1.53$

17. $s = \frac{1}{2}(a + b + c) = \frac{1}{2}(12 + 13 + 5) = 15$

$A = \sqrt{s(s - a)(s - b)(s - c)} = \sqrt{15 \cdot 3 \cdot 2 \cdot 10} = \sqrt{900} = 30$

19. $s = \frac{1}{2}(a + b + c) = \frac{1}{2}(2 + 2 + 2) = 3$

$A = \sqrt{s(s - a)(s - b)(s - c)} = \sqrt{3 \cdot 1 \cdot 1 \cdot 1} = \sqrt{3} \approx 1.73$

21. $s = \frac{1}{2}(a + b + c) = \frac{1}{2}(5 + 8 + 9) = 11$

$A = \sqrt{s(s - a)(s - b)(s - c)} = \sqrt{11 \cdot 6 \cdot 3 \cdot 2} = \sqrt{396} \approx 19.90$

23. $s = \frac{1}{2}(a + b + c) = \frac{1}{2}(10 + 8 + 5) = \frac{23}{2}$

$A = \sqrt{s(s - a)(s - b)(s - c)} = \sqrt{\frac{23}{2} \cdot \frac{3}{2} \cdot \frac{7}{2} \cdot \frac{13}{2}} = \sqrt{\frac{6279}{16}} = \sqrt{392.4375} \approx 19.81$

25. We let $a = 100$, $b = 50$, and $c = 75$. The area of the lot can be found using Heron's Formula:

$s = \frac{1}{2}(a + b + c) = \frac{1}{2}(100 + 50 + 75) = \frac{225}{2}$

$A = \sqrt{s(s - a)(s - b)(s - c)} = \sqrt{\frac{225}{2} \cdot \frac{25}{2} \cdot \frac{125}{2} \cdot \frac{75}{2}} = \sqrt{\frac{51,734,375}{16}}$

$= \sqrt{3,295,898.438} \approx 1815.4609$ square feet

If the price of the land is \$3 per square foot, the triangular lot costs $1815.4609 \times \$3 = \$5446.38$.

27. We know that $A = \frac{1}{2}ab \sin \gamma$. We use this formula to prove that $A = \frac{a^2 \sin \beta \sin \gamma}{2 \sin \alpha}$.

$A = \frac{1}{2}ab \sin \gamma = \frac{1}{2}a \sin \gamma \left[\frac{a \sin \beta}{\sin \alpha} \right] = \frac{a^2 \sin \beta \sin \gamma}{2 \sin \alpha}$

29. $A = \frac{a^2 \sin \beta \sin \gamma}{2 \sin \alpha}$, $\gamma = 180° - \alpha - \beta = 120°$

$A = \frac{2^2 \sin 20° \sin 120°}{2 \sin 40°} = \frac{4(0.3420)(0.8660)}{2(0.6428)} \approx 0.92$

31. $A = \frac{b^2 \sin \beta \sin \gamma}{2 \sin \beta}$, $\alpha = 180° - \beta - \gamma = 100°$

$A = \frac{25 \sin 100° \sin 10°}{2 \sin 70°} = \frac{25(0.9848)(0.1736)}{2(0.9397)} \approx 2.27$

33. $A = \frac{c^2 \sin \alpha \sin \beta}{2 \sin \gamma}$, $\beta = 180° - \alpha - \gamma = 40°$

$A = \frac{9 \sin 110° \sin 40°}{2 \sin 30°} = \frac{9(0.9397)(0.6428)}{2(0.5)} \approx 5.44$

35. $A = \dfrac{c^2 \sin \alpha \sin \beta}{2 \sin \gamma}, \gamma = 180° - \alpha - \beta = 100°$

$A = \dfrac{4 \sin 40° \sin 40°}{2 \sin 100°} = \dfrac{4(0.6428)(0.6428)}{2(0.9848)} \approx 0.84$

37. The area A is the sum of the area of a triangle and a sector. Thus,

$$A = \frac{1}{2}r \cdot r \sin(\pi - \theta) + \frac{1}{2}r^2\theta = \frac{1}{2}r^2[\sin(\pi - \theta) + \theta]$$

$$= \frac{1}{2}r^2[\sin \pi \cos \theta + \cos \pi \sin \theta + \theta] = \frac{1}{2}r^2[0 - \sin \theta + \theta] = \frac{1}{2}r^2(\theta - \sin \theta)$$

8.5 Sinusoidal Graphs; Simple Harmonic Motion

1. $y = 2 \sin x$
Comparing $y = 2 \sin x$ to $y = A \sin \omega x$, we find $A = 2$ and $\omega = 1$.
Thus, the amplitude is $|A| = |2| = 2$.
The period is $T = \dfrac{2\pi}{\omega} = \dfrac{2\pi}{1} = 2\pi$.

3. $y = -4 \cos 2x$
Comparing $y = -4 \cos 2x$ to $y = A \cos \omega x$, we find $A = -4$ and $\omega = 2$.
Thus, the amplitude is $|A| = |-4| = 4$.
The period is $T = \dfrac{2\pi}{\omega} = \dfrac{2\pi}{2} = \pi$.

5. $y = 6 \sin \pi x$
Comparing $y = 6 \sin x\pi$ to $y = A \sin \omega x$, we find $A = 6$ and $\omega = \pi$.
Thus, the amplitude is $|A| = |6| = 6$.
The period is $T = \dfrac{2\pi}{\omega} = \dfrac{2\pi}{\pi} = 2$.

7. $y = -\dfrac{1}{2} \cos \dfrac{3}{2}x$

Comparing $y = -\dfrac{1}{2} \cos \dfrac{3}{2}x$ to $y = A \cos \omega x$,

we find $A = -\dfrac{1}{2}$ and $\omega = \dfrac{3}{2}$.

Thus, the amplitude is $|A| = \left|-\dfrac{1}{2}\right| = \dfrac{1}{2}$.

The period is

$$T = \dfrac{2\pi}{\omega} = \dfrac{2\pi}{\dfrac{3}{2}} = \dfrac{2\pi}{1} \cdot \dfrac{2}{3} = \dfrac{4\pi}{3}.$$

9. $y = \dfrac{5}{3} \sin \left[\dfrac{-2\pi}{3}x\right]$

Comparing $y =$

$\dfrac{5}{3}\sin\left[\dfrac{-2\pi}{3}x\right] = \dfrac{-5}{3}\sin\left[\dfrac{2\pi}{3}x\right]$ to $y = A \sin$

ωx, we find $A = \dfrac{-5}{3}$ and $\omega = \dfrac{2\pi}{3}$.

Thus, the amplitude is $|A| = \left|\dfrac{-5}{3}\right| = \dfrac{5}{3}$.

The period is $T = \dfrac{2\pi}{\omega} = \dfrac{2\pi}{\dfrac{2\pi}{3}} = 3$.

11. $y = 2 \sin \dfrac{\pi}{2}x$
Amplitude: $|A| = |2| = 2$

Period: $T = \dfrac{2\pi}{\dfrac{\pi}{2}} = \dfrac{2\pi}{1} \cdot \dfrac{2}{\pi} = 4$

Hence, graph F is a sin graph with amplitude $= 2$ and period $= 4$.

13. $y = 2 \cos \frac{1}{2}x$

Amplitude: $|A| = |2| = 2$

Period: $T = \dfrac{2\pi}{\frac{1}{2}} = \dfrac{2\pi}{1} \cdot \dfrac{2}{1} = 4\pi$

Hence, graph A is a cos graph with amplitude 2 and period 4π.

15. $y = -3 \sin 2x$

Amplitude: $|A| = |-3| = 3$

Period: $T = \dfrac{2\pi}{2} = \pi$

Hence, graph H is a reflected sin graph with amplitude 3 and period π.

17. $y = -2 \cos \frac{1}{2}x$

Amplitude: $|A| = |-2| = 2$

Period: $T = \dfrac{2\pi}{\frac{1}{2}} = 4\pi$

Hence, graph C is a reflected cos graph with amplitude 2 and period 4π.

19. $y = 3 \sin 2x$

Amplitude: $|A| = |3| = 3$

Period: $T = \dfrac{2\pi}{2} = \pi$

Hence, graph J is a sin graph with amplitude 3 and period π.

21. A
23. D
25. B

27. $y = 5 \sin 4x$

Amplitude: $|A| = |5| = 5$

Period: $T = \dfrac{2\pi}{4} = \dfrac{\pi}{2}$

Use the amplitude to scale the y-axis and the period to scale the x-axis. Then, fill in the graph of the sine function.

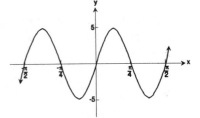

29. $y = 5 \cos \pi x$

Amplitude: $|A| = |5| = 5$

Period: $T = \dfrac{2\pi}{\pi} = 2$

Use the amplitude to scale the y-axis and the period to scale the x-axis. Then, fill in the graph of the cosine function.

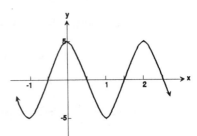

31. $y = -2 \cos 2\pi x$

Amplitude: $|A| = |-2| = 2$

Period: $T = \dfrac{2\pi}{2\pi} = 1$

Use the amplitude to scale the y-axis and the period to scale the x-axis. Then, fill in the graph of the reflected cosine function. $(A = -2 < 0)$

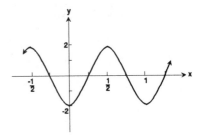

33. $y = -4 \sin \dfrac{1}{2}x$

Amplitude: $|A| = |-4| = 4$

Period: $T = \dfrac{2\pi}{\dfrac{1}{2}} = 4\pi$

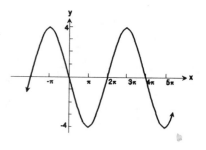

Use the amplitude to scale the y-axis and the period to scale the x-axis. Then, fill in the graph of the reflected sine function. $(A = -4 < 0)$

35. $y = \dfrac{3}{2}\sin\left(\dfrac{-2}{3}x\right) = \dfrac{-3}{2}\sin\left(\dfrac{2}{3}x\right)$

Amplitude: $|A| = \left|\dfrac{-3}{2}\right| = \dfrac{3}{2}$

Period: $T = \dfrac{2\pi}{\dfrac{2}{3}} = \dfrac{2\pi}{1} \cdot \dfrac{3}{2} = 3\pi$

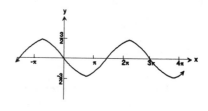

Use the amplitude to scale the y-axis and the period to scale the x-axis. Then, fill in the graph of the reflected sine function. $(A = \dfrac{-3}{2} < 0)$

37. Since the graph starts at $(0, 5)$, we have a cos graph with $A = 5$. Since the period is 8, we have

$$8 = \dfrac{2\pi}{\omega} \text{ so}$$
$$8\omega = 2\pi$$
$$\omega = \dfrac{2\pi}{8} = \dfrac{\pi}{4}$$

Hence, the function is $y = 5 \cos \dfrac{\pi}{4}x$.

39. Since the graph starts at $(0, -3)$, we have a reflected cosine function with $A = -3$. Since the period is 4π, we have

$$4\pi = \dfrac{2\pi}{\omega}, \text{ so}$$
$$4\pi\omega = 2\pi$$
$$\omega = \dfrac{2\pi}{4\pi} = \dfrac{1}{2}$$

Hence, the function is $y = -3 \cos \dfrac{1}{2}x$.

41. Since the graph starts at $(0, 0)$, we have the sine function. Since it increases first to $\dfrac{3}{4}$, we have $A = \dfrac{3}{4}$. Since the period is 1, we have

$$1 = \dfrac{2\pi}{\omega}, \text{ so}$$
$$\omega = 2\pi$$

Hence, the function is $y = \dfrac{3}{4} \sin 2\pi x$.

43. Since the graph starts at $(0, 0)$, we have the sine function. Since it decreases first to -1, we have a reflected sine function with $A = -1$. The period is $\dfrac{4\pi}{3}$ so

$$\dfrac{4\pi}{3} = \dfrac{2\pi}{\omega}$$
$$4\pi\omega = 6\pi$$
$$\omega = \dfrac{6\pi}{4\pi} = \dfrac{6}{4} = \dfrac{3}{2}$$

Hence, the function is $y = -1 \sin \dfrac{3}{2}x$

$$y = -\sin \dfrac{3}{2}x$$

45. Since the graph starts at $(0, -2)$, we have the reflected cosine function with $A = -2$. The period is $\dfrac{4}{3}$, so

$$\dfrac{4}{3} = \dfrac{2\pi}{\omega}$$
$$4\omega = 6\pi$$
$$\omega = \dfrac{6\pi}{4} = \dfrac{3\pi}{2}$$

Hence, the function is $y = -2 \cos \dfrac{3\pi}{2}x$.

47. Amplitude $= 3$

Period $= 4 = \dfrac{2\pi}{\omega}$

$$\omega = \dfrac{\pi}{2}$$

sin function

$$y = 3 \sin \dfrac{\pi}{2}x$$

49. Amplitude $= 4$

Period $= \dfrac{2\pi}{3} = \dfrac{2\pi}{\omega}$

$$\omega = 3$$

reverse of cos function

$$y = -4 \cos 3x$$

51. $y = 4 \sin(2x - \pi)$ compared to
$y = A \sin(\omega x - \phi)$

Amplitude: $|A| = |4| = 4$

Period: $T = \dfrac{2\pi}{\omega} = \dfrac{2\pi}{2} = \pi$

Phase Shift: $\dfrac{\phi}{\omega} = \dfrac{\pi}{2}$

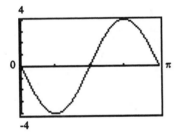

53. $y = 2 \cos \left[3x + \dfrac{\pi}{2} \right]$ compared to
$y = A \cos (\omega x - \phi)$.

$$y = 2 \cos \left[3x - \left[-\dfrac{\pi}{2} \right] \right]$$

Amplitude: $|A| = |2| = 2$

Period: $T = \dfrac{2\pi}{\omega} = \dfrac{2\pi}{3}$

Phase Shift: $\dfrac{\phi}{\omega} = \dfrac{-\dfrac{\pi}{2}}{3} = -\dfrac{\pi}{6}$

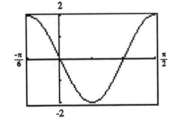

55. $y = -3 \sin\left[2x + \dfrac{\pi}{2}\right]$ compared to

$y = A \sin(\omega x - \phi)$

$y = -3 \sin\left[2x - \left[-\dfrac{\pi}{2}\right]\right]$

Amplitude: $|A| = |-3| = 3$

Period: $T = \dfrac{2\pi}{\omega} = \dfrac{2\pi}{2} = \pi$

Phase Shift: $\dfrac{\phi}{\omega} = \dfrac{-\dfrac{\pi}{2}}{2} = -\dfrac{\pi}{4}$

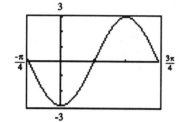

57. $y = 4 \sin(\pi x + 2)$

$y = A \sin(\omega x - \phi)$

$A = 4,\ \omega = \pi,\ \phi = -2$

Amplitude: 4

Period: $\dfrac{2\pi}{\omega} = \dfrac{2\pi}{\pi} = 2$

Phase Shift: $\dfrac{\phi}{\omega} = \dfrac{-2}{\pi}$

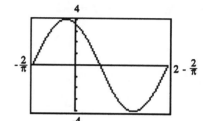

59. $y = 3 \cos(\pi x - 2)$

$y = A \cos(\omega x - \phi)$

$A = 3,\ \omega = \pi,\ \phi = 2$

Amplitude: 3

Period: $\dfrac{2\pi}{\omega} = \dfrac{2\pi}{\pi} = 2$

Phase Shift: $\dfrac{\phi}{\omega} = \dfrac{2}{\pi}$

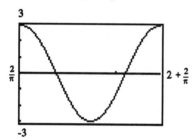

61. $y = 3 \sin\left[-2x + \dfrac{\pi}{2}\right]$

$y = -3 \sin\left[2x - \dfrac{\pi}{2}\right]$

$y = A \cos(\omega x - \phi)$

Amplitude: 3

Period: $\dfrac{2\pi}{\omega} = \dfrac{2\pi}{2} = \pi$

Phase Shift: $\dfrac{\phi}{\omega} = \dfrac{\dfrac{\pi}{2}}{2} = \dfrac{\pi}{4}$

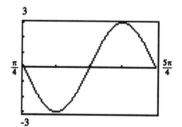

63. $d = 5 \cos \pi t$

65. $d = 6 \cos 2t$

67. $d = 5 \sin \pi t$

69. $d = 6 \sin 2t$

71. (a) Simple harmonic
    (b) 5 m

    (c) $\dfrac{2\pi}{3}$ sec

    (d) $\dfrac{3}{2\pi}$ oscillations/sec

73. (a) Simple harmonic
    (b) 6 m
    (c) 2 sec

    (d) $\dfrac{1}{2}$ oscillation/sec

75. (a) Simple harmonic
    (b) 3 m
    (c) 4π sec
    (d) $\dfrac{1}{4\pi}$ oscillation/sec

77. (a) Simple harmonic
    (b) 2 m
    (c) 1 sec
    (d) 1 oscillation/sec

79. $I = 220 \sin 60\pi t,\ t \geq 0$

Period $= \dfrac{2\pi}{60\pi} = \dfrac{1}{30}$

Amplitude $= 220$

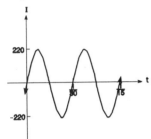

81. $I = 120 \sin\left[30\pi t - \dfrac{\pi}{3}\right],\ t \geq 0$

Period $= \dfrac{2\pi}{30\pi} = \dfrac{1}{15}$

Amplitude $= 120$

Phase shift compares to $\dfrac{\pi}{3}$

so $\dfrac{2\pi}{\dfrac{1}{15}} = \dfrac{\dfrac{\pi}{3}}{x}$

$x = \dfrac{\pi}{45} \cdot \dfrac{1}{2\pi}$

$x = \dfrac{1}{90}$

The phase shift is $\dfrac{1}{90}$.

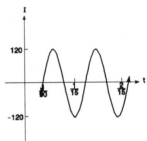

83. $V = 220 \sin 120\pi t$
    (a) Amplitude $= 220$

    Period $= \dfrac{2\pi}{120\pi} = \dfrac{1}{60}$

    (b) and (e)

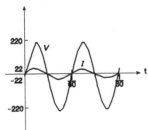

(c) $V = IR$
    $220 \sin 120\pi t = 10\,I$
    $22 \sin 120\pi t = I$

(d) Amplitude $= 22$

    Period $= \dfrac{2\pi}{120\pi} = \dfrac{1}{60}$

85. (a) $P = \dfrac{\left(V_0 \sin 2\pi\ ft\right)^2}{R} = \dfrac{V_0^2 \sin^2 2\pi\ ft}{R}$

    (b) $P = \dfrac{V_0^2}{R} \cdot \dfrac{1}{2}(1 - \cos 4\pi\ ft)$

1.

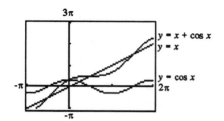

3.

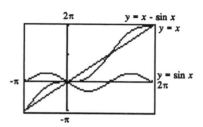

5.

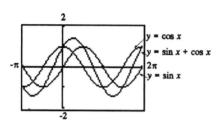

7.

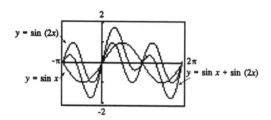

9.

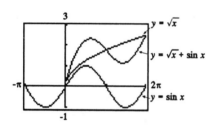

11.

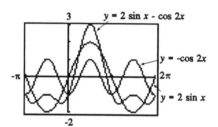

13.

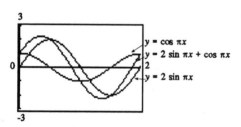

15.

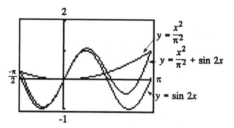

17.

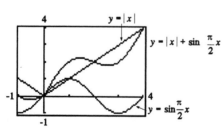

19.

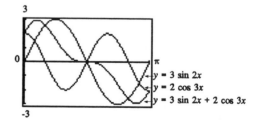

21.

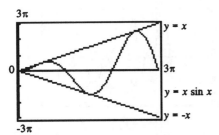

Touches: $(0, 0)$; $(\pi, -\pi)$; $(2\pi, 2\pi)$; $(3\pi, -3\pi)$
Turning Points: $(0.86, 0.56)$, $(3.42, -3.28)$, $(6.43, 6.36)$

23.

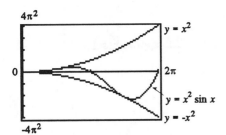

Touches: $(0, 0)$, $\left[\dfrac{\pi}{2}, \dfrac{\pi^2}{4}\right]$, $\left[\dfrac{3\pi}{2}, -\dfrac{9\pi^2}{4}\right]$
Turning Points: $(2, 28, 3.94)$, $(5.08, -24.08)$

25.

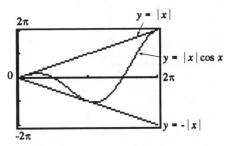

Touches: $(0, 0)$, $(\pi, -\pi)$, $(2\pi, 2\pi)$
Turning Points: $(0.86, 0.56)$, $(3.42, -3.28)$

27.

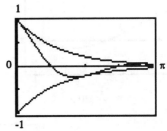

Touches $(0, 0)$, $(1.57, -0.20)$, $(3.14, 0.04)$
Turning Points: $(1.33, -0.23)$, $(2.90, 0.04)$

29.

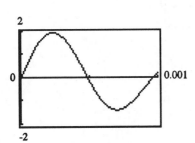

31. (a)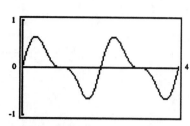

(b)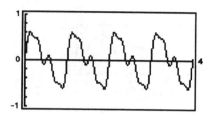

(c)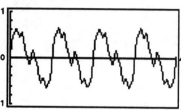

33.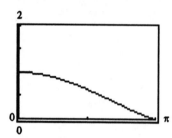

As x approaches 0, $\dfrac{\sin x}{x}$ approaches 1.

35.

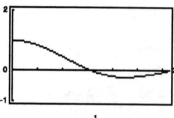

$$y = \frac{1}{x} \sin x$$

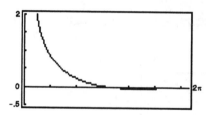

$$y = \frac{1}{x^2} \sin x$$

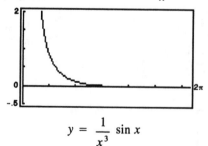

$$y = \frac{1}{x^3} \sin x$$

As x gets larger, the graph of $y = \dfrac{1}{x^n} \sin x$ gets closer to $y = 0$.

1. $\alpha = 90° - 20° = 70°$

$\sin 20° = \dfrac{b}{10}$, so $b = 10 \sin 20° \approx 3.42$

$a^2 = c^2 - b^2 = 10^2 - (3.42)^2 \approx 88.3$

$a = \sqrt{88.3} \approx 9.4$

3. $a^2 = c^2 - b^2 = 5^2 - 2^2 = 21$

$a = 4.58$

$\cos \alpha = \dfrac{2}{5}$, so $\alpha = 66.4°$

$\sin \beta = \dfrac{2}{5}$, so $\beta = 23.6°$

5. $\dfrac{\sin \alpha}{a} = \dfrac{\sin \beta}{b}$

$\dfrac{\sin 50°}{1} = \dfrac{\sin 30°}{b}$

$b = \dfrac{\sin 30°}{\sin 50°} \approx 0.65$

$\gamma = 180° - \alpha - \beta = 100°$

$c^2 = a^2 + b^2 - 2ab \cos \gamma$

$c^2 = 1 + 0.4260 - 2(1)(0.6527)\cos 100° \approx 1.6527$

$c \approx 1.29$

7. $\dfrac{\sin \alpha}{a} = \dfrac{\sin \gamma}{c}$

$\dfrac{\sin 100°}{5} = \dfrac{\sin \gamma}{2}$

$\sin \gamma = \dfrac{2 \sin 100°}{5} \approx 0.3939$

$\gamma = 23.2°$

$\beta = 180° - \alpha - \gamma = 56.8°$

$b^2 = a^2 + c^2 - 2ac \cos \beta$

$b^2 = 25 + 4 - 2(5)(2)\cos 56.8°$

$b^2 \approx 18.0487$

$b \approx 4.25$

9. $\dfrac{\sin \alpha}{a} = \dfrac{\sin \gamma}{c}$

$\dfrac{\sin \alpha}{3} = \dfrac{\sin 100°}{1}$

$\sin \alpha = 3 \sin 100° = 2.8191$
Impossible! No Triangle.

11. $b^2 = a^2 + c^2 - 2ac \cos \beta$

$b^2 = 9 + 1 - 2(3)(1) \cos 100°$

$b^2 = 10 - 6(-0.17365) \approx 11.0419$

$b \approx 3.32$

$\cos \alpha = \dfrac{b^2 + c^2 - a^2}{2bc}$

$\cos \alpha = \dfrac{11.0419 + 1 - 9}{2(3.3229)(1)} = \dfrac{3.0419}{6.6458} \approx 0.4577$

$\alpha \approx 62.8°$

$\gamma \approx 180° - \alpha - \beta \approx 17.2°$

13. $\cos \alpha = \dfrac{b^2 + c^2 - a^2}{2bc}$

$\cos \alpha = \dfrac{9 + 1 - 4}{2(3)(1)} = 1$

$\alpha = 0°$

No Triangle.

15. $c^2 = a^2 + b^2 - 2ab \cos \gamma$

$c^2 = 1 + 9 - 2(1)(3) \cos 40°$

$\quad = 10 - 6(0.76604) = 5.4037$

$c \approx 2.32$

$\cos \alpha = \dfrac{b^2 + c^2 - a^2}{2bc}$

$\cos \alpha = \dfrac{9 + 5.4037 - 1}{2(3)(2.3246} = \dfrac{13.4037}{13.9476} \approx 0.9610$

$\alpha \approx 16.1°$

$\beta \approx 180° - \alpha - \gamma \approx 123.9°$

17. $\dfrac{\sin \alpha}{a} = \dfrac{\sin \beta}{b}$

$\dfrac{\sin 80°}{5} = \dfrac{\sin \beta}{3}$

$\sin \beta = \dfrac{3 \sin 80°}{5} \approx 0.5909$

$\beta = 36.2°$

$\gamma = 180° - \alpha - \beta \approx 63.8°$

$c^2 = a^2 + b^2 - 2ab \cos \gamma$

$c^2 = 25 + 9 - 2(5)(3)(0.4415) \approx 20.755$

$c \approx 4.56$

19. $\cos \alpha = \dfrac{b^2 + c^2 - a^2}{2bc}$

$\cos \alpha = \dfrac{\dfrac{1}{4} + \dfrac{16}{9} - 1}{2 \cdot \dfrac{1}{2} \cdot \dfrac{4}{3}} = \dfrac{\dfrac{9 + 64 - 36}{36}}{\dfrac{4}{3}} = \dfrac{\dfrac{37}{36}}{\dfrac{4}{3}} = \dfrac{37}{48} \approx 0.7708$

$\alpha = 39.6°$

$\dfrac{\sin \alpha}{a} = \dfrac{\sin \beta}{b}$

$\dfrac{\sin 39.6°}{1} = \dfrac{\sin \beta}{\dfrac{1}{2}}$

$\sin \beta = \dfrac{1}{2} \sin 39.6° \approx 0.3187$

$\beta = 18.5°$

$\gamma = 180° - \alpha - \beta = 121.9°$

21.

$$\frac{\sin \alpha}{a} = \frac{\sin \beta}{b}$$

$$\frac{\sin 10°}{3} = \frac{\sin \beta}{4}$$

$$\sin \beta = \frac{4 \sin 10°}{3} \approx 0.2315 \quad \text{(Two Triangles)}$$

$$\beta_1 \approx 13.4° \qquad\qquad \beta_2 = 180° - 13.4° = 166.6°$$

$$\gamma_1 = 180° - \alpha - \beta_1 = 156.6° \qquad \gamma_2 = 180° - \alpha - \beta_2 \approx 3.4°$$

$$\frac{\sin 10°}{3} = \frac{\sin 156.6°}{c} \qquad\qquad \frac{\sin 10°}{3} = \frac{\sin 3.4°}{c}$$

$$c = \frac{3 \sin 156.6°}{\sin 10°} \qquad\qquad c = \frac{3 \sin 3.4°}{\sin 10°}$$

$$c_1 \approx 6.86 \qquad\qquad c \approx 1.02$$

23.

$$a^2 = b^2 + c^2 - 2bc \cos \alpha$$

$$a^2 = 16 + 25 - 2(4)(5)\cos 70°$$

$$a^2 = 41 - 40(5.3420) \approx 27.3192$$

$$a \approx 5.23$$

$$\cos \beta = \frac{a^2 + c^2 - b^2}{2ac}$$

$$\cos \beta = \frac{27.3192 + 25 - 16}{2(5.2268)(5)} \approx \frac{36.3192}{52.268} \approx 0.6949$$

$$\beta = 46°$$

$$\gamma \approx 180° - \alpha - \gamma \approx 64°$$

25.

$$A = \frac{1}{2}ab \sin \gamma$$

$$A = \frac{1}{2} \cdot 2 \cdot 3 \cdot \sin 40°$$

$$\approx 1.93$$

27.

$$A = bc \sin \frac{1}{2}\alpha$$

$$A = \frac{1}{2} \cdot 4 \cdot 10 \cdot \sin 70°$$

$$\approx 18.79$$

29.

$$s = \frac{1}{2}(a + b + c)$$

$$s = \frac{1}{2}(4 + 3 + 5) = 6$$

$$A = \sqrt{s(s - a)(s - b)(s - c)}$$

$$= \sqrt{6(2)(3)(1)} = \sqrt{36} = 6$$

31.

$$s = \frac{1}{2}(a + b + c)$$

$$s = \frac{1}{2}(4 + 2 + 5) = \frac{11}{2}$$

$$A = \sqrt{s(s - a)(s - b)(s - c)}$$

$$= \sqrt{\frac{11}{2}\left[\frac{3}{2}\right]\left[\frac{7}{2}\right]\left[\frac{1}{2}\right]}$$

$$= \sqrt{\frac{231}{16}} = \frac{\sqrt{231}}{4} \approx 3.80$$

33.

$$A = \frac{a^2 \sin \beta \sin \gamma}{2 \sin \alpha}$$

$$\gamma = 180° - \alpha - \beta = 100°$$

$$A = \frac{1^2 \sin 30° \sin 100°}{2 \sin 50°} = \frac{(.5)(0.9848)}{(1.5321)} \approx 0.32$$

35. $y = 4 \cos x$
Amplitude = 4
Period = 2π

37. $y = -8 \sin \dfrac{\pi}{2}x$
Amplitude = 8
Period = 4

39. $y = 4 \sin 3x$ compared to $y = A \sin(\omega x - \phi)$ so $A = 4$, $\omega = 3$, and $\phi = 0$

Amplitude: $|A| = |4| = 4$

Period: $T = \dfrac{2\pi}{\omega} = \dfrac{2\pi}{3}$

Phase Shift: $\dfrac{\phi}{\omega} = \dfrac{0}{3} = 0$

Use amplitude to scale y-axis and the period to scale x-axis.

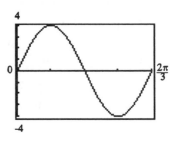

41. $y = -2 \sin\left[\dfrac{\pi}{2}x + \dfrac{1}{2}\right]$ compare to $y = A \sin(\omega x - \phi)$, so $A = -2$, $\omega = \dfrac{\pi}{2}$, and $\phi = -\dfrac{1}{2}$

Amplitude: $|A| = |-2| = 2$

Period: $T = \dfrac{2\pi}{\omega} = \dfrac{2\pi}{\dfrac{\pi}{2}} = 4$

Phase Shift: $\dfrac{\phi}{\omega} = \dfrac{-\dfrac{1}{2}}{\dfrac{\pi}{2}} = -\dfrac{1}{2} \cdot \dfrac{2}{\pi} = -\dfrac{1}{\pi}$

Since $A < 0$, graph is reflected about x-axis.

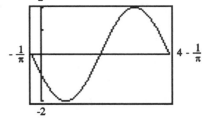

43. $y = \dfrac{1}{2} \sin\left[\dfrac{3}{2}x - \pi\right]$ compare to $y = A \sin(\omega x - \phi)$ so $A = \dfrac{1}{2}$, $\omega = \dfrac{3}{2}$, and $\theta = \pi$

Amplitude: $|A| = \left|\dfrac{1}{2}\right| = \dfrac{1}{2}$

Period: $T = \dfrac{2\pi}{\omega} = \dfrac{2\pi}{\dfrac{3}{2}} = 2\pi \cdot \dfrac{2}{3} = \dfrac{4\pi}{3}$

Phase Shift: $\dfrac{\phi}{\omega} = \dfrac{\pi}{\dfrac{3}{2}} = \pi \cdot \dfrac{2}{3} = \dfrac{2\pi}{3}$

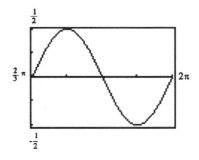

45. $y = -\dfrac{2}{3}\cos(\pi x - 6)$ compare to $y = A\sin(\omega x - \phi)$ so

$A = -\dfrac{2}{3}$, $\omega = \pi$, and $\theta = 6$

Amplitude: $\;|A| = \left|-\dfrac{2}{3}\right| = \dfrac{2}{3}$

Period: $\;T = \dfrac{2\pi}{\omega} = \dfrac{2\pi}{\pi} = 2$

Phase Shift: $\;\dfrac{\phi}{\omega} = \dfrac{6}{\pi}$

Since $A < 0$, graph is reflected about x-axis.

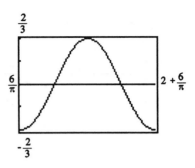

(Problems 47 and 49 could have other answers besides those given.)

47. Since the graph starts one cycle at $(0, 5)$, it is a cosine function with the amplitude is 5 so $A = 5$. The period is 8π.

$$\dfrac{2\pi}{\omega} = 8\pi$$
$$8\pi\omega = 2\pi$$
$$\omega = \dfrac{2\pi}{8\pi} = \dfrac{1}{4}$$

There is no phase shift.

Hence, we get $y = 5\cos\dfrac{1}{4}x$ or $y = 5\cos\dfrac{x}{4}$

49. Since the graph goes through $(0, -6)$, we see that we have a reflected cosine function with $A = -6$ and period $= 8$, so

$$\dfrac{2\pi}{\omega} = 8$$
$$8\omega = 2\pi$$
$$\omega = \dfrac{2\pi}{8} = \dfrac{\pi}{4}$$

Hence, $y = -6\cos\dfrac{\pi}{4}x$

51.

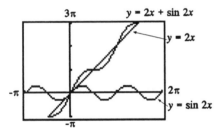

53.

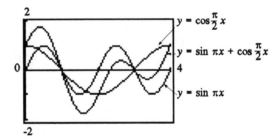

55.

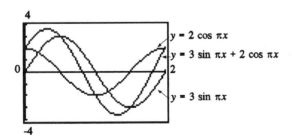

57.

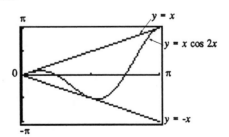

Touches: $(0, 0)$, $\left[\dfrac{\pi}{2}, -\dfrac{\pi}{2}\right]$, (π, π)

Turning Points: $(0.43, 0.28)$, $(1.71, -1.64)$

59.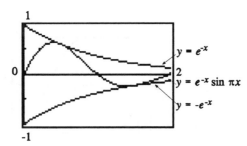

Touches: $(0.50, 0.60)$, $(1.50, -0.22)$
Turning Points: $(0.40, 0.63)$, $(1.40, -0.23)$

61.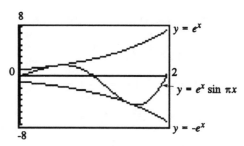

Touches: $(0, 50, 1.64)$, $(1.50, -4.48)$
Turning Points: $(0.59, 1.73)$, $(1.59, -4.71)$

63. (a) Simple harmonic
  (b) 6 ft
  (c) π sec
  (d) $\dfrac{1}{\pi}$ oscillation/sec

65. (a) Simple harmonic
  (b) 2 ft
  (c) 2 sec
  (d) $\dfrac{1}{2}$ oscillation/sec

67. $E = 120 \sin 120\pi t, \ t \geq 0$

  (a) The amplitude is the maximum value of which is 120.

  (b) Period $= \dfrac{2\pi}{120\pi} = \dfrac{1}{60}$

(c)

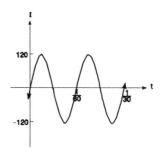

69.
$$\cot 65° = \frac{b}{500}$$
$$500(0.4663) = b$$
$$233.15 = b$$
$$\cot 25° = \frac{a + b}{500}$$
$$500(\cot 25°) = a + b$$
$$500(2.1445) = a + 233.15$$
$$1072.25 - 233.15 = a$$
$$839.1 = a$$
$$839 \text{ feet}$$

71.
$$\tan 25° = \frac{b}{50}$$
$$50(0.4663) = b$$
$$23.315 = b$$
$$23.32 \text{ feet}$$

73.

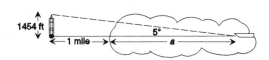

$$1454 = 0.2754 \text{ mi}$$

$$\cot 5° = \frac{a + 1}{0.2754}$$

$$0.2754(\cot 5°) = a + 1$$

$$0.2754(11.43) = a + 1$$

$$3.147822 = a + 1$$

$$3.15 = a + 1$$

$$3.15 - 1 = a$$

$$2.15 \text{ mi}$$

75.

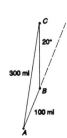

Let $\beta = 180° - 20° = 160°$

Then, $\dfrac{\sin \beta}{b} = \dfrac{\sin \gamma}{c}$

$$\frac{\sin 160°}{300} = \frac{\sin \gamma}{100}$$

$$\sin \gamma = \frac{100 \sin 160°}{300} \approx 0.1140$$

$$\gamma = 6.55°$$

Therefore, $\alpha = 180° - \beta - \gamma \approx 13.45°$

Now, we can find a, the distance from city B to city C, using the Law of Cosines:

$$a^2 = b^2 + c^2 - 2bc \cos \alpha$$

$$a^2 = 90,000 + 10,000 - 2(300)(100)\cos 13.45°$$

$$a^2 \approx 41,645.6$$

$$a \approx 204.1$$

The distance from city B to city C is 204.1 miles.

77. Draw a line perpendicular to the shore

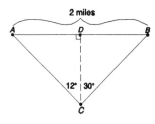

(a) $\angle ACB = 12° + 30° = 42°$

$\angle ABC = 180° - 90° - 30° = 60°$

Using the Law of Sines, we can find $\overline{AC}$:

$$\frac{\sin 60°}{\overline{AC}} = \frac{\sin 42°}{2}$$

$$\overline{AC} = \frac{2 \sin 60°}{\sin 42°} \approx 2.59 \text{ miles}$$

(b) $\angle BAC = 180° - 90° - 12° = 78°$

Using the Law of Sines:

$$\frac{\sin 78°}{\overline{BC}} = \frac{\sin 42°}{2}$$

$$\overline{BC} = \frac{2 \sin 78°}{\sin 42°} \approx 2.92 \text{ miles}$$

(c) Using the Law of Sines:

$$\frac{\sin 90°}{2.92} = \frac{\sin 60°}{\overline{CD}}$$

$$\overline{CD} = \frac{2.92 \sin 60°}{\sin 90°} \approx 2.53 \text{ miles}$$

79. See Figure. The boat has traveled a total distance of $(18 \text{ mi/hr})(4 \text{ hr}) = 72$ miles.

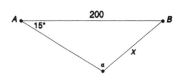

(a) To find the distance to the British West Indies, we use the Law of Cosines since we know two sides and the included angle.

$$x^2 = 200^2 + 72^2 - 2(200)(72)\cos 15° = 17365.3$$

$$x \approx 131.8 \text{ miles}$$

The sailboat is 131.8 miles from the British West Indies.

(b)   We now know three sides of the triangle. We use the Law of Cosines to find the angle α opposite 200.

$$200^2 = 72^2 + 131.8^2 - 2(72)(131.8)\cos \alpha$$

$$\cos \alpha = \frac{40000 - 5184 - 17365.3}{-2(72)(131.8)} \approx -0.9195$$

$$\alpha = 156.9°$$

The sailboat should turn through an angle of $\theta = 180° - 156.9° = 23.1°$

(c)   The extra $(72 + 131.8) - 200 = 3.8$ miles will take $\dfrac{3.8 \text{ miles}}{18 \text{ mph}} \approx 0.2$ hours or 12 minutes.

81.   The approximate area of the lake is $A_1 + A_2 + A_3$.

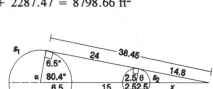

$$A_1 = \frac{1}{2}(125)(100) \sin 50° \approx 4787.78 \text{ ft}^2$$

$$A_2 = \frac{1}{2}(50)(70) \sin 100° \approx 1723.41 \text{ ft}^2$$

To find A_3, we need to find a and b. To find a, we use the Law of Cosines.

$$a^2 = 125^2 + 100^2 - 2(125)(100)\cos 50° \approx 9555.31$$

$$a \approx 97.75$$

To find b, we use the Law of Cosines:

$$b^2 = 50^2 + 70^2 - 2(50)(70)\cos 100° \approx 8615.54$$

$$b \approx 92.82$$

Since $S = \frac{1}{2}(a + b + c)$, we have $S = \frac{1}{2}(97.75 + 92.82 + 50) = 120.285$

Now, $A_3 = \sqrt{S(S - a)(S - b)(S - c)}$

$$= \sqrt{120.285(120.285 - 97.75)(120.285 - 92.82)(120.285 - 50)} \approx 2287.47 \text{ ft}^2$$

So, the approximate area of the lake is $4787.78 + 1723.41 + 2287.47 = 8798.66 \text{ ft}^2$

83. $$\frac{6.5}{2.5} = \frac{26.5 + x}{2.5 + x}$$

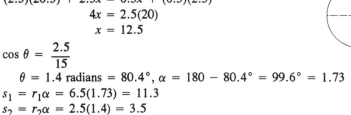

$$(2.5)(26.5) + 2.5x = 6.5x + (6.5)(2.5)$$

$$4x = 2.5(20)$$

$$x = 12.5$$

$$\cos \theta = \frac{2.5}{15}$$

$\theta = 1.4$ radians $= 80.4°$, $\alpha = 180 - 80.4° = 99.6° = 1.73$

$s_1 = r_1\alpha = 6.5(1.73) = 11.3$

$s_2 = r_2\alpha = 2.5(1.4) = 3.5$

Length of belt $= 2(11.3 + 24 + 3.5) = 77.6$ inches

POLAR COORDINATES; VECTORS

9.1 Polar Coordinates

1.

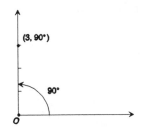

3.

5.

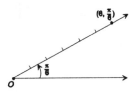

7.

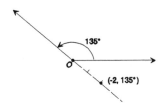

9.

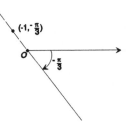

11.

13.

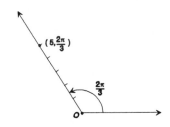

(a) $r > 0, -2\pi < \theta < 0$

$$\left[5, \frac{2\pi}{3} - 2\pi\right] = \left[5, \frac{-4\pi}{3}\right]$$

(b) $r < 0, 0 \le \theta < 2\pi$

$$\left[-5, \frac{2\pi}{3} + \pi\right] = \left[-5, \frac{5\pi}{3}\right]$$

(c) $r > 0, 2\pi \le \theta < 4\pi$

$$\left[5, \frac{2\pi}{3} + 2\pi\right] = \left[5, \frac{8\pi}{3}\right]$$

15.

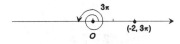

(a) $(2, 3\pi - \pi - 4\pi) = (2, -2\pi)$

(b) $(-2, 3\pi - 2\pi) = (-2, \pi)$

(c) $(2, 3\pi - \pi) = (2, 2\pi)$

17.

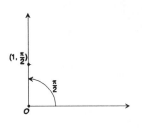

(a) $\left[1, \dfrac{\pi}{2} - 2\pi\right] = \left[1, \dfrac{-3\pi}{2}\right]$

(b) $\left[-1, \dfrac{\pi}{2} + \pi\right] = \left[-1, \dfrac{3\pi}{2}\right]$

(c) $\left[1, \dfrac{\pi}{2} + 2\pi\right] = \left[1, \dfrac{5\pi}{2}\right]$

19.

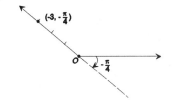

(a) $\left[3, \dfrac{-\pi}{4} - \pi\right] = \left[3, \dfrac{-5\pi}{4}\right]$

(b) $\left[-3, \dfrac{-\pi}{4} + 2\pi\right] = \left[-3, \dfrac{7\pi}{4}\right]$

(c) $\left[3, \dfrac{-\pi}{4} + \pi + 2\pi\right] = \left[3, \dfrac{11\pi}{4}\right]$

21. $x = r \cos \theta = 3 \cos \dfrac{\pi}{2} = 3 \cdot 0 = 0$

$y = r \sin \theta = 3 \sin \dfrac{\pi}{2} = 3 \cdot 1 = 3$

The rectangular coordinates of the point $\left[3, \dfrac{\pi}{2}\right]$ are $(0, 3)$.

23. $x = r \cos \theta = -2 \cos 0 = -2 \cdot 1 = -2$
$y = r \sin \theta = -2 \sin 0 = -2 \cdot 0 = 0$
The rectangular coordinates of the point $(-2, 0)$ are $(-2, 0)$.

25. $x = r \cos \theta = 6 \cos 150° = 6 \cdot \dfrac{-\sqrt{3}}{2} = -3\sqrt{3}$

$y = r \sin \theta = 6 \sin 150° = 6 \cdot \dfrac{1}{2} = 3$

The rectangular coordinates of the point $(6, 150°)$ are $\left(-3\sqrt{3}, 3\right)$.

27. $x = r \cos \theta = -2 \cos \dfrac{3\pi}{4} = -2 \cdot \dfrac{-\sqrt{2}}{2} = \sqrt{2}$

$y = r \sin \theta = -2 \sin \dfrac{3\pi}{4} = -2 \cdot \dfrac{\sqrt{2}}{2} = -\sqrt{2}$

The rectangular coordinates of the point $\left[-2, \dfrac{3\pi}{4}\right]$ are $\left(\sqrt{2}, -\sqrt{2}\right)$.

29. $x = r \cos \theta = -1 \cos \dfrac{-\pi}{3} = -1 \cdot \dfrac{1}{2} = \dfrac{-1}{2}$

$y = r \sin \theta = -1 \sin \dfrac{-\pi}{3} = -1 \cdot \dfrac{-\sqrt{3}}{2} = \dfrac{\sqrt{3}}{2}$

The rectangular coordinates of the point $\left[-1, \dfrac{-\pi}{3}\right]$ are $\left[\dfrac{-1}{2}, \dfrac{\sqrt{3}}{2}\right]$.

31. $x = r \cos \theta = -2 \cos -180° = -2 \cdot -1 = 2$
$y = r \sin \theta = -2 \sin -180° = -2 \cdot 0 = 0$
The rectangular coordinates of the point $(-2, -180°)$ are $(2, 0)$.

33. $x = r \cos \theta = 7.5 \cos 110° = -2.57$
$y = r \sin \theta = 7.5 \sin 110° = 7.05$

35. $x = 6.3 \cos 3.8 = -4.98$
$y = 6.3 \sin 3.8 = -3.86$

37. Since the point $(3, 0)$ lies on the positive x-axis, then $r = \sqrt{x^2 + y^2} = \sqrt{9 + 0} = 3$
and $\theta = \tan^{-1} \dfrac{y}{x} = \tan^{-1} \dfrac{0}{3} = 0$.
Thus, polar coordinates are $(3, 0)$.

39. Since the point $(-1, 0)$ lies on the negative x-axis, then $r = -\sqrt{1 + 0} = -1$ and
$\theta = \tan^{-1} \dfrac{0}{-1} = 0$.
Thus, polar coordinates are
$(-1, 0) = (1, \pi)$.

41. Since the point $(1, -1)$ lies in quadrant IV, then $r = \sqrt{1 + 1} = \sqrt{2}$ and
$\theta = \tan^{-1}(-1) = \dfrac{-\pi}{4}$.

Thus, polar coordinates are $\left[\sqrt{2}, \dfrac{-\pi}{4}\right]$.

43. Since the point $\left(\sqrt{3}, 1\right)$ lies in quadrant I, then

$r = \sqrt{3 + 1} = 2$ and $\theta = \tan^{-1} \dfrac{1}{\sqrt{3}} = \dfrac{\pi}{6}$.

Thus, polar coordinates are $\left[2, \dfrac{\pi}{6}\right]$.

45. $x = 1.3, y = -2.1$

$r = \sqrt{x^2 + y^2} = \sqrt{6.1} = 2.47$

$\alpha = \tan^{-1}\left|\dfrac{y}{x}\right| = \tan^{-1}(1.615) = 1.02$

$\theta = -\alpha = -1.02$
Polar coordinates of (x, y) are $(2.47, -1.02)$.

47. $x = 8.3, y = 4.2$

$r = \sqrt{x^2 + y^2} = \sqrt{86.53} = 9.3$

$\alpha = \tan\left|\dfrac{y}{x}\right| = \tan^{-1}(0.506) = 0.47$

$\theta = \alpha = 0.47$
Polar coordinates of (x, y) are $(9.3, 0.47)$.

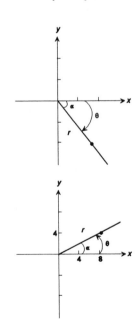

49.
$$2x^2 + 2y^2 = 3$$
$$2(x^2 + y^2) = 3$$
$$2r^2 = 3$$
$$r^2 = \frac{3}{2}$$

51.
$$x^2 = 4y$$
$$(r \cos \theta)^2 = 4r \sin \theta$$
$$r^2 \cos^2 \theta - 4r \sin \theta = 0$$

53.
$$2xy = 1$$
$$2(r \cos \theta)(r \sin \theta) = 1$$
$$2r^2 \cos \theta \sin \theta = 1$$
$$r^2 \sin 2\theta = 1$$

55.
$$x = 4$$
$$r \cos \theta = 4$$

57.
$$r = \cos \theta$$
$$r^2 = r \cos \theta$$
$$x^2 + y^2 = x$$
$$x^2 + y^2 - x = 0$$
$$(x^2 - x) + y^2 = 0$$
$$\left[x^2 - x + \frac{1}{4} \right] + y^2 = 0 + \frac{1}{4}$$
$$\left[x - \frac{1}{2} \right]^2 + y^2 = \frac{1}{4}$$

This is the equation of a circle with center at $\left[\frac{1}{2}, 0 \right]$ and its radius $\frac{1}{2}$.

59.
$$r^2 = \cos \theta$$
$$r^3 = r \cos \theta$$
$$(x^2 + y^2)\left(\pm \sqrt{x^2 + y^2} \right) = x$$
$$(x^2 + y^2)^{3/2} - x = 0$$

61.
$$r = 2$$
$$\pm \sqrt{x^2 + y^2} = 2$$
$$x^2 + y^2 = 4$$

This is the equation of a circle with center $(0, 0)$ and radius 2.

63.
$$r = \frac{4}{1 - \cos \theta}$$
$$r(1 - \cos \theta) = 4$$
$$r - r \cos \theta = 4$$
$$\pm \sqrt{x^2 + y^2} - x = 4$$
$$\pm \sqrt{x^2 + y^2} = 4 + x$$
$$x^2 + y^2 = 16 + 8x + x^2$$
$$y^2 = 8(x + 2)$$

65. $P_1 = (r_1, \theta_1)$ and $P_2 = (r_2, \theta_2)$ or $P_1 = (r_1, \cos \theta_1, r_1 \sin \theta_1)$ and $P_2 = (r_2, \cos \theta_2, r_2, \sin \theta_2)$

$$d = \sqrt{(r_2 \cos \theta_2 - r_1 \cos \theta_1)^2 + (r_2 \sin \theta_2 - r_1 \sin \theta_1)^2}$$
$$= \sqrt{r_2^2 \cos^2 \theta_2 - 2r_1 r_2 \cos \theta_2 \cos \theta_1 + r_1^2 \cos^2 \theta_1 + r_2^2 \sin^2 \theta_2 - 2r_1 r_2 \sin \theta_2 \sin \theta_1 + r_1^2 \sin^2 \theta}$$
$$= \sqrt{r_2^2 (\cos^2 \theta_2 + \sin^2 \theta_2) + r_1^2 (\cos^2 \theta_1 + \sin^2 \theta_1) - 2r_1 r_2 (\cos \theta_2 \cos \theta_1 + \sin \theta_2 \sin \theta_1)}$$
$$= \sqrt{r_1^2 + r_2^2 - 2r_1 r_2 \cos(\theta_2 - \theta_1)}$$

9.2 Polar Equations and Graphs

1. $r = 4$

   This is of the form $r = a$, $a > 0$. Thus, by Table 7, the graph of $r = 4$ is a circle, center at the pole and radius 4.

   If we convert the polar equation to a rectangular equation, then
   $$r = 4$$
   $$r^2 = 16$$
   $$x^2 + y^2 = 16$$
   Circle: Center at the pole and radius 4.

   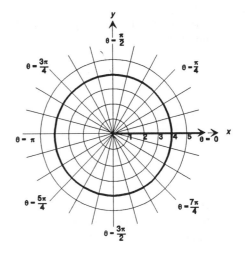

3. $\theta = \dfrac{\pi}{3}$

   This is of the form $\theta = \alpha$. Thus, by Table 7, the graph of $\theta = \dfrac{\pi}{3}$ is a line passing through the pole making an angle of $\dfrac{\pi}{3}$ with the polar axis.

   If we convert the polar equation to a rectangular equation, then
   $$\theta = \frac{\pi}{3}$$
   $$\tan \theta = \tan \frac{\pi}{3} = \sqrt{3}$$
   $$\frac{y}{x} = \sqrt{3}$$
   $$y = \sqrt{3}\, x$$

   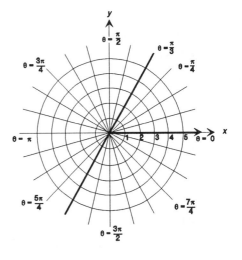

   This is the graph of a line passing through the pole making an angle of $\dfrac{\pi}{3}$ with the polar axis.

5. $r \sin \theta = 4$

    This is of the form $r \sin \theta = b$. Thus, by Table 7, the graph of $r \sin \theta = 4$ is a horizontal line. Since $y = r \sin \theta$, we can write the rectangular equation as
    $$y = 4$$
    We conclude that the graph of $r \sin \theta = 4$ is a horizontal line 4 units above the pole.

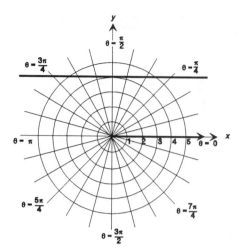

7. $r \cos \theta = -2$

    This is of the form $r \cos \theta = a$. Thus, by Table 7, the graph of $r \cos \theta = -2$ is a vertical line.

    Since $x = r \cos \theta$, we can write the rectangular equation as
    $$x = -2$$
    which is the graph of a vertical line 2 units to the left of the pole.

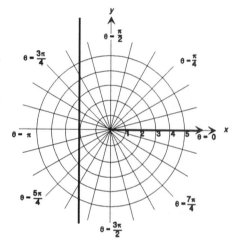

9. $r = 2 \cos \theta$

    This is of the form $r = 2a \cos \theta$, $a > 0$. Thus, by Table 7, the graph of $r = 2 \cos \theta$ is a circle, passing through the pole, tangent to the line $\theta = \dfrac{\pi}{2}$, center on polar axis.

    If we convert to a rectangular equation,
    $$r = 2 \cos \theta$$
    $$r^2 = 2r \cos \theta$$
    $$x^2 + y^2 = 2x$$
    $$x^2 - 2x + y^2 = 0$$
    $$(x - 1)^2 + y^2 = 1$$
    This is the equation of a circle, center at $(1, 0)$ in rectangular coordinates and radius 1.

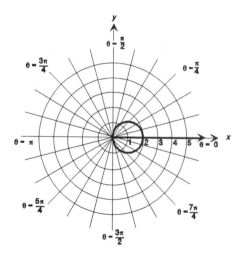

11. $r = -4 \sin \theta$ is the graph of a circle which passes through the pole, is tangent to the polar axis with center on line $\theta = \dfrac{\pi}{2}$. Converting to rectangular coordinates, we have:

$$r = -4 \sin \theta$$
$$r^2 = -4r \sin \theta$$
$$x^2 + y^2 = -4y$$
$$x^2 + y^2 + 4y = 0$$
$$x^2 + (y + 2)^2 = 4$$

This is the equation of a circle, center at $(0, -2)$ in rectangular coordinates or $\left[2, \dfrac{3\pi}{2}\right]$ in polar coordinates, and radius 2.

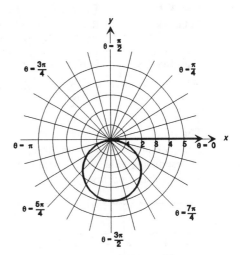

13. We convert the polar equation to a rectangular equation.

$$r \sec \theta = 4$$
$$r \, \dfrac{1}{\cos \theta} = 4$$
$$r = 4 \cos \theta$$
$$r^2 = 4r \cos \theta$$
$$x^2 + y^2 = 4x$$
$$x^2 - 4x + y^2 = 0$$
$$(x - 2)^2 + y^2 = 4$$

This is the equation of a circle, center $(2, 0)$ in rectangular coordinates, $(2, 0)$ in polar coordinates, and radius 2.

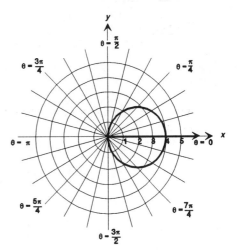

15. We convert to a rectangular equation.

$$r \csc \theta = -2$$
$$r \, \dfrac{1}{\sin \theta} = -2$$
$$r = -2 \sin \theta$$
$$r^2 = -2r \sin \theta$$
$$x^2 + y^2 = -2y$$
$$x^2 + y^2 + 2y = 0$$
$$x^2 + (y + 1)^2 = 1$$

This is the equation of a circle, center $(0, -1)$ in rectangular coordinates, $\left[1, \dfrac{3\pi}{2}\right]$ in polar coordinates, and radius 1.

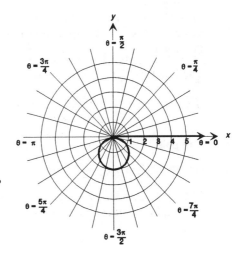

17. $r = 2$; E   19. $r = 2 \cos \theta$; F   21. $r = 1 + \cos \theta$; H   23. $\theta = \dfrac{3\pi}{4}$; D

25. $r = 4$; D   27. $r = 3 \sin \theta$; F   29. $r \cos \theta$; A

Section 9.2 Polar Equations and Graphs

31. $r = 2 + 2 \cos \theta$.   The graph will be a cardiod.   We check for symmetry first.
    Polar axis: Replace θ by $-\theta$.   The result is $r = 2 + 2 \cos(-\theta) = 2 + 2 \cos \theta$
                                   Thus, the graph is symmetric with respect to the polar axis.

The line $\theta = \dfrac{\pi}{2}$: Replace θ by $\pi - \theta$.

$$r = 2 + 2 \cos(\pi - \theta) = 2 + 2(\cos \pi \cos \theta + \sin \pi \sin \theta)$$
$$= 2 + 2(-\cos \theta + 0) = 2 - 2 \cos \theta$$
The test fails.

The pole: Replace r by $-r$. $-r = 2 + 2 \cos \theta$.   The test fails.

Next, we identify points on the graph by assigning values to the angle θ and calculating the corresponding values of r.   Due to the symmetry with respect to the polar axis, we only need to assign values to θ from 0 to π.

θ	0	$\dfrac{\pi}{6}$	$\dfrac{\pi}{3}$	$\dfrac{\pi}{2}$	$\dfrac{2\pi}{3}$	$\dfrac{5\pi}{6}$	π
$r = 2 + 2 \cos \theta$	4	$2 + \sqrt{3} \approx 3.7$	3	2	1	$2 - \sqrt{3} \approx 0.3$	0
(r, θ)	$(4, 0)$	$\left(3.7, \dfrac{\pi}{6}\right)$	$\left(3, \dfrac{\pi}{3}\right)$	$\left(2, \dfrac{\pi}{2}\right)$	$\left(1, \dfrac{2\pi}{3}\right)$	$\left(0.3, \dfrac{5\pi}{6}\right)$	$(0, \pi)$

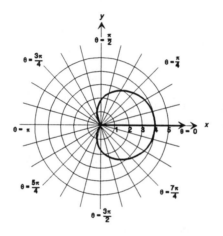

33. $r = 3 - 3 \sin \theta$ The graph will be a cardiod.   We check for symmetry first.
    Polar axis: $r = 3 - 3 \sin(-\theta) = 3 + 3 \sin \theta$.   The test fails.

The line $\theta = \dfrac{\pi}{2}$: $r = 3 - 3 \sin(\pi - \theta) = 3 - 3(\sin \pi \cos \theta - \cos \pi \sin \theta)$

$$= 3 - 3[0 - (-\sin \theta] = 3 - 3 \sin \theta$$

Thus, the graph is symmetric with respect to the line $\theta = \dfrac{\pi}{2}$.

The pole: $-r = 3 - 3 \sin \theta$.   The test fails.

Due to the symmetry with respect to the line $\theta = \dfrac{\pi}{2}$, we only need to assign values to θ from

$\dfrac{-\pi}{2}$ to $\dfrac{\pi}{2}$.

θ	$\dfrac{-\pi}{2}$	$\dfrac{-\pi}{3}$	$\dfrac{-\pi}{6}$	0	$\dfrac{\pi}{6}$	$\dfrac{\pi}{3}$	$\dfrac{\pi}{2}$
$r = 3 - 3\sin\theta$	6	$3 + \dfrac{3\sqrt{3}}{2} \approx 5.6$	$\dfrac{9}{2}$	3	$\dfrac{3}{2}$	$3 - \dfrac{3\sqrt{3}}{2} \approx 0.4$	0
(r, θ)	$\left(6, \dfrac{-\pi}{2}\right)$	$\left(5.6, \dfrac{-\pi}{3}\right)$	$\left(\dfrac{9}{2}, \dfrac{-\pi}{6}\right)$	$(3, 0)$	$\left(\dfrac{3}{2}, \dfrac{\pi}{6}\right)$	$\left(0.4, \dfrac{\pi}{3}\right)$	$\left(0, \dfrac{\pi}{2}\right)$

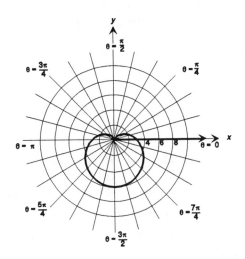

35. $r = 2 + \sin\theta$ The graph will be a limacon without inner loop. We check for symmetry.

Polar axis: $r = 2 + \sin(-\theta) = 2 - \sin\theta$. The test fails.

The line $\theta = \dfrac{\pi}{2}$: $r = 2 + \sin(\pi - \theta) = 2 + (\sin\pi\cos\theta - \cos\pi\sin\theta) = 2 + \sin\theta$

Thus, the graph is symmetric with respect to the line $\theta = \dfrac{\pi}{2}$.

The pole: $-r = 2 + \sin\theta$. The test fails.

Due to the symmetry with respect to the line $\theta = \dfrac{\pi}{2}$, we only need to assign values to θ from

$\dfrac{-\pi}{2}$ to $\dfrac{\pi}{2}$.

θ	$\dfrac{-\pi}{2}$	$\dfrac{-\pi}{3}$	$\dfrac{-\pi}{6}$	0	$\dfrac{\pi}{6}$	$\dfrac{\pi}{3}$	$\dfrac{\pi}{2}$
$r = 2 + \sin\theta$	1	$2 - \dfrac{\sqrt{3}}{2} \approx 1.1$	$\dfrac{3}{2}$	2	$\dfrac{5}{2}$	$2 + \dfrac{\sqrt{3}}{2} \approx 2.9$	3
(r, θ)	$\left(1, \dfrac{-\pi}{2}\right)$	$\left(1.1, \dfrac{-\pi}{3}\right)$	$\left(\dfrac{3}{2}, \dfrac{-\pi}{6}\right)$	$(2, 0)$	$\left(\dfrac{5}{2}, \dfrac{\pi}{6}\right)$	$\left(2.9, \dfrac{\pi}{3}\right)$	$\left(3, \dfrac{\pi}{2}\right)$

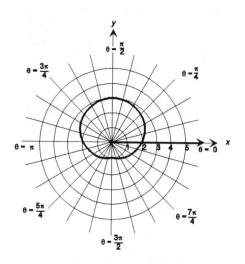

37. $r = 4 - 2 \cos \theta$ The graph will be a limacon without inner loop. We check for symmetry.

Polar axis: $r = 4 - 2 \cos(-\theta) = 4 - 2 \cos \theta$

Thus, the graph is symmetric with respect to the polar axis.

The line $\theta = \dfrac{\pi}{2}$: $r = 4 = -2 \cos(\pi - \theta) = 4 - 2(\cos \pi \cos \theta + \sin \pi \sin \theta)$

$= 4 - 2(-\cos \theta) = 4 + 2 \cos \theta$

The test fails.

The pole: $-r = 4 - 2 \cos \theta$. The test fails.

Due to the symmetry with respect to the polar axis, we only need to assign values to θ from 0 to π.

θ	0	$\dfrac{\pi}{6}$	$\dfrac{\pi}{4}$	$\dfrac{\pi}{3}$	$\dfrac{\pi}{2}$	$\dfrac{2\pi}{3}$
$r = 4 - 2 \cos \theta$	2	$4 - \sqrt{3} \approx 2.3$	$4 - \sqrt{2} \approx 2.6$	3	4	5
(r, θ)	$(2, 0)$	$\left[2.3, \dfrac{\pi}{6}\right]$	$\left[2.6, \dfrac{\pi}{4}\right]$	$\left[3, \dfrac{\pi}{3}\right]$	$\left[4, \dfrac{\pi}{2}\right]$	$\left[5, \dfrac{2\pi}{3}\right]$

θ	$\dfrac{5\pi}{6}$	π
$r = 4 - 2 \cos \theta$	$4 + \sqrt{3} \approx 5.7$	6
(r, θ)	$\left[5.7, \dfrac{5\pi}{6}\right]$	$(6, \pi)$

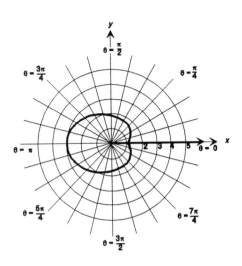

39. $r = 1 + 2 \sin \theta$ The graph will be a limacon with inner loop. We check for symmetry.
Polar axis: $r = 1 + 2 \sin(-\theta) = 1 - 2 \sin \theta$. The test fails.

The line $\theta = \dfrac{\pi}{2}$: $r = 1 + 2 \sin(\pi - \theta) = 1 + 2 \sin \theta$

Thus, the graph is symmetric with respect to the line $\theta = \dfrac{\pi}{2}$.

The pole: $-r = 1 + 2 \sin \theta$. The test fails.

Due to the symmetry with respect to the line $\theta = \dfrac{\pi}{2}$, we only need to assign values to θ from

$\dfrac{-\pi}{2}$ to $\dfrac{\pi}{2}$.

θ	$\dfrac{-\pi}{2}$	$\dfrac{-\pi}{3}$	$\dfrac{-\pi}{6}$	0	$\dfrac{\pi}{6}$	$\dfrac{\pi}{3}$	$\dfrac{\pi}{2}$
$r = 1 + 2 \sin \theta$	-1	$1 - \sqrt{3} \approx -0.7$	0	1	2	$1 + \sqrt{3} \approx 2.7$	3
(r, θ)	$\left(-1, \dfrac{-\pi}{2}\right)$	$\left(-0.7, \dfrac{-\pi}{3}\right)$	$\left(0, \dfrac{-\pi}{6}\right)$	$(1, 0)$	$\left(2, \dfrac{\pi}{6}\right)$	$\left(2.7, \dfrac{\pi}{3}\right)$	$\left(3, \dfrac{\pi}{2}\right)$

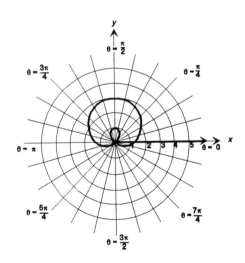

41. $r = 2 - 3 \cos \theta$. The graph will be a limacon with inner loop. We check for symmetry.

Polar axis: $r = 2 - 3 \cos(-\theta) = 2 - 3 \cos \theta$

Thus, the graph is symmetric with respect to the polar axis.

The line $\theta = \dfrac{\pi}{2}$: $r = 2 - 3 \cos(\pi - \theta) = 2 + 3 \cos \theta$. The test fails.

The pole: $-r = 2 - 3 \cos \theta$. The test fails.

Due to the symmetry with respect to the polar axis, we only need to assign values to θ from 0 to π.

θ	0	$\dfrac{\pi}{6}$	$\dfrac{\pi}{3}$	$\dfrac{\pi}{2}$	$\dfrac{2\pi}{3}$	$\dfrac{5\pi}{6}$	π
$r = 2 - 3 \cos \theta$	-1	$2 - \dfrac{3\sqrt{3}}{2} \approx -0.6$	$\dfrac{1}{2}$	2	$\dfrac{7}{2}$	$2 + \dfrac{3\sqrt{3}}{2} \approx 4.6$	5
(r, θ)	$(-1, 0)$	$\left(-0.6, \dfrac{\pi}{6}\right)$	$\left(\dfrac{1}{2}, \dfrac{\pi}{3}\right)$	$\left(2, \dfrac{\pi}{3}\right)$	$\left(\dfrac{7}{2}, \dfrac{2\pi}{3}\right)$	$\left(4.6, \dfrac{5\pi}{6}\right)$	$(5, \pi)$

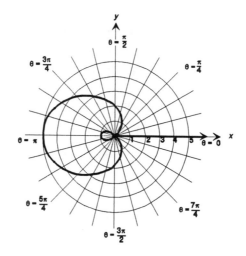

43. $r = 3 \cos 2\theta$ This graph will be a rose with 4 petals. We check for symmetry.

Polar axis: $r = 3 \cos 2(-\theta) = 3 \cos(-2\theta) = 3 \cos 2\theta$

Thus, the graph is symmetric with respect to the polar axis.

The line $\theta = \dfrac{\pi}{2}$: $r = 3 \cos 2(\pi - \theta) = 2 \cos 2\theta$

Thus, the graph is symmetric with respect to the line $\theta = \dfrac{\pi}{2}$. Consequently, since the graph is symmetric with respect to both the polar axis and the line $\theta = \dfrac{\pi}{2}$, it must be symmetric with respect to the pole. Due to the fact that $r = 3 \cos 2\theta$ has period π and is symmetric with respect to the polar axis, the line $\theta = \dfrac{\pi}{2}$, and the pole, we only consider values of θ from 0 to $\dfrac{\pi}{2}$.

θ	0	$\dfrac{\pi}{6}$	$\dfrac{\pi}{4}$	$\dfrac{\pi}{3}$	$\dfrac{\pi}{2}$
$r = 3 \cos 2\theta$	3	$\dfrac{3}{2}$	0	$\dfrac{-3}{2}$	-3
(r, θ)	$(3, 0)$	$\left(\dfrac{3}{2}, \dfrac{\pi}{6}\right)$	$\left(0, \dfrac{\pi}{4}\right)$	$\left(\dfrac{-3}{2}, \dfrac{\pi}{3}\right)$	$\left(-3, \dfrac{\pi}{2}\right)$

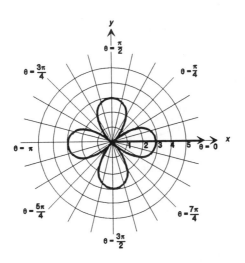

45. $r = 4 \sin 3\theta$ The graph will be a rose with 3 petals. We check for symmetry.

Polar axis: $r = 4 \sin 3(-\theta) = -4 \sin 3\theta$. The test fails.

The line $\theta = \dfrac{\pi}{2}$: $r = -4 \sin 3(\pi - \theta) = 4 \sin 3\theta$

Thus, the graph is symmetric with respect to $\theta = \dfrac{\pi}{2}$.

The pole: $-r = 4 \sin 3\theta$. The test fails.

Due to the symmetry with respect to the line $\theta = \dfrac{\pi}{2}$, we only need to assign values to θ from

$\dfrac{-\pi}{2}$ to $\dfrac{\pi}{2}$.

θ	$\dfrac{-\pi}{2}$	$\dfrac{-\pi}{3}$	$\dfrac{-\pi}{4}$	$\dfrac{-\pi}{6}$	0	$\dfrac{\pi}{6}$
$r = 4 \sin 3\theta$	4	0	$-2\sqrt{2} \approx -2.8$	-4	0	4
(r, θ)	$\left(4, \dfrac{-\pi}{2}\right)$	$\left(0, \dfrac{-\pi}{3}\right)$	$\left(-2.8, \dfrac{-\pi}{4}\right)$	$\left(-4, \dfrac{-\pi}{6}\right)$	$(0, 0)$	$\left(4, \dfrac{\pi}{6}\right)$

θ	$\dfrac{\pi}{4}$	$\dfrac{\pi}{3}$	$\dfrac{\pi}{2}$
$r = 4 \sin 3\theta$	$2\sqrt{2} \approx 2.8$	0	-4
(r, θ)	$\left(2.8, \dfrac{\pi}{4}\right)$	$\left(0, \dfrac{\pi}{3}\right)$	$\left(-4, \dfrac{\pi}{2}\right)$

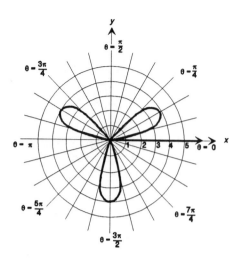

47. $r^2 = 9 \cos 2\theta$ The graph will be a lemniscate. We check for symmetry.
   Polar axis: $r^2 = 9 \cos 2(-\theta) = 9 \cos 2\theta$
   Thus, the graph is symmetric with respect to the polar axis.

   The line $\theta = \dfrac{\pi}{2}$: $r^2 = 9 \cos 2(\pi - \theta) = 9 \cos 2\theta$

   Thus, the graph is symmetric with respect to the line $\theta = \dfrac{\pi}{2}$.

   The pole: $(-r)^2 = 9 \cos 2\theta$
   $r^2 = 9 \cos 2\theta$
   Thus, the graph is symmetric with respect to the pole.

Due to the fact that $r^2 = 9 \cos 2\theta$ has period π and is symmetric with respect to the polar axis, the

line $\theta = \dfrac{\pi}{2}$, and the pole, we only consider values of θ from 0 to $\dfrac{\pi}{2}$.

θ	0	$\dfrac{\pi}{6}$	$\dfrac{\pi}{4}$
$r^2 = 9 \cos 2\theta$	9	$\dfrac{9}{2}$	0
$r = \pm\sqrt{9 \cos 2\theta}$	± 3	$\dfrac{\pm 3\sqrt{2}}{2} \approx \pm 2.1$	0
(r, θ)	$(3, 9)$ $(-3, 9)$	$\left(2.1, \dfrac{\pi}{6}\right)$ $\left(-2.1, \dfrac{\pi}{6}\right)$	$(0, 0)$

Note there are no points on the graph for $\dfrac{\pi}{4} < \theta < \dfrac{3\pi}{4}$ since $\cos 2\pi < 0$ for such values.

Chapter 9 Polar Coordinates; Vectors

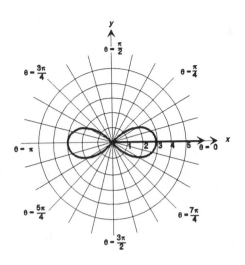

49. $r = 2^{\theta}$ The graph will be a spiral. We check for symmetry.

Polar axis: $r = 2^{-\theta}$. The test fails.

The line $\theta = \dfrac{\pi}{2}$: $r = 2^{\pi - \theta}$. The test fails.

The pole: $-r = 2^{\theta}$. The test fails.

There is no number θ for which $r = 0$. Hence, the graph does not pass through the pole. We observe that r is positive for all θ, that r increases as θ increases, that $r \to 0$ as $\theta \to -\infty$, and that $r \to \infty$ as $\theta \to \infty$.

θ	0	$\dfrac{\pi}{4}$	$\dfrac{\pi}{2}$	π	$\dfrac{3\pi}{2}$	2π
$r = 2^{\theta}$	1	1.7	3.0	8.8	26.2	77.9
(r, θ)	$(1, 0)$	$\left(1.7, \dfrac{\pi}{4}\right)$	$\left(3.0, \dfrac{\pi}{2}\right)$	$(8.8, \pi)$	$\left(26.2, \dfrac{3\pi}{2}\right)$	$(77.9, 2\pi)$

θ	$\dfrac{-\pi}{4}$	$\dfrac{-\pi}{2}$	$-\pi$	$\dfrac{-3\pi}{2}$
$r = 2^{\theta}$	0.6	0.3	0.1	0.0
(r, θ)	$\left(0.6, \dfrac{-\pi}{4}\right)$	$\left(0.3, \dfrac{-\pi}{2}\right)$	$(0.1, -\pi)$	$\left(0.0, \dfrac{-3\pi}{2}\right)$

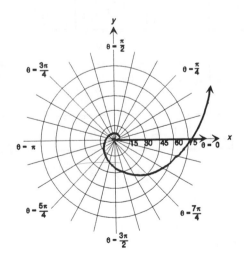

51. $r = 1 - \cos \theta$
The graph will be a cardioid.

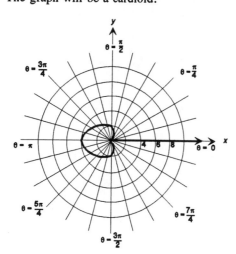

53. $r = 1 - 3 \cos \theta$
The graph will be a limacon with inner loop.

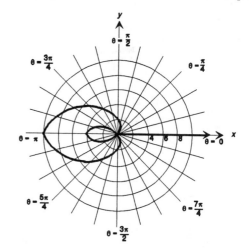

55. $r = \dfrac{2}{1 - \cos \theta}$ We check for symmetry.

Polar axis: $r = \dfrac{2}{1 - \cos(-\theta)} = \dfrac{2}{1 - \cos \theta}$
Thus, the graph is symmetric with respect to the polar axis.

The line $\theta = \dfrac{\pi}{2}$: $r = \dfrac{2}{1 - \cos(\pi - \theta)} = \dfrac{2}{1 + \cos \theta}$. The test fails.

The pole: $-r = \dfrac{2}{1 - \cos \theta}$. The test fails.

Due to symmetry with respect to the polar axis, we only need to assign values to θ from 0 to π.
Since $1 - \cos 0 = 0$, $\theta \neq 0 \pm 2k\pi$, k any integer.

θ	$\dfrac{\pi}{6}$	$\dfrac{\pi}{4}$	$\dfrac{\pi}{3}$	$\dfrac{\pi}{2}$	$\dfrac{2\pi}{3}$
$r = \dfrac{2}{1 - \cos \theta}$	$\dfrac{2}{1 - \dfrac{\sqrt{3}}{2}} \approx 14.9$	$\dfrac{2}{1 - \dfrac{\sqrt{2}}{2}} \approx 6.8$	4	2	$\dfrac{4}{3}$
(r, θ)	$\left[14.9, \dfrac{\pi}{6}\right]$	$\left[6.8, \dfrac{\pi}{4}\right]$	$\left[4, \dfrac{\pi}{3}\right]$	$\left[2, \dfrac{\pi}{2}\right]$	$\left[\dfrac{4}{3}, \dfrac{2\pi}{3}\right]$

θ	$\dfrac{3\pi}{4}$	$\dfrac{5\pi}{6}$	π
$r = \dfrac{2}{1 - \cos \theta}$	$\dfrac{2}{1 + \dfrac{\sqrt{2}}{2}} \approx 1.2$	$\dfrac{2}{1 + \dfrac{\sqrt{3}}{2}} \approx 1.1$	1
(r, θ)	$\left[1.2, \dfrac{3\pi}{4}\right]$	$\left[1.1, \dfrac{5\pi}{6}\right]$	$(1, \pi)$

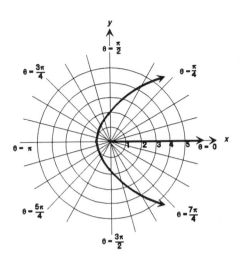

57. $r = \dfrac{1}{3 - 2 \cos \theta}$ We check for symmetry.

Polar axis: $\quad r = \dfrac{1}{3 - 2 \cos(-\theta)} = \dfrac{1}{3 - 2 \cos \theta}$

Thus, the graph is symmetric with respect to the polar axis.

The line $\theta = \dfrac{\pi}{2}$: $\quad r = \dfrac{1}{3 - 2 \cos(\pi - \theta)} = \dfrac{1}{3 + 2 \cos \theta}$. The test fails.

The pole: $\quad -r = \dfrac{1}{3 - 2 \cos \theta}$. The test fails.

Due to symmetry with respect to the polar axis, we only need to assign values to θ from 0 to π.

θ	0	$\dfrac{\pi}{6}$	$\dfrac{\pi}{4}$	$\dfrac{\pi}{3}$	$\dfrac{\pi}{2}$
$r = \dfrac{1}{3 - 2 \cos \theta}$	1	$\dfrac{1}{3 - \sqrt{3}} \approx 0.8$	$\dfrac{1}{3 - \sqrt{2}} \approx 0.6$	$\dfrac{1}{2}$	$\dfrac{1}{3}$
(r, θ)	$(1, 0)$	$\left(0.8, \dfrac{\pi}{6}\right)$	$\left(0.6, \dfrac{\pi}{4}\right)$	$\left(\dfrac{1}{2}, \dfrac{\pi}{3}\right)$	$\left(\dfrac{1}{3}, \dfrac{\pi}{2}\right)$
θ	$\dfrac{2\pi}{3}$	$\dfrac{3\pi}{4}$	$\dfrac{5\pi}{6}$	π	
$r = \dfrac{1}{3 - 2 \cos \theta}$	$\dfrac{1}{4}$	$\dfrac{1}{3 + \sqrt{2}} \approx 0.2$	$\dfrac{1}{3 + \sqrt{3}} \approx 0.2$	$\dfrac{1}{5}$	
(r, θ)	$\left(\dfrac{1}{4}, \dfrac{2\pi}{3}\right)$	$\left(0.2, \dfrac{3\pi}{4}\right)$	$\left(1.1, \dfrac{5\pi}{6}\right)$	$\left(\dfrac{1}{5}, \pi\right)$	

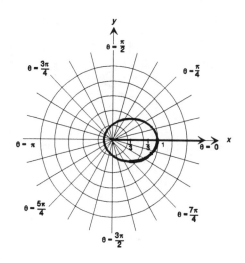

59. $r = \theta, \theta \geq 0$

The tests for symmetry with respect to the pole, polar axis, and the line $\theta = \dfrac{\pi}{2}$ fail. We observe that r increases as θ increases.

θ	0	$\dfrac{\pi}{4}$	$\dfrac{\pi}{2}$	π	$\dfrac{3\pi}{2}$	2π
$r = \theta$	0	$\dfrac{\pi}{4} \approx 0.8$	$\dfrac{\pi}{2} \approx 1.6$	$\pi \approx 3.1$	$\dfrac{3\pi}{2} \approx 4.7$	$2\pi \approx 6.3$
(r, θ)	$(0, 0)$	$\left(0.8, \dfrac{\pi}{4}\right)$	$\left(1.6, \dfrac{\pi}{2}\right)$	$(3.1, \pi)$	$\left(4.7, \dfrac{3\pi}{2}\right)$	$(6.3, 2\pi)$

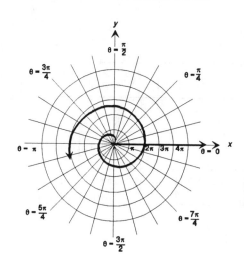

61. $r = \csc \theta - 2, 0 < \theta < \pi$

$r = \dfrac{1}{\sin \theta} - 2$

The graph is symmetric with respect to the line $\theta = \dfrac{\pi}{2}$. We only need to assign values to θ from 0

to $\dfrac{\pi}{2}$. We observe that the points (r, θ) where $\theta = k\pi$, k any integer, are undefined.

θ	$\dfrac{\pi}{6}$	$\dfrac{\pi}{4}$	$\dfrac{\pi}{3}$	$\dfrac{\pi}{2}$
$r = \dfrac{1}{\sin \theta} - 2$	0	$\sqrt{2} - 2 \approx -0.6$	$\dfrac{2\sqrt{3}}{3} - 2 \approx -0.8$	-1
(r, θ)	$\left(0, \dfrac{\pi}{6}\right)$	$\left(-0.6, \dfrac{\pi}{4}\right)$	$\left(-0.8, \dfrac{\pi}{3}\right)$	$\left(-1, \dfrac{\pi}{2}\right)$

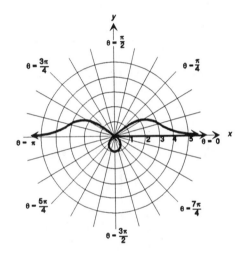

63. $r = \tan \theta$ Check for symmetry.

Polar axis: $r = \tan(-\theta) = -\tan \theta$. The test fails.

The line $\theta = \dfrac{\pi}{2}$: $r = \tan(\pi - \theta) = \dfrac{\tan \pi - \tan \theta}{1 + \tan \pi \tan \theta} = -\tan \theta$. The test fails.

The pole: $-r = \tan \theta$. The test fails.

θ	0	$\dfrac{\pi}{6}$	$\dfrac{\pi}{4}$	$\dfrac{\pi}{3}$	$\dfrac{2\pi}{3}$
$r = \tan\theta$	0	$\dfrac{1}{\sqrt{3}} \approx 0.6$	1	$\sqrt{3} \approx 1.7$	$-\sqrt{3} \approx -1.7$
(r, θ)	$(0, 0)$	$\left[0.6, \dfrac{\pi}{6}\right]$	$\left[1, \dfrac{\pi}{4}\right]$	$\left[1.7, \dfrac{\pi}{3}\right]$	$\left[-1.7, \dfrac{2\pi}{3}\right]$
θ	$\dfrac{3\pi}{4}$	$\dfrac{5\pi}{6}$	π	$\dfrac{7\pi}{6}$	$\dfrac{5\pi}{4}$
$r = \tan\theta$	-1	$\dfrac{-1}{\sqrt{3}} \approx -0.6$	0	$\dfrac{1}{\sqrt{3}} \approx 0.6$	1
(r, θ)	$\left[-1, \dfrac{3\pi}{4}\right]$	$\left[-0.6, \dfrac{5\pi}{6}\right]$	$(0, \pi)$	$\left[0.6, \dfrac{7\pi}{6}\right]$	$\left[1, \dfrac{5\pi}{4}\right]$
θ	$\dfrac{4\pi}{3}$	$\dfrac{5\pi}{3}$	$\dfrac{7\pi}{4}$	$\dfrac{11\pi}{6}$	2π
$r = \tan\theta$	$\sqrt{3} \approx 1.7$	$-\sqrt{3} \approx -1.7$	-1	$\dfrac{1}{-\sqrt{3}} \approx -0.6$	0
(r, θ)	$\left[1.7, \dfrac{4\pi}{3}\right]$	$\left[-1.7, \dfrac{5\pi}{3}\right]$	$\left[-1, \dfrac{7\pi}{4}\right]$	$\left[-0.6, \dfrac{11\pi}{6}\right]$	$(0, 2\pi)$

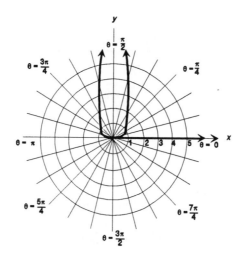

65.   We convert the polar equation to a rectangular equation.

$$r \sin\theta = a$$
$$y = a$$

Thus, the graph of $r \sin\theta = a$ is a horizontal line a units above the pole, if $a > 0$, and a units below the pole, if $a < 0$.

67.   We convert the polar equation to a rectangular equation.

$$r = 2a \sin\theta, \ a > 0$$
$$r^2 = 2ar \sin\theta$$
$$x^2 + y^2 = 2ay$$
$$x^2 + y^2 - 2ay = 0$$
$$x^2 + (y - a)^2 = a^2$$

Circle: radius a, center at rectangular coordinates $(0, a)$.

69. We convert the polar equation to a rectangular equation.

$$r = 2\,a\cos\theta, \; a > 0$$
$$r^2 = 2\,ar\cos\theta$$
$$x^2 + y^2 = 2ax$$
$$x^2 - 2ax + y^2 = 0$$
$$(x - a)^2 + y^2 = a^2$$

Circle: radius a, center at rectangular coordinates $(a, 0)$.

71. (a) $r^2 = \cos\theta$: $r^2 = \cos(\pi - \theta)$ $(-r^2) = \cos(-\theta)$
$\qquad\qquad\qquad\qquad\quad r^2 = -\cos\theta\qquad\qquad\quad r^2 = \cos\theta$
$\qquad\qquad\qquad$ Not equivalent; test fails   New test works

   (b) $r^2 = \sin\theta$: $r^2 = \sin(\pi - \theta)$ $(-r^2) = \sin(-\theta)$
$\qquad\qquad\qquad\qquad\quad r^2 = \sin\theta\qquad\qquad\qquad r^2 = -\sin\theta$
$\qquad\qquad\qquad$ Test works   New test fails

9.3   The Complex Plane; Demoivre's Theorem

1. $r = \sqrt{x^2 + y^2} = \sqrt{(1)^2 + (1)^2} = \sqrt{2}$ and $\tan\theta = \dfrac{y}{x} = 1$

Thus, $\theta = 45°$ and $r = \sqrt{2}$, so the polar form of $z = 1 + i$ is

$z = r(\cos\theta + i\sin\theta) = \sqrt{2}\,(\cos 45° + i\sin 45°)$

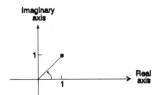

3. $r = \sqrt{x^2 + y^2} = \sqrt{\left(\sqrt{3}\right)^2 + (-1)^2} = \sqrt{4} = 2$ and $\tan\theta = \dfrac{y}{x} = \dfrac{-1}{\sqrt{3}}$

Thus, $\theta = 330°$ and $r = 2$, so the polar form of $z = \sqrt{3} - i$ is
$z = r(\cos\theta + i\sin\theta) = 2(\cos 330° + i\sin 330°)$

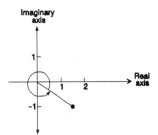

5. $r = \sqrt{x^2 + y^2} = \sqrt{(0)^2 + (-3)^2} = \sqrt{9} = 3$ and $\tan\theta = \dfrac{y}{x} = \dfrac{-3}{0}$
undefined.

Thus, $\theta = 270°$ and $r = 3$, so the polar form of $z = -3i$ is
$z = r(\cos\theta + i\sin\theta) = 3(\cos 270° + i\sin 270°)$

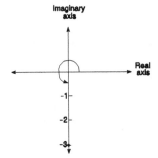

7. $r = \sqrt{x^2 + y^2} = \sqrt{(4)^2 + (-4)^2} = \sqrt{32} = 4\sqrt{2}$,

and $\tan\theta = \dfrac{y}{x} = \dfrac{-4}{4} = -1$

Thus, $\theta = 315°$ and $r = 4\sqrt{2}$, so the polar form of $z = 4 - 4i$ is

$z = r(\cos\theta + i\sin\theta) = 4\sqrt{2}\,(\cos 315° + i\sin 315°)$

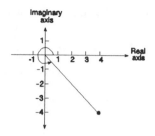

9. $r = \sqrt{x^2 + y^2} = \sqrt{(3)^2 + (-4)^2} = \sqrt{25} = 5$

$\tan\theta = \dfrac{y}{x} = \dfrac{-4}{3} \approx -1.3333$

$\theta = 306.9°$

The polar form of $z = 3 - 4i$ is $z = 5(\cos 306.9° + i\sin 306.9°)$

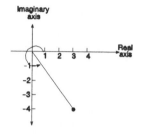

11. $r = \sqrt{x^2 + y^2} = \sqrt{(-2)^2 + (3)^2} = \sqrt{13}$

$\tan\theta = \dfrac{y}{x} = \dfrac{-3}{2} \approx -1.5$

$\theta = 123.7°$

The polar form of $z = -2 + 3i$ is $z = \sqrt{13}\,(\cos 123.7° + i\sin 123.7°)$

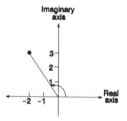

13. $2(\cos 120° + i\sin 120°) = 2\left[-\dfrac{1}{2} + \dfrac{\sqrt{3}}{2}i\right] = -1 + \sqrt{3}\,i$

15. $4\left[\cos\dfrac{7\pi}{4} + i\sin\dfrac{7\pi}{4}\right] = 4\left[\dfrac{\sqrt{2}}{2} + -\dfrac{\sqrt{2}}{2}i\right] = 2\sqrt{2} - 2\sqrt{2}\,i$

17. $3\left[\cos\dfrac{3\pi}{2} + i\sin\dfrac{3\pi}{2}\right] = 3(0 + -i) = -3i$

19. $0.2(\cos 100° + i\sin 100°) = 0.2(-0.1736 + 0.9848i) = -0.035 + 0.197i$

21. $2\left[\cos\dfrac{\pi}{18} + i\sin\dfrac{\pi}{18}\right] = 2(0.9848 + 0.1736i) = 1.97 + 0.347i$

23. $zw = 2 \cdot 4[\cos(40° + 20°) + i\sin(40° + 20°) = 8(\cos 60° + i\sin 60°)$

$\dfrac{z}{w} = \dfrac{2}{4}[\cos(40° - 20°) + i\sin(40° - 20°)]$

$\dfrac{z}{w} = \dfrac{1}{2}(\cos 20° + i\sin 20°)$

25.
$$zw = 3 \cdot 4[\cos(130° + 270°) + i\sin(130° + 270°)]$$
$$zw = 12[\cos(400° - 360°) + i\sin(400° - 360°)]$$
$$zw = 12(\cos 40° + i\sin 40°)$$
$$\frac{z}{w} = \frac{3}{4}[\cos(130° - 270°) + i\sin(130° - 270°)]$$
$$\frac{z}{w} = \frac{3}{4}[\cos(-140°) + i\sin(-140°)]$$
$$\frac{z}{w} = \frac{3}{4}(\cos 220° + i\sin 220°)$$

27.
$$zw = 4\left[\cos\left(\frac{\pi}{8} + \frac{\pi}{10}\right) + i\sin\left(\frac{\pi}{8} + \frac{\pi}{10}\right)\right]$$
$$zw = 4\left[\cos\frac{9\pi}{40} + i\sin\frac{9\pi}{40}\right]$$
$$\frac{z}{w} = \frac{2}{2}\left[\cos\left(\frac{\pi}{8} - \frac{\pi}{10}\right) + i\sin\left(\frac{\pi}{8} - \frac{\pi}{10}\right)\right]$$
$$\frac{z}{w} = \cos\frac{\pi}{40} + i\sin\frac{\pi}{40}$$

29.
$$r = \sqrt{x^2 + y^2} = \sqrt{8} = 2\sqrt{2}$$
$$\tan\theta = \frac{y}{x} = 1, \theta = 45°$$
$$z = 2\sqrt{2}(\cos 45° + i\sin 45°)$$
$$r = \sqrt{x^2 + y^2} = \sqrt{4} = 2$$
$$\tan\theta = \frac{y}{x} = \frac{-1}{\sqrt{3}}, \theta = 330°$$
$$w = 2(\cos 330° + i\sin 330°)$$
$$zw = 2\sqrt{2} \cdot 2[\cos(45° + 330°) + i\sin(45° + 330°)]$$
$$zw = 4\sqrt{2}[\cos(375° - 360°) + i\sin(375° - 360°)]$$
$$zw = 4\sqrt{2}(\cos 15° + i\sin 15°)$$
$$\frac{z}{w} = \frac{2\sqrt{2}}{2}[\cos(45° - 330°) + i\sin(45° - 330°)]$$
$$\frac{z}{w} = \sqrt{2}(\cos 75° + i\sin 75°)$$

31.
$$[4(\cos 40° + i\sin 40°)]^3 = 4^3[\cos(3 \cdot 40°) + i\sin(3 \cdot 40°)] = 64(\cos 120° + i\sin 120°)$$
$$= 64\left(\frac{-1}{2} + \frac{\sqrt{3}}{2}i\right) = -32 + 32\sqrt{3}\,i = 32(-1 + \sqrt{3}\,i)$$

33.
$$\left[2\left(\cos\frac{\pi}{10} + i\sin\frac{\pi}{10}\right)\right]^5 = 2^5\left[\cos\left(5 \cdot \frac{\pi}{10}\right) + i\sin\left(5 \cdot \frac{\pi}{10}\right)\right]$$
$$= 32\left[\cos\frac{\pi}{2} + i\sin\frac{\pi}{2}\right] = 32i$$

35. $\left[\sqrt{3}(\cos 10° + i \sin 10°)\right]^6 = \sqrt{3^6}\left[\cos(6 \cdot 10°) + i \sin(6 \cdot 10°)\right]$

$$= 27(\cos 60° + i \sin 60°) = 27\left[\frac{1}{2} + \frac{\sqrt{3}}{2}i\right] = \frac{27}{2}\left(1 + \sqrt{3}\,i\right)$$

37. $\left[\sqrt{5}\left(\cos\frac{3\pi}{16} + i \sin\frac{3\pi}{16}\right)\right]^4 = \sqrt{5^4}\left[\cos\left(4 \cdot \frac{3\pi}{16}\right) + i \sin\left(4 \cdot \frac{3\pi}{16}\right)\right]$

$$= 25\left(\cos\frac{3\pi}{4} + i \sin\frac{3\pi}{4}\right) = 25\left[\frac{-\sqrt{2}}{2} + \frac{\sqrt{2}}{2}i\right]$$

$$= \frac{25\sqrt{2}}{2}(-1 + i)$$

39. $(1 - i)^5 = \left[\sqrt{2}\left(\cos\frac{7\pi}{4} + i \sin\frac{7\pi}{4}\right)\right]^5 = 4\sqrt{2}\left(\cos\frac{35\pi}{4} + i \sin\frac{35\pi}{4}\right)$

$$= 4\sqrt{2}\left[\frac{-\sqrt{2}}{2} + i\frac{\sqrt{2}}{2}\right] = -4 + 4i = 4(-1 + i)$$

41.
$$r = \sqrt{x^2 + y^2} = \sqrt{3}$$
$$\tan \theta = \frac{y}{x} = \frac{-1}{\sqrt{2}}, \theta = 324.7°$$
$$\sqrt{2} - i = \sqrt{3}(\cos 324.7 + i \sin 324.7)$$
$$\left(\sqrt{2} - i\right)^6 = \left[\sqrt{3}\right]^6\left[\cos(6 \cdot 324.7) + i \sin(6 \cdot 324.7)\right]$$
$$= 27(\cos 1948.2° + i \sin 1948.2°) = -23 + 14.15i$$

43. $1 + i = \sqrt{2}(\cos 45° + i \sin 45°)$

The three complex cube roots of $1 + i = \sqrt{2}(\cos 45° + i \sin 45°)$ are:

$$z_k = \sqrt[3]{\sqrt{2}}\left[\cos\left(\frac{45°}{3} + \frac{360°k}{3}\right) + i \sin\left(\frac{45°}{3} + \frac{360°k}{3}\right)\right].$$

Thus, $z_0 = \sqrt[6]{2}(\cos 15° + i \sin 15°)$

$$z_1 = \sqrt[6]{2}(\cos 135° + i \sin 135°)$$

$$z_2 = \sqrt[6]{2}(\cos 255° + i \sin 255°)$$

45. $4 - 4\sqrt{3}\,i = 8(\cos 300° + i \sin 300°)$

The four complex cube roots of $4 - 4\sqrt{3}\,i = 8(\cos 300° + i \sin 300°)$ are:

$$z_k = \sqrt[4]{8}\left[\cos\left(\frac{300°}{4} + \frac{360°k}{4}\right) + i \sin\left(\frac{300°}{4} + \frac{360°k}{4}\right)\right], k = 0, 1, 2, 3$$

Thus, $z_0 = \sqrt[4]{8}(\cos 75° + i \sin 75°)$

$$z_1 = \sqrt[4]{8}(\cos 165° + i \sin 165°)$$

$$z_2 = \sqrt[4]{8}(\cos 255° + i \sin 255°)$$

$$z_3 = \sqrt[4]{8}(\cos 345° + i \sin 345°)$$

47. $-16i = 16(\cos 270° = i \sin 270°)$

$$z_k = \sqrt[4]{16}\left[\cos\left(\frac{270°}{4} + \frac{360°k}{4}\right) + i \sin\left(\frac{270°}{4} + \frac{360°k}{4}\right)\right], \; k = 0, 1, 2, 3$$

Thus, $z_0 = 2(\cos 67.5° + i \sin 67.5°)$
$z_1 = 2(\cos 157.5° + i \sin 157.5°)$
$z_2 = 2(\cos 247.5° + i \sin 247.5°)$
$z_3 = 2(\cos 337.5° + i \sin 377.5°)$

49. $i = 1(\cos 90° = i \sin 90°), k = 0, 1, 2, 3, 4$

$$z_k = \sqrt[5]{1}\left[\cos\left(\frac{90°}{5} + \frac{360°k}{5}\right) + i \sin\left(\frac{90°}{5} + \frac{360°k}{5}\right)\right]$$

Thus, $z_0 = \cos 18° + i \sin 18°$
$z_1 = \cos 90° + i \sin 90°$
$z_2 = \cos 162° + i \sin 162°$
$z_3 = \cos 234° + i \sin 234°$
$z_4 = \cos 306° + i \sin 306°$

51. $z = 1 + 0i = 1(\cos 0° + i \sin 0°) \quad k = 0, 1, 2, 3$

$$z_k = \sqrt[4]{1}\left[\cos\left(\frac{0°}{4} + \frac{360°k}{4}\right) + i \sin\left(\frac{0°}{4} + \frac{360°k}{4}\right)\right]$$

Thus, $z_0 = \cos 0° + i \sin 0° = 1$
$z_1 = \cos 90° + i \sin 90° = i$
$z_2 = \cos 180° + i \sin 180° = -1$
$z_3 = \cos 270° + i \sin 270° = -i$
$1, i, -1, -i$

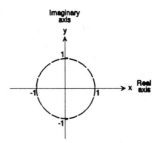

53. Let $w = r(\cos \theta + i \sin \theta)$ be a complex number. If $w \neq 0$, there are n distinct complex nth roots of w, given by the formula:

$$z_k = \sqrt[n]{r}\left[\cos\left(\frac{\theta}{n} + \frac{2k\pi}{n}\right) + i \sin\left(\frac{\theta}{n} + \frac{2k\pi}{n}\right)\right], \text{ where } k = 0, 1, 2, \dots, n - 1$$

$|z_k| = \sqrt[n]{r}$ for all k.

55. Looking at the formula for the number of distinct complex nth roots of the complex number $w = r(\cos \theta + i \sin \theta)$,

$$z_k = \sqrt[n]{r}\left[\cos\left(\frac{\theta}{n} + \frac{2k\pi}{n}\right) + i \sin\left(\frac{\theta}{n} + \frac{2k\pi}{n}\right)\right]$$

where $k = 0, 1, 2, \dots, n - 1$, we see that the z_k are spaced apart by an angle of $\frac{2\pi}{n}$.

Problems 1–8 use the diagram at right.

1. **v + w:**
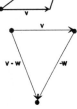

3. **3v:**

5. **v − w:**

7. **3v + u − 2w:**
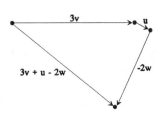

In Problems 9-16, refer to the figure at right.

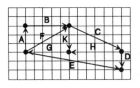

9. We see that **A + B = F**, so **x + B = F** implies **x = A**.

11. From the figure, **F + C + D = E**. Therefore, **C = −F + E − D**.

13. We see that **H + G + E = D**, so that **E = −G − H + D**.

15. Notice that **A + B + K + G = 0**, so **x** must be **0**.

17. If $\|\mathbf{v}\| = 4$, then $\|3\mathbf{v}\| = |3| \|\mathbf{v}\| = (3)(4) = 12$.

For Problems 19-26, use the fact that if $P = (x_1, y_1)$ and $Q = (x_2, y_2)$, then, $\overrightarrow{PQ}$ represents the vector

$\mathbf{v} = (x_2 − x_1)\mathbf{i} + (y_2 − y_1)\mathbf{j}$, *where $\mathbf{i}$ and $\mathbf{j}$ are unit vectors in the positive x-direction and positive y-direction respectively.*

19. $P = (0, 0)$, $Q = (3, 4)$
Then, $\mathbf{v} = (3 − 0)\mathbf{i} + (4 − 0)\mathbf{j} = 3\mathbf{i} + 4\mathbf{j}$

21. $P = (3, 2)$, $Q = (5, 6)$
Then, $\mathbf{v} = (5 − 3)\mathbf{i} + (6 − 2)\mathbf{j} = 2\mathbf{i} + 4\mathbf{j}$

23. $P = (−2, −1)$, $Q = (6, −2)$
Then, $\mathbf{v} = (6 − (−2))\mathbf{i} + (−2 − (−1))\mathbf{j}$
$= 8\mathbf{i} − \mathbf{j}$

25. $P = (1, 0)$, $Q = (0, 1)$
Then, $\mathbf{v} = −\mathbf{i} + \mathbf{j}$

27. If $\mathbf{v} = 3\mathbf{i} - 4\mathbf{j}$,

$\quad$ then $\|\mathbf{v}\| = \sqrt{3^2 + (-4)^2} = 5$

29. For $\mathbf{v} = \mathbf{i} - \mathbf{j}$,

$\quad \|\mathbf{v}\| = \sqrt{1^2 + (-1)^2} = \sqrt{2}$

31. For $\mathbf{v} = -2\mathbf{i} + 3\mathbf{j}$, $\|\mathbf{v}\| = \sqrt{4 + 9} = \sqrt{13}$

33. $\mathbf{v} = 3\mathbf{i} - 5\mathbf{j}$ and $\mathbf{w} = -2\mathbf{i} + 3\mathbf{j}$

$\quad$ So $2\mathbf{v} + 3\mathbf{w} = (6\mathbf{i} - 10\mathbf{j}) + (6\mathbf{i} + 9\mathbf{j})$

$$= -\mathbf{j}$$

35. For $\mathbf{v} = 3\mathbf{i} - 5\mathbf{j}$, $\mathbf{w} = -2\mathbf{i} + 3\mathbf{j}$, we have

$$\|\mathbf{v} - \mathbf{w}\| = \|5\mathbf{i} - 8\mathbf{j}\|$$
$$= \sqrt{25 + 64} = \sqrt{89}$$

37. For $\mathbf{v} = 3\mathbf{i} - 5\mathbf{j}$ and $\mathbf{w} = -2\mathbf{i} + 3\mathbf{j}$,

$$\|\mathbf{v}\| - \|\mathbf{w}\| = \sqrt{9 + 25} - \sqrt{4 + 9}$$
$$= \sqrt{34} - \sqrt{13}$$

39. For $\mathbf{v} = 5\mathbf{i}$, $\|\mathbf{v}\| = \sqrt{25} = 5$, and a unit vector in the same direction as $\mathbf{v}$ is:

$$\mathbf{u} = \frac{1}{\|\mathbf{v}\|}\mathbf{v} = \frac{1}{5}\mathbf{v} = \frac{1}{5}(5\mathbf{i}) = \mathbf{i}$$

41. For $\mathbf{v} = 3\mathbf{i} - 4\mathbf{j}$, $\|\mathbf{v}\| = \sqrt{9 + 16} = 5$, and a unit vector in the same direction as $\mathbf{v}$ is:

$$\mathbf{u} = \frac{1}{\|\mathbf{v}\|}\mathbf{v} = \frac{1}{5}(3\mathbf{i} - 4\mathbf{j}) = \frac{3}{5}\mathbf{i} - \frac{4}{5}\mathbf{j}$$

$\quad$ Notice: $\|\mathbf{u}\| = \sqrt{\frac{9}{25} + \frac{16}{25}} = \sqrt{\frac{25}{25}} = \sqrt{1} = 1$, so $\mathbf{u}$ is indeed a unit vector.

43. For $\mathbf{v} = \mathbf{i} - \mathbf{j}$, $\|\mathbf{v}\| = \sqrt{1 + 1} = \sqrt{2}$, and a unit vector in the same direction as $\mathbf{v}$ is:

$$\mathbf{u} = \frac{1}{\|\mathbf{v}\|}\mathbf{v} = \frac{1}{\sqrt{2}}(\mathbf{i} - \mathbf{j}) = \frac{\sqrt{2}}{2}\mathbf{i} - \frac{\sqrt{2}}{2}\mathbf{j}$$

45. Let $\mathbf{v} = a\mathbf{i} + b\mathbf{j}$. We want $\|\mathbf{v}\| = 4$ and $a = 2b$.

$\quad$ Now $\|\mathbf{v}\| = \sqrt{a^2 + b^2} = \sqrt{(2b)^2 + b^2} = \sqrt{4b^2 + b^2} = \sqrt{5b^2}$
$\quad$ We want

$$\sqrt{5b^2} = 4$$
$$5b^2 = 16$$
$$b^2 = \frac{16}{3}$$
$$b = \sqrt{\frac{16}{5}} = \frac{4}{\sqrt{5}} = \frac{\pm 4\sqrt{5}}{5}$$

$\quad$ Then, $a = 2b = \frac{\pm 8\sqrt{5}}{5}$

$\quad$ and $\mathbf{v} = \frac{8\sqrt{5}}{5}\mathbf{i} + \frac{4\sqrt{5}}{5}\mathbf{j}$ or $\mathbf{v} = \frac{-8\sqrt{5}}{5}\mathbf{i} - \frac{4\sqrt{5}}{5}\mathbf{j}$

47. Given: $\quad \mathbf{v} = 2\mathbf{i} - \mathbf{j}$
$\quad\quad\quad\quad \mathbf{w} = x\mathbf{i} + 3\mathbf{j}$
$\quad\quad\quad\quad \|\mathbf{v} + \mathbf{w}\| = 5$, solve for x.

We have: $\mathbf{v} + \mathbf{w} = (2 + x)\mathbf{i} + 2\mathbf{j}$, so that

$$\|\mathbf{v} + \mathbf{w}\| = \sqrt{(2 + x)^2 + 4} = \sqrt{4 + 4x + x^2 + 4} = \sqrt{x^2 + 4x + 8}$$

Therefore, $\|\mathbf{v} + \mathbf{w}\| = 5$ implies

$$\sqrt{x^2 + 4x + 8} = 5$$
$$\sqrt{x^2 + 4x + 8} = 25$$
$$\sqrt{x^2 + 4x - 17} = 0$$

so by the quadratic formula,

$$x = \frac{-4 \pm \sqrt{16 - 4(-17)}}{2} = \frac{-4 \pm \sqrt{84}}{2} = \frac{-4 \pm 2\sqrt{21}}{2}$$

or $\quad x = -2 + \sqrt{21} \approx 2.58 \quad$ or $\quad x = -2 - \sqrt{21} \approx -6.58$

$$\left\{ -2 + \sqrt{21}, \ -2 - \sqrt{21} \right\}$$

49. Let: $\mathbf{v}_a$ = Velocity of aircraft relative to the air
    $\mathbf{v}_w$ = Wind velocity
    $\mathbf{v}_g$ = Velocity of aircraft relative to ground
    Then $\mathbf{v}_g = \mathbf{v}_a + \mathbf{v}_w$
    So, we need to find expressions for $\mathbf{v}_a$ and $\mathbf{v}_w$
    (a) A unit vector in the easterly direction is $\mathbf{i}$. Therefore, $\mathbf{v}_a = 500\mathbf{i}$.
    (b) A vector in the northwesterly direction is $-\mathbf{i} + \mathbf{j}$, with length

    $\|-\mathbf{i} + \mathbf{j}\| = \sqrt{2}$. Therefore, a *unit* vector in the northwesterly direction is:

    $$\mathbf{u} = \frac{1}{\sqrt{2}}(-\mathbf{i} + \mathbf{j}) = \frac{-\sqrt{2}}{2}\mathbf{i} + \frac{\sqrt{2}}{2}\mathbf{j}$$

    From this, $\mathbf{v}_w = 60\mathbf{u} = -30\sqrt{2}\,\mathbf{i} + 30\sqrt{2}\,\mathbf{j}$

    (c) Now, $\mathbf{v}_g = \mathbf{v}_a + \mathbf{v}_w = (500 - 30\sqrt{2})\mathbf{i} + 30\sqrt{2}\,\mathbf{j}$
    The *speed* of the aircraft is $\|\mathbf{v}_g\|$:

    $$\|\mathbf{v}_g\| = \sqrt{\left(500 - 30\sqrt{2}\right)^2 + \left(30\sqrt{2}\right)^2}$$

    or $\|\mathbf{v}_g\| = \sqrt{500^2 - 2(30)(500)\sqrt{2} + 2(30)^2 + 2(30)^2}$

    $$= \sqrt{250{,}000 - 30{,}000\sqrt{2} + 3{,}600}$$
    $$= \sqrt{253{,}600 - 42{,}426.4} = \sqrt{211{,}173.6}$$
    $$= 460 \text{ kilometers per hour}$$

51. Refer to the figure and the notation used in the solution to Problem 49 (above).

    We will use the basic equation: $\mathbf{v}_g = \mathbf{v}_a + \mathbf{v}_w$ (1)
    If there was no wind, the plane's velocity would be simply $\mathbf{v}_a$. Here we are given information about the velocity of the airplane relative to the ground, $\mathbf{v}_g$, and the speed of the wind, $\mathbf{v}_w$, and we want to find $\mathbf{v}_a$, the plane's velocity in still air. By (1),
    $\mathbf{v}_a = \mathbf{v}_g - \mathbf{v}_w$
    (a) What is $\mathbf{v}_g$? We are told that, relative to the ground, the plane has a northwesterly direction and a speed of 250 mph.

    From the solution to Problem 49, a *unit* vector in the northwesterly direction is

    $$\mathbf{u} = \frac{-\sqrt{2}}{2}\mathbf{i} + \frac{\sqrt{2}}{2}\mathbf{j}$$

Therefore, $\mathbf{v}_g = 250(\mathbf{u}) = -125\sqrt{2}\,\mathbf{i} + 125\sqrt{2}\,\mathbf{j}$

(b) What is $\mathbf{v}_w$? An easterly wind is a wind *from* the east, so $\mathbf{v}_w = -50\mathbf{i}$.

(c) Therefore, $\mathbf{v}_a = \mathbf{v}_g - \mathbf{v}_w = \left(-125\sqrt{2} + 50\right)\mathbf{i} + 125\sqrt{2}\,\mathbf{j}$,
and the speed *would have* been

$$\|\mathbf{v}_a\| = \sqrt{\left(-125\sqrt{2} + 50\right)^2 + \left(125\sqrt{2}\right)^2} = \sqrt{31250 - 12500\sqrt{2} + 2500 + 31250} \approx 217.5$$

or 218 mph

53. Referring to the figure and using the grid,

$\mathbf{F}_1 = -3\mathbf{i}$
$\mathbf{F}_2 = -\mathbf{i} + 4\mathbf{j}$
$\mathbf{F}_3 = 4\mathbf{i} - 2\mathbf{j}$
$\mathbf{F}_4 = -4\mathbf{j}$

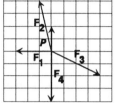

We will add an unknown force, $\mathbf{X} = a\mathbf{i} + b\mathbf{j}$. The object at P will not move
provided the sum of all the forces acting upon it is $\mathbf{0}$; i.e., we want

$\mathbf{F}_1 + \mathbf{F}_2 + \mathbf{F}_3 + \mathbf{F}_3 + \mathbf{X} = \mathbf{0}$

or, $-3\mathbf{i} + (-\mathbf{i} + 4\mathbf{j}) + (4\mathbf{i} - 2\mathbf{j}) + (-4\mathbf{j}) + (a\mathbf{i} + b\mathbf{j}) = \mathbf{0}$

or, $-2\mathbf{j} + (a\mathbf{i} + b\mathbf{j}) = \mathbf{0}$

$a\mathbf{i} + (-2 + b)\mathbf{j} = \mathbf{0}$

This means $a = 0$

$-2 + b = 0, b = 2$

Therefore, $\mathbf{X} = 2\mathbf{j}$

9.5 The Dot Product

1. $\mathbf{v} = \mathbf{i} - \mathbf{j}, \mathbf{w} = \mathbf{i} + \mathbf{j}$

(a) $\mathbf{v} \cdot \mathbf{w} = 1 \cdot 1 + (-1) \cdot 1 = 1 - 1 = 0$

(b) $\cos \theta = \dfrac{\mathbf{v} \cdot \mathbf{w}}{\|\mathbf{v}\|\,\|\mathbf{w}\|} = \dfrac{0}{\|\mathbf{v}\|\,\|\mathbf{w}\|} = 0$

(Note: If $\cos \theta = 0$, then $\theta = 90°$, so $\mathbf{v}$ and $\mathbf{w}$ are perpendicular.)

3. $\mathbf{v} = 2\mathbf{i} + \mathbf{j}, \mathbf{w} = \mathbf{i} + 2\mathbf{j}$

(a) $\mathbf{v} \cdot \mathbf{w} = 2 + 2 = 4$

(b) $\|\mathbf{v}\| = \sqrt{4 + 1} = \sqrt{5}$ $\|\mathbf{w}\| = \sqrt{1 + 4} = \sqrt{5}$ $\cos \theta = \dfrac{\mathbf{v} \cdot \mathbf{w}}{\|\mathbf{v}\|\,\|\mathbf{w}\|} = \dfrac{4}{\sqrt{5}\sqrt{5}} = \dfrac{4}{5}$

5. $\mathbf{v} = \sqrt{3}\mathbf{i} - \mathbf{j}, \mathbf{w} = \mathbf{i} + \mathbf{j}$

(a) $\mathbf{v} \cdot \mathbf{w} = \sqrt{3} - 1$

(b) $\|\mathbf{v}\| = \sqrt{3 + 1} = 2$ $\|\mathbf{w}\| = \sqrt{1 + 1} = \sqrt{2}$ $\cos \theta = \dfrac{\sqrt{3} - 1}{2\sqrt{2}} = \dfrac{\sqrt{6} - \sqrt{2}}{4}$

7. $\mathbf{v} = 3\mathbf{i} + 4\mathbf{j}\ \mathbf{w} = 4\mathbf{i} + 3\mathbf{j}$

(a) $\mathbf{v} \cdot \mathbf{w} = 12 + 12 = 24$

(b) $\|\mathbf{v}\| = \sqrt{9 + 16} = 5$ $\|\mathbf{w}\| = \sqrt{16 + 9} = 5$ $\cos \theta = \dfrac{24}{5 \cdot 5} = \dfrac{24}{25}$

9. $\mathbf{v} = 4\mathbf{i}$, $\mathbf{w} = \mathbf{j}$

   (a) $\mathbf{v} \cdot \mathbf{w} = 4 \cdot 0 + 0 \cdot 1 = 0$ (Because $\mathbf{v} = 4\mathbf{i} = 4\mathbf{i} + 0\mathbf{j}$ and $\mathbf{w} = 0\mathbf{i} + 1\mathbf{j}$)

   (b) $\cos \theta = \dfrac{\mathbf{v} \cdot \mathbf{w}}{\|\mathbf{v}\| \, \|\mathbf{w}\|} = 0$

11. Given: $\mathbf{v} = a\mathbf{i} - \mathbf{j}$, $\mathbf{w} = 2\mathbf{i} + 3\mathbf{j}$

    We want $\theta = \dfrac{\pi}{2}$. Then $\cos \theta = \cos \dfrac{\pi}{2} = 0$.

    But $\cos \theta = \dfrac{\mathbf{v} \cdot \mathbf{w}}{\|\mathbf{v}\| \, \|\mathbf{w}\|} = 0$ only if $\mathbf{v} \cdot \mathbf{w} = 0$.

    Therefore, we want $\mathbf{v} \cdot \mathbf{w} = 2a - 3 = 0$, or $2a = 3$; $a = \dfrac{3}{2}$

13. $\mathbf{v} = 2\mathbf{i} - 3\mathbf{j}$, $\mathbf{w} = \mathbf{i} - \mathbf{j}$

$$\mathbf{v_1} = \text{proj}_{\mathbf{w}}\mathbf{v} = \frac{2 + 3}{\left(\sqrt{2}\right)^2}(\mathbf{i} - \mathbf{j}) = \frac{5}{2}(\mathbf{i} - \mathbf{j})$$

$$\mathbf{v_2} = \mathbf{v} - \mathbf{v_1} = (2\mathbf{i} - 3\mathbf{j}) - \frac{5}{2}(\mathbf{i} - \mathbf{j}) = \frac{-1}{2}\mathbf{i} - \frac{1}{2}\mathbf{j}$$

15. $\mathbf{v} = \mathbf{i} - \mathbf{j}$, $\mathbf{w} = \mathbf{i} + 2\mathbf{j}$

$$\mathbf{v_1} = \text{proj}_{\mathbf{w}}\mathbf{v} = \frac{1 - 2}{\left(\sqrt{5}\right)^2}(\mathbf{i} + 2\mathbf{j}) = \frac{-1}{5}(\mathbf{i} + 2\mathbf{j})$$

$$\mathbf{v_2} = \mathbf{v} - \mathbf{v_1} = (\mathbf{i} - \mathbf{j}) + \frac{1}{5}(\mathbf{i} + 2\mathbf{j}) = \mathbf{i} + \frac{1}{5}\mathbf{i} - \mathbf{j} + \frac{2}{5}\mathbf{j} = \frac{6}{5}\mathbf{i} - \frac{3}{5}\mathbf{j}$$

17. $\mathbf{v} = 3\mathbf{i} + \mathbf{j}$, $\mathbf{w} = -2\mathbf{i} - \mathbf{j}$

$$\mathbf{v_1} = \text{proj}_{\mathbf{w}}\mathbf{v} = \frac{-7}{\left(\sqrt{5}\right)^2}(-2\mathbf{i} - \mathbf{j}) = \frac{7}{5}(2\mathbf{i} + \mathbf{j})$$

$$\mathbf{v_2} = \mathbf{v} - \mathbf{v_1} = (3\mathbf{i} + \mathbf{j}) - \frac{7}{5}(2\mathbf{i} + \mathbf{j}) = 3\mathbf{i} - \frac{14}{5}\mathbf{i} + \mathbf{j} - \frac{7}{5}\mathbf{j} = \frac{1}{5}\mathbf{i} - \frac{2}{5}\mathbf{j}$$

19. Let $\mathbf{v_a}$ = velocity of plane in still air

    $\mathbf{v_w}$ = velocity of the wind

    $\mathbf{v_g}$ = velocity of the plane relative to the ground

    Then, $\mathbf{v_g} = \mathbf{v_a} + \mathbf{v_w}$

    (a) To derive an expression for $\mathbf{v_a}$, we first need to find a unit vector in the direction southwest. The vector $-\mathbf{i} - \mathbf{j}$ points in the desired direction, so a unit vector in a southwesterly direction is:

$$\mathbf{u} = \frac{1}{\|-\mathbf{i} - \mathbf{j}\|}(-\mathbf{i} - \mathbf{j}) = \frac{\sqrt{2}}{2}\mathbf{i} - \frac{\sqrt{2}}{2}\mathbf{j}$$

    Hence, $\mathbf{v_a} = 550\left(\dfrac{-\sqrt{2}}{2}\mathbf{i} - \dfrac{\sqrt{2}}{2}\mathbf{j}\right)$ or $\mathbf{v_a} = -275\sqrt{2}\mathbf{i} - 275\sqrt{2}\mathbf{j}$

    (b) The wind is *from* the west, so $\mathbf{v_w} = 80\mathbf{i}$

    (c) We know $\mathbf{v_g} = \mathbf{v_a} + \mathbf{v_w}$, so $\mathbf{v_g} = -275\sqrt{2}\mathbf{i} - 275\sqrt{2}\mathbf{j} + 80\mathbf{i}$, or $\mathbf{v_g} = (80 - 275\sqrt{2})\mathbf{i} - 275\sqrt{2}\mathbf{j}$. Note that both components of $\mathbf{v_g}$ are negative, so $\mathbf{v_g}$ points into the third quadrant, somewhere between W and S.

    (d) The speed of the jet, relative to the earth, is

$$\|\mathbf{v_g}\| = \sqrt{\left(80 - 275\sqrt{2}\right)^2 + \left(275\sqrt{2}\right)^2} = \sqrt{6400 - 44000\sqrt{2} + 151250 + 151250}$$

$$= \sqrt{246674.6} \approx 496.7 \text{ miles per hour}$$

(e) To find the direction of v_g, pick a simple vector in a known direction such as $-i$ which points due west, and then determine the angle between v_g and $-i$.

We use the formula $\cos \theta = \dfrac{v_g \cdot (-i)}{\|v_g\| \; \| i \|}$

Now $v_g = (80 - 275\sqrt{2})i - 275\sqrt{2}j$,

so $v_g \cdot (-i) = (80 - 275\sqrt{2})(-1) + (-275\sqrt{2})(0) = 275\sqrt{2} - 80 \approx 308.91$

In part (d), we found $\|v_g\| \approx 496.7$

Finally, $\|-i\| = 1$.

Therefore, $\cos \theta \approx \dfrac{308.91}{496.7} = .6219$

Then $\theta - \cos^{-1}(.6219) \approx 51.5°$

The angle between v_g and due west is $51.5°$; i.e., the plane's direction is $51.5°$ south of west.

21. For convenience, let the positive x-axis point downstream so that the velocity of the current, v_c, is:
$$v_c = 3i$$
Let v_w = velocity of the boat in the water and v_g = velocity of the boat relative to the land.

Then, $v_g = v_w + v_c$

Here is what we are given.
(1) $v_c = 3i$
(2) The *speed* of the boat in water is given:
$$\|v_w\| = 20,$$
but the *direction* of v_w is what we want to find.
(3) We know the *direction* of the boat relative to the land, j (directly *across* the stream), but we do not know $\|v_g\|$.

To get started, let $v_w = ai + bj$

By (2), $\|v_w\| = \sqrt{a^2 + b^2} = 20$, so that $a^2 + b^2 = 400$.

We know the direction of v_g, so we let $v_g = kj$

Since $v_g = v_w + v_c$, we have: $\quad kj = ai + bj + 3i$
$$kj = (a + 3)i + bj$$

So, $\quad a + 3 = 0$ and $b = k$
$$a = -3$$

Now, $\qquad\quad a^2 + b^2 = 400$
$$9 + b^2 = 400$$
$$b^2 = 391$$
$$b \approx 19.77$$

and $\qquad\qquad k = b \approx 19.77$

We now have $v_w = -3i + 19.77j$ and $v_g = 19.77j$

Since the i component of v_w is negative, the boat is headed slightly *upstream*.

Let's find the angle between v_w and j:
$$\cos \theta = \frac{v_w \cdot j}{\|v_w\| \; \|j\|} = \frac{19.77}{20}, \text{ since by (2), } \|v_w\| = 20 \approx .9885$$

So, $\theta \approx \cos^{-1}(.9885) \approx 8.7°$.

Thus, the heading of the boat needs to be $8.7°$ upstream from the line directly across the stream.

Finally, the speed of the boat, relative to the land is $\|\mathbf{v}_g\| = \|19.77\mathbf{j}\| = 19.77$, and the time needed to cross the stream is given by:

$$\text{time} = \frac{\text{distance}}{\text{rate}} = \frac{0.5 \text{ kilometer}}{19.77 \text{ kilometer/hr}} \approx .0253 \text{ hr} \approx 1.5 \text{ min.}$$

23. $\mathbf{v}_m = 2(\cos\theta\mathbf{i} + \sin\theta\mathbf{j})$
$\mathbf{v}_a = k\mathbf{j}$
$\mathbf{v}_R = \mathbf{i}$
$\mathbf{v}_m + \mathbf{v}_R = \mathbf{v}_a$
$k^2 + 1 = 4$
$k = \sqrt{3}$
$2\cos\theta\mathbf{i} + 2\sin\theta\mathbf{j} + \mathbf{i} = k\mathbf{j}$
$2\cos\theta + 1 = 0, \quad 2\sin\theta = \sqrt{3}$

$$\cos\theta = \frac{-1}{2} \quad \sin\theta = \frac{\sqrt{3}}{2}$$

$$\theta = 120°$$

The swimmer should head at an angle of 60° to the shore.

$$\text{time} = \frac{\text{distance}}{\text{velocity}} = \frac{\frac{1}{2} \text{ km}}{\sqrt{3} \text{ km/hour}} = .289 \text{ hours} = 17.32 \text{ minutes}$$

25. $\mathbf{F} = 3\dfrac{(2\mathbf{i} + \mathbf{j})}{\sqrt{5}} \qquad \overrightarrow{AB} = 2\mathbf{j}$

$$\mathbf{W} = \mathbf{F} \cdot \overrightarrow{AB} = \left(\frac{6\mathbf{i}}{\sqrt{5}} + \frac{3}{\sqrt{5}}\mathbf{j}\right) \cdot 2\mathbf{j} = \frac{6}{\sqrt{5}}\mathbf{j} = \sqrt{\left(\frac{6}{\sqrt{5}}\right)^2} = \frac{6}{\sqrt{5}} \approx 2.68 \text{ ft-lb}$$

27. $\mathbf{F} = 20\cos(30\mathbf{i} - \sin 30\mathbf{j}) \qquad \overrightarrow{AB} = 100\mathbf{i}$

$$\mathbf{W} = \mathbf{F} \cdot \overrightarrow{AB} = 2000 \cdot \frac{\sqrt{3}}{2} = 1732 \text{ ft.-lb.}$$

29. Let $\mathbf{u} = a_1\mathbf{i} + b_1\mathbf{j}$
$\mathbf{v} = a_2\mathbf{i} + b_2\mathbf{j}$
$\mathbf{w} = a_3\mathbf{i} + b_3\mathbf{j}$
Then, $\mathbf{u} \cdot (\mathbf{v} + \mathbf{w}) = (a_1\mathbf{i} + b_1\mathbf{j}) \cdot [a_2\mathbf{i} + b_2\mathbf{j} + a_3\mathbf{i} + b_3\mathbf{j}] = (a_1\mathbf{i} + b_1\mathbf{j}) \cdot [(a_2 + a_3)\mathbf{i} + (b_2 + b_3)\mathbf{j}]$
$= a_1(a_2 + a_3) + b_1(b_2 + b_3)$
On the other hand, $\mathbf{u} \cdot \mathbf{v} + \mathbf{u} \cdot \mathbf{w} = (a_1\mathbf{i} + b_1\mathbf{j}) \cdot (a_2\mathbf{i} + b_2\mathbf{j}) + (a_1\mathbf{i} + b_1\mathbf{j} \cdot (a_3\mathbf{i} + b_3\mathbf{j})$
$= a_1a_2 + b_1b_2 + a_1a_3 + b_1b3 = a_1a_2 + a_1a_3 + b_1b_2 + b_1b_3$
$= a_1(a_2 + a_3) + b_1(b_2 + b_3)$
Thus, $\mathbf{u} \cdot (\mathbf{v} + \mathbf{w}) = \mathbf{u} \cdot \mathbf{v} + \mathbf{u} \cdot \mathbf{w}$

31. Let $\mathbf{v} = a\mathbf{i} + b\mathbf{j}$
Since $\mathbf{v}$ is a unit vector, we know:

$$\|\mathbf{v}\| = \sqrt{a^2 + b^2} = 1 \text{ or } a^2 + b^2 = 1$$

If α is the angle between $\mathbf{v}$ and $\mathbf{i}$, then $\cos\alpha = \dfrac{\mathbf{v} \cdot \mathbf{i}}{\|\mathbf{v}\| \|\mathbf{i}\|}$ or $\cos\alpha = \dfrac{(a\mathbf{i} + b\mathbf{j}) \cdot (\mathbf{i})}{1 \cdot 1}$

$$\cos\alpha = a$$

Then, since $a^2 + b^2 = 1$, we have: $\qquad \cos^2 \alpha + b^2 = 1$
$$b^2 = 1 - \cos^2 \alpha$$
$$b^2 = \sin^2 \alpha$$
$$b = \sin \alpha$$

Hence, $\mathbf{v} = \cos \alpha \mathbf{i} + \sin \alpha \mathbf{j}$

33. Let $\mathbf{v} = a\mathbf{i} + b\mathbf{j}$
$$\text{proj}_i \mathbf{v} = \frac{\mathbf{v} \cdot \mathbf{i}}{\|\mathbf{i}\|^2}\mathbf{i} = \mathbf{v}(\mathbf{v} \cdot \mathbf{i})\mathbf{i}$$
$$\mathbf{v} \cdot \mathbf{i} = a, \quad \mathbf{v} \cdot \mathbf{j} = b, \text{ so } \mathbf{v} = (\mathbf{v} \cdot \mathbf{i}) + (\mathbf{v} \cdot \mathbf{j})\mathbf{j}$$

35. $(\mathbf{v} - \alpha\mathbf{w})\mathbf{w} = \mathbf{v} \cdot \mathbf{w} - \alpha\mathbf{w} \cdot \mathbf{w} = \mathbf{v} \cdot \mathbf{w} - \alpha\|\mathbf{w}\|^2 = \mathbf{v} \cdot \mathbf{w} - \dfrac{\mathbf{v} \cdot \mathbf{w}}{\|\mathbf{w}\|^2}\|\mathbf{w}\|^2 = 0$

37. If $\mathbf{F}$ is orthogonal to $\overrightarrow{\mathbf{AB}}$, then because
$$\text{work} = \mathbf{F} \cdot \overrightarrow{\mathbf{AB}} = (\mathbf{i} + \mathbf{j})(\mathbf{i} - \mathbf{j}) = 1 - 1 = 0$$

⑨ Chapter Review

1. $x = r \cos \theta \qquad\qquad y = r \sin \theta$

$x = 3 \cos \dfrac{\pi}{6} \qquad\quad\; y = 3 \sin \dfrac{\pi}{6}$

$x = \dfrac{3\sqrt{3}}{2} \qquad\qquad\; y = \dfrac{3}{2}$

$\left[\dfrac{3\sqrt{3}}{2}, \dfrac{3}{2}\right]$

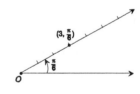

3. $x = r \cos \theta \qquad\qquad y = r \sin \theta$

$x = -2 \cos \dfrac{4\pi}{3} \qquad y = -2 \sin \dfrac{4\pi}{3}$

$x = 1 \qquad\qquad\qquad y = \sqrt{3}$

$\left(1, \sqrt{3}\right)$

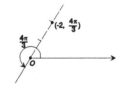

5. $x = r \cos \theta \qquad\qquad y = r \sin \theta$

$x = -3 \cos \dfrac{-\pi}{2} \qquad y = -3 \sin \dfrac{-\pi}{2}$

$x = 0 \qquad\qquad\qquad y = 3$

$(0, 3)$

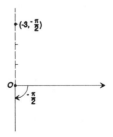

7. Since the point $(-3, 3)$ lies in quadrant II, then $r = -\sqrt{x^2 + y^2} = -\sqrt{9 + 9} = -\sqrt{18} = -3\sqrt{2}$ and $\theta = \tan^{-1} \dfrac{y}{x} = \tan^{-1}(-1) = \dfrac{-\pi}{4}$. Hence, two pairs of polar coordinates are $\left[-3\sqrt{2}, \dfrac{-\pi}{4} \right]$ and $\left[3\sqrt{2}, \dfrac{3\pi}{4} \right]$.

9. Since the point $(0, -2)$ lies on the negative y-axis, then $r = |y| = |-2| = 2$ and $\theta = \dfrac{-\pi}{2}$. Hence, two pairs of polar coordinates are $\left[2, \dfrac{-\pi}{2} \right]$ and $\left[-2, \dfrac{\pi}{2} \right]$.

11. Since the point $(3, 4)$ lies in quadrant I, then $r = \sqrt{9 + 16} = \sqrt{25} = 5$ and $\theta = \tan^{-1} \dfrac{y}{x} = \tan^{-1} \dfrac{4}{3} \approx 0.93$. Hence, two pairs of polar coordinates are $(5, 0.93)$ and $(-5, 4.07)$.

13.
$$3x^2 + 3y^2 = 6y$$
$$3(x^2 + y^2) = 6y$$
$$3r^2 = 6(r \sin \theta)$$
$$3r^2 - 6r \sin \theta = 0$$

15.
$$2x^2 - y^2 = \dfrac{y}{x}$$
$$2x^2 - 3y^2 + 2y^2 = \dfrac{y}{x}$$
$$2(x^2 + y)^2 - 3y^2 = \tan \theta$$
$$2r^2 - 3(r \sin \theta)^2 = \tan \theta$$
$$2r^2 - 3r^2 \sin^2 \theta - \tan \theta = 0$$
$$r^2(2 - 3 \sin^2 \theta) - \tan \theta = 0$$

17.
$$x(x^2 + y^2) = 4$$
$$r \cos \theta (r^2) = 4$$
$$r^3 \cos \theta = 4$$

19.
$$r(r) = (2 \sin \theta)r$$
$$r^2 = 2r \sin \theta$$
$$x^2 + y^2 = 2y$$
$$x^2 + y^2 - 2y = 0$$

21.
$$r = 5$$
$$r^2 = 25$$
$$x^2 + y^2 = 25$$

23.
$$r \cos \theta + 3r \sin \theta = 6$$
$$x + 3y = 6$$

25. $r = 4 \cos \theta$

We test for symmetry:

With respect to the Pole: Replace r by $-r$. The test fails, so the graph may not be symmetric with respect to the pole.

With respect to the Polar Axis: Replace θ by $-\theta$. The result is $r = 4 \cos(-\theta) = 4 \cos \theta$. Thus, the graph is symmetric with respect to the polar axis.

With respect to the Line $\theta = \dfrac{\pi}{2}$: Replace θ by $\pi - \theta$. The result is $r = 4 \cos(\pi - \theta) = 4(\cos \pi \cos \theta + \sin \pi \sin \theta) = -4 \cos \theta$. The test fails.

Due to the periodicity of $\cos \theta$, we only consider values of θ between 0 and 2π.

θ	0	$\dfrac{\pi}{6}$	$\dfrac{\pi}{3}$	$\dfrac{\pi}{2}$
$r = 4\cos\theta$	4	$4\left[\dfrac{\sqrt{3}}{2}\right] = 2\sqrt{3}$	$4\left[\dfrac{1}{2}\right] = 2$	$4(0) = 0$
(r, θ)	$(4, 0)$	$\left[2\sqrt{3}, \dfrac{\pi}{6}\right]$	$\left[2, \dfrac{\pi}{3}\right]$	$\left[0, \dfrac{\pi}{2}\right]$

θ	$\dfrac{2\pi}{3}$	$\dfrac{5\pi}{6}$	π
$r = 4\cos\theta$	$4\left[\dfrac{-1}{2}\right] = -2$	$4\left[\dfrac{-\sqrt{3}}{2}\right] = -2\sqrt{3}$	$4(-1) = -4$
(r, θ)	$\left[-2, \dfrac{2\pi}{3}\right]$	$\left[-2\sqrt{3}, \dfrac{5\pi}{6}\right]$	$(-4, \pi)$

The remaining points on the graph can be found by using symmetry with respect to the polar axis.

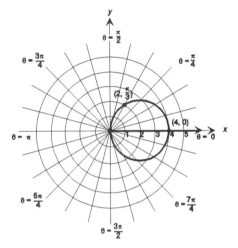

27. $r = 3 - 3\sin\theta$ This is a cardiod.
Test for symmetry:

With respect to the Pole: Test fails.

With respect to the Line $\theta = \dfrac{\pi}{2}$: $r = 3 - 3\sin(\pi - \theta) = 3 - 3\sin\theta$

The graph is symmetric with respect to the line $\theta = \dfrac{\pi}{2}$.

θ	$\dfrac{-\pi}{2}$	$\dfrac{-\pi}{3}$	$\dfrac{-\pi}{6}$	0	$\dfrac{\pi}{6}$	$\dfrac{\pi}{3}$	$\dfrac{\pi}{2}$
$r = 3 - 3\sin\theta$	6	$3 + \dfrac{3\sqrt{3}}{2} \approx 5.6$	$\dfrac{9}{2}$	3	$\dfrac{3}{2}$	$3 - \dfrac{3\sqrt{3}}{2} \approx 0.4$	0
(r, θ)	$\left[6, \dfrac{-\pi}{2}\right]$	$\left[5.6, \dfrac{-\pi}{3}\right]$	$\left[\dfrac{9}{2}, \dfrac{-\pi}{6}\right]$	$(3, 0)$	$\left[\dfrac{3}{2}, \dfrac{\pi}{6}\right]$	$\left[0.4, \dfrac{\pi}{3}\right]$	$\left[0, \dfrac{\pi}{2}\right]$

Due to the periodicity of $\sin\theta$, we only consider values of θ between 0 and 2π.

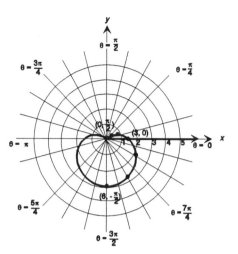

29. $r = 4 - \cos \theta$ This is a limacon without inner loop.

Test for symmetry:

    With respect to the Pole: Test fails.

    With respect to the Polar Axis: $r = 4 - \cos(-\theta) = 4 - \cos \theta$. The graph is symmetric with respect to the polar axis.

    With respect to the Line $\theta = \dfrac{\pi}{2}$: The result is $r = 4 - \cos(\pi - \theta) = 4 + \cos \theta$. Test fails.

θ	0	$\dfrac{\pi}{6}$	$\dfrac{\pi}{3}$	$\dfrac{\pi}{2}$
$r = 4 - \cos \theta$	3	$4 - \dfrac{\sqrt{3}}{2} = \dfrac{8 - \sqrt{3}}{2}$	$4 - \dfrac{1}{2} = \dfrac{7}{2}$	$4 - 0 = 4$
(r, θ)	$(3, 0)$	$\left(\dfrac{8 - \sqrt{3}}{2}, \dfrac{\pi}{6} \right)$	$\left(\dfrac{7}{2}, \dfrac{\pi}{3} \right)$	$\left(4, \dfrac{\pi}{2} \right)$

θ	$\dfrac{2\pi}{3}$	$\dfrac{5\pi}{6}$	π
$r = 4 - \cos \theta$	$4 + \dfrac{1}{2} = \dfrac{9}{2}$	$4 + \dfrac{\sqrt{3}}{2} = \dfrac{8 + \sqrt{3}}{2}$	$4 - (-1) = -5$
(r, θ)	$\left(\dfrac{9}{2}, \dfrac{2\pi}{3} \right)$	$\left(\dfrac{8 + \sqrt{3}}{2}, \dfrac{5\pi}{6} \right)$	$(5, \pi)$

The remaining points on the graph can be found by using symmetry with respect to the polar axis.

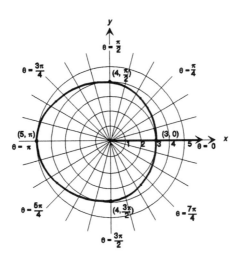

31. $r = \sqrt{x^2 + y^2} = \sqrt{2}$, $\tan \theta = \dfrac{-1}{-1} = 1$

$\theta = 225°$

Thus, $-1 - i = \sqrt{2}\,(\cos 225° + i \sin 225°)$

33. $r = \sqrt{x^2 + y^2} = \sqrt{(4)^2 + (-3)^2} = 5$

$\tan \theta = \dfrac{-3}{4}$, $\theta = 323.1°$

Thus,
$$4 - 3i = 5(\cos 323.1° + i \sin 323.1°)$$

35. $2(\cos 150° + i \sin 150°) = 2\left[\dfrac{-\sqrt{3}}{2} + \dfrac{1}{2}i\right] = -\sqrt{3} + i$

37. $3\left[\cos \dfrac{2\pi}{3} + i \sin \dfrac{2\pi}{3}\right] = 3\left[\dfrac{-1}{2} + \dfrac{\sqrt{3}}{2}i\right] = \dfrac{-3}{2} + \dfrac{3\sqrt{3}}{2}i$

39. $0.1(\cos 350° + i \sin 350°) = 0.1(0.9848 + -0.1736i) = 0.098 - 0.017i$

41. $zw = \cos(80° + 50°) + i \sin(80° + 50°)$
$zw = \cos 130° + i \sin 130°$

$\dfrac{z}{w} = \cos(80° - 50°) + i \sin(80° - 50°)$

$\dfrac{z}{w} = \cos 30° + i \sin 30°$

43. $zw = 6\left[\cos\left[\dfrac{9\pi}{5} + \dfrac{\pi}{5}\right] + i \sin\left[\dfrac{9\pi}{5} + \dfrac{\pi}{5}\right]\right] = 6(\cos 2\pi + i \sin 2\pi) = 6(\cos 0 + i \sin 0)$

$\dfrac{z}{w} = \dfrac{3}{2}\left[\cos \dfrac{8\pi}{5} + i \sin \dfrac{8\pi}{5}\right]$

45. $zw = \cos(10° + 355°) + i \sin(10° + 355°)$
$zw = 5(\cos 5° + i \sin 5°)$

$\dfrac{z}{w} = 5[\cos(10° - 355°) + i \sin(10° - 355°)]$

$\dfrac{z}{w} = 5(\cos(-345°) + i \sin(-345°)) = 5(\cos 15° + i \sin 15°)$

47. $[3(\cos 20° + i \sin 20°)]^3 = 3^3[3 \cdot 20°) + i \sin(3 \cdot 20°)] = 27(\cos 60° + i \sin 60°)$

$$= 27\left[\frac{1}{2} + i\frac{\sqrt{3}}{2}\right] = \frac{27}{2}\left(1 + \sqrt{3}\,i\right)$$

49. $\left[\sqrt{2}\left[\cos \frac{5\pi}{8} + i \sin \frac{5\pi}{8}\right]\right]^4 = \sqrt{2^4}\left[\cos\left[4 \cdot \frac{5\pi}{8}\right] + i \sin\left[4 \cdot \frac{5\pi}{8}\right]\right]$

$$= 4\left[\cos \frac{5\pi}{2} + i \sin \frac{5\pi}{2}\right] = 4\left[\cos\frac{\pi}{2} + i \sin\frac{\pi}{2}\right] = 4(0 + i)$$
$$= 4i$$

51. $\left(1 - \sqrt{3}\,i\right) = 2(\cos 300° + i \sin 300°)$

$\left(1 - \sqrt{3}\,i\right)^6 = 2^6[\cos(6 \cdot 300°) + i \sin(6 \cdot 300°)]$
$\qquad = 64(\cos 1800° + i \sin 1800°) = 64(1 + 0i) = 64$

53. $\quad (3 + 4i) = 5(\cos 53.1° + i \sin 53.1°)$
$(3 + 4i)^4 = 5^4[\cos(4 \cdot 53.1°) + i \sin(4 \cdot 53.1°)]] = 625(\cos 212.4° + i \sin 212.4°)$
$\qquad = -527.1 - 335.8i$

55. $z_k = \sqrt[3]{27} = 3\left[\cos\left[\frac{0°}{4} + \frac{360°k}{4}\right] + i \sin\left[\frac{0°}{4} + \frac{360°k}{4}\right]\right], \ k = 0, 1, 2$
Thus, $z_0 = 3(\cos 0° + i \sin 0°) = 3$
$\qquad z_1 = 3(\cos 120° + i \sin 120°)$
$\qquad z_2 = 3(\cos 240° + i \sin 240°)$

57. $P = (1, -2)$ and $Q = (3, -6)$, so if $\mathbf{v}$ is represented by the directed line segment $\overrightarrow{PQ}$, then
$$\mathbf{v} = (3 - 1)\mathbf{i} + (-6 - (-2))\mathbf{j} = 2\mathbf{i} - 4\mathbf{j} \text{ and } \|\mathbf{v}\| = \sqrt{2^2 + (-4)^2} = \sqrt{20} = 2\sqrt{5}$$

59. For $P = (0, -2)$, $Q = (-1, 1)$, we have:
$$\mathbf{v} = (-1 - 0)\mathbf{i} + (1 - (-2))\mathbf{j} = -\mathbf{i} + 3\mathbf{j} \text{ and } \|\mathbf{v}\| = \sqrt{(-1)^2 + 3^2} = \sqrt{10}$$

61. $4\mathbf{v} - 3\mathbf{w} = 4(-2\mathbf{i} + \mathbf{j}) - 3(4\mathbf{i} - 3\mathbf{j}) = -8\mathbf{i} + 4\mathbf{j} - 12\mathbf{i} + 9\mathbf{j} = -20\mathbf{i} + 13\mathbf{j}$

63. $\|\mathbf{v}\| = \|-2\mathbf{i} + \mathbf{j}\| = \sqrt{(-2)^2 + 1^2} = \sqrt{5}$

65. $\|\mathbf{v}\| + \|\mathbf{w}\| = \|-2\mathbf{i} + \mathbf{j}\| = \|4\mathbf{i} - 3\mathbf{j}\| = \sqrt{4 + 1} + \sqrt{16 + 9} = \sqrt{5} + 5 \approx 7.24$

67. From Problem 63, $\|\mathbf{v}\| = \sqrt{5}$. Then a unit vector having the same direction as $\mathbf{v}$ is:
$$\mathbf{u} = \frac{1}{\|\mathbf{v}\|}\mathbf{v} = \frac{1}{\sqrt{5}}(-2\mathbf{i} + \mathbf{j}) = \frac{\sqrt{5}}{5}(-2\mathbf{i} + \mathbf{j}) \text{ or } \mathbf{u} = \frac{-2\sqrt{5}}{5}\mathbf{i} + \frac{\sqrt{5}}{5}\mathbf{j}$$

Note: $\|\mathbf{u}\| = \sqrt{\left[\frac{-2\sqrt{5}}{5}\right]^2 + \left[\frac{\sqrt{5}}{5}\right]^2} = \sqrt{\frac{4(5)}{25} + \frac{5}{25}} = \sqrt{\frac{25}{25}} = 1$

69. $\mathbf{v} = -2\mathbf{i} + \mathbf{j}$, $\mathbf{w} = 4\mathbf{i} - 3\mathbf{j}$
$\mathbf{v} \cdot \mathbf{w} = (-2)(4) + (1)(-3) = -11$

$$\cos \theta = \frac{\mathbf{v} \cdot \mathbf{w}}{\|\mathbf{v}\|\ \|\mathbf{w}\|} = \frac{-11}{\sqrt{5} \cdot \sqrt{25}} = \frac{-11\sqrt{5}}{25}$$

71. $\mathbf{v} = \mathbf{i} - 3\mathbf{j}$, $\mathbf{w} = -\mathbf{i} + \mathbf{j}$
$\mathbf{v} \cdot \mathbf{w} = (1)(-1) + (-3)(1) = -4$

$$\cos \theta = \frac{\mathbf{v} \cdot \mathbf{w}}{\|\mathbf{v}\|\ \|\mathbf{w}\|} = \frac{-4}{\sqrt{10} \cdot \sqrt{2}} = \frac{-4}{2\sqrt{5}} = \frac{-2\sqrt{5}}{5}$$

73. $\mathbf{v} = 2\mathbf{i} + 3\mathbf{j}$, $\mathbf{w} = 3\mathbf{i} + \mathbf{j}$

$$\text{proj}_{\mathbf{w}}\mathbf{v} = \frac{6 + 3}{\left(\sqrt{10}\right)^2}(3\mathbf{i} + \mathbf{j}) = \frac{9}{10}(3\mathbf{i} + \mathbf{j})$$

75. For $\mathbf{v} = 3\mathbf{i} - 4\mathbf{j}$ and $\mathbf{w} = 12\mathbf{i} - 5\mathbf{j}$, we have:

$\mathbf{v} \cdot \mathbf{w} = (3\mathbf{i} - 4\mathbf{j}) \cdot (12\mathbf{i} - 5\mathbf{j}) = (3)(12) + (-4)(-5) = 56$

$\|\mathbf{v}\| = \sqrt{9 + 16} = 5$ and $\|\mathbf{w}\| = \sqrt{144 + 25} = 13$

Therefore, if θ is the angle between $\mathbf{v}$ and $\mathbf{w}$, $\cos \theta = \dfrac{\mathbf{v} \cdot \mathbf{w}}{\|\mathbf{v}\|\ \|\mathbf{w}\|} = \dfrac{56}{5 \cdot 13}$ or $\cos \theta = .86154$.

From this, $\theta = \cos^{-1}(.86154) \approx 30.5°$.

77. Let $\mathbf{v}_c$ = velocity of the current
$\mathbf{v}_s$ = velocity of the swimmer relative to the water
and $\mathbf{v}_\ell$ = velocity of the swimmer relative to the land
Then, $\mathbf{v}_\ell = \mathbf{v}_c + \mathbf{v}_s$
We are given *two* of the three vectors;
$\mathbf{v}_c = 2\mathbf{i}$, if we let $\mathbf{i}$ point directly downstream,
and $\mathbf{v}_s = 5\mathbf{j}$, since the swimmer swims perpendicular to the current, at 5 miles per hour.

Then, $\mathbf{v}_\ell = \mathbf{v}_c + \mathbf{v}_s = 2\mathbf{i} + 5\mathbf{j}$

The swimmer's *actual* speed is: $\|\mathbf{v}_\ell\| = \sqrt{4 + 25} = \sqrt{29} \approx 5.39$ miles per hour
Since it takes $1/5.39 = 0.18553$ hours to cross the stream, the swimmer will end up $2(0.18553) = 0.4$ miles downstream.

79. $15(\cos \theta \mathbf{i} + \sin \theta \mathbf{j}) + 5\mathbf{i} = 2\mathbf{j}$
$a^2 = (15)^2 - (5)^2$
$a^2 = 225 - 25$
$a^2 = 200$
$a = \sqrt{200}$

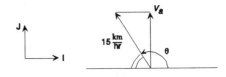

$15 \cos \theta \mathbf{i} + 5\mathbf{i} + 15 \sin \theta \mathbf{j} = \sqrt{200}$
So, $15 \cos \theta + 5 = 0$

$$\cos \theta = \frac{-1}{3} \text{ and } \sin \theta = \frac{\sqrt{200}}{15} = \frac{10\sqrt{2}}{15} = \frac{2\sqrt{2}}{3}$$
$$\theta = 109.5°$$

The person should head at an angle of 70.5° to the shore.

ANALYTIC GEOMETRY

10.2 The Parabola

1. **B**   3. **E**   5. **H**   7. **C**   9. **F**   11. **G**   13. **D**   15. **B**

17. We want an equation of the parabola with focus at (4, 0) and vertex at (0, 0). The focus and vertex both lie on the horizontal line $y = 0$ (i.e., the x-axis). The distance, a, from (4, 0) to (0, 0) is $a = 4$. Also, the focus is to the right of the vertex, so the parabola opens to the right. Since the vertex is at (0, 0), the equation of the parabola is:
$$y^2 = 4ax \qquad \text{(from Table 1)}$$
$$\text{or} \quad y^2 = 16x$$
Letting $x = 4$, we find $y^2 = 64$ or $y = \pm 8$. The points (4, 8) and (4, −8) define the latus rectum.

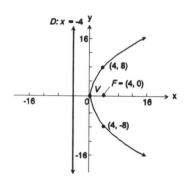

19. The focus, (0, −3), and the vertex, (0, 0), both lie on the vertical line $x = 0$ (the y-axis). We have $a = 3$, and since (0, −3) is *below* (0, 0), the parabola opens down. Therefore, from Table 1,
$$x^2 = -4ay$$
$$\text{or} \quad x^2 = -12y \ (a = 3)$$
Letting $y = -3$, we find $x^2 = 36$ or $x = \pm 6$. The points (−6, −3) and (6, −3) define the latus rectum.

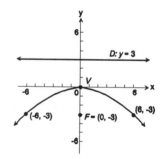

21. The vertex must be midway between the focus, (−2, 0), and the directrix, the line $x = 2$. Therefore, the vertex is the point (0, 0). The distance from the focus to the vertex is $a = 2$, and the parabola opens to the left, since (−2, 0) is to the *left* of the vertex. Therefore, we have:
$$y^2 = -4ax$$
$$\text{or} \quad y^2 = -8x \qquad (a = 2)$$
Letting $x = -2$, we find $y^2 = 16$ or $y = \pm 4$. The points (−2, 4) and (−2, −4) define the latus rectum.

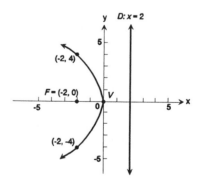

23. The directrix, $y = -\frac{1}{2}$, lies $\frac{1}{2}$ unit **below** the vertex, $(0, 0)$, so

the focus must be $\frac{1}{2}$ unit **above** the vertex, at the point $\left[0, \frac{1}{2}\right]$.

Therefore, $a = \frac{1}{2}$ and the parabola opens upward, so we have:

$$x^2 = 4ay \quad \text{or} \quad x^2 = 2y \quad \left[a = \frac{1}{2}\right]$$

Letting $y = \frac{1}{2}$, we find $x^2 = 1$ or $x = \pm 1$.

The points $\left[1, \frac{1}{2}\right]$ and $\left[-1, \frac{1}{2}\right]$ define the latus rectum.

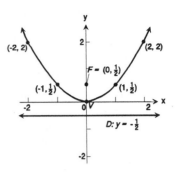

25. Here, the vertex, $(2, -3)$, and the focus, $(2, -5)$, both lie on the vertical line, $x = 2$. The distance between the vertex and the focus is $a = 2$, and the parabola opens down, since the focus is **below** the vertex. From Table 2, we have:

$(x - h)^2 = -4a(y - k)$

$(x - 2)^2 = -4a(y - (-3))$, since $(h, k) = (2, -3)$

$(x - 2)^2 = -8(y + 3)$, since $a = 2$

Letting $y = -5$, we find $(x - 2)^2 = 16$ or $x - 2 = \pm 4$ so that $x = 6$ or $x = -2$. The points $(-2, -5)$ and $(6, -5)$ define the latus rectum.

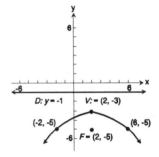

27. Since the axis of symmetry, the y-axis, is vertical, the parabola opens up or down. The point $(2, 3)$ is **above** the vertex, $(0, 0)$, so the parabola opens **up**. Therefore, from Table 1,

$$x^2 = 4ay$$

But here we don't know what a is. But the point $(2, 3)$ must satisfy the equation, since it lies on the graph.

Therefore,

$\quad x^2 = 4ay$

$\quad 4 = 4a(3) \qquad$ (using $x = 2$, $y = 3$)

$\quad 1 = 3a$

$\quad a = \frac{1}{3}$

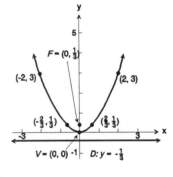

The equation of the parabola is: $x^2 = 4\left[\frac{1}{3}\right]y \quad \text{or} \quad x^2 = \frac{4}{3}y$

To help sketch the graph, note that the focus is at $(0, a) = \left[0, \frac{1}{3}\right]$. Letting $y = \frac{1}{3}$, we

find $x^2 = \frac{4}{9}$ or $x = \pm\frac{2}{3}$. The points $\left[\frac{-2}{3}, \frac{1}{3}\right]$ and $\left[\frac{2}{3}, \frac{1}{3}\right]$ define the latus rectum.

29. The directrix, $y = 2$, is horizontal, so the parabola opens up or down. Also, the axis of symmetry must be vertical, and it contains the focus, $(-3, 4)$, so it must be the line $x = -3$. The vertex lies on the axis of symmetry, midway between the focus and the directrix, so the vertex must be the point $(-3, 3)$.

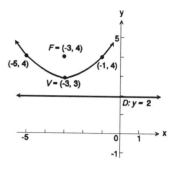

The focus, then, is one unit **above** the vertex, so we know that $a = 1$ and the parabola opens up. From Table 2, we have:
$$(x - h)^2 = 4a(y - k)$$
$$(x - (-3))^2 = 4a(y - 3), \text{ since } (h, k) = (-3, 3)$$
$$(x + 3)^2 = 4(y - 3), \text{ since } a = 1$$
Letting $y = 4$, we find $(x + 3)^2 = 4$ or $x + 3 = \pm 2$, so that $x = -1$ or $x = -5$. The points $(-1, 4)$ and $(-5, 4)$ define the latus rectum.

31. Here, the directrix, $x = 1$, is vertical, so the axis of symmetry is horizontal, and the parabola opens to the left or right. Since the focus, $(-3, -2)$ is on the axis of symmetry, the equation of the axis of symmetry must be $y = -2$. Again, the vertex is on the axis, midway between focus and directrix, so the vertex is $(-1, -2)$. The parabola opens to the left, since the focus is to the left of the vertex. Finally, $a = 2$.

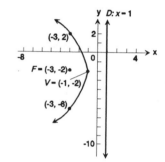

From 2, $(y - k)^2 = -4a(x - h)$
or $(y + 2)^2 = -8(x + 1)$

Letting $x = -3$, $(y + 2)^2 = 16$ or $y + 2 = \pm 4$ so that $y = 2$ or $y = -6$. The points $(-3, -6)$ or $(-3, 2)$ define the latus rectum.

33. The equation $x^2 = 4y$ is in the form:
$$x^2 = 4ay$$
where $4a = 4$,
or $a = 1$
Thus, by Table 1, we have: Vertex: $(0, 0)$
Focus: $(0, 1)$
Directrix: $y = -1$
Letting $y = 1$, we find $x^2 = 4$ or $x = \pm 2$. The points $(-2, 1)$ and $(2, 1)$ define the latus rectum.

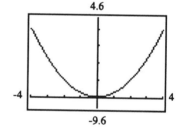

35. The equation $y^2 = -16x$ is in the form:
$$y^2 = -4ax$$
where $-4a = -16$
or $a = 4$
By Table 1, we have: Vertex: $(0, 0)$
Focus: $(-4, 0)$
Directrix: $x = 4$
Letting $x = -4$, we have $y^2 = 64$ or $y = \pm 8$. The points $(-4, 8)$ and $(-4, -8)$ define the latus rectum.

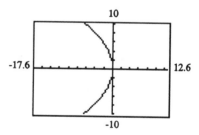

Enter $y_1 = \sqrt{-16x}$ and $y_2 = -\sqrt{-16x}$ to graph $y^2 = -16x$ on a graphing utility.

37. The equation $(y - 2)^2 = 8(x + 1)$ is in the form:

$$(y - k)^2 = 4a(x - h),$$

where: (1) $4a = 8$, or $a = 2$
         (2) $y - k = y - 2$, or $k = 2$
         (3) $x - h = x + 1$, or $-h = 1$, or $h = -1$

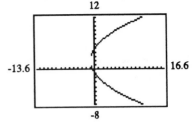

Thus, the vertex is at $(h, k) = (-1, 2)$. From Table 2, the parabola opens to the right, so the focus is $a = 2$ units to the **right** of the vertex. The focus is $(1, 2)$. The directrix is a vertical line 2 units to the **left** of the vertex: $x = -3$.

Letting $x = 1$, we have $(y - 2)^2 = 16$ or $y - 2 = \pm 4$, so that $y = 6$ or $y = -2$. The points $(1, 6)$ and $(1, -2)$ define the latus rectum.

Enter $y_1 = 2 + \sqrt{8(x + 1)}$ and $y_2 = 2 - \sqrt{8(x + 1)}$ to graph the parabola on a graphing utility.

39. The equation $(x - 3)^2 = -(y + 1)$ is in the form $(x - h)^2 = -4a(y - k)$, where:

   (1) $-4a = -1$, or $a = \dfrac{1}{4}$

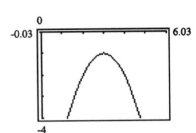

   (2) $x - h = x - 3$
            $-h = -3$
              $h = 3$

and   (3) $y - k = y + 1$
                $-k = 1$
                $k = -1$

So, from Table 2, we have: Vertex: $(h, k) = (3, -1)$

The parabola opens **down**, so the focus is **below** the vertex: Focus: $\left(3, -\dfrac{5}{4}\right)$

The directrix is a horizontal line $a = \dfrac{1}{4}$ unit **above** the vertex: Directrix: $y = -\dfrac{3}{4}$

Letting $y = \dfrac{-5}{4}$, we have $(x - 3)^2 = \dfrac{1}{4}$ or $(x - 3) = \pm\dfrac{1}{2}$ so that $x = \dfrac{7}{2}$ or $x = \dfrac{5}{2}$.

The points $\left(\dfrac{7}{2}, \dfrac{-5}{2}\right)$ and $\left(\dfrac{5}{2}, \dfrac{-5}{2}\right)$ define the latus rectum.

Enter $y_1 = -1 - (x - 3)^2$ to graph the parabola on a graphing utility.

41. The equation $(y + 3)^2 = 8(x - 2)$ is in the form:

$$(y - k)^2 = 4a(x - h)$$

where: (1) $4a = 8$, or $a = 2$
          (2) $y - k = y + 3$, or $k = -3$
          (3) $x - h = x - 2$, or $h = 2$

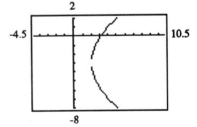

Thus, from Table 2, we have: Vertex: $(h, k) = (2, -3)$
The parabola opens to the right, so the focus is $a = 2$ units to the right of the vertex:

     Focus: $(4, -3)$

The directrix is the vertical line 2 units to the left of the vertex:

     Directrix: $x = 0$

Letting $x = 4$, we have $(y + 3)^2 = 16$ or $y + 3 = \pm 4$ so that $y = 1$ or $y = -7$. The points $(4, 1)$ and $(4, -7)$ define the latus rectum.

Enter $y_1 = -3 + \sqrt{8(x - 2)}$ and $y_2 = -3 - \sqrt{8(x - 2)}$ to graph the parabola on a graphing utility.

43.

$$y^2 - 4y + 4x + 4 = 0$$
$$y^2 - 4y = -4x - 4$$
$$y^2 - 4y + 4 = -4x$$
$$(y - 2)^2 = -4(x + 0)$$

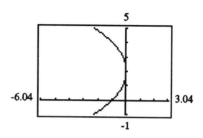

The equation is now in the form: $(y - k)^2 = -4a(x - h)$,
where: (1) $-4a = -4$, or $a = 1$
           (2) $y - k = y - 2$, or $k = 2$
           (3) $x - h = x + 0$, or $h = 0$
Thus, from Table 2, we have: Vertex: $(h, k) = (0, 2)$
The parabola opens to the left, so the focus is $a = 1$ unit to
the left of the vertex: Focus: $(-1, 2)$
The directrix is the vertical line 1 unit to the right of the vertex:
      Directrix: $x = 1$
Letting $x = -1$, we have $(y - 2)^2 = 4$ or $y - 2 = \pm2$ so that $y = 4$ or $y = 0$. The points $(-1, 4)$
and $(-1, 0)$ define the latus rectum.

Graph $y_1 = 2 + \sqrt{-4x}$ and $y_2 = 2 - \sqrt{-4x}$

45.

$$x^2 + 8x = 4y - 8$$
$$x^2 + 8x + 16 = 4y + 8$$
$$(x + 4)^2 = 4(y + 2)$$

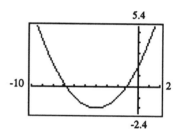

The equation is now in the form $(x - h)^2 = 4a(y - k)$, where:
     (1) $4a = 4$, or $a = 1$
     (2) $x - h = x + 4$, or $h = -4$
     (3) $y - k = y + 2$, or $y = -2$
Thus, from Table 2, we have:
     Vertex: $(h, k) = (-4, -2)$
The parabola opens up, so the focus is $a = 1$ unit above the
vertex:
     Focus: $(-4, -1)$
The directrix is the horizontal line 1 unit below the vertex:
     Directrix: $y = -3$
Letting $y = -1$, we have $(x + 4)^2 = 4$ or $x + 4 = \pm2$ so that $x = -2$ or $x = -6$. The points
$(-2, -1)$ and $(-6, -1)$ define the latus rectum.
Graph $y = -2 + (x + 4)^2/4$.

47.   For $y^2 + 2y - x = 0$, complete the square:

$$y^2 + 2y - x = 0$$
$$y^2 + 2y + \underline{} = x + \underline{}$$
$$y^2 + 2y + 1 = x + 1$$
$$(y + 1)^2 = (x + 1)$$

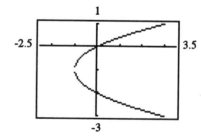

This is in the form: $(y - k)^2 = 4a(x - h)$,

where: (1) $4a = 1$, or $a = \dfrac{1}{4}$

           (2) $x - h = x + 1$
                 $h = -1$

and   (3) $y - k = y + 1$
                 $-k = 1$
                 $k = -1$

From Table 2, we have: Vertex: $(h, k) = (-1, -1)$

The parabola opens to the right, and has: Focus: $\left[-\dfrac{3}{4}, -1\right]$

The directrix is $x = -\dfrac{5}{4}$

Letting $x = \frac{-3}{4}$, we have $(y + 1)^2 = \frac{7}{4}$ or $y + 1 = \pm\frac{\sqrt{7}}{2}$ so that $y = \frac{\sqrt{7}}{2} - 1$ or

$y = \frac{-\sqrt{7}}{2} - 1$. The points $\left[\frac{-3}{4}, \frac{\sqrt{7}}{2} - 1\right]$ and $\left[\frac{-3}{4}, \frac{-\sqrt{7}}{2} - 1\right]$ define the latus rectum.

Graph $y_1 = -1 + \sqrt{x + 1}$ and $y_2 = -1 - \sqrt{x + 1}$.

49. We complete the square:
$$x^2 - 4x + \underline{} = y + 4 + \underline{}$$
$$x^2 - 4x + 4 = y + 4 + 4$$
$$(x - 2)^2 = y + 8$$
which is in the form: $(x - h)^2 = 4a(y - k)$

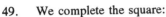

where: (1) $4a = 1,\ a = \frac{1}{4}$

(2) $x - h = x - 2$
$h = 2$

and (3) $y - k = y + 8$
$k = -8$

We have: Vertex: $(2, -8)$

Focus: $\left[2, -7\frac{3}{4}\right]$ $=$ $\left[2, \frac{-31}{4}\right]$

Directrix: $y = -8\frac{1}{4}$ or $y = \frac{-33}{4}$

Letting $y = \frac{-31}{4}$, we have $(x - 2)^2 = \frac{1}{4}$ or $x - 2 = \pm\frac{1}{2}$ so that $x = \frac{5}{2}$ or $x = \frac{3}{2}$.

The points $\left[\frac{5}{2}, \frac{-31}{4}\right]$ and $\left[\frac{3}{2}, \frac{-31}{4}\right]$ define the latus rectum.
Graph $y = -8 + (x - 2)^2$.

51. $(y - 1)^2 = c(x - 0)$
$(y - 1)^2 = cx$
$(2 - 1)^2 = c(1)$
$1 = c$
$(y - 1)^2 = x$

53. $(y - 1)^2 = c(x - 2)$
$(0 - 1)^2 = c(1 - 2)$
$1 = c(-1)$
$c = -1$
$(y - 1)^2 = -(x - 2)$

55. $(x - 0)^2 = c(y - 1)$
$x^2 = c(y - 1)$
$2^2 = c(2 - 1)$
$4 = c$
$x^2 = 4(y - 1)$

57. $(y - 0)^2 = c(x - (-2))$
$y^2 = c(x + 2)$
$1^2 = c(0 + 2)$
$1 = 2c$
$\frac{1}{2} = c$
$y^2 = \frac{1}{2}(x + 2)$

59. Situate the parabola so that its vertex is at $(0, 0)$, and it opens up. Then, we know:
$$x^2 = 4ay$$
Since the parabola is 10 feet across and 4 feet deep, the points $(-5, 4)$ and $(5, 4)$ must satisfy the equation:
$$x^2 = 4ay$$
$$25 = 4a(4) \qquad (x = 5, y = 4)$$
$$25 = 16a$$
$$a = \frac{25}{16} \approx 1.5625$$
But a is, by definition, the distance from the vertex to the focus.

Therefore, the receiver (at the focus) is $\frac{25}{16} = 1.5625$ feet, or 18.75 inches, from the base of the dish, along the axis of symmetry.

61. The light is placed at the focus of the parabola. $x^2 = 4y$. Since $F = (0, 1)$, the bulb should be placed 1 inch from the vertex.

63. Situate the vertex at $(0, 0)$: $x^2 = cy$
Point on parabola is $(300, 80)$:
$$300^2 = c \cdot 80$$
$$c = 1125$$
$$x^2 = 1125y$$
When $x = 150$, $\quad (150)^2 = 1125h$
$$h = \frac{150^2}{1125}$$
$$= 20 \text{ feet}$$

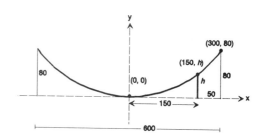

65. Situate the parabola so that its vertex is at $(0, 0)$ and it opens up. We know:
$$x^2 = 4ay$$
The light source is located 2 feet from the base, so $a = 2$.
Since the width is 5 feet, $x = \pm 2.5$ feet. Thus, we have:
$$(2.5)^2 = 4(2)y$$
$$6.25 = 8y$$
$$y = 0.78125 \text{ feet}$$
Thus, the depth of the search light is 0.78125 feet.

67. Situate the parabola so that its vertex is at $(0, 0)$ and it opens up. We know:
$$x^2 = 4ay$$
The parabola is 20 feet wide, so $x = \pm 10$ feet, and 6 feet deep, so $y = 6$ feet. Thus, we have:
$$(10)^2 = 4a(6)$$
$$100 = 24a$$
$$a \approx 4.17 \text{ feet}$$
Therefore, the heat source will be concentrated 4.17 feet from the base, along the axis of symmetry.

69. Situate the parabola so that its vertex is at $(0, 0)$ and it opens down. Then, we know
$$x^2 = -4ay$$
Since the bridge has a span of 120 feet and a maximum height of 25 feet, two points on the parabola are $(-60, -25)$ and $(60, -25)$. See figure.
Since the form of the equation of this parabola is:
$$x^2 = -4ay$$

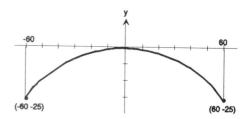

Letting $x = 60$ and $y = -25$, we can find a.
$$60^2 = -4a(-25)$$
$$a = 36$$
So, the equation of the parabola is:
$$x^2 = -144y$$
So the height of the bridge 10 feet from the center is:
$$10^2 = -144y$$
$$y = -0.69 \text{ feet}$$
Therefore, the height of the bridge 10 feet from center is $25 - 0.69 = 24.31$ feet.

The height of the bridge 30 feet from center is $25 - 6.25 = 18.75$ feet.

The height of the bridge 50 feet from center is $25 - 17.36 = 7.64$ feet.

71. $\quad Ax^2 + Ey = 0 \quad A \neq 0, E \neq 0$
$$Ax^2 = -Ey$$
$$x^2 = \frac{-E}{A}y$$

Parabola; vertex at $(0, 0)$ and axis of symmetry the y-axis; focus at $\left[0, \dfrac{-E}{4A}\right]$;

directrix the line $y = \dfrac{E}{4A}$. The parabola opens up if $\dfrac{-E}{A} > 0$ and down if $\dfrac{-E}{A} < 0$.

73. $\quad Ax^2 + Dx + Ey + F = 0 \qquad A \neq 0$

(a) If $E \neq 0$, then
$$Ax^2 + Dx + Ey + F = 0$$
$$Ax^2 + Dx = -Ey - F$$
$$A\left[x^2 + \frac{D}{A}x + \frac{D^2}{4A^2}\right] = -Ey - F + \frac{D^2}{4A}$$
$$\left[x + \frac{D}{2A}\right]^2 = \frac{1}{A}\left[-Ey - F + \frac{D^2}{4A}\right]$$
$$\left[x + \frac{D}{2A^2}\right] = \frac{-E}{A}\left[y + \left[\frac{F}{E} - \frac{D^2}{4AE}\right]\right]$$
$$\left[x + \frac{D}{2A}\right]^2 = \frac{-E}{A}\left[y - \frac{D^2 - 4AF}{4AE}\right]$$

Parabola; vertex at $\left[\dfrac{-D}{2A}, \dfrac{D^2 - 4AF}{4AE}\right]$, axis of symmetry parallel to y-axis.

(b) If $E = 0$, then
$$Ax^2 + Dx + F = 0$$
$$x = \frac{-D \pm \sqrt{D^2 - 4AF}}{2A}$$
If $D^2 - 4AF = 0$, then
$$x = \frac{-D}{2A}$$
Vertical line

(c) If $E = 0$, then
$$Ax^2 + Dx + F = 0$$
$$x = \frac{-D \pm \sqrt{D^2 - 4AF}}{2A}$$
If $D^2 - 4AF > 0$, then
$$x = \frac{-D + \sqrt{D^2 - 4AF}}{2A} \text{ or}$$
$$x = \frac{-D - \sqrt{D^2 - 4AF}}{2A}$$
Two vertical lines

(d) If $E = 0$, then
$$Ax^2 + Dx + F = 0$$
$$x = \frac{-D \pm \sqrt{D^2 - 4AF}}{2A}$$
If $D^2 - 4AF < 0$, there is no real solution. Hence, the graph contains no points.

1. C   3. B   5. C   7. D

In Problems 9–17, write the equation in the form shown in Table 3 in the text, so that you can identify (h, k), a, and b, and also tell whether the major axis is parallel to the x-axis or the y-axis.

9. $\dfrac{x^2}{25} + \dfrac{y^2}{4} = 1$

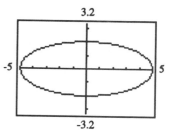

This is in the proper form, and we see that $h = 0$ and $k = 0$, so the center of the ellipse is at the origin. Since the larger denominator is associated with x^2, we see from Table 3 that the major axis is parallel to the x-axis, and:

$a^2 = 25$, so $a = 5$
$b^2 = 4$, so $b = 2$

We will be able to locate the vertices using $a = 5$, but to find the foci we will need c:

$c^2 = a^2 - b^2 = 21$

$c = \sqrt{21}$

From Table 3, the vertices are located at $(h \pm a, k) = (0 \pm 5, 0)$:

Vertices: $(-5, 0)$, $(5, 0)$

and the foci are at $(h \pm c, k)$:

Foci: $\left(-\sqrt{21}, 0\right), \left(\sqrt{21}, 0\right)$

(Notice that, since the major axis is parallel to the x-axis, we find the vertices by moving left and right of the center (h, k), a distance $a = 5$. Similarly, the foci are $c = \sqrt{21}$ units to the left and right of the center.)

Remember that we use b to locate the end-points of the ***minor axis***, which passes through the center and is perpendicular to the major axis.

Graph $y_1 = \sqrt{4\left(1 + \dfrac{x^2}{25}\right)}$ and $y_2 = -\sqrt{4\left(1 + \dfrac{x^2}{25}\right)}$

11. $\dfrac{x^2}{9} + \dfrac{y^2}{25} = 1$

This is in the proper form, and the fact that the larger denominator appears in the y^2-term means that the major axis is parallel to the y-axis. By Table 3,

(1) $h = 0$, $k = 0$, so the center is $(0, 0)$
(2) $a^2 = 25$, so $a = 5$
(3) $b^2 = 9$, so $b = 3$, and
(4) $c^2 = a^2 - b^2 = 16$, so $c = 4$

Therefore, we have: Vertices: $(h, k \pm a)$: $(0, -5)$ and $(0, 5)$

Foci: $(h, k \pm c)$: $(0, -4)$ and $(0, 4)$

Use b to find the endpoints of the ***minor*** axis.

Graph $y_1 = 5\sqrt{1 - \dfrac{x^2}{9}}$ and $y_2 = -5\sqrt{1 - \dfrac{x^2}{9}}$

13. $4x^2 + y^2 = 16$ We must get a 1 on the right-hand side.

$\dfrac{4x^2}{16} + \dfrac{y^2}{16} = 1$ Divide both sides by 16. But we can't have the 4 in the numerator, so we simplify.

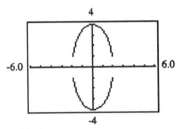

$\dfrac{x^2}{4} + \dfrac{y^2}{16} = 1$

Now we have the proper form, with $h = 0$, $k = 0$. By Table 3, the major axis is parallel to the y-axis, and:

    $a^2 = 16$, $a = 4$

    $b^2 = 4$, $b = 2$

    $c^2 = a^2 - b^2 = 12$, $c = \sqrt{12} = 2\sqrt{3}$

Also: Foci: $(h, k \pm c)$: $\left(0, -2\sqrt{3}\right), \left(0, 2\sqrt{3}\right)$

          Vertices: $(h, k \pm a)$: $(0, -4), (0, 4)$

Graph $y_1 = \sqrt{16 - 4x^2}$ and $y_2 = -\sqrt{16 - 4x^2}$

15. $4y^2 + x^2 = 8$

$\dfrac{4y^2}{8} + \dfrac{x^2}{8} = 1$

Obtain a 1 on the right.

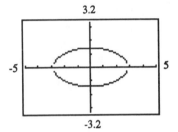

$\dfrac{y^2}{2} + \dfrac{x^2}{8} = 1$

Simplify.

We know: (1) Major axis is parallel to the x-axis.

          (2) $a^2 = 8$, so $a = \sqrt{8} = 2\sqrt{2}$

          (3) $b^2 = 2$, so $b = \sqrt{2}$

          (4) $c^2 = a^2 - b^2 = 6$, $c = \sqrt{6}$

          (5) $h = 0$, $k = 0$, so the center is at $(0, 0)$

          (6) Foci: $(h \pm c, k)$: $\left(-\sqrt{6}, 0\right), \left(\sqrt{6}, 0\right)$

          (7) Vertices: $(h \pm a, k)$: $\left(-2\sqrt{2}, 0\right), \left(2\sqrt{2}, 0\right)$

Graph $y_1 = \dfrac{\sqrt{8 - x^2}}{2}$ and $y_2 = -\dfrac{\sqrt{8 - x^2}}{2}$

17. $x^2 + y^2 = 16$

This is the equation of a circle, with center at $(0, 0)$ and radius $= 4$.

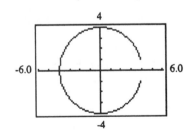

    Vertices: $(\pm 4, 0)$

    Foci: $(0, 0)$

Graph $y_1 = \sqrt{16 - x^2}$ and $y_2 = -\sqrt{16 - x^2}$

*For Problems 19–28, in order to find the **equation** of an ellipse, it is necessary to find a, b, h and k, and to know whether the major axis is parallel to the x-axis or the y-axis. Once we have these pieces of information, we can use Table 3 to write down the equation. We will use the following facts:*

*(1)   (h, k) are the coordinates of the **center** of the ellipse.*
(2)   a = the distance from the center to either vertex (half the length of the major axis)
*(3)   b = the distance from the center to either endpoint of the minor axis (half the length of the minor axis), **or**, if c is known, then $b^2 = a^2 - c^2$*
(4)   c = the distance from the center to either focus.
(5)   Since $b^2 = a^2 - c^2$, if any two of the three numbers a, b, and c are known, we can find the third.
(6)   The center, the foci, and the vertices all lie on the major axis.
Use these facts to find a, b, h, and k, and to determine if the major axis is vertical or horizontal.

19.   We are given: Center: $C(0, 0)$
                                    Focus: $F(3, 0)$
                                    Vertex: $V(5, 0)$

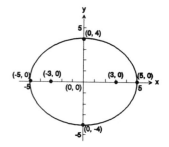

Please refer to the paragraph above.
By fact (1) above, $h = 0$, $k = 0$
By fact (2), $a = d(C, V) = 5$
By fact (4), $c = d(C, F) = 3$
By fact (3), $b^2 = a^2 - c^2 = 25 - 9 = 16$
By fact (6), since C, F and V all lie on the horizontal line, $y = 0$, the
        major axis is parallel to the x-axis.

Then from Table 1, the equation is: $\dfrac{(x - h)^2}{a^2} + \dfrac{(y - k)^2}{b^2} = 1$ or $\dfrac{x^2}{25} + \dfrac{y^2}{16} = 1$

21.   We are given: Center: $C(0, 0)$
                                    Focus: $F(0, -4)$
                                    Vertex: $V(0, 5)$

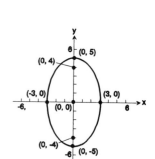

The numbering below refers to the paragraph preceding the solution to Problem 19.
By (1), $h = 0$, $k = 0$
By (2), $a = d(C, V) = 5$
By (4), $c = d(C, F) = 4$
By (3), $b^2 = a^2 - c^2 = 9$
By (6), the major axis is the vertical line $x = 0$.

From Table 1, the equation is: $\dfrac{(x - h)^2}{b^2} + \dfrac{(y - k)^2}{a^2} = 1$ or $\dfrac{x^2}{9} + \dfrac{y^2}{25} = 1$

23.   We are given: Foci: $F_1(-2, 0)$ and $F_2(2, 0)$
                                    Length of Major Axis = 6.

First of all, the center is the midpoint between F_1 and F_2: $C(0, 0)$.
The numbers below refer to the paragraph preceding the solution to Problem 19.
By (1), $h = 0$, $k = 0$
By (2), $a =$ half the length of the major axis, or $a = 3$
By (4), $c = d(F_1, C) = 2$
By (3), $b^2 = a^2 - c^2 = 5$
By (6), the major axis is the horizontal line, $y = 0$.

Then, by Table 3, we have: $\dfrac{(x - h)^2}{a^2} + \dfrac{(y - k)^2}{b^2} = 1$ or $\dfrac{x^2}{9} + \dfrac{y^2}{5} = 1$

25. We are given: Foci: $F_1(0, -3)$ and $F_2(0, 3)$
                      x-intercepts: ± 2

The center is $C(0, 0)$, and the major axis is the vertical line $x = 0$, (the y-axis). Therefore, the minor axis lies on the x-axis, so the x-intercepts are the endpoints of the minor axis. Thus, we have $b = 2$. We know $c = d(C, F) = 3$, and $b^2 = a^2 - c^2$, or $a^2 = b^2 + c^2 = 13$. From Table 3, we have:

$$\frac{(x - h)^2}{b^2} + \frac{(y - k)^2}{a^2} = 1 \quad \text{or} \quad \frac{x^2}{4} + \frac{y^2}{13} = 1$$

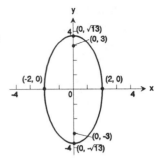

27. Here we are given: Center: $C(0, 0)$
                             Vertex: $V(0, 4)$
                         and $b = 1$

Therefore, $h = 0$, $k = 0$, and $a = d(C, V) = 4$, so we have all we need to write down the equation. Since the major axis is parallel to the y-axis ($x = 0$), we have:

$$\frac{(x - h)^2}{b^2} + \frac{(y - k)^2}{a^2} = 1 \quad \text{or} \quad \frac{x^2}{1} + \frac{y^2}{16} = 1$$

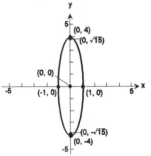

29. $\dfrac{(x + 1)^2}{4} + (y - 1)^2 = 1$ **31.** $(x - 1)^2 + \dfrac{y^2}{4} = 1$

33. $\dfrac{(x - 3)^2}{4} + \dfrac{(y + 1)^2}{9} = 1$

This is in the form: $\dfrac{(x - h)^2}{b^2} + \dfrac{(y - k)^2}{a^2} = 1$

where $h = 3$, $k = -1$, $a = 3$, $b = 2$.
Since the larger denominator belongs to the y^2-term, we see from Table 3 that the major axis is parallel to the y-axis, and we have:
      Center: (h, k), or $(3, -1)$
        Foci: $(h, k \pm c)$, where $c^2 = a^2 - b^2 = 5$,
          or $c = \sqrt{5}$, giving the points $(3, -1 - \sqrt{5})$ and $(3, -1 + \sqrt{5})$
    Vertices: $(h, k \pm a)$, or $(3, -4)$ and $(3, 2)$

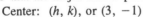

Graph $y_1 = -1 + 3\sqrt{1 - \dfrac{(x - 3)^2}{4}}$ and $y_2 = -1 - 3\sqrt{1 - \dfrac{(x - 3)^2}{4}}$

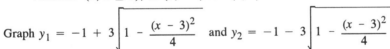

35. $(x + 5)^2 + 4(y - 4)^2 = 16$
To put this into the form listed in Table 3, we need to obtain a 1 on the right-hand-side:

$$\frac{(x + 5)^2}{16} + \frac{4(y - 4)^2}{16} = 1$$

Now get rid of the 4 in the numerator of the y^2-term (multiply top and bottom by $\dfrac{1}{4}$).

$$\frac{(x + 5)^2}{16} + \frac{(y - 4)^2}{4} = 1$$

We now have the proper form: $\dfrac{(x - h)^2}{a^2} + \dfrac{(y - k)^2}{b^2} = 1$

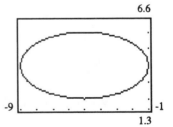

where $h = -5$, $k = 4$, $a = 4$, $b = 2$. Also, $c^2 = a^2 - b^2 = 12$, so $c = \sqrt{12}$.

Since the larger denominator is associated with the x^2 term, the major axis is parallel to the x-axis, and, from Table 3, we have:

Center: (h, k), or $(-5, 4)$

Foci: $(h \pm c, k)$, or $\left(-5 - 2\sqrt{3},\ 4\right)$ and $\left(-5 + 2\sqrt{3},\ 4\right)$

Vertices: $(h \pm a, k)$, or $(-9, 4)$ and $(-1, 4)$

Graph $y_1 = 4 + 2\sqrt{1 - \dfrac{(x + 5)^2}{16}}$ and $y_2 = 4 - 2\sqrt{1 - \dfrac{(x + 5)^2}{16}}$

37. $x^2 + 4x + 4y^2 - 8y + 4 = 0$

Here we start by completing the square in both x and y:

$x^2 + 4x + \underline{\quad} + 4(y^2 - 2y + \underline{\quad}) = -4$

$x^2 + \underline{4x + 4} + 4(\underline{y^2 - 2y + 1}) = -4 + 4 + 4$

$(4)(1) = 4$

$(x + 2)^2 + 4(y - 1)^2 = 4$

$\dfrac{(x + 2)^2}{4} + \dfrac{(y - 1)^2}{1} = 1$

This is in the form: $\dfrac{(x - h)^2}{a^2} + \dfrac{(y - k)^2}{b^2} = 1$

where $h = -2$, $k = 1$, $a = 2$, $b = 1$, and $c^2 = a^2 - b^2 = 3$, so $c = \sqrt{3}$.

From Table 3, we have:

Center: (h, k), or $(-2, 1)$

Foci: $(h \pm c, k)$, or $\left(-2 - \sqrt{3},\ 1\right)$ and $\left(-2 + \sqrt{3},\ 1\right)$

Vertices: $(h \pm a, k)$, or $(-4, 1)$ and $(0, 1)$

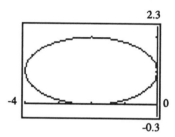

Graph $y_1 = 1 + \sqrt{1 + \dfrac{(x + 2)^2}{4}}$ and $y_2 = 1 - \sqrt{1 + \dfrac{(x + 2)^2}{4}}$

39.
$$2x^2 + 3y^2 - 8x + 6y + 5 = 0$$
$$2x^2 - 8x + 3y^2 + 6y = -5$$
$$2(x^2 - 4x + \underline{\quad}) + 3(y^2 + 2y + \underline{\quad}) = -5$$
$$2(x^2 - 4x + 4) + 3(y^2 + 2y + 1) = -5 + 8 + 3$$
$$2(x - 2)^2 + 3(y + 1)^2 = 6$$
$$\dfrac{(x - 2)^2}{3} + \dfrac{(y + 1)^2}{2} = 1$$

This is in the form: $\dfrac{(x - h)^2}{a^2} + \dfrac{(y - k)^2}{b^2} = 1$, where

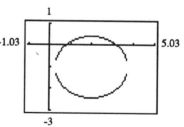

$h = 2$, $k = -1$, $a^2 = 3$, so $a = \sqrt{3}$, $b^2 = 2$, so $b = \sqrt{2}$, and $c^2 = a^2 - b^2 = 1$, so that $c = 1$. Then we have:

Center: (h, k), or $(2, -1)$

Foci: $(h \pm c, k)$, or $(1, -1)$ and $(3, -1)$

Vertices: $(h \pm a, k)$, or $\left(2 - \sqrt{3},\ -1\right)$ and $\left(2 + \sqrt{3},\ -1\right)$

Graph $y_1 = -1 + \sqrt{2\left[1 - \dfrac{(x - 2)^2}{3}\right]}$ and $y_2 = -1 - \sqrt{2\left[1 - \dfrac{(x - 2)^2}{3}\right]}$

41.

$$9x^2 + 4y^2 - 18x + 16y - 11 = 0$$
$$9x^2 - 18x + 4y^2 + 16y = 11$$
$$9(x^2 - 2x + \underline{\quad}) + 4(y^2 + 4y + \underline{\quad}) = 11$$
$$9(x^2 - 2x + 1) + 4(y^2 + 4y + 4) = 11 + 9 + 16$$
$$9(x - 1)^2 + 4(y + 2)^2 = 36$$

$$\frac{(x - 1)^2}{4} + \frac{(y + 2)^2}{9} = 1$$

This is in the form: $\dfrac{(x - h)^2}{b^2} + \dfrac{(y - k)^2}{a^2} = 1$

where $h = 1$, $k = -2$, $a^2 = 9$, so $a = 3$, $b^2 = 4$, so $b = 2$, and $c^2 = a^2 - b^2 = 5$, so that

$c = \sqrt{5}$.

Then we have: Center: (h, k), or $(1, -2)$

Foci: $(h, k \pm c)$, or $(1, -2 + \sqrt{5})$ and $(1, -2 - \sqrt{5})$
Vertices: $(h, k \pm a)$, or $(1, 1)$ and $(1, -5)$

Graph $y_1 = -2 + 3\sqrt{1 - \dfrac{(x - 1)^2}{4}}$ and $y_2 = -2 - 3\sqrt{1 - \dfrac{(x - 1)^2}{4}}$

43.

$$4x^2 + y^2 + 4y = 0$$
$$4x^2 + (y^2 + 4y + \underline{\quad}) = 0$$
$$4x^2 + (y^2 + 4y + 4) = 4$$
$$4x^2 + (y + 2)^2 = 4$$

$$x^2 + \frac{(y + 2)^2}{4} = 1$$

This is in the form: $\dfrac{(x - h)^2}{b^2} + \dfrac{(y - k)^2}{a^2} = 1$, where

$h = 0$, $k = -2$, $a^2 = 4$, so $a = 2$, $b^2 = 1$, so $b = 1$, and $c^2 = a^2 - b^2 = 3$, so that $c = \sqrt{3}$.
Then we have:

Center: $(0, -2)$

Foci: $(h, k \pm c)$, or $(0, -2 + \sqrt{3})$ and $(0, -2 - \sqrt{3})$
Vertices: $(h, k \pm a)$, or $(0, 0)$ and $(0, -4)$

Graph $y_1 = -2 + \sqrt{4(1 - x^2)}$ and $y_2 = -2 - \sqrt{4(1 - x^2)}$

45. We are given: Center: $C(2, -2)$
Vertex: $V(7, -2)$
Focus: $F(4, -2)$

From the center, $h = 2$, $k = -2$

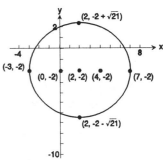

Also, $a = d(C, V) = 5$, and $c = d(C, F) = 2$. Therefore,

$b^2 = a^2 - c^2 = 21$, so $b = \sqrt{21}$. Finally, C, V, and F all lie on
the horizontal line $y = -2$, so the major axis is parallel to the
x-axis.

From Table 3, the equation is of the form:

$$\frac{(x - h)^2}{a^2} + \frac{(y - k)^2}{b^2} = 1 \quad \text{or} \quad \frac{(x - 2)^2}{25} + \frac{(y + 2)^2}{21} = 1$$

47. We are given: Vertices: $V_1(4, 3)$ and $V_2(4, 9)$
                                       Focus: $F(4, 8)$

First of all, the center is the midpoint between V_1 and V_2:

   $C(4, 6)$, and $h = 4$, $k = 6$.

Then, $a = d(C, V) = 3$, $c = d(C, F) = 2$

and $b^2 = a^2 - c^2 = 5$, so $b = \sqrt{5}$

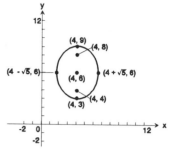

Now V_1, V_2, F, and C all lie on the vertical line $x = 4$, so the major
axis is parallel to the y-axis, and the equation is of the form:

$$\frac{(x - h)^2}{b^2} + \frac{(y - k)^2}{a^2} = 1 \quad \text{or} \quad \frac{(x - 4)^2}{5} + \frac{(y - 6)^2}{9} = 1$$

49. We are given: Foci: $F_1(5, 1)$ and $F_2(-1, 1)$
                                       Length of major axis: 8

Then the center is midway between F_1 and F_2: $C(2, 1)$, so $h = 2$,
$k = 1$. The length of the major axis is $2a$, so $a = 4$, and
$c = d(C, F_1) = 3$. Therefore, $b^2 = a^2 - c^2 = 7$. The major
axis (the line $y = 1$) is parallel to the x-axis, so we have:

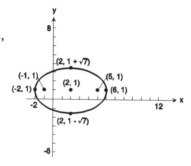

$$\frac{(x - h)^2}{a^2} + \frac{(y - k)^2}{b^2} = 1 \quad \text{or} \quad \frac{(x - 2)^2}{16} + \frac{(y - 1)^2}{7} = 1$$

51. We are given: Center: $C(1, 2)$
                                       Focus: $F(4, 2)$
                              Contains the point: $(1, 3)$

From the center $h = 1$, $k = 2$. Also, $c = d(C, F) = 3$, so
$c^2 = 9$. The major axis is parallel to the x-axis, so we have:

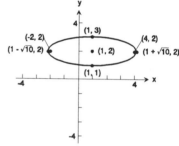

$$\frac{(x - h)^2}{a^2} + \frac{(y - k)^2}{b^2} = 1 \quad \text{or} \quad \frac{(x - 1)^2}{a^2} + \frac{(y - 2)^2}{b^2} = 1$$

But the point $(1, 3)$ must satisfy the equation since it lies on the
graph. Therefore:

$$\frac{0}{a^2} + \frac{1}{b^2} = 1 \text{ or } b^2 = 1, \text{ so } b = 1 \text{ and } a^2 = b^2 + c^2 = 10.$$

Thus, $\dfrac{(x - 1)^2}{10} + \dfrac{(y - 2)^2}{1} = 1$

53. We are given: Center: $C(1, 2)$
                                       Vertex: $V(4, 2)$
                              Contains the point: $(1, 3)$

From the center $h = 1$, $k = 2$. Also, $a = d(C, V) = 3$, so $a^2 = 9$.
The major axis is parallel to the x-axis, so we have:

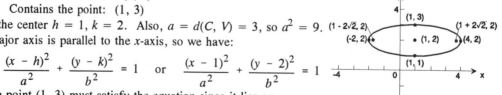

$$\frac{(x - h)^2}{a^2} + \frac{(y - k)^2}{b^2} = 1 \quad \text{or} \quad \frac{(x - 1)^2}{a^2} + \frac{(y - 2)^2}{b^2} = 1$$

But the point $(1, 3)$ must satisfy the equation since it lies on
the graph. Therefore:

$$\frac{0}{a^2} + \frac{1}{b^2} = 1 \text{ or } b^2 = 1, \text{ so } b = 1 \text{ and } c^2 = a^2 - b^2 = 8.$$

Thus, $\dfrac{(x - 1)^2}{9} + \dfrac{(y - 2)^2}{1} = 1$

55.

$$y = \sqrt{16 - 4x^2}$$
$$y^2 = 16 - 4x^2, \qquad y \geq 0$$
$$4x^2 + y^2 = 16, \qquad y \geq 0$$
$$\frac{x^2}{4} + \frac{y^2}{16} = 1, \qquad y \geq 0$$

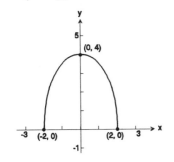

57.

$$y = -\sqrt{64 - 16x^2}$$
$$y^2 = 64 - 16x^2, \qquad y \leq 0$$
$$16x^2 + y^2 = 64, \qquad y \leq 0$$
$$\frac{x^2}{4} + \frac{y^2}{64} = 1, \qquad y \leq 0$$

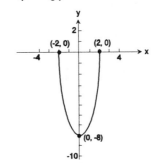

59. The center of the ellipse will be at $(0, 0)$ due to the nice positioning of the axes. The length of the major axis is 20, so we know $a = 10$. The length of *half* the minor axis is 6, i.e., $b = 6$. Finally, the major axis is horizontal, so we have:

$$\frac{(x - h)^2}{a^2} + \frac{(y - k)^2}{b^2} = 1 \quad \text{or} \quad \frac{x^2}{100} + \frac{y^2}{36} = 1$$

61. Assume the half-ellipse formed by the whispering galley is centered at $(0, 0)$. Since the hall is 100 feet long, we have $2a = 100$, so $a = 50$. The distance from the center of the ellipse to the foci is 25 feet, so $c = 25$. Since $b^2 = a^2 - c^2$, $b^2 = 50^2 - 25^2 = 1875$; therefore, $b = \sqrt{1875} = 43.3$. So the ceiling will be 43.3 feet high at the center.

63. Situate the semielliptical arch so that the x-axis coincides with the water and the y-axis passes through the center of the arch. Since the bridge has a span of 120 feet, the length of the major axis is 120, so $2a = 120$ or $a = 60$. The length of half the minor axis is 25, so $b = 25$. Therefore, we have:

$$\frac{x^2}{3600} + \frac{y^2}{625} = 1$$

When $x = 10$ feet from center:

$$\frac{10^2}{3600} + \frac{y^2}{625} = 1$$
$$\frac{y^2}{625} = 1 - \frac{100}{3600}$$
$$y^2 = 625 \cdot \frac{3500}{3600}$$
$$y \approx 24.65 \text{ feet}$$

When $x = 30$ feet from center:

$$\frac{30^2}{3600} + \frac{y^2}{625} = 1$$
$$y \approx 21.65 \text{ feet}$$

When $x = 50$ feet from center:

$$\frac{50^2}{3600} + \frac{y^2}{625} = 1$$
$$y \approx 13.82 \text{ feet}$$

65. First we need to find an equation for the ellipse. Let the center of the ellipse be the origin, $(0, 0)$, with the x-axis at ground-level. Then the major axis has length 40, so we know $a = 20$, and the length of **half** the minor axis is 15, i.e., $b = 15$. Since the major axis is horizontal, we have:

$$\frac{x^2}{a^2} + \frac{y^2}{b^2} = 1 \text{ or } \frac{x^2}{400} + \frac{y^2}{225} = 1$$

We wish to find y when $x = 0, \pm 10, \pm 20$, so we solve for y:

$$\frac{x^2}{400} + \frac{y^2}{225} = 1$$

$$\frac{y^2}{225} = 1 - \frac{x^2}{400}$$

$$\frac{y^2}{225} = \frac{400 - x^2}{400}$$

$$y^2 = 225 \left[\frac{400 - x^2}{400} \right]$$

$$y^2 = \frac{225}{400}(400 - x^2)$$

$$y = \sqrt{\frac{225}{400}} \sqrt{400 - x^2}$$

$$\text{or} \quad y = \frac{15}{20}\sqrt{400 - x^2} = \frac{3}{4}\sqrt{400 - x^2}$$

x	$y = \frac{3}{4}\sqrt{400 - x^2}$
0	$\frac{3}{4}\sqrt{400} = \frac{3}{4}(20) = 15$
± 10	$\frac{3}{4}\sqrt{400 - 100} \approx 12.99$
± 20	$\frac{3}{4}\sqrt{400 - 400} = 0$

67. If the mean distance is 93 million miles, then $a = 93$. Therefore, the length of the major axis is 186. So, the perihelion is $186 - 94.5 = 91.5$ million miles. The distance from the center of the ellipse to the sun (the focus) is $93 - 91.5 = 1.5$ million miles; therefore, $c = 1.5$. Since $b^2 = a^2 - c^2$, we have:

$$b^2 = 93^2 - 1.5^2$$
$$b^2 = 8646.75$$

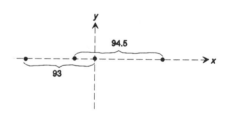

So, the equation of the earth's orbit around the sun is: $\dfrac{x^2}{93^2} + \dfrac{y^2}{8646.75} = 1$

69. The mean distance is $507 - 23.2 = 483.8$ million miles. Therefore, the perihelion is $483.3 - 23.2 = 406.6$ million miles. So, since $a = 483.8$ and $c = 23.2$, we can find b using $b^2 = a^2 - c^2$. Therefore, $b^2 = 483.8^2 - 23.2^2 = 233524.2$. So, the equation for the orbit of Pluto is

$$\frac{x^2}{483.8^2} + \frac{y^2}{233524.2} = 1.$$

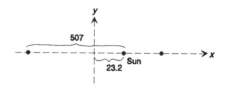

Chapter 10 Analytic Geometry

71. If the *x*-axis is placed along the 100 ft. portion and the *y*-axis along the 50 ft. portion, one equation for the ellipse is:

$$\frac{x^2}{(50)^2} + \frac{y^2}{(25)^2} = 1$$

When $x = 40$, then $\dfrac{(40)^2}{(50)^2} + \dfrac{y^2}{(25^2)} = 1$

$$\frac{y^2}{(25)^2} = 1 - \left[\frac{4}{5}\right]^2 = \frac{9}{25}$$

$$y^2 = (25)(9)$$

$$y = (5)(3) = 15$$

The width 10 feet from the side is 30 feet.

73. (a) $Ax^2 + Cy^2 + F = 0$
$Ax^2 + Cy^2 = -F$

If *A* and *C* are the same sign and *F* is of the opposite sign, then the equation takes the form $x^2/(-F/A) + y^2/(-F/C) = 1$, where $-F/A$ and $-F/C$ are positive. This is the equation of an ellipse with center at (0, 0).

(b) If $A = C$, the equation may be written as $x^2 + y^2 = -F/A$. This is the equation of a circle with center at (0, 0) and radius equal to $\sqrt{-F/A}$.

75. (a) $e = \dfrac{c}{a}$ is close to zero when *c* is close to zero. Since $c^2 = a^2 - b^2$, *c* is close to zero when $a^2 \approx b^2$. Hence, the ellipse is close to a circle.

(b) $e = \dfrac{c}{a} = \dfrac{1}{2}$ when $c = 1$ and $a = 2$. Thus, $b^2 = a^2 - c^2 = 4 - 1 = 3$. Hence, the ellipse is oval.

(c) $e = \dfrac{c}{a}$ is close to 1 when $c \approx a$ or $c^2 \approx a^2$. Hence, $a^2 - c^2$ will be close to zero. Since $b^2 = a^2 - c^2$, then b^2 is close to zero. Thus, the ellipse is elongated with the length of the minor axis small in comparison to the major axis.

10.4 The Hyperbola

1. B   3. A   5. B   7. C

For Problems 9–18, refer to Table 4. There we see that in order to find the equation of a hyperbola, we need to determine, h, k, a, b, and c to decide whether the transverse axis is horizontal or vertical.

We will use the following facts:
*(1) (h, k) are the coordinates of the **center** of the hyperbola, which is midway between the vertices, and also midway between the foci.*
(2) a = the distance from the center to either vertex.
(3) $b^2 = c^2 - a^2$, where c = the distance from the center to either focus.
(4) The center, the vertices and the foci all lie on the transverse axis.

9. We are given: Center: $C(0, 0)$
                      Focus: $F_2(3, 0)$
                      Vertex: $V_2(1, 0)$

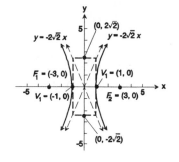

Please refer to the paragraph above.
By (1) we have $h = 0$, $k = 0$
By (2), $a = d(C, V_2) = 1$
By (3), $c = d(C, F_2) = 3$, and

$$b^2 = c^2 - a^2 = 8, \text{ so that } b = \sqrt{8} = 2\sqrt{2}$$

By (4), since C, F_2 and V_2 lie on the horizontal line $y = 0$, the transverse axis is parallel to the x-axis.

Then by Table 4, the equation is: $\dfrac{(x - h)^2}{a^2} - \dfrac{(y - k)^2}{b^2} = 1$ or $\dfrac{x^2}{1} - \dfrac{y^2}{8} = 1$

As an aid in sketching the graph, we locate the asymptotes of the hyperbola. First, plot the points that lie on the conjugate axis (perpendicular to the transverse axis) a distance b from the center: $\left(0, -2\sqrt{2}\right)$ and $\left(0, 2\sqrt{2}\right)$. These two points, together with the vertices, determine a rectangle whose diagonals are the asymptotes of the hyperbola.

11. Here we are given: Center: $C(0, 0)$
                        Focus: $F_1(0, -6)$
                        Vertex: $V_2(0, 4)$

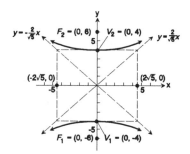

Please refer to the paragraph before the solution to Problem 9.
By (1), $h = 0$, $k = 0$
By (2), $a = d(C, V_2) = 4$
By (3), $c = d(C, F_1) = 6$, and $b^2 = c^2 - a^2 = 20$, so that

$$b = \sqrt{20} = 2\sqrt{5}$$

By (4), the transverse axis is the vertical line $x = 0$.

Hence, from Table 4, we have: $\dfrac{(y - k)^2}{a^2} - \dfrac{(x - h)^2}{b^2} = 1$ or $\dfrac{y^2}{16} - \dfrac{x^2}{20} = 1$

13. We are given: Foci: $F_1(-5, 0)$ and $F_2(5, 0)$
                     Vertex: $V_2(3, 0)$

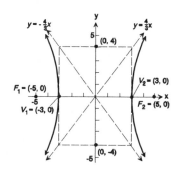

Refer to the paragraph before the solution to Problem 9.

By (1), the center is the midpoint between F_1 and F_2: $C(0, 0)$, so
     $h = 0$, $k = 0$.
By (2), $a = d(C, V_2) = 3$
By (3), $c = d(C, F_1) = 5$, and $b^2 = c^2 - a^2 = 16$
     $b = 4$
By (4), the transverse axis is the horizontal line $y = 0$.
Therefore, the equation is:

$$\frac{(x - h)^2}{a^2} - \frac{(y - k)^2}{b^2} = 1 \quad \text{or} \quad \frac{x^2}{9} - \frac{y^2}{16} = 1$$

15. We are given: Vertices: $V_1(0, -6)$ and $V_2(0, 6)$

Asymptote: $y = 2x$

The center is the midpoint between V_1 and V_2: $C(0, 0)$. Then $a = d(C, V_1) = 6$. The transverse axis is the vertical line $x = 0$. From Table 4, we see that the asymptotes of a hyperbola with a vertical transverse axis are:

$$y - k = \pm\frac{a}{b}(x - h)$$

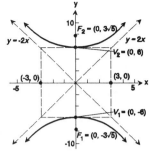

Here, $h = 0$, $k = 0$, so one asymptote would be: $y = \frac{a}{b}x$

Comparing this with the given asymptote, $y = 2x$, we find:

$$\frac{a}{b} = 2$$
$$a = 2b$$
$$b = \frac{1}{2}a$$
$$b = 3 \quad \text{(since } a = 6\text{)}$$

Then the equation of the hyperbola is: $\dfrac{(y - k)^2}{a^2} - \dfrac{(x - h)^2}{b^2} = 1$ or $\dfrac{y^2}{36} - \dfrac{x^2}{9} = 1$

17. Here we are given: Foci: $F_1(-4, 0)$ and $F_2(4, 0)$

Asymptote: $y = -x$

The center is the midpoint between F_1 and F_2: $C(0, 0)$. The transverse axis is the **horizontal** line $y = 0$.

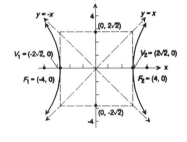

By Table 4, the asymptotes are:

$$y - k = \pm\frac{b}{a}(x - h) \quad \text{or} \quad y = \pm\frac{b}{a}x \quad \text{(since } h = 0, k = 0\text{)}$$

Comparing this with the given asymptote, $y = -x$, we see:

$$-\frac{b}{a} = -1$$

$$b = a$$

Now $c = d(C, F_1) = 4$, and
$$b^2 = c^2 - a^2$$
$$a^2 + b^2 = c^2$$
$$a^2 + b^2 = 16 \qquad (c = 4)$$
$$a^2 + a^2 = 16 \qquad (b = a)$$
$$2a^2 = 16$$
$$a^2 = 8$$
$$a = \sqrt{8} = 2\sqrt{2}$$
$$b = \sqrt{8} = 2\sqrt{2} \qquad (b = a)$$

The equation is: $\dfrac{(x - h)^2}{a^2} - \dfrac{(y - k)^2}{b^2} = 1$ or $\dfrac{x^2}{8} - \dfrac{y^2}{8} = 1$

19. $\dfrac{x^2}{16} - \dfrac{y^2}{4} = 1$

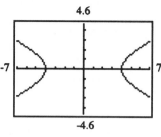

This is in the form: $\dfrac{(x-h)^2}{a^2} - \dfrac{(y-k)^2}{b^2} = 1$

with $h = 0$, $k = 0$, $a = 4$, and $b = 2$.

From Table 4, we have:

$\qquad c^2 = a^2 + b^2 = 20$, so $c = \sqrt{20} = 2\sqrt{5}$, and

$\qquad$ Center: $(h, k) = (0, 0)$

$\qquad$ Transverse axis: Parallel to x-axis, and contains the Center: $y = 0$

$\qquad\qquad$ Foci: $(h \pm c, k)$: $\left(-2\sqrt{5},\, 0\right)$ and $\left(2\sqrt{5},\, 0\right)$

$\qquad\qquad$ Vertices: $(h \pm a, k)$: $(-4, 0)$ and $(4, 0)$

$\qquad$ Asymptotes: $y - k = \pm\dfrac{b}{a}(x - h)$, or $y = \pm\dfrac{2}{4}x = \pm\dfrac{1}{2}x$

$\qquad$ (Lines through $(0, 0)$ with slopes $\dfrac{1}{2}$ and $-\dfrac{1}{2}$.)

Graph $y_1 = 2\sqrt{\dfrac{x^2}{16} - 1}$ and $y_2 = -2\sqrt{\dfrac{x^2}{16} - 1}$

21. $4x^2 - y^2 = 16$

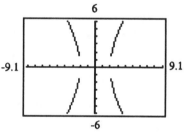

$\qquad \dfrac{4x^2}{16} - \dfrac{y^2}{16} = 1$ $\quad$ Obtain a 1 on the right-hand side.

$\qquad \dfrac{x^2}{4} - \dfrac{y^2}{16} = 1$ $\quad$ Simplify.

This is a hyperbola with transverse axis parallel to the x-axis (since the x^2-term is the positive one), with $h = 0$, $k = 0$, $a^2 = 4$ and $b^2 = 16$. Then $c^2 = a^2 + b^2 = 20$.

$\qquad$ Therefore, $\qquad a = 2$

$\qquad\qquad\qquad\qquad b = 4$

$\qquad\qquad\qquad\qquad c = \sqrt{20} = 2\sqrt{5}$

$\qquad\qquad$ Center: $(h, k) = (0, 0)$

$\qquad$ Transverse axis: $y = 0$

$\qquad\qquad$ Foci: $(h \pm c, k)$: $(-2\sqrt{5},\, 0)$ and $(2\sqrt{5},\, 0)$

$\qquad\qquad$ Vertices: $(h \pm a, k)$: $(-2, 0)$ and $(2, 0)$

$\qquad$ Asymptotes: $y - k = \pm\dfrac{b}{a}(x - h)$, or $y = \pm 2x$

(Lines through $(0, 0)$ with slopes 2 and -2.)

Graph $y_1 = 4\sqrt{\dfrac{x^2}{4} - 1}$ and $y_2 = -4\sqrt{\dfrac{x^2}{4} - 1}$

23. $y^2 - 9x^2 = 9$

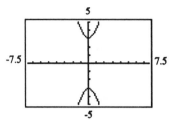

$$\frac{y^2}{9} - \frac{x^2}{1} = 1$$

This is a hyperbola with transverse axis parallel to the y-axis (since the y^2 term is positive), with $h = 0$, $k = 0$, $a^2 = 9$ and $b^2 = 1$.
Therefore, $a = 3$
$\qquad\qquad b = 1$

$c^2 = a^2 + b^2 = 10$, so $c = \sqrt{10}$
$\qquad\qquad$ Center: $(h, k) = (0, 0)$
$\qquad$ Transverse axis: $x = 0$

$\qquad\qquad$ Foci: $(h, k \pm c)$: $\left(0, -\sqrt{10}\right)$ and $\left(0, \sqrt{10}\right)$
$\qquad\qquad$ Vertices: $(h, k \pm a)$: $(0, -3)$ and $(0, 3)$

$\qquad$ Asymptotes: $y - k = \pm\dfrac{a}{b}(x - h)$, or $y = \pm 3x$

Graph $y_1 = 3\sqrt{1 + x^2}$ and $y_2 = -3\sqrt{1 + x^2}$

25. $y^2 - x^2 = 25$

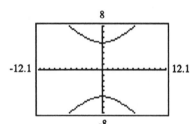

$$\frac{y^2}{25} - \frac{x^2}{25} = 1$$

This is a hyperbola with transverse axis parallel to the y-axis (since the y^2-term is positive), with $h = 0$, $k = 0$, $a^2 = 25$, $b^2 = 25$. Therefore, $c^2 = a^2 + b^2 = 50$, and we have:
$\qquad a = 5$
$\qquad b = 5$
$\qquad c = \sqrt{50} = 5\sqrt{2}$
$\qquad\qquad$ Center: $(h, k) = (0, 0)$
$\qquad$ Transverse axis: $x = 0$

$\qquad\qquad$ Foci: $(h, k \pm c)$: $\left(0, -5\sqrt{2}\right)$ and $\left(0, 5\sqrt{2}\right)$
$\qquad\qquad$ Vertices: $(h, k \pm a)$: $(0, -5)$ and $(0, 5)$

$\qquad$ Asymptotes: $y - k = \pm\dfrac{a}{b}(x - h)$, or $y = \pm x$

Graph $y_1 = \sqrt{25 + x^2}$ and $-\sqrt{25 + x^2}$

27. $x^2 - y^2 = 1$ $\qquad\qquad\qquad\qquad$ **29.** $\dfrac{y^2}{36} - \dfrac{x^2}{9} = 1$

31. We are given:$\qquad$ Center: $C(4, -1)$
$\qquad\qquad\qquad\qquad\quad$ Focus: $F_2(7, -1)$
$\qquad\qquad\qquad\qquad\quad$ Vertex: $V_2(6, -1)$

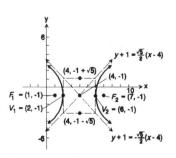

Please refer to the paragraph before the solution to Problem 9.
By (1), $h = 4$, $k = -1$
By (2), $a = d(C, V_2) = 2$
By (3), $c = d(C, F_2) = 3$, and
$\qquad\qquad b^2 = c^2 - a^2 = 5$, so that $b = \sqrt{5}$
By (4), the Transverse axis is the horizontal line $y = -1$

Then, by Table 4, we have: $\dfrac{(x - h)^2}{a^2} - \dfrac{(y - k)^2}{b^2} = 1$ or $\dfrac{(x - 4)^2}{4} - \dfrac{(y + 1)^2}{5} = 1$

33. We are given: Center: $C(-3, -4)$

                   Focus: $F_1(-3, -8)$

                  Vertex: $V_2(-3, -2)$

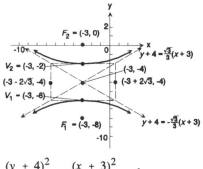

Refer to the paragraph before the solution to Problem 9.

By (1), $h = -3$, $k = -4$

By (2), $a = d(C, V_2) = 2$

By (3), $c = d(C, F_1) = 4$

$$b^2 = c^2 - a^2 = 12$$

$$b = \sqrt{12} = 2\sqrt{3}$$

By (4), the transverse axis is the vertical line $x = -3$

Then, by Table 4, we have: $\dfrac{(y - k)^2}{a^2} - \dfrac{(x - h)^2}{b^2} = 1$ or $\dfrac{(y + 4)^2}{4} - \dfrac{(x + 3)^2}{12} = 1$

35. We are given: Foci: $F_1(3, 7)$ and $F_2(7, 7)$

                   Vertex: $V_1(6, 7)$

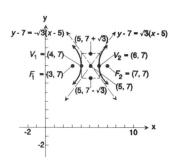

Refer to the paragraph before the solution to Problem 9.

By (1) the Center is midway between F_1 and F_2: $C(5, 7)$, so

    $h = 5$, $k = 7$.

By (2), $a = d(C, V_1) = 1$

By (3), $c = d(C, F_1) = 2$

    $b^2 = c^2 - a^2 = 3$

       $b = \sqrt{3}$

By (4), the transverse axis is the *horizontal* line $y = 7$

By Table 4, $\dfrac{(x - h)^2}{a^2} - \dfrac{(y - k)^2}{b^2} = 1$ or $\dfrac{(x - 5)^2}{1} - \dfrac{(y - 7)^2}{3} = 1$

37. We are given: Vertices: $V_1(-1, -1)$ and $V_2(3, -1)$

                   Asymptote: $\dfrac{(x - 1)}{2} = \dfrac{(y + 1)}{3}$

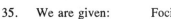

Refer to the paragraph before the solution to Problem 9.

By (1) the center is $C(1, -1)$.

By (2), $a = d(C, V_1) = 2$

By (4), the transverse axis is the *horizontal* line

    $y = -1$.

We still need b, and we don't know c. From Table 4, the asymptotes would be:

$$y - k = \pm \frac{b}{a}(x - h), \text{ or } y + 1 = \pm \frac{b}{a}(x - 1)$$

Compare that formula with the *given* asymptote:

$$\frac{(x - 1)}{2} = \frac{(y + 1)}{3}, \text{ or } y + 1 = \frac{3}{2}(x - 1)$$

We see that $\dfrac{b}{a} = \dfrac{3}{2}$ or $b = \dfrac{3a}{2}$

$$b = \frac{6}{2} \quad \text{(since } a = 2\text{)}$$

$$b = 3$$

Therefore, the equation is: $\dfrac{(x - h)^2}{a^2} - \dfrac{(y - k)^2}{b^2} = 1$ or $\dfrac{(x - 1)^2}{4} - \dfrac{(y + 1)^2}{9} = 1$

             Chapter 10 Analytic Geometry

39. $\dfrac{(x-2)^2}{4} - \dfrac{(y+3)^2}{9} = 1$

This is in the form found in Table 4. Since the x^2-term is positive, the transverse axis is parallel to the x-axis. We have $h = 2$, $k = -3$, $a^2 = 4$, and $b^2 = 9$. Therefore,

$a = 2$
$b = 3$
$c^2 = a^2 + b^2 = 13$, so that

$c = \sqrt{13}$

Center: $(h, k) = (2, -3)$

Foci: $(h \pm c, k)$: $\left(2 - \sqrt{13}, -3\right)$ and $\left(2 + \sqrt{13}, -3\right)$
Vertices: $(h \pm a, k)$: $(0, -3)$ and $(4, -3)$

Asymptotes: $y - k = \pm\dfrac{b}{a}(x - h)$, or $y + 3 = \pm\dfrac{3}{2}(x - 2)$

(Lines through $(3, -2)$ with slopes $\dfrac{3}{2}$ and $-\dfrac{3}{2}$.)

Graph $y_1 = -3 + 3\sqrt{\dfrac{(x-2)^2}{4} - 1}$ and $y_2 = -3 - 3\sqrt{\dfrac{(x-2)^2}{4} - 1}$

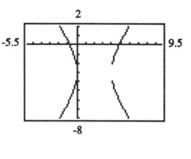

41. $(y - 2)^2 - 4(x + 2)^2 = 4$

$\dfrac{(y-2)^2}{4} - \dfrac{(x+2)^2}{1} = 1$

This is a hyperbola with $h = -2$, $k = 2$; the transverse axis is parallel to the y-axis, and $a^2 = 4$, $b^2 = 1$. Then:

$a = 2$
$b = 1$
$c^2 = a^2 + b^2 = 5$, so

$c = \sqrt{5}$

Center: (h, k): $(-2, 2)$

Foci: $(h, k \pm c)$: $(-2, 2 - \sqrt{5})$ and $(-2, 2 + \sqrt{5})$
Vertices: $(h, k \pm a)$: $(-2, 0)$ and $(-2, 4)$

Asymptotes: $y - k = \pm\dfrac{a}{b}(x - h)$, or $y - 2 = \pm2(x + 2)$

(Lines through $(-2, 2)$ with slopes 2 and -2.)

Graph $y_1 = 2 + 2\sqrt{1 + (x + 2)^2}$ and $y_2 = 2 - 2\sqrt{1 + (x + 2)^2}$

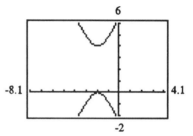

43. $(x + 1)^2 - (y + 2)^2 = 4$

$\dfrac{(x+1)^2}{4} - \dfrac{(y+2)^2}{4} = 1$

This is a hyperbola with $h = -1$, $k = -2$; the transverse axis is parallel to the x-axis, and $a^2 = 4$, $b^2 = 4$. Then:

$a = 2$
$b = 2$

$c^2 = a^2 + b^2 = 8$, so $c = 2\sqrt{2}$

Center: $(h, k) = (-1, -2)$

Foci: $(h \pm c, k)$: $\left(-1 - 2\sqrt{2}, -2\right)$ and $\left(-1 + 2\sqrt{2}, -2\right)$
Vertices: $(h \pm a, k)$: $(-3, -2)$ and $(1, -2)$

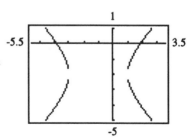

Asymptotes: $y - k = \pm \dfrac{b}{a}(x - h)$, or $y + 2 = \pm(x + 1)$

(Lines through $(-1, -2)$ with slopes 1 and -1.)

Graph $y_1 = -2 + 2\sqrt{\dfrac{(x+1)^2}{4} - 1}$ and $y_2 = -2 - 2\sqrt{\dfrac{(x+1)^2}{4} - 1}$

45.
$$x^2 - y^2 - 2x - 2y - 1 = 0$$
$$(x^2 - 2x + 1) - (y^2 + 2y + 1) = 1 + 1 - 1$$
$$(x - 1)^2 - (y + 1)^2 = 1$$

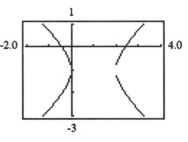

This is a hyperbola with $h = 1$, $k = -1$; the transverse axis is parallel to the x-axis, and $a^2 = 1$, $b^2 = 1$. Then:

$a = 1$
$b = 1$
$c^2 = a^2 + b^2 = 2$, so

$c = \sqrt{2}$

Center: $(h, k) = (1, -1)$

Foci: $(h \pm c, k)$: $\left(1 - \sqrt{2}, -1\right)$ and $\left(1 + \sqrt{2}, -1\right)$
Vertices: $(h \pm a, k)$: $(0, -1)$ and $(2, -1)$
Asymptotes: $y + 1 = \pm(x - 1)$
(Lines through $(1, -1)$ with slopes 1 and -1.)

Graph $y_1 = -1 + \sqrt{(x - 1)^2 - 1}$ and $y_2 = -1 - \sqrt{(x - 1)^2 - 1}$

47.
$$y^2 - 4x^2 - 4y - 8x - 4 = 0$$
$$(y^2 - 4y + 4) - 4(x^2 + 2x + 1) = 4 + 4 - 4$$
$$(y - 2)^2 - 4(x + 1)^2 = 4$$
$$\dfrac{(y - 2)^2}{4} - (x + 1)^2 = 1$$

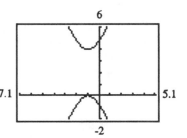

This is a hyperbola with $h = -1$, $k = 2$; the transverse axis is parallel to the y-axis, and $a^2 = 4$, $b^2 = 1$. Then:

$a = 2$
$b = 1$

$c^2 = a^2 + b^2 = 5$, so $c = \sqrt{5}$

Center: $(h, k) = (-1, 2)$

Foci: $(h, k \pm c)$: $\left(-1, 2 - \sqrt{5}\right)$ and $\left(-1, 2 + \sqrt{5}\right)$
Vertices: $(h, k \pm a)$: $(-1, 0)$ and $(-1, 4)$
Asymptotes: $y - 2 = \pm 2(x + 1)$
(Lines through $(-1, 2)$ with slopes 2 and -2.)

Graph $y_1 = 2 + 2\sqrt{1 + (x + 1)^2}$ and $y_2 = 2 - 2\sqrt{1 + (x + 1)^2}$

49.

$$4x^2 - y^2 - 24x - 4y + 16 = 0$$
$$4x^2 - 24x - y^2 - 4y = -16$$
$$4(x^2 - 6x + \underline{}) - 1(y^2 + 4y + \underline{}) = -16$$
$$4(x^2 - 6x + 9) - 1(y^2 + 4y + 4) = -16 + 36 - 4$$
$$4(x - 3)^2 - (y + 2)^2 = 16$$
$$\frac{(x - 3)^2}{4} - \frac{(y + 2)^2}{16} = 1$$

This is now in a form we can recognize: A hyperbola with transverse axis parallel to the x-axis (since the x^2-term is positive), with center at $C(3, -2)$, and $a^2 = 4$, and $b^2 = 16$. Then $c^2 = a^2 + b^2 = 20$, and we have:

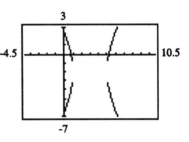

$a = 2$

$b = 4$

$c = \sqrt{20} = 2\sqrt{5}$

Center: $C(3, -2)$

Foci: $(h \pm c, k)$: $\left(3 - 2\sqrt{5},\ -2\right)$ and $\left(3 + 2\sqrt{5},\ -2\right)$

Vertices: $(h \pm a, k)$: $(1, -2)$ and $(5, -2)$

Asymptotes: $y - k = \pm\frac{b}{a}(x - h)$, or $y + 2 = \pm 2(x - 3)$

Graph $y_1 = -2 + 4\sqrt{\dfrac{(x - 3)^2}{4} - 1}$ and $y_2 = -2 - 4\sqrt{\dfrac{(x - 3)^2}{4} - 1}$

51.

$$y^2 - 4x^2 - 16x - 2y - 19 = 0$$
$$y^2 - 2y - 4x^2 - 16x = 19$$
$$(y^2 - 2y + \underline{}) - 4(x^2 + 4x + \underline{}) = 19$$
$$(y^2 - 2y + 1) - 4(x^2 + 4x + 4) = 19 + 1 - 16$$
$$(y - 1)^2 - 4(x + 2)^2 = 4$$
$$\frac{(y - 1)^2}{4} - \frac{(x + 2)^2}{1} = 1$$

This is the equation of a hyperbola with transverse axis parallel to the y-axis, with center at $C(-2, 1)$, with $a^2 = 4$ and $b^2 = 1$. Then $c^2 = a^2 + b^2 = 5$, and we have:

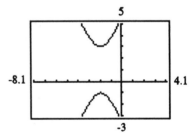

$a = 2$

$b = 1$

$c = \sqrt{5}$

Center: $C(-2, 1)$

Foci: $(h, k \pm c)$: $\left(-2, 1 - \sqrt{5}\right)$ and $\left(-2, 1 + \sqrt{5}\right)$

Vertices: $(h, k \pm a)$: $(-2, -1)$ and $(-2, 3)$

Asymptotes: $y - k = \pm\frac{a}{b}(x - h)$, or $y - 1 = \pm 2(x + 2)$

Graph $y_1 = 1 + 2\sqrt{1 + (x + 2)^2}$ and $y_2 = 1 - 2\sqrt{1 + (x + 2)^2}$

53.
$$y = \sqrt{16 + 4x^2}$$
$$y^2 = 16 + 4x^2, \qquad y \geq 0$$
$$y^2 - 4x^2 = 16, \qquad y \geq 0$$
$$\frac{y^2}{16} - \frac{x^2}{4} = 1, \qquad y \geq 0$$

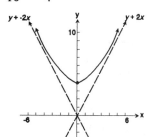

55.
$$y = -\sqrt{-25 + x^2}$$
$$y^2 = -25 + x^2, \qquad y \leq 0$$
$$y^2 - x^2 = -25, \qquad y \leq 0$$
$$\frac{y^2}{25} - \frac{x^2}{25} = -1, \qquad y \leq 0$$
$$\frac{x^2}{25} - \frac{y^2}{25} = 1, \qquad y \leq 0$$

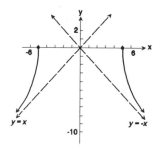

57. (a) Set up a rectangular coordinate system so that the two stations lie on the x-axis and the origin is midway between them. The ship lies on a hyperbola whose foci are the locations of the two stations. Since the time difference is .00038 seconds and the speed of the signal is 186,000 miles per second, the difference of the distances from the ship to each station (foci) is:

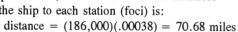

$$\text{distance} = (186,000)(.00038) = 70.68 \text{ miles}$$

The difference of the distances from the ship to each station, 70.68, equals $2a$, so $a = 35.34$ and the vertex of the corresponding hyperbola is at (35.34, 0). Since the focus is at (100, 0), following this hyperbola, the ship would reach shore 64.66 miles from the master station.

(b) The ship should follow a hyperbola with vertex at (80, 0). For this hyperbola, $a = 80$, so the constant difference of the distances from the ship to each station is 160. The time difference the ship should look for is:

$$\text{time} = \frac{160}{186,000} = 0.00086 \text{ seconds}$$

(c) We need to find the equation of the hyperbola with vertex at (80, 0) and a focus at (100, 0). The form of the equation of this hyperbola is:

$$\frac{x^2}{a^2} - \frac{y^2}{b^2} = 1$$

where $a = 80$. Since $c = 100$ and $b^2 = c^2 - a^2$, we have $b^2 = 100^2 - 80^2 = 3600$. So, the equation of the hyperbola is:

$$\frac{x^2}{6400} - \frac{y^2}{3600} = 1$$

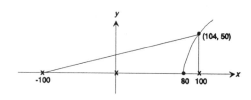

Since the ship is 50 miles off shore, we have $y = 50$. Solve the above equation for x.

$$\frac{x^2}{6400} - \frac{50^2}{3600} = 1$$

$$\frac{x^2}{6400} = 1 + \frac{2500}{3600} = \frac{6100}{3600}$$

$$x^2 = 6400 \cdot \frac{6100}{3600}$$

$$x \approx 104 \text{ miles}$$

The ship is at the position (104, 50).

59. (a) Set up a rectangular coordinate system so the two devices lie on the x-axis and the origin is midway between them. The devices serve as foci to the hyperbola so $c = \dfrac{2000}{2} = 1000$.

Since the explosion occurs 200 feet from point B, the vertex of the hyperbola will be (800, 0); therefore, $a = 800$. Since $b^2 = c^2 - a^2$, we have $b^2 = 1000^2 - 800^2 = 360{,}000$. Therefore, the equation of the hyperbola is:

$$\frac{x^2}{800^2} - \frac{y^2}{360{,}000} = 1$$

If $x = 1000$ feet, then we can find y.

$$\frac{1000^2}{800^2} - \frac{y^2}{360{,}000} = 1$$

$$y^2 = 360{,}000 \left[\frac{360{,}000}{640{,}000} \right]$$

$$y = 450 \text{ feet}$$

Therefore, the second detonation should occur 450 feet above Point B.

61. By definition of the eccentricity, e,

$$e = \frac{c}{a}, \text{ or } c = ae$$

Therefore, if $e \approx 1$, then $c \approx a$, and $b^2 = c^2 - a^2 \approx 0$, so that b is close to 0.

Assume, for the sake of simplicity, that we have a hyperbola, centered at (0, 0), with transverse axis lying along the x-axis. The asymptotes are:

$$y = \pm \frac{b}{a} x,$$

i.e., lines through the origin with slope $\pm \dfrac{b}{a}$. Now, if $e \approx 1$, we have $b \approx 0$, so the slopes of the asymptotes are nearly 0. Hence, the asymptotes are nearly horizontal, so the hyperbola is very narrow.

On the other hand, if e is very large, we have:

$$c = ae \text{ and } \quad b^2 = c^2 - a^2$$
$$= e^2 a^2 - a^2$$
$$= (e^2 - 1)a^2, \text{ and}$$
$$b = \left(\sqrt{e^2 - 1} \right) a$$

If e is much larger than 1, then b will be much larger than a.

In this case, a hyperbola with horizontal transverse axis will have asymptotes with slopes

$$\pm \frac{b}{a} = \pm \frac{\left(\sqrt{e^2 - 1} \right) a}{a} = \pm \sqrt{e^2 - 1} > 1.$$

Thus, the asymptotes will be nearly vertical, producing a *wide* hyperbola. As an example, look at the graph of the hyperbola in Problem 9. There, $e = \dfrac{c}{a} = \dfrac{4}{1} = 4$, and the asymptotes have slopes $\pm\sqrt{e^2 - 1} = \pm\sqrt{15} \approx \pm 3.9$. As you can see, the hyperbola is very wide.

63. $\dfrac{x^2}{4} - y^2 = 1 \quad (a^2 = 4,\ b^2 = 1)$

is a hyperbola with *horizontal* transverse axis, centered at $(0, 0)$ with asymptotes

$$y - k = \pm\dfrac{b}{a}(x - h),\ \text{or}\ y = \pm\dfrac{1}{2}x$$

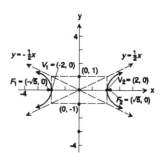

$y^2 - \dfrac{x^2}{4} = 1 \quad (a^2 = 1,\ b^2 = 4)$

is a hyperbola with *vertical* transverse axis, also is centered at $(0, 0)$ and has asymptotes

$$y - k = \pm\dfrac{a}{b}(x - h),\ \text{or}\ y = \pm\dfrac{1}{2}x$$

Since the two hyperbolas have the same asymptotes, they are conjugate.

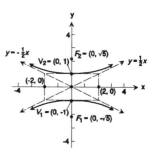

65. $Ax^2 + Cy^2 + F = 0 \quad A \neq 0,\ C \neq 0,\ F \neq 0$
$Ax^2 + Cy^2 = -F$
Since A and C are of opposite signs and $F \neq 0$, this equation may be written as
$x^2/(-F/A) + y^2/(-F/C) = 1$, where $-F/A$ and $-F/C$ are opposite in sign. This is the equation of a hyperbola with center at $(0, 0)$. The transverse axis is the x-axis if $-F/A > 0$. The transverse axis is the y-axis if $-F/A < 0$.

10.5 Rotation of Axes; General Form of a Conic

1. $x^2 + 4x + y + 3 = 0$
   Here $A = 1$, $B = 0$, and $C = 0$, so that $B^2 - 4AC = 0$.
   Since $B^2 - 4AC = 0$, the equation defines a parabola.

3. $6x^2 + 3y^2 - 12x + 6y = 0$
   Here $A = 6$, $B = 0$, and $C = 3$, so that $B^2 - 4AC = -72$.
   Since $B^2 - 4AC < 0$, the equation defines an ellipse.

5. $3x^2 - 2y^2 + 6x + 4 = 0$
   Here $A = 3$, $B = 0$, and $C = -2$, so that $B^2 - 4AC = 24$.
   Since $B^2 - 4AC > 0$, the equation defines a hyperbola.

7. $2y^2 - x^2 - y + x = 0$
   Here $A = -1$, $B = 0$, and $C = 2$, so that $B^2 - 4AC = 8 > 0$.
   The equation defines a hyperbola.

9. $x^2 + y^2 - 8x + 4y = 0$

Here $A = 1$, $B = 0$, and $C = 1$, so that $B^2 - 4AC = -4 < 0$.
The equation defines an ellipse, specifically a circle.

For Problems 11–12, we use the formulas $\cot 2\theta = \dfrac{A - C}{B}$, $x = x'\cos\theta - y'\sin\theta$, *and*
$y = x'\sin\theta + y'\cos\theta.$

11. $A = 1$, $B = 4$, $C = 1$, $\cot 2\theta = 0$ so that $\theta = \dfrac{\pi}{4}$

$x = \dfrac{\sqrt{2}}{2}(x' - y');\ y = \dfrac{\sqrt{2}}{2}(x' + y')$

13. $A = 5$, $B = 6$, $C = 5$, $\cot 2\theta = 0$ so that $\theta = \dfrac{\pi}{4}$

$x = \dfrac{\sqrt{2}}{2}(x' - y');\ y = \dfrac{\sqrt{2}}{2}(x' + y')$

15. $A = 13$, $B = -6\sqrt{3}$, $C = 7$, $\cot 2\theta = \dfrac{6}{-6\sqrt{3}} = \dfrac{-1}{\sqrt{3}}$; $\cos 2\theta = \dfrac{-1}{2}$

$\sin\theta = \sqrt{\dfrac{1 + \dfrac{1}{2}}{2}} = \dfrac{\sqrt{3}}{2}$; $\cos\theta = \sqrt{\dfrac{1 - \dfrac{1}{2}}{2}} = \dfrac{1}{2}$

$x = \dfrac{1}{2}x' - \dfrac{\sqrt{3}}{2}y' = \dfrac{1}{2}\left(x' - \sqrt{3}y'\right)$

$y = \dfrac{\sqrt{3}}{2}x' + \dfrac{1}{2}y' = \dfrac{1}{2}\left(\sqrt{3}x' + y'\right)$

17. $A = 4$, $B = -4$, $C = 1$, $\cot 2\theta = \dfrac{-3}{4}$; $\cos 2\theta = \dfrac{-3}{5}$

$\sin\theta = \sqrt{\dfrac{1 + \dfrac{3}{5}}{2}} = \dfrac{2}{\sqrt{5}} = \dfrac{2\sqrt{5}}{5}$; $\cos\theta = \sqrt{\dfrac{1 - \dfrac{3}{5}}{2}} = \dfrac{1}{\sqrt{5}} = \dfrac{\sqrt{5}}{5}$

$x = \dfrac{\sqrt{5}}{5}x' - \dfrac{2\sqrt{5}}{2}y' = \dfrac{\sqrt{5}}{5}(x' - 2y')$ $\qquad$ $y = \dfrac{2\sqrt{5}}{5}x' + \dfrac{\sqrt{5}}{5}y' = \dfrac{\sqrt{5}}{5}(2x' + y')$

19. $A = 25$, $B = -36$, $C = 40$, $\cot 2\theta = \dfrac{-15}{-36} = \dfrac{5}{12}$; $\cos 2\theta = \dfrac{5}{13}$

$\sin\theta = \sqrt{\dfrac{1 - \dfrac{5}{13}}{2}} = \dfrac{2}{\sqrt{13}} = \dfrac{2\sqrt{13}}{13}$; $\cos\theta = \sqrt{\dfrac{1 + \dfrac{5}{13}}{2}} = \dfrac{3}{\sqrt{13}} = \dfrac{3\sqrt{13}}{13}$

$x = \dfrac{3\sqrt{13}}{13}x' - \dfrac{2\sqrt{13}}{13}y' = \dfrac{\sqrt{13}}{13}(3x' - 2y')$ $\qquad$ $y = \dfrac{2\sqrt{13}}{13}xa' + \dfrac{3\sqrt{13}}{13}y' = \dfrac{\sqrt{13}}{13}(2x' + 3y')$

21.

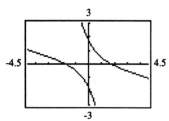

$\theta = 45°$ (see Problem 11)

$x^2 + 4xy + y^2 - 3 = 0$

$$\left[\frac{\sqrt{2}}{2}(x' - y')\right]^2 + 4\left[\frac{\sqrt{2}}{2}(x' - y')\right]\left[\frac{\sqrt{2}}{2}(x' + y')\right] + \left[\frac{\sqrt{2}}{2}(x' + y')\right]^2 - 3 = 0$$

$$\frac{1}{2}(x'^2 - 2x'y' + y'^2) + \frac{4}{2}(x'^2 - y'^2) + \frac{1}{2}(x'^2 + 2x'y' + y'^2) = 3$$

$$6x'^2 - 2y'^2 = 6$$

$$x'^2 - \frac{y'^2}{3} = 1$$

Hyperbola; center at origin; transverse axis the x'-axis; vertices at $(\pm 1, 0)$.

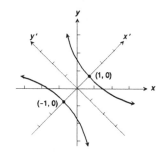

23.

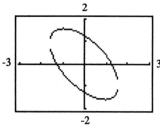

$\theta = 45°$ (see Problem 13)

$5x^2 + 6xy + 5y^2 - 8 = 0$

$$5\left[\frac{\sqrt{2}}{2}(x' - y')\right]^2 + 6\left[\frac{\sqrt{2}}{2}(x' - y')\right]\left[\frac{\sqrt{2}}{2}(x' + y')\right] + 5\left[\frac{\sqrt{2}}{2}(x' + y')\right]^2 - 8 = 0$$

$$\frac{5}{2}(x'^2 - 2x'y' + y'^2) + \frac{6}{2}(x'^2 - y'^2) + \frac{5}{2}(x'^2 + 2x'y' + y'^2) = 8$$

$$16x'^2 + 4y'^2 = 16$$

$$x'^2 + \frac{y'^2}{4} = 1$$

Ellipse; center at $(0, 0)$; major axis the y'-axis; vertices at $(0, \pm 2)$.

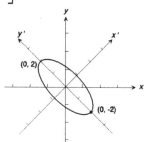

25.

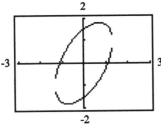

$\theta = 60°$ (see Problem 15)

$13x^2 - 6\sqrt{3}\,xy + 7y^2 - 16 = 0$

$13\left[\frac{1}{2}(x' - \sqrt{3}\,y')\right]^2 - 6\sqrt{3}\left[\frac{1}{2}(x' - \sqrt{3}\,y')\right]\left[\frac{1}{2}(\sqrt{3}\,x' + y')\right] + 7\left[\frac{1}{2}(\sqrt{3}\,x' + y')\right]^2 = 16$

$13(x'^2 - 2\sqrt{3}\,x'y'^2 + 3y'^2) - 6\sqrt{3}(\sqrt{3}\,x'^2 - 2x'y' - \sqrt{3}\,y'^2) + 7(3x'^2 + 2\sqrt{3}\,x'y' + y'^2) = 64$

$16x'^2 + 64y'^2 = 64$

$\dfrac{x'^2}{4} + \dfrac{y'^2}{1} = 1$

Ellipse; center at $(0, 0)$; major axis is the x'-axis; vertices at $(\pm 2, 0)$.

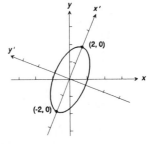

27.

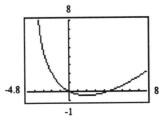

$\theta = 63°$ (see Problem 17)

$4x^2 - 4xy + y^2 - 8\sqrt{5}\,x - 16\sqrt{5}\,y = 0$

$4\left[\frac{\sqrt{5}}{5}(x' - 2y')\right]^2 - 4\left[\frac{\sqrt{5}}{5}(x' - 2y')\right]\left[\frac{\sqrt{5}}{5}(2x' + y')\right] + \left[\frac{\sqrt{5}}{5}(2x' + y')\right]^2$

$\qquad - 8\sqrt{5}\left[\frac{\sqrt{5}}{5}(x' - 2y')\right] - 16\sqrt{5}\left[\frac{\sqrt{5}}{5}(2x' + y')\right] = 0$

$\frac{4}{5}(x'^2 - 4x'y' + 4y'^2) - \frac{4}{5}(2x'^2 - 3x'y' - 2y'^2)$

$\qquad + \frac{1}{5}(4x'^2 + 4x'' + y'^2) - 8(x' - 2y')$

$\qquad - 16(2x' + y' = 0$

$5y'^2 - 40x' = 0$

$y'^2 = 8x'$

Parabola; vertex at $(0, 0)$; focus at $(2, 0)$.

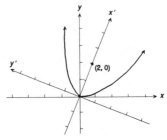

29.

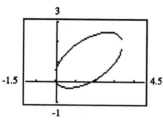

$\theta \approx 34°$ (see Problem 19)

$25x^2 - 36xy + 40y^2 - 12\sqrt{13}\,x - 8\sqrt{13}\,y = 0$

$$25\left[\frac{\sqrt{13}}{13}(3x' - 2y')\right]^2 - 36\left[\frac{\sqrt{13}}{13}(3x' - 2y')\right]\left[\frac{\sqrt{13}}{13}(2x' + 3y')\right]$$

$$+ 40\left[\frac{\sqrt{13}}{13}(2x' + 3y')\right]^2 - 12\sqrt{13}\left[\frac{\sqrt{13}}{13}(3x' - 2y')\right] - 8\sqrt{13}\left[\frac{\sqrt{13}}{13}(2x' + 3y')\right] = 0$$

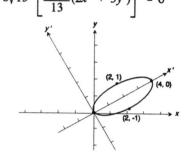

$$\frac{25}{13}\left(9x'^2 - 12x'y' + 4y'^2\right) - \frac{36}{13}\left(6x'^2 + 5x'y' - 6y'^2\right)$$

$$+ \frac{40}{13}\left(4x'^2 + 12x'y' + 9y'^2\right) - 12(3x' - 2y')$$

$$- 8(2x' + 3y') = 0$$

$$\frac{169}{13}x'^2 + \frac{676}{13}y'^2 - 52x' = 0$$

$$13x'^2 + 52y'^2 - 52x' = 0$$

$$x'^2 - 4x' + 4y'^2 = 0$$

$$(x' - 2)^2 + 4y'^2 = 4$$

$$\frac{(x' - 2)^2}{4} + y'^2 = 1$$

Ellipse; center at $(2, 0)$; major axis the x'-axis; vertices at $(4, 0)$ and $(0, 0)$.

31.

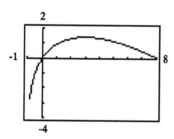

$A = 16$, $B = 24$, $C = 9$, $\cot 2\theta = \dfrac{7}{24}$; $\cos 2\theta = \dfrac{7}{25}$

$$\sin\theta = \sqrt{\frac{1 - \dfrac{7}{25}}{2}} = \frac{3}{5}; \quad \cos\theta = \sqrt{\frac{1 + \dfrac{7}{25}}{2}} = \frac{4}{5}; \theta \approx 37°$$

$x = \dfrac{4}{5}x' - \dfrac{3}{5}y' = \dfrac{1}{5}(4x' - 3y')$ $\qquad y = \dfrac{3}{5}x' + \dfrac{4}{5}y' = \dfrac{1}{5}(3x' + 4y')$

$16x^2 + 24xy + 9y^2 - 130x + 90y = 0$

$$16\left[\frac{1}{5}(4x' - 3y')\right]^2 + 24\left[\frac{1}{5}(4x' - 3y')\right]\left[\frac{1}{5}(3x' + 4y')\right]$$

$$9\left[\frac{1}{5}(3x' + 4y')\right]^2 - 130\left[\frac{1}{5}(4x' - 3y')\right] + 90\left[\frac{1}{5}(3x' + 4y')\right] = 0$$

$$\frac{16}{25}(16x'^2 - 24x'y' + 9y'^2) + \frac{24}{25}(12x'^2 + 7x'y' - 12y'^2)$$

$$+ \frac{9}{25}(9x'^2 + 24x'y' + 16y'^2) - 26(4x' - 3y')$$

$$+ 18(3x' + 4y') = 0$$

$$\frac{625}{25}x'^2 - 50x' + 150y' = 0$$

$$x'^2 - 2x' = -6y'$$

$$(x' - 1)^2 = -6\left[y' - \frac{1}{6}\right]$$

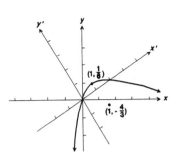

Parabola; vertex at $\left[1, \frac{1}{6}\right]$; focus at $\left[1, \frac{-4}{3}\right]$.

33.　$B^2 - 4AC = 9 + 8 = 17 > 0$; hyperbola　35.　$B^2 - 4AC = 49 - 12 = 37 > 0$; hyperbola

37.　$B^2 - 4AC = 144 - 144 = 0$; parabola　39.　$B^2 - 4AC = 144 - 160 = -16 < 0$; ellipse

41.　$B^2 - 4AC = 4 - 12 = -8 < 0$; ellipse

43.　Refer to Equation (6).
$A' = A \cos^2 \theta + B \sin \theta \cos \theta + C \sin^2 \theta$
$B' = B(\cos^2 \theta - \sin^2 \theta) + 2(C - A)(\sin \theta \cos \theta)$
$C' = A \sin^2 \theta - B \sin \theta \cos \theta + C \cos^2 \theta$
$D' = D \cos \theta + E \sin \theta$
$E' = -D \sin \theta + E \cos \theta$
$F' = F$

45.　$B'^2 - 4A'C'$
$= [B(\cos^2 \theta - \sin^2 \theta) + 2(C - A)\sin \theta \cos \theta]^2 - 4[A \cos^2 \theta + B \sin \theta \cos \theta + C \sin^2 \theta)]$
　　$[A \sin^2 \theta - B \sin \theta \cos \theta + C \sin^2 \theta]$
$= B^2(\cos^4 \theta - 2 \sin^2 \theta \cos^2 \theta + \sin^4 \theta) + 4B(C - A) \sin \theta \cos \theta(\cos^2 \theta - \sin^2 \theta)$
　　$+ 4(C - A)^2 \sin^2 \theta \cos^2 \theta - 4[A^2 \sin^2 \theta \cos^2 \theta - AB \sin \theta \cos^3 \theta + AC \cos^4 \theta$
　　$- B^2 \sin^2 \theta \cos^2 \theta + BC \sin \theta \cos^3 \theta + AB \sin^3 \theta \cos \theta + AC \sin^4 \theta - BC \sin^3 \theta \cos \theta$
　　$+ C^2 \sin^2 \theta \cos^2 \theta]$
$= B^2[\cos^4 \theta - 2 \sin^2 \theta \cos^2 \theta + \sin^4 \theta + 4 \sin^2 \theta \cos^2 \theta] + BC[4 \sin \theta \cos \theta(\cos^2 \theta - \sin^2 \theta)$
　　$- 4 \sin \theta \cos \theta(\cos^2 \theta - \sin^2 \theta)] - AB[4 \sin \theta \cos \theta(\cos^2 \theta - \sin^2 \theta) - 4 \sin \theta \cos^3 \theta$
　　$+ 4 \sin^3 \theta \cos \theta] + 4C^2[\sin^2 \theta \cos^2 \theta - \sin^2 \theta \cos^2 \theta] - 4AC[2 \sin^2 \theta \cos^2 \theta$
　　$+ \cos^4 \theta \sin^4 \theta] + 4A^2[\sin^2 \theta \cos^2 \theta - \sin^2 \theta \cos^2 \theta]$
$= B^2[\cos^4 \theta + 2 \sin^2 \theta \cos^2 \theta + \sin^4 \theta] - 4AC[\cos^4 \theta + 2 \sin^2 \theta \cos^2 \theta + \sin^4 \theta]$
$= B^2[\sin^2 \theta + \cos^2 \theta]^2 - 4AC[\cos^2 \theta + \sin^2 \theta]^2$
$= B^2 - 4AC$

47.　Refer to Equation (5).
$d^2 = (y_2 - y_1)^2 + (x_2 - x_1)^2$
$= [x_2' \sin \theta + y_2' \cos \theta - x_1' \sin \theta - y_1' \cos \theta]^2 + [x_2' \cos \theta - y_2' \sin \theta - x_1' \cos \theta + y_1' \sin \theta]^2$
$= [(x_2' - x_1') \sin \theta + (y_2' - y_1')\cos \theta]^2 + [(x_2' - x_1') \cos \theta - (y_2' - y_1')^2 \sin \theta]^2$
$= (x_2' - x_1')^2 \sin^2 \theta + 2(x_2' - x_1')^2(y_2' - y_1')\sin \theta \cos \theta + (y_2' - y_1')^2 \cos^2 \theta + (x_2' - x_1')^2 \cos^2 \theta$
　　$- 2(x_2' - x_1')^2(y_2' - y_1')\sin \theta \cos \theta + (y_2' - y_1')^2 \sin^2 \theta$
$= (x_2' - x_1')^2(\sin^2 \theta + \cos^2 \theta) + (y_2' - y_1')^2(\cos^2 \theta + \sin^2 \theta)$
$= (x_2' - x_1')^2 + (y_2' - y_1')^2$

For Problems 1–18, use Formulas 4 and 5.

1. $e = 1; p = 1$; parabola; directrix is perpendicular to the polar axis 1 unit to the right of the pole.

3. $r = \dfrac{4}{2\left(1 - \dfrac{3}{2}\sin\theta\right)} = \dfrac{2}{1 - \dfrac{3}{2}\sin\theta}$; $ep = 2$, $e = \dfrac{3}{2}$; $p = \dfrac{4}{3}$

   Hyperbola; directrix is parallel to the polar axis $\dfrac{4}{3}$ units below the pole.

5. $r = \dfrac{3}{4\left(1 - \dfrac{1}{2}\cos\theta\right)} = \dfrac{\dfrac{3}{4}}{1 - \dfrac{1}{2}\cos\theta}$; $ep = \dfrac{3}{4}$, $e = \dfrac{1}{2}$; $p = \dfrac{3}{2}$

   Ellipse; directrix is perpendicular to the polar axis $\dfrac{3}{2}$ units to the left of the pole.

7.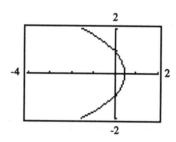

   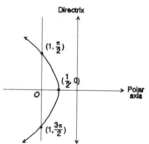

   $r = \dfrac{1}{1 + \cos\theta}$

   $ep = 1$, $e = 1$, $p = 1$

   Parabola; directrix is perpendicular to the polar axis 1 unit to the right of the pole; vertex is $\left(\dfrac{1}{2}, 0\right)$.

9.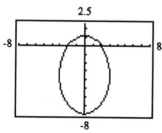

   $r = \dfrac{8}{4\left(1 + \dfrac{3}{4}\sin\theta\right)} = \dfrac{2}{1 + \dfrac{3}{4}\sin\theta}$; $ep = 2$, $e = \dfrac{3}{4}$, $p = \dfrac{8}{3}$

Ellipse; directrix parallel to the polar axis $\dfrac{8}{3}$ units above the pole.

Vertices are at $\left[\dfrac{8}{7}, \dfrac{\pi}{2}\right]$ and $\left[8, \dfrac{3\pi}{2}\right]$.

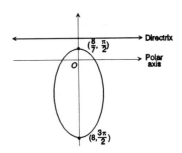

11.

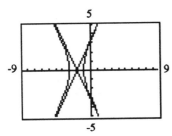

$$r = \frac{9}{3(1 - 2\cos\theta)} = \frac{3}{1 - 2\cos\theta}; \; ep = 3, \; e = 2, \; p = \frac{3}{2}$$

Hyperbola, directrix is perpendicular to the polar axis $\dfrac{3}{2}$ units to the left of the pole. Vertices are at $(1, \pi)$ and $(-3, 0)$.

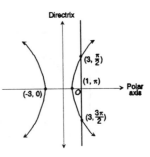

13.

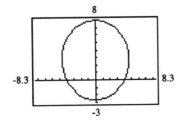

$$r = \frac{8}{2 - \sin\theta}; \; r = \frac{8}{2\left[1 - \dfrac{1}{2}\sin\theta\right]} = \frac{4}{1 - \dfrac{1}{2}\sin\theta}$$

$$e = \frac{1}{2}; \; ep = 4, \text{ so } p = \frac{4}{\dfrac{1}{2}} = 8$$

The conic is an ellipse; major axis is perpendicular to the directrix. The directrix is parallel to the polar axis at a distance 8 units below the pole.

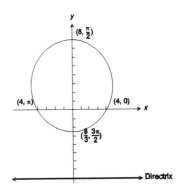

15.

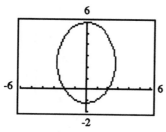

$$r = \frac{6}{(3 - 2 \sin \theta)} = \frac{2}{1 - \frac{2}{3} \sin \theta}$$

$ep = 2$, $e = \frac{2}{3}$, $p = 3$

Ellipse; directrix parallel to the polar axis 3 units below the pole.

Vertices are at $\left[6, \frac{\pi}{2}\right]$ and $\left[\frac{6}{5}, \frac{3\pi}{2}\right]$.

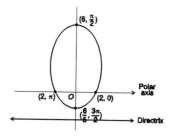

17.

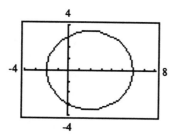

$$r = \frac{6 \sec \theta}{2 \sec \theta - 1} = \frac{6}{2 - \cos \theta} = \frac{3}{1 - \frac{1}{2} \cos \theta}$$

$ep = 3$, $e = \frac{1}{2}$, $p = 6$

Ellipse; directrix is perpendicular to the polar axis 6 units to the left of the pole. Vertices are at $(6, 0)$ and $(2, \pi)$.

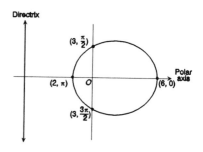

19. $r = \dfrac{1}{1 + \cos \theta}$

$r + r \cos \theta = 1$

$r = 1 - r \cos \theta$

$r^2 = (1 - r \cos \theta)^2$

$x^2 + y^2 = (1 - x)^2$

$x^2 + y^2 = 1 - 2x + x^2$

$y^2 + 2x - 1 = 0$

21. $r = \dfrac{8}{4 + 3 \sin \theta}$

$4r + 3r \sin \theta = 8$

$4r = 8 - 3r \sin \theta$

$16r^2 = (8 - 3r \sin \theta)^2$

$16(x^2 + y^2) = (8 - 3y)^2$

$16x^2 + 16y^2 = 64 - 48y + 9y^2$

$16x^2 + 7y^2 + 48y - 64 = 0$

23. $r = \dfrac{9}{3 - 6 \cos \theta}$

$3r - 6r \cos \theta = 9$

$3r = 9 + 6r \cos \theta$

$r = 3 + 2r \cos \theta$

$r^2 = (3 + 2r \cos \theta)^2$

$x^2 + y^2 = (3 + 2x)^2$

$x^2 + y^2 = 9 + 12x + 4x^2$

$3x^2 - y^2 + 12x + 9 = 0$

25. $r = \dfrac{8}{2 - \sin \theta}$

$r(2 - \sin \theta) = 8$

$2r - r \sin \theta = 8$

$2r = 8 + r \sin \theta$

$4r^2 = (8 + r \sin \theta)^2$

$4(x^2 + y^2) = (8 + y)^2$

$4x^2 + 4y^2 = 64 + 16y + y^2$

$4x^2 + 3y^2 - 16y - 64 = 0$

27. $r(3 - 2 \sin \theta) = 6$

$3r - 2r \sin \theta = 6$

$3r = 6 + 2r \sin \theta$

$9r^2 = (6 + 2r \sin \theta)^2$

$9(x^2 + y^2) = (6 + 2y)^2$

$9x^2 + 9y^2 = 36 + 24y + 4y^2$

$9x^2 + 5y^2 - 24y - 36 = 0$

29. $r = \dfrac{6 \sec \theta}{2 \sec \theta - 1}$

$r = \dfrac{6}{2 - \cos \theta}$

$2r - r \cos \theta = 6$

$2r = 6 + r \cos \theta$

$4r^2 = (6 + r \cos \theta)^2$

$4(x^2 + y^2) = (6 + x)^2$

$4x^2 + 4y^2 = 36 + 12x + x^2$

$3x^2 + 4y^2 - 12x - 36 = 0$

31. $r = \dfrac{ep}{1 + e \sin \theta}$

$e = 1, p = 1$

$r = \dfrac{1}{1 + \sin \theta}$

33. $r = \dfrac{ep}{1 - e \cos \theta}$

$e = \dfrac{4}{5}, p = 3$

$r = \dfrac{\dfrac{12}{5}}{1 - \dfrac{4}{5} \cos \theta} = \dfrac{12}{5 - 4 \cos \theta}$

35. $r = \dfrac{ep}{1 - e \sin \theta}$

$e = 6, p = 2$

$r = \dfrac{12}{1 - 6 \sin \theta}$

37. $d(F, P) = e \cdot d(D, P)$

$d(D, P) = p - r \cos \theta$

$\therefore r = e(p - r \cos \theta)$

$r = ep - er \cos \theta$

$r + er \cos \theta = ep$

$r = \dfrac{ep}{1 + e \cos \theta}$

39. $d(FP) = e \cdot d(D, P)$

$d(D, P) = p + r \sin \theta$

$\therefore r = e(p + r \sin \theta)$

$r = ep + er \sin \theta$

$r - er \sin \theta = ep$

$r = \dfrac{ep}{1 - e \sin \theta}$

1.

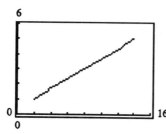

$x = 3t + 2, y = t + 1$
$x = 3(y - 1) + 2$
$x = 3y - 1$
$x - 3y + 1 = 0$

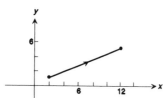

3.

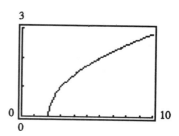

$x = t + 2, y = \sqrt{t}$
$y = \sqrt{x - 2}$

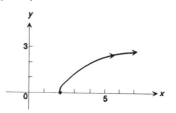

5.

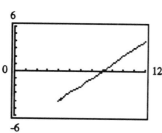

$x = t^2 + 4, y = t^2 - 4$
$x = (y + 4) + 4$
$x = y + 8$

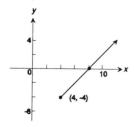

7.

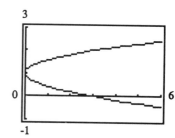

$x = 3t^2, y = t + 1$
$x = 3(y - 1)^2$

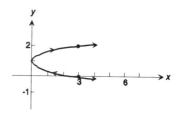

9.

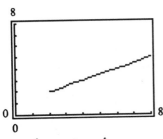

$x = 2e^t, y = 1 + e^t$

$y = 1 + \dfrac{x}{2}$

$2y = 2 + x$

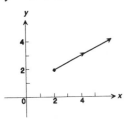

11.

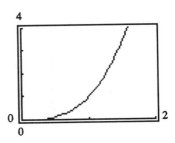

$x = \sqrt{t}, y = t^{3/2}$
$y = (x^2)^{3/2}$
$y = x^3$

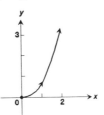

13.

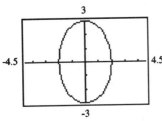

$x = 2\cos t, y = 3\sin t; \ 0 \le t \le 2\pi$

$\left[\dfrac{x}{2}\right]^2 + \left[\dfrac{y}{3}\right]^2 = \cos^2 t + \sin^2 t$

$\dfrac{x^2}{4} + \dfrac{y^2}{9} = 1$

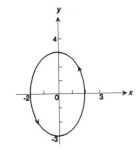

15.

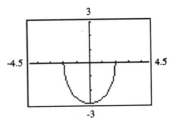

$x = 2\cos t, y = 3\sin t; \ -\pi \le t \le 0$

$\left[\dfrac{x}{2}\right]^2 + \left[\dfrac{y}{3}\right]^2 = \cos^2 t + \sin^2 t$

$\dfrac{x^2}{4} + \dfrac{y^2}{9} = 1$

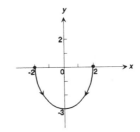

17.

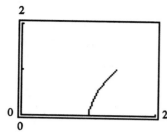

$x = \sec t, y = \tan t$
$1 + \tan^2 t = \sec^2 t$
$1 + y^2 = x^2$
$x^2 - y^2 = 1$

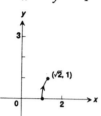

19.

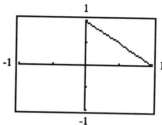

$x = \sin^2 t, y = \cos^2 t$
$\sin^2 t + \cos^2 t = 1$
$x + y = 1$

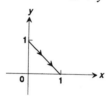

21. $x = t, y = t^3$

$x = \sqrt[3]{t}, y = t$

23. $x = t, y = t^{2/3}$
$x = t^{3/2}, y = t$

25. $x = 2 \cos \omega t, y = -3 \sin \omega t$

$\dfrac{2\pi}{\omega} = 2, \omega = \pi$

$x = 2 \cos \pi t, y = -3 \sin \pi t,$
$0 \le t \le 2$

27. $x = -2 \sin \omega t, y = 3 \cos \omega t$

$\dfrac{2\pi}{\omega} = 1, \omega = 2\pi$

$x = -2 \sin 2\pi t,$
$y = 3 \cos 2\pi t, 0 \le t \le 1$

29.

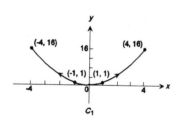

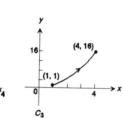

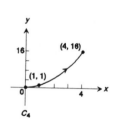

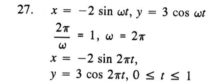

31. $x = (x_2 - x_1)t + x_1$
$y = (y_2 - y_1)t + y_1$

$\dfrac{x - x_1}{x_2 - x_1} = t$

$y = (y_2 - y_1) \left[\dfrac{x - x_1}{x_2 - x_1} \right] + y_1$

$y - y_1 = \left[\dfrac{y_2 - y_1}{x_2 - x_1} \right] (x - x_1)$

This is the equation of a line. Its orientation is from (x_1, y_1) to (x_2, y_2).

33.

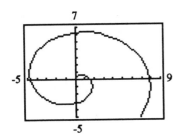

35.

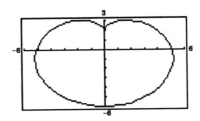

10 Chapter Review

For Problems 1–20, use the following rules:

   I.   *If only **one** variable is squared, the equation represents a parabola.*

   II.   *If **both** variables are squared, the equation is either an ellipse or a hyperbola.*

      A.   *If both the x^2-term and the y-term are positive, the graph is an ellipse.*

      B.   *If one of the two squared terms is negative, the graph is a hyperbola.*

1. $y^2 = -16x$.

This is a parabola, with equation of the form: $y^2 = -4ax$

where $a = 4$. By Table 3:   Vertex: $(0, 0)$

                                Focus: $(-4, 0)$

                         Directrix: $x = 4$

3. $\dfrac{x^2}{25} - y^2 = 1$

This is a hyperbola in the form: $\dfrac{(x - h)^2}{a^2} - \dfrac{(y - k)^2}{b^2} = 1$

where $a = 5$, $b = 1$, $h = 0$, $k = 0$.

By Table 4, $c^2 = a^2 + b^2 = 26$, so $c = \sqrt{26}$, and we have:

    Center: $(0, 0)$

      Foci: $(h \pm c, k)$: $\left(-\sqrt{26}, 0\right)$ and $\left(\sqrt{26}, 0\right)$

   Vertices: $(h \pm a, k)$: $(-5, 0)$ and $(5, 0)$

   Asymptotes: $y - k = \pm\dfrac{b}{a}(x - h)$ or $y = \pm\dfrac{1}{5}x$

5. $\dfrac{y^2}{25} + \dfrac{x^2}{16} = 1$

This is an ellipse since both variables are squared, and both terms are positive. The equation is already in the form:

$$\dfrac{(y - k)^2}{a^2} + \dfrac{(x - h)^2}{b^2} = 1$$

where $h = 0$, $k = 0$, $a = 5$, $b = 4$

By Table 3, we have $c^2 = a^2 - b^2 = 9$, so that $c = 3$, and:

Center: (h, k), or $(0, 0)$

Foci: $(h, k \pm c)$, or $(0, -3)$ and $(0, 3)$

Vertices: $(h, k \pm a)$, or $(0, -5)$ and $(0, 5)$

7. $x^2 + 4y = 4$ is a *parabola*.

$$x^2 + 4y = 4$$
$$x^2 = -4y + 4$$
$$x^2 = -4(y - 1)$$

This is in the form $(x - h)^2 = -4a(y - k)$, where

(1) $a = 1$

(2) $x - h = x$
     $h = 0$

(3) $y - k = y - 1$
     $k = 1$

From Table 2, we have: Vertex: $(0, 1)$

Focus: $(0, 0)$

Directrix: $y = 2$

9. $4x^2 - y^2 = 8$

This is a hyperbola, since it consists of a *difference* of squared terms:

$$4x^2 - y^2 = 8$$
$$\frac{x^2}{2} - \frac{y^2}{8} = 1$$

From Table 4, $a^2 = 2$, so $a = \sqrt{2}$, and $b^2 = 8$, so $b = \sqrt{8} = 2\sqrt{2}$.

Also, $c^2 = a^2 + b^2 = 10$, so $c = \sqrt{10}$. Then we have:

Transverse axis: horizontal: $y = 0$

Center: $(0, 0)$

Foci: $\left(-\sqrt{10}, 0\right)$ and $\left(\sqrt{10}, 0\right)$

Vertices: $\left(-\sqrt{2}, 0\right)$ and $\left(\sqrt{2}, 0\right)$

Asymptotes: $y - k = \pm \frac{b}{a}(x - h)$, or $y = \pm 2x$

11. $x^2 - 4x = 2y$ is a parabola:

$$x^2 - 4x + \underline{\quad} = 2y + \underline{\quad}$$
$$x^2 - 4x + 4 = 2y + 4$$
$$(x - 2)^2 = 2(y + 2)$$

We have: (1) $4a = 2$, or $a = \frac{1}{2}$

(2) $h = 2$

(3) $k = -2$

and: Vertex: $(2, -2)$

Focus: $\left(2, -\frac{3}{2}\right)$

Directrix: $y = -\frac{5}{2}$

13. $y^2 - 4y - 4x^2 + 8x = 4$

Complete the square:

$$(y^2 - 4y + \underline{}) - 4(x^2 - 2x + \underline{}) = 4$$
$$(y^2 - 4y + 4) - 4(x^2 - 2x + 1) = 4 + 4 - 4$$
$$(y - 2)^2 - 4(x - 1)^2 = 4$$
$$\frac{(y - 2)^2}{4} - \frac{(x - 1)^2}{1} = 1$$

This is a hyperbola with vertical transverse axis, center at $(1, 2)$, $a^2 = 4$, $b^2 = 1$, and $c^2 = a^2 + b^2 = 5$.

Then,
$$a = 2$$
$$b = 1$$
$$c = \sqrt{5}$$

Center: $(h, k) = (1, 2)$

Foci: $(h, k \pm c) = \left(1, 2 - \sqrt{5}\right)$ and $\left(1, 2 + \sqrt{5}\right)$

Vertices: $(h, k \pm a) = (1, 0)$ and $(1, 4)$

Asymptotes: $y - 2 = \pm 2(x - 1)$

(Lines through $(1, 2)$ with slopes 2 and -2.)

15. $4x^2 + 9y^2 - 16x - 18y = 11$

Since both variables are squared, this is either an ellipse or a hyperbola. Since both squared terms are positive, the graph *must* be an ellipse. We start by completing the square, to put the equation in a recognizable form:

$$4x^2 - 16x + 9y^2 - 18y = 11$$
$$4(x^2 - 4x + \underline{}) + 9(y^2 - 2y + \underline{}) = 11$$
$$4(x^2 - 4x + 4) + 9(y^2 - 2y + 1) = 11 + 16 + 9$$
$$4(x - 2)^2 + 9(y - 1)^2 = 36$$
$$\frac{4(x - 2)^2}{36} + \frac{9(y - 1)^2}{36} = 1$$
$$\frac{(x - 2)^2}{9} + \frac{(y - 1)^2}{4} = 1$$

We now have the form: $\dfrac{(x - h)^2}{a^2} + \dfrac{(y - k)^2}{b^2} = 1$

where $h = 2$, $k = 1$, $a = 3$, $b = 2$. By Table 3, $c^2 = a^2 - b^2 = 5$, so $c = \sqrt{5}$, and we have:

Center: $(h, k) = (2, 1)$

Foci: $(h \pm c, k)$: $\left(2 - \sqrt{5}, 1\right)$ and $\left(2 + \sqrt{5}, 1\right)$

Vertices: $(h \pm a, k)$: $(-1, 1)$ and $(5, 1)$

17. $4x^2 - 16x + 16y + 32 = 0$ is a *parabola*.

$$4x^2 - 16x = -16y - 32$$
$$4(x^2 - 4x + \underline{}) = -16y - 32 + \underline{}$$
$$4(x^2 - 4x + 4) = -16y - 32 + 16$$

$$(4)(4) = 16$$
$$4(x - 2)^2 = -16y - 16$$
$$4(x - 2)^2 = -16(y + 1)$$
$$(x - 2)^2 = -4(y + 1)$$

We have: (1) $a = 1$
          (2) $h = 2$
          (3) $k = -1$
and: Vertex: $(2, -1)$
          Focus: $(2, -2)$
          Directrix: $y = 0$

19. $9x^2 + 4y^2 - 18x + 8y = 23$
     Both variables are squared, and the squared terms are both positive, so this is an ellipse.
$$9x^2 - 18x + 4y^2 + 8y = 23$$
$$9(x^2 - 2x + \underline{\quad}) + 4(y^2 + 2y + \underline{\quad}) = 23$$
$$9(x^2 - 2x + 1) + 4(y^2 + 2y + 1) = 23 + 9 + 4$$
$$9(x - 1)^2 + 4(y + 1)^2 = 36$$
$$\frac{(x - 1)^2}{4} + \frac{(y + 1)^2}{9} = 1$$

This is in the form: $\dfrac{(x - h)^2}{b^2} + \dfrac{(y - k)^2}{a^2} = 1$
where $h = 1, k = -1, a = 3, b = 2.$

From Table 3, $c^2 = a^2 - b^2 = 5$, so $c = \sqrt{5}$, and we have:
    Center: $(h, k) = (1, -1)$

       Foci: $(h, k \pm c)$, or $\left(1, -1 - \sqrt{5}\right)$ and $\left(1, -1 + \sqrt{5}\right)$
    Vertices: $(h, k \pm a)$, or $(1, -4)$ and $(1, 2)$

21. We are given: Type of graph: Parabola
                      Focus: $F(-2, 0)$
                      Directrix: $x = 2$

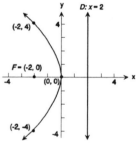

From Table 2, we see that to find the equation of a parabola, we need to know the coordinates of the vertex, (h, k), the distance from the vertex to the focus (a), and whether the parabola opens up, down, left, or right.

Here, the directrix, $x = 2$, is a vertical line, so the axis of symmetry is a horizontal line (which passes through the focus, $(-2, 0)$, i.e., $y = 0$.

The vertex is on the axis, midway between the focus and the directrix, at $V(0, 0)$. Then we have:
    $a = d(V, F) = 2$
Finally, the focus is to the *left* of the vertex, so the parabola opens to the left. By Table 2, the parabola has an equation of the form:
$$(y - k)^2 = -4a(x - h), \text{ or}$$
$$y^2 = -4ax, \text{ since } h = 0, k = 0$$
$$y^2 = -8x, \text{ since } a = 2$$

23. We are given: Type of graph: Hyperbola
                      Center: $C(0, 0)$
                      Focus: $F_2(0, 4)$
                      Vertex: $V_1(0, -2)$

As we see from Table 4, to determine the equation of a hyperbola, we need to know the orientation of the transverse axis, the coordinates of the center, (h, k) and the constants a and b.

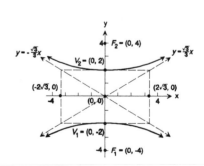

We know $h = 0, k = 0$
       $c = d(C, F_2) = 4$

$$a = d(C, V_1) = 2$$
and $$b^2 = c^2 - a^2$$
$$= 16 - 4 = 12$$
or $$b = \sqrt{12} = 2\sqrt{3}$$

The center, focus, and vertex all lie on the **vertical** line $x = 0$, so the transverse axis is parallel to the y-axis. Therefore,

$$\frac{(y - k)^2}{a^2} - \frac{(x - h)^2}{b^2} = 1, \quad \text{or} \quad \frac{y^2}{4} - \frac{x^2}{12} = 1$$

As a further aid in graphing the hyperbola, the asymptotes are:

$$y - k = \pm\frac{a}{b}(x - h), \quad \text{or}$$

$$y = \pm\frac{1}{\sqrt{3}}x$$

$$y = \pm\frac{\sqrt{3}}{3}x$$

(Lines through $(0, 0)$ with slopes $\dfrac{\sqrt{3}}{3}$ and $\dfrac{-\sqrt{3}}{3}$.)

25. We are given: Type of Graph: Ellipse
    Foci: $F_1(-3, 0)$, $F_2(3, 0)$
    Vertex: $V_2(4, 0)$

From Table 3, we see that we need the center, (h, k), a (the distance from the center to a vertex), and b, and we need to know whether the major axis is vertical or horizontal.

In this problem, the foci both lie on the **horizontal** line $y = 0$. The center is the midpoint between F_1 and F_2: $C(0, 0)$, so $h = 0$, $k = 0$.

Also: $$c = d(C, F_1) = 3$$
$$a = d(C, V_2) = 4$$
and $$b^2 = a^2 - c^2 = 7, \text{ so}$$
$$b = \sqrt{7}$$

Therefore, the equation is of the form: $\dfrac{(x - h)^2}{a^2} + \dfrac{(y - k)^2}{b^2} = 1$, or $\dfrac{x^2}{16} + \dfrac{y^2}{7} = 1$

27. We are given: Type of graph: Parabola
    Vertex: $V(2, -3)$
    Focus: $F(2, -4)$

From Table 2, we need to know the vertex, $V(h, k)$; the distance, a, from the vertex to the focus; and which direction the parabola opens.

Since the focus is **below** the vertex, the parabola opens **down**. Also, $a = d(V, F) = 1$.

Therefore, the equation will be of the form:
$$(x - h)^2 = -4a(y - k)$$
or $$(x - 2)^2 = -4(y + 3)$$

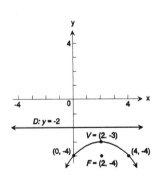

29.   We are given: Type of graph: Hyperbola

                                   Center: $C(-2, -3)$

                                    Focus: $F_1(-4, -3)$

                                  Vertex: $V_1(-3, -3)$

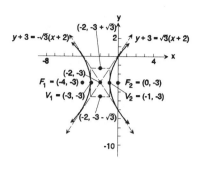

The transverse axis is the *horizontal* line $y = -3$, and:

$a = d(C, V_1) = 1$

$c = d(C, F_1) = 2$

$b^2 = c^2 - a^2 = 3$, so

$b = \sqrt{3}$

        Foci: $(h \pm c, k)$: $(-4, -3)$ and $(0, -3)$

    Vertices: $(h \pm a, k)$: $(-3, -3)$ and $(-1, -3)$

Asymptotes: $y - k = \pm\dfrac{b}{a}(x - h)$ or $y + 3 = \pm\sqrt{3}\,(x + 2)$

The equation is: $\dfrac{(x - h)^2}{a^2} - \dfrac{(y - k)^2}{b^2} = 1$, or $\dfrac{(x + 2)^2}{1} - \dfrac{(y + 3)^2}{3} = 1$

31.   We are given: Type of graph: Ellipse

                                   Foci: $F_1(-4, 2)$ and $F_2(-4, 8)$

                                 Vertex: $V_2(-4, 10)$

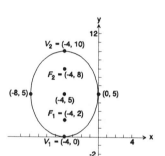

As we see in Table 3, we need to know the center, (h, k); the distance, a, from the center to either vertex; b, the distance from the center to either endpoint of the minor axis; and whether the major axis is vertical or horizontal. The center is the midpoint between F_1 and F_2: $C(-4, 5)$.

Then $a = d(C, V_2) = 5$, and $c = d(C, F_1) = 3$.

Therefore, $b^2 = a^2 - c^2 = 16$, so that $b = 4$.

Finally, F_1, F_2 and V_2 all lie on the *vertical* line $x = -4$, so the major axis is parallel to the y-axis. By Table 3, the equation of the ellipse is:

$$\frac{(x - h)^2}{b^2} + \frac{(y - k)^2}{a^2} = 1, \quad \text{or} \quad \frac{(x + 4)^2}{16} + \frac{(y - 5)^2}{25} = 1$$

33.   We are given: Center: $C(-1, 2)$

                                          $a = 3$

                                          $c = 4$

Transverse axis parallel to the x-axis.

Since the conic section has a transverse axis, it must be a hyperbola. For hyperbolas, $b^2 = c^2 - a^2$, so $b^2 = 7$, or $b = \sqrt{7}$.

From Table 4, the equation is of the form:

$$\frac{(x - h)^2}{a^2} - \frac{(y - k)^2}{b^2} = 1$$

where (h, k) are the coordinates of the center. Therefore, we have:

$$\frac{(x + 1)^2}{9} - \frac{(y - 2)^2}{7} = 1$$

As an aid to graphing, the vertices are at $(h \pm a, k)$, or $(-4, 2)$ and $(2, 2)$, and the asymptotes are:

$$y - 2 = \pm\frac{\sqrt{7}}{3}(x + 1)$$

(Lines through $(-1, 2)$ with slopes $\dfrac{\sqrt{7}}{3} \approx .88$ and $-\dfrac{\sqrt{7}}{3} \approx -.88$)

35. We are given: Vertices: $V_1(0, 1)$ and $V_2(6, 1)$
                      Asymptote: $3y + 2x - 9 = 0$

Since this conic section has asymptotes, it must be a hyperbola.
The center is midway between V_1 and V_2: $C(3, 1)$, and the
transverse axis is the **horizontal** line $y = 1$. Also,
$a = d(C, V_1) = 3$. But we need to know b to determine the
equation.

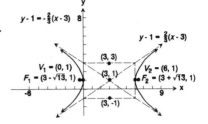

From Table 4, the asymptotes are:

$$y - k = \pm\frac{b}{a}(x - h), \quad \text{or} \quad y - 1 = \pm\frac{b}{a}(x - 3)$$

We must put the given asymptote into this form:

$$3y + 2x - 9 = 0$$
$$3y = -2x + 9$$
$$y = -\frac{2}{3}x + 3$$
$$y - 1 = -\frac{2}{3}x + 2$$
$$y - 1 = -\frac{2}{3}x + \frac{2 \cdot 3}{3}$$
$$y - 1 = -\frac{2}{3}(x - 3)$$

Therefore, $-\dfrac{b}{a} = -\dfrac{2}{3}$
$$3b = 2a$$
$$3b = 6 \quad \text{(since } a = 3)$$
$$b = 2$$

The equation of the hyperbola is: $\dfrac{(x - 3)^2}{9} - \dfrac{(y - 1)^2}{4} = 1$

37. $y^2 + 4x + 3y - 8 = 0$
Here $A = 0$ and $C = 1$ so that $AC = 0$. The equation is a parabola.

39. $x^2 + 2y^2 + 4x - 8y + 2 = 0$
Here $A = 1$ and $C = 2$ so that $AC = 2$. The equation is a ellipse.

41. $9x^2 - 12xy + 4y^2 + 8x + 12y = 0$
Here $A = 9$ and $B = -12$, and $C = 4$ so that $B^2 - 4AC = 144 - 144 = 0$. The equation is a
parabola.

43. $4x^2 + 10xy + 4y^2 - 9 = 0$
Here $A = 4$ and $B = 10$, and $C = 4$ so that $B^2 - 4AC = 100 - 64 = 36$. The equation is a
hyperbola.

45. $x^2 - 2xy + 3y^2 + 2x + 4y - 1 = 0$
Here $A = 1$ and $B = -2$, and $C = 3$ so that $B^2 - 4AC = 4 - 12 = -8$. The equation is an
ellipse.

47. $A = 2$, $B = 5$, $C = 2$, $\cot 2\theta = 0$ so that $\theta = 45°$

$$x = \frac{\sqrt{2}}{2}(x' - y'); \; y = \frac{\sqrt{2}}{2}(x' + y')$$

$$2\left[\frac{\sqrt{2}}{2}(x' - y')\right]^2 + 5\left[\frac{\sqrt{2}}{2}(x' - y')\right]\left[\frac{\sqrt{2}}{2}(x' + y')\right] + 2\left[\frac{\sqrt{2}}{2}(x' + y')\right]^2 - \frac{9}{2} = 0$$

$$2 \cdot \frac{1}{2}(x'^2 - 2x'y'^2 + y'^2) + 5 \cdot \frac{1}{2}(x'^2 - y'^2) + 2 \cdot \frac{1}{2}(x'^2 + 2x'y' + y'^2) = \frac{9}{2}$$

$$\frac{9}{2}x'^2 - \frac{1}{2}y'^2 = \frac{9}{2}$$

$$x'^2 - \frac{-y'^2}{9} = 1$$

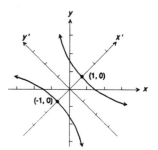

Hyperbola; center at origin; transverse axis is the x'-axis; vertices at $(\pm 1, 0)$.

49. $A = 6$, $B = 4$, $C = 9$, $\cot 2\theta = \frac{-3}{4}$; $\cos 2\theta = \frac{-3}{5}$

$$\sin \theta = \sqrt{\frac{1 + \frac{3}{5}}{2}} = \frac{2}{\sqrt{5}} = \frac{2\sqrt{5}}{5}; \; \cos \theta = \sqrt{\frac{1 - \frac{3}{5}}{2}} = \frac{1}{\sqrt{5}} = \frac{\sqrt{5}}{5}$$

$\theta = 63°$

$$x = \frac{\sqrt{5}}{5}(x' - 2y'), \; y = \frac{\sqrt{5}}{5}(2x' + y')$$

$$6\left[\frac{\sqrt{5}}{5}(x' - 2y')\right]^2 + 4\left[\frac{\sqrt{5}}{5}(x' - 2y')\right]\left[\frac{\sqrt{5}}{5}(2x' + y')\right] + 9\left[\frac{\sqrt{5}}{5}(2x' + y')\right]^2 = 20$$

$$\frac{6}{5}(x'^2 - 4x'y' + 4y'^2) + \frac{4}{5}(2x'^2 - 3x'y' - 2y'^2)$$

$$+ \frac{9}{5}(4x'^2 + 4x'y' + y'^2) = 20$$

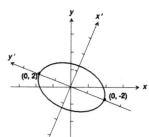

$$10x'^2 + 5y'^2 = 20$$

$$\frac{x'^2}{2} + \frac{y'^2}{4} = 1$$

Ellipse; center at origin; major axis is the y'-axis; vertices at $(0, \pm 2)$.

51. $A = 4$, $B = -12$, $C = 9$, $\cot 2\theta = \frac{5}{12}$; $\cos 2\theta = \frac{5}{13}$

$$\sin \theta = \sqrt{\frac{1 - \frac{5}{13}}{2}} = \frac{2}{\sqrt{13}} = \frac{2\sqrt{13}}{13}; \; \cos \theta = \sqrt{\frac{1 + \frac{5}{13}}{2}} = \frac{3}{\sqrt{13}} = \frac{3\sqrt{13}}{13}$$

$\theta \approx 34°$

$$x = \frac{3\sqrt{13}}{13}x' - \frac{2\sqrt{13}}{13}y' = \frac{\sqrt{13}}{13}(3x' - 2y') \qquad y = \frac{2\sqrt{13}}{13}x' + \frac{3\sqrt{13}}{13}y' = \frac{\sqrt{13}}{13}(2x' + 3y')$$

$$4\left[\frac{\sqrt{13}}{13}(3x' - 2y')\right]^2 - 12\left[\frac{\sqrt{13}}{13}(3x' - 2y')\right]\left[\frac{\sqrt{13}}{13}(2x' + 3y')\right]$$

$$+ 9\left[\frac{\sqrt{13}}{13}(2x' + 3y')\right]^2 - 12\left[\frac{\sqrt{13}}{13}(3x' - 2y')\right] + 8\left[\frac{\sqrt{13}}{13}(2x' + 3y')\right] = 0$$

$$\frac{4}{13}(9x'^2 - 12x'y'^2 + 4y'^2) - \frac{12}{13}(6x'^2 + 5x'y' - 6y'^2)$$

$$+ \frac{9}{13}(4x'^2 + 12x'y' + 9y'^2) + 52\frac{\sqrt{13}}{13}x' = 0$$

$$13y'^2 + 4\sqrt{13}\,x' = 0$$

$$y'^2 = \frac{-4\sqrt{13}}{13}x'$$

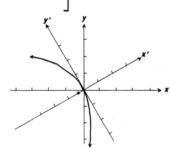

Parabola; vertex at the origin; focus on the x'-axis at $\left[\dfrac{-\sqrt{13}}{13}, 0\right]$.

53. $r = \dfrac{4}{1 - \cos\theta}$

$ep = 4, e = 1, p = 4$

Parabola; directrix is perpendicular to the polar axis 4 units to the left of the pole.

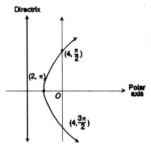

55. $r = \dfrac{6}{2 - \sin\theta} = \dfrac{3}{1 - \dfrac{1}{2}\sin\theta}$

$ep = 3, e = \dfrac{1}{2}, p = 6$

Ellipse; directrix parallel to the polar axis 6 units below the pole.

Vertices are at $\left[6, \dfrac{\pi}{2}\right]$ and $\left[2, \dfrac{3\pi}{2}\right]$.

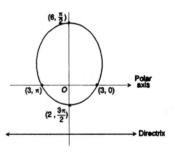

57. $r = \dfrac{8}{4 + 8\cos\theta} = \dfrac{2}{1 + 2\cos\theta}$

$ep = 2, e = 2, p = 1$

Hyperbola, directrix is perpendicular to the polar axis 1 unit to the right

of the pole. Vertices are at $\left[\dfrac{2}{3}, 0\right]$ and $(-2, \pi)$.

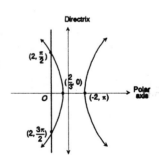

59. $r = \dfrac{4}{1 - \cos \theta}$

$r(1 - \cos \theta) = 4$

$r - r \cos \theta = 4$

$r = 4 + r \cos \theta$

$r^2 = (4 + r \cos \theta)^2$

$x^2 + y^2 = (4 + x)^2$

$x^2 + y^2 = 16 + 8x + x^2$

$y^2 - 8x - 16 = 0$

61. $r = \dfrac{8}{4 + 8 \cos \theta}$

$r(4 + 8 \cos \theta) = 8$

$4r = 8 - 8r \cos \theta$

$r = 2(1 - r \cos \theta)$

$r^2 = 4(1 - r \cos \theta)^2$

$x^2 + y^2 = 4(1 - x)^2$

$x^2 + y^2 = 4(1 - 2x + x^2)$

$3x^2 - y^2 - 8x + 4 = 0$

63.

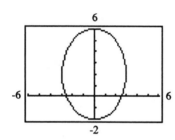

$x = 4t - 2,$
$y = 1 - t$
$x = 4(1 - y) - 2$
$x = 4 - 4y - 2$
$x + 4y = 2$

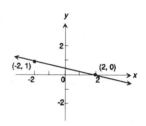

65.

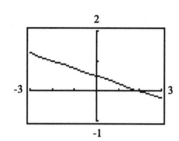

$x = 3 \sin t, \; y = 4 \cos t + 2$

$\dfrac{x}{3} = \sin t; \; \dfrac{y - 2}{4} = \cos t$

$\sin^2 t + \cos^2 t = 1$

$\dfrac{x^2}{9} + \dfrac{(y - 2)^2}{16} = 1$

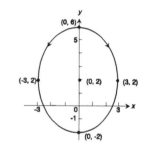

67.

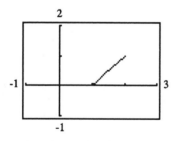

$x = \sec^2 t, \; y = \tan^2 t$

$1 + \tan^2 t = \sec^2 t$

$1 + y = x$

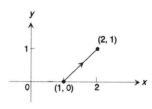

69. We start with the ellipse and determine its foci and vertices:

$$4x^2 + 9y^2 = 36$$

$$\dfrac{x^2}{9} + \dfrac{y^2}{4} = 1$$

This is an ellipse centered at $(0, 0)$. The larger denominator is associated with the x^2 term, so we know that the major axis is horizontal, and $a^2 = 9$, $b^2 = 4$, and $c^2 = a^2 - b^2 = 5$, so $c = \sqrt{5}$.

From Table 3: Vertices of the ellipse: $(h \pm a, k)$, or $(-3, 0)$ and $(3, 0)$

Foci of the ellipse: $(h \pm c, k)$, or $\left(-\sqrt{5}, 0\right)$ and $\left(\sqrt{5}, 0\right)$

Therefore, Foci of the hyperbola: $F_1(-3, 0)$ and $F_2(3, 0)$

Vertices of the hyperbola: $V_1\left(-\sqrt{5}, 0\right)$ and $V_2\left(\sqrt{5}, 0\right)$

The center of the hyperbola is midway between F_1 and F_2: $C(0, 0)$, and the transverse axis is the **horizontal** line, $y = 0$.

Finally, $a = d(C, V_1) = \sqrt{5}$
               $c = d(C, F_1) = 3$, and
               $b^2 = c^2 - a^2 = 9 - 5 = 4$, so
               $b = 2$

By Table 3, the equation of the **hyperbola** is:

$$\frac{(x - h)^2}{a^2} - \frac{(y - k)^2}{b^2} = 1, \text{ or } \frac{x^2}{5} - \frac{y^2}{4} = 1$$

71.   Let (x, y) be any point in this collection of points.

The distance from (x, y), to $(3, 0) = \sqrt{(x - 3)^2 + y^2}$

The distance (x, y) to the line $x = \dfrac{16}{3}$ is $\left| x - \dfrac{16}{3} \right|$

Therefore, we have

$$\sqrt{(x - 3)^2 + y^2} = \frac{3}{4}\left| x - \frac{16}{3} \right|$$

$$(x - 3)^2 + y^2 = \frac{9}{16}\left[x - \frac{16}{3} \right]^2 \quad \text{square both sides}$$

$$x^2 - 6x + 9 + y^2 = \frac{9}{16}\left[x^2 - \frac{32}{3}x + \frac{256}{9} \right]$$

$$16x^2 - 96x + 144 + 16y^2 = 9x^2 - 96x + 256$$

$$7x^2 + 16y^2 = 122$$

$$\frac{7x^2}{112} + \frac{16y^2}{112} = 1$$

$$\frac{x^2}{16} + \frac{y^2}{7} = 1$$

Thus, the set of points is an ellipse.

73.   Situate the parabola so that its vertex is at $(0, 0)$ and it opens up. We know
$$x^2 = 4ay$$
Since the light source is located at the focus, and the light source is one foot from the base, we have $a = 1$. Since the diameter of the opening is 2 feet across, we have $x = \dfrac{1}{2}(2) = 1$. Thus, we have:

$$x^2 = 4ay$$
$$1^2 = 4(1)y$$
$$y = 0.25 \text{ feet}$$

The mirror should be 0.25 feet or 3 inches deep.

75.   Situate the ellipse so that its center is at $(0, 0)$. Since the bridge has a span of 60 feet, the length of the major axis is 60; thus, $a = 30$. The maximum height of the bridge is 20 feet; thus, $b = 20$. The equation of the ellipse is therefore:

$$\frac{x^2}{900} + \frac{y^2}{400} = 1$$

The height, y, of the arch a distance $x = 5$ feet from center is:

$$\frac{5^2}{900} + \frac{y^2}{400} = 1$$

$$y^2 = 400 \cdot \frac{875}{900}$$

$$y \approx 19.72 \text{ feet}$$

Similarly, when $x = 10$ feet, $y \approx 18.86$ feet. When $x = 20$ feet, $y \approx 14.91$ feet.

77. (a) Set up a rectangular coordinate system so that the
          two stations lie on the x-axis and the origin is
          midway between them. The ship lies on a
          hyperbola whose foci are the locations of the two
          stations. Since the time difference is .00032
          seconds and the speed of the signal is 186,000
          miles per second, the difference of the distances
          from the ship to each station is:

distance $= (186,000)(.00032) = 59.52$ miles

The difference of the distances from the ship to each station, 59.52, equals $2a$, so $a = 29.76$
and the vertex of the corresponding hyperbola is at (29.76, 0). Since the focus is at (75, 0),
the ship would reach shore $75 - 29.76 = 45.24$ miles from the master station.

(b) The ship should follow a hyperbola with vertex at (60, 0). For this hyperbola, $a = 60$, so the
     constant difference of the distances from the ship to each station is 120. The time difference
     the ship should look for is:

$$\text{time} = \frac{120}{186,000} = 0.000645 \text{ seconds}$$

(c) We need to find the equation of the hyperbola with vertex at (60, 0) and a focus at (75, 0).
     The form of the equation of this hyperbola is:

$$\frac{x^2}{a^2} - \frac{y^2}{b^2} = 1$$

where $a = 60$. Since $c = 75$ and $b^2 = c^2 - a^2$, we have $b^2 = 75^2 - 60^2 = 2025$. So, the
equation of the hyperbola is:

$$\frac{x^2}{3600} - \frac{y^2}{2025} = 1$$

Since the ship is 20 miles off shore, we have
$y = 20$. Solve the above equation for x.

$$\frac{x^2}{3600} - \frac{20^2}{2025} = 1$$

$$\frac{x^2}{3600} = \frac{2425}{2025}$$

$$x^2 = 3600 \cdot \frac{2425}{2025}$$

$$x \approx 66 \text{ miles}$$

The ship is at the position (66, 20).

SYSTEMS OF EQUATIONS AND INEQUALITIES

11.1 Systems of Linear Equations: Substitution; Elimination

1. $\left.\begin{array}{r} 2x - y = 5 \\ 5x + 2y = 8 \end{array}\right\}$ $\quad x = 2, y = -1 \quad$ $\begin{array}{rl} 2(2) - (-1) = 5 & 5 = 5 \\ 5(2) + 2(-1) = 8 & 8 = 8 \end{array}$

Therefore, $x = 2$, $y = -1$ is a solution to the system of equations, since they satisfy **both** equations.

3. $\left.\begin{array}{r} 3x - 4y = 4 \\ \dfrac{1}{2}x - 3y = \dfrac{-1}{2} \end{array}\right\}$ $\quad x = 2, y = \dfrac{1}{2} \quad$ $\begin{array}{rl} 3(2) - 4\left(\dfrac{1}{2}\right) = 4 & 4 = 4 \\ \dfrac{1}{2}(2) - 3\left(\dfrac{1}{2}\right) = \dfrac{-1}{2} & \dfrac{-1}{2} = \dfrac{-1}{2} \end{array}$

Each equation is satisfied, so $x = 2$, $y = \dfrac{1}{2}$ is a solution of the system.

5. $\left.\begin{array}{r} x^2 - y^2 = 3 \\ xy = 2 \end{array}\right\}$ $\quad x = 2, y = 1 \quad$ $\begin{array}{rl} 2^2 - 1^2 = 3 & 3 = 3 \\ (2)(1) = 2 & 2 = 2 \end{array}$

7. $\left.\begin{array}{r} \dfrac{x}{1 + x} + 3y = 6 \\ x + 9y^2 = 36 \end{array}\right\}$ $\quad x = 0, y = 2 \quad$ $\begin{array}{rl} \dfrac{0}{1 + 0} + 3(2) = 6 & 6 = 6 \\ 0 + 9(2)^2 = 36 & 36 = 36 \end{array}$

9. $\left.\begin{array}{r} 3x + 3y + 2z = 4 \\ x - y - z = 0 \\ 2y - 3z = -8 \end{array}\right\}$ $x = 1, y = -1, z = 2 \quad$ $\begin{array}{rl} 3(1) + 3(-1) + 2(2) = 4 & 4 = 4 \\ 1 - (-1) - 2 = 0 & 0 = 0 \\ 2(-1) - 3(2) = -8 & -8 = -8 \end{array}$

Therefore, $x = 1$, $y = 2$, $z = 2$ is a solution to the system.

11. $\begin{cases} x + y = 8 & (1) \\ x - y = 4 & (2) \end{cases}$

We can use substitution.

Step 1: We can easily solve either equation for *x or y*. Let us begin by solving equation (1) for x:
$$x + y = 8$$
$$x = -y + 8$$

Step 2: Now substitute $x = -y + 8$ into (2): $\qquad \begin{array}{r} x - y = 4 \\ -y + 8 - y = 4 \\ -2y + 8 = 4 \end{array}$

Step 3: Solve the last equation for y:
$$-2y + 8 = 4$$
$$-2y = -4$$
$$y = 2$$

Step 4: We can now find x by using the fact that $x = -y + 8$ (Step 1) and $y = 2$ (Step 3):
$$x = -y + 8$$
$$x = -(2) + 8, \text{ since } y = 2$$
$$x = 6$$
The solution of the system is $x = 6$, $y = 2$.

13. $\begin{cases} 5x - y = 13 & (1) \\ 2x + 3y = 12 & (2) \end{cases}$

We use elimination.
$$\begin{cases} 15x - 3y = 39 & (1) \quad \text{Multiply both sides by 3.} \\ 2x + 3y = 12 & (2) \end{cases}$$
$$\begin{cases} 15x - 3y = 39 & (1) \\ 17x = 51 & (2) \quad \text{Replace (2) by (1) + (2)} \end{cases}$$
$$\begin{cases} 15x - 3y = 39 & (1) \\ x = 3 & (2) \end{cases}$$
$$\begin{cases} 45 - 3y = 39 & (1) \quad \text{Back-substitute; } x = 3 \\ x = 3 & (2) \end{cases}$$
$$\begin{cases} -3y = -6 & (1) \\ x = 3 & (2) \end{cases}$$
$$\begin{cases} y = 2 & (1) \\ x = 3 & (2) \end{cases}$$

The solution is $x = 3$, $y = 2$.

15. $\begin{cases} 3x = 24 & (1) \\ x + 2y = 0 & (2) \end{cases}$

We use substitution.
Step 1: Since equation (1) contains only one variable, it is best to start there:
$$3x = 24$$
$$x = 8$$
Step 2: Substitute $x = 8$ into (2): $\qquad x + 2y = 0$
$$8 + 2y = 0$$
Step 3: Solve for y: $\qquad\qquad\qquad 8 + 2y = 0$
$$2y = -8$$
$$y = -4$$
We already know x, so the solution of the system is $x = 8$, $y = -4$.

17. $\begin{cases} 3x - 6y = 2 & (1) \\ 5x + 4y = 1 & (2) \end{cases}$

We use substitution.
Step 1: It is harder to decide which equation to solve for which variable. Since the coefficient of x in equation (1) divides all other constants in equation (1), I choose to solve for x, in order to avoid fractions for as long as possible:
$$3x - 6y = 2$$
$$3x = 6y + 2$$
$$x = 2y + \frac{2}{3}$$

Step 2: Substitute $x = 2y + 8$ into (2):

$$5x + 4y = 1$$

$$5\left(2x + \frac{2}{3}\right) + 4y = 1$$

$$10y + \frac{10}{3} + 4y = 12$$

$$14y + \frac{10}{3} = 1$$

$$42y + 10 = 3$$

Step 3: Solve for y:

$$42y + 10 = 3$$

$$42y = -7$$

$$y = -\frac{1}{6}$$

Step 4: Determine x:

$$x = 2y + \frac{2}{3} \quad \text{(Step 1)}$$

$$x = 2\left(-\frac{1}{6}\right) + \frac{2}{3}$$

$$x = \frac{1}{3}$$

The solution is $x = \dfrac{1}{3}$, $y = -\dfrac{1}{6}$.

19. $\begin{cases} 2x + y = 1 & (1) \\ 4x + 2y = 3 & (2) \end{cases}$

We use substitution.

Step 1: It is easiest to solve equation (1) for y:

$$2x + y = 1$$
$$y = -2x + 1$$

Step 2: Substitute $y = -2x + 1$ into (2):

$$4x + 2y = 3$$
$$4x + 2(-2x + 1) = 3$$
$$4x - 4x + 2 = 3$$
$$0x = 1$$

This has no solution, so the system is inconsistent.

21. $\begin{cases} 2x - y = 0 & (1) \\ 3x + 2y = 7 & (2) \end{cases}$

We use elimination.

$\begin{cases} 4x - 2y = 0 & (1) \\ 3x + 2y = 7 & (2) \end{cases}$ Multiply both sides by 2.

$\begin{cases} 4x - 2y = 0 & (1) \\ 7x = 7 & (2) \end{cases}$ Replace (2) by (1) + (2).

$\begin{cases} 4x - 2y = 0 & (1) \\ x = 1 & (2) \end{cases}$

$\begin{cases} 4 - 2y = 0 & (1) \\ x = 1 & (2) \end{cases}$ Back-substitute; $x = 1$.

$\begin{cases} -2y = -4 & (1) \\ x & (2) \end{cases}$

$\begin{cases} y = 2 & (1) \\ x = 1 & (2) \end{cases}$

The solution is $x = 1$, $y = 2$.

23. $\begin{cases} x + 2y = 4 & (1) \\ 2x + 4y = 8 & (2) \end{cases}$

We use substitution.

Step 1:   Solve for x in (1):

$$x + 2y = 4$$
$$x = -2y + 4$$

Step 2:   Substitute $x = -2y + 4$ into (2):

$$2x + 4y = 8$$
$$2(-2y + 4) + 4y = 8$$
$$-4y + 8 + 4y = 8$$
$$0y = 0$$

This is an identity (i.e., true for *any* value of y), so *any* value of y is a solution, and we let $x = 4 - 2y$, y any real number.

25. $\begin{cases} 2x - 3y = -1 & (1) \\ 10x + y = 11 & (2) \end{cases}$

We use elimination.

$\begin{cases} -10x + 15y = 5 & (1) \\ 10x + y = 11 & (2) \end{cases}$ Multiply both sides by -5.

$\begin{cases} -10x + 15y = 5 & (1) \\ 16y = 16 & (2) \end{cases}$ Replace (2) by (1) + (2).

$\begin{cases} -10x + 15y = 5 & (1) \\ y = 1 & (2) \end{cases}$

$\begin{cases} -10x + 15(1) = 5 & (1) \\ y = 1 & (2) \end{cases}$ Back-substitute; $y = 1$.

$\begin{cases} -10x + 15 = 5 & (1) \\ y = 1 & (2) \end{cases}$

$\begin{cases} x = 1 \\ y = 1 \end{cases}$

The solution is $x = 1$, $y = 1$.

27. $\begin{cases} 2x + 3y = 6 & (1) \\ x - y = \dfrac{1}{2} & (2) \end{cases}$

We use substitution.

Step 1:   Solve for (2) for x:

$$x - y = \frac{1}{2}$$
$$x = y + \frac{1}{2}$$

Step 2:   Substitute $x = y + \dfrac{1}{2}$ into (1):

$$2x + 3y = 6$$
$$2\left(y + \frac{1}{2}\right) + 3y = 6$$
$$2y + 1 + 3y = 6$$
$$5y + 1 = 6$$

Step 3:   Solve for y:

$$5y + 1 = 6$$
$$5y = 5$$
$$y = 1$$

Step 4: Determine x:

$$x = y + \frac{1}{2} \quad \text{(Step 1)}$$

$$x = 1 + \frac{1}{2} \quad (y = 1)$$

$$x = \frac{3}{2}$$

The solution is $x = \frac{3}{2}$, $y = 1$.

29. $\begin{cases} \frac{1}{2}x + \frac{1}{3}y = 3 & (1) \\ \frac{1}{4}x - \frac{2}{3}y = -1 & (2) \end{cases}$

Multiply equation (1) by 6.
Multiply equation (2) by 12.
This eliminates the fractions.

$$\begin{cases} 3x + 2y = 18 & (1) \\ 3x - 8y = -12 & (2) \end{cases}$$

We use elimination.

$$\begin{cases} 3x + 2y = 18 \\ \quad\quad 10y = 30 \quad \text{Replace (2) by (1) - (2).} \end{cases}$$

$$\begin{cases} 3x + 2y = 18 \\ \quad\quad y = 3 \end{cases}$$

$$\begin{cases} 3x + 2(3) = 18 \\ \quad\quad\quad y = 3 \end{cases}$$

$$\begin{cases} x = 4 \\ y = 3 \end{cases}$$

The solution is $x = 4$, $y = 3$.

31. $\begin{cases} 3x - 5y = 3 & (1) \\ 15x + 5y = 21 & (2) \end{cases}$

We use elimination.

$$\begin{cases} 3x - 5y = 3 & (1) \\ \quad\quad 18x = 24 & (2) \quad \text{Replace (2) by (1) + (2).} \end{cases}$$

$$\begin{cases} 3x - 5y = 3 & (1) \\ \quad\quad x = \frac{4}{3} & (2) \end{cases}$$

$$\begin{cases} 4 - 5y = 3 & (1) \quad \text{Back-substitution; } x = \frac{4}{3}. \\ \quad\quad x = \frac{4}{3} & (2) \end{cases}$$

$$\begin{cases} 5y = 1 & (1) \\ x = \frac{4}{3} & (2) \end{cases}$$

$$\begin{cases} y = \frac{1}{5} & (1) \\ x = \frac{4}{3} & (2) \end{cases}$$

The solution is $x = \frac{4}{3}$, $y = \frac{1}{5}$.

33.
$$\begin{cases} x - y = 6 & (1) \\ 2x - 3z = 16 & (2) \\ 2y + z = 4 & (3) \end{cases}$$

We use substitution.

Step 1: We can solve for x in equation (1): $\qquad x - y = 6$
$$x = y + 6$$

Step 2: Substitute this expression for x into equations (2) and (3):
$$\begin{cases} 2x - 3z = 16 & (2) \\ 2y + z = 4 & (3) \end{cases}$$
$$\begin{cases} 2(y + 6) - 3z = 16 & (2) \\ 2y + z = 4 & (3) \end{cases}$$
$$\begin{cases} 2y - 3z + 12 = 16 & (2) \\ 2y + z = 4 & (3) \end{cases}$$
$$\begin{cases} 2y - 3z + = 4 & (2) \\ 2y + z = 4 & (3) \end{cases}$$

Now we must return to Step 1 and solve (2) or (3) for y or z.

It is easiest to solve for z in equation (3): $\qquad 2y + z = 4$
$$z = 4 - 2y$$

Now substitute $z = 4 - 2y$ into (2): $\qquad 2y - 3z = 4$
$$2y - 3(4 - 2y) = 4$$
$$2y - 12 + 6y = 4$$
$$8y = 16$$

Step 3: $\quad y = 2$

Step 4: From $z = 4 - 2y$, we have: $\qquad z = 4 - 2(2)$
$$z = 0,$$

and from $\quad x = y + 6 \quad$ (Step 1)
we have $\quad x = 2 + 6$
$$x = 8$$

The solution is: $x = 8$, $y = 2$, and $z = 0$.

35.
$$\begin{cases} x - 2y + 3z = 7 & (1) \\ 2x + y + z = 4 & (2) \\ -3x + 2y - 2z = -10 & (3) \end{cases}$$

We use substitution.

Step 1A: Solve equation (1) for x: $\qquad x - 2y + 3z = 7$
$$x = 7 + 2y - 3z$$

Step 2A: Substitute $x = 7 + 2y - 3z$ into (2) and (3):
$$\begin{cases} 2x + y + z = 4 & (2) \\ -3x + 2y - 2z = -10 & (3) \end{cases}$$
$$\begin{cases} 2(7 + 2y - 3z) + y + z = 4 & (2) \\ -3(7 + 2y - 3z) + 2y - 2z = -10 & (3) \end{cases}$$
$$\begin{cases} 14 + 4y - 6z + y + z = 4 & (2) \\ -21 - 6y + 9z + 2y - 2z = -10 & (3) \end{cases}$$
$$\begin{cases} 5y - 5z = -10 & (2) \\ -4y + 7z = 11 & (3) \end{cases}$$

Step 1B: We must now solve for y or z in (2) or (3). If we work with (2), we can avoid fractions on this step:
$$5y - 5z = -10 \qquad (2)$$
$$-5z = -5y - 10$$
$$z = y + 2$$

Step 2B: Substitute this into (3):

$$-4y + 7z = 11$$
$$-4y + 7(y + 2) = 11$$
$$-4y + 7y + 14 = 11$$
$$3y = -3$$

Step 3: $y = -1$

Step 4: Determine z:

$$z = y + 2 \quad \text{(Step 1B)}$$
$$z = -1 + 2 \quad (y = -1)$$
$$z = 1$$

Determine x:

$$x = 7 + 2y - 3z \quad \text{(Step 1A)}$$
$$x = 7 + 2(-1) - 3(1) \quad (y = -1), z = 1)$$
$$x = 2$$

The solution is $x = 2$, $y = -1$, $z = 1$.

37. $\begin{cases} x - y - z = 1 & (1) \\ 2x + 3y + z = 2 & (2) \\ 3x + 2y \quad = 0 & (3) \end{cases}$

We use elimination.

Since z is already missing from equation (3), we try to eliminate it from either (1) or (2).

$\begin{cases} x - y - z = 1 & (1) \\ 3x + 2y \quad = 3 & (2) \quad \text{Replace (2) by (1) + (2).} \\ 3x + 2y \quad = 0 & (3) \end{cases}$

$\begin{cases} x - y - z = 1 & (1) \\ -3x - 2y \quad = -3 & (2) \quad \text{Multiply both sides by } -1. \\ 3x + 2y \quad = 0 & (3) \end{cases}$

$\begin{cases} x - y - z = 1 & (1) \\ -3x - 2y \quad = -3 & (2) \\ 0x + 0y \quad = -3 & (3) \quad \text{Replace (3) by (2) + (3).} \end{cases}$

(3) has no solution, so the original system is inconsistent.

39. $\begin{cases} x - y - z = 1 & (1) \\ -x + 2y - 3z = -4 & (2) \\ 3x - 2y - 7z = 0 & (3) \end{cases}$

We use substitution.

Step 1A: Solve for x in (1):

$$x - y - z = 1$$
$$x = 1 + y + z$$

Step 2A: Substitute into (2) and (3):

$$\begin{cases} -x + 2y - 3z = -4 & (2) \\ 3x - 2y - 7z = 0 & (3) \end{cases}$$

$$\begin{cases} -(1 + y + z) + 2y - 3z = -4 & (2) \\ 3(1 + y + z) - 2y - 7z = 0 & (3) \end{cases}$$

$$\begin{cases} y - 4z = -3 & (2) \\ y - 4z = -3 & (3) \end{cases}$$

Step 1B: Solve (2) for y:

$$y - 4z = -3$$
$$y = -3 + 4z$$

Step 2B: Substitute this into (3):

$$y - 4z = -3$$
$$-3 + 4z - 4z = -3$$
$$0 \cdot z = 0$$

This equation is satisfied by all values of z, so z can be any real number.

Then, from Step 1B, $y = 4z - 3$, and from Step 1A,
$$x = 1 + y + z$$
$$x = 1 + (4z - 3) + z$$
$$x = 5z - 2$$

Thus, we can write the solution as: $x = 5z - 2$; $y = 4z - 3$; where z is *any* real number.

We could also express our answer in terms of the following:

$$x = \frac{5}{4}y + \frac{7}{4}; \ z = \frac{1}{4}y + \frac{3}{4}, \text{ where } y \text{ is any real number.}$$

Also, the solution is: $y = \frac{4}{5}x - \frac{7}{5}; \ z = \frac{1}{5}x + \frac{2}{5}$, where x is any real number.

41. $\begin{cases} 2x - 2y + 3z = 6 & (1) \\ 4x - 3y + 2z = 0 & (2) \\ -2x + 3y - 7z = 1 & (3) \end{cases}$

We use elimination.

We can eliminate x from equation (3) by adding (1) and (3):
$$\begin{cases} 2x - 2y + 3z = 6 & (1) \\ 4x - 3y + 2z = 0 & (2) \\ \phantom{2x - {}} y - 4z = 7 & (3) \quad \text{Replace (3) by (1) + 3.} \end{cases}$$

We need to eliminate the *same* variable, x, from either (1) or (2):
$$\begin{cases} -4x + 4y - 6z = -12 & (1) \quad \text{Multiply both sides by } -2. \\ 4x - 3y + 2z = 0 & (2) \\ \phantom{-4x + 4{}} y - 4z = 7 & (3) \end{cases}$$
$$\begin{cases} -4x + 4y - 6z = -12 & (1) \\ \phantom{-4x + 4{}} y - 4z = -12 & (2) \quad \text{Replace (2) by (1) + (2).} \\ \phantom{-4x + 4{}} y - 4z = 7 & (3) \end{cases}$$

We can see that (2) and (3) are contradictory: if we subtract (3) from (2), we obtain $0 = -19$.

Thus, the system is inconsistent.

43. $\begin{cases} x + y - z = 6 & (1) \\ 3x - 2y + z = -5 & (2) \\ x + 3y - 2z = 14 & (3) \end{cases}$

We use substitution.

Step 1A: I choose to solve equation (2) for z: $\quad 3x - 2y + z = -5$
$$z = -5 - 3x + 2y$$

Step 2A: Now substitute this expression for z into (1) and (3):
$$\begin{cases} x + y - z = 6 & (1) \\ x + 3y - 2z = 14 & (3) \end{cases}$$
$$\begin{cases} x + y - (-5 - 3x + 2y) = 6 & (1) \\ x + 3y - 2(-5 - 3x + 2y) = 14 & (3) \end{cases}$$
$$\begin{cases} 4x - y = 1 & (1) \\ 7x - y = 4 & (3) \end{cases}$$

Step 1B: Now I solve for y in (1). (Solving equation (3) for y would be equally easy.)
$$4x - y = 1$$
$$-y = 1 - 4x$$
$$y = -1 + 4x$$

Step 2B: Substitute this into (3): $\qquad 7x - y = 4$
$$7x - (-1 + 4x) = 4$$
$$3x + 1 = 4$$

Step 3: $\quad x = 1$

Step 4:   We now find y and z:

$$y = -1 + 4x \qquad \text{(Step 1B)}$$
$$y = -1 + 4(1) \qquad (x = 1)$$
$$y = 3$$

Also, $z = -5 - 3x + 2y \qquad \text{(Step 1A)}$
$$z = -5 - 3(1) + 2(3) \qquad (x = 1, y = 3)$$
$$z = -2$$

So, the solution is: $x = 1$, $y = 3$, $z = -2$.

45. $\begin{cases} x + 2y - z = -3 & (1) \\ 2x - 4y + z = -7 & (2) \\ -2x + 2y - 3z = 4 & (3) \end{cases}$

We can eliminate z by adding (1) and (2):

$\begin{cases} x + 2y - z = -3 & (1) \\ 3x - 2y = -10 & (2) \quad \text{Replace (2) by (1) + (2).} \\ -2x + 2y - 3z = 4 & (3) \end{cases}$

Now eliminate z from (1) or (3):

$\begin{cases} -3x - 6y + 3z = 9 & (1) \quad \text{Multiply both sides by } -3. \\ 3x - 2y = -10 & (2) \\ -2x + 2y - 3z = 4 & (3) \end{cases}$

$\begin{cases} -3x - 6y + 3z = 9 & (1) \\ 3x - 2y = -10 & (2) \\ -5x - 4y = 13 & (3) \quad \text{Replace (3) by (1) + (3).} \end{cases}$

$\begin{cases} -3x - 6y + 3z = 9 & (1) \\ -6x + 4y = 20 & (2) \quad \text{Multiply both sides by } -2. \\ -5x - 4y = 13 & (3) \end{cases}$

$\begin{cases} -3x - 6y + 3z = 9 & (1) \\ -6x + 4y = 20 & (2) \\ -11x = 33 & (3) \quad \text{Replace (3) by (2) + (3).} \end{cases}$

$\begin{cases} -3x - 6y + 3z = 9 & (1) \\ -6x + 4y = 20 & (2) \\ x = -3 & (3) \end{cases}$

$\begin{cases} 9 - 6y + 3z = 9 & (1) \quad \text{Back-substitute; } x = -3. \\ 18 + 4y = 20 & (2) \quad \text{Back-substitute; } x = -3. \\ x = -3 & (3) \end{cases}$

$\begin{cases} -6y + 3z = 0 & (1) \\ y = \dfrac{1}{2} & (2) \\ x = -3 & (3) \end{cases}$

$\begin{cases} -3 + 3z = 0 & (1) \quad \text{Back-substitute; } y = \dfrac{1}{2}. \\ y = \dfrac{1}{2} & (2) \\ x = -3 & (3) \end{cases}$

$\begin{cases} z = 1 & (1) \\ y = \dfrac{1}{2} & (2) \\ x = -3 & (3) \end{cases}$

The solution is $x = -3$, $y = \dfrac{1}{2}$, $z = 1$.

47. $\begin{cases} \dfrac{1}{x} + \dfrac{1}{y} = 8 & (1) \\[2mm] \dfrac{3}{x} - \dfrac{5}{y} = 0 & (2) \end{cases}$

$\begin{cases} \dfrac{1}{x} + \dfrac{1}{y} = 8 & (1) \\[2mm] 3\left(\dfrac{1}{x}\right) - 5\left(\dfrac{1}{y}\right) = 0 & (2) \end{cases}$

Now let $u = \dfrac{1}{x},\ v = \dfrac{1}{y}$: $\begin{cases} u + v = 8 & (1) \\ 3u - 5v = 0 & (2) \end{cases}$

Step 1: Solve (1) for v:
$$u + v = 8$$
$$v = 8 - u$$

Step 2: Substitute into (2):
$$3u - 5v = 0$$
$$3u - 5(8 - u) = 0$$
$$3u - 40 + 5u = 0$$

Step 3: Solve for u:
$$8u = 40$$
$$u = 5$$

Step 4: $v = 8 - u$ (Step 1)
$$v = 8 - 5 \quad (u = 5)$$
$$v = 3$$

So $u = 5,\ v = 3$.

But we are supposed to find x and y.

We have: $u = \dfrac{1}{x},\ v = \dfrac{1}{y}$

$$5 = \dfrac{1}{x},\ 3 = \dfrac{1}{y}$$

$$x = \dfrac{1}{5},\ y = \dfrac{1}{3}$$

49. $\begin{cases} y_1 = \sqrt{2}x - 20\sqrt{7} \\ y_2 = -0.1x + 20 \end{cases}$

Graph y_1 and y_2.
Using the ZOOM and TRACE features on the graphing calculator, the point of intersection is $x = 48.15,\ y = 15.18$.

51. $\begin{cases} \sqrt{2}x + \sqrt{3}y + \sqrt{6} = 0 \\ \sqrt{3}x - \sqrt{2}y + 60 = 0 \end{cases}$

Solve each equation for y.
$$\sqrt{2}x + \sqrt{3}y + \sqrt{6} = 0$$
$$\sqrt{3}y = -\sqrt{2}x - \sqrt{6}$$

$$y_1 = \dfrac{-\sqrt{2}x - \sqrt{6}}{\sqrt{3}}$$

$$\sqrt{3}x - \sqrt{2}y + 60 = 0$$
$$-\sqrt{2}y = -\sqrt{3}x - 60$$

$$y_2 = \dfrac{\sqrt{3}x + 60}{\sqrt{2}}$$

Graph y_1 and y_2.
Using the ZOOM and TRACE features on the graphing calculator, the point of intersection is $x = -21.47,\ y = 16.12$.

53.
$$\begin{cases} \sqrt{3}x + \sqrt{2}y = \sqrt{0.3} \\ 100x - 95y = 20 \end{cases}$$

Solve each equation for y.

$$\sqrt{3}x + \sqrt{2}y = \sqrt{0.3}$$

$$y_1 = \frac{-\sqrt{3}x + \sqrt{0.3}}{\sqrt{2}}$$

$$100x - 95y = 20$$

$$y_2 = \frac{-100x + 20}{-95}$$

Graph y_1 and y_2.
Using the ZOOM and TRACE features on the graphing calculator, the point of intersection is $x = 0.26$, $y = 0.06$.

In Problems 55–64, start by giving variable names (x, y, etc.) to the unknowns. Then translate each statement about the unknowns into an equation involving the variables. (The number of equations and the number of variables must be equal.) Solve the equations for the unknowns. Finally, be sure you have answered the original question.

55. Let the two numbers be x and y.
$$\begin{cases} x + y = 81 & \text{(Their sum is 81.)} \\ 2x - 3y = 62 & \text{(Twice one minus three times the other is 62.)} \end{cases}$$

Solve the first equation for x:
$$x + y = 81$$
$$x = 81 - y$$

Substitute this into the second equation:
$$2x - 3y = 62$$
$$2(81 - y) - 3y = 62$$
$$-5y = -100$$
$$y = 20$$

Now find x: We know $x = 81 - y$, and $y = 20$,
$$\text{so } x = 81 - 20$$
$$x = 61$$

The two numbers are 20 and 61.

57. Denote the width and length of the rectangle by w and ℓ. Then:
$$(1) \qquad 2w + 2\ell = 90 \quad \text{(the perimeter is 90)}$$
$$(2) \qquad \ell = 2w \quad \text{(length is twice the width)}$$

Substitute $\ell = 2w$ into (1):
$$2w + 2\ell = 90$$
$$2w + 2(2w) = 90$$
$$6w = 90$$
$$w = 15$$

Then $\ell = 2w$, or $\ell = 30$.
The room is 15 feet by 30 feet.

59. Let x = cost of one cheeseburger, in cents
   y = cost of a shake, in cents

Then: $4x + 2y = 790$ \qquad (1)
$$2y = x + 15$$

From (2), we have:
$$y = \frac{1}{2}x + \frac{15}{2}$$

Substitute this into equation (1):

$$4x + 2y = 790$$

$$4x + 2\left(\frac{1}{2}x + \frac{15}{2}\right) = 790$$

$$4x + x + 15 = 790$$

$$5x = 775$$

$$x = 155$$

Then, since $y = \frac{1}{2}x + \frac{15}{2}$, we have: $y = \frac{155}{2} + \frac{15}{2}$

$$y = 85$$

Therefore, a cheeseburger costs $1.55, and a shake costs $0.85.

61. Here, we really have only one unknown:

x = Number of pounds of cashews to use

Since we are using 30 pounds of peanuts, the mixture will contain $30 + x$ pounds.

The basic formula is: Revenue = (Number of pounds) · (Price per pound)

Revenue from peanuts alone = $(30)(1.50) = 45$ dollars

Revenue from cashews alone = $x(5) = 5x$ dollars

Revenue from mixture = $(30 + x) \cdot 3 = 90 + 3x$ dollars

So we want: $90 + 3x = 45 + 5x$

$$45 = 2x$$

$$x = \frac{45}{2} = 22\frac{1}{2}$$

The manager should use 22.5 pounds of cashews.

63. Here, the unknown quantities are speeds.

Let x = average windspeed

y = average air speed of the Piper

both in miles per hour.

Going *with* the wind, the plane has a groundspeed of $x + y$. Then:

(rate) · (time) = distance

$(x + y)(3) = 600$, or

$x + y = 200$ \qquad (1)

Against the wind, the speed of the plane will be $(y - x)$.

(rate) · (time) = distance

$(y - x)(4) = 600$, or

$y - x = 150$ \qquad (2)

We have: $\begin{cases} x + y = 200 & (1) \\ y - x = 150 & (2) \end{cases}$

Solve (2) for y: $y - x = 150$

$$y = 150 + x$$

Then from (1): $x + y = 200$

$$x + (150 + x) = 200$$

$$2x = 50$$

$$x = 25$$

Finally, $y = 150 + x = 175$

Thus, the wind speed is $x = 25$ mph, and the air speed of the plane is $y = 175$ mph.

65. Let x = one design of a set of dishes.
and y = second design of dishes

$$x + y = 200$$
$$25x + 45y = 7400$$
$$x = 200 - y$$
$$25(200 - y) + 45y = 7400$$
$$5000 - 25y + 45y = 7400$$
$$5000 + 20y = 7400$$
$$20y = 2400$$
$$y = 120$$
$$x = 200 - 120 = 80$$

80 $25 sets of dishes should be ordered and 120 $45 sets of dishes should be ordered.

67. In order to determine the size of the refund, we must determine the cost-per-package for bacon and eggs. We will use the general principle:

Total cost for an Item = (Cost per package) · (Number of Packages)

Let x = Cost-per-package of bacon
and y = Cost-per-carton of eggs

Now translate the sentences into equations:
  (1)   Three packages of bacon and two cartons of eggs cost $7.45:
$$3x + 2y = 7.45 \quad (1)$$
  (2)   Two packages of bacon and three cartons of eggs cost $6.45:
$$2x + 3y = 6.45 \quad (2)$$

Solve for x and y:

$$\begin{cases} 3x + 2y = 7.45 & (1) \\ 2x + 3y = 6.45 & (2) \end{cases}$$

$$\begin{cases} 6x + 4y = 14.90 & (1) \quad \text{Multiply both sides by 2.} \\ -6x - 9y = -19.35 & (2) \quad \text{Multiply both sides by } -3. \end{cases}$$

$$\begin{cases} 6x + 4y = 14.90 & (1) \\ -5y = -4.45 & (2) \quad \text{Replace (2) by (1) + (2).} \end{cases}$$

$$\begin{cases} 6x + 4y = 14.90 & (1) \\ y = .89 & (2) \end{cases}$$

$$\begin{cases} 6x + 3.56 = 14.90 & (1) \quad \text{Back-substitute; } y = .89. \\ y = .89 & (2) \end{cases}$$

$$\begin{cases} 6x = 11.34 & (1) \\ y = .89 & (2) \end{cases}$$

$$\begin{cases} x = 1.89 & (1) \\ y = .89 & (2) \end{cases}$$

We now know that bacon costs $1.89 per package, and eggs cost $.89 per carton. But the question is, how much money will be refunded on two packages of bacon and two cartons of eggs?
Refund = 2(1.89) + 2(.89) = $5.56

69. Let the numbers be x, y, and z, from smallest to largest ($x \le y \le z$).
Then we have:

$$\begin{cases} x + y + z = 48 & (1) \\ y + z = 3x & (2) \\ x + y = z + 6 & (3) \end{cases}$$

$$\begin{cases} x + y + z = 48 & (1) \\ -3x + y + z = 0 & (2) \\ x + y - z = 6 & (3) \end{cases}$$

We can eliminate z twice by adding (3) to (1) and (3) to (2):

$$\begin{cases} 2x + 2y = 54 & (1) \\ -2x + 2y = 6 & (2) \\ x + y - z = 6 & (3) \end{cases} \quad \begin{array}{l} \text{Replace (1) by (1) + (3).} \\ \text{Replace (2) by (2) + (3).} \end{array}$$

$$\begin{cases} 4y = 60 & (1) \\ -2x + 2y = 6 & (2) \\ x + y - z = 6 & (3) \end{cases} \quad \text{Replace (1) by (1) + (2).}$$

$$\begin{cases} y = 15 & (1) \\ -2x + 2y = 6 & (2) \\ x + y - z = 6 & (3) \end{cases}$$

$$\begin{cases} y = 15 & (1) \\ -2x + 30 = 6 & (2) \\ x + 15 - z = 6 & (3) \end{cases} \quad \begin{array}{l} \text{Back-substitute; } y = 15. \\ \text{Back-substitute; } y = 15. \end{array}$$

$$\begin{cases} y = 15 & (1) \\ -2x = -24 & (2) \\ x - z = -9 & (3) \end{cases}$$

$$\begin{cases} y = 15 & (1) \\ x = 12 & (2) \\ x - z = -9 & (3) \end{cases}$$

$$\begin{cases} y = 15 & (1) \\ x = 12 & (2) \\ 12 - z = -9 & (3) \end{cases} \quad \text{Back-substitute; } x = 12.$$

$$\begin{cases} y = 15 & (1) \\ x = 12 & (2) \\ z = 21 & (3) \end{cases}$$

The three numbers are: 12, 15, and 21.

71. $$\begin{cases} I_2 = I_1 + I_3 \\ 5 - 3I_1 - 5I_2 = 0 \\ 10 - 5I_2 - 7I_3 = 0 \end{cases}$$

$$\begin{cases} I_2 = I_1 + I_3 \\ 5 - 3I_1 - 5(I_1 + I_3) = 0 \\ 10 - 5(I_1 + I_3) - 7I_3 = 0 \end{cases}$$

$$\begin{cases} I_2 = I_1 + I_3 \\ -8I_1 - 5I_3 + 5 = 0 \\ -5I_1 - 12I_3 + 10 = 0 \end{cases}$$

$$\begin{cases} I_2 = I_1 + I_3 \\ 40I_1 + 25I_3 = 25 \\ -40I_1 - 96I_3 = -80 \end{cases}$$

Chapter 11 Systems of Equations and Inequalities

$$-71I_3 = -55$$

$$I_3 = \frac{55}{71}$$

$$-8I_1 - 5\left(\frac{55}{71}\right) + 5 = 0$$

$$-8I_1 = -5 + \frac{275}{71}$$

$$-8I_1 = \frac{-355 + 275}{71}$$

$$-8I_1 = \frac{-80}{71}$$

$$I_1 = \frac{10}{71}$$

$$I_2 = I_1 + I_3$$

$$I_2 = \frac{10}{71} + \frac{55}{71}$$

$$I_2 = \frac{65}{71}$$

$$I_1 = \frac{10}{71}, \ I_2 = \frac{65}{71}, \ I_3 = \frac{55}{71}$$

73. Let x = Number of orchestra seats,
$\quad\quad y$ = Number of main seats,
and $\ z$ = Number of balcony seats.

Then $x + y + z = 500$, since there are a total of 500 seats.

If all seats are sold, the revenue is \$17,100: $50x + 35y + 25z = 17,100$

Finally, if we sell only half of the orchestra seats, revenue is \$14,600:

$$50\left(\frac{1}{2}x\right) + 35y + 25z = 14,600$$

$$\begin{cases} x + y + z = 500 & (1) \\ 50x + 35y + 25z = 17,100 & (2) \\ 25x + 35y + 25z = 14,600 & (3) \end{cases}$$

$$\begin{cases} x + y + z = 500 & (1) \\ 10x + 7y + 5z = 3420 & (2) \quad \text{Divide both sides by 5.} \\ -5x - 7y - 5z = -2920 & (3) \quad \text{Divide both sides by } -5. \end{cases}$$

$$\begin{cases} x + y + z = 500 & (1) \\ 10x + 7y + 5z = 3420 & (2) \\ 5x = 500 & (3) \quad \text{Replace (3) by (2) + (3).} \end{cases}$$

$$\begin{cases} x + y + z = 500 & (1) \\ 10x + 7y + 5z = 3420 & (2) \\ x = 100 & (3) \end{cases}$$

$$\begin{cases} 100 + y + z = 500 & (1) \quad \text{Back-substitute; } x = 100. \\ 1000 + 7y + 5z = 3420 & (2) \quad \text{Back-substitute; } x = 100. \\ x = 100 & (3) \end{cases}$$

$$\begin{cases} y + z = 400 & (1) \\ 7y + 5z = 2420 & (2) \\ x = 100 & (3) \end{cases}$$

$$\begin{cases} -5y - 5z = -2000 & (1) \\ 7y + 5z = 2420 & (2) \\ x = 100 & (3) \end{cases} \quad \text{Multiply both sides by } -5.$$

$$\begin{cases} 2y = 420 & (1) \\ 7y + 5z = 2420 & (2) \\ x = 100 & (3) \end{cases} \quad \text{Replace (1) by (1) + (2).}$$

$$\begin{cases} y = 210 & (1) \\ 7y + 5z = 2420 & (2) \\ x = 100 & (3) \end{cases}$$

$$\begin{cases} y = 210 & (1) \\ 1470 + 5z = 2420 & (2) \\ x = 100 & (3) \end{cases} \quad \text{Back-substitute; } y = 210.$$

$$\begin{cases} y = 210 & (1) \\ 5z = 950 & (2) \\ x = 100 & (3) \end{cases}$$

$$\begin{cases} y = 210 & (1) \\ z = 190 & (2) \\ x = 100 & (3) \end{cases}$$

Thus, there are: $x = 100$ orchestra seats,
$y = 210$ main seats, and
$z = 190$ balcony seats.

79. We have $y = x^2 + bx + c$.

If this passes through $(1, 2)$, then:

$$y = x^2 + bx + c$$
$$2 = 1^2 + b(1) + c \quad (x = 1, y = 2)$$
$$2 = 1 + b + c$$
$$b + c = 1 \qquad (1)$$

If $(-1, 3)$ lies on the graph, then:

$$y = x^2 + bx + c$$
$$3 = (-1)^2 + b(-1) + c$$
$$3 = 1 - b + c$$
$$-b + c = 2 \qquad (2)$$

Let's solve (2) for c:

$$-b + c = 2$$
$$c = 2 + b$$

Substitute this into (1):

$$b + c = 1$$
$$b + (2 + b) = 1$$
$$2b = -1$$
$$b = -\frac{1}{2}$$

Now find c:

$$c = 2 + b$$
$$c = 2 + \left(-\frac{1}{2}\right)$$
$$c = \frac{3}{2}$$

The solution is $b = -\frac{1}{2}$, $c = \frac{3}{2}$.

81. We have $y = ax^2 + bx + c$.

From $(-1, 4)$:

$$4 = a(-1)^2 + b(-1) + c$$
$$4 = a - b + c \qquad (1)$$

From $(2, 3)$:

$$3 = a(2)^2 + b(2) + c$$
$$3 = 4a + 2b + c \qquad (2)$$

From $(0, 1)$:

$$1 = a(0)^2 + b(0) + c$$
$$c = 1 \qquad (3)$$

Substitute $c = 1$ into (1) and (2):

$$\begin{cases} a - b + c = 4 & (1) \\ 4a + 2b + c = 3 & (2) \end{cases}$$

$$\begin{cases} a - b + 1 = 4 & (1) \\ 4a + 2b + 1 = 3 & (2) \end{cases}$$

$$\begin{cases} a - b = 3 & (1) \\ 4a + 2b = 2 & (2) \end{cases}$$

Now solve for (1) for a:

$$a - b = 3$$
$$a = 3 + b$$

Then, from (2):

$$4a + 2b = 2$$
$$4(3 + b) + 2b = 2$$
$$6b + 12 = 2$$
$$6b = -10$$
$$b = -\frac{10}{6}$$

or

$$b = -\frac{5}{3}$$

Finally,

$$a = 3 + b$$
$$a = 3 - \frac{5}{3}$$
$$a = \frac{4}{3}$$

The solution is $a = \dfrac{4}{3}$, $b = -\dfrac{5}{3}$, $c = 1$.

83. Solve:

$$\begin{cases} y = m_1 x + b_1 & (1) \\ y = m_2 x + b_2 & (2) \end{cases}$$

From (1), $y = m_1 x + b_1$. Substitute this into (2):

$$y = m_2 x + b_2 \qquad (2)$$
$$m_1 x + b_1 = m_2 x + b_2$$
$$m_1 x - m_2 x = b_2 - b_1$$
$$(m_1 - m_2)x = b_2 - b_1$$
$$x = \frac{b_2 - b_1}{m_1 - m_2} = \frac{b_1 - b_2}{m_2 - m_1}$$

$$(\text{Note } m_1 - m_2 \neq 0)$$

Then:
$$y = m_1x + b_1$$
$$y = m_1\left(\frac{b_2 - b_1}{m_1 - m_2}\right) + b_1$$
$$y = \frac{m_1b_2 - m_1b_1}{m_1 - m_2} + \frac{b_1(m_1 - m_2)}{m_1 - m_2} \quad \text{(common denominator)}$$
$$y = \frac{m_1b_2 - m_1b_1 + m_1b_1 - m_2b_1}{m_1 - m_2}$$
$$y = \frac{m_1b_2 - m_2b_1}{m_1 - m_2} = \frac{m_2b_1 - m_1b_2}{m_2 - m_1}$$

85. We have: $\begin{cases} y = m_1x + b_1 & (1) \\ y = m_2x + b_2 & (2) \end{cases}$

Equation (1) is already solved for y, so we can substitute this into (2):
$$y = m_2x + b_2 \quad (2)$$
$$m_1x + b_1 = m_2x + b_2$$
$$m_1x - m_2x = b_2 - b_1$$
$$mx - mx = b - b$$
$$0 \cdot x = 0$$

This is solved by *every* value of x. So the solution is:
$y = mx + b$, where x can be *any* real number.

11.2 Systems of Linear Equations: Matrices

1. $\begin{cases} x - 5y = 5 \\ 4x + 3y = 6 \end{cases}$ becomes $\begin{bmatrix} 1 & -5 & | & 5 \\ 4 & 3 & | & 6 \end{bmatrix}$

3. $\begin{cases} 2x + 3y - 6 = 0 \\ 4x - 6y + 2 = 0 \end{cases}$

First, put the constants on the right-hand side: $\begin{cases} 2x + 3y = 6 \\ 4x - 6y = -2 \end{cases}$

This can be represented as: $\begin{bmatrix} 2 & 3 & | & 6 \\ 4 & -6 & | & -2 \end{bmatrix}$

5. $\begin{cases} 0.01x - 0.03y = 0.06 \\ 0.13x + 0.10y = 0.20 \end{cases}$ becomes $\begin{bmatrix} 0.01 & -0.03 & | & 0.06 \\ 0.13 & 0.10 & | & 0.20 \end{bmatrix}$

7. $\begin{cases} x - y + z = 10 \\ 3x + 2y = 5 \\ x + y + 2z = 2 \end{cases}$ becomes $\begin{bmatrix} 1 & -1 & 1 & | & 10 \\ 3 & 2 & 0 & | & 5 \\ 1 & 1 & 2 & | & 2 \end{bmatrix}$

9. $\begin{cases} x + y - z = 2 \\ 3x - 2y = 2 \end{cases}$ becomes $\begin{bmatrix} 1 & 1 & -1 & | & 2 \\ 3 & -2 & 0 & | & 2 \end{bmatrix}$

For Problems 11–20, we will show the augmented matrix that you should see on your graphing utility after each row operation is performed.

11. $\begin{bmatrix} 1 & -3 & -5 & | & -2 \\ 2 & -5 & -4 & | & 5 \\ -3 & 5 & 4 & | & 6 \end{bmatrix} \rightarrow \begin{bmatrix} 1 & -3 & -5 & | & -2 \\ 0 & 1 & 6 & | & 9 \\ -3 & 5 & 4 & | & 6 \end{bmatrix} \rightarrow \begin{bmatrix} 1 & -3 & -5 & | & -2 \\ 0 & 1 & 6 & | & 9 \\ 0 & -4 & -11 & | & 0 \end{bmatrix} \rightarrow \begin{bmatrix} 1 & -3 & -5 & | & -2 \\ 0 & 1 & 6 & | & 9 \\ 0 & 0 & 13 & | & 36 \end{bmatrix}$

         ↑   ↑   ↑

    (a) $R_2 = -2r_1 + r_2$ (b) $R_3 = 3r_1 + r_3$ (c) $R_3 = 4r_2 + r_3$

13. $\begin{bmatrix} 1 & -3 & 4 & | & 3 \\ 2 & -5 & 6 & | & 6 \\ -3 & 3 & 4 & | & 6 \end{bmatrix} \rightarrow \begin{bmatrix} 1 & -3 & 4 & | & 3 \\ 0 & 1 & -2 & | & 0 \\ -3 & 3 & 4 & | & 6 \end{bmatrix} \rightarrow \begin{bmatrix} 1 & -3 & 4 & | & 3 \\ 0 & 1 & -2 & | & 0 \\ 0 & -6 & 16 & | & 15 \end{bmatrix} \rightarrow \begin{bmatrix} 1 & -3 & 4 & | & 3 \\ 0 & 1 & -2 & | & 0 \\ 0 & 0 & 4 & | & 15 \end{bmatrix}$

         ↑   ↑   ↑

    (a) $R_2 = -2r_1 + r_2$ (b) $R_3 = 3r_1 + r_3$ (c) $R_3 = 6r_2 + r_3$

15. $\begin{bmatrix} 1 & -3 & 2 & | & -6 \\ 2 & -5 & 3 & | & -4 \\ -3 & -6 & 4 & | & 6 \end{bmatrix} \rightarrow \begin{bmatrix} 1 & -3 & 2 & | & -6 \\ 0 & 1 & -1 & | & 8 \\ -3 & -6 & 4 & | & 6 \end{bmatrix} \rightarrow \begin{bmatrix} 1 & -3 & 2 & | & -6 \\ 0 & 1 & -1 & | & 8 \\ 0 & -15 & 10 & | & -12 \end{bmatrix} \rightarrow \begin{bmatrix} 1 & -3 & 2 & | & -6 \\ 0 & 1 & -1 & | & 8 \\ 0 & 0 & -5 & | & 108 \end{bmatrix}$

         ↑   ↑   ↑

    (a) $R_2 = -2r_1 + r_2$ (b) $R_3 = 3r_1 + r_3$ (c) $R_3 = 15r_2 + r_3$

17. $\begin{bmatrix} 1 & -3 & 1 & | & -2 \\ 2 & -5 & 6 & | & -2 \\ -3 & 1 & 4 & | & 6 \end{bmatrix} \rightarrow \begin{bmatrix} 1 & -3 & 1 & | & -2 \\ 0 & 1 & 4 & | & 2 \\ -3 & 1 & 4 & | & 6 \end{bmatrix} \rightarrow \begin{bmatrix} 1 & -3 & 1 & | & -2 \\ 0 & 1 & 4 & | & 2 \\ 0 & -8 & 7 & | & 0 \end{bmatrix} \rightarrow \begin{bmatrix} 1 & -3 & 1 & | & -2 \\ 0 & 1 & 4 & | & 2 \\ 0 & 0 & 39 & | & 16 \end{bmatrix}$

         ↑   ↑   ↑

    (a) $R_2 = -2r_1 + r_2$ (b) $R_3 = 3r_1 + r_3$ (c) $R_3 = 8r_2 + r_3$

19. $\begin{bmatrix} 1 & -3 & -2 & | & 3 \\ 2 & -5 & 2 & | & -1 \\ -3 & -2 & 4 & | & 6 \end{bmatrix} \rightarrow \begin{bmatrix} 1 & -3 & -2 & | & 3 \\ 0 & 1 & 6 & | & -7 \\ -3 & -2 & 4 & | & 6 \end{bmatrix} \rightarrow \begin{bmatrix} 1 & -3 & -2 & | & 3 \\ 0 & 1 & 6 & | & -7 \\ 0 & -11 & -2 & | & 15 \end{bmatrix} \rightarrow \begin{bmatrix} 1 & -3 & -2 & | & 3 \\ 0 & 1 & 6 & | & -7 \\ 0 & 0 & 64 & | & -62 \end{bmatrix}$

         ↑   ↑   ↑

    (a) $R_2 = -2r_1 + r_2$ (b) $R_3 = 3r_1 + r_3$ (c) $R_3 = 11r_2 + r_3$

21. $\begin{cases} x = 5 \\ y = -1 \end{cases}$ consistent $x = 5, y = -1$ 23. $\begin{cases} x = 1 \\ y = 2 \\ 0 = 3 \end{cases}$ inconsistent

25. $\begin{cases} x + 2z = -1 \\ y - 4z = -2 \\ \quad\; 0 = 0 \end{cases}$ consistent $x = -1 - 2z, y = -2 + 4z, z$ any real number

27. $\begin{cases} x_1 = 1 \\ x_2 + x_4 = 2 \\ x_3 + 2x_4 = 3 \end{cases}$ consistent $x_1 = 1, x_2 = 2 - x_4, x_3 = 3 - 2x_4, \quad x_4$ any real number.

29. $\begin{cases} x_1 + 4x_4 = 2 \\ x_2 + x_3 + 3x_4 = 3 \\ \qquad\qquad 0 = 0 \end{cases}$ consistent $x_1 = 2 - 4x_4, x_2 = 3 - x_3 - 3x_4, x_3, x_4$ any real numbers.

31. $\begin{cases} x + y = 8 \\ x - y = 4 \end{cases}$ can be represented as: $\begin{bmatrix} 1 & 1 & | & 8 \\ 1 & -1 & | & 4 \end{bmatrix}$

Since we already have a 1 in the first row, first column, we proceed to obtain 0's below it:

$$\begin{bmatrix} 1 & 1 & | & 8 \\ 1 & -1 & | & 4 \end{bmatrix} \rightarrow \begin{bmatrix} 1 & 1 & | & 8 \\ 0 & -2 & | & -4 \end{bmatrix} \rightarrow \begin{bmatrix} 1 & 1 & | & 8 \\ 0 & 1 & | & 2 \end{bmatrix}$$

$$\uparrow \qquad\qquad \uparrow$$

$$R_2 = -1r_1 + r_2 \quad R_2 = -\frac{1}{2}r_2$$

That means: $\begin{cases} x + y = 8 \\ \quad\ y = 2 \end{cases}$

$\begin{cases} x + 2 = 8 \\ \quad\ y = 2 \end{cases}$ Back-substitute; $y = 2$.

$\begin{cases} x = 6 \\ y = 2 \end{cases}$

The solution is $x = 6$, $y = 2$.

33. $\begin{cases} x - 5y = -13 \\ 3x + 2y = 12 \end{cases}$

$$\begin{bmatrix} 1 & -5 & | & -13 \\ 3 & 2 & | & 12 \end{bmatrix} \rightarrow \begin{bmatrix} 1 & -5 & | & -13 \\ 0 & 17 & | & 51 \end{bmatrix} \rightarrow \begin{bmatrix} 1 & -5 & | & -13 \\ 0 & 1 & | & 3 \end{bmatrix}$$

$$y = 3 \qquad x - 5y = -13$$
$$x = 5(3) - 13$$
$$x = 2$$
$$x = 2, y = 3$$

35. $\begin{cases} 3x - 6y = 24 \\ 5x + 4y = 12 \end{cases}$ becomes: $\begin{bmatrix} 3 & -6 & | & 24 \\ 5 & 4 & | & 12 \end{bmatrix}$

$$\rightarrow \begin{bmatrix} 1 & -2 & | & 8 \\ 5 & 4 & | & 12 \end{bmatrix} \rightarrow \begin{bmatrix} 1 & -2 & | & 8 \\ 0 & 14 & | & -28 \end{bmatrix} \rightarrow \begin{bmatrix} 1 & -2 & | & 8 \\ 0 & 1 & | & -2 \end{bmatrix} \rightarrow \begin{bmatrix} 1 & 0 & | & 4 \\ 0 & 1 & | & -2 \end{bmatrix}$$

$$\uparrow \qquad\qquad \uparrow \qquad\qquad \uparrow \qquad\qquad \uparrow$$

$$R_1 = \frac{1}{3}r_1 \qquad R_2 = -5r_1 + r_2 \qquad R_2 = \frac{1}{14}r_2 \qquad R_1 = 2r_2 + r_1$$

The solution is $x = 4$, $y = -2$.

37. $\begin{cases} 2x + y = 1 \\ 4x + 2y = 6 \end{cases}$ becomes: $\begin{bmatrix} 2 & 1 & | & 1 \\ 4 & 2 & | & 6 \end{bmatrix} \rightarrow \begin{bmatrix} 1 & \frac{1}{2} & | & \frac{1}{2} \\ 4 & 2 & | & 6 \end{bmatrix} \rightarrow \begin{bmatrix} 1 & \frac{1}{2} & | & \frac{1}{2} \\ 0 & 0 & | & 4 \end{bmatrix}$

$$\uparrow \qquad\qquad \uparrow$$

$$R_1 = \frac{1}{2}r_1 \qquad R_2 = -4r_1 + r_2$$

The system is inconsistent.

39. $\begin{cases} 2x - 4y = -2 \\ 3x + 2y = 3 \end{cases}$ becomes: $\begin{bmatrix} 2 & -4 & | & -2 \\ 3 & 2 & | & 3 \end{bmatrix}$

$\rightarrow \begin{bmatrix} 1 & -2 & | & -1 \\ 3 & 2 & | & 3 \end{bmatrix} \rightarrow \begin{bmatrix} 1 & -2 & | & -1 \\ 0 & 8 & | & 6 \end{bmatrix} \rightarrow \begin{bmatrix} 1 & -2 & | & -1 \\ 0 & 1 & | & \frac{3}{4} \end{bmatrix} \rightarrow \begin{bmatrix} 1 & 0 & | & \frac{1}{2} \\ 0 & 1 & | & \frac{3}{4} \end{bmatrix}$

$\quad\quad\quad\uparrow \quad\quad\quad\quad\uparrow \quad\quad\quad\quad\uparrow \quad\quad\quad\quad\uparrow$

$\quad R_1 = \frac{1}{2}r_1 \quad R_2 = -3r_1 + r_2 \quad R_2 = \frac{1}{8}r_2 \quad R_1 = 2r_2 + r_1$

The solution is $x = \dfrac{1}{2}$, $y = \dfrac{3}{4}$.

41. $\begin{cases} x + 2y = 4 \\ 2x + 4y = 8 \end{cases}$ becomes: $\begin{bmatrix} 1 & 2 & | & 4 \\ 2 & 4 & | & 8 \end{bmatrix} \rightarrow \begin{bmatrix} 1 & 2 & | & 4 \\ 0 & 0 & | & 0 \end{bmatrix}$

$\quad\quad\quad\quad\quad\quad\quad\quad\quad\quad\quad\quad\quad\quad\uparrow$

$\quad\quad\quad\quad\quad\quad\quad\quad\quad\quad\quad\quad R_2 = -2r_1 + r_2$

Therefore, the system is equivalent to the single equation: $x + 2y = 4$

(a) In terms of y, $\quad x = -2y + 4$

where y can be any real number.

(b) In terms of x, $\quad 2y = -x + 4$,

or $\quad\quad\quad\quad y = -\dfrac{1}{2}x + 2$,

where x can be any real number.

43. $\begin{cases} 2x + 3y = 6 \\ x - y = \dfrac{1}{2} \end{cases}$ becomes: $\begin{bmatrix} 2 & 3 & | & 6 \\ 1 & -1 & | & \frac{1}{2} \end{bmatrix}$

$\rightarrow \begin{bmatrix} 1 & -1 & | & \frac{1}{2} \\ 2 & 3 & | & 6 \end{bmatrix} \rightarrow \begin{bmatrix} 1 & -1 & | & \frac{1}{2} \\ 0 & 5 & | & 5 \end{bmatrix} \rightarrow \begin{bmatrix} 1 & -1 & | & \frac{1}{2} \\ 0 & 1 & | & 1 \end{bmatrix} \rightarrow \begin{bmatrix} 1 & 0 & | & \frac{3}{2} \\ 0 & 1 & | & 1 \end{bmatrix}$

$\quad\uparrow \quad\quad\quad\quad\uparrow \quad\quad\quad\quad\uparrow \quad\quad\quad\quad\uparrow$

Interchange r_1 and r_2 $\quad R_2 = -2r_1 + r_2 \quad R_2 = \dfrac{1}{5}r_2 \quad R_1 = r_2 + r_1$

The solution is $x = \dfrac{3}{2}$, $y = 1$.

45. $\begin{cases} 3x - 5y = 3 \\ 15x + 5y = 21 \end{cases}$ becomes: $\begin{bmatrix} 3 & -5 & | & 3 \\ 15 & 5 & | & 21 \end{bmatrix}$

$\rightarrow \begin{bmatrix} 1 & -\frac{5}{3} & | & 1 \\ 15 & 5 & | & 21 \end{bmatrix} \rightarrow \begin{bmatrix} 1 & -\frac{5}{3} & | & 1 \\ 0 & 30 & | & 6 \end{bmatrix} \rightarrow \begin{bmatrix} 1 & -\frac{5}{3} & | & 1 \\ 0 & 1 & | & \frac{1}{5} \end{bmatrix}$

$\quad\uparrow \quad\quad\quad\quad\uparrow \quad\quad\quad\quad\uparrow$

$\quad R_1 = \frac{1}{3}r_1 \quad R_2 = -15r_1 + r_2 \quad R_2 = \dfrac{1}{30}r_2$

Thus, we have:

$$\begin{cases} x - \dfrac{5}{3}y = 1 \\[2mm] \qquad\quad y = \dfrac{1}{5} \end{cases}$$

$$\begin{cases} x - \dfrac{5}{3}\left(\dfrac{1}{5}\right) = 1 \qquad \text{(Back-substitution)} \\[2mm] \qquad\qquad\quad y = \dfrac{1}{5} \end{cases}$$

$$\begin{cases} x = \dfrac{4}{3} \\[2mm] y = \dfrac{1}{5} \end{cases}$$

The solution is $x = \dfrac{4}{3}$, $y = \dfrac{1}{5}$

47. $\begin{cases} x - y \quad = 6 \\ 2x \quad - 3z = 16 \\ \quad 2y + z = 4 \end{cases}$ becomes: $\begin{bmatrix} 1 & -1 & 0 & | & 6 \\ 2 & 0 & -3 & | & 16 \\ 0 & 2 & 1 & | & 4 \end{bmatrix}$ (We already have a 1 in row 1, column 1.)

$\rightarrow \begin{bmatrix} 1 & -1 & 0 & | & 6 \\ 0 & 2 & -3 & | & 4 \\ 0 & 2 & 1 & | & 4 \end{bmatrix}$ (Use the 1 to get 0's below it.)

$\uparrow$
$R_2 = -2r_1 + r_2$

$\rightarrow \begin{bmatrix} 1 & -1 & 0 & | & 6 \\ 0 & 1 & -\dfrac{3}{2} & | & 2 \\ 0 & 2 & 1 & | & 4 \end{bmatrix}$ (We need a 1 in row 2, column 2.)

$\uparrow$
$R_2 = \dfrac{1}{2}r_2$

$\rightarrow \begin{bmatrix} 1 & -1 & 0 & | & 6 \\ 0 & 1 & -\dfrac{3}{2} & | & 2 \\ 0 & 0 & 4 & | & 0 \end{bmatrix}$ (Obtain a 0 below the 1 in row 2.)

$\uparrow$
$R_3 = -2r_2 + r_3$

$\rightarrow \begin{bmatrix} 1 & -1 & 0 & | & 6 \\ 0 & 1 & -\dfrac{3}{2} & | & 2 \\ 0 & 0 & 1 & | & 0 \end{bmatrix}$

$\uparrow$
$R_3 = \dfrac{1}{4}r_3$

Thus, we have:

$$\begin{cases} x - y = 6 \\ y - \dfrac{3}{2}z = 2 \\ \quad\quad z = 0 \end{cases}$$

$$\begin{cases} x - y = 6 \\ \quad y = 2 \\ \quad\quad z = 0 \end{cases} \quad \text{(Back-substitution; } z = 0)$$

$$\begin{cases} x - 2 = 6 \\ \quad y = 2 \\ \quad\quad z = 0 \end{cases}$$

The solution is $x = 8$, $y = 2$, $z = 0$.

49. $\begin{cases} x - 2y + 3z = 7 \\ 2x + y + z = 4 \\ -3x + 2y - 2z = -10 \end{cases}$ becomes: $\left[\begin{array}{ccc|c} 1 & -2 & 3 & 7 \\ 2 & 1 & 1 & 4 \\ -3 & 2 & -2 & -10 \end{array}\right]$

$\rightarrow \left[\begin{array}{ccc|c} 1 & -2 & 3 & 7 \\ 0 & 5 & -5 & -10 \\ 0 & -4 & 7 & 11 \end{array}\right]$ (Use the 1 in row 1 to get 0's below it.)

$\uparrow$

$R_2 = -2r_1 + r_2$
$R_3 = 3r_1 + r_3$

To obtain a 1 in row 2, column 2, we can either multiply row 2 by $\dfrac{1}{5}$, or add row 3 to row 2:

$\rightarrow \left[\begin{array}{ccc|c} 1 & -2 & 3 & 7 \\ 0 & 1 & -1 & -2 \\ 0 & -4 & 7 & 11 \end{array}\right] \rightarrow \left[\begin{array}{ccc|c} 1 & 0 & 1 & 3 \\ 0 & 1 & -1 & -2 \\ 0 & 0 & 3 & 3 \end{array}\right]$

$\uparrow$ $\qquad\qquad\qquad$ $\uparrow$

$R_2 = \dfrac{1}{5}r_2 \qquad\qquad R_1 = 2r_2 + r_1$

$\qquad\qquad\qquad\qquad R_3 = 4r_2 + r_3$

The zero *above* the 1 in row 2 will eliminate the need to do back-substitution at the end of the problem.

$\rightarrow \left[\begin{array}{ccc|c} 1 & 0 & 1 & 3 \\ 0 & 1 & -1 & -2 \\ 0 & 0 & 1 & 1 \end{array}\right] \rightarrow \left[\begin{array}{ccc|c} 1 & 0 & 0 & 2 \\ 0 & 1 & 0 & -1 \\ 0 & 0 & 1 & 1 \end{array}\right]$

$\uparrow$ $\qquad\qquad\qquad$ $\uparrow$

$R_3 = \dfrac{1}{3}r_3 \qquad\qquad R_1 = -1r_3 + r_1$

$\qquad\qquad\qquad\qquad R_2 = r_3 + r_2$

The solution is $x = 2$, $y = -1$, $z = 1$.

51. $\begin{cases} 2x - 2y - 2z = 2 \\ 2x + 3y + z = 2 \\ 3x + 2y = 0 \end{cases}$ becomes: $\begin{bmatrix} 2 & -2 & -2 & | & 2 \\ 2 & 3 & 1 & | & 2 \\ 3 & 2 & 0 & | & 0 \end{bmatrix}$

$$\rightarrow \begin{bmatrix} 2 & -2 & -2 & | & 2 \\ 0 & 5 & 3 & | & 0 \\ 1 & 4 & 2 & | & -2 \end{bmatrix} \rightarrow \begin{bmatrix} 1 & 4 & 2 & | & -2 \\ 0 & 5 & 3 & | & 0 \\ 2 & -2 & -2 & | & 2 \end{bmatrix} \rightarrow \begin{bmatrix} 1 & 4 & 2 & | & -2 \\ 0 & 5 & 3 & | & 0 \\ 0 & -10 & -6 & | & 6 \end{bmatrix}$$

$\uparrow$ $\uparrow$ $\uparrow$

$R_2 = -r_1 + r_2$ $R_1 \leftrightarrow R_3$ $R_3 = -2r_1 + r_3$

$R_3 = -r_1 + r_3$

$$\rightarrow \begin{bmatrix} 1 & 4 & 2 & | & -2 \\ 0 & 1 & \dfrac{3}{5} & | & 0 \\ 0 & -10 & -6 & | & 6 \end{bmatrix} \rightarrow \begin{bmatrix} 1 & 0 & \dfrac{22}{5} & | & -2 \\ 0 & 5 & \dfrac{3}{5} & | & 0 \\ 0 & 0 & 0 & | & 6 \end{bmatrix}$$

$\uparrow$ $\uparrow$

$R_2 = \dfrac{1}{5} r_2$ $R_1 = -4r_2 + r_1$

$R_3 = 10r_2 + r_3$

No solution. Inconsistent.

53. $\begin{cases} -x + y + z = -1 \\ -x + 2y - 3z = -4 \\ 3x - 2y - 7z = 0 \end{cases}$ becomes: $\begin{bmatrix} -1 & 1 & 1 & | & -1 \\ -1 & 2 & -3 & | & -4 \\ 3 & -2 & -7 & | & 0 \end{bmatrix}$

$$\rightarrow \begin{bmatrix} 1 & -1 & -1 & | & 1 \\ -1 & 2 & -3 & | & -4 \\ 3 & -2 & -7 & | & 0 \end{bmatrix} \rightarrow \begin{bmatrix} 1 & -1 & -1 & | & 1 \\ 0 & 1 & -4 & | & -3 \\ 0 & 1 & -4 & | & -3 \end{bmatrix} \rightarrow \begin{bmatrix} 1 & 0 & -5 & | & -2 \\ 0 & 1 & -4 & | & -3 \\ 0 & 0 & 0 & | & 0 \end{bmatrix} \rightarrow \begin{matrix} x - 5z = -2 \\ y - 4z = -3 \end{matrix}$$

$\uparrow$ $\uparrow$ $\uparrow$

$R_1 = -r_1$ $R_2 = r_1 + r_2$ $R_1 = r_2 + r_1$

                   $R_3 = -3r_1 + r_3$ $R_3 = -r_2 + r_3$

Hence, $x = 5z - 2$; $y = 4z - 3$ where z is any real number.

55. $\begin{cases} 2x - 2y + 3z = 6 \\ 4x - 3y + 2z = 0 \\ -2x + 3y - 7z = 1 \end{cases}$ becomes: $\begin{bmatrix} 2 & -2 & 3 & | & 6 \\ 4 & -3 & 2 & | & 0 \\ -2 & 3 & -7 & | & 1 \end{bmatrix}$

$$\rightarrow \begin{bmatrix} 1 & -1 & \dfrac{3}{2} & | & 3 \\ 4 & -3 & 2 & | & 0 \\ -2 & 3 & -7 & | & 1 \end{bmatrix} \rightarrow \begin{bmatrix} 1 & -1 & \dfrac{3}{2} & | & 3 \\ 0 & 1 & -4 & | & -12 \\ 0 & 1 & -4 & | & 7 \end{bmatrix} \rightarrow \begin{bmatrix} 1 & 0 & -\dfrac{5}{2} & | & -9 \\ 0 & 1 & -4 & | & -12 \\ 0 & 0 & 0 & | & 19 \end{bmatrix}$$

$\uparrow$ $\uparrow$ $\uparrow$

$R_1 = \dfrac{1}{2} r_1$ $R_2 = -4r_1 + r_2$ $R_1 = r_2 + r_1$

                  $R_3 = 2r_1 + r_3$ $R_3 = -r_2 + r_3$

The system is inconsistent.

57. $\begin{cases} x + y - z = 6 \\ 3x - 2y + z = -5 \\ x + 3y - 2z = 14 \end{cases}$ becomes: $\begin{bmatrix} 1 & 1 & -1 & | & 6 \\ 3 & -2 & 1 & | & -5 \\ 1 & 3 & -2 & | & 14 \end{bmatrix}$

$\rightarrow \begin{bmatrix} 1 & 1 & -1 & | & 6 \\ 0 & -5 & 4 & | & -23 \\ 0 & 2 & -1 & | & 8 \end{bmatrix} \rightarrow \begin{bmatrix} 1 & 1 & -1 & | & 6 \\ 0 & 1 & -\frac{4}{5} & | & \frac{23}{5} \\ 0 & 2 & -1 & | & 8 \end{bmatrix} \rightarrow \begin{bmatrix} 1 & 0 & -\frac{1}{5} & | & \frac{7}{5} \\ 0 & 1 & -\frac{4}{5} & | & \frac{23}{5} \\ 0 & 0 & \frac{3}{5} & | & -\frac{6}{5} \end{bmatrix} \rightarrow \begin{bmatrix} 1 & 0 & -\frac{1}{5} & | & \frac{7}{5} \\ 0 & 1 & -\frac{4}{5} & | & \frac{23}{5} \\ 0 & 0 & 1 & | & -2 \end{bmatrix}$

$\uparrow$ $\uparrow$ $\uparrow$ $\uparrow$

$R_2 = -3r_1 + r_2$ $R_2 = \dfrac{-1}{5}r_2$ $R_1 = -r_2 + r_1$ $R_3 = \dfrac{5}{3}r_3$

$R_3 = -r_1 + r_3$ $R_3 = -2r_2 + r_3$

$\rightarrow \begin{bmatrix} 1 & 0 & 0 & | & 1 \\ 0 & 1 & 0 & | & 3 \\ 0 & 0 & 1 & | & -2 \end{bmatrix}$

$\uparrow$

$R_1 = \dfrac{1}{5}r_3 + r_1$

$R_2 = \dfrac{4}{5}r_3 + r_2$

The solution is $x = 1$, $y = 3$, $z = -2$.

59. $\begin{cases} x + 2y - z = -3 \\ 2x - 4y + z = -7 \\ -2x + 2y - 3z = 4 \end{cases}$ becomes: $\begin{bmatrix} 1 & 2 & -1 & | & -3 \\ 2 & -4 & 1 & | & -7 \\ -2 & 2 & -3 & | & 4 \end{bmatrix}$

$\rightarrow \begin{bmatrix} 1 & 2 & -1 & | & -3 \\ 0 & -8 & 3 & | & -1 \\ 0 & 6 & -5 & | & -2 \end{bmatrix} \rightarrow \begin{bmatrix} 1 & 2 & -1 & | & -3 \\ 0 & -2 & -2 & | & -3 \\ 0 & 6 & -5 & | & -2 \end{bmatrix}$ This will make the fractions in row 2 easier to work with.

$\uparrow$ $\uparrow$

$R_2 = -2r_1 + r_2$ $R_2 = r_3 + r_2$

$R_3 = 2r_1 + r_3$

$\rightarrow \begin{bmatrix} 1 & 2 & -1 & | & -3 \\ 0 & 1 & 1 & | & \frac{3}{2} \\ 0 & 6 & -5 & | & -2 \end{bmatrix} \rightarrow \begin{bmatrix} 1 & 0 & -3 & | & -6 \\ 0 & 1 & 1 & | & \frac{3}{2} \\ 0 & 0 & -11 & | & -11 \end{bmatrix}$

$\uparrow$ $\uparrow$

$R_2 = -\dfrac{1}{2}r_2$ $R_1 = -2r_2 + r_1$

$R_3 = -6r_2 + r_3$

$$\rightarrow \begin{bmatrix} 1 & 0 & -3 & | & -6 \\ 0 & 1 & 1 & | & \dfrac{3}{2} \\ 0 & 0 & 1 & | & 1 \end{bmatrix} \rightarrow \begin{bmatrix} 1 & 0 & 0 & | & -3 \\ 0 & 1 & 0 & | & \dfrac{1}{2} \\ 0 & 0 & 1 & | & 1 \end{bmatrix}$$

$$\uparrow \qquad\qquad\qquad \uparrow$$

$$R_3 = -\dfrac{1}{11}r_3 \qquad R_1 = 3r_3 + r_1$$

$$R_2 = -1r_3 + r_2$$

The solution is $x = -3$, $y = \dfrac{1}{2}$, $z = 1$.

61. $\begin{cases} 3x + y - z = \dfrac{2}{3} \\ 2x - y + z = 1 \\ 4x + 2y \quad = \dfrac{8}{3} \end{cases}$ becomes: $\begin{bmatrix} 3 & 1 & -1 & | & \dfrac{2}{3} \\ 2 & -1 & 1 & | & 1 \\ 4 & 2 & 0 & | & \dfrac{8}{3} \end{bmatrix}$

$$\rightarrow \begin{bmatrix} 1 & 2 & -2 & | & -\dfrac{1}{3} \\ 2 & -1 & 1 & | & 1 \\ 4 & 2 & 0 & | & \dfrac{8}{3} \end{bmatrix} \rightarrow \begin{bmatrix} 1 & 2 & -2 & | & -\dfrac{1}{3} \\ 0 & -5 & 5 & | & \dfrac{5}{3} \\ 0 & -6 & 8 & | & \dfrac{12}{3} \end{bmatrix}$$

$$\uparrow \qquad\qquad\qquad\qquad \uparrow$$

$$R_1 = -1r_2 + r_1 \qquad R_2 = -2r_1 + r_2$$

$$R_3 = -4r_1 + r_3$$

$$\rightarrow \begin{bmatrix} 1 & 2 & -2 & | & -\dfrac{1}{3} \\ 0 & 1 & -1 & | & -\dfrac{1}{3} \\ 0 & -6 & 8 & | & 4 \end{bmatrix} \rightarrow \begin{bmatrix} 1 & 0 & 0 & | & \dfrac{1}{3} \\ 0 & 1 & -1 & | & -\dfrac{1}{3} \\ 0 & 0 & 2 & | & 2 \end{bmatrix}$$

$$\uparrow \qquad\qquad\qquad\qquad \uparrow$$

$$R_2 = -\dfrac{1}{5}r_2 \qquad R_1 = -2r_2 + r_1$$

$$R_3 = 6r_2 + r_3$$

$$\rightarrow \begin{bmatrix} 1 & 0 & 0 & | & \dfrac{1}{3} \\ 0 & 1 & -1 & | & -\dfrac{1}{3} \\ 0 & 0 & 1 & | & 1 \end{bmatrix} \rightarrow \begin{bmatrix} 1 & 0 & 0 & | & \dfrac{1}{3} \\ 0 & 1 & 0 & | & \dfrac{2}{3} \\ 0 & 0 & 1 & | & 1 \end{bmatrix}$$

$$\uparrow \qquad\qquad\qquad\qquad \uparrow$$

$$R_3 = \dfrac{1}{2}r_3 \qquad R_2 = r_3 + r_2$$

The solution is $x = \dfrac{1}{3}$, $y = \dfrac{2}{3}$, $z = 1$.

63. $\begin{cases} x + y + z + w = 4 \\ 2x - y + z = 0 \\ 3x + 2y + z - w = 6 \\ x - 2y - 2z + 2w = -1 \end{cases}$ becomes: $\begin{bmatrix} 1 & 1 & 1 & 1 & | & 4 \\ 2 & -1 & 1 & 0 & | & 0 \\ 3 & 2 & 1 & -1 & | & 6 \\ 1 & -2 & -2 & 2 & | & -1 \end{bmatrix}$

$\rightarrow \begin{bmatrix} 1 & 1 & 1 & 1 & | & 4 \\ 0 & -3 & -1 & -2 & | & -8 \\ 0 & -1 & -2 & -4 & | & -6 \\ 0 & -3 & -3 & 1 & | & -5 \end{bmatrix} \rightarrow \begin{bmatrix} 1 & 1 & 1 & 1 & | & 4 \\ 0 & -1 & -2 & -4 & | & -6 \\ 0 & -3 & -1 & -2 & | & -8 \\ 0 & -3 & -3 & 1 & | & -5 \end{bmatrix}$

$\uparrow$
$R_2 = -2r_1 + r2$
$R_3 = -3r_1 + r_3$
$R_4 = -1r_1 + r_4$

$\uparrow$
Interchange r_2 and r_3

$\rightarrow \begin{bmatrix} 1 & 1 & 1 & 1 & | & 4 \\ 0 & 1 & 2 & 4 & | & 6 \\ 0 & -3 & -1 & -2 & | & -8 \\ 0 & -3 & -3 & 1 & | & -5 \end{bmatrix} \rightarrow \begin{bmatrix} 1 & 0 & -1 & -3 & | & -2 \\ 0 & 1 & 2 & 4 & | & 6 \\ 0 & 0 & 5 & 10 & | & 10 \\ 0 & 0 & 3 & 13 & | & 13 \end{bmatrix}$

$\uparrow$
$R_2 = -1r_2$

$\uparrow$
$R_1 = -1r_2 + r_1$
$R_3 = 3r_2 + r_3$
$R_4 = 3r_2 + r_4$

$\rightarrow \begin{bmatrix} 1 & 0 & -1 & -3 & | & -2 \\ 0 & 1 & 2 & 4 & | & 6 \\ 0 & 0 & 1 & 2 & | & 2 \\ 0 & 0 & 3 & 13 & | & 13 \end{bmatrix} \rightarrow \begin{bmatrix} 1 & 0 & 0 & -1 & | & 0 \\ 0 & 1 & 0 & 0 & | & 2 \\ 0 & 0 & 1 & 2 & | & 2 \\ 0 & 0 & 0 & 7 & | & 7 \end{bmatrix}$

$\uparrow$
$R_3 = \dfrac{1}{5}r_3$

$\uparrow$
$R_1 = r_3 + r_1$
$R_2 = -2r_3 + r_2$
$R_4 = -3r_3 + r_4$

$\rightarrow \begin{bmatrix} 1 & 0 & 0 & -1 & | & 0 \\ 0 & 1 & 0 & 0 & | & 2 \\ 0 & 0 & 1 & 2 & | & 2 \\ 0 & 0 & 0 & 1 & | & 1 \end{bmatrix} \rightarrow \begin{bmatrix} 1 & 0 & 0 & 0 & | & 0 \\ 0 & 1 & 0 & 0 & | & 2 \\ 0 & 0 & 1 & 0 & | & 0 \\ 0 & 0 & 0 & 1 & | & 1 \end{bmatrix}$

$\uparrow$
$R_4 = \dfrac{1}{7}r_4$

$\uparrow$
$R_1 = r_4 + r_1$
$R_3 = -2r_4 + r_3$

The solution is $x = 1, y = 2, z = 0, w = 1$.

65. $\begin{cases} x + 2y + z = 1 \\ 2x - y + 2z = 2 \\ 3x + y + 3z = 3 \end{cases}$ becomes: $\begin{bmatrix} 1 & 2 & 1 & | & 1 \\ 2 & -1 & 2 & | & 2 \\ 3 & 1 & 3 & | & 3 \end{bmatrix}$

$\rightarrow \begin{bmatrix} 1 & 2 & 1 & | & 1 \\ 0 & -5 & 0 & | & 0 \\ 0 & -5 & 0 & | & 0 \end{bmatrix}$

$\uparrow$

$R_2 = -2r_1 + r_2$
$R_3 = -3r_1 + r_3$

The system is equivalent to two equations: $\begin{cases} x + 2y + z = 1 \quad (1) \\ \quad -5y \quad = 0 \quad (2) \end{cases}$

From (2) we have: $y = 0$, and back-substitution into (1) yields: $x + z = 1$

We can write the solution as:

$\quad y = 0; x = -z + 1$; where z is any real number,

or

$\quad y = 0; z = -x + 1$; where x is any real number.

67. $\begin{cases} x - y + z = 5 \\ 3x + 2y - 2z = 0 \end{cases}$ becomes: $\begin{bmatrix} 1 & -1 & 1 & | & 5 \\ 3 & 2 & -2 & | & 0 \end{bmatrix}$

$\rightarrow \begin{bmatrix} 1 & -1 & 1 & | & 5 \\ 0 & 5 & -5 & | & -15 \end{bmatrix} \rightarrow \begin{bmatrix} 1 & -1 & 1 & | & 5 \\ 0 & 1 & -1 & | & -3 \end{bmatrix} \rightarrow \begin{bmatrix} 1 & 0 & 0 & | & 2 \\ 0 & 1 & -1 & | & -3 \end{bmatrix}$

$\uparrow \qquad\qquad\qquad \uparrow \qquad\qquad\qquad \uparrow$

$R_2 = -3r_1 + r_2 \qquad R_2 = \frac{1}{5}r_2 \qquad\qquad R_1 = r_2 + r_1$

This represents the system: $\begin{cases} x = 2 \\ y - z = -3 \end{cases}$

The solution is:

$\quad x = 2; y = z - 3$; where z is any real number,

or

$\quad x = 2; z = y + 3$; where y is any real number.

69. $\begin{cases} 2x + 3y - z = 3 \\ x - y - z = 0 \\ -x + y + z = 0 \\ x + y + 3z = 5 \end{cases}$ becomes: $\begin{bmatrix} 2 & 3 & -1 & | & 3 \\ 1 & -1 & -1 & | & 0 \\ -1 & 1 & 1 & | & 0 \\ 1 & 1 & 3 & | & 5 \end{bmatrix}$

$\rightarrow \begin{bmatrix} 1 & -1 & -1 & | & 0 \\ 2 & 3 & -1 & | & 3 \\ -1 & 1 & 1 & | & 0 \\ 1 & 1 & 3 & | & 5 \end{bmatrix} \rightarrow \begin{bmatrix} 1 & -1 & -1 & | & 0 \\ 0 & 5 & 1 & | & 3 \\ 0 & 0 & 0 & | & 0 \\ 0 & 2 & 4 & | & 5 \end{bmatrix} \rightarrow \begin{bmatrix} 1 & -1 & -1 & | & 0 \\ 0 & 5 & 1 & | & 3 \\ 0 & 2 & 4 & | & 5 \\ 0 & 0 & 0 & | & 0 \end{bmatrix}$

$\uparrow \qquad\qquad\qquad \uparrow \qquad\qquad\qquad\qquad \uparrow$

Interchange rows $\quad R_2 = -2r_1 + r_2 \quad$ Interchange r_3 and r_4
one and two $\qquad R_3 = r_1 + r_3$
$\qquad\qquad\qquad R_4 = -1r_1 + r_4$

$$\rightarrow \begin{bmatrix} 1 & -1 & -1 & 0 \\ 0 & 1 & -7 & -7 \\ 0 & 2 & 4 & 5 \\ 0 & 0 & 0 & 0 \end{bmatrix} \rightarrow \begin{bmatrix} 1 & 0 & -8 & -7 \\ 0 & 1 & -7 & -7 \\ 0 & 1 & 18 & 19 \\ 0 & 0 & 0 & 0 \end{bmatrix} \rightarrow \begin{bmatrix} 1 & 0 & -8 & -7 \\ 0 & 1 & -7 & -7 \\ 0 & 0 & 1 & \frac{19}{18} \\ 0 & 0 & 0 & 0 \end{bmatrix}$$

$$\uparrow \qquad\qquad \uparrow \qquad\qquad \uparrow$$
$$R_2 = -2r_3 + r_2 \qquad R_1 = r_2 + r_1 \qquad R_3 = \frac{1}{18}r_3$$
$$R_3 = -2r_2 + r_3$$

Because of the unusual fraction, we will do back-substitution. We have:

$$\begin{cases} x - 8z = -7 \\ y - 7z = -7 \\ \qquad z = \dfrac{19}{18} \end{cases}$$

$$\begin{cases} x - 8\left(\dfrac{19}{18}\right) = -7 \\ y - 7\left(\dfrac{19}{18}\right) = -7 \\ \qquad\qquad z = \dfrac{19}{18} \end{cases}$$

$$\begin{cases} x = \dfrac{-7 \cdot 18 + 8 \cdot 19}{18} = \dfrac{26}{18} = \dfrac{13}{9} \\ y = \dfrac{-7 \cdot 18 + 7 \cdot 19}{18} = \dfrac{7}{18} \\ z = \dfrac{19}{18} \end{cases}$$

Thus, the solution is $x = \dfrac{13}{9}, y = \dfrac{7}{18}, z = \dfrac{19}{18}$.

71. $\begin{cases} 4x + y + z - w = 4 \\ x - y + 2z + 3w = 3 \end{cases}$

$$\begin{bmatrix} 4 & 1 & 1 & -1 & 4 \\ 1 & -1 & 2 & 3 & 3 \end{bmatrix} \rightarrow \begin{bmatrix} 1 & -1 & 2 & 3 & 3 \\ 4 & 1 & 1 & -1 & 4 \end{bmatrix}$$

$$\uparrow$$
Interchange rows
$$\rightarrow \begin{bmatrix} 1 & -1 & 2 & 3 & 3 \\ 0 & 5 & -7 & -13 & -8 \end{bmatrix}$$

$$\uparrow$$
$$R_2 = -4r_1 + r_2$$

This is equivalent to the system: $\begin{cases} x - y + 2z - 3w = 3 & (1) \\ \quad 5y - 7z - 13w = -8 & (2) \end{cases}$

From (2): $\quad 5y = 7z + 13w - 8$

$$y = \frac{7}{5}z + \frac{13}{5}w - \frac{8}{5}$$

Then from (1): $x = y - 2z - 3w + 3$

or $x = \left(\dfrac{7}{5}z + \dfrac{13}{5}w - \dfrac{8}{5}\right) - 2z - 3w + 3$

$x = -\dfrac{3}{5}z - \dfrac{2}{5}w + \dfrac{7}{5}$

The solution is: $x = -\dfrac{3}{5}z - \dfrac{2}{5}w + \dfrac{7}{5}$

$y = \dfrac{7}{5}z + \dfrac{13}{5}w - \dfrac{8}{5}$

where z and w are any real numbers.

73.   We have $y = ax^2 + bx + c$, and each of the three points must satisfy this equation.

$(1, 2)$: $2 = a + b + c$
$(-2, -7)$: $-7 = 4a - 2b + c$
$(-2, -3)$: $-3 = 4a + 2b + c$

We have three equations in three unknowns which can be represented by:

$$\begin{bmatrix} 1 & 1 & 1 & | & 2 \\ 4 & -2 & 1 & | & -7 \\ 4 & 2 & 1 & | & -3 \end{bmatrix} \rightarrow \begin{bmatrix} 1 & 1 & 1 & | & 2 \\ 0 & -6 & -3 & | & -15 \\ 0 & -2 & -3 & | & -11 \end{bmatrix} \rightarrow \begin{bmatrix} 1 & 1 & 1 & | & 2 \\ 0 & 1 & \frac{1}{2} & | & \frac{5}{2} \\ 0 & -2 & -3 & | & -11 \end{bmatrix}$$

$$\uparrow \qquad\qquad\qquad \uparrow$$

$$R_2 = -4r_1 + r_2 \qquad R_2 = -\dfrac{1}{6}r_2$$

$$R_3 = -4r_1 + r_3$$

$$\rightarrow \begin{bmatrix} 1 & 0 & \frac{1}{2} & | & -\frac{1}{2} \\ 0 & 1 & \frac{1}{2} & | & \frac{5}{2} \\ 0 & 0 & -2 & | & -6 \end{bmatrix} \rightarrow \begin{bmatrix} 1 & 0 & \frac{1}{2} & | & -\frac{1}{2} \\ 0 & 1 & \frac{1}{2} & | & \frac{5}{2} \\ 0 & 0 & 1 & | & 3 \end{bmatrix} \rightarrow \begin{bmatrix} 1 & 0 & 0 & | & -2 \\ 0 & 1 & 0 & | & 1 \\ 0 & 0 & 1 & | & 3 \end{bmatrix}$$

$$\uparrow \qquad\qquad\qquad \uparrow \qquad\qquad\qquad \uparrow$$

$$R_1 = -1r_2 + r_1 \qquad R_3 = -\dfrac{1}{2}r_3 \qquad R_1 = -\dfrac{1}{2}r_3 + r_1$$

$$R_3 = 2r_2 + r_3 \qquad\qquad\qquad\qquad R_2 = -\dfrac{1}{2}r_3 + r_2$$

The solution is $a = -2$, $b = 1$, $c = 3$, so the parabola is $y = -2x^2 + x + 3$.

You can verify that each of the given points satisfies this equation.

75. $f(x) = ax^3 + bx^2 + cx + d$

$f(-3) = -112$ implies $-27a + 9b - 3c + d = -112$
$f(-1) = -2$ implies $-a + b - c + d = -2$
$f(1) = 4$ implies $a + b + c + d = 4$
and $f(2) = 13$ implies $-8a + 4b + 2c + d = 13$

Thus, we want to find the solution to a system of four equations in four unknowns.

$$\begin{bmatrix} -27 & 9 & -3 & 1 & | & -112 \\ -1 & 1 & -1 & 1 & | & -2 \\ 1 & 1 & 1 & 1 & | & 4 \\ 8 & 4 & 2 & 1 & | & 13 \end{bmatrix} \rightarrow \begin{bmatrix} 1 & 1 & 1 & 1 & | & 4 \\ -1 & 1 & -1 & 1 & | & -2 \\ -27 & 9 & -3 & 1 & | & -112 \\ 8 & 4 & 2 & 1 & | & 13 \end{bmatrix} \rightarrow \begin{bmatrix} 1 & 1 & 1 & 1 & | & 4 \\ 0 & 2 & 0 & 2 & | & 2 \\ 0 & 36 & 24 & 28 & | & -4 \\ 0 & -4 & -6 & -7 & | & -19 \end{bmatrix}$$

$\uparrow$
Interchange r_3 and r_1

$\uparrow$
$R_2 = r_1 + r_2$
$R_3 = 27r_1 + r_3$
$R_4 = -8r_1 + r_4$

$$\rightarrow \begin{bmatrix} 1 & 1 & 1 & 1 & | & 4 \\ 0 & 1 & 0 & 1 & | & 1 \\ 0 & 9 & 6 & 7 & | & -1 \\ 0 & -4 & -6 & -7 & | & -19 \end{bmatrix} \rightarrow \begin{bmatrix} 1 & 0 & 1 & 0 & | & 3 \\ 0 & 1 & 0 & 1 & | & 1 \\ 0 & 0 & 6 & -2 & | & -10 \\ 0 & 0 & -6 & -3 & | & -15 \end{bmatrix}$$

Now can we get a 1 in row 3 column 3, *and* avoid fractions?

$\uparrow$
$R_2 = \dfrac{1}{2}r_2$
$R_3 = \dfrac{1}{4}r_3$

$\uparrow$
$R_1 = -1r_2 + r_1$
$R_3 = -9r_2 + r_3$
$R_4 = 4r_2 + r_4$

$$\rightarrow \begin{bmatrix} 1 & 0 & 1 & 0 & | & 3 \\ 0 & 1 & 0 & 1 & | & 1 \\ 0 & 0 & 3 & -1 & | & -5 \\ 0 & 0 & -2 & -1 & | & -5 \end{bmatrix} \rightarrow \begin{bmatrix} 1 & 0 & 1 & 0 & | & 3 \\ 0 & 1 & 0 & 1 & | & 1 \\ 0 & 0 & 1 & -2 & | & -10 \\ 0 & 0 & -2 & -1 & | & -5 \end{bmatrix}$$

$\uparrow$
$R_3 = \dfrac{1}{2}r_3$
$R_4 = \dfrac{1}{3}r_4$

$\uparrow$
$R_3 = r_4 + r_3$

$$\rightarrow \begin{bmatrix} 1 & 0 & 0 & 2 & | & 13 \\ 0 & 1 & 0 & 1 & | & 1 \\ 0 & 0 & 1 & -2 & | & -10 \\ 0 & 0 & 0 & -5 & | & -25 \end{bmatrix} \rightarrow \begin{bmatrix} 1 & 0 & 0 & 2 & | & 13 \\ 0 & 1 & 0 & 1 & | & 1 \\ 0 & 0 & 1 & -2 & | & -10 \\ 0 & 0 & 0 & 1 & | & 5 \end{bmatrix} \rightarrow \begin{bmatrix} 1 & 0 & 0 & 0 & | & 3 \\ 0 & 1 & 0 & 0 & | & -4 \\ 0 & 0 & 1 & 0 & | & 0 \\ 0 & 0 & 0 & 1 & | & 5 \end{bmatrix}$$

$\uparrow$
$R_1 = -1r_3 + r_1$

$R_4 = 2r_3 + r_4$

$\uparrow$
$R_4 = -\dfrac{1}{5}r_4$

$\uparrow$
$R_1 = -2r_4 + r_1$

$R_2 = -1r_4 + r_2$
$R_3 = 2r_4 + r_3$

So we have: $a = 3$, $b = -4$, $c = 0$, $d = 5$.

The function is: $f(x) = 3x^3 - 4x^2 + 5$

77. Let x, y, and z represent the number of liters of 15%, 25% and 50% solutions which will be mixed. Then,

$$x + y + z = 100 \quad (1)$$

Also, in x liters of 15% solution, there will be $.15x$ liters of H_2SO_4, y liters of 25% solution contain $.25y$ liters of H_2SO_4, and the z liters contain $.50z$ liters of H_2SO_4. Meanwhile, our final 100 liter mixture is 40% H_2SO_4, so it contains $.40(100) = 40$ liters of H_2SO_4.

Thus, $.15x + .25y + .50z = 40$ (2)

We have 2 equations in three unknowns:

$$\begin{bmatrix} 1 & 1 & 1 & | & 100 \\ 0.15 & 0.25 & 0.50 & | & 40 \end{bmatrix} \rightarrow \begin{bmatrix} 1 & 1 & 1 & | & 100 \\ 0 & 0.10 & 0.35 & | & 25 \end{bmatrix} \rightarrow \begin{bmatrix} 1 & 1 & 1 & | & 100 \\ 0 & 1 & 3.5 & | & 250 \end{bmatrix}$$

$$\uparrow \qquad\qquad\qquad \uparrow$$

$$R_2 = -.15r_1 + r_2 \qquad R_2 = 10r_2$$

$$\rightarrow \begin{bmatrix} 1 & 0 & -2.5 & | & -150 \\ 0 & 1 & 3.5 & | & 250 \end{bmatrix}$$

$$\uparrow$$

$$R_1 = -1r_2 + r_1$$

This gives: $\begin{cases} x - 2.5z = -150 \\ y + 3.5z = 250 \end{cases}$

so $x = 2.5z - 150$
$y = -3.5z + 250$
where z can be any real number.

But, we require $x \geq 0$, $y \geq 0$, and $z \geq 0$.

Since $x \geq 0$, we have: $2.5z - 150 \geq 0$
$2.5 \geq 150$
$z \geq 60$

Also, $y \geq 0$ implies: $-3.5z + 250 \geq 0$
$-3.5 \geq -250$
$z \leq 71.43$

Some possible solutions are given below:

z (50%)	$x = 2.5z - 150$ (15%)	$y = -3.5z + 250$ (25%)	40%
60	0	40	100
64	10	26	100
68	20	12	100
70	25	5	100

79. $x =$ price of hamburger, $y =$ price of fries, $z =$ price of colas

$\begin{cases} 8x + 6y + 6z = 26.10 \\ 10x + 6y + 8z = 31.60 \end{cases}$

$$\begin{bmatrix} 8 & 6 & 6 & | & 26.10 \\ 10 & 6 & 8 & | & 31.60 \end{bmatrix}$$

$$\begin{bmatrix} 4 & 3 & 3 & | & 13.05 \\ 5 & 3 & 4 & | & 15.80 \end{bmatrix}$$

$$\begin{bmatrix} 4 & 3 & 3 & | & 13.05 \\ 1 & 0 & 1 & | & 12.75 \end{bmatrix}$$

$$\begin{bmatrix} 1 & 0 & 1 & | & 2.75 \\ 4 & 3 & 3 & | & 13.05 \end{bmatrix}$$

$$\begin{bmatrix} 1 & 0 & 1 & | & 2.75 \\ 0 & 3 & -1 & | & 2.05 \end{bmatrix}$$

$$\begin{bmatrix} 1 & 0 & 1 & \bigg| & 2.75 \\ 0 & 1 & -\dfrac{1}{3} & \bigg| & \dfrac{2.05}{3} \end{bmatrix}$$

$x = 2.75 - z$, z any real number

$y = \dfrac{2.05}{3} + \dfrac{1}{3}z$, z any real number

$y = 0.68 + \dfrac{1}{3}z$, z any real number

There is not sufficient information:

x	\$2.15	\$2.00	\$1.85
y	\$0.88	\$0.93	\$0.98
z	\$0.60	\$0.75	\$0.90

81. Let x = amount in Treasury bills, y = amount in corporate bonds, z = amount in junk bonds

(a) $\begin{cases} x + y + z = 20000 \\ .07x + .09y + .11z = 2000 \end{cases}$

$$\begin{bmatrix} 1 & 1 & 1 & \big| & 20000 \\ .07 & .09 & .11 & \big| & 2000 \end{bmatrix}$$

$$\rightarrow \begin{bmatrix} 1 & 1 & 1 & \big| & 20{,}000 \\ 7 & 9 & 11 & \big| & 200{,}000 \end{bmatrix}$$

$$\rightarrow \begin{bmatrix} 1 & 1 & 1 & \big| & 20{,}000 \\ 0 & 2 & 4 & \big| & 60{,}000 \end{bmatrix}$$

$$\rightarrow \begin{bmatrix} 1 & 1 & 1 & \big| & 20{,}000 \\ 0 & 1 & 2 & \big| & 30{,}000 \end{bmatrix}$$

$$\rightarrow \begin{bmatrix} 1 & 0 & -1 & \big| & -10{,}000 \\ 0 & 1 & 2 & \big| & 30{,}000 \end{bmatrix}$$

$x = -10{,}000 + z$, $y = 30{,}000 - 2z$, z any real number

Amount Invested At

7%	9%	11%
0	10,000	10,000
1,000	8,000	11,000
2,000	6,000	12,000
3,000	4,000	13,000
4,000	2,000	14,000
5,000	0	15,000

(b) $\begin{cases} x + y + z = 25000 \\ .07x + .09y + .11z = 2000 \end{cases}$

$$\begin{bmatrix} 1 & 1 & 1 & \big| & 25000 \\ .07 & .09 & .11 & \big| & 2000 \end{bmatrix}$$

$$\rightarrow \begin{bmatrix} 1 & 1 & 1 & \big| & 25{,}000 \\ 7 & 9 & 11 & \big| & 200{,}000 \end{bmatrix}$$

$$\rightarrow \begin{bmatrix} 1 & 1 & 1 & \big| & 25{,}000 \\ 0 & 2 & 4 & \big| & 25{,}000 \end{bmatrix}$$

$$\rightarrow \begin{bmatrix} 1 & 1 & 1 & 25{,}000 \\ 0 & 1 & 2 & 12{,}500 \end{bmatrix}$$

$$\rightarrow \begin{bmatrix} 1 & 0 & -1 & 12{,}500 \\ 0 & 1 & 2 & 12{,}500 \end{bmatrix}$$

$x = 12{,}500 + z$, $y = 12{,}500 - 2z$, z any real number

Amount Invested At		
7%	9%	11%
12,500	12,500	0
14,500	8,500	2,000
16,500	4,500	4,000
18,750	0	6,250

(c) $\begin{cases} x + y + z = 30{,}000 \\ .07x + .09y + .11z = 2{,}000 \end{cases}$

$$\begin{bmatrix} 1 & 1 & 1 & 30{,}000 \\ .07 & .09 & .11 & 2{,}000 \end{bmatrix}$$

$$\rightarrow \begin{bmatrix} 1 & 1 & 1 & 30{,}000 \\ 7 & 9 & 11 & 200{,}000 \end{bmatrix}$$

$$\rightarrow \begin{bmatrix} 1 & 1 & 1 & 30{,}000 \\ 0 & 2 & 4 & -10{,}000 \end{bmatrix}$$

$$\rightarrow \begin{bmatrix} 1 & 1 & 1 & 30{,}000 \\ 0 & 1 & 2 & -5{,}000 \end{bmatrix}$$

$$\rightarrow \begin{bmatrix} 1 & 0 & -1 & 35{,}000 \\ 0 & 1 & 2 & -5{,}000 \end{bmatrix}$$

$x = 35{,}000 + z$, $y = -5{,}000 - 2z$, z any real number

All the money invested at 7% provides $30{,}000(.07) = \$2100$, more than what is required.

83. $\begin{cases} I_1 + I_2 = I_3 \\ 16 - 8 - 9I_3 - 3I_1 = 0 \\ 16 - 4 - 9I_3 - 9I_2 = 0 \\ 8 - 4 - 9I_2 + 3I_1 = 0 \end{cases}$

$\begin{aligned} I_1 + I_2 - I_3 &= 0 \\ -3I_1 \qquad - 9I_3 &= -8 \\ -9I_2 - 9I_3 \quad\;\; &= -12 \\ 3I_1 - 9I_2 \qquad &= -4 \end{aligned}$

$$\begin{bmatrix} 1 & 1 & -1 & 0 \\ -3 & 0 & -9 & -8 \\ 0 & -9 & -9 & -12 \\ 3 & -9 & 0 & -4 \end{bmatrix}$$

$$\begin{bmatrix} 1 & 1 & -1 & | & 0 \\ 0 & 3 & -12 & | & -8 \\ 0 & -9 & -9 & | & -12 \\ 0 & -12 & 3 & | & -4 \end{bmatrix}$$

$$\begin{bmatrix} 1 & 0 & 3 & | & \dfrac{8}{3} \\ 0 & 1 & -4 & | & -\dfrac{8}{3} \\ 0 & 0 & -45 & | & -36 \\ 0 & 0 & -45 & | & -36 \end{bmatrix}$$

$$\begin{bmatrix} 1 & 0 & 3 & | & \dfrac{8}{3} \\ 0 & 1 & -4 & | & -\dfrac{8}{3} \\ 0 & 0 & 1 & | & \dfrac{36}{45} \\ 0 & 0 & 0 & | & 0 \end{bmatrix}$$

$$I_3 = \frac{36}{45} = \frac{4}{5}$$

$$I_2 = \frac{-8}{3} + 4\left(\frac{4}{5}\right) = \frac{-40 + 48}{15} = \frac{8}{15}$$

$$I_1 = \frac{8}{3} - 3\left(\frac{4}{5}\right) = \frac{40 - 36}{15} = \frac{4}{15}$$

$$I_1 = \frac{4}{15}, \ I_2 = \frac{8}{15}, \ I_3 = \frac{4}{5}$$

85. $\begin{cases} I_1 = I_3 + I_2 \\ 24 - 6I_1 - 3I_3 = 0 \\ 12 + 24 - 6I_1 - 6I_2 = 0 \end{cases}$

$$\begin{aligned} I_1 - I_2 - I_3 &= 0 \\ -6I_1 \qquad\quad -3I_3 &= -24 \\ 6I_1 - 6I_2 \qquad\quad &= -36 \end{aligned}$$

$$\begin{bmatrix} 1 & -1 & -1 & | & 0 \\ -6 & 0 & -3 & | & -24 \\ -6 & -6 & 0 & | & -36 \end{bmatrix}$$

$$\begin{bmatrix} 1 & -1 & -1 & | & 0 \\ 0 & -6 & -9 & | & -24 \\ 0 & -12 & -6 & | & -36 \end{bmatrix}$$

$$\begin{bmatrix} 1 & -1 & -1 & | & 0 \\ 0 & 1 & \frac{3}{2} & | & 4 \\ 0 & -12 & -6 & | & -36 \end{bmatrix}$$

$$\begin{bmatrix} 1 & -1 & -1 & | & 0 \\ 0 & 1 & \frac{3}{2} & | & 4 \\ 0 & 2 & 1 & | & 6 \end{bmatrix}$$

$$\begin{bmatrix} 1 & 0 & \frac{1}{2} & | & 4 \\ 0 & 1 & \frac{3}{2} & | & 4 \\ 0 & 0 & -2 & | & -2 \end{bmatrix}$$

$$\begin{bmatrix} 1 & 0 & \frac{1}{2} & | & 4 \\ 0 & 1 & \frac{3}{2} & | & 4 \\ 0 & 0 & 1 & | & 1 \end{bmatrix}$$

$$\begin{bmatrix} 1 & 0 & 0 & | & \frac{7}{2} \\ 0 & 1 & 0 & | & \frac{5}{2} \\ 0 & 0 & 1 & | & 1 \end{bmatrix}$$

$$I_1 = 3.5, I_2 = 2.5, I_3 = 1$$

89. $\begin{cases} a_1 x + b_1 y = c_1 \\ a_2 x + b_2 y = c_2 \end{cases}$ becomes:

$$\begin{bmatrix} a_1 & b_1 & | & c_1 \\ a_2 & b_2 & | & c_2 \end{bmatrix}$$

If $a_1 \neq 0$, we can divide row one by a_1 to obtain a 1 in the top left corner:

$$\rightarrow \begin{bmatrix} 1 & \frac{b_1}{a_1} & | & \frac{c_1}{a_1} \\ a_2 & b_2 & | & c_2 \end{bmatrix}$$

$$\uparrow$$

$$R_1 = \frac{1}{a_1} r_1, \text{ provided } a_1 \neq 0$$

Our next move would depend on whether a_2 is zero or not. If $a_2 \neq 0$, we continue:

$$\rightarrow \left[\begin{array}{cc|c} 1 & \dfrac{b_1}{a_1} & \dfrac{c_1}{a_1} \\[3mm] 0 & \dfrac{-a_2 b_1}{a_1} + b_2 & \dfrac{-a_2 c_1}{a_1} + c_2 \end{array}\right] \rightarrow \left[\begin{array}{cc|c} 1 & \dfrac{b_1}{a_1} & \dfrac{c_1}{a_1} \\[3mm] 0 & \dfrac{a_1 b_2 - a_2 b_1}{a_1} & \dfrac{a_1 c_2 - a_2 c_1}{a_1} \end{array}\right]$$

$$\uparrow \qquad\qquad\qquad\qquad\qquad \uparrow$$
$$R_2 = -a_2 r_1 + r_2 \qquad\qquad\qquad \text{Simplifying}$$

Now recall that $D = a_1 b_2 - a_2 b_1$ so we have:

$$\begin{cases} x + \dfrac{b_1}{a_1} y = \dfrac{c_1}{a_1} & (1) \\[4mm] \dfrac{D}{a_1} y = \dfrac{a_1 c_2 - a_2 c_1}{a_1} & (2) \end{cases}$$

Solve for (2) for y: $\quad y = \dfrac{a_1 c_2 - a_2 c_1}{D}$

Then use back-substitution to find x:

$$x + \frac{b_1}{a_1}\left(\frac{a_1 c_2 - a_2 c_1}{D}\right) = \frac{c_1}{a_1}$$

$$x = \frac{c_1}{a_1} - \frac{b_1(a_1 c_2 - a_2 c_1)}{a_1 D}$$

We now need to get a common denominator $(a_1 D)$ to simplify x:

$$x = \frac{c_1 D}{a_1 D} - \frac{b_1 a_1 c_2 - b_1 a_2 c_1}{a_1 D} = \frac{c_1(a_1 b_2 - a_2 b_1) - b_1 a_1 c_2 + b_1 a_2 c_1}{a_1 D}$$

$$= \frac{c_1 a_1 b_2 - c_1 a_2 b_1 - b_1 a_1 c_2 + b_1 a_2 c_1}{a_1 D} = \frac{c_1 a_1 b_2 - b_1 a_1 c_2}{a_1 D} = \frac{c_1 b_2 - b_1 c_2}{D}$$

and our solution is:

$$x = \frac{1}{D}(c_1 b_2 - b_1 c_2)$$

$$y = \frac{1}{D}(a_1 c_2 - a_2 c_1), \text{ provided } a_1 \neq 0, \ a_2 \neq 0, \text{ as desired.}$$

But what if a_2 *is* zero?

Then we have: $\quad \left[\begin{array}{cc|c} 1 & \dfrac{b_1}{a_1} & \dfrac{c_1}{a_1} \\[3mm] 0 & b_2 & c_2 \end{array}\right]$

Also, $D = a_1 b_2 - a_2 b_1$
$\qquad\quad = a_1 b_2 \quad$ (since $a_2 = 0$)

Therefore, b_2 *cannot* be 0, and we continue:

$$\rightarrow \left[\begin{array}{cc|c} 1 & \dfrac{b_1}{a_1} & \dfrac{c_1}{a_1} \\[2mm] 0 & 1 & \dfrac{c_2}{b_2} \end{array}\right] \rightarrow \left[\begin{array}{cc|c} 1 & 0 & -\dfrac{b_1c_2}{a_1b_2} + \dfrac{c_1}{a_1} \\[2mm] 0 & 1 & \dfrac{c_2}{b_2} \end{array}\right] \rightarrow \left[\begin{array}{cc|c} 1 & 0 & \dfrac{b_2c_1 - b_1c_2}{a_1b_2} \\[2mm] 0 & 1 & \dfrac{c_2}{b_2} \end{array}\right] \quad \text{Simplifying}$$

$$\uparrow \qquad\qquad \uparrow$$
$$R_2 = \frac{1}{b_2}r_2 \qquad R_1 = \frac{-b_1}{a_1}r_2 + r_1$$

The solution is:

$$x = \frac{b_2c_1 - b_1c_2}{a_1b_2} = \frac{1}{D}(b_2c_1 - b_1c_2)$$

$$y = \frac{c_2}{b_2} = \frac{a_1c_2}{a_1b_2} = \frac{1}{D}(a_1c_2) \qquad (D = a_1b_2 \text{ if } a_2 = 0)$$

This takes care of the case $a_1 \neq 0$, $a_2 = 0$.

Finally, what if $a_1 = 0$? Then

$D = a_1b_2 - a_2b_1 = -a_2b_1$, so $a_2 \neq 0$, $b_1 \neq 0$, and we want to show:

$$x = \frac{1}{D}(c_1b_2 - c_2b_1) = \frac{1}{-a_2b_1}(c_1b_2 - c_2b_1)$$

$$y = \frac{1}{D}(a_1c_2 - a_2c_1) = \frac{1}{-a_2b_1}(-a_2c_1) = \frac{c_1}{b_1}$$

Since $a_1 = 0$, we start with:

$$\left[\begin{array}{cc|c} 0 & b_1 & c_1 \\ a_2 & b_2 & c_2 \end{array}\right] \rightarrow \left[\begin{array}{cc|c} a_2 & b_2 & c_2 \\ 0 & b_1 & c_1 \end{array}\right] \rightarrow \left[\begin{array}{cc|c} 1 & \dfrac{b_2}{a_2} & \dfrac{c_2}{a_2} \\[2mm] 0 & 1 & \dfrac{c_1}{b_1} \end{array}\right]$$

$$\uparrow \qquad\qquad\qquad \uparrow$$
$$\text{Interchange rows} \quad R_1 = \frac{1}{a_2}r_1 \quad \text{(since } a_2 \neq 0)$$

$$R_2 = \frac{1}{b_1}r_2 \quad \text{(since } b_1 \neq 0)$$

Therefore, $y = \dfrac{c_1}{b_1}$, as desired, and by back-substitution,

$$x + \frac{b_2}{a_2}y = \frac{c_2}{a_2}$$

$$x + \frac{b_2}{a_2}\left(\frac{c_1}{b_1}\right) = \frac{c_2}{a_2}$$

$$x = \frac{c_2}{a_2} - \frac{c_1b_2}{a_2b_1}$$

$$x = \frac{c_2b_1 - c_1b_2}{a_2b_1}$$

or $$x = \frac{-1}{a_2b_1}(c_1b_2 - c_2b_1)$$

as desired.

11.3 Systems of Linear Equations: Determinants

For Problems 1–10, to answer (b) enter matrix A, then evaluate the determinant of A to obtain the solution.

1. (a) $\begin{vmatrix} 3 & 1 \\ 4 & 2 \end{vmatrix} = (3)(2) - (4)(1) = 2$

3. (a) $\begin{vmatrix} 6 & 4 \\ -1 & 3 \end{vmatrix} = (6)(3) - (-1)(4) = 18 + 4 = 22$

5. (a) $\begin{vmatrix} -3 & -1 \\ 4 & 2 \end{vmatrix} = (-3)(2) - (4)(-1) = -2$

7. (a) $\begin{vmatrix} 3 & 4 & 2 \\ 1 & -1 & 5 \\ 1 & 2 & -2 \end{vmatrix} = 3 \begin{vmatrix} -1 & 5 \\ 2 & -2 \end{vmatrix} - 4 \begin{vmatrix} 1 & 5 \\ 1 & -2 \end{vmatrix} + 2 \begin{vmatrix} 1 & -1 \\ 1 & 2 \end{vmatrix}$

   $= 3[(-1)(-2) - (2)(5)] - 4[(1)(-2) - (1)(5)] + 2[(1)(2) - (1)(-1)]$
   $= 3[2 - 10] - 4[-2 - 5] + 2[2 + 1]$
   $= (3)(-8) - (4)(-7) + (2)(3)$
   $= -24 + 28 + 6$
   $= 10$

9. (a) $\begin{vmatrix} 4 & -1 & 2 \\ 6 & -1 & 0 \\ 1 & -3 & 4 \end{vmatrix} = 4 \begin{vmatrix} -1 & 0 \\ -3 & 4 \end{vmatrix} - (-1) \begin{vmatrix} 6 & 0 \\ 1 & 4 \end{vmatrix} + 2 \begin{vmatrix} 6 & -1 \\ 1 & -3 \end{vmatrix}$

   $= 4[(-1)(4) - (-3)(0)] - (-1)[(6)(4) - (1)(0)] + 2[(6)(-3) - (1)(-1)]$
   $= 4[-4] - (-1)[24] + 2[-18 + 1]$
   $= -16 + 24 - 34$
   $= -26$

For Problems 11–40, to answer (b), enter matrix A as the entries of D, matrix B as the entries of D_x, matrix C as the entries of D_y, and matrix D as the entries of D_z (when solving a 3 × 3 system). Then x = det (B)/det (A), y = det (C)/det (A) and z = det (D)/det (A).

11. (a) $\begin{cases} x + y = 8 \\ x - y = 4 \end{cases}$

   Here, $D = \begin{vmatrix} 1 & 1 \\ 1 & -1 \end{vmatrix} = -1 - 1 = -2$

   Since $D \neq 0$, we proceed to find D_x and D_y.

   To obtain D_x, replace the first column in D by the constants on the right-hand-side of the original system of equations:

   $$D_x = \begin{vmatrix} 8 & 1 \\ 4 & -1 \end{vmatrix} = -8 - 4 = -12$$

To obtain D_y, replace the second column in D by the constants:

$$D_y = \begin{vmatrix} 1 & 8 \\ 1 & 4 \end{vmatrix} = 4 - 8 = -4$$

Then by Cramer's Rule,

$$x = \frac{D_x}{D} = \frac{-12}{-2} = 6 \qquad y = \frac{D_y}{D} = \frac{-4}{-2} = 2$$

(b)   Enter matrix $A = \begin{bmatrix} 1 & 1 \\ 1 & -1 \end{bmatrix}$, matrix $B = \begin{bmatrix} 8 & 1 \\ 4 & -1 \end{bmatrix}$ and matrix $C = \begin{bmatrix} 1 & 8 \\ 1 & 4 \end{bmatrix}$.

Then $x = \det(B)/\det(A) = 6$ and $y = \det(C)/\det(A) = 2$.

13.   (a) $\begin{cases} 5x - y = 13 \\ 2x + 3y = 12 \end{cases}$

Here, $D = \begin{vmatrix} 5 & -1 \\ 2 & 3 \end{vmatrix} = 15 - (-2) = 17$

Since $D \neq 0$, we find D_x and D_y.

To obtain D_x, replace the first column in D by the constants:

$$D_x = \begin{vmatrix} 13 & -1 \\ 12 & 3 \end{vmatrix} = 39 - (-12) = 51$$

Similarly, $D_y = \begin{vmatrix} 5 & 13 \\ 2 & 12 \end{vmatrix} = 60 - 26 = 34$

Then by Cramer's Rule,

$$x = \frac{D_x}{D} = \frac{51}{17} = 3 \quad \text{and} \quad y = \frac{D_y}{D} = \frac{34}{17} = 2$$

(b)   Enter matrix $A = \begin{bmatrix} 5 & -1 \\ 2 & 3 \end{bmatrix}$, matrix $B = \begin{bmatrix} 13 & -1 \\ 12 & 3 \end{bmatrix}$ and matrix $C = \begin{bmatrix} 5 & 13 \\ 2 & 12 \end{bmatrix}$.

Then $x = \det(B)/\det(A) = 3$ and $y = \det(C)/\det(A) = 2$.

15.   (a) $\begin{cases} 3x \quad\quad = 24 \\ x + 2y = 0 \end{cases}$

This is *easily* solved by inspection, but, to use Cramer's Rule:

$$D = \begin{vmatrix} 3 & 0 \\ 1 & 2 \end{vmatrix} = 6$$

$$D_x = \begin{vmatrix} 24 & 0 \\ 0 & 2 \end{vmatrix} = 48$$

and $D_y = \begin{vmatrix} 3 & 24 \\ 1 & 0 \end{vmatrix} = -24$

so that $x = \frac{D_x}{D} = \frac{48}{6} = 8$ and $y = \frac{D_y}{D} = \frac{-24}{6} = -4$

(b)   Enter matrix $A = \begin{bmatrix} 3 & 0 \\ 1 & 2 \end{bmatrix}$, matrix $B = \begin{bmatrix} 24 & 0 \\ 0 & 2 \end{bmatrix}$ and matrix $C = \begin{bmatrix} 3 & 24 \\ 1 & 0 \end{bmatrix}$.

Then $x = \det(B)/\det(A) = 8$ and $y = \det(C)/\det(A) = -4$.

17.   (a) $\begin{cases} 3x - 6y = 24 \\ 5x + 4y = 12 \end{cases}$

Here, $D = \begin{vmatrix} 3 & -6 \\ 5 & 4 \end{vmatrix} = 12 - (-30) = 42$

$$D_x = \begin{vmatrix} 24 & -6 \\ 12 & 4 \end{vmatrix} = 96 - (-72) = 168$$

and $D_y = \begin{vmatrix} 3 & 24 \\ 5 & 12 \end{vmatrix} = 36 - 120 = -84$

Therefore,

$$x = \frac{D_x}{D} = \frac{168}{42} = 4 \quad \text{and} \quad y = \frac{D_y}{D} = \frac{-84}{42} = -2$$

(b)   Enter matrix $A = \begin{bmatrix} 3 & -6 \\ 5 & 4 \end{bmatrix}$, matrix $B = \begin{bmatrix} 24 & -6 \\ 12 & 4 \end{bmatrix}$ and matrix $C = \begin{bmatrix} 3 & 24 \\ 5 & 12 \end{bmatrix}$.

Then $x = \det(B)/\det(A) = 4$ and $y = \det(C)/\det(A) = -2$.

19.   (a) $\begin{cases} 3x - 2y = 4 \\ 6x - 4y = 0 \end{cases}$

$$D = \begin{vmatrix} 3 & -2 \\ 6 & -4 \end{vmatrix} = -12 - (-12) = 0$$

Since $D = 0$, we cannot use Cramer's Rule. It is not applicable.

(b)   Enter matrix $A = \begin{bmatrix} 3 & -2 \\ 6 & -4 \end{bmatrix}$. Since $\det(A) = 0$, the system cannot be solved using Cramer's Rule.

21.   (a) $\begin{cases} 2x - 4y = -2 \\ 3x + 2y = 3 \end{cases}$

Here, $D = \begin{vmatrix} 2 & -4 \\ 3 & 2 \end{vmatrix} = 4 - (-12) = 16$

$$D_x = \begin{vmatrix} -2 & -4 \\ 3 & 2 \end{vmatrix} = -4 - (-12) = 8$$

and $D_y = \begin{vmatrix} 2 & -2 \\ 3 & 3 \end{vmatrix} = 6 - (-6) = 12$

By Cramer's Rule,

$$x = \frac{D_x}{D} = \frac{8}{16} = \frac{1}{2} \quad \text{and} \quad y = \frac{D_y}{D} = \frac{12}{16} = \frac{3}{4}$$

(b)   Enter matrix $A = \begin{bmatrix} 2 & -4 \\ 3 & 2 \end{bmatrix}$, matrix $B = \begin{bmatrix} -2 & -4 \\ 3 & 2 \end{bmatrix}$ and matrix $C = \begin{bmatrix} 2 & -2 \\ 3 & 3 \end{bmatrix}$.

Then $x = \det(B)/\det(A) = 0.5$ and $y = \det(C)/\det(A) = 0.75$.

23.   (a) $\begin{cases} 2x - 3y = -1 \\ 10x + 10y = 5 \end{cases}$

Here, $D = \begin{vmatrix} 2 & -3 \\ 10 & 10 \end{vmatrix} = 20 - (-30) = 50$

$$D_x = \begin{vmatrix} -1 & -3 \\ 5 & 10 \end{vmatrix} = -10 - (-15) = 5$$

and $D_y = \begin{vmatrix} 2 & -1 \\ 10 & 5 \end{vmatrix} = 10 - (-10) = 20$

By Cramer's Rule,

$$x = \frac{D_x}{D} = \frac{5}{50} = \frac{1}{10} \quad \text{and} \quad y = \frac{D_y}{D} = \frac{20}{50} = \frac{2}{5}$$

(b)   Enter matrix $A = \begin{bmatrix} 2 & -3 \\ 10 & 10 \end{bmatrix}$, matrix $B = \begin{bmatrix} -1 & -3 \\ 5 & 10 \end{bmatrix}$ and matrix $C = \begin{bmatrix} 2 & -1 \\ 10 & 5 \end{bmatrix}$.

Then $x = \det(B)/\det(A) = 0.1$ and $y = \det(C)/\det(A) = 0.4$.

25.   (a) $\begin{cases} 2x + 3y = 6 \\ x - y = \dfrac{1}{2} \end{cases}$

Here, $D = \begin{vmatrix} 2 & 3 \\ 1 & -1 \end{vmatrix} = -2 - 3 = -5$

$$D_x = \begin{vmatrix} 6 & 3 \\ \dfrac{1}{2} & -1 \end{vmatrix} = -6 - \frac{3}{2} = -\frac{15}{2}$$

and $D_y = \begin{vmatrix} 2 & 6 \\ 1 & \dfrac{1}{2} \end{vmatrix} = 1 - 6 = -5$

By Cramer's Rule,

$$x = \frac{D_x}{D} = \frac{\dfrac{-15}{2}}{-5} = \frac{3}{2} \quad \text{and} \quad y = \frac{D_y}{D} = \frac{-5}{-5} = 1$$

(b)   Enter matrix $A = \begin{bmatrix} 2 & 3 \\ 1 & -1 \end{bmatrix}$, matrix $B = \begin{bmatrix} 6 & 3 \\ 0.5 & -1 \end{bmatrix}$ and matrix $C = \begin{bmatrix} 2 & 6 \\ 1 & 0.5 \end{bmatrix}$.

Then $x = \det(B)/\det(A) = 1.5$ and $y = \det(C)/\det(A) = 1$.

27.   (a) $\begin{cases} 3x - 5y = 3 \\ 15x + 5y = 21 \end{cases}$

Here, $D = \begin{vmatrix} 3 & -5 \\ 15 & 5 \end{vmatrix} = 15 - (-5)(15) = 90$

$$D_x = \begin{vmatrix} 3 & -5 \\ 21 & 5 \end{vmatrix} = 15 - (-105) = 120$$

and $D_y = \begin{vmatrix} 3 & 3 \\ 15 & 21 \end{vmatrix} = (3)(21) - (3)(15) = (3)(21 - 15) = 18$

By Cramer's Rule,

$$x = \frac{D_x}{D} = \frac{120}{90} = \frac{12}{9} = \frac{4}{3} \quad \text{and} \quad y = \frac{D_y}{D} = \frac{18}{90} = \frac{1}{5}$$

(b)   Enter matrix $A = \begin{bmatrix} 3 & -5 \\ 15 & 5 \end{bmatrix}$, matrix $B = \begin{bmatrix} 3 & -5 \\ 21 & 5 \end{bmatrix}$ and matrix $C = \begin{bmatrix} 3 & 3 \\ 15 & 21 \end{bmatrix}$.

Then $x = \det(B)/\det(A) = 1.33$ and $y = \det(C)/\det(A) = 0.2$.

29. **(a)**
$$\begin{cases} x + y - z = 6 \\ 3x - 2y + z = -5 \\ x + 3y - 2z = 14 \end{cases}$$

Here,

$$D = \begin{vmatrix} 1 & 1 & -1 \\ 3 & -2 & 1 \\ 1 & 3 & -2 \end{vmatrix} = 1\begin{vmatrix} -2 & 1 \\ 3 & -2 \end{vmatrix} - 1\begin{vmatrix} 3 & 1 \\ 1 & -2 \end{vmatrix} + (-1)\begin{vmatrix} 3 & -2 \\ 1 & 3 \end{vmatrix}$$
$$= 1(4 - 3) - (1)(-6 - 1) + (-1)(9 - (-2))$$
$$= 1 - (-7) + (-11)$$
$$= -3$$

To obtain D_x, replace the first column in D by the column of constants:

$$D_x = \begin{vmatrix} 6 & 1 & -1 \\ -5 & -2 & 1 \\ 14 & 3 & -2 \end{vmatrix} = 6\begin{vmatrix} -2 & 1 \\ 3 & -2 \end{vmatrix} - 1\begin{vmatrix} -5 & 1 \\ 14 & -2 \end{vmatrix} + (-1)\begin{vmatrix} -5 & -2 \\ 14 & 3 \end{vmatrix}$$
$$= 6(4 - 3) - 1(10 - 14) + (-1)(-15 - (-28))$$
$$= 6 - (-4) + (-13)$$
$$= -3$$

Similarly, $D_y = \begin{vmatrix} 1 & 6 & -1 \\ 3 & -5 & 1 \\ 1 & 14 & -2 \end{vmatrix} = 1\begin{vmatrix} -5 & 1 \\ 14 & -2 \end{vmatrix} - 6\begin{vmatrix} 3 & 1 \\ 1 & -2 \end{vmatrix} + (-1)\begin{vmatrix} 3 & -5 \\ 1 & 14 \end{vmatrix}$

$$= 1(10 - 14) - 6(-6 - 1) + (-1)(42 - (-5))$$
$$= -4 - (-42) + (-47)$$
$$= -9$$

Finally, $D_z = \begin{vmatrix} 1 & 1 & 6 \\ 3 & -2 & -5 \\ 1 & 3 & 14 \end{vmatrix} = 1\begin{vmatrix} -2 & -5 \\ 3 & 14 \end{vmatrix} - 1\begin{vmatrix} 3 & -5 \\ 1 & 14 \end{vmatrix} + 6\begin{vmatrix} 3 & -2 \\ 1 & 3 \end{vmatrix}$

$$= 1(-28 - (-15)) - 1(42 - (-5)) + 6(9 - (-2))$$
$$= -13 - 47 + 66$$
$$= 6$$

$$x = \frac{D_x}{D} = \frac{-3}{-3} = 1, \quad y = \frac{D_y}{D} = \frac{-9}{-3} = 3, \quad \text{and} \quad z = \frac{D_z}{D} = \frac{6}{-3} = -2$$

(b) Enter matrix $A = \begin{bmatrix} 1 & 1 & -1 \\ 3 & -2 & 1 \\ 1 & 3 & -2 \end{bmatrix}$, matrix $B = \begin{bmatrix} 6 & 1 & -1 \\ -5 & -2 & 1 \\ 14 & 3 & -2 \end{bmatrix}$, matrix $C = \begin{bmatrix} 1 & 6 & -1 \\ 3 & -5 & 1 \\ 1 & 14 & -2 \end{bmatrix}$, and

matrix $D = \begin{vmatrix} 1 & 1 & 6 \\ 3 & -2 & -5 \\ 1 & 3 & 14 \end{vmatrix}$

Then $x = \det(B)/\det(A) = 1$, and $y = \det(C)/\det(A) = 3$, and $z = \det(D)/\det(A) = -2$.

31.　(a)　$\begin{cases} x + 2y - z = -3 \\ 2x - 4y + z = -7 \\ -2x + 2y - 3z = 4 \end{cases}$

$$D = \begin{vmatrix} 1 & 2 & -1 \\ 2 & -4 & 1 \\ -2 & 2 & -3 \end{vmatrix} = 1 \begin{vmatrix} -4 & 1 \\ 2 & -3 \end{vmatrix} - 2 \begin{vmatrix} 2 & 1 \\ -2 & -3 \end{vmatrix} + (-1) \begin{vmatrix} 2 & -4 \\ -2 & 2 \end{vmatrix}$$

$$= 1(12 - 2) - 2(-6 - (-2)) + (-1)(4 - 8)$$
$$= 10 - (-8) + 4$$
$$= 22$$

$$D_x = \begin{vmatrix} -3 & 2 & -1 \\ -7 & -4 & 1 \\ 4 & 2 & -3 \end{vmatrix} = -3 \begin{vmatrix} -4 & 1 \\ 2 & -3 \end{vmatrix} - 2 \begin{vmatrix} -7 & 1 \\ 4 & -3 \end{vmatrix} + (-1) \begin{vmatrix} -7 & -4 \\ 4 & 2 \end{vmatrix}$$

$$= -3(12 - 2) - 2(21 - 4) + (-1)(-14 - (-16))$$
$$= -30 - 34 + (-2)$$
$$= -66$$

$$D_y = \begin{vmatrix} 1 & -3 & -1 \\ 2 & -7 & 1 \\ -2 & 4 & -3 \end{vmatrix} = 1 \begin{vmatrix} -7 & 1 \\ 4 & -3 \end{vmatrix} - (-3) \begin{vmatrix} 2 & 1 \\ -2 & -3 \end{vmatrix} + (-1) \begin{vmatrix} 2 & -7 \\ -2 & 4 \end{vmatrix}$$

$$= 1(21 - 4) - (-3)(-6 - (-2)) + (-1)(8 - 14)$$
$$= 17 - 12 + 6$$
$$= 11$$

and

$$D_z = \begin{vmatrix} 1 & 2 & -3 \\ 2 & -4 & -7 \\ -2 & 2 & 4 \end{vmatrix} = 1 \begin{vmatrix} -4 & -7 \\ 2 & 4 \end{vmatrix} - 2 \begin{vmatrix} 2 & -7 \\ -2 & 4 \end{vmatrix} + (-3) \begin{vmatrix} 2 & -4 \\ -2 & 2 \end{vmatrix}$$

$$= 1(-16 - (-14)) - 2(8 - 14) + (-3)(4 - 8)$$
$$= -2 - (-12) + 12$$
$$= 22$$

By Cramer's Rule,

$$x = \frac{D_x}{D} = \frac{-66}{22} = -3, \ y = \frac{D_y}{D} = \frac{11}{22} = \frac{1}{2}, \text{ and } z = \frac{D_z}{D} = \frac{22}{22} = 1$$

(b)　Enter matrix $A = \begin{bmatrix} 1 & 2 & -1 \\ 2 & -4 & 1 \\ -2 & 2 & -3 \end{bmatrix}$, matrix $B = \begin{bmatrix} -3 & 2 & -1 \\ -7 & -4 & 1 \\ 4 & 2 & -3 \end{bmatrix}$, matrix $C = \begin{bmatrix} 1 & -3 & -1 \\ 2 & -7 & 1 \\ -2 & 4 & -3 \end{bmatrix}$, and

matrix $D = \begin{bmatrix} 1 & 2 & -3 \\ 2 & -4 & -7 \\ -2 & 2 & 4 \end{bmatrix}$.

Then $x = \det (B)/\det (A) = -3$, and $y = \det (C)/\det (A) = 0.5$, and $z = \det (D)/\det (A) = 1$.

33. (a)
$$\begin{cases} x - 2y + 3z = 1 \\ 3x + y - 2z = 0 \\ 2x - 4y + 6z = 2 \end{cases}$$

$$D = \begin{vmatrix} 1 & -2 & 3 \\ 3 & 1 & -2 \\ 2 & -4 & 6 \end{vmatrix} = 1 \begin{vmatrix} 1 & -2 \\ -4 & 6 \end{vmatrix} - (-2) \begin{vmatrix} 3 & -2 \\ 2 & 6 \end{vmatrix} + 3 \begin{vmatrix} 3 & 1 \\ 2 & -4 \end{vmatrix}$$

$$= 1(-2) - (-2)(22) + 3(-14)$$
$$= 0$$

Since $D = 0$, Cramer's Rule cannot be applied.

(b) Enter matrix $A = \begin{bmatrix} 1 & -2 & 3 \\ 3 & 1 & -2 \\ 2 & -4 & 6 \end{bmatrix}$. Since det $(A) = 0$, Cramer's Rule does not apply.

35. (a)
$$\begin{cases} x + 2y - z = 0 \\ 2x - 4y + z = 0 \\ -2x + 2y - 3z = 0 \end{cases}$$

$$D = \begin{vmatrix} 1 & 2 & -1 \\ 2 & -4 & 1 \\ -2 & 2 & -3 \end{vmatrix} = 1 \begin{vmatrix} -4 & 1 \\ 2 & -3 \end{vmatrix} - 2 \begin{vmatrix} 2 & 1 \\ -2 & -3 \end{vmatrix} + (-1) \begin{vmatrix} 2 & -4 \\ -2 & 2 \end{vmatrix}$$

$$= 1(10) - 2(-4) + (-1)(-4)$$
$$= 22$$

$$D_x = \begin{vmatrix} 0 & 2 & -1 \\ 0 & -4 & 1 \\ 0 & 2 & -3 \end{vmatrix} = 0 \begin{vmatrix} -4 & 1 \\ 2 & -3 \end{vmatrix} - 2 \begin{vmatrix} 0 & -1 \\ 0 & -3 \end{vmatrix} + (-1) \begin{vmatrix} 0 & -4 \\ 0 & -3 \end{vmatrix}$$

$$= 0(12 - 2) - 2(0 - 0) + (-1)(0 - 0)$$
$$= 0$$

(We could have used (12) in the text, which states that if any row or column contains only 0's, the value of the determinant is 0.)

$$D_y = \begin{vmatrix} 1 & 0 & -1 \\ 2 & 0 & 1 \\ -2 & 0 & -3 \end{vmatrix} = 0$$

Similarly, $D_z = 0$
Therefore,

$$x = \frac{D_x}{D} = \frac{0}{22} = 0 \quad \text{and} \quad y = 0, z = 0$$

(b) Enter matrix $A = \begin{bmatrix} 1 & 2 & -1 \\ 2 & -4 & 1 \\ -2 & 2 & -3 \end{bmatrix}$, matrix $B = \begin{bmatrix} 0 & 2 & -1 \\ 0 & -4 & 1 \\ 0 & 2 & -3 \end{bmatrix}$, matrix $C = \begin{bmatrix} 1 & 0 & -1 \\ 2 & 0 & 1 \\ -2 & 0 & -3 \end{bmatrix}$, and

matrix $D = \begin{bmatrix} 1 & 2 & 0 \\ 2 & -4 & 0 \\ -2 & 2 & 0 \end{bmatrix}$.

Then $x = $ det $(B)/$det $(A) = 0$, $y = $ det $(C)/$det $(A) = 0$, and $z = $ det $(D)/$det $(A) = 0$.

37. (a)
$$\begin{cases} x - 2y + 3z = 0 \\ 3x + y - 2z = 0 \\ 2x - 4y + 6z = 0 \end{cases}$$

$$D = \begin{vmatrix} 1 & -2 & 3 \\ 3 & 1 & -2 \\ 2 & -4 & 6 \end{vmatrix} = 1 \begin{vmatrix} 1 & -2 \\ -4 & 6 \end{vmatrix} - (-2) \begin{vmatrix} 3 & -2 \\ 2 & 6 \end{vmatrix} + 3 \begin{vmatrix} 3 & 1 \\ 2 & -4 \end{vmatrix}$$

$$= 1(6 - 8) - (-2)(18 - (-4)) + 3(-12 - 2)$$
$$= -2 - (-44) + (-42)$$
$$= 0$$

Since $D = 0$, Cramer's Rule cannot be applied.

(b) Enter matrix A $= \begin{bmatrix} 1 & -2 & 3 \\ 3 & 1 & -2 \\ 2 & -4 & 6 \end{bmatrix}$.

Since det (A) $= 0$, Cramer's Rule cannot be used to solve the system.

39. (a)
$$\begin{cases} \dfrac{1}{x} + \dfrac{1}{y} = 8 \\ \dfrac{3}{x} - \dfrac{5}{y} = 0 \end{cases}$$

$$\begin{cases} \dfrac{1}{x} + \dfrac{1}{y} = 8 \\ 3\left(\dfrac{1}{x}\right) - 5\left(\dfrac{1}{y}\right) = 0 \end{cases}$$

Let $u = \dfrac{1}{x}$ and $v = \dfrac{1}{y}$. Then we have: $u + v = 8$

$3u - 5v = 0$. Now we solve for u and v:

Here, $D = \begin{vmatrix} 1 & 1 \\ 3 & -5 \end{vmatrix} = -5 - 3 = -8$

$D_u = \begin{vmatrix} 8 & 1 \\ 0 & -5 \end{vmatrix} = -40 - 0 = -40$

and $D_v = \begin{vmatrix} 1 & 8 \\ 3 & 0 \end{vmatrix} = 0 - 24 = -24$

Therefore, by Cramer's Rule,

$$u = \frac{D_u}{D} = \frac{-40}{-8} = 5 \quad \text{and} \quad v = \frac{D_v}{D} = \frac{-24}{-8} = 3$$

But, we are supposed to find x and y.

Since $u = \dfrac{1}{x}$, we have:

$$5 = \frac{1}{x}$$
$$5x = 1$$
$$x = \frac{1}{5}$$

Also, $v = \dfrac{1}{y}$

$$3 = \frac{1}{y}$$
$$3y = 1$$
$$y = \frac{1}{3}$$

(b) Enter matrix A $= \begin{bmatrix} 1 & 1 \\ 3 & -5 \end{bmatrix}$, matrix B $= \begin{bmatrix} 8 & 1 \\ 0 & -5 \end{bmatrix}$ and matrix C $= \begin{bmatrix} 1 & 8 \\ 3 & 0 \end{bmatrix}$.

Then $u =$ det (B)/det (A) $= 5$ and $v =$ det (C)/det (A) $= 3$. Since $u = \dfrac{1}{x}$, $x = \dfrac{1}{5}$ and $v = \dfrac{1}{y}$,

so $y = \dfrac{1}{3}$.

41. Since $\begin{vmatrix} x & x \\ 4 & 3 \end{vmatrix} = 3x - 4x = -x$, we have: $-x = 5$

$$x = -5$$

43. $\begin{vmatrix} x & 1 & 1 \\ 4 & 3 & 2 \\ -1 & 2 & 5 \end{vmatrix} = x\begin{vmatrix} 3 & 2 \\ 2 & 5 \end{vmatrix} - 1\begin{vmatrix} 4 & 2 \\ -1 & 5 \end{vmatrix} + 1\begin{vmatrix} 4 & 3 \\ -1 & 2 \end{vmatrix}$

$$= x(15 - 4) - 1(20 - (-2)) + 1(8 - (-3))$$
$$= 11x - 22 + 11$$
$$= 11x - 11$$

so we have: $11x - 11 = 2$
$$11x = 13$$
$$x = \frac{13}{11}$$

45. $\begin{vmatrix} x & 2 & 3 \\ 1 & x & 0 \\ 6 & 1 & -2 \end{vmatrix} = x\begin{vmatrix} x & 0 \\ 1 & -2 \end{vmatrix} - 2\begin{vmatrix} 1 & 0 \\ 6 & -2 \end{vmatrix} + 3\begin{vmatrix} 1 & x \\ 6 & 1 \end{vmatrix}$

$$= x(-2x - 0) - 2(-2 - 0) + 3(1 - 6x)$$
$$= -2x^2 + 4 + 3 - 18x$$
$$= -2x^2 - 18x + 7$$

so we have: $-2x^2 - 18x + 7 = 7$
$$-2x^2 - 18x = 0$$
$$-2x(x + 9) = 0$$

so $x = 0$ or $x = -9$

47. Let $D = \begin{vmatrix} x & y & z \\ u & v & w \\ 1 & 2 & 3 \end{vmatrix} = 4$

Then, $\begin{vmatrix} 1 & 2 & 3 \\ u & v & w \\ x & y & z \end{vmatrix} = -4$, because the value of a determinant changes sign if any two rows are interchanged.

49. We try to use row operations to put $\begin{vmatrix} x & y & z \\ -3 & -6 & -9 \\ u & v & w \end{vmatrix}$ into the form $\begin{vmatrix} x & y & z \\ u & v & w \\ 1 & 2 & 3 \end{vmatrix}$

Since we know that the value of the determinant on the right is 4.

$$\begin{vmatrix} x & y & z \\ -3 & -6 & -9 \\ u & v & w \end{vmatrix} = -3 \begin{vmatrix} x & y & z \\ 1 & 2 & 3 \\ u & v & w \end{vmatrix} \quad \text{by (14)}$$

$$= (-3)(-1) \begin{vmatrix} x & y & z \\ u & v & w \\ 1 & 2 & 3 \end{vmatrix} \quad \text{by (11)}$$

Therefore, $\begin{vmatrix} x & y & z \\ -3 & -6 & -9 \\ u & v & w \end{vmatrix} = (-3)(-1)(4) = 12$

51. Let $D = \begin{vmatrix} x & y & z \\ u & v & w \\ 1 & 2 & 3 \end{vmatrix} = 4$

Now, $\begin{vmatrix} 1 & 2 & 3 \\ x-3 & y-6 & z-9 \\ 2u & 2v & 2w \end{vmatrix} = 2 \begin{vmatrix} 1 & 2 & 3 \\ x-3 & y-6 & z-9 \\ u & v & w \end{vmatrix} \quad \text{by (14)}$

$$= 2(-1) \begin{vmatrix} x-3 & y-6 & z-9 \\ 1 & 2 & 3 \\ u & v & w \end{vmatrix} \quad \text{by (11)}$$

$$= 2(-1)(-1) \begin{vmatrix} x-3 & y-6 & z-9 \\ u & v & w \\ 1 & 2 & 3 \end{vmatrix} \quad \text{by (11)}$$

Note that in this last determinant, row one can be obtained from D by the operation
$$R_1 = -3r_3 + r_1$$

By (15), that operation leaves the value of the determinant unchanged, so that
$$\begin{vmatrix} x-3 & y-6 & z-9 \\ u & v & w \\ 1 & 2 & 3 \end{vmatrix} = \begin{vmatrix} x & y & z \\ u & v & w \\ 1 & 2 & 3 \end{vmatrix} = 4$$

Therefore, $\begin{vmatrix} 1 & 2 & 3 \\ x-3 & y-6 & z-9 \\ 2u & 2v & 2w \end{vmatrix} = (2)(-1)(-1)(4) = 8$

53. Let $D = \begin{vmatrix} x & y & z \\ u & v & w \\ 1 & 2 & 3 \end{vmatrix} = 4$

$\begin{vmatrix} 1 & 2 & 3 \\ 2x & 2y & 2z \\ u-1 & v-2 & w-3 \end{vmatrix} = 2\begin{vmatrix} 1 & 2 & 3 \\ x & y & z \\ u-1 & v-2 & w-3 \end{vmatrix}$ by (14)

$= 2(-1)\begin{vmatrix} x & y & z \\ 1 & 2 & 3 \\ u-1 & v-2 & w-3 \end{vmatrix}$ by (11)

$= 2(-1)(-1)\begin{vmatrix} x & y & z \\ u-1 & v-2 & w-3 \\ 1 & 2 & 3 \end{vmatrix}$

Note, in this last determinant, row two can be obtained from D by the operation
$$R_2 = -1r_3 + r_2$$
which leaves the value of the determinant unchanged by (15). In other words,

$\begin{vmatrix} x & y & z \\ u-1 & v-2 & w-3 \\ 1 & 2 & 3 \end{vmatrix} = \begin{vmatrix} x & y & z \\ u & v & w \\ 1 & 2 & 3 \end{vmatrix} = 4$

and $\begin{vmatrix} 1 & 2 & 3 \\ 2x & 2y & 2z \\ u-1 & v-2 & w-3 \end{vmatrix} = 2(-1)(-1)4 = 8$

55. $\begin{vmatrix} x & y & 1 \\ x_1 & y_1 & 1 \\ x_2 & y_2 & 1 \end{vmatrix} = x(y_1 - y_2) - y(x_1 - x_2) + (x_1 y_2 - x_2 y_1) = 0$

$$x(y_1 - y_2) + y(x_2 - x_1) = x_2 y_1 - x_1 y_2$$
$$y(x_2 - x_1) = x_2 y_1 - x_1 y_2 + x(y_2 - y_1)$$
$$y(x_2 - x_1) - y_1(x_2 - x_1) = x_2 y_1 - x_1 y_2 + x(y_2 - y_1) - y_1(x_2 - x_1)$$
$$(x_2 - x_1)(y - y_1) = x(y_2 - y_1) + x_2 y_1 - x_1 y_2 - y_1(x_2 - x_1)$$
$$(x_2 - x_1)(y - y_1) = (y_2 - y_1)x - (y_2 - y_1)x_1$$
$$(x_2 - x_1)(y - y_1) = (y_2 - y_1)(x - x_1)$$
$$(y - y_1) = \frac{y_2 - y_1}{x_2 - x_1}(x - x_1)$$

57.

$$\begin{vmatrix} x^2 & x & 1 \\ y^2 & y & 1 \\ z^2 & z & 1 \end{vmatrix} = x^2 \begin{vmatrix} y & 1 \\ z & 1 \end{vmatrix} - x \begin{vmatrix} y^2 & 1 \\ z^2 & 1 \end{vmatrix} + 1 \begin{vmatrix} y^2 & y \\ z^2 & z \end{vmatrix}$$

$$= x^2(y - z) - x(y^2 - z^2) + 1(y^2z - yz^2)$$
$$= x^2(y - z) - x(y - z)(y + z) + yz(y - z)$$
$$= (y - z)[x^2 - x(y + z) + yz]$$
$$= (y - z)[x^2 - xy - xz + yz]$$
$$= (y - z)[x(x - y) - z(x + y)]$$
$$= (y - z)(x - y)(x - z) \qquad \text{as desired.}$$

59. Generally, below is a 3×3 determinant.

$$\begin{vmatrix} a_{13} & a_{12} & a_{11} \\ a_{23} & a_{22} & a_{21} \\ a_{33} & a_{32} & a_{31} \end{vmatrix} = a_{13}(a_{22}a_{31} - a_{32}a_{21}) - a_{12}(a_{23}a_{31} - a_{33}a_{21}) + a_{11}(a_{23}a_{32} - a_{33}a_{22})$$

$$= a_{13}a_{22}a_{31} - a_{13}a_{32}a_{21} - a_{12}a_{23}a_{31} + a_{12}a_{33}a_{21} + a_{11}a_{23}a_{32} - a_{11}a_{33}a_{22}$$
$$= a_{11}a_{22}a_{33} + a_{11}a_{32}a_{23} + a_{12}a_{21}a_{33} - a_{12}a_{31}a_{23} - a_{13}a_{21}a_{32} + a_{13}a_{31}a_{22}$$
$$= [a_{11}(a_{22}a_{33} - a_{32}a_{23}) - a_{12}(a_{21}a_{33} - a_{31}a_{23}) + a_{13}(a_{21}a_{32} - a_{31}a_{22})]$$

$$= - \begin{bmatrix} a_{11} & a_{12} & a_{13} \\ a_{21} & a_{22} & a_{23} \\ a_{31} & a_{32} & a_{33} \end{bmatrix}$$

As an example, let $A = \begin{vmatrix} 1 & 3 & 2 \\ -1 & 4 & -3 \\ 2 & 1 & 6 \end{vmatrix}$

Then, $A = 1 \begin{vmatrix} 4 & -3 \\ 1 & 6 \end{vmatrix} - 3 \begin{vmatrix} -1 & -3 \\ 2 & 6 \end{vmatrix} + 2 \begin{vmatrix} -1 & 4 \\ 2 & 1 \end{vmatrix} = 1(24 - (-3)) - 3(-6 - (-6)) + 2(-1 - 8)$

$$= 27 - 0 - 18 = 9$$

Interchange columns one and three:

$B = \begin{vmatrix} 2 & 3 & 1 \\ -3 & 4 & -1 \\ 6 & 1 & 2 \end{vmatrix} = 2 \begin{vmatrix} 4 & -1 \\ 1 & 2 \end{vmatrix} - 3 \begin{vmatrix} -3 & -1 \\ 6 & 2 \end{vmatrix} + 1 \begin{vmatrix} -3 & 4 \\ 6 & 1 \end{vmatrix}$

$$= 2(8 - (-1)) - 3(-6 - (-6)) + 1(-3 - 24) = 18 - 0 - 27 = -9$$

Therefore, $B = -9 = (-1)9 = (-1)A$

61. To give a proof of a theorem, we cannot simply show an example. Instead, let D represent *any* 3 by 3 determinant in which the entries in column one equal those in column three. Then D will be of the form:

$$D = \begin{vmatrix} a & d & a \\ b & e & b \\ c & f & c \end{vmatrix}$$

where a, b, c, d, e, f can be *any* real numbers.

Then, $D = \begin{vmatrix} a & d & a \\ b & e & b \\ c & f & c \end{vmatrix} = a \begin{vmatrix} e & b \\ f & c \end{vmatrix} - d \begin{vmatrix} b & b \\ c & c \end{vmatrix} + a \begin{vmatrix} b & e \\ c & f \end{vmatrix}$

$$= a(ce - bf) - d(bc - bc) + a(bf - ce)$$
$$= a(ce - bf) - \quad 0 \quad + a(bf - ce)$$
$$= ace - abf + abf - ace$$
$$= 0$$

11.4 Matrix Algebra

<u>Historical Problem</u>

1. **(a)** Using the correspondence $a + bi \leftrightarrow \begin{bmatrix} a & b \\ -b & a \end{bmatrix}$, we have: $2 - 5i \leftrightarrow \begin{bmatrix} 2 & -5 \\ 5 & 2 \end{bmatrix}$

$$1 + 3i \leftrightarrow \begin{bmatrix} 1 & 3 \\ -3 & 1 \end{bmatrix}$$

(b) $\begin{bmatrix} 2 & -5 \\ 5 & 2 \end{bmatrix} \begin{bmatrix} 1 & 3 \\ -3 & 1 \end{bmatrix} = \begin{bmatrix} 17 & 1 \\ -1 & 17 \end{bmatrix}$ **(c)** Now $\begin{bmatrix} 17 & 1 \\ -1 & 17 \end{bmatrix} \leftrightarrow 17 + i$

(d) On the other hand, we have $(2 - 5i)(1 + 3i) = 2 - 15i^2 + 6i - 5i = 17 + i$

<u>Exercises</u>

In Problems 1–16, we are using: $A = \begin{bmatrix} 0 & 3 & -5 \\ 1 & 2 & 6 \end{bmatrix}$; $B = \begin{bmatrix} 4 & 1 & 0 \\ -2 & 3 & -2 \end{bmatrix}$; $C = \begin{bmatrix} 4 & 1 \\ 6 & 2 \\ -2 & 3 \end{bmatrix}$

To answer (b), enter the matrices A, B, and C into your graphing utility and perform the indicated operation. Answer should agree with (a).

1. **(a)** $A + B = \begin{bmatrix} 0 & 3 & -5 \\ 1 & 2 & 6 \end{bmatrix} + \begin{bmatrix} 4 & 1 & 0 \\ -2 & 3 & -2 \end{bmatrix} = \begin{bmatrix} 0+4 & 3+1 & -5+0 \\ 1+(-2) & 2+3 & 6+(-2) \end{bmatrix} = \begin{bmatrix} 4 & 4 & -5 \\ -1 & 5 & 4 \end{bmatrix}$

3. **(b)** $4A = 4\begin{bmatrix} 0 & 3 & -5 \\ 1 & 2 & 6 \end{bmatrix} = \begin{bmatrix} 4\cdot0 & 4\cdot3 & 4(-5) \\ 4\cdot1 & 4\cdot2 & 4\cdot6 \end{bmatrix} = \begin{bmatrix} 0 & 12 & -20 \\ 4 & 8 & 24 \end{bmatrix}$

5. **(a)** $3A - 2B = \begin{bmatrix} 0 & 9 & -15 \\ 3 & 6 & 18 \end{bmatrix} - \begin{bmatrix} 8 & 2 & 0 \\ -4 & 6 & -4 \end{bmatrix} = \begin{bmatrix} -8 & 7 & -15 \\ 7 & 0 & 22 \end{bmatrix}$

7. **(a)** $AC = \begin{bmatrix} 0 & 3 & -5 \\ 1 & 2 & 6 \end{bmatrix} \begin{bmatrix} 4 & 1 \\ 6 & 2 \\ -2 & 3 \end{bmatrix} = \begin{bmatrix} 0\cdot4 + 3\cdot6 + (-5)(-2) & 0\cdot1 + 3\cdot2 + (-5)3 \\ 1\cdot4 + 2\cdot6 + 6(-2) & 1\cdot1 + 2\cdot2 + 6\cdot3 \end{bmatrix} = \begin{bmatrix} 28 & -9 \\ 4 & 23 \end{bmatrix}$

9. **(a)** $CA = \begin{bmatrix} 4 & 1 \\ 6 & 2 \\ -2 & 3 \end{bmatrix} \begin{bmatrix} 0 & 3 & -5 \\ 1 & 2 & 6 \end{bmatrix} = \begin{bmatrix} 4\cdot0 + 1\cdot1 & 4\cdot3 + 1\cdot2 & 4(-5) + 1\cdot6 \\ 6\cdot0 + 2\cdot1 & 6\cdot3 + 2\cdot2 & 6(-5) + 2\cdot6 \\ (-2)\cdot0 + 3\cdot1 & (-2)\cdot3 + 3\cdot2 & (-2)(-5) + 3\cdot6 \end{bmatrix} = \begin{bmatrix} 1 & 14 & -14 \\ 2 & 22 & -18 \\ 3 & 0 & 28 \end{bmatrix}$

11. **(a)** $C(A + B) = \begin{bmatrix} 4 & 1 \\ 6 & 2 \\ -2 & 3 \end{bmatrix} \begin{bmatrix} 4 & 4 & -5 \\ -1 & 5 & 4 \end{bmatrix} = \begin{bmatrix} 15 & 21 & -16 \\ 22 & 34 & -22 \\ -11 & 7 & 22 \end{bmatrix}$

13. (a) $AC - 3I_2 = \begin{bmatrix} 0 & 3 & -5 \\ 1 & 2 & 6 \end{bmatrix} \begin{bmatrix} 4 & 1 \\ 6 & 2 \\ -2 & 3 \end{bmatrix} - 3\underbrace{\begin{bmatrix} 1 & 0 \\ 0 & 1 \end{bmatrix}}_{I_2} = \begin{bmatrix} 28 & -9 \\ 4 & 23 \end{bmatrix} - \begin{bmatrix} 3 & 0 \\ 0 & 3 \end{bmatrix} = \begin{bmatrix} 25 & -9 \\ 4 & 20 \end{bmatrix}$

15. (a) $CA - CB = \begin{bmatrix} 4 & 1 \\ 6 & 2 \\ -2 & 3 \end{bmatrix} \begin{bmatrix} 0 & 3 & -5 \\ 1 & 2 & 6 \end{bmatrix} - \begin{bmatrix} 4 & 1 \\ 6 & 2 \\ -2 & 3 \end{bmatrix} \begin{bmatrix} 4 & 1 & 0 \\ -2 & 3 & -2 \end{bmatrix} = \begin{bmatrix} 1 & 14 & -14 \\ 2 & 22 & -18 \\ 3 & 0 & 28 \end{bmatrix} - \begin{bmatrix} 14 & 7 & -2 \\ 20 & 12 & -4 \\ -14 & 7 & -6 \end{bmatrix} = \begin{bmatrix} -13 & 7 & -12 \\ -18 & 10 & -14 \\ 17 & -7 & 34 \end{bmatrix}$

17. (a) $\begin{bmatrix} 2 & -2 \\ 1 & 0 \end{bmatrix} \begin{bmatrix} 2 & 1 & 4 & 6 \\ 3 & -1 & 3 & 2 \end{bmatrix} = \begin{bmatrix} 2 \cdot 2 + (-2)3 & 2 \cdot 1 + (-2)(-1) & 2 \cdot 4 + (-2)3 & 2 \cdot 6 + (-2)2 \\ 1 \cdot 2 + 0 \cdot 3 & 1 \cdot 1 + 0(-1) & 1 \cdot 4 + 0 \cdot 3 & 1 \cdot 6 + 0 \cdot 2 \end{bmatrix}$

$= \begin{bmatrix} -2 & 4 & 2 & 8 \\ 2 & 1 & 4 & 6 \end{bmatrix}$

(b) Enter matrix A $= \begin{bmatrix} 2 & -2 \\ 1 & 0 \end{bmatrix}$ and matrix B $= \begin{bmatrix} 2 & 1 & 4 & 6 \\ 3 & -1 & 3 & 2 \end{bmatrix}$.
Then enter [A] [B] to obtain the answer in (a).

19. (a) $\begin{bmatrix} 1 & 0 & 1 \\ 2 & 4 & 1 \\ 3 & 6 & 1 \end{bmatrix} \begin{bmatrix} 1 & 3 \\ 6 & 2 \\ 8 & -1 \end{bmatrix} = \begin{bmatrix} 9 & 2 \\ 34 & 13 \\ 47 & 20 \end{bmatrix}$

(b) Enter matrix A $= \begin{bmatrix} 1 & 0 & 1 \\ 2 & 4 & 1 \\ 3 & 6 & 1 \end{bmatrix}$ and matrix B $= \begin{bmatrix} 1 & 3 \\ 6 & 2 \\ 8 & -1 \end{bmatrix}$.
Then enter [A] [B] to obtain the answer in (a).

21. $A = \begin{bmatrix} 2 & 1 \\ 1 & 1 \end{bmatrix}$

Step 1: Form $\begin{bmatrix} 2 & 1 & | & 1 & 0 \\ 1 & 1 & | & 0 & 1 \end{bmatrix} = \left[A \mid I_2\right]$

Step 2: $\begin{bmatrix} 2 & 1 & | & 1 & 0 \\ 1 & 1 & | & 0 & 1 \end{bmatrix} \rightarrow \underset{\uparrow}{\begin{bmatrix} 1 & 1 & | & 0 & 1 \\ 2 & 1 & | & 1 & 0 \end{bmatrix}} \rightarrow \underset{\uparrow}{\begin{bmatrix} 1 & 1 & | & 0 & 1 \\ 0 & -1 & | & 0 & -2 \end{bmatrix}}$

Interchange $\quad R_2 = -2r_1 + r2$
rows one and two

$\rightarrow \underset{\uparrow}{\begin{bmatrix} 1 & 0 & | & 1 & -1 \\ 0 & -1 & | & 1 & -2 \end{bmatrix}} \rightarrow \underset{\uparrow}{\begin{bmatrix} 1 & 0 & | & 1 & -1 \\ 0 & 1 & | & -1 & 2 \end{bmatrix}}$

$R_2 = r_2 + r_1 \qquad R_2 - (1)r_2$

Step 3: We have now achieved the form $\left[I_2 | A^{-1}\right]$,

so $A^{-1} = \begin{bmatrix} 1 & -1 \\ -1 & 2 \end{bmatrix}$.

Check: $A \cdot A^{-1} = \begin{bmatrix} 2 & 1 \\ 1 & 1 \end{bmatrix} \begin{bmatrix} 1 & -1 \\ -1 & 2 \end{bmatrix} = \begin{bmatrix} 1 & 0 \\ 0 & 1 \end{bmatrix} = I_2!$

23. $A = \begin{bmatrix} 6 & 5 \\ 2 & 2 \end{bmatrix}$

Step 1: $[A | I_2] = \begin{bmatrix} 6 & 5 & | & 1 & 0 \\ 2 & 2 & | & 0 & 1 \end{bmatrix}$

Step 2: $\begin{bmatrix} 6 & 5 & | & 1 & 0 \\ 2 & 2 & | & 0 & 1 \end{bmatrix} \rightarrow \begin{bmatrix} 2 & 2 & | & 0 & 1 \\ 6 & 5 & | & 1 & 0 \end{bmatrix} \rightarrow \begin{bmatrix} 2 & 2 & | & 0 & 1 \\ 6 & -1 & | & 1 & -3 \end{bmatrix} \rightarrow \begin{bmatrix} 2 & 0 & | & 2 & -5 \\ 0 & -1 & | & 1 & -3 \end{bmatrix} \rightarrow \begin{bmatrix} 1 & 0 & | & 1 & -\frac{5}{2} \\ 0 & 1 & | & -1 & 3 \end{bmatrix}$

$\uparrow$ Interchange rows $\uparrow R_2 = -3r_1 + r_2$ $\uparrow R_1 = 2r_2 + R_1$ $\uparrow R_1 = \frac{1}{2}r_1$

$R_2 = (-1)r_2$

Step 3: We have: $A^{-1} = \begin{bmatrix} 1 & -\dfrac{5}{2} \\ -1 & 3 \end{bmatrix}$

25. $A = \begin{bmatrix} 2 & 1 \\ a & a \end{bmatrix}$, where $a \neq 0$.

$\begin{bmatrix} 2 & 1 & | & 1 & 0 \\ a & a & | & 0 & 1 \end{bmatrix} \rightarrow \begin{bmatrix} 1 & \frac{1}{2} & | & \frac{1}{2} & 0 \\ a & a & | & 0 & 1 \end{bmatrix} \rightarrow \begin{bmatrix} 1 & \frac{1}{2} & | & \frac{1}{2} & 0 \\ 0 & \frac{1}{2}a & | & -\frac{1}{2}a & 1 \end{bmatrix} \rightarrow \begin{bmatrix} 1 & \frac{1}{2} & | & \frac{1}{2} & 0 \\ 0 & 1 & | & -1 & \frac{2}{a} \end{bmatrix} \rightarrow \begin{bmatrix} 1 & 0 & | & 1 & -\frac{1}{a} \\ 0 & 1 & | & -1 & \frac{2}{a} \end{bmatrix}$

$\uparrow R_1 = \frac{1}{2}r_1$ $\uparrow R_2 = -ar_1 + r_2$ $\uparrow R_2 = \left(\frac{2}{a}\right)r_2$ $\uparrow R_1 = -\frac{1}{2}r_2 + r_1$

Therefore, $A^{-1} = \begin{bmatrix} 1 & -\dfrac{1}{a} \\ -1 & \dfrac{2}{a} \end{bmatrix}$

27. $A = \begin{bmatrix} 1 & -1 & 1 \\ 0 & -2 & 1 \\ -2 & -3 & 0 \end{bmatrix}$

$\begin{bmatrix} 1 & -1 & 1 & | & 1 & 0 & 0 \\ 0 & -2 & 1 & | & 0 & 1 & 0 \\ -2 & -3 & 0 & | & 0 & 0 & 1 \end{bmatrix} \rightarrow \begin{bmatrix} 1 & -1 & 1 & | & 1 & 0 & 0 \\ 0 & -2 & 1 & | & 0 & 1 & 0 \\ 0 & -5 & 2 & | & 2 & 0 & 1 \end{bmatrix} \rightarrow \begin{bmatrix} 1 & -1 & 1 & | & 1 & 0 & 0 \\ 0 & 1 & -\frac{1}{2} & | & 0 & -\frac{1}{2} & 0 \\ 0 & -5 & 2 & | & 2 & 0 & 1 \end{bmatrix}$

$\uparrow R_3 = 2r_1 + r_3$ $\uparrow R_2 = -\frac{1}{2}r_2$

$$\rightarrow \begin{bmatrix} 1 & 0 & \frac{1}{2} & | & 1 & -\frac{1}{2} & 0 \\ 0 & 1 & -\frac{1}{2} & | & 0 & -\frac{1}{2} & 0 \\ 0 & 0 & -\frac{1}{2} & | & 2 & -\frac{5}{2} & 1 \end{bmatrix} \rightarrow \begin{bmatrix} 1 & 0 & 0 & | & 3 & -3 & 1 \\ 0 & 1 & 0 & | & -2 & 2 & -1 \\ 0 & 0 & -\frac{1}{2} & | & 2 & -\frac{5}{2} & 1 \end{bmatrix} \rightarrow \begin{bmatrix} 1 & 0 & 0 & | & 3 & -3 & 1 \\ 0 & 1 & 0 & | & -2 & 2 & -1 \\ 0 & 0 & 1 & | & -4 & 5 & -2 \end{bmatrix}$$

$\uparrow$ $\uparrow$ $\uparrow$

$R_1 = r_2 + r_1$ $R_1 = r_3 + r_1$ $R_3 = -2r_3$

$R_3 = 5r_2 + r_3$ $R_2 = (-1)r_3 + r_2$

We have $A^{-1} = \begin{bmatrix} 3 & -3 & 1 \\ -2 & 2 & -1 \\ -4 & 5 & -2 \end{bmatrix}$

29. $A = \begin{bmatrix} 1 & 1 & 1 \\ 3 & 2 & -1 \\ 3 & 1 & 2 \end{bmatrix}$

Then, $\begin{bmatrix} 1 & 1 & 1 & | & 1 & 0 & 0 \\ 3 & 2 & -1 & | & 0 & 1 & 0 \\ 3 & 1 & 2 & | & 0 & 0 & 1 \end{bmatrix} \rightarrow \begin{bmatrix} 1 & 1 & 1 & | & 1 & 0 & 0 \\ 0 & -1 & -4 & | & -3 & 1 & 0 \\ 0 & -2 & -1 & | & -3 & 0 & 1 \end{bmatrix}$

$\uparrow$

$R_2 = -3r_1 + r_2$

$R_3 = -3r_1 + r_3$

$$\rightarrow \begin{bmatrix} 1 & 1 & 1 & | & 1 & 0 & 0 \\ 0 & 1 & 4 & | & 3 & -1 & 0 \\ 0 & -2 & -1 & | & -3 & 0 & 1 \end{bmatrix} \rightarrow \begin{bmatrix} 1 & 0 & -3 & | & -2 & 1 & 0 \\ 0 & 1 & 4 & | & 3 & -1 & 0 \\ 0 & 0 & 7 & | & -3 & -2 & 1 \end{bmatrix}$$

$\uparrow$ $\uparrow$

$R_2 = (-1)r_2$ $R_1 = (-1)r_2 + r_1$

$R_3 = 2r_2 + r_3$

$$\rightarrow \begin{bmatrix} 1 & 0 & -3 & | & -2 & 1 & 0 \\ 0 & 1 & 4 & | & 3 & -1 & 0 \\ 0 & 0 & 1 & | & \frac{3}{7} & -\frac{2}{7} & \frac{1}{7} \end{bmatrix} \rightarrow \begin{bmatrix} 1 & 0 & 0 & | & -\frac{5}{7} & \frac{1}{7} & \frac{3}{7} \\ 0 & 1 & 0 & | & \frac{9}{7} & \frac{1}{7} & -\frac{4}{7} \\ 0 & 0 & 1 & | & \frac{3}{7} & -\frac{2}{7} & \frac{1}{7} \end{bmatrix}$$

$\uparrow$ $\uparrow$

$R_3 = \left(\frac{1}{7}\right)r_3$ $R_1 = 3r_3 + r_1$

$R_2 = -4r_3 + r_2$

Thus, $A^{-1} = \begin{bmatrix} -\dfrac{5}{7} & \dfrac{1}{7} & \dfrac{3}{7} \\[2mm] \dfrac{9}{7} & \dfrac{1}{7} & -\dfrac{4}{7} \\[2mm] \dfrac{3}{7} & -\dfrac{2}{7} & \dfrac{1}{7} \end{bmatrix}$

31. Let $A = \begin{bmatrix} 2 & 1 \\ 1 & 1 \end{bmatrix}$, $X = \begin{bmatrix} x \\ y \end{bmatrix}$, $B = \begin{bmatrix} 8 \\ 5 \end{bmatrix}$.

Then $2x + y = 8$
        $x + y = 5$

can be written compactly as $A \cdot X = B$.

From Problem 21, $A^{-1} = \begin{bmatrix} 1 & -1 \\ -1 & 2 \end{bmatrix}$ and $X = A^{-1}B = \begin{bmatrix} 1 & -1 \\ -1 & 2 \end{bmatrix}\begin{bmatrix} 8 \\ 5 \end{bmatrix} = \begin{bmatrix} 3 \\ 2 \end{bmatrix}$;

or in other words, $x = 3$ and $y = 2$.

33. Here, $A = \begin{bmatrix} 2 & 1 \\ 1 & 1 \end{bmatrix}$, $X = \begin{bmatrix} x \\ y \end{bmatrix}$, $B = \begin{bmatrix} 0 \\ 5 \end{bmatrix}$.

From Problem 21, $A^{-1} = \begin{bmatrix} 1 & -1 \\ -1 & 2 \end{bmatrix}$, so $X = A^{-1}B = \begin{bmatrix} 1 & -1 \\ -1 & 2 \end{bmatrix}\begin{bmatrix} 0 \\ 5 \end{bmatrix} = \begin{bmatrix} -5 \\ 10 \end{bmatrix}$; or $x = -5$ and $y = 10$.

35. $A = \begin{bmatrix} 6 & 5 \\ 2 & 2 \end{bmatrix}$, $X = \begin{bmatrix} x \\ y \end{bmatrix}$, $B = \begin{bmatrix} 7 \\ 2 \end{bmatrix}$.

From Problem 23, $A^{-1} = \begin{bmatrix} 1 & -\dfrac{5}{2} \\[2mm] -1 & 3 \end{bmatrix}$, so $X = A^{-1}B = \begin{bmatrix} 1 & -\dfrac{5}{2} \\[2mm] -1 & 3 \end{bmatrix}\begin{bmatrix} 7 \\ 2 \end{bmatrix} = \begin{bmatrix} 2 \\ -1 \end{bmatrix}$; or $x = 2$ and $y = -1$.

37. $A = \begin{bmatrix} 6 & 5 \\ 2 & 2 \end{bmatrix}$, $X = \begin{bmatrix} x \\ y \end{bmatrix}$, $B = \begin{bmatrix} 13 \\ 5 \end{bmatrix}$.

From Problem 23, $A^{-1} = \begin{bmatrix} 1 & -\dfrac{5}{2} \\[2mm] -1 & 3 \end{bmatrix}$, so $X = A^{-1}B = \begin{bmatrix} 1 & -\dfrac{5}{2} \\[2mm] -1 & 3 \end{bmatrix}\begin{bmatrix} 13 \\ 5 \end{bmatrix} = \begin{bmatrix} \dfrac{1}{2} \\[2mm] 2 \end{bmatrix}$; or $x = \dfrac{1}{2}$ and $y = 2$.

39. $A = \begin{bmatrix} 2 & 1 \\ a & a \end{bmatrix}$, $X = \begin{bmatrix} x \\ y \end{bmatrix}$, $B = \begin{bmatrix} -3 \\ -a \end{bmatrix}$, where $a \neq 0$.

From Problem 25, $A^{-1} = \begin{bmatrix} 1 & -\dfrac{1}{a} \\[2mm] -1 & \dfrac{2}{a} \end{bmatrix}$, so $X = A^{-1}B = \begin{bmatrix} 1 & -\dfrac{1}{a} \\[2mm] -1 & \dfrac{2}{a} \end{bmatrix}\begin{bmatrix} -3 \\ -a \end{bmatrix} = \begin{bmatrix} -2 \\ 1 \end{bmatrix}$; or $x = -2$ and $y = 1$.

41. $A = \begin{bmatrix} 2 & 1 \\ a & a \end{bmatrix}$, $X = \begin{bmatrix} x \\ y \end{bmatrix}$, $B = \begin{bmatrix} \dfrac{7}{a} \\ 5 \end{bmatrix}$.

From Problem 25, $A^{-1} = \begin{bmatrix} 1 & -\dfrac{1}{a} \\ -1 & \dfrac{2}{a} \end{bmatrix}$, so $X = A^{-1}B = \begin{bmatrix} 1 & -\dfrac{1}{a} \\ -1 & \dfrac{2}{a} \end{bmatrix}\begin{bmatrix} \dfrac{7}{a} \\ 5 \end{bmatrix} = \begin{bmatrix} \dfrac{2}{a} \\ \dfrac{3}{a} \end{bmatrix}$; or $x = \dfrac{2}{a}$ and $y = \dfrac{3}{a}$.

43. $A = \begin{bmatrix} 1 & -1 & 1 \\ 0 & -2 & 1 \\ -2 & -3 & 0 \end{bmatrix}$, $X = \begin{bmatrix} x \\ y \\ z \end{bmatrix}$, $B = \begin{bmatrix} 0 \\ -1 \\ -5 \end{bmatrix}$.

By Problem 27, $A^{-1} = \begin{bmatrix} 3 & -3 & 1 \\ -2 & 2 & -1 \\ -4 & 5 & -2 \end{bmatrix}$, so $X = A^{-1}B = \begin{bmatrix} 3 & -3 & 1 \\ -2 & 2 & -1 \\ -4 & 5 & -2 \end{bmatrix}\begin{bmatrix} 0 \\ -1 \\ -5 \end{bmatrix} = \begin{bmatrix} -2 \\ 3 \\ 5 \end{bmatrix}$;

or $x = -2$, $y = 3$, and $z = 5$.

45. $A = \begin{bmatrix} 1 & -1 & 1 \\ 0 & -2 & 1 \\ -2 & -3 & 0 \end{bmatrix}$, $X = \begin{bmatrix} x \\ y \\ z \end{bmatrix}$, $B = \begin{bmatrix} 2 \\ 2 \\ \dfrac{1}{2} \end{bmatrix}$.

By Problem 27, $A^{-1} = \begin{bmatrix} 3 & -3 & 1 \\ -2 & 2 & -1 \\ -4 & 5 & -2 \end{bmatrix}$, so $X = A^{-1}B = \begin{bmatrix} 3 & -3 & 1 \\ -2 & 2 & -1 \\ -4 & 5 & -2 \end{bmatrix}\begin{bmatrix} 2 \\ 2 \\ \dfrac{1}{2} \end{bmatrix} = \begin{bmatrix} \dfrac{1}{2} \\ -\dfrac{1}{2} \\ 1 \end{bmatrix}$;

or $x = \dfrac{1}{2}$, $y = -\dfrac{1}{2}$, and $z = 1$.

47. $A = \begin{bmatrix} 1 & 1 & 1 \\ 3 & 2 & -1 \\ 3 & 1 & 2 \end{bmatrix}$, $X = \begin{bmatrix} x \\ y \\ z \end{bmatrix}$, $B = \begin{bmatrix} 9 \\ 8 \\ 1 \end{bmatrix}$.

By Problem 29, $A^{-1} = \dfrac{1}{7}\begin{bmatrix} -5 & 1 & 3 \\ 9 & 1 & -4 \\ 3 & -2 & 1 \end{bmatrix}$, so $X = A^{-1}B = \dfrac{1}{7}\begin{bmatrix} -5 & 1 & 3 \\ 9 & 1 & -4 \\ 3 & -2 & 1 \end{bmatrix}\begin{bmatrix} 9 \\ 8 \\ 1 \end{bmatrix} = \dfrac{1}{7}\begin{bmatrix} -34 \\ 85 \\ 12 \end{bmatrix} = \begin{bmatrix} -\dfrac{34}{7} \\ \dfrac{85}{7} \\ \dfrac{12}{7} \end{bmatrix}$;

or $x = -\dfrac{34}{7}$, $y = \dfrac{85}{7}$, $z = \dfrac{12}{7}$

49. $A = \begin{bmatrix} 1 & 1 & 1 \\ 3 & 2 & -1 \\ 3 & 1 & 2 \end{bmatrix}$, $X = \begin{bmatrix} x \\ y \\ z \end{bmatrix}$, $B = \begin{bmatrix} 2 \\ 7 \\ 3 \\ \dfrac{10}{3} \end{bmatrix}$.

By Problem 29, $A^{-1} = \dfrac{1}{7}\begin{bmatrix} -5 & 1 & 3 \\ 9 & 1 & -4 \\ 3 & -2 & 1 \end{bmatrix}$, so $X = A^{-1}B = \dfrac{1}{7}\begin{bmatrix} -5 & 1 & 3 \\ 9 & 1 & -4 \\ 3 & -2 & 1 \end{bmatrix}\begin{bmatrix} 2 \\ 7 \\ 3 \\ \dfrac{10}{3} \end{bmatrix} = \dfrac{1}{7}\begin{bmatrix} \dfrac{7}{3} \\ 7 \\ \dfrac{14}{3} \end{bmatrix} = \begin{bmatrix} \dfrac{1}{3} \\ 1 \\ \dfrac{2}{3} \end{bmatrix}$;

or $x = \dfrac{1}{3}$, $y = 1$, $z = \dfrac{2}{3}$

51. $A = \begin{bmatrix} 4 & 2 \\ 2 & 1 \end{bmatrix}$. We start with $[A\,|\,I_2]$ and put it into reduced echelon form. If I_2 does *not* appear to the left of the vertical bar, then A has no inverse.

$$[A\,|\,I_2] = \begin{bmatrix} 4 & 2 & | & 1 & 0 \\ 2 & 1 & | & 0 & 1 \end{bmatrix} \rightarrow \begin{bmatrix} 4 & 2 & | & 1 & 0 \\ 0 & 0 & | & -\dfrac{1}{2} & 1 \end{bmatrix} \rightarrow \begin{bmatrix} 1 & \dfrac{1}{2} & | & \dfrac{1}{4} & 0 \\ 0 & 0 & | & -\dfrac{1}{2} & 1 \end{bmatrix}$$

$$\uparrow \qquad\qquad\qquad \uparrow$$
$$R_2 = -\dfrac{1}{2}r_1 + r_2 \quad R_1 = -\dfrac{1}{4}r_1$$

This is in reduced echelon form, but the identity matrix I_2 does not appear on the left. Thus, A has no inverse.

53. $$[A\,|\,I_2] = \begin{bmatrix} 15 & 3 & | & 1 & 0 \\ 10 & 2 & | & 0 & 1 \end{bmatrix} \rightarrow \begin{bmatrix} 1 & \dfrac{1}{5} & | & \dfrac{1}{15} & 0 \\ 10 & 2 & | & 0 & 1 \end{bmatrix} \rightarrow \begin{bmatrix} 1 & \dfrac{1}{5} & | & \dfrac{1}{15} & 0 \\ 0 & 0 & | & -\dfrac{2}{3} & 1 \end{bmatrix}$$

$$\uparrow \qquad\qquad\qquad \uparrow$$
$$R_1 = \dfrac{1}{15}r_1 \qquad R_2 = -10r_1 + r_2$$

We cannot obtain I_2 on the left, so A has no inverse.

55. $$\begin{bmatrix} -3 & 1 & -1 & | & 1 & 0 & 0 \\ 1 & -4 & -7 & | & 0 & 1 & 0 \\ 1 & 2 & 5 & | & 0 & 0 & 1 \end{bmatrix} \rightarrow \begin{bmatrix} 1 & 2 & 5 & | & 0 & 0 & 1 \\ 1 & -4 & -7 & | & 0 & 1 & 0 \\ -3 & 1 & -1 & | & 1 & 0 & 0 \end{bmatrix} \rightarrow \begin{bmatrix} 1 & 2 & 5 & | & 0 & 0 & 1 \\ 0 & -6 & -12 & | & 0 & 1 & -1 \\ 0 & 7 & 14 & | & 1 & 0 & 3 \end{bmatrix}$$

$$\uparrow \qquad\qquad\qquad\qquad \uparrow$$

Interchange rows $\qquad R_2 = -r_1 + r_2$
one and three $\qquad\qquad R_3 = 3r_1 + r_3$

$$\rightarrow \begin{bmatrix} 1 & 2 & 5 & 0 & 0 & 1 \\ 0 & 1 & 2 & 0 & -\frac{1}{6} & \frac{1}{6} \\ 0 & 1 & 2 & \frac{1}{7} & 0 & \frac{3}{7} \end{bmatrix} \rightarrow \begin{bmatrix} 1 & 2 & 5 & 0 & 0 & 1 \\ 0 & 1 & 2 & 0 & -\frac{1}{6} & \frac{1}{6} \\ 0 & 0 & 0 & \frac{1}{7} & \frac{1}{6} & \frac{11}{42} \end{bmatrix}$$

$$\uparrow \qquad\qquad\qquad\qquad \uparrow$$

$$R_2 = -\frac{1}{6}r_2 \qquad\qquad R_3 = -r_2 + r_3$$

$$R_3 = \frac{1}{7}r_3$$

The left side is not the identity matrix, I_3, so A has no inverse.

57. $\begin{bmatrix} 0.01 & 0.05 & -0.01 \\ 0.01 & -0.02 & 0.01 \\ -0.02 & 0.01 & 0.03 \end{bmatrix}$

59. $\begin{bmatrix} 0.02 & -0.04 & -0.01 & 0.01 \\ -0.02 & 0.05 & 0.03 & -0.03 \\ -0.02 & 0.01 & -0.04 & 0.00 \\ -0.02 & 0.06 & 0.07 & 0.06 \end{bmatrix}$

61. $x = 4.57, y = -6.44, z = -24.07$

63. $x = -1.19, y = -6.63, z = 8.27$

65. (a) Since we want to use a 2 by 3 matrix to represent the data, we can let rows represent stainless steel and aluminum, respectively, while the columns can represent 10-gallon, 5-gallon, and 1-gallon containers:

	Production		
	10 g.	5 g.	1 g.
Stainless steel	500	350	400
Aluminum	700	500	850

$$\left. \right\} \begin{bmatrix} 500 & 350 & 400 \\ 700 & 500 & 850 \end{bmatrix}$$

This could have been represented by a 3 by 2 matrix, by letting rows represent size of container, and columns represent type of material.

$$\begin{bmatrix} 500 & 700 \\ 350 & 500 \\ 400 & 850 \end{bmatrix}$$

(b) Now we are given the following information:

	Pounds of Material
10-gal.	15
5-gal.	8
1-gal.	3

$$\left. \right\} \begin{bmatrix} 15 \\ 8 \\ 3 \end{bmatrix}$$

(c)   From (a) and (b): $\begin{bmatrix} 500 & 350 & 400 \\ 700 & 500 & 850 \end{bmatrix} \begin{bmatrix} 15 \\ 8 \\ 3 \end{bmatrix} = \begin{bmatrix} 11{,}500 \\ 17{,}050 \end{bmatrix}$

The first row represents stainless steel containers (11,500 pounds), and the second row represents aluminum (17,050 pounds).

(d)

	Stainless Steel	Aluminum
Cost per pound	0.10	0.05

$\Big\} [0.10 \quad 0.05]$

(e) $[0.10 \quad 0.05] \begin{bmatrix} 11{,}500 \\ 17{,}050 \end{bmatrix} = [2002.50]$; i.e., total cost of material = \$2002.50.

Note: Since the first entry in (c) was for stainless steel, the first entry, a_{11}, in (d) had to be for stainless steel.

67.   Let $A = \begin{bmatrix} a & b \\ c & d \end{bmatrix}$.

We are assuming that $D = ad - bc \neq 0$, and we wish to show that A has no inverse. We start, as usual, with

$$[A \mid I_2] = \begin{bmatrix} a & b & 1 & 0 \\ c & d & 0 & 1 \end{bmatrix}.$$

Our first step in putting this into echelon form would be to multiply row one by $\dfrac{1}{a}$, to get a 1 in the first row, first column. But that will be impossible if $a = 0$. Thus, we need to consider two cases:

Case 1:   If $a \neq 0$, proceed as usual:

$$\begin{bmatrix} a & b & 1 & 0 \\ c & d & 0 & 1 \end{bmatrix} \rightarrow \begin{bmatrix} 1 & \dfrac{b}{a} & \dfrac{1}{a} & 0 \\ c & d & 0 & 1 \end{bmatrix} \rightarrow \begin{bmatrix} 1 & \dfrac{b}{a} & \dfrac{1}{a} & 0 \\ 0 & d - \dfrac{cb}{a} & -\dfrac{c}{a} & 1 \end{bmatrix}$$

$$\uparrow \qquad\qquad\qquad \uparrow$$
$$R_1 = \dfrac{1}{a} r_1 \qquad R_2 = -cr_1 + r_2$$

$$= \begin{bmatrix} 1 & \dfrac{b}{a} & \dfrac{1}{a} & 0 \\ 0 & \dfrac{ad - bc}{a} & -\dfrac{c}{a} & 1 \end{bmatrix} \left(\text{since } d - \dfrac{cb}{a} = \dfrac{ad - bc}{a} \right)$$

$$\rightarrow \begin{bmatrix} 1 & \dfrac{b}{a} & \dfrac{1}{a} & 0 \\ 0 & 1 & \dfrac{-c}{ad - bc} & \dfrac{a}{ad - bc} \end{bmatrix} \rightarrow \begin{bmatrix} 1 & 0 & \dfrac{1}{a} + \dfrac{bc}{a(ad - bc)} & \dfrac{-b}{ad - bc} \\ 0 & 1 & \dfrac{-c}{ad - bc} & \dfrac{a}{ad - bc} \end{bmatrix}$$

$$\uparrow \qquad\qquad\qquad\qquad \uparrow$$
$$R_2 = \left(\dfrac{a}{ad - bc} \right) r_2 \qquad R_1 = \left(\dfrac{-b}{a} \right) r_2 + r_1$$

Now some algebra:

$$\frac{1}{a} + \frac{bc}{a(ad-bc)} = \frac{ad-bc}{a(ad-bc)} + \frac{bc}{a(ad-bc)} = \frac{ad}{a(ad-bc)} = \frac{d}{ad-bc}$$

Thus, our inverse is: $A^{-1} = \begin{bmatrix} \dfrac{d}{ad-bc} & \dfrac{-b}{ad-bc} \\ \dfrac{-c}{ad-bc} & \dfrac{a}{ad-bc} \end{bmatrix} = \dfrac{1}{D}\begin{bmatrix} d & -b \\ -c & a \end{bmatrix}$, since $D = ad - bc$.

Case 2: But what if $a = 0$? Then, $ad - bc \neq 0$, so we know $b \neq 0$ and $c \neq 0$. Also, we will have:

$$\left[A\,|\,I_2\right] = \begin{bmatrix} 0 & b & | & 1 & 0 \\ c & d & | & 0 & 1 \end{bmatrix}$$

So $\qquad \begin{bmatrix} 0 & b & | & 1 & 0 \\ c & d & | & 0 & 1 \end{bmatrix} \rightarrow \begin{bmatrix} c & d & | & 0 & 1 \\ 0 & b & | & 1 & 0 \end{bmatrix} \qquad \rightarrow \begin{bmatrix} 1 & \dfrac{d}{c} & | & 0 & \dfrac{1}{c} \\ 0 & b & | & 1 & 0 \end{bmatrix}$

$\qquad\qquad\qquad \uparrow \qquad\qquad\qquad\qquad \uparrow$

Interchange rows $R_1 = \dfrac{1}{c}r_1$ (since $c \neq 0$)

$$= \begin{bmatrix} 1 & \dfrac{d}{c} & | & 0 & \dfrac{1}{c} \\ 0 & 1 & | & \dfrac{1}{b} & 0 \end{bmatrix} \qquad \rightarrow \begin{bmatrix} 1 & 0 & | & \dfrac{-d}{bc} & \dfrac{1}{c} \\ 0 & 1 & | & \dfrac{1}{b} & 0 \end{bmatrix}$$

$\qquad\qquad \uparrow \qquad\qquad\qquad\qquad\qquad \uparrow$

$R_2 = \dfrac{1}{b}r_2$ (since $b \neq 0$) $\quad R_1 = \left(-\dfrac{d}{c}\right)r_2 + r_1$

Therefore, $A^{-1} = \begin{bmatrix} \dfrac{-d}{bc} & \dfrac{1}{c} \\ \dfrac{1}{b} & 0 \end{bmatrix}$.

That looks odd, but if we get a common denominator, we have:

$$A^{-1} = \begin{bmatrix} \dfrac{-d}{bc} & \dfrac{b}{bc} \\ \dfrac{c}{bc} & \dfrac{0}{bc} \end{bmatrix} = \frac{1}{-bc}\begin{bmatrix} d & -b \\ -c & 0 \end{bmatrix} = \frac{1}{D}\begin{bmatrix} d & -b \\ -c & a \end{bmatrix},$$

since $a = 0$, and $D = ad - bc = -bc$!

11.5 Partial Fraction Decomposition

1.   The expression $\dfrac{x}{x^2-1}$ is proper, since the degree of the numerator is less than that of the denominator.

3. $\dfrac{x^2 + 5}{x^2 - 4}$ is improper, so we do the division:

$$
\begin{array}{r}
1 \quad \leftarrow \text{Quotient} \\
x^2 - 4\overline{\smash{\big)}\, x^2 + 0x + 5} \\
\underline{x^2 \qquad\;\; - 4} \\
9 \quad \leftarrow \text{Remainder}
\end{array}
$$

Then, $\dfrac{x^2 + 5}{x^2 - 4} = 1 + \dfrac{9}{x^2 - 4}$.

5. $\dfrac{5x^3 + 2x - 1}{x^2 - 4}$ is improper, so we must do long division:

$$
\begin{array}{r}
5x \qquad\qquad\quad \\
x^2 - 4\overline{\smash{\big)}\, 5x^3 + 0x^2 + 2x - 1} \\
\underline{5x^3 \qquad\quad -20x} \\
22x - 1
\end{array}
$$

Thus, $\dfrac{5x^3 + 2x - 1}{x^2 - 4} = 5x + \dfrac{22x - 1}{x^2 - 4}$

7. Here, $\dfrac{x(x - 1)}{(x + 4)(x - 3)} = \dfrac{x^2 - x}{x^2 + x - 12}$ is improper:

$$
\begin{array}{r}
1 \qquad\qquad \\
x^2 + x - 12\overline{\smash{\big)}\, x^2 - x + 0} \\
\underline{x^2 + x - 12} \\
-2x + 12
\end{array}
$$

Hence, $\dfrac{x(x - 1)}{(x + 4)(x - 3)} = 1 + \dfrac{-2x + 12}{x^2 + x - 12} = 1 + \dfrac{-2(x - 6)}{(x + 4)(x - 3)}$

In Problems 9–38, we will use the following steps:

Step 1: Perform long division, if necessary, to obtain a **proper** fraction, $\dfrac{p(x)}{q(x)}$, and put it in lowest terms.

Step 2: Factor the denominator, $q(x)$, completely into linear and irreducible quadratic factors, and identify which of the four cases in the text apply.

Step 3: Write out the partial fraction expansion, based on the factors of $q(x)$.

Step 4: Solve for the unknown coefficients in Step 3.

Step 5: Write the final decomposition.

9. $\dfrac{4}{x(x - 1)}$

Step 1: Already *proper*.

Step 2: Done: $q(x) = x(x - 1)$.
This is Case 1 (only non-repeated linear factors).

Step 3: $\dfrac{4}{x(x - 1)} = \dfrac{A}{x} + \dfrac{B}{x - 1}$

Step 4: Multiply both sides by $x(x - 1)$: $4 = A(x - 1) + Bx$
Let $x = 1$: then $4 = A(0) + B$, or $B = 4$
Let $x = 0$: then $4 = A(-1) + B \cdot 0$, or $A = -4$

Step 5: Therefore, $\dfrac{4}{x(x - 1)} = \dfrac{-4}{x} + \dfrac{4}{x - 1}$

11. $\dfrac{1}{x(x^2 + 1)}$

Step 1: Already proper.

Step 2: $q(x) = x(\underbrace{x^2 + 1}_{\text{cannot be factored}})$

This is Case 1 (nonrepeated linear) and Case 3 (nonrepeated irreducible quadratic).

Step 3:
$$\frac{1}{x(x^2+1)} = \frac{A}{x} + \frac{Bx+C}{x^2+1}$$

(Remember, irreducible quadratic factors in the denominator require *first degree* numerators in the decomposition.)

Step 4: Multiply by $x(x^2+1)$: $1 = A(x^2+1) + (Bx+C)x$

Let $x = 0$: then $1 = A$.

We still have two unknowns, B and C, so we need two equations:

Let $-x = 1$: $1 = A(1+1) + (B \cdot 1 + C) \cdot 1$

or $\quad 1 = 2A + B + C$

or $\quad B + C = -1$

Let $x = -1$: $1 = 2A + (-B + C)(-1)$

or $\quad 1 = 2 + B - C$ (since $A = 1$)

or $\quad B - C = -1$ (since $A = 1$)

We have $\quad B + C = -1$

$\underline{\quad\quad\quad B - C = -1}$

Add: $2B \quad\quad = -2$

$\quad\quad\quad B = -1$

Then, from $B + C = -1$, we get:

$-1 + C = -1$

$C = 0$

Step 5:
$$\frac{1}{x(x^2+1)} = \frac{1}{x} + \frac{-x}{x^2+1}$$

13. $\dfrac{1}{(x-1)(x-2)}$

Step 1: This is proper.

Step 2: $q(x) = (x-1)(x-2)$. Case 1 only.

Step 3: $\dfrac{x}{(x-1)(x-2)} = \dfrac{A}{x-1} + \dfrac{B}{x-2}$

Step 4: Multiply by $(x-1)(x-2)$: $x = A(x-2) + B(x-1)$

Let $x = 2$: $2 = B$

Let $x = 1$: $1 = -A$, or $A = -1$

Step 5: $\dfrac{x}{(x-1)(x-2)} = \dfrac{-1}{x-1} + \dfrac{2}{x-2}$

15. $\dfrac{x^2}{(x-1)^2(x+1)}$

Step 1: This is proper.

Step 2: $q(x) = (x-1)^2(x+1)$.

This involves both Case 1 and Case 2 (a repeated linear factor).

Step 3: For the factor $(x-1)^2$, we need two terms, $\dfrac{A}{x-1} + \dfrac{B}{(x-1)^2}$, since it is a linear factor

raised to the *second* power.

$$\frac{x^2}{(x-1)^2(x+1)} = \frac{A}{x-1} + \frac{B}{(x-1)^2} + \frac{C}{x+1}$$

Step 4: Multiply by $(x-1)^2(x+1)$: $x^2 = A(x-1)(x+1) + B(x+1) + C(x-1)^2$

Since we have three unknowns, we need three equations, so we will choose three values of x starting with the *zeros* of $q(x)$:

Let $x = 1$: $\quad 1 = 2B$, or $B = \dfrac{1}{2}$

Let $x = -1$: $1 = 4C$, or $C = \dfrac{1}{4}$

Now pick a third value for x:

Let $x = 0$: $0 = -A + B + C$

$0 = -A + \dfrac{1}{2} + \dfrac{1}{4}$, since we know B and C.

$A = \dfrac{3}{4}$

Step 5:
$$\frac{x^2}{(x - 1)^2(x + 1)} = \frac{\dfrac{3}{4}}{x - 1} + \frac{\dfrac{1}{2}}{(x - 1)^2} + \frac{\dfrac{1}{4}}{x + 1}$$

17. $\dfrac{1}{x^3 - 8}$

Step 1:   This is proper.

Step 2:   We need to do some factoring:
$$q(x) = (x - 2)(x^2 + 2x + 4)$$

Can this be factored?

To determine if $x^2 + 2x + 4$ can be factored or if it is irreducible, check its discriminant:

$b^2 - 4ac = 4 - 4(4) = 4 - 16 = -12 < 0$

Therefore, $x^2 + 2x + 4$ has no real zeros; i.e., it is irreducible over the reals.

We have Case 1 and Case 3.

Step 3:
$$\frac{1}{x^3 - 8} = \frac{1}{(x - 2)(x^2 + 2x + 4)} = \frac{A}{x - 2} + \frac{Bx + C}{x^2 + 2x + 4}$$

Step 4:   Multiply both sides by $x^3 - 8 = (x - 2)(x^2 + 2x + 4)$:

$1 = A(x^2 + 2x + 4) + (Bx + C)(x - 2)$ (1)

We only have one zero for $q(x)$, $x = 2$:

Let $x = 2$: $1 = 12A$, or $A = \dfrac{1}{12}$

But we need three equations. Another approach is to expand (1) and collect like terms:

$1 = Ax^2 + 2Ax + 4A + Bx^2 - 2Bx + Cx - 2C$

or

$1 = (A + B)x^2 + (2A - 2B + C)x + (4A - 2C)$

There is no x^2 on the left-hand side, so we must have $A + B = 0$. Also, there is no x-term on the left, so

$2A - 2B + C = 0$

Finally, the constant terms must be equal:

$1 = 4A - 2C$

But, we already know $A = \dfrac{1}{12}$, so from $A + B = 0$, we have $B = -\dfrac{1}{12}$, and from $1 = 4A -$

$2C$ we have:

$2C = 4A - 1$

$2C = \dfrac{4}{12} - 1$

$2C = \dfrac{-8}{12}$

$C = \dfrac{-4}{12} = \dfrac{-1}{3}$

Step 5: $$\frac{1}{x^3 - 8} = \frac{\frac{1}{12}}{x - 2} + \frac{\frac{-1}{12}x - \frac{1}{3}}{x^2 + 2x + 4} = \frac{\frac{1}{12}}{x - 2} + \frac{\frac{-1}{12}(x + 4)}{x^2 + 2x + 4}$$

19. $$\frac{x^2}{(x - 1)^2(x + 1)^2}$$

Step 1: This is proper.

Step 2: $q(x) = (x - 1)^2(x + 1)^2$. This is case 2.

Step 3: $$\frac{x^2}{(x - 1)^2(x + 1)^2} = \frac{A}{x - 1} + \frac{B}{(x - 1)^2} + \frac{C}{x + 1} + \frac{D}{(x + 1)^2}$$

Step 4: Multiply by $(x - 1)^2(x + 1)^2$:
$$x^2 = A(x - 1)(x + 1)^2 + B(x + 1)^2 + C(x - 1)^2(x + 1) + D(x - 1)^2$$

Let $x = 1$: $\quad 1 = 4B, \quad B = \dfrac{1}{4}$

Let $x = -1$: $\quad 1 = 4D, \quad D = \dfrac{1}{4}$

Let $x = 0$: $\quad 0 = -A + B + C + D$

$\qquad$ or $\qquad A - C = \dfrac{1}{2}$

Let $x = 2$: $\quad 4 = 9A + 9B + 3C + D$

$\qquad$ or $\qquad -9A - 3C = -\dfrac{3}{2}$

$\qquad$ or $\qquad 3A + C = \dfrac{1}{2}$

We have two equations in two unknowns:

$$A - C = \frac{1}{2}$$

$$3A + C = \frac{1}{2}$$

Add: $\quad 4A \qquad = 1$

$$A = \frac{1}{4}$$

and $\qquad A - C = \dfrac{1}{2}$

$$\frac{1}{4} - C = \frac{1}{2}$$

$$C = -\frac{1}{4}$$

Step 5: $$\frac{x^2}{(x - 1)^2(x + 1)^2} = \frac{\frac{1}{4}}{x - 1} + \frac{\frac{1}{4}}{(x - 1)^2} + \frac{\frac{-1}{4}}{x + 1} + \frac{\frac{1}{4}}{(x + 1)^2}$$

21. $\dfrac{x - 3}{(x + 2)(x + 1)^2}$

Steps 1 and 2:   Proper, with $q(x) = (x + 2)(x + 1)^2$, Case 1 and Case 2.

Step 3: $\dfrac{x - 3}{(x + 2)(x + 1)^2} = \dfrac{A}{x + 2} + \dfrac{B}{(x + 1)} + \dfrac{C}{(x + 1)^2}$

Step 4:   Clear of fractions:

$$x - 3 = A(x + 1)^2 + B(x + 2)(x + 1) + C(x + 2)$$

Let $x = -1$: $-4 = C$

Let $x = -2$: $-5 = A$

Let $x = 0$: $-3 = A + 2B + 2C$

or $-3 = -5 + 2B - 8$

$10 = 2B$

$B = 5$

Step 5: $\dfrac{x - 3}{(x + 2)(x + 1)^2} = \dfrac{-5}{x + 2} + \dfrac{5}{x + 1} + \dfrac{-4}{(x + 1)^2}$

23. $\dfrac{x + 4}{x^2(x^2 + 4)}$

Step 1:   This is proper.

Step 2: $q(x) = x^2(x^2 + 4)$.

Be careful to note the difference between x^2 and $x^2 + 4$:

The factor x^2 is a linear factor (x), repeated:

$x^2 = x \cdot x$, so it is Case 2.

The factor $x^2 + 4$ is a quadratic that cannot be factored as a product of linear factors (Case 3).

Step 3: $\dfrac{x + 4}{x^2(x^2 + 4)} = \dfrac{A}{x} + \dfrac{B}{x^2} + \dfrac{Cx + D}{x^2 + 4}$

Step 4: $x + 4 = Ax(x^2 + 4) + B(x^2 + 4) + (Cx + D)x^2$.   (1)

There is only one zero, $x = 0$:

Let $x = 0$: $4 = 4B$, or $B = 1$.

Let's expand equation (1) and combine like terms:

$$x + 4 = Ax^3 + 4Ax + Bx^2 + 4B + Cx^3 + Dx^2$$
$$= (A + C)x^3 = (B + D)x^2 + (4A)x + 4B$$

Now equate coefficients of like powers of x on the left and right:

For x^3: $0 = A + C$

For x^2: $0 = B + D$

For x^1: $1 = 4A$, or $A = \dfrac{1}{4}$

For x^0: $4 = 4B$, or $B = 1$

Then: $A + C = 0$ $B + D = 0$

$\dfrac{1}{4} + C = 0$ $1 + D = 0$

$C = \dfrac{-1}{4}$ $D = -1$

Step 5: $\dfrac{x + 4}{x^2(x^2 + 4)} = \dfrac{\frac{1}{4}}{x} + \dfrac{1}{x^2} + \dfrac{-\frac{1}{4}x - 1}{x^2 + 4} = \dfrac{\frac{1}{4}}{x} + \dfrac{1}{x^2} - \dfrac{\frac{1}{4}(x + 4)}{x^2 + 4}$

25. $\dfrac{x^2 + 2x + 3}{(x + 1)(x^2 + 2x + 4)}$

Step 1:   This is proper.

Step 3: $q(x) = (x + 1)\underbrace{(x^2 + 2x + 4)}$

$b^2 - 4ac = 4 - 16 = -12 < 0$

We have Case 1 and Case 3.

Step 3: $\dfrac{x^2 + 2x + 3}{(x + 1)(x^2 + 2x + 4)} = \dfrac{A}{x + 1} = \dfrac{Bx + C}{x^2 + 2x + 4}$

Step 4: $x^2 + 2x + 3 = A(x^2 + 2x + 4) + (Bx + C)(x + 1)$

Let $x = -1$: $2 = 3A$; $A = \dfrac{2}{3}$

Let $x = 0$: $3 = 4A + C$; $C = 3 - 4A = 3 - \dfrac{8}{3} = \dfrac{1}{3}$

Let $x = 1$: $6 = 7A + 2B + 2C$

$2B = 6 - 7A - 2C$

$2B = 6 - \dfrac{14}{3} - \dfrac{2}{3} = \dfrac{2}{3}$; $B = \dfrac{1}{3}$

Step 5: $\dfrac{x^2 + 2x + 3}{(x + 1)(x^2 + 2x + 4)} = \dfrac{\frac{2}{3}}{x + 1} + \dfrac{\frac{1}{3}(x + 1)}{x^2 + 2x + 4}$

27. $\dfrac{x}{(3x - 2)(2x + 1)}$

Step 1: This is proper.

Step 2: $q(x) = (3x - 2)(2x + 1)$, Case 1.

Step 3: $\dfrac{x}{(3x - 2)(2x + 1)} = \dfrac{A}{3x - 2} + \dfrac{B}{2x + 1}$

Step 4: $x = A(2x + 1) + B(3x - 2)$

The two zeros are $x = -\dfrac{1}{2}$ and $x = \dfrac{2}{3}$

Let $x = -\dfrac{1}{2}$: $-\dfrac{1}{2} = B\left(-\dfrac{7}{2}\right)$

$1 = 7B$

$B = \dfrac{1}{7}$

Let $x = \dfrac{2}{3}$: $\dfrac{2}{3} = A\left(\dfrac{7}{3}\right)$

$2 = 7A$

$A = \dfrac{2}{7}$

Step 5: $\dfrac{x}{(3x - 2)(2x + 1)} = \dfrac{\frac{2}{7}}{3x - 2} + \dfrac{\frac{1}{7}}{2x + 1}$

29. $\dfrac{x}{x^2 + 2x - 3}$

Step 1: Already proper.

Step 2: $q(x) = x^2 + 2x - 3 = (x + 3)(x - 1)$, Case 1.

Step 3: $\dfrac{x}{x^2 + 2x - 3} = \dfrac{A}{x + 3} + \dfrac{B}{x - 1}$

Step 4: $x = A(x - 1) + B(x + 3)$

Let $x = 1$: $\quad 1 = 4B, B = \dfrac{1}{4}$

Let $x = -3$: $\quad -3 = -4A, A = \dfrac{3}{4}$

Step 5: $\dfrac{x}{(x^2 + 2x - 3)} = \dfrac{\frac{3}{4}}{x + 3} + \dfrac{\frac{1}{4}}{x - 1}$

---

31. $\dfrac{x^2 + 2x + 3}{(x^2 + 4)^2}$

Steps 1 and 2: This is proper, and falls under Case 4 (repeated irreducible quadratic factor).

Step 3: $\dfrac{x^2 + 2x + 3}{(x^2 + 4)^2} = \dfrac{Ax + B}{x^2 + 4} + \dfrac{Cx + D}{(x^2 + 4)^2}$

Step 4: Clear of fractions:

$$x^2 + 2x + 3 = (Ax + B)(x^2 + 4) + (Cx + D)$$

We have no zeros of $q(x)$, so expand the right-hand side:

$$x^2 + 2x + 3 = Ax^3 + Bx^2 + 4Ax + 4B + Cx + D$$
$$x^2 + 2x + 3 = Ax^3 + Bx^2 + (4A + C)x + (4B + D)$$

x^3: $\quad 0 = A$

x^2: $\quad 1 = B$

x: $\quad 2 = 4A + C, C = 2$ (since $A = 0$)

Constant: $\quad 3 = 4B + D, D = 3 - 4B = -1$

Step 5: $\dfrac{x^2 + 2x + 3}{(x^2 + 4)^2} = \dfrac{1}{x^2 + 4} + \dfrac{2x - 1}{(x^2 + 4)^2}$

---

33. $\dfrac{7x + 3}{x^3 - 2x^2 - 3x}$

Step 1: This is proper.

Step 2: $q(x) = x^3 - 2x^2 - 3x = x(x^2 - 2x - 3) = x(x - 3)(x + 1)$

This is Case 1.

Step 3: $\dfrac{7x + 3}{x^3 - 2x^2 - 3x)} = \dfrac{A}{x} = \dfrac{B}{x - 3} + \dfrac{C}{x + 1}$

Step 4: Multiply both sides by $x^3 - 2x^2 - 3x = x(x - 3)(x + 1)$:

$$7x + 3 = A(x - 3)(x + 1) + Bx(x + 1) + Cx(x - 3)$$

Let $x = 0$: $\quad 3 = -3A; A = -1$

Let $x = 3$: $\quad 24 = 12B, B = 2$

Let $x = -1$: $-4 = 4C, C = -1$

Step 5: $\dfrac{7x + 3}{x^3 - 2x^2 - 3x} = \dfrac{-1}{x} + \dfrac{2}{x - 3} + \dfrac{-1}{x + 1}$

---

35. $\dfrac{x^2}{x^3 - 4x^2 + 5x - 2}$

Step 1: This is proper.

Step 2: Try to find a zero by synthetic division:

```
1| 1  -4   5   -2          -1| 1  -4   5   -2
        1  -5   10                -1   3   -2
   ─────────────────          ─────────────────
      1  -5  10  -12            1  -3   2   [0]
                                └─────────┘
                                x² - 3x + 2
```

Thus, $x - 1$ is a factor:
$$q(x) = x^3 - 4x^2 + 5x - 2 = (x - 1)(x^2 - 3x + 2) = (x - 1)(x - 1)(x - 2)$$
$$= (x - 1)^2(x - 2)$$

This involves Case 1 and Case 2.

Step 3: $$\frac{x^2}{x^3 - 4x^2 + 5x - 2} = \frac{A}{x - 1} + \frac{B}{(x - 1)^2} + \frac{C}{x - 2}$$

Step 4: Multiply by $x^3 - 4x^2 + 5x - 2 = (x - 1)^2(x - 2)$:
$$x^2 = A(x - 1)(x - 2) + B(x - 2) + C(x - 1)^2$$
Let $x = 1$: $\quad 1 = -B, B = -1$
Let $x = 2$: $\quad 4 = C$
Let $x = 0$: $\quad 0 = 2A - 2B + C$
$$2A = 2B - C$$
$$2A = -2 - 4$$
$$A = -3$$

Step 5: $$\frac{x^2}{x^3 - 4x^2 + 5x - 2} = \frac{-3}{x - 1} + \frac{-1}{(x - 1)^2} + \frac{4}{x - 2}$$

37. $$\frac{x^3}{(x^2 + 16)^3}$$

Step 1 and 2: Proper, Case 4

Step 3: $$\frac{x^3}{(x^2 + 16)^3} = \frac{Ax + B}{x^2 + 16} + \frac{Cx + D}{(x^2 + 16)^2} + \frac{Ex + F}{(x^2 + 16)^3}$$

Step 4: $$x^3 = (Ax + B)(x^2 + 16)^2 + (Cx + D)(x^2 + 16) + Ex + F \quad \text{or}$$
$$x^3 = (Ax + B)(x^4 + 32x^2 + 256) + Cx^3 + Dx^2 + 16Cx + 16D + Ex + F$$
$$x^3 = Ax^5 + 32Ax^3 + 256Ax + Bx^4 + 32Bx^2 + 256B + Cx^3 + Dx^2$$
$$+ 16Cx + 16D + Ex + F$$
$$x^3 = Ax^5 + Bx^4 + (32A + C)x^3 + (32B + D)x^2 + (256A + 16C + E)x$$
$$+ (256B + 16D + F)$$

Now we equate coefficients of like powers of x to obtain six equations in six unknowns.
$$x^5: 0 = A$$
$$x^4: 0 = B$$
$$x^3: 1 = 32A + C, C = 1$$
$$x^2: 0 = 32B + D, D = 0$$
$$x^1: 0 = 256A + 16C + E, E = -16C = -16$$
$$x^0: 0 = 256B + 16D + F, F = 0$$

Step 5: $$\frac{x^3}{(x^2 + 16)^3} = \frac{x}{(x^2 + 16)^2} + \frac{-16x}{(x^2 + 16)^3}$$

39. $$\frac{4}{2x^2 - 5x - 3}$$

Step 1: This is proper.

Step 2: $q(x) = 2x^2 - 5x - 3 = (2x + 1)(x - 3)$, Case 1.

Step 3: $$\frac{4}{2x^2 - 5x - 3} = \frac{A}{2x + 1} + \frac{B}{x - 3}$$

Step 4: $4 = A(x - 3) + B(2x + 1)$

The two zeros are $x = 3$ and $x = \frac{-1}{2}$

Let $x = 3$: $\quad\quad 4 = 7B$
$$B = \frac{4}{7}$$

$$\text{Let } x = \frac{-1}{2}: \qquad 4 = \frac{-7}{2}A$$

$$A = \frac{-8}{7}$$

Step 5: $\qquad \dfrac{4}{2x^2 - 5x - 3} = \dfrac{-\frac{8}{7}}{2x + 1} + \dfrac{\frac{4}{7}}{x - 3}$

41. $\dfrac{2x + 3}{x^4 - 9x^2}$

Step 1: This is proper.

Step 2: $q(x) = x^4 - 9x^2 = x^2(x - 3)(x + 3)$, Case 1 and 2.

Step 3: $\dfrac{2x + 3}{x^4 - 9x^2} = \dfrac{A}{x} + \dfrac{B}{x^2} + \dfrac{C}{x - 3} + \dfrac{D}{x + 3}$

Step 4: $2x + 3 = Ax(x - 3)(x + 3) + B(x - 3)(x + 3) + Cx^2(x - 3) + Dx^2(x - 3)$

Let $x = 0$: $\quad 3 = B(-9)$

$$B = \frac{-1}{3}$$

Let $x = 3$: $\quad 9 = C(9)(6)$

$$C = \frac{1}{6}$$

Let $x = -3$: $\; -3 = D(9)(-6)$

$$D = \frac{1}{18}$$

$2x + 3 = Ax(x - 3)(x + 3) - \dfrac{1}{3}(x - 3)(x + 3) + \dfrac{1}{6}x^2(x + 3) + \dfrac{1}{18}x^2(x - 3)$

Let $x = 1$: $\quad 5 = A(-2)(4) - \dfrac{1}{3}(-2)(4) + \dfrac{1}{6}(4) + \dfrac{1}{18}(-2)$

$$5 = A(-8) + \frac{8}{3} + \frac{2}{3} - \frac{1}{9}$$

$$\frac{45}{9} - \frac{30}{9} + \frac{1}{9} = A(-8)$$

$$\frac{16}{9} = A(-8)$$

$$A = \frac{-2}{9}$$

Step 5: $\qquad \dfrac{2x + 3}{x^4 - 9x^2} = \dfrac{-\frac{2}{9}}{x} - \dfrac{\frac{1}{3}}{x^2} + \dfrac{\frac{1}{6}}{x - 3} + \dfrac{\frac{1}{18}}{x + 3}$

11.6 Systems of Nonlinear Equations

<u>Historical Problem</u>

1. We wish to solve

$$\begin{cases} x^2 + y^2 = 100 & (1) \\ x = \frac{3}{4}y & (2) \end{cases}$$

This is readily done by substitution, since we know from (2) that $x = \frac{3}{4}y$. Substituting this expression for x into (1) produces:

$$x^2 + y^2 = 100$$

$$\left(\frac{3}{4}y\right)^2 + y^2 = 100$$

$$\frac{9}{16}y^2 + y^2 = 100$$

$$\frac{25}{16}y^2 = 100$$

$$y^2 = \frac{16}{25}(100)$$

$$y = \pm\sqrt{\frac{16(100)}{25}}$$

$$y = \pm\frac{(4)(10)}{5}, \quad \text{or} \quad y = \pm 8$$

Now determine x from (2): $x = \frac{3}{4}y$

If $y = -8$, $x = \frac{3}{4}(-8) = -6$

If $y = 8$, $x = \frac{3}{4}(8) = 6$

Thus, we have two solutions: $x = -6$ and $y = -8$;
$\qquad\qquad\qquad\qquad\qquad\qquad\quad x = 6$ and $y = 8$

Each solution checks, as you can verify.

<u>Exercises</u>

1. $\begin{cases} y = x^2 + 1 \\ y = x + 1 \end{cases}$

Graphically: Graph $y_1 = x^2 + 1$ and $y_2 = x + 1$.

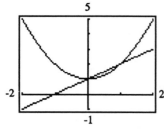

Algebraically:
$$x^2 + 1 = x + 1$$
$$x^2 - x = 0$$
$$x(x - 1) = 0$$
$$x = 0, \; x = 1$$
$$x = 0, \; y = 1; \; x = 1, \; y = 2$$

$(0.00, 1.00)$ and $(1.00, 2.00)$ are the intersection points.

3. $\begin{cases} y = \sqrt{36 - x^2} \\ y = 8 - x \end{cases}$

Graphically: Graph $y_1 = \sqrt{36 - x^2}$ and $y_2 = 8 - x$.

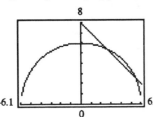

(2.58, 5.41) and (5.41, 2.58) are the intersection points.

Algebraiclly: $\sqrt{36 - x^2} = 8 - x$

$$36 - x^2 = 64 - 16x + x^2$$
$$2x^2 - 16x + 28 = 0$$
$$x^2 - 8x + 14 = 0$$
$$x = \frac{8 \pm \sqrt{64 - 56}}{2}$$
$$x = \frac{8 \pm 2\sqrt{2}}{2}$$
$$x = 4 \pm \sqrt{2}$$
$$x = 8 - x$$
$$x = 8 - (4 \pm \sqrt{2})$$
$$y = 4 \mp \sqrt{2}$$
$$\left(4 + \sqrt{2}, 4 - \sqrt{2}\right) \text{ and } \left(4 - \sqrt{2}, 4 + \sqrt{2}\right)$$

5. $\begin{cases} y = \sqrt{x} \\ y = 2 - x \end{cases}$

Graphically: Graph $y_1 = \sqrt{x}$ and $y_2 = 2 - x$.

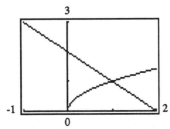

(1.00, 1.00) is the intersection point.

Algebraically: $\sqrt{x} = 2 - x$
$$x = 4 - 4x + x^2$$
$$x^2 - 5x + 4 = 0$$
If $x = 1$, then $y = 1$.
$$(1, 1)$$

7. $\begin{cases} x = 2y \\ x = y^2 - 2y \end{cases}$

Graphically: We must solve the second equation for y:
$$y^2 - 2y + 1 = x + 1$$
$$(y - 1)^2 = x + 1$$
$$y - 1 = \pm\sqrt{x + 1}$$
$$y = 1 \pm \sqrt{x + 1}$$
Graph $y_1 = \frac{x}{2}$, $y_2 = 1 + \sqrt{x + 1}$ and $y_3 = 1 - \sqrt{x + 1}$

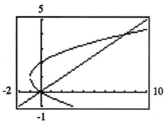

Graph (0.00, 0.00); (8.00, 4.00)
are the intersection points.

Algebraically:

$$2y = y^2 - 2y$$
$$y^2 - 4y = 0$$
$$y(y - 4) = 0$$
$$y = 0, y = 4$$

If $y = 0$, $x = 0$; if $y = 4$, $x = 8$

$(0, 0)$ and $(8, 4)$

9. $\begin{cases} x^2 + y^2 = 4 \\ x^2 + 2x + y^2 = 0 \end{cases}$

Graphically: Graph $y_1 = \sqrt{4 - x^2}$, $y_2 = -\sqrt{4 - x^2}$, $y_3 = \sqrt{-x^2 - 2x}$, $y_4 = -\sqrt{-x^2}$

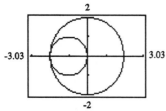

$(-2.00, 0.00)$ is the only
intersection point.

Algebraically:

Since $x^2 + y^2 = 4$, we use substitution:
$$2x + 4 = 0$$
$$2x = -4$$
$$x = -2$$
$$x^2 + y^2 = 4$$
$$(-2)^2 + y^2 = 4$$
$$4 + y^2 = 4$$
$$y^2 = 0$$
$$y = 0$$

11. $\begin{cases} y = 3x - 5 \quad (1) \\ x^2 + y^2 = 5 \quad (2) \end{cases}$

Graphically: Graph $y_1 = 3x - 5$, $y_2 = \sqrt{5 - x^2}$, and $y_3 = -\sqrt{5 - x^2}$

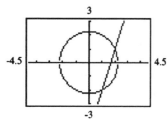

$(1.00, -2.00)$; $(2.00, 1.00)$ are
the intersection points.

Algebraically:

We cannot use elimination, since both x and
y are *squared* in one equation, but not in the
other. Instead, we use substitution:
$$y = 3x - 5 \qquad (1)$$
Now substitute this expression into (2):
$$x^2 + y^2 = 5$$
$$x^2 + (3x - 5)^2 = 5$$
$$x^2 + 9x^2 - 30x + 25 = 5$$
$$10x^2 - 30x + 20 = 0$$
$$10(x^2 - 3x + 2) = 0$$
$$10(x - 1)(x - 2) = 0$$
$$x - 1 = 0 \qquad x - 2 = 0$$
$$x = 1 \qquad x = 2$$
Now $y = 3x - 5$, so we have:
$$y = 3(1) - 5 \qquad \text{or} \qquad y = 3(2) - 5$$
$$y = -2 \qquad \text{or} \qquad y = 1$$
We have two possible solutions:
$$x = 1, x = -2$$
and $\qquad x = 2, y = 1$
Both solutions check, as you can verify.

13. $$\begin{cases} x^2 + y^2 = 4 & (1) \\ y^2 - x = 4 & (2) \end{cases}$$

Graphically: Graph $y_1 = \sqrt{4 - x^2}$, $y_2 = -\sqrt{4 - x^2}$, $y_3 = \sqrt{4 + x}$ and $y_4 = -\sqrt{4 + x}$.

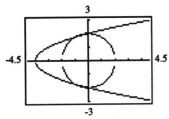

$(-1, 1.73)$; $(-1, -1.73)$; $(0.00, 2.00)$; $(0.00, -2.00)$ are the intersection points.

Algebraically:

$$\begin{cases} x^2 + y^2 = 4 & (1) \\ y^2 - x = 4 & (2) \end{cases}$$

$$\begin{cases} x^2 + y^2 = 4 & (1) \\ -y^2 + x = -4 & (2) \quad \text{Multiply by } -1. \end{cases}$$

$$\begin{cases} x^2 + y^2 = 4 & (1) \\ x^2 + x = 0 & (2) \quad \text{Replace (2) by (1) + (2).} \end{cases}$$

Now solve (2) for x:

$$x^2 + x = 0 \quad (2)$$
$$x(x + 1) = 0$$
$$x = 0 \quad \text{or} \quad x + 1 = 0$$
$$x = -1$$

Now use (1) to find y: If $x = 0$, then

$$x^2 + y^2 = 4 \quad (1)$$
$$0 + y^2 = 4$$
$$y^2 = 4$$
$$y = \pm 2$$

If $x = -1$, then

$$x^2 + y^2 = 4 \quad (1)$$
$$1 + y^2 = 4$$
$$y^2 = 3$$
$$y = \pm\sqrt{3}$$

We have four possible solutions:

$$x = 0 \text{ and } y = -2;$$
$$x = 0 \text{ and } y = 2;$$
$$x = -1 \text{ and } y = -\sqrt{3};$$
$$x = -1 \text{ and } y = \sqrt{3}$$

Let's check these:

For $x = 0$, $y = \pm 2$:

$$0^2 + (\pm 2)^2 = 0 + 4 = 4 \qquad (1)$$
$$(\pm 2)^2 - 0 = 4 - 0 = 4 \qquad (2)$$

For $x = -1$, $y = \pm\sqrt{3}$:

$$(-1)^2 + (\pm\sqrt{3})^2 = 1 + 3 = 4 \qquad (1)$$
$$(\pm\sqrt{3})^2 - (-1) = 3 + 1 = 4 \qquad (2)$$

Thus, all four possibilities check.

15. $\begin{cases} xy = 4 & (1) \\ x^2 + y^2 = 8 & (2) \end{cases}$

Graphically: Graph $y_1 = \dfrac{4}{x}$, $y_2 = \sqrt{8 - x^2}$ and $y_3 = -\sqrt{8 - x^2}$.

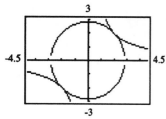

$(2.00, 2.00)$; $(-2.00, -2.00)$ are the intersection points.

Algebraically: Solve (1) for y: $xy = 4$ (1)

$$y = \frac{4}{x}$$

Substitute this into (2): $x^2 + y^2 = 8$

$$x^2 + \left[\frac{4}{x}\right]^2 = 8$$

$$x^2 + \frac{16}{x^2} = 8$$

$x^4 + 16 = 8x^2$ Multiply both sides by x^2.

$x^4 - 8x^2 + 16 = 0$ This is a disguised quadratic.

Let $u = x^2$ (so that $u^2 = x^4$): $u^2 - 8u + 16 = 0$

$$(u - 4)(u - 4) = 0$$

Therefore, $u = 4$, or $x^2 = 4$

$$x = \pm 2$$

If $x = 2$, then: $y = \dfrac{4}{x}$ (from (1))

$$y = \frac{4}{x} = 2$$

If $x = -2$, then: $y = \dfrac{4}{x}$

$$y = -2$$

We have two possible solutions: $x = 2$ and $y = 2$;

$x = -2$ and $y = -2$

both of which check.

17. $\begin{cases} x^2 + y^2 = 4 & (1) \\ y = x^2 - 9 & (2) \end{cases}$

Graphically: Graph $y = \sqrt{4 - x^2}$, $y_2 = -\sqrt{4 - x^2}$, $y_3 = x^2 - 9$.

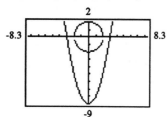

No solution; Inconsistent.

Algebraically: Let's substitute expression (2) into (1): $x^2 + (x^2 - 9)^2 = 4$
$$x^2 + x^4 - 18x^2 + 81 = 4$$
$$x^4 - 17x^2 + 77 = 0 \text{ This is quadratic.}$$

Use the quadratic formula: $x^2 = \dfrac{17 \pm \sqrt{289 - 4(77)}}{2}$

$$x^2 = \dfrac{17 \pm \sqrt{289 - 308}}{2} = \dfrac{17 \pm \sqrt{-19}}{2}$$

No real solution. Inconsistent.

19. $\begin{cases} y = x^2 - 4 & (1) \\ y = 6x - 13 & (2) \end{cases}$

Graphically: Graph $y_1 = x^2 - 4$ and $y_2 = 6x - 13$.

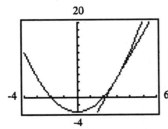

(3.00, 5.00) is the only intersection point.

Algebraically: Substituting equation (2) into equation (1): $6x - 13 = x^2 - 4$
$$x^2 - 6x + 9 = 0$$
$$(x - 3)^2 = 0$$
$$x = 3$$

Now, back-substitute this result into (2): $y = 6(3) - 13$
$$y = 5$$

We have **one** solution: $x = 3$ and $y = 5$

Now to check: $\begin{cases} 5 = (3)^2 - 4 = 5 & (1) \\ 5 = 6(3) - 13 = 5 & (2) \end{cases}$

The solution checks.

21. $\begin{cases} 2x^2 + y^2 = 18 & (1) \\ xy = 4 & (2) \end{cases}$

Solve equation (2) for y: $xy = 4$
$$y = \frac{4}{x}$$

Substitute this into (1): $2x^2 + y^2 = 18$

$$2x^2 + \left[\frac{4}{x}\right]^2 = 18$$

$$2x^2 + \frac{16}{x^2} = 18$$

$$2x^4 + 16 = 18x^2 \qquad \text{Multiply both sides by } x^2.$$
$$x^4 + 8 = 9x^2 \qquad \text{Divide all terms by 2.}$$
$$x^4 - 9x^2 + 8 = 0 \qquad \text{This is a disguised quadratic.}$$

Let $u = x^2$: $u^2 - 9u + 8 = 0$
$$(u - 1)(u - 8) = 0$$

Therefore, $u = 1$ or $u = 8$, so that $x^2 = 1$ or $x^2 = 8$, and

$$x = -1, 1, -\sqrt{8}, \sqrt{8}$$

If $x = -1$, then $y = \dfrac{4}{x} = -4$

If $x = 1$, then $y = \dfrac{4}{x} = 4$

If $x = -\sqrt{8} = -2\sqrt{2}$, then $y = \dfrac{4}{x} = \dfrac{4}{-2\sqrt{2}} = -\sqrt{2}$

If $x = \sqrt{8} = 2\sqrt{2}$, then $y = \sqrt{2}$

We have four *possible* solutions:

$x = -1$ and $y = -4$;

$x = 1$ and $y = 4$;

$x = -2\sqrt{2}$ and $y = -\sqrt{2}$;

and $x = 2\sqrt{2}$ and $y = \sqrt{2}$.

All four check, as you may verify.

23. $\begin{cases} y = 2x + 1 & (1) \\ 2x^2 + y^2 = 1 & (2) \end{cases}$

Solve for y in (1): $y = 2x + 1$

Substitute this expression for y into (2):

$$2x^2 + y^2 = 1$$
$$2x^2 + (2x + 1)^2 = 1$$
$$2x^2 + 4x^2 + 4x + 1 = 1$$
$$6x^2 + 4x = 0$$
$$2x + (3x + 2) = 0$$

So we have $x = 0$ or $x = -\dfrac{2}{3}$.

If $x = 0$, then:

$y = 2x + 1 \qquad$ from (1)

$y = 2(0) + 1$

$y = 1$

If $x = -\dfrac{2}{3}$, then:

$y = 2x + 1$

$y = 2\left[-\dfrac{2}{3}\right] + 1$

$y = -\dfrac{1}{3}$

We have two possible solutions: $x = 0$ and $y = 1$

and $x = -\dfrac{2}{3}$ and $y = -\dfrac{1}{3}$

Let's check these:

For $x = 0$, $y = 1$: $\begin{cases} 1 = 2(0) + 1 & (1) \\ 2(0)^2 + (1) = 1 & (2) \end{cases}$

For $x = -\dfrac{2}{3}$ and $y = -\dfrac{1}{3}$: $\qquad -\dfrac{1}{3} = 2\left[-\dfrac{2}{3}\right] + 1 \qquad\qquad (1)$

$$2\left[-\dfrac{2}{5}\right]^2 + \left[-\dfrac{1}{3}\right]^2 = \dfrac{8}{9} + \dfrac{1}{9} = 1 \quad (2)$$

Both solutions check.

25. $\begin{cases} x + y + 1 = 0 & (1) \\ x^2 + y^2 + 6y - x = -5 & (2) \end{cases}$

We could add (1) and (2) and eliminate the x-terms, but we would still have an x^2-term. Instead, let's solve (1) for x:

$$x + y + 1 = 0$$
$$x = -y - 1$$

Now substitute this expression for x into (2):
$$x^2 + y^2 + 6y - x = -5$$
$$(-y - 1)^2 + y^2 + 6y - (-y - 1) = -5$$
$$y^2 + 2y + 1 + y^2 + 6y + y + 1 = -5 \qquad \text{This is a quadratic.}$$
$$2y^2 + 9y + 7 = 0$$
$$(2y + 7)(y + 1) = 0$$

So we have $y = \dfrac{-7}{2}$ or $y = -1$.

If $y = \dfrac{-7}{2}$, then: $\qquad x = -y - 1 \qquad\qquad \text{from (1)}$

$$x = \frac{7}{2} - 1 = \frac{5}{2}$$

If $y = -1$, then: $\qquad x = -y - 1$
$$x = 1 - 1 = 0$$

We have two possible solutions: $x = \dfrac{5}{2}$ and $y = \dfrac{-7}{2}$; $x = 0$ and $y = -1$

You can verify that both of these solutions check in the original system of equations.

27. $\begin{cases} 4x^2 - 3xy + 9y^2 = 15 & (1) \\ \quad\ 2x + 3y = 5 & (2) \end{cases}$

Solve for x in (2): $\qquad 2x + 3y = 5$
$$2x = -3y + 5$$
$$x = \frac{-3}{2}y + \frac{5}{2}$$

Substitute this expression into (1):
$$4x^2 - 3xy + 9y^2 = 15$$
$$4\left[\frac{-3}{2}y + \frac{5}{2}\right]^2 - 3\left[\frac{-3}{2}y + \frac{5}{2}\right]y + 9y^2 = 15$$
$$4\left[\frac{9}{4}y^2 - \frac{15}{2}y + \frac{25}{4}\right] - 3\left[\frac{-3}{2}y^2 + \frac{5}{2}y\right] + 9y^2 = 15$$
$$9y^2 - 30y + 25 + \frac{9}{2}y^2 - \frac{15}{2}y + 9y^2 = 15$$
$$18y^2 - 60y + 50 + 9y^2 - 15y + 18y^2 = 30$$
$$45y^2 - 75y + 20 = 0$$
$$9y^2 - 15y + 4 = 0$$
$$(3y - 1)(3y - 4) = 0$$

So $y = \dfrac{1}{3}$ or $y = \dfrac{4}{3}$

If $y = \dfrac{1}{3}$, then: $\quad x = \dfrac{-3}{2}y + \dfrac{5}{2} \qquad \text{from (2)}$

$$x = \frac{-3}{2}\left[\frac{1}{3}\right] + \frac{5}{2} = 2$$

If $y = \dfrac{4}{3}$, then: $\quad x = \dfrac{-3}{2}y + \dfrac{5}{2}$

$$x = \frac{-3}{2}\left[\frac{4}{3}\right] + \frac{5}{2} = \frac{1}{2}$$

The two possible solutions are: $x = 2$ and $y = \dfrac{1}{3}$; $x = \dfrac{1}{2}$ and $y = \dfrac{4}{3}$

Both of these solutions check.

29.
$$\begin{cases} x^2 - 4y^2 + 7 = 0 & (1) \\ 3x^2 + y^2 - 31 = 0 & (2) \end{cases}$$

Here we can use the method of elimination:

$$\begin{cases} x^2 - 4y^2 + 7 = 0 & (1) \\ 12x^2 + 4y^2 - 124 = 0 & (2) \quad \text{Multiply both sides by 4.} \end{cases}$$

$$\begin{cases} x^2 - 4y^2 + 7 = 0 & (1) \\ 13x^2 - 117 = 0 & (2) \quad \text{Replace (2) by (1) + (2).} \end{cases}$$

Now solve (2) for x:

$$13x^2 - 117 = 0$$
$$13x^2 = 117$$
$$x^2 = 9$$

so, $x = 3$ or $x = -3$

If $x = 3$, then:
$$x^2 - 4y^2 + 7 = 0 \quad (1)$$
$$9 - 4y^2 + 7 = 0$$
$$-4y^2 = -16$$
$$y^2 = 4$$
$$y = \pm 2$$

If $x = -3$, then:
$$x^2 - 4y^2 + 7 = 0$$
$$9 - 4y^2 + 7 = 0$$
$$-4y^2 = -16$$
$$y^2 = 4$$
$$y = \pm 2$$

We have four possible solutions:
$$x = 3, y = 2;$$
$$x = 3, y = -2;$$
$$x = -3, y = 2;$$
$$x = -3, y = -2$$

We can check all four at once. If $x = \pm 3$, and $y = \pm 2$:

$$\begin{cases} x^2 - 4y^2 + 7 = 9 - 16 + 7 = 0 & (1) \\ 3x^2 + y^2 - 31 = 27 + 4 - 31 = 0 & (2) \end{cases}$$

All four solutions check.

31.
$$\begin{cases} 7x^2 - 3y^2 + 5 = 0 & (1) \\ 3x^2 + 5y^2 - 12 = 0 & (2) \end{cases}$$

Let's solve (2) for x^2:

$$3x^2 + 5y^2 - 12 = 0$$
$$3x^2 = -5y^2 + 12$$
$$x^2 = \frac{-5}{3}y^2 + 4$$

Substitute this expression for x^2 into (1):

$$7x^2 - 3y^2 + 5 = 0$$
$$7\left[\frac{-5}{3}y^2 + 4\right] - 3y^2 + 5 = 0$$
$$\frac{-35}{3}y^2 + 28 - 3y^2 + 5 = 0$$
$$-35y^2 + 84 - 9y^2 + 15 = 0 \qquad \text{Multiply by 3.}$$
$$-44y^2 = -99$$
$$y^2 = \frac{99}{44} = \frac{9}{4}$$
$$y = \pm\frac{3}{2}$$

If $y = \dfrac{3}{2}$, then: $\qquad x^2 = \dfrac{-5}{3}y^2 + 4 \qquad\qquad$ from (2)

$$x^2 = \dfrac{-5}{3}\left[\dfrac{9}{4}\right] + 4$$

$$x^2 = \dfrac{1}{4}$$

$$x = \pm\dfrac{1}{2}$$

If $y = \dfrac{-3}{2}$, then: $\qquad x^2 = \dfrac{-5}{3}y^2 + 4$

$$x^2 = \dfrac{-5}{3}\left[\dfrac{9}{4}\right] + 4, \text{ as above}$$

$$x = \pm\dfrac{1}{2}$$

We have four solutions: $\quad x = \dfrac{1}{2}, y = \dfrac{3}{2}; x = -\dfrac{1}{2}, y = \dfrac{3}{2};$

$$x = \dfrac{1}{2}, y = \dfrac{-3}{2}; x = \dfrac{-1}{2}, y = \dfrac{-3}{2}$$

(All of them check.)

33. $\begin{cases} x^2 + 2xy = 10 & (1) \\ 3x^2 - xy = 2 & (2) \end{cases}$

$\begin{cases} x^2 + 2xy = 10 & (1) \\ 6x^2 - 2xy = 4 & (2) \quad \text{Multiply by 2.} \end{cases}$

$\begin{cases} x^2 + 2xy = 10 & (1) \\ 7x^2 = 14 & (2) \quad \text{Replace (2) by (1) + (2).} \end{cases}$

From (2): $\qquad x^2 = 2$

$$x = \pm\sqrt{2}$$

If $x = \sqrt{2}$, then: $\qquad x^2 + 2xy = 10 \qquad\qquad (1)$

$$2 + 2\sqrt{2}\,y = 10$$

$$2\sqrt{2}\,y = 8$$

$$y = \dfrac{8}{2\sqrt{2}} = 2\sqrt{2}$$

If $x = -\sqrt{2}$, then: $\qquad x^2 + 2xy = 10$

$$2 - 2\sqrt{2}\,y = 10$$

$$-2\sqrt{2}\,y = 8$$

$$y = \dfrac{8}{-2\sqrt{2}} = -2\sqrt{2}$$

The two possible solutions are: $\quad x = \sqrt{2}, y = 2\sqrt{2}$ and $x = -\sqrt{2}, y = -2\sqrt{2}$
Both solutions check.

35. $\begin{cases} 2x^2 + y^2 = 2 & (1) \\ x^2 - 2y^2 + 8 = 0 & (2) \end{cases}$

$\begin{cases} 4x^2 + 2y^2 = 4 & (1) \quad \text{Multiply equation (1) by 2.} \\ x^2 - 2y^2 = -8 & (2) \end{cases}$

Adding equation (1) and (2): $\qquad 5x^2 = -4$

$$x^2 = \frac{-4}{5}$$

No solution. The system is inconsistent.

37. $\begin{cases} x^2 + 2y^2 = 16 & (1) \\ 4x^2 - y^2 = 24 & (2) \end{cases}$

$\begin{cases} x^2 + 2y^2 = 16 & (1) \\ 8x^2 - 2y^2 = 48 & (2) \quad \text{Multiply by 2.} \end{cases}$

$\begin{cases} x^2 + 2y^2 = 16 & (1) \\ \qquad 9x^2 = 64 & (2) \quad \text{Replace (2) by (1) + (2).} \end{cases}$

From (2): $\qquad x^2 = \dfrac{64}{9}$

$$x = \pm\frac{8}{3}$$

If $x = \pm\dfrac{8}{3}$, then:

$$x^2 + 2y^2 = 16 \qquad (1)$$

$$\frac{64}{9} + 2y^2 = 16$$

$$2y^2 = \frac{144}{9} - \frac{64}{9} = \frac{80}{9}$$

$$y^2 = \frac{40}{9}$$

$$\text{and} \quad y = \pm\sqrt{\frac{40}{9}} = \pm\frac{-2\sqrt{10}}{3}$$

We have four solutions, all of which check:

$$x = \frac{8}{3},\ y = \frac{2\sqrt{10}}{3};\ x = \frac{8}{3},\ y = \frac{-2\sqrt{10}}{3};$$

$$x = \frac{-8}{3},\ y = \frac{2\sqrt{10}}{3};\ x = \frac{-8}{3},\ y = \frac{-2\sqrt{10}}{3}$$

39. $\begin{cases} \dfrac{5}{x^2} - \dfrac{2}{y^2} + 3 = 0 & (1) \\[2mm] \dfrac{3}{x^2} + \dfrac{1}{y^2} - 7 = 0 & (2) \end{cases}$

$\begin{cases} \dfrac{5}{x^2} - \dfrac{2}{y^2} + 3 = 0 & (1) \\[2mm] \dfrac{6}{x^2} + \dfrac{2}{y^2} - 14 = 0 & (2) \quad \text{Multiply both sides by 2.} \end{cases}$

$\begin{cases} \dfrac{5}{x^2} - \dfrac{2}{y^2} + 3 = 0 & (1) \\[2mm] \dfrac{11}{x^2} \qquad - 11 = 0 & (2) \quad \text{Replace (2) by (1) + (2).} \end{cases}$

Chapter 11 Systems of Equations and Inequalities

Now, from (2):
$$\frac{11}{x^2} = 11$$
$$11 = 11x^2$$
$$x^2 = 1$$
$$x = \pm 1$$

If $x = 1$ or $x = -1$, we have:
$$\frac{5}{x^2} - \frac{2}{y^2} + 3 = 0 \quad (9((1)$$
$$\frac{5}{1} - \frac{2}{y^2} + 3 = 0 \quad (\text{since } x = \pm 1)$$
$$8 = \frac{2}{y^2}$$
$$8y^2 = 2$$
$$y^2 = \frac{1}{4}$$
$$y = \pm \frac{1}{2}$$

We have four solutions, all of which can be checked:
$$x = 1, y = \frac{1}{2}; \; x = 1, y = \frac{-1}{2}; \; x = -1, y = \frac{1}{2}; \; x = -1, y = \frac{-1}{2}$$

41.
$$\begin{cases} \dfrac{1}{x^4} + \dfrac{6}{y^4} = 6 & (1) \\[2mm] \dfrac{2}{x^4} - \dfrac{2}{y^4} = 19 & (2) \end{cases}$$

Let's make a substitution: let $u = \dfrac{1}{x^4}$ and $v = \dfrac{1}{y^4}$. Then we have:

$$\begin{cases} u + 6v = 6 & (1) \\ 2u - 2v = 19 & (2) \end{cases}$$

$$\begin{cases} u + 6v = 6 & (1) \\ 6u - 6v = 57 & (2) \end{cases} \quad \text{Multiply each term by 3.}$$

$$\begin{cases} u + 6v = 6 & (1) \\ 7u = 63 & (2) \end{cases} \quad \text{Replace (2) by (1) + (2).}$$

From (2): $7u = 63$
$$u = 9$$

Then, from (1): $u + 6v = 6$
$$9 + 6v = 6$$
$$6v = -3$$
$$v = \frac{-1}{2}$$

Recall that $u = \dfrac{1}{x^4}$ and $v = \dfrac{1}{y^4}$. Thus:
$$\frac{1}{y^4} = v = \frac{-1}{2}$$
$$2 = -y^4$$
$$y^4 = -2$$

This is impossible, since an even power can never be negative, so we have no real solution. The system is inconsistent.

43.
$$\begin{cases} x^2 - 3xy + 2y^2 = 0 & \text{(1)} \\ \quad x^2 + xy = 6 & \text{(2)} \end{cases}$$

Since both x and y appear in two terms of equation (1), we cannot use elimination. We can either solve for y in (2), or factor (1). Let's try the latter approach:

$$x^2 - 3xy + 2y^2 = 0 \qquad \text{(1)}$$
$$(x - y)(x - 2y) = 0$$

Therefore, either $x = y$ or $x = 2y$.

If $x = y$, then from (2):
$$x^2 + xy = 6$$
$$y^2 + y^2 = 6 \quad (x = y)$$
$$y^2 = 3$$
$$y = \pm\sqrt{3}$$

We know $x = y$, so we have two possible solutions: $x = \sqrt{3}$, $y = \sqrt{3}$ and $x = -\sqrt{3}$, $y = -\sqrt{3}$.

If $x = 2y$, then:
$$x^2 + xy = 6 \qquad \text{(2)}$$
$$4y^2 + 2y^2 = 6 \qquad (x = 2y)$$
$$y^2 = 1$$
$$y = \pm 1$$

Then, from $x = 2y$, we have two more possible solutions: $x = 2, y = 1$ and $x = -2, y = -1$
You can check that all four solutions are valid.

45.
$$\begin{cases} xy - x^2 = -3 & \text{(1)} \\ 3xy - 4y^2 = 2 & \text{(2)} \end{cases}$$

Solve equation (1) for y:
$$xy - x^2 = -3 \qquad \text{(1)}$$
$$xy = x^2 - 3$$
$$y = \frac{x^2 - 3}{x}, \text{ provided } x \neq 0.$$

We will check the possibility $x = 0$ later.

Substitute $y = \dfrac{x^2 - 3}{x}$ into (2):

$$3xy - 4y^2 = 2 \qquad \text{(2)}$$
$$3x\left[\frac{x^2 - 3}{x}\right] - 4\left[\frac{x^2 - 3}{x}\right]^2 = 2$$
$$3x^2 - 9 - \frac{4(x^2 - 3)^2}{x^2} = 2$$
$$3x^4 - 9x^2 - 4(x^4 - 6x^2 + 9) = 2x^2 \quad \text{Multiply by } x^2.$$
$$3x^4 - 9x^2 - 4x^4 + 24x^2 - 36 - 2x^2 = 0$$
$$-x^4 + 13x^2 - 36 = 0$$
$$x^4 - 13x^2 + 36 = 0$$
$$(x^2 - 4)(x^2 - 9) = 0$$
$$x^2 = 4 \quad \text{or} \quad x^2 = 9$$
$$x = \pm 2 \quad \text{or} \quad x = \pm 3$$

If $x = 2$, then $\quad y = \dfrac{x^2 - 3}{x} \qquad$ or $\qquad y = \dfrac{4 - 3}{2} = \dfrac{1}{2}$

If $x = -2$, then $\quad y = \dfrac{x^2 - 3}{x} \qquad$ or $\qquad y = \dfrac{4 - 3}{-2} = \dfrac{-1}{2}$

If $x = 3$, then $\quad y = \dfrac{x^2 - 3}{x} \qquad$ or $\qquad y = \dfrac{9 - 3}{3} = 2$

If $x = -3$, then $\quad y = \dfrac{x^2 - 3}{x} \qquad$ or $\qquad y = \dfrac{9 - 3}{-3} = -2$

Chapter 11 Systems of Equations and Inequalities

This gives us four possible solutions. What about the possibility $x = 0$? From (1) we would have $0 = -3$. Thus, there are no solutions for which $x = 0$. Each of the four solutions above checks:

$$x = 2, y = \frac{1}{2}; x = -2, y = \frac{-1}{2}; x = 3, y = 2; x = -3, y = -2$$

47. $\begin{cases} x^3 - y^3 = 26 & (1) \\ x - y = 2 & (2) \end{cases}$

Solve (2) for x: $x - y = 2$
$$x = y + 2$$
and substitute this into (1): $x^3 - y^3 = 26$
$$(y + 2)^3 - y^3 = 26$$
$$y^3 + 6y^2 + 12y + 8 - y^3 = 26$$
$$6y^2 + 12y - 18 = 0$$
$$y^2 + 2y - 3 = 0$$
$$(y + 3)(y - 1) = 0$$

so $y = -3$ or $y = 1$.
If $y = -3$, then $x = y + 2 = -1$
If $y = 1$, then $x = y + 2 = 3$.
The two solutions (both of which check) are $x = -1, y = -3$, and $x = 3, y = 1$.

49. $\begin{cases} y^2 + y + x^2 - x - 2 = 0 & (1) \\ y + 1 + \dfrac{x - 2}{y} = 0 & (2) \end{cases}$

$\begin{cases} y^2 + y + x^2 - x - 2 = 0 & (1) \\ -y^2 - y - x + 2 = 0 & (2) \quad \text{Multiply both sides by } -y. \end{cases}$

$\begin{cases} y^2 + y + x^2 - x - 2 = 0 & (1) \\ x^2 - 2x = 0 & (2) \quad \text{Replace (2) by (1) + (2).} \end{cases}$

From (2): $x - 2x = 0$
$$x(x - 2) = 0$$
$$x = 0 \quad \text{or} \quad x = 2$$

Now use (1):
If $x = 0$: $y^2 + y + x^2 - x - 2 = 0$ (1)
$$y^2 + y - 2 = 0$$
$$(y + 2)(y - 1) = 0$$
  and $y = -2 \quad \text{or} \quad y = 1$
If $x = 2$: $y^2 + y + x^2 - x - 2 = 0$ (1)
$$y^2 + y + 4 - 2 - 2 = 0$$
$$y^2 + y = 0$$
$$y(y + 1) = 0$$
  and $y = 0 \quad \text{or} \quad y = -1$

Thus, we have four possible solutions:
$$x = 0, y = -2; x = 0, y = 1; x = 2, y = 0; \text{ and } x = 2, y = -1$$
We see from equation (2) that y cannot be zero. That eliminates the third solution above. The others can be checked:
$$x = 0, y = -2$$
$$x = 0, y = 1$$
$$x = 2, y = -1$$

51. $\begin{cases} \log_x y = 3 \\ \log_x (4y) = 5 \end{cases}$

$\begin{cases} y = x^3 \\ 4y = x^5 \end{cases}$

$$4x^3 = x^5$$
$$x^5 - 4x^3 = 0$$
$$x^3(x^2 - 4) = 0$$
$$x = 0, x = -2, x = 2$$

0 and -2 are extraneous (the base of a logarithm must be positive). Thus, $x = 2$ and $y = 2^3 = 8$.

53. Graph $y_1 = x^{2/3}$ and $y_2 = e^{-x}$. The only intersection point is $(0.48, 0.61)$.

55. Graph $y_1 = \sqrt[3]{2 - x^2}$ and $y_2 = \dfrac{4}{x^3}$. The only intersection point is $(-1.64, -0.89)$.

57. Graph $y_1 = \sqrt[4]{12 - x^4} = (12 - x^4)^{1/4}$, $y_2 = -(12 - x^4)^{1/4}$, $y_3 = \sqrt{\dfrac{2}{x}}$, and $y_4 = -\sqrt{\dfrac{2}{x}}$.
The intersection points are $(0.58, 1.85)$; $(1.81, 1.05)$; $(0.58, -1.85)$ and $(1.81, -1.05)$.

59. Graph $y_1 = \dfrac{2}{x}$ and $y_2 = \ln x$. The only intersection point is $(2.34, 0.85)$.

61. Let x and y be the two numbers. Then we have:

$$\begin{cases} x - y = 2 & (1) \\ x^2 + y^2 = 10 & (2) \end{cases}$$

Solve (1) for x: $\quad x - y = 2$
$$x = y + 2$$
Substitute this into (2): $\quad (y + 2)^2 + y^2 = 10 \quad (2)$
$$y^2 + 4y + 4 + y^2 = 10$$
$$2y^2 + 4y - 6 = 0$$
$$2(y^2 + 2y - 3) = 0$$
$$2(y + 3)(y - 1) = 0$$
$$y = -3 \text{ or } y = 1$$
If $y = -3$, then $x = y + 2 = -3 + 2 = -1$
If $y = 1$, then $x = y + 2 = 1 + 2 = 3$
The two numbers are 3 and 1 or -3 and -1.
$$3^2 + 1^2 = 10 \text{ and } (-3)^2 + (-1)^2 = 10$$
so both solutions check.

63. Let x and y be the two numbers. Then we have:

$$\begin{cases} xy = 4 & (1) \\ x^2 + y^2 = 8 & (2) \end{cases}$$

Solve for y in (1): $\quad y = \dfrac{4}{x}$

Then from (2):
$$x^2 + y^2 = 8$$
$$x^2 + \left[\frac{4}{x}\right]^2 = 8$$
$$x^2 + \frac{16}{x^2} = 8$$
$$x^4 + 16 = 8x^2 \qquad \text{Multiply by } x^2.$$
$$x^4 - 8x^2 + 16 = 0$$
$$(x^2 - 4)(x^2 - 4) = 0$$
$$x^2 = 4$$
$$x = \pm 2$$

Recall that $y = \dfrac{4}{x}$.

If $x = 2$, then $y = 2$. If $x = -2$, then $y = -2$.

Thus, we have *two* pairs of numbers: 2 and 2, or -2 and -2. Both pairs solve the original problem.

65. Let x and y be the two numbers. Then:

$$\begin{cases} x - y = xy & (1) \\ \dfrac{1}{x} + \dfrac{1}{y} = 5 & (2) \end{cases}$$

$$\begin{cases} x - y = xy & (1) \\ y + x = 5xy & (2) \end{cases} \quad \text{Multiply by } xy.$$

$$\begin{cases} x - y = xy & (1) \\ 2x = 6xy & (2) \end{cases} \quad \text{Replace (2) by (1) + (2).}$$

From (2) we have:
$$2x - 6xy = 0$$
$$x - 3xy = 0$$
$$x(1 - 3y) = 0$$

So either $x = 0$ or $y = \dfrac{1}{3}$. But from the original equation (2), we see that $x = 0$ is impossible.
Thus, $y = \dfrac{1}{3}$.

From (1): $\quad x - y = xy \quad (1)$

$$x - \frac{1}{3} = \frac{1}{3}x \qquad \left[y = \frac{1}{3}\right]$$
$$\frac{2}{3}x = \frac{1}{3}$$
$$x = \frac{1}{2}$$

Thus, the two numbers are $\dfrac{1}{2}$ and $\dfrac{1}{3}$.

67. $\begin{cases} \dfrac{a}{b} = \dfrac{2}{3} \\ a + b = 10 \end{cases}$

$$\frac{10 - b}{b} = \frac{2}{3}$$
$$3(10 - b) = 2b$$
$$30 = 5b$$
$$b = 6$$
$$a = 4$$
$$a + b = 10, \; b - a = 2$$

Ratio of $a + b$ to $b - a$ is $\dfrac{10}{2} = 5$.

69. $\begin{cases} x + 2y = 0 & (1) \\ (x - 1)^2 + (y - 1)^2 = 5 & (2) \end{cases}$

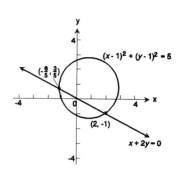

The line $x + 2y = 0$ can be rewritten as $y = \dfrac{-1}{2}x$,

so we see it is a line with slope $m = \dfrac{-1}{2}$ which passes through $(0, 0)$.

Also, $(x - 1)^2 + (y - 1)^2 = 5$ is a circle with center $C(1, 1)$ and radius $r = \sqrt{5} \approx 2.24$.

To find the points of intersection, substitute $y = \dfrac{-1}{2}x$ (from (1)) into (2):

$$(x - 1)^2 + (y - 1)^2 = 5 \quad (2)$$
$$(x - 1)^2 + \left[\frac{-1}{2}x - 1\right]^2 = 5$$
$$x^2 - 2x + 1 + \frac{1}{4}x^2 + x + 1 = 5$$
$$\frac{5}{4}x^2 - x - 3 = 0$$
$$5x^2 - 4x - 12 = 0 \quad \text{Multiply by 4.}$$
$$(5x + 6)(x - 2) = 0$$
$$\text{so } x = \frac{-6}{5}, \text{ or } x = 2$$

If $x = \dfrac{-6}{5}$, then $y = \dfrac{-1}{2}x = \dfrac{3}{5}$

If $x = 2$, then $y = \dfrac{-1}{2}x = -1$

Both solutions check, so we have two points of intersection: $\left[\dfrac{-6}{5}, \dfrac{3}{5}\right]$ and $(2, -1)$

71. $\begin{cases} (x - 1)^2 + (y + 2)^2 = 4 & (1) \\ y^2 + 4y - x + 1 = 0 & (2) \end{cases}$

Equation (1) is a circle with center at $C(1, -2)$ and radius $r = 2$. To graph the parabola, (2), let's complete the square and then plot points:

$$y^2 + 4y + \underline{} = x - 1 + \underline{}$$
$$y^2 + 4y + 4 = x - 1 + 4$$
$$(y + 2)^2 = x + 3$$
$$x = (y + 2)^2 - 3$$

y	$x = (y + 2)^2 - 3$	(x, y)
-4	1	$(1, -4)$
-3	-2	$(-2, -3)$
-2	-3	$(-3, -2)$
-1	-2	$(-2, -1)$
0	1	$(1, 0)$

To solve the system, let's use the standard form we obtained for the parabola:

$$\begin{cases} (x - 1)^2 + (y + 2)^2 = 4 & (1) \\ (y + 2)^2 - 3 = x & (2) \end{cases}$$

$$\begin{cases} (x - 1)^2 + 3 = 4 - x & (1) \quad \text{Replace (1) by (1) - (2)} \\ (y + 2)^2 - 3 = x & (2) \end{cases}$$

From (1): $\qquad\qquad (x - 1)^2 + 3 = 4 - x$

$$x - 2x + 1 + 3 = 4 - x$$
$$x^2 - x = 0$$
$$x(x - 1) = 0$$
$$x = 0 \text{ or } x = 1$$

If $x = 0$, then by (2): $\qquad (y + 2)^2 - 3 = x$

$$(y + 2)^2 - 3 = 0$$
$$(y + 2)^2 = 3$$
$$y + 2 = \pm\sqrt{3}$$
$$y = -2 \pm \sqrt{3}$$

If $x = 1$, then: $\qquad\qquad (y + 2)^2 - 3 = 1$

$$(y + 2)^2 = 4$$
$$y + 2 = \pm 2$$
$$y = -2 \pm 2$$

i.e., $y = 0$ or $y = -4$

We have *four* points of intersection: $\quad (0, -2 - \sqrt{3}\,), (0, -2 + \sqrt{3}\,), (1, 0),$ and $(1, -4)$

73.
$$\begin{cases} y = \dfrac{4}{x - 3} & (1) \\ x^2 - 6x + y^2 + 1 = 0 & (2) \end{cases}$$

To graph (1), plot points:

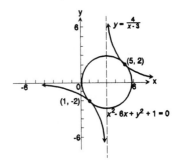

x	$y = \dfrac{4}{x - 3}$	(x, y)
6	$\dfrac{4}{3}$	$\left(6, \dfrac{4}{3}\right)$
4	$\dfrac{4}{1}$	$(4, 4)$
3	Undefined	
2	$\dfrac{4}{-1}$	$(2, -4)$
0	$\dfrac{4}{-3}$	$\left(0, -\dfrac{4}{3}\right)$

To graph (2), complete the square: $\quad x^2 - 6x + \underline{} + y^2 = -1 + \underline{}$

$$x^2 - 6x + 9 + y^2 = -1 + 9$$
$$(x - 3)^2 + y^2 = 8$$

This is a circle with center at $C(3, 0)$ and radius $r = \sqrt{8} \approx 2.83$.

Now solve the system:
$$\begin{cases} y = \dfrac{4}{x - 3} & (1) \\ x^2 - 6x + y^2 + 1 = 0 & (2) \end{cases}$$

We could substitute $y = \dfrac{4}{x - 3}$ into (2), but then we obtain a complicated denominator, $(x - 3)^2 = x^2 - 6x + 9$.

Instead, let's solve (1) for x: $y(x - 3) = 4$

$$c - 3 = \frac{4}{y}$$

$$x = \frac{4}{y} + 3$$

and substitute this into (2):

$$x^2 - 6x + y^2 + 1 = 0$$

$$\left[\frac{4}{y} + 3\right]^2 - 6\left[\frac{4}{y} + 3\right] + y^2 + 1 = 0$$

$$\frac{16}{y^2} + \frac{24}{y} + 9 - \frac{24}{y} - 18 + y^2 + 1 = 0$$

$$\frac{16}{y^2} + y^2 - 8 = 0$$

$$16 + y^4 - 8y^2 = 0 \quad \text{Multiply by } y^2.$$

$$y^4 - 8y^2 + 16 = 0$$

$$(y^2 - 4)(y^2 - 4) = 0$$

so $y^2 = 4$

$\quad y = \pm 2$

If $y = 2$, then $x = \dfrac{4}{y} + 3 = 5$.

If $y = -2$, then $x = \dfrac{4}{y} + 3 = 1$.

We have two points of intersection: (5, 2) and (1, −2)

75. $\ell = $ length, $w = $ width

$$\begin{cases} 2\ell + 2w = 16 & (1) \quad \text{(Perimeter)} \\ \ell w = 15 & (2) \quad \text{(Area)} \end{cases}$$

$2\ell + 2\left[\dfrac{15}{\ell}\right] = 16$ Substituting $w = \dfrac{15}{\ell}$ from (2) into equation (1).

$2\ell^2 - 16\ell + 30 = 0$ Multiply by ℓ.

$2(\ell^2 - 8\ell + 15) = 0$

$2(\ell - 5)(\ell - 3) = 0$

$\ell = 5 \qquad\qquad \ell = 3$

$w = \dfrac{15}{\ell} = 3 \qquad w = \dfrac{15}{3} = 5$

The dimensions are 3 inches $\times$ 5 inches, or 5 inches $\times$ 3 inches.

77. $\begin{cases} 2\pi r_1 + 2\pi r_2 = 12\pi \\ \pi r_1^2 + \pi r_2^2 = 20\pi \end{cases}$

$$2\pi(r_1 + r_2) = 12\pi \qquad (1)$$

$$r_1 + r_2 = \frac{12\pi}{2\pi}$$

$$r_1 + r_2 = 6$$

$$r_1 = 6 - r_2$$

$$\pi(r_1^2 + r_2^2) = 20\pi \qquad (2)$$

$$r_1^2 + r_2^2 = \frac{20\pi}{\pi}$$

$$r_1^2 + r_2^2 = 20$$

We substitute equation (1) into (2):

$$r_1^2 + r_2^2 = 20$$

$$(6 - r_2)^2 + r_2^2 = 20$$

$$36 - 12r_2 + r_2^2 + r_2^2 = 20$$

$$2r_2^2 - 12r_2 + 16 = 0$$

$$2(r_2^2 - 6r_2 + 8) = 0$$

$$2(r_2 - 2)(r_2 - 4) = 0$$

$$r_2 - 2 = 0 \qquad \text{or} \qquad r_2 - 4 = 0$$

$$r_2 = 2 \qquad \text{or} \qquad r_2 = 4$$

$$r_1 = 6 - r_2 \qquad\qquad r_1 = 6 - r_2$$

$$r_1 = 6 - 2 \qquad\qquad r_1 = 6 - 4$$

$$r_1 = 4 \qquad\qquad\quad r_1 = 2$$

The radius of each circle is 2 cm and 4 cm.

79. We know that rate $\times$ time = distance

The tortoise takes a total of $9 + 3 = 12$ *minutes* longer to complete the 21 meter race. In hours, 12 minutes is $\frac{12}{60}$ hour which is $\frac{1}{5}$ hour. The hare runs at a speed 0.5 meter per hour faster, which means the tortoise runs 0.5 meter per hour slower than the hare. Hence,

$$(r - 0.5)\left(t + \frac{1}{5}\right) = 21 \leftrightarrow \text{Tortoise}$$

$$rt = 21 \leftrightarrow \text{Hare}$$

$$r = \frac{21}{t}$$

$$\left(r - \frac{1}{2}\right)\left(t + \frac{1}{5}\right) = 21$$

$$\left(\frac{21}{t} - \frac{1}{2}\right)\left(t + \frac{1}{5}\right) = 21$$

$$21 + \frac{21}{5t} - \frac{t}{2} - \frac{1}{10} = 21$$

$$210t + 42 - 5t^2 - t = 210t$$

$$-5t^2 - t + 42 = 0$$

$$5t^2 + t - 42 = 0$$

$$(5t - 14)(t + 3) = 0$$

$$5t - 14 = 0 \qquad\qquad t = -3$$

$$t = \frac{14}{5} \qquad \text{Ignore because time cannot be negative}$$

$$= 2\frac{4}{5} \text{ hours}$$

$$r = \frac{21}{t} = \frac{21}{\frac{14}{5}} = \frac{105}{14} = 7.5 \frac{\text{meters}}{\text{hour}}$$

$$r - 0.5 = 7.5 - 0.5 = 7 \frac{\text{meters}}{\text{hour}}$$

The tortoise runs at an average speed of 7 meters per hour. The hare runs at an average speed of 7.5 meters per hour.

81. Let the piece of cardboard have width x, and length y. Then its area is $A = xy$. From the drawing, we see that the width of the box will be $w = x - 4$, its length will be $\ell = y - 4$, and its height will be $h = 2$. Therefore, its volume is

$$V = \ell \cdot w \cdot h = 2(x - 4)(y - 4)$$

We have: $\begin{cases} xy = 216 & (1) \\ 2(x - 4)(y - 4) = 224 & (2) \end{cases}$

Solve (1) for y: $y = \dfrac{216}{x}$

Then from (2): $2(x - 4)(y - 4) = 224$

$$(2x - 8)\left[\frac{216}{x} - 4\right] = 224$$

$$432 - 8x - \frac{1728}{x} + 32 = 224$$

$$432x - 8x^2 - 1728 + 32x = 224x \qquad \text{Multiply by } x.$$
$$-8x^2 + 240x - 1728 = 0$$
$$x^2 - 30x + 216 = 0 \qquad \text{Divide by } -8.$$

With some trial and error, this can be factored: $(x - 12)(x - 18) = 0$
so $x = 12$ or $x = 18$

If $x = 12$, then $y = \dfrac{216}{x} = 18$

If $x = 18$, then $y = \dfrac{216}{x} = 12$

Thus, the cardboard should be 12 centimeters by 18 centimeters.

83.
$$x^2 + y^2 = 4500$$
$$3x + 3y + (x - y) = 300$$
$$4x + 2y = 300$$
$$2y = 300 - 4x$$
$$y = 150 - 2x$$
$$x^2 + (150 - 2x)^2 = 4500$$
$$x^2 + 22{,}500 - 600x + 4x^2 = 4500$$
$$5x^2 - 600x + 22{,}500 = 4500$$
$$5x^2 - 600x + 18{,}000 = 0$$
$$5(x^2 - 120x + 3600) = 0$$
$$5(x - 60)^2 = 0$$
$$x = 60 \text{ ft.}$$
$$y = 150 - 2x$$
$$y = 150 - 2(60)$$
$$y = 30 \text{ ft.}$$

85. Let ℓ and w be the length and width of a rectangle. Then the area, A, and perimeter, P, are given by:

$$\begin{cases} A = \ell w & (1) \\ P = 2\ell + 2w & (2) \end{cases}$$

We solve for ℓ and w, treating A and P as constants:

From (2), $\qquad 2\ell = P - 2w$

$$\ell = \frac{P}{2} - w$$

Then from (1),

$$\ell = A \qquad (1)$$

$$\left(\frac{P}{2} - 3\right)w = A$$

$$\frac{P}{2}w - w^2 = A$$

$$-w^2 + \frac{P}{2}w - A = 0$$

$$w^2 - \frac{P}{2}w + A = 0$$

This is a quadratic, in w, with $a = 1$, $b = \dfrac{-P}{2}$ and $c = A$.

Then, $w = \dfrac{-b \pm \sqrt{b^2 - 4ac}}{2a} = \dfrac{\dfrac{P}{2} \pm \sqrt{\dfrac{P^2}{4} - 4A}}{2} = \dfrac{\dfrac{P}{2} \pm \sqrt{\dfrac{P^2 - 16A}{4}}}{2} = \dfrac{\dfrac{P}{2} - \dfrac{\sqrt{P^2 - 16A}}{2}}{2}$

or $\quad w = \dfrac{P \pm \sqrt{P^2 - 16A}}{4}$ Multiplying top and bottom by 2.

Recall that $\ell = \dfrac{P}{2} - w$.

If $w = \dfrac{P \pm \sqrt{P^2 - 16A}}{4}$, then we have

$$\ell = \frac{P}{2} - \frac{P}{4} + \frac{\sqrt{P^2 - 16A}}{4} \text{ or } \ell = \frac{P}{4} - \frac{\sqrt{P^2 - 16A}}{4} = \frac{P - \sqrt{P^2 - 16A}}{4}$$

This gives a length *smaller* than the width, so we reject this solution.

If $w = \dfrac{P - \sqrt{P^2 - 16A}}{4}$, then we obtain $\ell = \dfrac{P + \sqrt{P^2 - 16A}}{4}$, and these are the proper

formulas.

87. $\quad M^2 - 4(2M - 4) = 0$
$\qquad M^2 - 8M + 16 = 0$
$\qquad\quad (M - 4)^2 = 0$
$\qquad\qquad\quad M = 4$

Now, using the point-slope equation with slope 4 and the point, (2, 4),
$$y - 4 = 4(x - 2)$$
$$y = 4x - 8 + 4$$
$$y = 4x - 4$$

89. Refer to Problem 87 for the method to be used. We want the system

$$\begin{cases} y = x^2 + 2 & (1) \\ y = mx + b & (2) \end{cases}$$

to have one solution. Substitute $y = mx + b$ into the first equation:
$$mx + b = x^2 + 2$$
$$-x^2 + mx + b - 2 = 0$$
$$x^2 - mx + 2 - b = 0 \qquad \text{(a quadratic)}$$

Here $A = 1$, $B = -m$ and $C = 2 - b$. The quadratic formula will produce just one solution provided $B^2 - 4AC = 0$, i.e.,

$$(-m)^2 - 4(1)(2 - b) = 0$$
$$m^2 - 8 + 4b = 0 \qquad (1)$$

We also want $(1, 3)$ to lie on the line $y = mx + b$, so that $3 = m(1) + b$. We have two equations:

$$\begin{cases} m^2 - 8 + 4b = 0 & (1) \\ 3 = m + b & (2) \end{cases}$$

From (2), $b = 3 - m$, so, by (1):
$$m^2 - 8 + 4b = 0 \qquad (1)$$
$$m^2 - 8 + 4(3 - m) = 0$$
$$m^2 - 8 + 12 - 4m = 0$$
$$m^2 - 4m + 4 = 0$$
$$(m - 2)(m - 2) = 0$$

so that $m = 2$. Then $b = 3 - m - 1$, and the equation of the tangent line is: $y = mx + b$, or
$$y = 2x + 1$$

91.   Refer to Problems 87 and 89 to see the method used. The system

$$\begin{cases} 2x^2 + 3y^2 = 14 & (1) \\ y = mx + b & (2) \end{cases}$$

is to have just one solution. Substitute $y = mx + b$ into the first equation:

$$2x^2 + 3y^2 = 14$$
$$2x^2 + 3(mx + b)^2 = 14$$
$$2x^2 + 3(m^2x^2 + 2mbx + b^2) = 14$$
$$2x^2 + 3m^2x^2 + 6mbx + 3b^2 - 14) = 0$$
$$(2 + 3m^2)x^2 + (6mb)x + (3b^2 - 14) = 0 \quad \text{(a quadratic)}$$

Here $A = 2 + 3m^2$, $B = 6mb$ and $C = 3b^2 - 14$.
There will be **one** solution to the quadratic if $B^2 - 4AC = 0$, i.e.,

$$(6mb)^2 - 4(2 + 3m^2)(3b^2 - 14) = 0$$
$$36m^2b^2 - 4(6b^2 - 28 + 9m^2b^2 - 42m^2) = 0$$
$$36m^2b^2 - 24b^2 + 112 - 36m^2b^2 + 168m^2 = 0$$
$$-24b^2 + 112 + 168m^2 = 0$$
$$3b^2 - 14 - 21m^2 = 0 \quad \text{(1) Divide by } -8.$$

We also want $(1, 2)$ to lie on the line $y = mx + b$, i.e, we need
$$2 = m + b \quad (2)$$

We have two equations to solve:

$$\begin{cases} 3b^2 - 14 - 21m^2 = 0 & (1) \\ 2 = m + b & (2) \end{cases}$$

From (2), $b = 2 - m$, and by (1),
$$3b^2 - 14 - 21m^2 = 0$$
$$3(2 - m)^2 - 14 - 21m^2 = 0$$
$$3(4 - 4m + m^2) - 14 - 21m^2 = 0$$
$$12 - 12m + 3m^2 - 14 - 21m^2 = 0$$
$$-18m^2 - 12m - 2 = 0$$
$$9m^2 + 6m + 1 = 0$$
$$(3m + 1)(3m + 1) = 0$$

so $m = \dfrac{-1}{3}$ and $b = 2 - m = \dfrac{7}{3}$, and the equation of the tangent line is

$$y = mx + b, \text{ or}$$
$$y = \dfrac{-1}{3}x + \dfrac{7}{3}$$

93. We want the system
$$\begin{cases} x^2 - y^2 = 3 \\ y = mx + b \end{cases}$$
to have just *one* solution.

Substitute $y = mx + b$ into the first equation:
$$x^2 - y^2 = 3$$
$$x^2 - (mx + b)^2 = 3$$
$$x^2 - (m^2x^2 + 2mbx - b^2) = 3$$
$$x^2 - m^2x^2 + 2mbx - b^2 = 3$$
$$(1 - m^2)x^2 + (-2mb)x + (-b^2 - 3) = 0$$

Let $A = 1 - m^2$, $B = -2mb$, $C = -b^2 - 3$

We want $B^2 - 4AC = 0$, or
$$(-2mb)^2 - 4(1 - m^2)(-b^2 - 3) = 0$$
$$4m^2b^2 - 4(-b^2 - 3 + m^2b^2 + 3m^2) = 0$$
$$4b^2 + 12 - 12m^2 = 0$$
$$b^2 + 3 - 3m^2 = 0 \qquad (1)$$

We want $(2, 1)$ to lie on the line $y = mx + b$: $\quad 1 = 2m + b \qquad (2)$

This gives us the system:
$$\begin{cases} b^2 + 3 - 3m^2 = 0 & (1) \\ 1 = 2m + b & (2) \end{cases}$$

From (2): $\quad b = 1 - 2m$, and we have
$$b^2 + 3 - 3m^2 = 0 \qquad (1)$$
$$(1 - 2m)^2 + 3 - 3m^2 = 0$$
$$1 - 4m + 4m^2 + 3 - 3m^2 = 0$$
$$m^2 - 4m + 4 = 0$$
$$(m - 2)(m - 2) = 0$$
so $m = 2$, $b = 1 - 2m = -3$, and the tangent line is
$$y = mx + b, \text{ or}$$
$$y = 2x - 3$$

95. Solve for r_1 and r_2:
$$\begin{cases} r_1 + r_2 = \dfrac{-b}{a} & (1) \\[2mm] r_1r_2 = \dfrac{c}{a} & (2) \end{cases}$$

From (1),
$$r_1 = -r_2 - \frac{b}{a}, \text{ and}$$
$$r_1r_2 = \frac{c}{a} \qquad (2)$$
$$\left[-r_2 - \frac{b}{a}\right] r_2 = \frac{c}{a}$$
$$-r_2^2 - \frac{b}{a}r_2 - \frac{c}{a} = 0$$
$$ar_2^2 + br_2 + c = 0 \quad \text{Multiply by } -a.$$

By the quadratic formula,
$$r_2 = \frac{-b \pm \sqrt{b^2 - 4ac}}{2a}$$

Then,

$$r_1 = -r_2 - \frac{b}{a} = -\left[\frac{-b \pm \sqrt{b^2 - 4ac}}{2a}\right] - \frac{2b}{2a} = \frac{b \mp \sqrt{b^2 - 4ac} - 2b}{2a}$$

$$= \frac{-b \mp \sqrt{b^2 - 4ac}}{2a}$$

Thus, we have one pair of numbers:

$$\frac{-b + \sqrt{b^2 - 4ac}}{2a} \quad \text{and} \quad \frac{-b - \sqrt{b^2 - 4ac}}{2a}$$

11.7   Systems of Inequalities

1. $x \geq 0$:

Step 1:   Graph the line $x = 0$ (a vertical line through the origin, i.e., the y-axis). Use a solid line, since the inequality is nonstrict.

Step 2:   Choose a test point not on the line, say, the point $(2, 0)$, which is to the right of the line.

Step 3:   For this point we have $x = 2 \geq 0$, so that the inequality *is* satisfied.

Step 4:   Therefore, we shade to the right of the line $x = 0$.

3. $x \geq 4$:

Step 1:   Graph the line $x = 4$, using a solid line (for a nonstrict inequality). The x-intercept is 4.

Step 2:   Choose a test point say, $(5, 0)$, which is to the right of the line.

Step 3:   For this point $(5, 0)$, we have: $5 \geq 4$, so the inequality is satisfied.

Step 4:   Therefore, we shade the region to the right of the line $x = 4$.

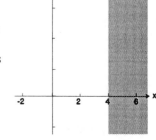

5. $2x + y \geq 6$:

Step 1:   Graph $2x + y = 6$, using a *solid* line. The x-intercept is $x = 3$, and the y-intercept is $y = 6$.

Step 2:   Choose a test point, say, $(0, 0)$, which is below the line.

Step 3:   For this point $(0, 0)$, we have: $2x + y = 0 < 6$, so the point does *not* satisfy the inequality.

Step 4:   Therefore, we shade the region *above* the line $2x + y = 6$.

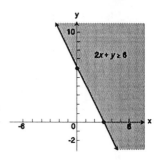

7.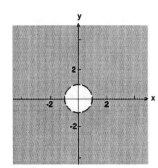

9.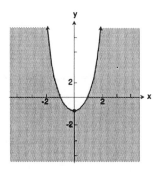

11.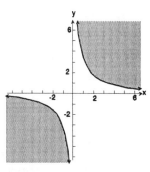

13. $\begin{cases} x + y \le 2 & (1) \\ 2x + y \ge 4 & (2) \end{cases}$

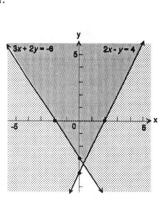

(a) $x + y \le 2$:

Step 1: Graph $x + y = 2$ with a solid line. The x-intercept is $x = 2$, and the y-intercept is $y = 2$.

Step 2: We will use $(0, 0)$, below the line, as a test point.

Step 3: At $(0, 0)$, $x + y = 0 < 2$, so the inequality is satisfied.

Step 4: Shade the region below the line $x + y = 2$.

(b) $2x + y \ge 4$:

Step 1: Graph $2x + y = 4$ with a solid line. The x-intercept is $x = 2$, and the y-intercept is $y = 4$.

Step 2: Use $(0, 0)$, below the line, as a test point.

Step 3: At $(0, 0)$, $2x + y = 0 < 4$, so the inequality is *not* satisfied.

Step 4: Shade the region above the line.

(c) Where the two shaded regions overlap is the graph of the system.

15. $\begin{cases} 2x - y \le 4 & (1) \\ 3x + 2y \ge -6 & (2) \end{cases}$

(a) $2x - y \le 4$:

Step 1: Graph $2x - y = 4$ with a solid line. The x-intercept is $x = 2$, and the y-intercept is $y = -4$.

Step 2: We will use $(0, 0)$ as a test point (above the line).

Step 3: At $(0, 0)$, $2x - y = 0 < 4$.

Step 4: Shade the region above the line $2x - y = 4$.

(b) $3x + 2y \ge -6$:

Step 1: Graph $3x + 2y = -6$ with a solid line. The x-intercept is $x = -2$, and the y-intercept is $y = -3$.

Step 2: Use $(0, 0)$, above the line.

Step 3: At $(0, 0)$, $3x + 2y = 0 > -6$.

Step 4: Shade the region above the line.

17. $\begin{cases} 2x - 3y \le 0 & (1) \\ 3x + 2y \le 6 & (2) \end{cases}$

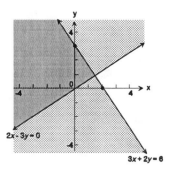

(a) $2x - 3y \le 0$:

    Step 1:   Graph $2x - 3y = 0$, a line through $(0, 0)$ with slope $= \dfrac{2}{3}$.

    Step 2:   Let's use $(1, 0)$ below the line.

    Step 3:   At $(1, 0)$, $2x - 3y = 2 > 0$, so the point does not satisfy the inequality.

    Step 4:   Shade the region *above* the line $2x - 3y = 0$.

(b) $3x + 2y \le 6$:

    Step 1:   Graph $3x + 2y = 6$ with a solid line. The x-intercept is $x = 2$, and the y-intercept is $y = 3$.

    Step 2:   We can use $(0, 0)$, below the line.

    Step 3:   At $(0, 0)$, $3x + 2y = 0 < 6$.

    Step 4:   Shade the region *below* the line $3x + 2y = 6$.

19. $\begin{cases} x^2 + y^2 \le 9 \\ x + y \ge 3 \end{cases}$

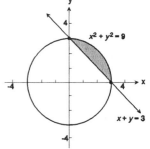

(a) $x^2 + y^2 \le 9$:

    Step 1:   Graph circle $x^2 + y^2 = 9$ with radius $= 3$.

    Step 2:   Use $(0, 0)$ inside the circle.

    Step 3:   At $(0, 0)$, $0 \le 9$, so the point does satisfy the inequality.

    Step 4:   Shade the region *inside* circle $x^2 + y^2 \le 9$.

(b) $x + y \ge 3$:

    Step 1:   Graph solid line $x + y = 3$. The x-intercept is $x = 3$ and the y-intercept is $y = 3$.

    Step 2:   Use $(3, 2)$ above the line.

    Step 3:   At $(3, 2)$, $5 \ge 3$, so the point does satisfy the inequality.

    Step 4:   Shade the region *above* the line $x + y = 3$.

21. $\begin{cases} y \ge x^2 - 4 \\ y \le x - 2 \end{cases}$

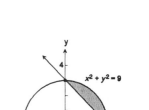

(a) $y \ge x^2 - 4$:

    Step 1:   Graph parabola $y = x^2 - 4$ with vertex $(0, -4)$ and going upward.

    Step 2:   Use $(0, 2)$ inside the parabola.

    Step 3:   At $(0, 2)$, $2 \ge -4$ so the point satisfies the inequality.

    Step 4:   Shade the region *inside* the prarabola $y = x^2 - 4$

(b) $y \le x - 2$:

    Step 1:   Graph the line $y = x - 2$ with x-intercept $x = 2$ and y-intercept $y = -2$.

    Step 2:   Use $(0, -3)$ below the line.

    Step 3:   At $(0, -3)$, $-3 \le -2$ so the point satisfies the inequality.

    Step 4:   Shade the region *below* the line $y = x - 2$.

23. $\begin{cases} xy \geq 4 \\ y \geq x^2 + 1 \end{cases}$

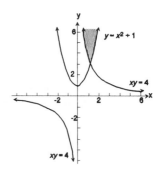

(a) $xy \geq 4$:
  Step 1: Graph $xy = 4$ using solid lines.
  Step 2: Use $(0, 0)$ between parts of the graph.
  Step 3: At $(0, 0)$, $0 \geq 4$ is false.
  Step 4: Shade the region *outside* the parts of the graph $xy = 4$.

(b) $y \geq x^2 + 1$:
  Step 1: Graph $y = x^2 + 1$. It is an upward parabola with vertex $(0, 1)$.
  Step 2: Use $(0, 2)$ inside the parabola.
  Step 3: At $(0, 2)$ $2 \geq 1$ is true.
  Step 4: Shade the region *inside* the parabola $y = x^2 + 1$.

25. $\begin{cases} x - 2y \leq 6 \quad (1) \\ 2x - 4y \geq 0 \quad (2) \end{cases}$

(a) $x - 2y \leq 6$:
  Step 1: Graph $x - 2y = 6$, using a solid line. The x-intercept is $x = 6$; the y-intercept is $y = -3$.
  Step 2: Use $(0, 0)$ above the line.
  Step 3: At $(0, 0)$, $x - 2y = 0 < 6$.
  Step 4: Shade the region *above* the line $x - 2y = 6$.

(b) $2x - 4y \geq 0$:
  Step 1: Graph $2x - 4y = 0$ with a solid line. This is a line through $(0, 0)$ with slope $= \dfrac{1}{2}$.
  Step 2: Let's use $(1, 0)$ below the line.
  Step 3: At $(1, 0)$, $2x - 4y = 2 > 0$.
  Step 4: Shade the region *below* $2x - 4y = 0$.

27. $\begin{cases} 2x + y \geq -2 \quad (1) \\ 2x + y \geq 2 \quad (2) \end{cases}$

(a) $2x + y \geq -2$:
  Step 1: Graph $2x + y = -2$, with a solid line. The x-intercept is $x = -1$; the y-intercept is $y = -2$.
  Step 2: Use $(0, 0)$ above the line.
  Step 3: At $(0, 0)$, $2x + y = 0 > -2$, so the inequality is satisfied.
  Step 4: Shade the region above the line $2x + y = -2$.

(b) $2x + y \geq 2$:
  Step 1: Graph $2x + y = 2$ with a solid line. The x-intercept is $x = 1$; the y-intercept is $y = 2$.
  Step 2: Use $(0, 0)$ *below* the line.
  Step 3: At $(0, 0)$, $2x + y = 0 < 2$, so $(0, 0)$ does *not* satisfy the inequality.
  Step 4: Therefore, shade the region *above* the line $2x + y = 2$.

29. $\begin{cases} 2x + 3y \geq 6 & (1) \\ 2x + 3y \leq 0 & (2) \end{cases}$

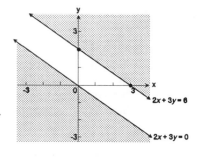

This system has no solution, because $2x + 3y$ cannot be greater than 6 **and** less than 0 at the same time.

(a) $2x + 3y \geq 6$:
  Step 1: Graph $2x + 3y = 6$, with a solid line. The x-intercept is $x = 3$; the y-intercept is $y = 2$.
  Step 2: Use $(0, 0)$ below the line as a test point.
  Step 3: At $(0, 0)$, $2x + 3y = 0 < 6$, so the inequality is **not** satisfied.
  Step 4: Shade the region **above** the line $2x + 3y = 6$.

(b) $2x + 3y \leq 0$:
  Step 1: Graph $2x + 3y = 0$ with a solid line. This is a line through $(0, 0)$ with slope $= \dfrac{-2}{3}$.
  Step 2: Use $(1, 0)$ above the line as a test point.
  Step 3: At $(1, 0)$, $2x + 3y = 2 > 0$.
  Step 4: Therefore, we shade the region **below** the line $2x + 3y = 0$.

(c)   Notice that the two shaded regions do **not** overlap. Thus, the system of inequalities has no solution.

31. $\begin{cases} x \geq 0 & (1) \\ y \geq 0 & (2) \\ 2x + y \leq 6 & (3) \\ x + 2y \leq 6 & (4) \end{cases}$

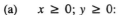

(a) $x \geq 0;\ y \geq 0$:
  These two inequalities require that our shaded region must be restricted to quadrant I.

(b) $2x + y \leq 6$:
  Step 1: Graph $2x + y = 6$ with a solid line. The x-intercept is $x = 3$; the y-intercept is $y = 6$.
  Step 2: Use $(0, 0)$ below the line as a test point.
  Step 3: At $(0, 0)$, $2x + y = 0 < 6$, so the inequality is satisfied.
  Step 4: Shade the region below the line $2x + y = 6$.

(c) $x + 2y \leq 6$:
  Step 1: Graph $x + 2y = 6$ with a solid line. The x-intercept is $x = 6$; the y-intercept is $y = 3$.
  Step 2: Let's use $(0, 0)$ **below** the line as a test point.
  Step 3: At $(0, 0)$, $x + 2y = 0 < 6$.
  Step 4: Therefore, shade the region **below** the line $x + 2y = 6$.

(d)   The graph is bounded.

(e)   To list the vertices, consult the graph. We see that there are four vertices:
  (1)   Intersection of x-axis and y-axis: $(0, 0)$.
  (2)   Intersection of $x + 2y = 6$ and y-axis: $(0, 3)$.
  (3)   Intersection of $x + 2y = 6$ and $2x + y = 6$. To find this point of intersection, it will be necessary to solve a system of equations:
  $$\begin{cases} x + 2y = 6 & (1) \\ 2x + y = 6 & (2) \end{cases}$$
  $$\begin{cases} -2x - 4y = -12 & (1) \quad \text{Multiply by } -2. \\ 2x + y = 6 & (2) \end{cases}$$
  $$\begin{cases} -3y = -6 & (1) \quad \text{Replace (1) by (1) + (2).} \\ 2x + y = 6 & (2) \end{cases}$$

$$\begin{cases} y = 2 & (1) \\ 2x + y = 6 & (2) \end{cases} \quad \text{Divide by } -3.$$

$$\begin{cases} y = 2 & (1) \\ 2x + 2 = 6 & (2) \end{cases} \quad \text{Back-substitution: } y = 2.$$

$$\begin{cases} y = 2 & (1) \\ x = 2 & (2) \end{cases}$$

Thus, the point of intersection is $(2, 2)$.

(4) The intersection of $2x + y = 6$ and the x-axis: $(3, 0)$

So, the four vertices are: $(0, 0)$, $(0, 3)$, $(2, 2)$, and $(3, 0)$.

33.  $\begin{cases} x \geq 0 & (1) \\ y \geq 0 & (2) \\ x + y \geq 2 & (3) \\ 2x + y \geq 4 & (4) \end{cases}$

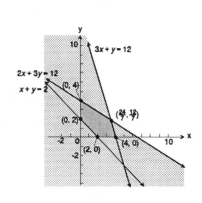

(a) $x \geq 0; \; y \geq 0$:

These inequalities require that the final shaded area be in quadrant I.

(b) $x + y \geq 2$:

Step 1: Graph $x + y = 2$ with a solid line. The x-intercept is $x = 2$; the y-intercept is $y = 2$.

Step 2: Use $(0, 0)$ below the line as a test point.

Step 3: At $(0, 0)$, $x + y = 0 < 2$, so $(0, 0)$ does *not* satisfy the inequality.

Step 4: Shade the region *above* the line $x + y = 2$.

(c) $2x + y \geq 4$:

Step 1: Graph $2x + y = 4$. The x-intercept is 2; the y-intercept is 4.

Step 2: Use $(0, 0)$ *below* the line as a test point.

Step 3: At $(0, 0)$, $2x + y = 0 < 4$.

Step 4: Therefore, we shade the region *above* the line $2x + y = 4$.

(d) The graph is *unbounded* since it extends forever toward the upper right.

(e) We only have two vertices:

(1) The intersection of $2x + y = 4$ and the y-axis: $(0, 4)$.

(2) The intersection of either $x + y = 2$ or $2x + y = 4$ and the x-axis: $(2, 0)$.

35. 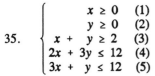 $\begin{cases} x \geq 0 & (1) \\ y \geq 0 & (2) \\ x + y \geq 2 & (3) \\ 2x + 3y \leq 12 & (4) \\ 3x + y \leq 12 & (5) \end{cases}$

(a) $x \geq 0; \; y \geq 0$:

Our graph will lie in quadrant I.

(b) $x + y \geq 2$:

Step 1: Graph $x + y = 2$. The x-intercept is 2; the y-intercept is 2.

Step 2: Use $(0, 0)$ below the line

Step 3: At $(0, 0)$, $x + y = 0 < 2$.

Step 4: Therefore, we shade the region *above* the line $x + y = 2$.

(c) $2x + 3y \leq 12$:

Step 1: Graph $2x + 3y = 12$. The x-intercept is $= 6$; the y-intercept is 4.

Step 2: Use $(0, 0)$ below the line.

Step 3: At $(0, 0)$, $2x + 3y = 0 < 12$.

Step 4: Therefore, shade *below* the line $2x + 3y = 12$.

(d) $3x + y \leq 12$:

Step 1: Graph $3x + y = 12$. The x-intercept is $x = 4$; the y-intercept is 12.

Step 2: Use $(0, 0)$ *below* the line as a test point.

Step 3: At $(0, 0)$, $3x + y = 0 < 12$.

Step 4: Shade the region *below* the line $3x + y = 12$.

(e) We see that the graph is bounded.

(f) We have five vertices:

(1) Intersection of $x + y = 2$ and the y-axis: $(0, 2)$.

(2) Intersection of $2x + 3y = 12$ and the y-axis: $(0, 4)$.

(3) Intersection of $2x + 3y = 12$ and $3x + y = 12$:

$$\begin{cases} 2x + 3y = 12 & (1) \\ 3x + y = 12 & (2) \end{cases}$$

$$\begin{cases} 2x + 3y = 12 & (1) \\ -9x - 3y = -36 & (2) \quad \text{Multiply by } -3. \end{cases}$$

$$\begin{cases} 2x + 3y = 12 & (1) \\ -7x = -24 & (2) \quad \text{Replace (2) by (1) + (2).} \end{cases}$$

$$\begin{cases} 2x + 3y = 12 & (1) \\ x = \dfrac{24}{7} & (2) \quad \text{Divide by } -7. \end{cases}$$

$$\begin{cases} \dfrac{48}{7} + 3y = 12 & (1) \quad \text{Back–substitution: } x = \dfrac{24}{7} \\ \phantom{\dfrac{48}{7} + 3y}x = \dfrac{24}{7} & (2) \end{cases}$$

$$\begin{cases} 3y = \dfrac{84}{7} - \dfrac{48}{7} & (1) \\ x = \dfrac{24}{7} & (2) \end{cases}$$

$$\begin{cases} y = \dfrac{12}{7} & (1) \\ x = \dfrac{24}{7} & (2) \end{cases}$$

The point of intersection is $\left(\dfrac{24}{7}, \dfrac{12}{7} \right)$.

(4) The intersection of $3x + y = 12$ and the x-axis: $(4, 0)$

(5) The intersection of $x + y = 2$ and the x-axis: $(2, 0)$

37. $\begin{cases} x \geq 0 & (1) \\ y \geq 0 & (2) \\ x + y \geq 2 & (3) \\ x + y \leq 8 & (4) \\ 2x + y \leq 10 & (5) \end{cases}$

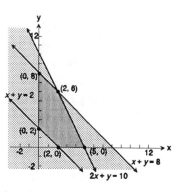

(a) $x \geq 0;\ y \geq 0$:

This places our final graph in quadrant I.

(b) $x + y \geq 2$:

Step 1:   Graph $x + y = 2$. The x-intercept is 2; the y-intercept is 2.

Step 2:   Use $(0, 0)$ *below* the line, as a test point.

Step 3:   At $(0, 0)$, $x + y = 0 < 2$, so the inequality is *not* satisfied.

Step 4:   Therefore, shade *above* the line $x + y = 2$.

(c) $x + y \leq 8$:

Step 1:   Graph $x + y = 8$. The x-intercept is 8; the y-intercept is 8.

Step 2:   Use $(0, 0)$ below the line.

Step 3:   At $(0, 0)$, $x + y = 0 < 8$.

Step 4:   Therefore, shade *below* the line $x + y = 8$.

(d) $2x + y \leq 10$:

Step 1:   Graph $2x + y = 10$. The x-intercept is $x = 5$; the y-intercept is $y = 10$.

Step 2:   Use $(0, 0)$ below the line as a test point.

Step 3:   At $(0, 0)$, $2x + y = 0 < 10$.

Step 4:   Shade the region below the line $2x + y = 10$.

(e)   The graph is bounded.

(f)   We have five vertices:

    (1)   Intersection of $x + y = 2$ and the y-axis: $(0, 2)$.

    (2)   Intersection of $x + y = 8$ and the y-axis: $(0, 8)$.

    (3)   Intersection of $x + y = 8$ and $2x + y = 10$:

$$\begin{cases} x + y = 8 & (1) \\ 2x + y = 10 & (2) \end{cases}$$

$$\begin{cases} x + y = 8 & (1) \\ -2x - y = -10 & (2) \end{cases}$$

$$\begin{cases} x + y = 8 & (1) \\ -x = -2 & (2) \end{cases} \quad \text{Replace (2) by (1) + (2).}$$

$$\begin{cases} x + y = 8 & (1) \\ x = 2 & (2) \end{cases}$$

$$\begin{cases} y = 6 & (1) \\ x = 2 & (2) \end{cases} \quad \text{Back-substitution:}\ x = 2.$$

The vertex is $(2, 6)$.

    (4)   The intersection of $2x + y = 10$ and the x-axis: $(5, 0)$

    (5)   The intersection of $x + y = 2$ and the x-axis: $(2, 0)$

39.
$$\begin{cases} x \ge 0 & (1) \\ y \ge 0 & (2) \\ x + 2y \ge 1 & (3) \\ x + 2y \le 10 & (4) \end{cases}$$

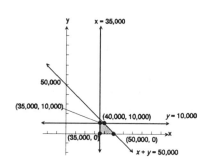

(a) $x \ge 0; \, y \ge 0$:

Our graph will lie in quadrant I.

(b) $x + 2y \ge 1$:

Step 1:   Graph $x + 2y = 1$. The x-intercept is 1; the y-intercept is $\dfrac{1}{2}$.

Step 2:   Use $(0, 0)$ below the line as a test point.

Step 3:   At $(0, 0)$, $x + 2y = 0 < 1$.

Step 4:   Therefore, we shade the region *above* the line $x + 2y = 1$.

(c) $x + 2y \le 10$:

Step 1:   Graph $x + 2y = 10$. The x-intercept is $= 10$; the y-intercept is 5.

Step 2:   Use $(0, 0)$ below the line.

Step 3:   At $(0, 0)$, $x + 2y = 0 < 10$.

Step 4:   Therefore, we shade the region *below* the line $x + 2y = 10$.

(d) We see that the region is bounded.

(e) From the graph, there are four vertices:

(1)   The intersection of $x + 2y = 1$ and the y-axis: $\left[0, \dfrac{1}{2}\right]$.

(2)   The intersection of $x + 2y = 10$ and the y-axis: $(0, 5)$.

(3)   The intersection of $x + 2y = 10$ and the x-axis: $(10, 0)$.

(4)   The intersection of $x + 2y = 1$ and the x-axis: $(1, 0)$

41. $\begin{cases} x \le 4 \\ x + y \le 6 \\ x \ge 0, \, y \ge 0 \end{cases}$

43. $\begin{cases} x \le 20 \\ y \ge 15 \\ x + y \le 50 \\ x \le y \\ x \ge 0 \end{cases}$

45. (a) $\begin{cases} x + y \le 50{,}000 \\ x \ge 35{,}000 \\ y \le 10{,}000 \\ x \ge 0 \\ y \ge 0 \end{cases}$

(b)

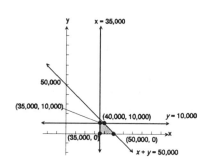

47. (a) x denotes the number of packages of the economy blend.
         y denotes the number of packages of the superior blend.

    $4x + 8y \leq 75(16)$ denotes the equation which states that the blends cannot exceed 75 pounds of A grade coffee or $(76)(16) = 1200$ ounces.

    $12x + 8y \leq 120(16)$ denotes the equation which states that the blends cannot exceed 120 pounds of B grade coffee or $(120)(16) = 1920$ ounces.

    We also must denote that the number of packages of both blends must be non-negative.

    The following equations are simplified:
          from $4x + 8y \leq 1200$
            to $x + 2y \leq 300$
    and   from $12x + 8y \leq 1920$
            to $3x + 2y \leq 480$

    We have the following system:

    $$x \geq 0, y \geq 0$$
    $$x + 2y \leq 300$$
    $$3x + 2y \leq 480$$

(b)

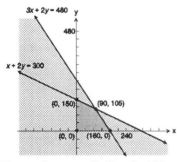

The vertices are $(0, 0)$, $(0, 150)$, $(90, 105)$, and $(160, 0)$.

49. (a)
$$\begin{cases} 30x + 20y \leq 1600 \\ 3x + 2y \leq 150 \\ x \geq 0 \\ y \geq 0 \end{cases}$$

(b)

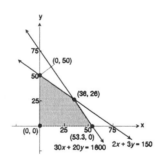

11.8 Linear Programming

In Problems 1-6, we have the same set of vertices: (0, 3), (0, 6), (5, 6), (5, 2), and (4, 0).

1. $z = x + y$

Vertex	Value of Objective Function ($z = x + y$)
(0, 3)	$z = 0 + 3 = 3$
(0, 6)	$z = 0 + 6 = 6$
(5, 6)	$z = 5 + 6 = 11$
(5, 2)	$z = 5 + 2 = 7$
(4, 0)	$z = 4 + 0 = 4$

The maximum value is 11, at (5, 6), and the minimum value is 3, at (0, 3).

3.

Vertex	Value of $z = x + 10y$
(0, 3)	$z = 0 + 10 \cdot 3 = 30$
(0, 6)	$z = 0 + 10 \cdot 6 = 60$
(5, 6)	$z = 5 + 10 \cdot 6 = 65$ ← Maximum value
(5, 2)	$z = 5 + 10 \cdot 2 = 25$
(4, 0)	$z = 4 + 10 \cdot 0 = 4$ ← Minimum value

5.

Vertex	Value of Objective Function ($z = 5x + 7y$)
(0, 3)	$z = 5 \cdot 0 + 7 \cdot 3 = 21$
(0, 6)	$z = 5 \cdot 0 + 7 \cdot 6 = 42$
(5, 6)	$z = 5 \cdot 5 + 7 \cdot 6 = 67$ ← Maximum value
(5, 2)	$z = 5 \cdot 5 + 7 \cdot 2 = 39$
(4, 0)	$z = 5 \cdot 4 + 7 \cdot 0 = 20$ ← Minimum value

7. Maximize $z = 2x + y$

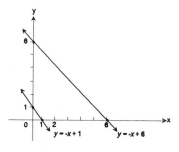

Subject to $x \geq 0, y \geq 0, x + y \leq 6, x + y \geq 1$

(a) We have two lines:
$y = -x + 6$ and $y = -x + 1$.

There is no point of intersection, since they are parallel lines. (Their slopes are both -1).

(b) Now evaluate $z = 2x + y$ at each vertex:

Vertex	Value of $z = 2x + y$
(0, 1)	$z = 2 \cdot 0 + 1 = 1$
(0, 6)	$z = 2 \cdot 0 + 6 = 6$
(6, 0)	$z = 2 \cdot 6 + 0 = 12$
(1, 0)	$z = 2 \cdot 1 + 0 = 2$

The maximum possible value for z is 12, at the point (6, 0).

9.   Minimize $z = 2x + 5y$

Subject to $x \geq 0$, $y \geq 0$, $x + y \geq 2$, $x \leq 5$, $y \leq 3$.
(a)   Graph the constraints:

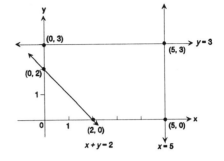

Vertex	Value of $z = 2x + 5y$
(0, 2)	$z = 10$
(0, 3)	$z = 15$
(5, 3)	$z = 25$
(5, 0)	$z = 10$
(2, 0)	$z = 4$

The minimum value of z is 4, at the point (2, 0).

11. $z = 3x + 5y$

Subject to $x \geq 0$, $y \geq 0$, $x + y \geq 2$, $2x + 3y \leq 12$,
$3x + 2y \leq 12$.

We have three lines:
(1) $y = -x + 2$

(2) $y = -\dfrac{2}{3}x + 4$

(3) $y = -\dfrac{3}{2}x + 6$

Let's find the points of intersection:

(1) and (2):
$$-x + 2 = -\dfrac{2}{3}x + 4$$
$$-3x + 6 = -2x + 12$$
$$-x = 6$$
$$x = -6 \text{ and } y = -x + 2$$
$$y = 8$$
But, $(-6, 8)$ is not in the feasible region ($x \geq 0$).

(1) and (3):
$$-x + 2 = -\dfrac{3}{2}x + 6$$
$$-2x + 4 = -3x + 12$$
$$x = 8$$
and $y = -x + 2 = -6$
But $(8, -6)$ is not in the feasible region.

(2) and (3):
$$-\dfrac{2}{3}x + 4 = -\dfrac{3}{2}x + 6$$
$$-4x + 24 = -9x + 36$$
$$5x = 12$$
$$x = \dfrac{12}{5}$$
and $y = -\dfrac{3}{2}x + 6$

$$y = -\dfrac{3}{2}\left[\dfrac{12}{5}\right] + 6 = \dfrac{12}{5}$$

This gives the point $\left[\dfrac{12}{5}, \dfrac{12}{5}\right]$.

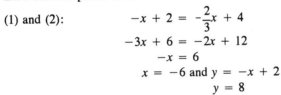

We now graph the constraints:

Vertex	Value of $z = 3x + 5y$
(0, 2)	$z = 10$
(0, 4)	$z = 20$
$\left(\dfrac{12}{5}, \dfrac{12}{5}\right)$	$z = \dfrac{96}{5} = 19.2$
(4, 0)	$z = 12$
(2, 0)	$z = 6$

The maximum value of z is 20, at the point (0, 4).

13. Minimize $z = 5x + 4y$

Subject to $x \geq 0$, $y \geq 0$, $x + y \geq 2$, $2x + 3y \leq 12$, $3x + y \leq 12$.

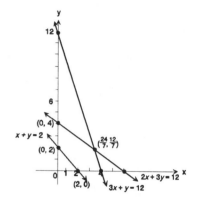

We have three lines to graph:
(1) $y = -x + 2$;
    y-intercept $= 2$; x-intercept $= 2$
(2) $y = -\dfrac{2}{3}x + 4$;
    y-intercept $= 4$; x-intercept $= 6$
(3) $y = -3x + 12$;
    y-intercept $= 12$; x-intercept $= 4$
The only intersection we need is:

$$(2) \text{ and } (3): \quad -\frac{2}{3}x + 4 = -3x + 12$$
$$-2x + 12 = -9x + 36$$
$$7x = 24$$
$$x = \frac{24}{7}$$

and $y = -3x + 12$; $y = \dfrac{12}{7}$

$$\left(\frac{24}{7}, \frac{12}{7}\right)$$

Vertex	Value of $z = 5x + 4y$
(0, 2)	$z = 8$
(0, 4)	$z = 16$
$\left(\dfrac{24}{7}, \dfrac{12}{7}\right)$	$z = \dfrac{168}{7} = 24$
(4, 0)	$z = 20$
(2, 0)	$z = 10$

The minimum value of z is 8, at the point (0, 2).

15. Maximize $z = 5x + 2y$

Subject to $x \geq 0, y \geq 0, x + y \leq 10, 2x + y \geq 10,$ $x + 2y \geq 10$.

We have:
(1) $y = -x + 10$;
   y-intercept $= 10$; x-intercept $= 10$
(2) $y = -2x + 10$;
   y-intercept $= 10$; x-intercept $= 5$
(3) $y = -\dfrac{1}{2}x + 5$;
   y-intercept $= 5$; x-intercept $= 10$

To find the intersection of (2) and (3):

$$-2x + 10 = -\dfrac{1}{2}x + 5$$
$$-4x + 20 = -x + 10$$
$$-3x = -10$$
$$x = \dfrac{10}{3}$$

and
$$y = -2x + 10$$
$$y = -2\left[\dfrac{10}{3}\right] + 10$$
$$y = \dfrac{10}{3}; \quad \left[\dfrac{10}{3}, \dfrac{10}{3}\right]$$

Vertex	Value of $z = 5x + 2y$
$(0, 10)$	$z = 20$
$(10, 0)$	$z = 50$
$\left[\dfrac{10}{3}, \dfrac{10}{3}\right]$	$z = \dfrac{70}{3} = 23\dfrac{1}{3}$

The maximum value of z is 50, at the point $(10, 0)$.

17. Let $x =$ Number of Downhill skis produced
   $y =$ Number of Cross-country skis produced

We want to maximize profit (which is $70 per Downhill ski, $50 per Cross-Country ski). Thus, total profit is:
   $P = 70x + 50y$
This is our objective function. Now we need to determine our constraints:
   $x \geq 0, y \geq 0$ Nonnegative constraints
We only have 40 hours manufacturing time available:
   $2x + y \leq 40$ Manufacturing time
Also, $x + y \leq 32$ Finishing time constraint

So we have two lines to graph:

(1) $y = -2x + 40$;
    y-intercept $= 40$; x-intercept $= 20$
(2) $y = -x + 32$;
    y-intercept $= 32$; x-intercept $= 32$

For the point of intersection of (1) and (2):
$$-2x + 40 = -x + 32$$
$$-x = -8$$
$$x = 8$$
and
$$y = -x + 32$$
$$y = 24; \quad (8, 24)$$

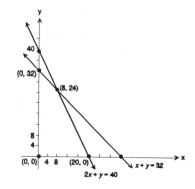

Vertex	Value of Profit: $P = 70x + 50y$
(0, 0)	$P = 0$
(0, 32)	$P = 50 \cdot 32 = 1600$
(8, 24)	$P = 70 \cdot 8 + 50 \cdot 24 = 1760$
(20, 0)	$P = 70 \cdot 20 = 1400$

The maximum profit is $1760, obtained when
$x = 8$ Downhill skis
$y = 24$ Cross-country skis
are produced.

19.   Maximize profit $P = 200x + 250y$

Let $x =$ acres of soybeans
      $y =$ acres of corn
$x \geq 0, y \geq 0$ Nonnegative constraints

$40x + 60y \leq 1800$ Cultivation Cost
$60x + 60y \leq 2400$ Labor Cost

For the point of intersection,
$$\frac{1800 - 40x}{60} = \frac{2400 - 60x}{60}$$
$$30 - \frac{2}{3}x = 40 - x$$
$$\frac{1}{3}x = 10$$
$$x = 30$$
$$y = \frac{1800 - 40(30)}{60}$$
$$y = 10$$

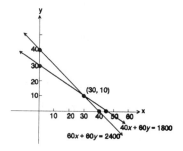

Vertex	Value of Profit $P = 200x + 250y$
(0, 0)	$P = 0$
(0, 30)	$P = 0 + 250(30) = 7500$
(30, 10)	$P = 200(30) + 250(10) = 8500$
(40, 0)	$P = 200(40) + 250(0) = 8000$

The maximum profit is $8500 obtained with 30 acres of soybeans and 10 acres of corn.

Chapter 11 Systems of Equations and Inequalities

21. Let x = Machine I
 y = Machine II

Minimize $C = 50x + 30y$

Subject to
$$x \geq 0, y \geq 0$$
$$60x + 40y \geq 0$$
$$70x + 20y \geq 200$$

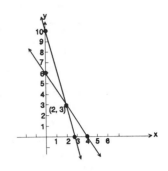

Vertex	Cost $C = 50x + 30y$
(0, 10)	$C = 50(0) + 30(10) = \$300$
(2, 3)	$C = 50(2) + 30(3) = \$190$
(4, 0)	$C = 50(4) + 30(0) = \$200$

The minimum cost is \$190 with 2 hours on Machine I and 3 hours on Machine II.

23. Let x = number of pounds of ground beef.
Let y = number of pounds of ground pork.

Minimize $C = .75x + .45y$

Subject to
$$x \geq 0, y \geq 0$$
$$.75x + .60y \geq .70(x + y)$$
$$.75x + .60y \geq .70x + .70y$$
$$.05x - .10y \geq 0$$
$$x \leq 200$$
$$y \geq 50$$
$$.05x \geq .10y$$

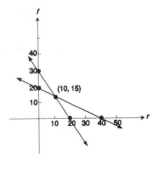

Vertex	Cost $C = .75x + .45y$
(100, 50)	$C = .75(100) + .45(50) = \$97.50$
(200, 100)	$C = .75(200) + .45(100) = \$195.00$
(200, 50)	$C = .75(200) + .45(50) = \$172.50$

The minimum cost is \$97.50 obtained with 100 lbs. of ground beef and 50 lbs of pork.

25. Maximize Profit $P = \$10r + \$12f$

Let r = number of racing skates
 f = number of figure skates

Subject to
$$r \geq 0, f \geq 0$$
$$6r + 4f \leq 120$$
$$1r + 2f \leq 40$$
$$\frac{120 - 4f}{6} = 40 - 2f$$
$$120 - 4f = 240 - 12f$$
$$8f = 120$$
$$f = 15$$
$$r = 40 - 2(15)$$
$$r = 10$$

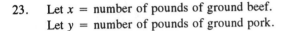

Vertex	Profit
(0, 20)	$P = 10(0) + 12(20) = \$240$
(10, 15)	$P = 10(10) + 12(15) = \$280$
(20, 0)	$P = 10(20) + 12(0) = \$200$

The maximum profit is \$280 which occurs when manufacturing 10 racing skates and 15 figure skates.

27. Minimize Cost $C = 9m + 4p$
    Let m = number of metal fasteners
    p = number of plastic fasteners
    Subject to:
    $$m \geq 2$$
    $$p \geq 2$$
    $$m + 2 \geq 6$$
    $$4m + 2p \leq 24$$

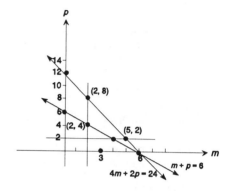

Vertex	Profit
(2, 8)	$C = 9(2) + 4(8) = \$50.00$
(2, 4)	$C = 9(2) + 4(4) = \$34.00$
(4, 2)	$C = 9(4) + 4(2) = \$44.00$
(5, 2)	$C = 9(5) + 4(2) = \$53.00$

The entrepreneur should order 2 metal fasteners and 4 plastic fasteners. The cost of the samples will be \$34.00.

29. Let x = Number of first class seats
    y = Number of coach seats

    The constraints are:
    $$8 \leq x \leq 16$$
    $$80 \leq y \leq 120$$

    (a) If the ratio of first class to coach cannot exceed $\dfrac{1}{12}$, then $\dfrac{x}{y} \leq \dfrac{1}{12}$.

    Let $y = 120$. Then $\dfrac{x}{120} \leq \dfrac{1}{12}$.
    $$12x \leq 120$$
    $$\text{or } x \leq 10$$
    This satisfies the first constraint. Thus, the number of seats that maximizes revenue is 10 first class and 120 coach.

    (b) If the ratio of first class to coach cannot exceed $\dfrac{1}{8}$, then $\dfrac{x}{y} \leq \dfrac{1}{8}$.

    Let $y = 120$. Then $\dfrac{x}{120} \leq \dfrac{1}{8}$.

    $8x \leq 120$ or $x \leq 15$ which satisifes the first constraint. Thus, the number of seats that maximizes revenue is 15 first class and 120 coach.

1. $\begin{cases} 2x - y = 5 & (1) \\ 5x + 2y = 8 & (2) \end{cases}$

   Step 1:   It is easiest to solve (1) for y:

   $$2x - y = 5$$
   $$-y = 5 - 2x$$
   $$y = -5 + 2x$$

   Step 2:   Substitute $y = -5 + 2x$ into (2):

   $$5x + 2y = 8$$
   $$5x + 2(-5 + 2x) = 8$$
   $$5x - 10 + 4x = 8$$
   $$9x = 18$$

   Step 3: $x = 2$

   Step 4:   Determine y:

   $$y = -5 + 2x \quad \text{(Step 1)}$$
   $$y = -5 + 2(2) \qquad (x = 2)$$
   $$y = -1$$

   The solution is $x = 2$, $y = -1$.

3. $\begin{cases} 3x - 4y = 4 & (1) \\ x - 3y = \dfrac{1}{2} & (2) \end{cases}$

   Step 1:   Let's solve for x in (2):

   $$x - 3y = \frac{1}{2}$$
   $$x = \frac{1}{2} + 3y$$

   Step 2:   Substitute $x = \dfrac{1}{2} + 3y$ into (1):

   $$3x - 4y = 4$$
   $$3\left(\frac{1}{2} + 3y \right) - 4y = 4$$
   $$\frac{3}{2} + 9y - 4y = 4$$
   $$3 + 18y - 8y = 8 \quad \text{Multiply by 2.}$$
   $$10y = 5$$

   Step 3: $y = \dfrac{5}{10}$

   $$y = \frac{1}{2}$$

   Step 4:   Determine x:

   $$x = \frac{1}{2} + 3y \qquad \text{(Step 1)}$$
   $$x = \frac{1}{2} + 3\left(\frac{1}{2} \right) \quad \left(y = \frac{1}{2} \right)$$
   $$x = 2$$

   The solution is $x = 2$, $y = \dfrac{1}{2}$.

5.
$$\begin{cases} x - 2y - 4 = 0 & (1) \\ 3x + 2y - 4 = 0 & (2) \end{cases}$$

or

$$\begin{cases} x - 2y = 4 & (1) \\ 3x + 2y = 4 & (2) \end{cases}$$

Step 1: It is easiest to solve for x in (1):
$$x - 2y = 4$$
$$x = 4 + 2y$$

Step 2: Substitute $x = 4 + 2y$ into (2):
$$3x + 2y = 4$$
$$3(4 + 2y) + 2y = 4$$
$$8y = -8$$

Step 3: $y = -1$

Step 4: Determine x:
$$x = 4 + 2y \qquad \text{(Step 1)}$$
$$x = 4 + 2(-1)$$
$$x = 2$$

The solution is $x = 2$, $y = -1$.

7.
$$\begin{cases} y = 2x - 5 & (1) \\ x = 3y + 4 & (2) \end{cases}$$

We can substitute $y = 2x - 5$ into (2):
$$x = 3y + 4$$
$$x = 3(2x - 5) + 4$$
$$x = 6x - 11$$
$$-5x = -11$$
$$x = \frac{11}{5}$$

Then we have:
$$y = 2x - 5$$
$$y = \frac{22}{5} - 5$$
$$y = \frac{-3}{5}$$

The solution is $x = \dfrac{11}{5}$, $y = \dfrac{-3}{5}$.

9.
$$\begin{cases} x - y + 4 = 0 & (1) \\ \dfrac{1}{2}x + \dfrac{1}{6}y + \dfrac{2}{5} = 0 & (2) \end{cases}$$

or

$$\begin{cases} x - y = 4 & (1) \\ 15x + 5y = -12 & (2) \quad \text{Multiply by 30.} \end{cases}$$

Step 1: Solve equation (1) for x:
$$x - y = -4$$
$$x = -4 + y$$

Step 2: Substitute $x = -4 + y$ into (2):
$$15x + 5y = -12$$
$$15(-4 + y) + 5y = -12$$
$$-60 + 20y = -12$$
$$20y = 48$$
$$y = \frac{48}{20}$$

Step 3: $y = \dfrac{12}{5}$

Step 4:   Determine x:

$$x = -4 + y \qquad \text{(Step 1)}$$

$$x = -4 + \frac{12}{5}$$

$$x = \frac{-20}{5} + \frac{12}{5}$$

$$x = \frac{-8}{5}$$

The solution is $x = \dfrac{-8}{5}$, $y = \dfrac{12}{5}$.

11. $\begin{cases} x - 2y - 8 = 0 & (1) \\ 2x + 2y - 10 = 0 & (2) \end{cases}$

$\begin{cases} x - 2y - 8 = 0 & (1) \\ 3x \qquad\quad - 18 = 0 & (2) \end{cases}$ Replace (2) by (1) + (2).

$\begin{cases} x - 2y - 8 = 0 & (1) \\ \qquad\qquad\quad x = 6 & (2) \end{cases}$

$\begin{cases} 6 - 2y - 8 = 0 & (1) \\ \qquad\qquad\quad x = 6 & (2) \end{cases}$ Back-substitute; $x = 6$.

$\begin{cases} -2y = 2 & (1) \\ \quad x = 6 & (2) \end{cases}$

$\begin{cases} y = -1 & (1) \\ x = 6 & (2) \end{cases}$

The solution is $x = 6$; $y = -1$.

13. $\begin{cases} y - 2x = 11 & (1) \\ 2y - 3x = 18 & (2) \end{cases}$

$\begin{cases} -2y + 4x = -22 & (1) \\ \quad 2y - 3x = 18 & (2) \end{cases}$ Multiply both sides by -2.

$\begin{cases} -2y + 4x = -22 & (1) \\ \qquad\qquad\; x = -4 & (2) \end{cases}$ Replace (2) by (1) + (2).

$\begin{cases} -2y - 16 = -22 & (1) \\ \qquad\qquad x = -4 & (2) \end{cases}$ Back-substitute; $x = -4$.

$\begin{cases} -2y = -6 & (1) \\ \quad x = -4 & (2) \end{cases}$

$\begin{cases} y = 3 & (1) \\ x = -4 & (2) \end{cases}$

The solution is $x = -4$; $y = 3$.

15. $\begin{cases} 2x + 3y - 13 = 0 & (1) \\ 3x - 2y \qquad = 0 & (2) \end{cases}$

$\begin{cases} 6x + 9y - 39 = 0 & (1) \\ -6x + 4y \qquad = 0 & (2) \end{cases}$ Multiply both sides by 3.
   Multiply both sides by -2.

$\begin{cases} 6x + 9y - 39 = 0 & (1) \\ \qquad 13y - 39 = 0 & (2) \end{cases}$ Replace (2) by (1) + (2).

$\begin{cases} 6x + 9y - 39 = 0 & (1) \\ \qquad\qquad\quad y = 3 & (2) \end{cases}$

$\begin{cases} 6x + 27 - 39 = 0 & (1) \\ \qquad\qquad\quad y = 3 & (2) \end{cases}$ Back-substitute; $y = 3$.

$$\begin{cases} x = 2 & (1) \\ y = 3 & (2) \end{cases}$$

The solution is $x = 2$, $y = 3$.

17. $$\begin{cases} 3x - 2y = 8 & (1) \\ x - \dfrac{2}{3}y = 12 & (2) \end{cases}$$

$$\begin{cases} 3x - 2y = 8 & (1) \\ -3x + 2y = -36 & (2) \end{cases} \quad \text{Multiply both sides by } -3.$$

$$\begin{cases} 3x - 2y = 8 & (1) \\ 0x + 0y = -28 & (2) \end{cases} \quad \text{Replace (2) by (1) + (2).}$$

The system is inconsistent.

19. $$\begin{cases} x + 2y - z = 6 & (1) \\ 2x - y + 3z = -13 & (2) \\ 3x - 2y + 3z = -16 & (3) \end{cases}$$

We can eliminate the variable y by adding (1) and (3):

$$\begin{cases} x + 2y - z = 6 & (1) \\ 2x - y + 3z = -13 & (2) \\ 4x \qquad + 2z = -10 & (3) \quad \text{Replace (3) by (1) + (3).} \end{cases}$$

Now we must eliminate the same variable, y, from either (1) or (2):

$$\begin{cases} x + 2y - z = 6 & (1) \\ 4x - 2y + 6z = -26 & (2) \quad \text{Multiply both sides by 2.} \\ 4x \qquad + 2z = -10 & (3) \end{cases}$$

$$\begin{cases} x + 2y - z = 6 & (1) \\ 5x \qquad + 5z = -20 & (2) \quad \text{Replace (2) by (1) + (2).} \\ 4x \qquad + 2z = -10 & (3) \end{cases}$$

Now attack (2) and (3):

$$\begin{cases} x + 2y - z = 6 & (1) \\ 20x \qquad + 20z = -80 & (2) \quad \text{Multiply both sides by 4.} \\ -20x \qquad - 10z = 50 & (3) \quad \text{Mutliply both sides by } -5. \end{cases}$$

$$\begin{cases} x + 2y - z = 6 & (1) \\ 20x \qquad + 20z = -80 & (2) \\ \qquad\qquad 10z = -30 & (3) \quad \text{Replace (3) by (2) + (3).} \end{cases}$$

$$\begin{cases} x + 2y - z = 6 & (1) \\ 20x \qquad + 20z = -80 & (2) \\ \qquad\qquad z = -3 & (3) \end{cases}$$

$$\begin{cases} x + 2y + 3 = 6 & (1) \quad \text{Back-substitute: } z = -3. \\ 20x \qquad - 60 = -80 & (2) \quad \text{Back-substitute; } z = -3. \\ \qquad\qquad z = -3 & (3) \end{cases}$$

$$\begin{cases} x + 2y = 3 & (1) \\ \qquad x = -1 & (2) \\ \qquad z = -3 & (3) \end{cases}$$

$$\begin{cases} -1 + 2y = 3 & (1) \quad \text{Back-substitute; } x = -1. \\ \qquad x = -1 & (2) \\ \qquad z = -3 & (3) \end{cases}$$

$$\begin{cases} y = 2 & (1) \\ x = -1 & (2) \\ z = -3 & (3) \end{cases}$$

The solution is $x = -1, y = 2, z = -3$.

21. $A + C = \begin{bmatrix} 1 & 0 \\ 2 & 4 \\ -1 & 2 \end{bmatrix} + \begin{bmatrix} 3 & -4 \\ 1 & 5 \\ 5 & -2 \end{bmatrix} = \begin{bmatrix} 4 & -4 \\ 3 & 9 \\ 4 & 0 \end{bmatrix}$

23. $6A = 6\begin{bmatrix} 1 & 0 \\ 2 & 4 \\ -1 & 2 \end{bmatrix} = \begin{bmatrix} 6 & 0 \\ 12 & 24 \\ -6 & 12 \end{bmatrix}$

25. $AB = \begin{bmatrix} 1 & 0 \\ 2 & 4 \\ -1 & 2 \end{bmatrix}\begin{bmatrix} 4 & -3 & 0 \\ 1 & 1 & -2 \end{bmatrix} = \begin{bmatrix} 4+0 & -3+0 & 0+0 \\ 8+4 & -6+4 & 0+(-8) \\ -4+2 & 3+2 & 0+(-4) \end{bmatrix} = \begin{bmatrix} 4 & -3 & 0 \\ 12 & -2 & -8 \\ -2 & 5 & -4 \end{bmatrix}$

27. $CB = \begin{bmatrix} 3 & -4 \\ 1 & 5 \\ 5 & -2 \end{bmatrix}\begin{bmatrix} 4 & -3 & 0 \\ 1 & 1 & -2 \end{bmatrix} = \begin{bmatrix} 8 & -13 & 8 \\ 9 & 2 & -10 \\ 18 & -17 & 4 \end{bmatrix}$

29. $A = \begin{bmatrix} 4 & 6 \\ 1 & 3 \end{bmatrix}$. We start with $[A \,|\, I_2]$, and proceed to put it into reduced echelon form.

$$[A \,|\, I_2] = \begin{bmatrix} 4 & 6 & | & 1 & 0 \\ 1 & 3 & | & 0 & 1 \end{bmatrix} \underset{\uparrow}{\rightarrow} \begin{bmatrix} 1 & 3 & | & 0 & 1 \\ 4 & 6 & | & 1 & 0 \end{bmatrix}$$

Interchange rows

$$\rightarrow \underset{\underset{R_2 = -4r^1 + r^2}{\uparrow}}{\begin{bmatrix} 1 & 3 & | & 0 & 1 \\ 0 & -6 & | & 1 & -4 \end{bmatrix}} \rightarrow \underset{\underset{R^1 = \frac{1}{2}r_2 + r_1}{\uparrow}}{\begin{bmatrix} 1 & 0 & | & \frac{1}{2} & -1 \\ 1 & -6 & | & 1 & -4 \end{bmatrix}} \rightarrow \underset{\underset{R^2 = -\frac{1}{6}r_2}{\uparrow}}{\begin{bmatrix} 1 & 0 & | & \frac{1}{2} & -1 \\ 0 & 1 & | & -\frac{1}{6} & \frac{2}{3} \end{bmatrix}}$$

Therefore, $A^{-1} = \begin{bmatrix} \frac{1}{2} & -1 \\ -\frac{1}{6} & \frac{2}{3} \end{bmatrix}$

31. $A = \begin{bmatrix} 1 & 3 & 3 \\ 1 & 2 & 1 \\ 1 & -1 & 2 \end{bmatrix}$

$$\begin{bmatrix} 1 & 3 & 3 & | & 1 & 0 & 0 \\ 1 & 2 & 1 & | & 0 & 1 & 0 \\ 1 & -1 & 2 & | & 0 & 0 & 1 \end{bmatrix} \underset{\underset{\substack{R_2 = -1r_1 + r_2 \\ R_3 = -1r_1 + r_3}}{\uparrow}}{\rightarrow} \begin{bmatrix} 1 & 3 & 3 & | & 1 & 0 & 0 \\ 0 & -1 & -2 & | & -1 & 1 & 0 \\ 0 & -4 & -1 & | & -1 & 0 & 1 \end{bmatrix} \underset{\underset{R_2 = -1r_2}{\uparrow}}{\rightarrow} \begin{bmatrix} 1 & 3 & 3 & | & 1 & 0 & 0 \\ 0 & 1 & 2 & | & 1 & -1 & 0 \\ 0 & -4 & -1 & | & -1 & 0 & 1 \end{bmatrix}$$

$$\rightarrow \begin{bmatrix} 1 & 0 & -3 & \bigm| & -2 & 3 & 0 \\ 0 & 1 & 2 & \bigm| & 1 & -1 & 0 \\ 0 & 0 & 7 & \bigm| & 3 & -4 & 1 \end{bmatrix} \rightarrow \begin{bmatrix} 1 & 0 & -3 & \bigm| & -2 & 3 & 0 \\ 0 & 1 & 2 & \bigm| & 1 & -1 & 0 \\ 0 & 0 & 1 & \bigm| & \dfrac{3}{7} & -\dfrac{4}{7} & \dfrac{1}{7} \end{bmatrix} \rightarrow \begin{bmatrix} 1 & 0 & 0 & \bigm| & -\dfrac{5}{7} & \dfrac{9}{7} & \dfrac{3}{7} \\ 0 & 1 & 0 & \bigm| & \dfrac{1}{7} & \dfrac{1}{7} & -\dfrac{2}{7} \\ 0 & 0 & 1 & \bigm| & \dfrac{3}{7} & -\dfrac{4}{7} & \dfrac{1}{7} \end{bmatrix}$$

$$\uparrow \qquad\qquad\qquad \uparrow \qquad\qquad\qquad \uparrow$$

$$R_1 = -3r_2 + r_1 \qquad R_3 = \frac{1}{7}r_3 \qquad R_1 = 3r_3 + r_1$$

$$R_3 = 4r_2 + r_3 \qquad\qquad\qquad\qquad R_2 = -2r_3 + r_2$$

Therefore, $A^{-1} = \begin{bmatrix} -\dfrac{5}{7} & \dfrac{9}{7} & \dfrac{3}{7} \\[2mm] \dfrac{1}{7} & \dfrac{1}{7} & -\dfrac{2}{7} \\[2mm] \dfrac{3}{7} & -\dfrac{4}{7} & \dfrac{1}{7} \end{bmatrix} = \dfrac{1}{7}\begin{bmatrix} -5 & 9 & 3 \\ 1 & 1 & -2 \\ 3 & -4 & 1 \end{bmatrix}$

33. $A = \begin{bmatrix} 4 & -8 \\ -1 & 2 \end{bmatrix}$

$$\begin{bmatrix} 4 & -8 & \bigm| & 1 & 0 \\ -1 & 2 & \bigm| & 0 & 1 \end{bmatrix} \rightarrow \begin{bmatrix} 1 & -2 & \bigm| & \dfrac{1}{4} & 0 \\ -1 & 2 & \bigm| & 0 & 1 \end{bmatrix} \rightarrow \begin{bmatrix} 1 & -2 & \bigm| & \dfrac{1}{4} & 0 \\ 0 & 0 & \bigm| & \dfrac{1}{4} & 1 \end{bmatrix}$$

$$\uparrow \qquad\qquad\qquad \uparrow$$

$$R_1 = \frac{1}{4}r_1 \qquad R_2 = r_1 + r_2$$

This did not take the form $\begin{bmatrix} I_2 | B \end{bmatrix}$, so A has no inverse; i.e., A is singular.

35. $\begin{cases} 3x - 2y = 1 \\ 10x + 10y = 5 \end{cases}$ becomes: $\begin{bmatrix} 3 & -2 & \bigm| & 1 \\ 10 & 10 & \bigm| & 5 \end{bmatrix}$

In two steps, we can get a 1 in row 1, column 1, **and** avoid fractions:

$$\rightarrow \begin{bmatrix} 3 & -2 & \bigm| & 1 \\ 1 & 16 & \bigm| & 2 \end{bmatrix} \rightarrow \begin{bmatrix} 1 & 16 & \bigm| & 2 \\ 3 & -2 & \bigm| & 1 \end{bmatrix} \rightarrow \begin{bmatrix} 1 & 16 & \bigm| & 2 \\ 0 & -50 & \bigm| & -5 \end{bmatrix}$$

$$\uparrow \qquad\qquad\qquad \uparrow \qquad\qquad\qquad \uparrow$$

$$R_2 = -3r_1 + r_2 \quad \text{Interchange rows} \quad R_2 = -3r_1 + r_2$$

$$\rightarrow \begin{bmatrix} 1 & 16 & \bigm| & 2 \\ 0 & 1 & \bigm| & \dfrac{1}{10} \end{bmatrix} \rightarrow \begin{bmatrix} 1 & 0 & \bigm| & \dfrac{20}{10} - \dfrac{16}{10} \\ 0 & 1 & \bigm| & \dfrac{1}{10} \end{bmatrix}$$

$$\uparrow \qquad\qquad\qquad \uparrow$$

$$R_2 = -\frac{1}{50}r_2 \qquad R_1 = -16r_2 + r_1$$

The solution is $x = \dfrac{2}{5}$, $y = \dfrac{1}{10}$.

37. $\begin{cases} 5x + 6y - 3z = 6 \\ 4x - 7y - 2z = -3 \\ 3x + y - 7z = 1 \end{cases}$ becomes: $\begin{bmatrix} 5 & 6 & -3 & | & 6 \\ 4 & -7 & -2 & | & -3 \\ 3 & 1 & -7 & | & 1 \end{bmatrix}$

$\rightarrow \begin{bmatrix} 1 & 13 & -1 & | & 9 \\ 4 & -7 & -2 & | & -3 \\ 3 & 1 & -7 & | & 1 \end{bmatrix} \rightarrow \begin{bmatrix} 1 & 13 & -1 & | & 9 \\ 0 & -59 & 2 & | & -39 \\ -1 & -38 & -4 & | & -26 \end{bmatrix}$ Let's get some smaller numbers:

$\uparrow$ $\uparrow$

$R_1 = -1r_2 + r_1$ $R_2 = -4r_1 + r_2$
                        $R_3 = -3r_1 + r_3$

$\rightarrow \begin{bmatrix} 1 & 13 & -1 & | & 9 \\ 0 & -59 & 2 & | & -39 \\ 0 & 19 & 2 & | & 13 \end{bmatrix} \rightarrow \begin{bmatrix} 1 & 13 & -1 & | & 9 \\ 0 & -2 & 8 & | & 0 \\ 0 & 19 & 2 & | & 13 \end{bmatrix}$

$\uparrow$ $\uparrow$

$R_3 = -\dfrac{1}{2}r_3$ $R_2 = 3r_3 + r_2$

$\rightarrow \begin{bmatrix} 1 & 13 & -1 & | & 9 \\ 0 & 1 & -4 & | & 0 \\ 0 & 19 & 2 & | & 13 \end{bmatrix}$ We got a 1 in row 2, column 2, *and* avoided fractions, by taking a couple of extra steps!

$\uparrow$

$R_2 = -\dfrac{1}{2}r_2$

$\rightarrow \begin{bmatrix} 1 & 0 & 51 & | & 9 \\ 0 & 1 & -4 & | & 0 \\ 0 & 0 & 78 & | & 13 \end{bmatrix} \rightarrow \begin{bmatrix} 1 & 0 & 51 & | & 9 \\ 0 & 1 & -4 & | & 0 \\ 0 & 0 & 1 & | & \dfrac{13}{78} = \dfrac{1}{6} \end{bmatrix} \rightarrow \begin{bmatrix} 1 & 0 & 0 & | & \dfrac{-51}{6} + \dfrac{54}{6} \\ 0 & 1 & 0 & | & \dfrac{4}{6} \\ 0 & 0 & 1 & | & \dfrac{1}{6} \end{bmatrix}$

$\uparrow$ $\uparrow$ $\uparrow$

$R_1 = -13r_2 + r_1$ $R_3 = \dfrac{1}{78}r_3$ $R_1 = -51r_3 + r_1$

$R_3 = -19r_2 + r_3$ $R_2 = 4r_3 + r_2$

The solution is $x = \dfrac{1}{2}$, $y = \dfrac{2}{3}$, $z = \dfrac{1}{6}$.

39. $\begin{cases} x \quad\quad - 2z = 1 \\ 2x + 3y \quad\quad = -3 \\ 4x - 3y - 4z = 3 \end{cases}$ becomes: $\begin{bmatrix} 1 & 0 & -2 & | & 1 \\ 2 & 3 & 0 & | & -3 \\ 4 & -3 & -4 & | & 3 \end{bmatrix}$

$\rightarrow \begin{bmatrix} 1 & 0 & -2 & | & 1 \\ 0 & 3 & 4 & | & -5 \\ 0 & -3 & 4 & | & -1 \end{bmatrix} \rightarrow \begin{bmatrix} 1 & 0 & -2 & | & 1 \\ 0 & 1 & \dfrac{4}{3} & | & -\dfrac{5}{3} \\ 0 & -3 & 4 & | & -1 \end{bmatrix} \rightarrow \begin{bmatrix} 1 & 0 & -2 & | & 1 \\ 0 & 1 & \dfrac{4}{3} & | & -\dfrac{5}{3} \\ 0 & 0 & 8 & | & -6 \end{bmatrix}$

$\uparrow$ $\uparrow$ $\uparrow$

$R_2 = -2r_1 + r_2$ $R_2 = \dfrac{1}{3}r_2$ $R_3 = 3r_2 + r_3$

$R_3 = -4r_1 + r_3$

$$\rightarrow \begin{bmatrix} 1 & 0 & -2 & | & 1 \\ 0 & 1 & \frac{4}{3} & | & -\frac{5}{3} \\ 0 & 0 & 1 & | & -\frac{3}{4} \end{bmatrix} \rightarrow \begin{bmatrix} 1 & 0 & 0 & | & -\frac{1}{2} \\ 0 & 1 & 0 & | & -\frac{2}{3} \\ 0 & 0 & 1 & | & -\frac{3}{4} \end{bmatrix}$$

$$\uparrow \qquad\qquad \uparrow$$

$$R_3 = -\frac{1}{8}r_3 \qquad R_1 = 2r_3 + r_1$$

$$R_2 = -\frac{4}{3}r_3 + r_2$$

The solution is $x = -\frac{1}{2}$, $y = -\frac{2}{3}$, $z = -\frac{3}{4}$.

41. $\begin{cases} x - y + z = 0 \\ x - y - 5z = 6 \\ 2x - 2y + z = 1 \end{cases}$ (Get the constants on the right-hand-side.)

$$\begin{bmatrix} 1 & -1 & 1 & | & 0 \\ 1 & -1 & -5 & | & 6 \\ 2 & -2 & 1 & | & 1 \end{bmatrix} \rightarrow \begin{bmatrix} 1 & -1 & 1 & | & 0 \\ 0 & 0 & -6 & | & 6 \\ 0 & 0 & -1 & | & 1 \end{bmatrix}$$

$$\uparrow$$

$$R_2 = -1r_1 + r_2$$
$$R_3 = -2r_1 + r_3$$

It is impossible to obtain a 1 in row 2 column 2 (we cannot use row 1, since that would mess up the 0's we have in column 1).

So we focus on row 2, column 3:

$$\rightarrow \begin{bmatrix} 1 & -1 & 1 & | & 0 \\ 0 & 0 & 1 & | & -1 \\ 0 & 0 & -1 & | & 1 \end{bmatrix} \rightarrow \begin{bmatrix} 1 & -1 & 0 & | & 1 \\ 0 & 0 & 1 & | & -1 \\ 0 & 0 & 0 & | & 0 \end{bmatrix}$$

$$\uparrow \qquad\qquad \uparrow$$

$$R_2 = -\frac{1}{6}r_2 \qquad R_1 = -r_2 + r_1$$

$$R_3 = r_2 + r_3$$

Thus, we have: $x - y = 1$
$$z = -1$$

We can write the solution as:

$z = -1$, $x = y + 1$, where y can be any real number, or

$z = -1$, $y = x - 1$, where x can be any real number.

43. $\begin{cases} x - y - z - t = 1 \\ 2x + y + z + 2t = 3 \\ x - 2y - 2z - 3t = 0 \\ 3x - 4y + z + 5t = -3 \end{cases}$ becomes: $\begin{bmatrix} 1 & -1 & -1 & -1 & | & 1 \\ 2 & 1 & 1 & 2 & | & 3 \\ 1 & -2 & -2 & -3 & | & 0 \\ 3 & -4 & 1 & 5 & | & -3 \end{bmatrix}$

$\rightarrow \begin{bmatrix} 1 & -1 & -1 & -1 & | & 1 \\ 0 & 3 & 3 & 4 & | & 1 \\ 0 & -1 & -1 & -2 & | & -1 \\ 0 & -1 & 4 & 8 & | & -6 \end{bmatrix} \rightarrow \begin{bmatrix} 1 & -1 & -1 & -1 & | & 1 \\ 0 & 1 & 1 & 0 & | & -1 \\ 0 & -1 & -1 & -2 & | & -1 \\ 0 & -1 & 4 & 8 & | & -6 \end{bmatrix}$

↑   ↑

$R_2 = -2r_1 + r_2$ $R_2 = 2r_3 + r_2$
$R_3 = -1r_2 + r_3$
$R_4 = -3r_1 + r_4$

$\rightarrow \begin{bmatrix} 1 & 0 & 0 & -1 & | & 0 \\ 0 & 1 & 1 & 0 & | & -1 \\ 0 & 0 & 0 & -2 & | & -2 \\ 0 & 0 & 5 & 8 & | & -7 \end{bmatrix}$ Now we need a 1 in row 3, column 3.

↑

$R_1 = r_2 + r_1$
$R_3 = r_2 + r_3$
$R_4 = r_2 + r_4$

$\rightarrow \begin{bmatrix} 1 & 0 & 0 & -1 & | & 0 \\ 0 & 1 & 1 & 0 & | & -1 \\ 0 & 0 & 5 & 8 & | & -7 \\ 0 & 0 & 0 & -2 & | & -2 \end{bmatrix} \rightarrow \begin{bmatrix} 1 & 0 & 0 & -1 & | & 0 \\ 0 & 1 & 1 & 0 & | & -1 \\ 0 & 0 & 1 & \frac{8}{5} & | & -\frac{7}{5} \\ 0 & 0 & 0 & 1 & | & 1 \end{bmatrix}$

↑   ↑

Interchange r_3 and r_4 $R_3 = \frac{1}{5}r_3$

$R_4 = -\frac{1}{2}r_4$

$\rightarrow \begin{bmatrix} 1 & 0 & 0 & -1 & | & 0 \\ 0 & 1 & 0 & -\frac{8}{5} & | & \frac{2}{5} \\ 0 & 0 & 1 & \frac{8}{5} & | & -\frac{7}{5} \\ 0 & 0 & 0 & 1 & | & 1 \end{bmatrix} \rightarrow \begin{bmatrix} 1 & 0 & 0 & 0 & | & 1 \\ 0 & 1 & 0 & 0 & | & 2 \\ 0 & 0 & 1 & 0 & | & -3 \\ 0 & 0 & 0 & 1 & | & 1 \end{bmatrix}$

↑   ↑

$R_2 = -1r_3 + r_2$ $R_1 = r_4 + r_1$

$R_2 = \frac{8}{5}r_4 + r_2$

$R_3 = -\frac{8}{5}r_4 + r_3$

The solution is: $x = 1$, $y = 2$, $z = -3$, $t = 1$

45. $\begin{vmatrix} 3 & 4 \\ 1 & 3 \end{vmatrix} = (3)(3) - (1)(4) = 9 - 4 = 5$

47. $\begin{vmatrix} 1 & 4 & 0 \\ -1 & 2 & 6 \\ 4 & 1 & 3 \end{vmatrix} = 1 \begin{vmatrix} 2 & 6 \\ 1 & 3 \end{vmatrix} - 4 \begin{vmatrix} -1 & 6 \\ 4 & 3 \end{vmatrix} + 0 \begin{vmatrix} -1 & 2 \\ 4 & 1 \end{vmatrix}$

$\qquad = 1(6 - 6) - 4(-3 - 24) + 0$

$\qquad = 0 - 4(-27) + 0$

$\qquad = 108$

49. $\begin{vmatrix} 2 & 1 & -3 \\ 5 & 0 & 1 \\ 2 & 6 & 0 \end{vmatrix} = 2 \begin{vmatrix} 0 & 1 \\ 6 & 0 \end{vmatrix} - 1 \begin{vmatrix} 5 & 1 \\ 2 & 0 \end{vmatrix} + (-3) \begin{vmatrix} 5 & 0 \\ 2 & 6 \end{vmatrix}$

$\qquad = 2(0 - 6) - 1(0 - 2) + (-3)(30 - 0)$

$\qquad = -12 + 2 - 90$

$\qquad = -100$

51. $\begin{cases} x - 2y = 4 \\ 3x + 2y = 4 \end{cases}$

Here, $D = \begin{vmatrix} 1 & -2 \\ 3 & 2 \end{vmatrix} = 2 - (-6) = 8$

To obtain D_x, replace the first column in D by the column of constants.

$$D_x = \begin{vmatrix} 4 & -2 \\ 4 & 2 \end{vmatrix} = 8 - (-8) = 16$$

To obtain D_y, replace the second column of D by the column of constants.

$$D_y = \begin{vmatrix} 1 & 4 \\ 3 & 4 \end{vmatrix} = 4 - 12 = -8$$

Then, $x = \dfrac{D_x}{D} = \dfrac{16}{8} = 2$ and $y = \dfrac{D_y}{D} = \dfrac{-8}{8} = -1$

53. $\begin{cases} 2x + 3y - 13 = 0 \\ 3x - 2y \quad\quad = 0 \end{cases}$

$\begin{cases} 2x + 3y = 13 \\ 3x - 2y = 0 \end{cases}$

Here, $\qquad D = \begin{vmatrix} 2 & 3 \\ 3 & -2 \end{vmatrix} = -4 - 9 = -13$

$\qquad\qquad D_x = \begin{vmatrix} 13 & 3 \\ 0 & -2 \end{vmatrix} = -26 - 0 = -26$

$\qquad\qquad D_y = \begin{vmatrix} 2 & 13 \\ 3 & 0 \end{vmatrix} = 0 - 39 = -39$

By Cramer's Rule, $x = \dfrac{D_x}{D} = \dfrac{-26}{-13} = 2$ and $y = \dfrac{D_y}{D} = \dfrac{-39}{-13} = 3$

55.
$$\begin{cases} x + 2y - z = 6 \\ 2x - y + 3z = -13 \\ 3x - 2y + 3z = -16 \end{cases}$$

$$D = \begin{vmatrix} 1 & 2 & -1 \\ 2 & -1 & 3 \\ 3 & -2 & 3 \end{vmatrix} = 1\begin{vmatrix} -1 & 3 \\ -2 & 3 \end{vmatrix} - 2\begin{vmatrix} 2 & 3 \\ 3 & 3 \end{vmatrix} + (-1)\begin{vmatrix} 2 & -1 \\ 3 & -2 \end{vmatrix}$$

$$= 1(-3 + 6) - 2(6 - 9) + (-1)(-4 + 3)$$
$$= 3 + 6 + 1$$
$$= 10$$

$$D_x = \begin{vmatrix} 6 & 2 & -1 \\ -13 & -1 & 3 \\ -16 & -2 & 3 \end{vmatrix} = 6\begin{vmatrix} -1 & 3 \\ -2 & 3 \end{vmatrix} - 2\begin{vmatrix} -13 & 3 \\ -16 & 3 \end{vmatrix} + (-1)\begin{vmatrix} -13 & -1 \\ -16 & -2 \end{vmatrix}$$

$$= 6(-3 + 6) - 2(-39 + 48) + (-1)(26 - 16)$$
$$= 18 - 18 - 10$$
$$= -10$$

$$D_y = \begin{vmatrix} 1 & 6 & -1 \\ 2 & -13 & 3 \\ 3 & -16 & 3 \end{vmatrix} = 1\begin{vmatrix} -13 & 3 \\ -16 & 3 \end{vmatrix} - 6\begin{vmatrix} 2 & 3 \\ 3 & 3 \end{vmatrix} + (-1)\begin{vmatrix} 2 & -13 \\ 3 & -16 \end{vmatrix}$$

$$= 1(-39 + 48) - 6(6 - 9) + (-1)(-32 + 39)$$
$$= 9 + 18 - 7$$
$$= 20$$

$$D_z = \begin{vmatrix} 1 & 2 & 6 \\ 2 & -1 & -13 \\ 3 & -2 & -16 \end{vmatrix} = 1\begin{vmatrix} -1 & -13 \\ -2 & -16 \end{vmatrix} - 2\begin{vmatrix} 2 & -13 \\ 3 & -16 \end{vmatrix} + 6\begin{vmatrix} 2 & -1 \\ 3 & -2 \end{vmatrix}$$

$$= 1(16 - 26) - 2(-32 + 39) + 6(-4 + 3)$$
$$= -10 - 14 - 6$$
$$= -30$$

By Cramer's Rule, $x = \dfrac{D_x}{D} = \dfrac{-10}{10} = -1$, $y = \dfrac{D_y}{D} = \dfrac{20}{10} = 2$, and $y = \dfrac{D_z}{D} = \dfrac{-30}{10} = -3$

57. $\dfrac{6}{x(x - 4)}$

Step 1: This is a proper fraction.

Step 2: $q(x) = x(x - 4)$

This is Case 1 (nonrepeated linear factors).

Step 3: $\dfrac{6}{x(x - 4)} = \dfrac{A}{x} + \dfrac{B}{x - 4}$

Step 4: Multiply both sides by $x(x - 4)$: $6 = A(x - 4) + Bx$

Let $x = 0$: $6 = -4A$, or $A = -\dfrac{6}{4} = -\dfrac{3}{2}$

Let $x = 4$: $6 = 4B$, or $B = \dfrac{6}{4} = \dfrac{3}{2}$

Step 5: $\dfrac{6}{x(x - 4)} = \dfrac{-\dfrac{3}{2}}{x} + \dfrac{\dfrac{3}{2}}{x - 4}$

59.

$$\dfrac{x - 4}{x^2(x - 1)}$$

Step 1:   This is a proper fraction.

Step 2: $q(x) = x^2(x - 1)$

Note that x^2 is a *repeated linear* factor: $x^2 = (x)(x)$.

Step 3: $\dfrac{x - 4}{x^2(x - 1)} = \dfrac{A}{x} + \dfrac{B}{x^2} + \dfrac{C}{x - 1}$

Step 4:   Multiply by $x^2(x - 1)$: $x - 4 = Ax(x - 1) + B(x - 1) + Cx^2$

Let $x = 0$: $-4 = -B$, or $B = 4$

Let $x = 1$: $-3 = C$

Now choose any value of x, say

$$x = 2: \quad -2 = 2A + B + 4C$$
$$-2 = 2A + 4 - 12$$
$$2A = -2 - 4 + 12$$
$$A = 3$$

Step 5: $\dfrac{x - 4}{x^2(x - 1)} = \dfrac{3}{x} + \dfrac{4}{x^2} - \dfrac{3}{x - 1}$

61.

$$\dfrac{x}{(x^2 + 9)(x + 1)}$$

Step 1:   This is proper.

Step 2: $q(x) = (x^2 + 9)(x + 1)$

cannot be factored

This is Case 1 and Case 3.

Step 3: $\dfrac{x}{(x^2 + 9)(x + 1)} = \dfrac{Ax + B}{x^2 + 9} + \dfrac{C}{x + 1}$

Step 4: $x = (Ax + B)(x + 1) + C(x^2 + 9) \qquad (1)$

Now $q(x)$ only has one zero, $x = -1$:

Let $x = -1$: $-1 = 10C = -\dfrac{1}{10}$

To find the coefficients A and B, let's expand (1):

$$x = Ax^2 + Ax + Bx + B + Cx^2 + 9C$$
$$x = (A + C)x^2 + (A + B)x + (B + 9C)$$

Now equate coefficients of like powers of x:

For x^2: $0 = A + C$

For x: $1 = A + B$

For the constant: $0 = B + 9C$

But we know $C = -\dfrac{1}{10}$, so

$$0 = A + C$$
$$0 = A - \dfrac{1}{10}$$
$$A = \dfrac{1}{10}$$

and $0 = B + 9C$

$$0 = B - \dfrac{9}{10}$$
$$B = \dfrac{9}{10}$$

Step 5:
$$\frac{x}{(x^2 + 9)(x + 1)} = \frac{\frac{1}{10}x + \frac{9}{10}}{x^2 + 9} = \frac{-\frac{1}{10}}{x + 1}$$

63. $\dfrac{x^3}{\left(x^2 + 4\right)^2}$

Step 1: This is a proper fraction.

Step 2: $q(x) = (x^2 + 4)^2$

This is Case 4 (a *repeated* irreducible quadratic factor).

Step 3: $\dfrac{x^3}{\left(x^2 + 4\right)^2} = \dfrac{Ax + B}{x^2 + 4} + \dfrac{Cx + D}{\left(x^2 + 4\right)^2}$

Step 4: Clear of fractions:
$$x^3 = (Ax + B)(x^2 + 4) + Cx + D$$

Since $q(x)$ has <u>no</u> real zeros, we choose to expand the right-hand-side and equate coefficients of like powers of x:
$$x^3 = Ax^3 + Bx^2 + 4Ax + 4B + Cx + D$$
$$x^3 = Ax^3 + Bx^2 + (4A + C)x + (4B + D)$$

Coefficient of x^3: $1 = A$
Coefficient of x^2: $0 = B$
Coefficient of x: $0 = 4A + C$
Constant term: $0 = 4B + D$

Since $A = 1$, we have:
$$0 = 4A + C$$
$$0 = 4 + C$$
$$C = -4$$

Since $B = 0$, we have:
$$0 = 4B + D$$
$$0 = 0 + D$$
$$D = 0$$

Step 5: $\dfrac{x^3}{\left(x^2 + 4\right)^2} = \dfrac{x}{x^2 + 4} + \dfrac{-4x}{(x^2 + 4)^2}$

65. $\dfrac{x^2}{(x^2 + 1)(x^2 - 1)}$

Step 1: This fraction is proper.

Step 2: $q(x) = (x^2 + 1)(x^2 - 1) = (x^2 + 1)(x + 1)(x - 1)$

This is Case 1 and Case 3.

Step 3: $\dfrac{x^2}{(x^2 + 1)(x^2 - 1)} = \dfrac{Ax + B}{x^2 + 1} + \dfrac{C}{x + 1} + \dfrac{D}{x - 1}$

Step 4: $x^2 = (Ax + B)(x + 1)(x - 1) + C(x^2 + 1)(x - 1) + D(x^2 + 1)(x + 1)$

Let $x = 1$: $1 = 4D$ *or* $D = \dfrac{1}{4}$

Let $x = -1$: $1 = 4C$ *or* $C = -\dfrac{1}{4}$

Now we need two more equations:
Let $x = 0$: $0 = -B - C + D$
$$0 = -B + \frac{1}{4} + \frac{1}{4}$$
$$B = \frac{1}{2}$$

Finally, let $x = 2$:
$$4 = (2A + B)(3) + C(5) + D(15)$$
$$4 = 6A + 3B + 5C + 15D$$
$$4 = 6A + 3\left(\frac{1}{2}\right) + 5\left(-\frac{1}{4}\right) + 15\left(\frac{1}{4}\right)$$
$$4 = 6A + \frac{6}{4} - \frac{5}{4} + \frac{15}{4}$$
$$4 = 6A + \frac{16}{4}$$
$$0 = 6A$$
$$A = 0$$

Step 5:
$$\frac{x^2}{(x^2 + 1)(x^2 - 1)} = \frac{\frac{1}{2}}{x^2 + 1} + \frac{-\frac{1}{4}}{x + 1} + \frac{\frac{1}{4}}{x - 1}$$

67. $\begin{cases} 2x + y + 3 = 0 \quad (1) \\ x^2 + y^2 = 5 \quad (2) \end{cases}$

We solve (1) for y:
$$2x + y + 3 = 0$$
$$y = -2x - 3$$

Substitute this expression for y into (2):
$$x^2 + y^2 = 5$$
$$x^2 + (-2x - 3)^2 = 5$$
$$x^2 + 4x^2 + 12x + 9 = 5$$
$$5x^2 + 12x + 4 = 0$$
$$(5x + 2)(x + 2) = 0$$
$$\text{so } x = \frac{-2}{5} \text{ or } x = -2$$

If $x = \frac{-2}{5}$, then, from (1): $\quad y = -2x - 3$
$$y = \frac{4}{5} - 3 = \frac{-11}{5}$$

If $x = -2$, then: $\quad y = -2x - 3 = 1$

Thus, we have two possible solutions: $\quad x = \frac{-2}{5}, \; y = \frac{-11}{5}$ and $x = -2, \; y = 1$

Both of these check.

69. $\begin{cases} 2xy + y^2 = 10 \quad (1) \\ 3y^2 - xy = 2 \quad (2) \end{cases}$

$\begin{cases} 2xy + y^2 = 10 \quad (1) \\ -2xy + 6y^2 = 4 \quad (2) \quad \text{Multiply by 2.} \end{cases}$

$\begin{cases} 2xy + y^2 = 10 \quad (1) \\ 7y^2 = 14 \quad (2) \quad \text{Replace (2) by (1) + (2).} \end{cases}$

From (2): $\quad y^2 = 2$
$$y = \pm\sqrt{2}$$

If $y = \sqrt{2}$, then:

$$2xy + y^2 = 10 \qquad (1)$$
$$2\sqrt{2}x + 2 = 10$$
$$2\sqrt{2}x = 8$$
$$x = \frac{8}{2\sqrt{2}} = 2\sqrt{2}$$

If $y = -\sqrt{2}$, then:

$$-2\sqrt{2}x + 2 = 10$$
$$-2\sqrt{2}x = 8$$
$$x = -2\sqrt{2}$$

We have two possible solutions: $x = 2\sqrt{2}$, $y = \sqrt{2}$; and $x = -2\sqrt{2}$, $y = -\sqrt{2}$ and they both check.

71. $\begin{cases} x^2 + y^2 = 6y & (1) \\ x^2 = 3y & (2) \end{cases}$

From (2): $\quad y = \dfrac{1}{3}x^2$

Then, by (1):

$$x^2 + y^2 = 6y$$
$$x^2 + \left(\frac{1}{3}x^2\right)^2 = 6\left(\frac{1}{3}x^2\right)$$
$$x^2 + \frac{x^4}{9} = 2x^2$$
$$9x^2 + x^4 = 18x^2 \qquad \text{Multiply by 9.}$$
$$x^4 - 9x^2 = 0$$
$$x^2(x^2 - 9) = 0$$
$$x^2(x - 3)(x + 3) = 0$$

Thus, $x = 0$, $x = 3$, or $x = -3$.

If $x = 0$, then $y = \dfrac{x^2}{3} = 0$

If $x = 3$, then $y = \dfrac{x^2}{3} = 3$

If $x = -3$, then $y = \dfrac{x^2}{3} = 3$

All of these solutions check: $\quad x = 0, y = 0$; $x = 3, y = 3$; and $x = -3, y = 3$

73. $\begin{cases} 3x^2 + 4xy + 5y^2 = 8 & (1) \\ x^2 + 3xy + 2y^2 = 0 & (2) \end{cases}$

Equation (2) can be factored: $\qquad x^2 + 3xy + 2y^2 = 0$
$$(x + y)(x + 2y) = 0$$

So $x = -y$ or $x = -2y$.

From (1), if $x = -y$, then: $\quad 3(-y)^2 + 4(-y)y + 5y^2 = 8$
$$3y^2 - 4y^2 + 5y^2 = 8$$
$$4y^2 = 8$$
$$y^2 = 2$$
$$y = \pm\sqrt{2}$$

Since $x = -y$, if $y = \sqrt{2}$, $x = -\sqrt{2}$, and if $y = -\sqrt{2}$, $x = \sqrt{2}$.

This gives two possible solutions: $x = \sqrt{2}, y = -\sqrt{2}; x = -\sqrt{2}, y = \sqrt{2}$, both of which check.

On the other hand, if $x = -2y$, then from (1): $3(-2y)^2 + 4(-2y)y + 5y^2 = 8$

$$12y^2 - 8y^2 + 5y^2 = 8$$
$$9y^2 = 8$$
$$y^2 = \frac{8}{9}$$
$$y = \pm\sqrt{\frac{8}{9}} = \pm\frac{2\sqrt{2}}{3}$$

Then, since $x = -2y$, we have:

If $y = \dfrac{2\sqrt{2}}{3}$, then $x = \dfrac{-4\sqrt{2}}{3}$

If $y = \dfrac{-2\sqrt{2}}{3}$, then $x = \dfrac{4\sqrt{2}}{3}$

Let's check these two possible solutions:

For $x = \dfrac{4\sqrt{2}}{3}, y = \dfrac{-2\sqrt{2}}{3}$

$$\begin{cases} 3x^2 + 4xy + 5y^2 = 3\left(\dfrac{32}{9}\right) + 4\left(\dfrac{-16}{9}\right) + 5\left(\dfrac{8}{9}\right) = \dfrac{72}{9} = 8 \\ x^2 + 3xy + 2y^2 = \left(\dfrac{32}{9}\right) + 3\left(\dfrac{-16}{9}\right) + 2\left(\dfrac{8}{9}\right) = \dfrac{0}{9} = 0 \end{cases}$$

The solution $x = \dfrac{-4\sqrt{2}}{3}, y = \dfrac{2\sqrt{2}}{3}$ also checks.

Thus, we have four solutions: $x = \sqrt{2}, y = -\sqrt{2}; x = -\sqrt{2}, y = \sqrt{2};$

$$x = \frac{4\sqrt{2}}{3}, y = \frac{-2\sqrt{2}}{3}; \text{ and } x = \frac{-4\sqrt{2}}{3}, y = \frac{2\sqrt{2}}{3}$$

75. $\begin{cases} x^2 - 3x + y^2 + y = -2 & (1) \\ \dfrac{x^2 - x}{y} + y + 1 = 0 & (2) \end{cases}$

$\begin{cases} x^2 - 3x + y^2 + y + 2 = 0 & (1) \\ x^2 - x + y^2 + y = 0 & (2) \quad \text{Multiply by } y. \end{cases}$

We can now eliminate y:

$\begin{cases} x^2 - 3x + y^2 + y + 2 = 0 & (1) \\ -x^2 + x - y^2 - y = 0 & (2) \quad \text{Multiply by } -1. \end{cases}$

$\begin{cases} x^2 - 3x + y^2 + y + 2 = 0 & (1) \\ -2x + 2 = 0 & (2) \quad \text{Replace (2) by (1) + (2).} \end{cases}$

From (2): $x = 1$. Then from (1): $x^2 - 3x + y^2 + y + 2 = 0$

$$1 - 3 + y^2 + y + 2 = 0$$
$$y^2 + y = 0$$
$$y(y + 1) = 0$$

So, $y = 0$ or $y = -1$.

Thus, we have two possible solutions: $x = 1, y = 0$ or $x = 1, y = -1$.

From the original equation (2), we see that y *cannot* be 0. But the other solution checks. Thus, we have just one solution:

$$x = 1, y = -1.$$

77. $\begin{cases} -2x + y \leq 2 & (1) \\ x + y \geq 2 & (2) \end{cases}$

(a) $-2x + y \leq 2$:

   Step 1: Graph $-2x + y = 2$ with a solid line (since the inequality is nonstrict). The x-intercept is $x = -1$; the y-intercept is $y = 2$.

   Step 2: Let's use $(0, 0)$, which lies *below* the line, as a test point.

   Step 3: At $(0, 0)$, $-2x + y = 0 \leq 2$, so the inequality is *satisfied*.

   Step 4: Therefore, we shade the region *below* the line $-2x + y = 2$.

(b) $x + y \geq 2$:

   Step 1: Graph $x + y = 2$ with a solid line. The x-intercept is 2; the y-intercept is 2.

   Step 2: Use $(0, 0)$ *below* the line as a test point.

   Step 3: At $(0, 0)$, $x + y = 0 < 2$, so the inequality is *not* satisfied.

   Step 4: Shade the region *above* the line $x + y = 2$.

(c) We see that the overlapping shaded region is unbounded.

(d) There is only one vertex, the intersection of $-2x + y = 2$ and $x + y = 2$:

$$\begin{cases} -2x + y = 2 & (1) \\ x + y = 2 & (2) \end{cases}$$

$$\begin{cases} -2x + y = 2 & (1) \\ -x - y = -2 & (2) \end{cases} \quad \text{Multiply by } -1.$$

$$\begin{cases} -2x + y = 2 & (1) \\ -3x = 0 & (2) \end{cases} \quad \text{Replace (2) by (1) + (2).}$$

$$\begin{cases} -2x + y = 2 & (1) \\ x = 0 & (2) \end{cases}$$

$$\begin{cases} y = 2 & (1) \\ x = 0 & (2) \end{cases}$$

The vertex is $(0, 2)$.

79. $\begin{cases} x \geq 0 \\ y \geq 0 \\ x + y \leq 4 \\ 2x + 3y \leq 6 \end{cases}$

(a) $x \geq 0; y \geq 0$:

These inequalities require that our graph be located in quadrant I.

(b) $x + y \leq 4$:

   Step 1: Graph $x + y = 4$. The x-intercept is $x = 4$; the y-intercept is $y = 4$.

   Step 2: Use $(0, 0)$ below the line as a test point.

   Step 3: At $(0, 0)$, $x + y = 0 < 4$.

   Step 4: Therefore, we shade *below* the line $x + y = 4$.

(c) $2x + 3y \leq 6$:

   Step 1: Graph $2x + 3y = 6$. The x-intercept is 3; the y-intercept is 2.

   Step 2: Use $(0, 0)$ below the line as a test point.

   Step 3: At $(0, 0)$, $2x + 3y = 0 < 6$.

   Step 4: Shade *below* the line $2x + 3y = 6$.

(d) We see that the graph is bounded.

(e) There are *three* vertices:

   (1) The intersection of $x = 0$ and $y = 0$: $(0, 0)$.

   (2) The intersection of $2x + 3y = 6$ and the y-axis: $(0, 2)$.

   (3) The intersection of $2x + 3y = 6$ and the x-axis: $(3, 0)$.

81. $\begin{cases} x \geq 0 & (1) \\ y \geq 0 & (2) \\ 2x + y \leq 8 & (3) \\ x + 2y \geq 2 & (4) \end{cases}$

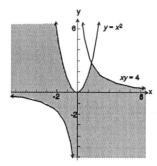

(a) $x \geq 0$; $y \geq 0$:
The graph will lie in quadrant I.

(b) $2x + y \leq 8$:
Step 1:   Graph $2x + y = 8$. The x-intercept is $x = 4$; the y-intercept is 8.
Step 2:   Use $(0, 0)$ below the line as a test point.
Step 3:   At $(0, 0)$, $2x + y = 0 < 8$.
Step 4:   Therefore, we shade *below* the line $2x + y = 8$.

(c) $x + 2y \geq 2$:
Step 1:   Graph $x + 2y = 2$. The x-intercept is 2; the y-intercept is 1.
Step 2:   Use $(0, 0)$ below the line as a test point.
Step 3:   At $(0, 0)$, $x + 2y = 0 < 2$, so the inequality is *not* satisfied.
Step 4:   Therefore, we shade *above* the line $x + 2y = 2$.

(d) The graph is bounded.

(e) There are *four* vertices, all are either x-intercepts or y-intercepts.
$(0, 1)$, $(0, 8)$, $(4, 0)$, and $(2, 0)$.

83.

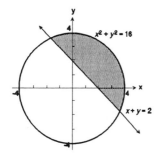

85.

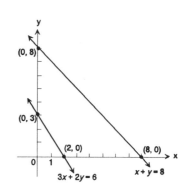

87. Maximize $z = 3x + 4y$
Subject to $x \geq 0$, $y \geq 0$, $3x + 2y \geq 6$, $x + y \leq 8$.
We graph the two lines:

(1) $y = -\dfrac{3}{2}x + 3$; y-intercept $= 3$; x-intercept $= 2$;

(2) $y = -x + 8$; y-intercept $= 8$; x-intercept $= 8$.

Vertex	Value of $z = 3x + 4y$
$(0, 3)$	$z = 12$
$(0, 8)$	$z = 32$
$(8, 0)$	$z = 24$
$(2, 0)$	$z = 6$

The maximum value of z is 32, at the point $(0, 8)$.

89. Minimize $z = 3x + 5y$
Subject to $x \geq 0, y \geq 0, x + y \geq 1, 3x + 2y \leq 12,$
$x + 3y \leq 12.$
We graph the three lines:

(1) $y = -x + 1$; y-intercept $= 1$; x-intercept $= 1$

(2) $y = -\dfrac{3}{2}x + 6$; y-intercept $= 6$; x-intercept $= 4$

(3) $y = -\dfrac{1}{3}x + 4$; y-intercept $= 4$; x-intercept $= 12$

We need the point of intersection of (2) and (3):

$$-\frac{3}{2}x + 6 = -\frac{1}{3}x + 4$$
$$-9x + 36 = -2x + 24$$
$$-7x = -12$$
$$x = \frac{12}{7}$$

and $y = -\dfrac{1}{3}x + 4 = -\dfrac{4}{7} + \dfrac{28}{7} = \dfrac{24}{7};\ \left(\dfrac{12}{7}, \dfrac{24}{7}\right)$

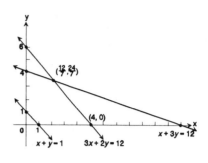

Vertex	Value of $z = 3x + 5y$
$(0, 1)$	$z = 5$
$(0, 4)$	$z = 20$
$\left(\dfrac{12}{7}, \dfrac{24}{7}\right)$	$z = \dfrac{36}{7} + \dfrac{120}{7} = \dfrac{156}{7} \approx 22.3$
$(4, 0)$	$z = 12$
$(1, 0)$	$z = 3$

The minimum value of z is 3, at $(1, 0)$.

91. Maximize $z = 5x + 4y$
Subject to $x \geq 0, y \geq 0, x + 2y \geq 2, 3x + 4y \leq 12, y \geq x$
We have three lines:

(1) $y = -\dfrac{1}{2}x + 1$; y-intercept $= 1$; x-intercept $= 2$

(2) $y = -\dfrac{3}{4}x + 3$; y-intercept $= 3$; x-intercept $= 4$

(3) $y = x$ We need the following points of intersection:

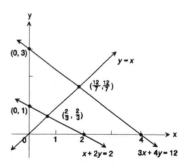

(1) and (3): $x = -\dfrac{1}{2}x + 1$
$$2x = -x + 2$$
$$3x = 2$$
$$x = \frac{2}{3}$$

and $y = x$
$$y = \frac{2}{3}$$
$$\left[\frac{2}{3}, \frac{2}{3}\right]$$

$$\text{(2) and (3):} \quad x = -\frac{3}{4}x + 3$$

$$4x = -3x + 12$$

$$7x = 12$$

$$x = \frac{12}{7}$$

$$\text{and} \quad y = x$$

$$y = \frac{12}{7}$$

$$\left[\frac{12}{7}, \frac{12}{7}\right]$$

Vertex	Value of $z = 5x + 4y$
$(0, 1)$	$z = 4$
$(0, 3)$	$z = 12$
$\left[\dfrac{12}{7}, \dfrac{12}{7}\right]$	$z = \dfrac{108}{7}$
$\left[\dfrac{2}{3}, \dfrac{2}{3}\right]$	$z = \dfrac{18}{3} = 6$

The maximum value of z is $\dfrac{108}{7}$, when $x = \dfrac{12}{7}$, and $y = \dfrac{12}{7}$.

93. $\begin{cases} 2x + 5y = 5 & (1) \\ 4x + 10y = A & (2) \end{cases}$

For the system of equations to have infinitely many solutions, all solutions must satisfy both equations. This means that both equations must be equal.

$$2x + 5y = 5 \rightarrow y = \frac{5 - 2x}{5}$$

$$4x + 10y = A \rightarrow y = \frac{A - 4x}{10}$$

Thus, $\quad \dfrac{5 - 2x}{5} = \dfrac{A - 4x}{10}$

$$50 - 20x = 5A - 20x$$

$$50 = 5A$$

$$10 = A$$

Solve by elimination:

$\begin{cases} 2x + 5y = 5 & (1) \\ 4x + 10y = A & (2) \end{cases}$

$\begin{cases} -4x - 10y = -10 & (1) \quad \text{Multiply by } -2. \\ 4x + 10y = A & (2) \end{cases}$

$\begin{cases} -4x - 10y = -10 & (1) \\ 0 = A - 10 & (2) \quad \text{Replace (2) by (1) + (2).} \end{cases}$

If $A - 10 = 0$, the system will have infinitely many solutions of the form:

$$-4x - 10y = -10$$

$$-4x = 10y - 10$$

$$x = \frac{-5}{2}y + \frac{5}{2}$$

where y is any real number.

Therefore, we need $A = 10$.

95. We are given $y = ax^2 + bx + c$. If $(0, 1)$ satisfies this equation, then:
$$1 = 0a + 0b + c,$$
or $\quad 1 = c$

From $(1, 0)$, we have: $\qquad 0 = a + b + c$
$$\text{or} \qquad a + b = -c$$
$$a + b = -1 \qquad \text{since } c = 1$$

From $(-2, 1)$: $\qquad\qquad 1 = 4a - 2b + c$
$$4a - 2b = 1 - c$$
$$4a - 2b = 0 \qquad \text{since } c = 1$$

We now have two equations in the two unknowns a and b.

$$\begin{cases} a + b = -1 & (1) \\ 4a - 2b = 0 & (2) \end{cases}$$

$$\begin{cases} 2a - 2b = -2 & (1) \\ 4a - 2b = 0 & (2) \end{cases} \quad \text{Multiply by 2.}$$

$$\begin{cases} 2a + 2b = -2 & (1) \\ \quad\;\; 6a = -2 & (2) \end{cases} \quad \text{Replace (2) by (1) + (2).}$$

From (2): $\quad a = \dfrac{-1}{3}$

Then from (1),
$$2a + 2b = -2$$
$$2\left[\dfrac{-1}{3}\right] + 2b = -2$$
$$2b = -2 + \dfrac{2}{3}$$
$$b = -1 + \dfrac{1}{3}$$
$$b = -\dfrac{2}{3}$$

Therefore, $y = ax^2 + bx + c = -\dfrac{1}{3}x^2 - \dfrac{2}{3}x + 1$.

97. Let $\quad x =$ blend of \$3.00 coffee
$\qquad\quad y =$ blend of \$6.00 coffee

$$\begin{cases} x + y = 100 \\ 3.00x + 6.00y = 3.90(x + y) \end{cases}$$

$$\begin{cases} x + y = 100 \\ 3x + 6y = 3.9x + 3.9y \end{cases}$$

$$\begin{cases} x + y = 100 \\ .9x - 2.1y = 0 \end{cases}$$

$$\begin{cases} x + y = 100 \\ 9x - 21y = 0 \end{cases}$$

$$\begin{cases} y = 100 - x \\ 9x - 21(100 - x) = 0 \end{cases}$$

$$\begin{cases} y = 100 - x \\ 9x - 2100 + 21x) = 0 \end{cases}$$

$$\begin{cases} y = 100 - x \\ 30x = 2100 \end{cases}$$

$$\begin{cases} y = 100 - x \\ x = 70 \end{cases}$$

$$\begin{cases} y = 30 \\ x = 70 \end{cases}$$

The desired blend consists of 70 lbs of \$3.00 coffee and 30 lbs of \$6.00 coffee.

99. Let x = number of small boxes
         y = number of medium boxes
         z = number of large boxes

(1) $x + 2y + 2z = 15$
(2) $x + y + 2z = 10$
(3) $y + 3z = 11$
(1)–(2) $y = 5$
(3) $5 + 3z = 11$
         $3z = 6$
          $z = 2$
(1) $x + 2(5) + 2(2) = 15$
        $x + 10 + 4 = 15$
             $x = 1$

You should buy 1 small box, 5 medium boxes, and 2 large boxes.

101. Let w = width of the plot
         ℓ = length of the plot
         d = length of the diagonal
         p = perimeter of the rectangle.

Now w, ℓ, d form a right triangle with hypotenuse = d, so we have
$$w^2 + \ell^2 = d^2 \qquad (1)$$
Also $2w + 2\ell = p \qquad (2)$

So we have two equations:
$$\begin{cases} w^2 + \ell^2 = (26)^2 & (1) \\ 2w + 2\ell = 68 & (2) \end{cases}$$
$$\begin{cases} w^2 + \ell^2 = 676 & (1) \\ w + \ell = 34 & (2) \quad \text{Divide by 2.} \end{cases}$$
Solve for ℓ in (2): $w + \ell = 34$
                  $\ell = 34 - w$
Substitute this into (1): $w^2 + \ell^2 = 676$
                $w^2 + (34 - w)^2 = 676$
        $w^2 + 1156 - 68w + w^2 = 676$
              $2w^2 - 68w + 480 = 0$
               $w^2 - 34w + 240 = 0$
             $(w - 24)(w - 10) = 0$
so $w = 24$ or $w = 10$
Finally, from (2), $\ell = 34 - w$, so: If $w = 24$, then $\ell = 10$;
                              If $w = 10$, then $\ell = 24$
Thus, the rectangle is 10 feet by 24 feet.

103. Let x and y be the legs of the triangle. We have:
$$\begin{cases} x + y + 6 = 14 & (1) \quad \text{The perimeter.} \\ x^2 + y^2 = 36 & (2) \quad \text{Pythagorean Theorem.} \end{cases}$$
From (1): $x + y = 8$
             $y = -x + 8$
Then, by (2): $x^2 + y^2 = 36$
              $x^2 + (-x + 8)^2 = 36$
        $x^2 + x^2 - 16x + 64 = 36$
            $2x^2 - 16x + 28 = 0$
             $x^2 - 8x + 14 = 0$

so $x = \dfrac{8 \pm \sqrt{64 - 4(14)}}{2} = \dfrac{8 \pm \sqrt{8}}{2}$ or $x = 4 \pm \sqrt{2}$

If $x = 4 + \sqrt{2}$, then $y = 8 - x = 4 - \sqrt{2}$

If $x = 4 - \sqrt{2}$, then $y = 8 - x = 4 + \sqrt{2}$

In either case, the lengths of the legs are $4 + \sqrt{2}$ inches and $4 - \sqrt{2}$ inches.

105.

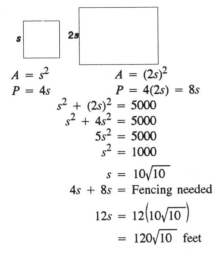

$A = s^2$

$P = 4s$

$A = (2s)^2$

$P = 4(2s) = 8s$

$$s^2 + (2s)^2 = 5000$$
$$s^2 + 4s^2 = 5000$$
$$5s^2 = 5000$$
$$s^2 = 1000$$
$$s = 10\sqrt{10}$$

$4s + 8s =$ Fencing needed

$$12s = 12\left(10\sqrt{10}\right)$$
$$= 120\sqrt{10} \text{ feet}$$

107. Let $\quad x =$ the amount Katy receives.

$\qquad y =$ the amount Mike receives.

$\qquad z =$ the amount Danny receives.

and $\quad w =$ the amount Colleen receives.

We are told: $x + y + z + w = 45$, and

$$y = 2x$$
$$w = x$$
$$z = \frac{x}{2}$$

Therefore, by substitution:

$$x + y + z + w = 45$$
$$x + 2x + \frac{x}{2} + x = 45$$
$$2x + 4x + x + 2x = 90 \qquad \text{Multiply by 2.}$$
$$9x = 90$$
$$x = 10$$

Katy receives $x = 10$ dollars.

Mike receives $y = 2x = 20$ dollars.

Danny receives $z = \dfrac{x}{2} = 5$ dollars.

Colleen receives $w = x = 10$ dollars.

109. Let $\quad x = $ # of hours for Katy to do the job alone,

$\qquad y = $ # of hours for Mike to do the job alone,

$\qquad z = $ # of hours for Danny to do the job alone.

We have:

	Fraction of the job done in one hour
Katy	$\dfrac{1}{x}$
Mike	$\dfrac{1}{y}$
Danny	$\dfrac{1}{z}$
Katy & Mike	$\dfrac{1}{x} + \dfrac{1}{y}$
Mike & Danny	$\dfrac{1}{y} + \dfrac{1}{z}$
Danny & Katy	$\dfrac{1}{x} + \dfrac{1}{z}$

But we are told that, working together, Katy and Mike can do the job in 1 hour and 20 minutes $\left(1\dfrac{2}{6} = \dfrac{8}{6} = \dfrac{4}{3} \text{ hours}\right)$.

Therefore, together, they complete $\dfrac{1}{\left(\dfrac{4}{3}\right)} = \dfrac{3}{4}$ of the job per hour:

$$\dfrac{1}{x} + \dfrac{1}{y} = \dfrac{3}{4} \quad (1)$$

Similarly, Mike and Danny do the job in 1 hour and 36 minutes

$\left(1\dfrac{36}{60} = 1\dfrac{6}{10} = \dfrac{16}{10} = \dfrac{8}{5} \text{ hr.}\right)$

Therefore, $\qquad \dfrac{1}{y} + \dfrac{1}{z} = \dfrac{1}{\left(\dfrac{8}{5}\right)} = \dfrac{5}{8} \quad (2)$

Finally, Danny and Katy take 2 hours 40 minutes $\left(2\dfrac{4}{6} = \dfrac{16}{6} = \dfrac{8}{3} \text{ hr.}\right)$ to do the job:

$$\dfrac{1}{x} + \dfrac{1}{z} = \dfrac{1}{\left(\dfrac{8}{3}\right)} = \dfrac{3}{8}$$

We have:
$$
\begin{cases}
\dfrac{1}{x} + \dfrac{1}{y} = \dfrac{3}{4} & (1) \\[2mm]
\dfrac{1}{y} + \dfrac{1}{z} = \dfrac{5}{8} & (2) \\[2mm]
\dfrac{1}{x} + \dfrac{1}{z} = \dfrac{3}{8} & (3)
\end{cases}
$$

Let $u = \dfrac{1}{x}, \ v = \dfrac{1}{y}, \ w = \dfrac{1}{z}$:

$$
\begin{cases}
u + v = \dfrac{3}{4} & (1) \\[2mm]
v + w = \dfrac{5}{8} & (2) \\[2mm]
u + w = \dfrac{3}{8} & (3)
\end{cases}
$$

Chapter 11 Systems of Equations and Inequalities

$$\begin{cases} u + v = \dfrac{3}{4} & (1) \\[2mm] v + w = \dfrac{5}{8} & (2) \\[2mm] -v + w = -\dfrac{3}{8} & (3) \end{cases} \quad \text{Replace (3) by (3) } - \text{ (1).}$$

$$\begin{cases} u + v = \dfrac{3}{4} & (1) \\[2mm] v + w = \dfrac{5}{8} & (2) \\[2mm] 2w = \dfrac{1}{4} & (3) \end{cases} \quad \text{Replace (3) by (3) } + \text{ (2).}$$

$$\begin{cases} u + v = \dfrac{3}{4} & (1) \\[2mm] v + w = \dfrac{5}{8} & (2) \\[2mm] w = \dfrac{1}{8} & (3) \end{cases} \quad \text{Divide by 2.}$$

$$\begin{cases} u + v = \dfrac{3}{4} & (1) \\[2mm] v + \dfrac{1}{8} = \dfrac{5}{8} & (2) \\[2mm] w = \dfrac{1}{8} & (3) \end{cases} \quad \text{Back-substitution; } w = \dfrac{1}{8}.$$

$$\begin{cases} u + v = \dfrac{3}{4} & (1) \\[2mm] v = \dfrac{1}{2} & (2) \\[2mm] w = \dfrac{1}{8} & (3) \end{cases}$$

$$\begin{cases} u = \dfrac{3}{4} & (1) \quad \text{Back-substitution; } v = \dfrac{1}{2}. \\[2mm] v = \dfrac{1}{2} & (2) \\[2mm] w = \dfrac{1}{8} & (3) \end{cases}$$

Finally, $\dfrac{1}{x} = u = \dfrac{1}{4}$, so $x = 4$; $\dfrac{1}{y} = v = \dfrac{1}{2}$, so $y = 2$; and $\dfrac{1}{z} = w = \dfrac{1}{8}$, so $z = 8$.

Katy can do the job in $x = 4$ hours;

Mike can do it in $y = 2$ hours;

Danny can do it in $z = 8$ hours.

111. Let x = Number of gasoline engines produced per week, and
y = Number of diesel engines produced per week

Then the objective function, or *cost*, is given by:

$C = 450x + 550y$ (in dollars)

Our constraints are:

(1) $x \geq 0, y \geq 0$

(2) $x \geq 20$ (must deliver at least 20 gasoline engines)

(3) $y \geq 15$ (must deliver at least 15 diesel engines)

(4) $x \leq 60$ (factory cannot make more than 60 gasoline engines)

(5) $x \leq 40$ (factory cannot make more than 40 diesel engines)

(6) $x + y \geq 50$ (must produce a total of 50 engines)

Equations (2)–(5) bound a rectangle with vertices (20, 15), (20, 40), (60, 40), and (60, 15). We will need two points of intersection:

(6) and (2): $\begin{cases} x + y = 50 \\ x \quad\quad = 20 \end{cases}$

Then $y = 30$

The point (20, 30) *is* in the feasible region.

(6) and (3): $\begin{cases} x + y = 50 \\ x \quad\quad = 15 \end{cases}$

Then $x = 35$

The point (35, 15) is also in the feasible region.

Vertex	Value of Cost: $C = 450x + 550y$
(20, 30)	$C = 25{,}500$
(20, 40)	$C = 31{,}000$
(60, 40)	$C = 49{,}000$
(60, 15)	$C = 35{,}250$
(35, 15)	$C = 24{,}000$

The minimum possible cost is $24,000, obtained by producing $x = 35$ gasoline engines, and $y = 15$ diesel engines.

Since the factory is *obligated* to deliver 20 gasoline engines and 15 diesel engines, their excess capacity is 15 gasoline engines (and no excess diesel engines).

Chapter 12

SEQUENCES; INDUCTION; COUNTING; PROBABILITY

12.1 Sequences

1. $a_1 = 1$
$a_2 = 2$
$a_3 = 3$
$a_4 = 4$
$a_5 = 5$

3. $a_1 = \dfrac{1}{1+2} = \dfrac{1}{3}$

$a_2 = \dfrac{2}{2+2} = \dfrac{2}{4} = \dfrac{1}{2}$

$a_3 = \dfrac{3}{3+2} = \dfrac{3}{5}$

$a_4 = \dfrac{4}{4+2} = \dfrac{4}{6} = \dfrac{2}{3}$

$a_5 = \dfrac{5}{5+2} = \dfrac{5}{7}$

5. $a_1 = (1)(1^2) = 1$
$a_2 = (-1)(2^2) = -4$
$a_3 = (1)(3^2) = 9$
$a_4 = (-1)(4^2) = -16$
$a_5 = (1)(5^2) = 25$

7. $a_1 = \dfrac{2}{4} = \dfrac{1}{2}$

$a_2 = \dfrac{4}{10} = \dfrac{2}{5}$

$a_3 = \dfrac{8}{28} = \dfrac{2}{7}$

$a_4 = \dfrac{16}{82} = \dfrac{8}{41}$

$a_5 = \dfrac{32}{244} = \dfrac{8}{61}$

9. $a_1 = \dfrac{-1}{(2)(3)} = \dfrac{-1}{6}$

$a_2 = \dfrac{1}{(3)(4)} = \dfrac{1}{12}$

$a_3 = \dfrac{-1}{(4)(5)} = \dfrac{-1}{20}$

$a_4 = \dfrac{1}{(5)(6)} = \dfrac{1}{30}$

$a_5 = \dfrac{-1}{(6)(7)} = \dfrac{-1}{42}$

11. $a_1 = \dfrac{1}{e}$

$a_2 = \dfrac{2}{e^2}$

$a_3 = \dfrac{3}{e^3}$

$a_4 = \dfrac{4}{e^4}$

$a_5 = \dfrac{5}{e^5}$

13. $\dfrac{n}{n+1}$

15. $\dfrac{1}{2^{n-1}}$

17. $(-1)^{n+1}$

19. $(-1)^{n+1} n$

21. $a_1 = 2$
$a_2 = 3 + 2 = 5$
$a_3 = 3 + 5 = 8$
$a_4 = 3 + 8 = 11$
$a_5 = 3 + 11 = 14$

23. $a_1 = -2$
$a_2 = 1 + -2 = -1$
$a_3 = 2 + -1 = 1$
$a_4 = 3 + 1 = 4$
$a_5 = 4 + 4 = 8$

25. $a_1 = 5$
$a_2 = 2 \cdot 5 = 10$
$a_3 = 2 \cdot 10 = 20$
$a_4 = 2 \cdot 20 = 40$
$a_5 = 2 \cdot 40 = 80$

27. $a_1 = 3$
$a_2 = \dfrac{3}{1} = 3$
$a_3 = \dfrac{3}{2}$
$a_4 = \dfrac{\frac{3}{2}}{3} = \dfrac{1}{2}$
$a_5 = \dfrac{\frac{1}{2}}{4} = \dfrac{1}{8}$

29. $a_1 = 1$
$a_2 = 2$
$a_3 = 1 \cdot 2 = 2$
$a_4 = 2 \cdot 2 = 4$
$a_5 = 2 \cdot 4 = 8$

31. $a_1 = A$
$a_2 = A + d$
$a_3 = A + 2d$
$a_4 = A + 3d$
$a_5 = A + 4d$

33. $a_1 = \sqrt{2}$
$a_2 = \sqrt{2 + \sqrt{2}}$
$a_3 = \sqrt{2 + \sqrt{2 + \sqrt{2}}}$
$a_4 = \sqrt{2 + \sqrt{2 + \sqrt{2 + \sqrt{2}}}}$
$a_5 = \sqrt{2 + \sqrt{2 + \sqrt{2 + \sqrt{2 + \sqrt{2}}}}}$

35. $\displaystyle\sum_{k=1}^{n} (k + 2) = 3 + 4 + 5 + \cdots + (n + 2)$

37. $\displaystyle\sum_{k=1}^{n} \dfrac{k^2}{2} = \dfrac{1}{2} + 2 + \dfrac{9}{2} + \cdots + \dfrac{n^2}{2}$

39. $\displaystyle\sum_{k=0}^{n} \dfrac{1}{3^k} = 1 + \dfrac{1}{3} + \dfrac{1}{9} + \cdots + \dfrac{1}{3^n}$

41. $\displaystyle\sum_{k=0}^{n-1} \dfrac{1}{3^{k+1}} = \dfrac{1}{3} + \dfrac{1}{9} + \dfrac{1}{27} + \cdots + \dfrac{1}{3^n}$

43. $\displaystyle\sum_{k=2}^{n} (-1)^k \ln k = \ln 2 - \ln 3 + \ln 4 = \cdots + (-1)^n \ln n$

45. $1 + 2 + 3 + \cdots + n = \displaystyle\sum_{k=1}^{n} k$

47. $\dfrac{1}{2} + \dfrac{2}{3} + \dfrac{3}{4} + \cdots + \dfrac{n}{n + 1} = \displaystyle\sum_{k=1}^{n} \dfrac{k}{k + 1}$

49. $1 - \dfrac{1}{3} + \dfrac{1}{9} - \dfrac{1}{27} + \cdots + (-1)^n \dfrac{1}{3^n} = \displaystyle\sum_{k=0}^{n} (-1)^k \left(\dfrac{1}{3^k} \right)$

51. $3 + \dfrac{3^2}{2} + \dfrac{3^3}{3} + \cdots + \dfrac{3^n}{n} = \displaystyle\sum_{k=1}^{n} \dfrac{3k}{k}$

53. $a + (a + d) + (a + 2d) + \cdots + (a + nd) = \displaystyle\sum_{k=1}^{n} a + (k - 1)d = \displaystyle\sum_{k=0}^{n} (a + kd)$

55. $a_1 = 1$, $a_2 = 1$, $a_3 = 2$, $a_4 = 3$, $a_5 = 5$, $a_6 = 8$, $a_7 = 13$, $a_8 = 21$, $a_{n+2} = a_{n+1} + a_n$
$a_8 = a_6 + a_7$
$a_8 = 8 + 13$
$a_8 = 21$

59. 1, 1, 2, 3, 5, 8, 13
The Fibonacci sequence.

12.2 Arithmetic Sequences

1. $d = a_{n+1} - a_n$
$d = (n + 5) - (n + 4)$
$d = n + 5 - n - 4$
$d = 1$
$a_1 = 5$
$a_2 = 6$
$a_3 = 7$
$a_4 = 8$

3. $d = [2(n + 1) - 5] - (2n - 5)$
$d = 2n + 2 - 5 - 2n + 5$
$d = 2$
$a_1 = -3$
$a_2 = -1$
$a_3 = 1$
$a_4 = 3$

5. $d = [6 - 2(n + 1)] - (6 - 2n)$
$d = 6 - 2n - 2 - 6 + 2n$
$d = -2$
$a_1 = 4$
$a_2 = 2$
$a_3 = 0$
$a_4 = -2$

7. $d = \left[\dfrac{1}{2} - \dfrac{1}{3}(n + 1)\right] - \dfrac{1}{2} - \dfrac{1}{3}n$
$d = \dfrac{1}{2} - \dfrac{1}{3}n - \dfrac{1}{3} - \dfrac{1}{2} + \dfrac{1}{3}n$
$d = -\dfrac{1}{3}$
$a_1 = \dfrac{1}{6}$
$a_2 = -\dfrac{1}{6}$
$a_3 = -\dfrac{1}{2}$
$a_4 = -\dfrac{5}{6}$

9. $d = (\ln 3^{n+1}) - (\ln 3^n)$
$d = (n + 1)\ln 3 - n(\ln 3)$
$d = \ln 3(n + 1 - n)$
$d = \ln 3$
$a_1 = \ln 3$
$a_2 = 2 \ln 3$
$a_3 = 3 \ln 3$
$a_4 = 4 \ln 3$

11. $a_n = 2 + (n - 1)3$
$a_n = 2 + 3n - 3$
$a_n = 3n - 1$
$a_5 = 3(5) - 1 = 14$

13. $a_n = 5 + (n - 1) - 3$
$a_n = 5 - 3n + 3$
$a_n = 8 - 3n$
$a_5 = 5 + 4(-3) = -7$

15. $a_n = 0 + (n - 1)\dfrac{1}{2}$
$a_n = \dfrac{1}{2}n - \dfrac{1}{2}$
$a_n = \dfrac{1}{2}(n - 1)$
$a_5 = 4\left(\dfrac{1}{2}\right) = 2$

17. $a_n = \sqrt{2} + (n - 1)\sqrt{2}$
$a_n = \sqrt{2}(1 + n - 1)$
$a_n = n\sqrt{2} = \sqrt{2}n$
$a_5 = 5\sqrt{2}$

19. The first term of the arithmetic sequence is $a = 2$, and the common difference is 2. The nth term obeys the formula:
$$a_n = 2 + (n - 1)2$$
Hence, the 12th term is:
$$a_{12} = 2 + (11)2$$
$$a_{12} = 2 + 22$$
$$a_{12} = 24$$

21. $a = 1, d = -3$
$a_n = 1 + (n - 1) - 3$
$a_n = 1 - 3n + 3$
$a_n = 4 - 3n$
$a_{10} = 1 + 9(-3)$
$a_{10} = 1 - 27 = -26$
Also, $a_{10} = 4 - 3(10)$
$a_{10} = 4 - 30$
$a_{10} = -26$

23. $a = a, d = b$
$a_n = a + (n - 1)b$
$a_8 = a + 7b$

25. $a_8 = a + 7d = 8$
$a_{20} = a + 19d = 44$
$-12d = -36$
$d = 3$
$a = 8 - 7d$
$a_1 = 8 - 7(3)$
$a_1 = 8 - 21$
$a_1 = -13$
$a_n = -16 + 3n$

27. $a_9 = a + 8d = -5$
$a_{15} = a + 14d = 31$
$-6d = -36$
$d = 6$
$a = -5 - 8d$
$a_1 = -5 - 8(6)$
$a_1 = -53$
$a_n = -59 + 6n$

29. $a_{15} = a + 14d = 0$
$a_{40} = a + 39d = -50$
$-25d = 50$
$d = -2$
$a = -14d$
$a_1 = -14(-2)$
$a_1 = 28$
$a_n = 30 - 2n$

31. $a_{14} = a + 13d = -1$
$a_{18} = a + 17d = -9$
$-4d = 8$
$d = -2$
$a = -1 - 13d$
$a_1 = -1 - 13(-2)$
$a_1 = 25$
$a_n = 27 - 2n$

33. $S_n = \dfrac{n}{2}(1 + (2n - 1))$

$\quad = \dfrac{n}{2} \cdot 2n$

$\quad = n^2$

35. $S_n = \dfrac{n}{2}(7 + (2 + 5n))$

$\quad = \dfrac{n}{2}(9 + 5n)$

37. $S_{35} = \dfrac{35}{2}(2 + 70)$

$\quad = \dfrac{35}{2} \cdot 72$

$\quad = 1260$

39. $S_{11} = \dfrac{12}{2}(5 + 49)$

$\quad = \dfrac{12}{2} \cdot 54$

$\quad = 324$

41.
```
sum seq(3.45N+4.
12,N,1,20,1)
            806.9
```

43.
```
sum seq(2.4N+.4,
N,1,15,1)
             294
```

45.
```
sum seq(2.58N+2.
32,N,1,25,1)
            896.5
```

47.
$$2x + 1 - (x + 3) = d$$
$$5x + 2 - (2x + 1) = d$$
$$x - 2 = d$$
$$\underline{3x + 1 = d}$$
$$-2x - 3 = 0$$
$$-2x = 3$$
$$x = \frac{-3}{2}$$

49. The total number of seats, s, is
$$S = \underbrace{25 + 26 + 27 + \cdots}$$

This is the sum of an arithmetic sequence with $d = 1$ and $n = 30$.
$$S_n = \frac{n}{2}[2a + (n - 1)d]$$
$$S_{30} = \frac{30}{2}[2(25) + 29(1)]$$
$$S_{30} = 15(50 + 29) = 1185 \text{ seats}$$

51. The lighter colored tile has 20 tiles in the bottom row and 1 tile at the top. The number of tiles decreases by one as we move up the triangle. This is an arithmetic sequence with $a = 20$ and $d = -1$. The number of terms to be added is 20.
$$S = \frac{20}{2}(40 + 19 \cdot (-1)) = \frac{20}{2} \cdot 21 = 210 \text{ tiles}$$

The darker tile has 19 tiles in the bottom row and 1 tile at the top. The number of tiles decreases by one as we move up the triangle. This is an arithmetic sequence with $a_1 = 19$ and $d = -1$. The number of terms to be added is 19.
$$S = \frac{19}{2}(38 + 18(-1)) = \frac{19}{2} \cdot 20 = 190 \text{ tiles}$$

12.3 Geometric Sequences; Geometric Series

1. $r = \dfrac{3^{n+1}}{3^n}$

$r = 3^{n+1-n}$

$r = 3$

$a_1 = 3$

$a_2 = 9$

$a_3 = 27$

$a_4 = 81$

3. $r = \dfrac{-3\left(\dfrac{1}{2}\right)^{n+1}}{-3\left(\dfrac{1}{2}\right)^n}$

$r = \left(\dfrac{1}{2}\right)^{n+1-n}$

$r = \dfrac{1}{2}$

$a_1 = -\dfrac{3}{2}$

$a_2 = -\dfrac{3}{4}$

$a_3 = -\dfrac{3}{8}$

$a_4 = -\dfrac{3}{16}$

5. $r = \dfrac{\dfrac{2^{(n+1)-1}}{4}}{\dfrac{2^{n-1}}{4}}$

$r = \dfrac{2^n}{2^{n-1}}$

$r = 2^{n-(n-1)}$

$r = 2$

$a_1 = \dfrac{1}{4}$

$a_2 = \dfrac{1}{2}$

$a_3 = 1$

$a_4 = 2$

7. $r = \dfrac{2^{\frac{n+1}{3}}}{2^{\frac{n}{3}}}$

$r = 2^{\frac{n+1}{3} - \frac{n}{3}}$

$r = 2^{1/3}$

$a_1 = 2^{1/3}$

$a_2 = 2^{2/3}$

$a_3 = 2$

$a_4 = 2^{4/3}$

9. $r = \dfrac{\dfrac{3^{(n-1)+1}}{2^{n+1}}}{\dfrac{3^{n-1}}{2^n}}$

$r = \dfrac{3^n}{2(3^{n-1})}$

$r = \dfrac{1}{2} 3^{n-(n-1)}$

$r = \dfrac{3}{2}$

$a_1 = \dfrac{1}{2}$

$a_2 = \dfrac{3}{4}$

$a_3 = \dfrac{9}{8}$

$a_4 = \dfrac{27}{16}$

11. Arithmetic

$d = [(n+1) + 2] - (n + 2)$

$d = n + 3 - n - 2$

$d = n + 3 - n - 2$

$d = 1$

13. Neither

15. Arithmetic

$d = \left[3 - \dfrac{2}{3}(n + 1)\right] - \left(3 - \dfrac{2}{3}n\right)$

$d = 3 - \dfrac{2}{3}n - \dfrac{2}{3} - 3 + \dfrac{2}{3}n$

$d = \dfrac{-2}{3}$

17. Neither

19. Geometric

$r = \dfrac{\left(\dfrac{2}{3}\right)^{n+1}}{\left(\dfrac{2}{3}\right)^n}$

$r = \dfrac{2^{n+1-n}}{3}$

$r = \dfrac{2}{3}$

21. Geometric

$r = \dfrac{-\left(2^{(n+1)-1}\right)}{-\left(2^{n-1}\right)}$

$r = 2^{n-(n-1)}$

$r = 2$

23. Geometric

$r = \dfrac{3^{\frac{n+1}{2}}}{3^{\frac{n}{2}}}$

$r = 3^{\frac{n+1}{2} - \frac{n}{2}}$

$r = 3$

$r = 3^{1/2}$

25. $a_n = ar^{n-1}$

$a_5 = 3^4(2) = 2 \cdot 81 = 162$

$a_n = 2 \cdot 3^{n-1}$

27. $a_5 = -1^4(5) = 1(5) = 5$

$a_n = (-1)^{n-1}(5)$

29. $a_5 = \left(\dfrac{1}{2}\right)^4 (0) = 0$

$a_n = \left(\dfrac{1}{2}\right)^{n-1} (0)$

$a_n = 0$

31. $a_5 = \sqrt{2}^4\left(\sqrt{2}\right) = \sqrt{2}^5 = 4\sqrt{2}$

$a_n = \sqrt{2}^{n-1}\sqrt{2}$

$a_n = \left(\sqrt{2}\right)^n$

33. The first term of this geometric sequence is $a = 1$, and the common ratio is $\dfrac{1}{2} \cdot \dfrac{\frac{1}{2}}{1} = \dfrac{1}{2}$.

The nth term obeys the formula: $a_n = \left(\dfrac{1}{2}\right)^{n-1}(1) = \left(\dfrac{1}{2}\right)^{n-1}$

$$a_7 = \left(\dfrac{1}{2}\right)^8 = \dfrac{1}{64}$$

35. $a = 1,\ r = -1$
$a_n = (-1)^{n-1}(1)$
$a_n = (-1)^{n-1}$
$a_9 = (-1)^8 = 1$

37. $a = 0.4,\ r = 0.1$
$a_n = (0.1)^{n-1}(0.4)$
$a_8 = (0.1)^7(0.4)$
$a_8 = (0.0000001)(0.4)$
$a_8 = 0.00000004$

39. The sequence $\dfrac{2^{n-1}}{4}$ is a geometric sequence with $a = \dfrac{1}{4}$ and $r = 2$.

$$S_n = \sum_{k=1}^{n} a_k = a_1 + a_2 + \dots + a_n = a\left(\dfrac{1 - r^n}{1 - r}\right)$$

In this sequence, $S_n = \displaystyle\sum_{k=1}^{n} \dfrac{2^{k-1}}{4} = \dfrac{1}{4} + \dfrac{2}{4} + \dfrac{2^2}{4} + \dfrac{2^3}{4} + \dots + \dfrac{2^{n-1}}{4} = \dfrac{1}{4}\left[\dfrac{1 - 2^n}{1 - 2}\right] = -\dfrac{1}{4}(1 - 2^n)$

41. Geometric sequence: $a = \dfrac{2}{3},\ r = \dfrac{2}{3}$

$$S_n = \sum_{k=1}^{n}\left(\dfrac{2}{3}\right)^k = \dfrac{2}{3} + \left(\dfrac{2}{3}\right)^2 + \left(\dfrac{2}{3}\right)^3 + \dots + \left(\dfrac{2}{3}\right)^n$$

$$= \dfrac{2}{3}\left[\dfrac{1 - \left(\dfrac{2}{3}\right)^n}{1 - \left(\dfrac{2}{3}\right)}\right] = \dfrac{2}{3}\left[\dfrac{1 - \left(\dfrac{2}{3}\right)^n}{\dfrac{1}{3}}\right] = 2\left[1 - \left(\dfrac{2}{3}\right)^n\right]$$

43. Geometric sequence: $a = -1,\ r = 2$

$$S_n = \sum_{k=1}^{n} -\left(2^{k-1}\right) = -1 - 2 - 4 - 8 - \dots - \left(2^{n-1}\right) = -1\left(\dfrac{1 - 2^n}{1 - 2}\right) = 1 - 2^n$$

45.
```
(1/4)sum seq(2^N
,N,0,14,1)
          8191.75
```

47.
```
sum seq((2/3)^N,
N,1,15,1)
       1.995432683
```

49.
```
-1sum seq(2^N,N,
0,14,1)
           -32767
```

51. This geometric series has first term $a = 1$ and common ratio $r = \frac{1}{3}$. Since $|r| < 1$, its sum is:
$$1 + \frac{1}{3} + \frac{1}{9} + \cdots = \frac{1}{1 - \frac{1}{3}} = \frac{3}{2}$$

53. This geometric series has first terms $a = 8$ and common ratio $r = \frac{1}{2}$. Since $|r| < 1$, its sum is:
$$8 + 4 + 2 + \cdots = \frac{8}{1 - \frac{1}{2}} = 16$$

55. This geometric series has first terms $a = 2$ and common ratio $r = \frac{-1}{4}$. Since $|r| < 1$, its sum is:
$$2 + \frac{1}{2} + \frac{1}{8} - \frac{1}{32} + \cdots = \frac{2}{1 + \frac{1}{4}} = \frac{8}{5}$$

57. This geometric series has first terms $a = 5$ and common ratio $r = \frac{1}{4}$. Since $|r| < 1$, its sum is:
$$\sum_{k=1}^{\infty} 5\left(\frac{1}{4}\right)^{k-1} = \frac{5}{1 - \frac{1}{4}} = \frac{20}{3}$$

59. This geometric series has first term $a = 6$ and common ratio $r = \frac{-2}{3}$. Since $|r| < 1$, its sum is:
$$\sum_{k=1}^{\infty} 6\left(\frac{-2}{3}\right)^{k-1} = \frac{6}{1 + \frac{2}{3}} = \frac{18}{5}$$

61. The ratio of successive terms must be the same. Thus,
$$\frac{x + 2}{x} = \frac{x + 3}{x + 2}$$
$$x^2 + 4x + 4 = x^2 + 3x$$
$$x = -4$$

63. (a) $a = 2$, geometric sequence
$a_{10} = (.9)^9(2) = 0.775$ feet

(b) $2(.9)^{n-1} < 1$ since $a_1 = 2$
$.9^{n-1} < .5$
$(n - 1) \log .9 < \log .5$
$$n - 1 < \frac{\log .5}{\log .9}$$
$n - 1 < 6.58$
$n < 7.58$
So, on the 8th swing of the pendulum, the arc is less than 1 foot.

(c) The sum of a geometric sequence with $n = 15$, $a = 2$, and $d = .9$ is
$$S_{15} = 2\left(\frac{1 - .9^{15}}{1 - .9}\right) = 15.88 \text{ feet}$$

(d) We must find the sum of an infinite geometric sequence with $a = 2$ and $d = 0.9$.
$$S_n = \frac{2}{1 - .9} = 20 \text{ feet}$$

65. **Begin:**   **After 1 year:**   **After 2 years:**
18,000   18000 + .05(18000) $18000(1 + .05) + 18000(1 + .05)(.05) = 18000(1 + .05)^2$

After 4 years, the salary is $18,000(1 + .05)^4 = \$21,879.11$

67. With option 1, your total salary is $\$2,000,000(7) + \$100,000(7) = \$14,700,000$.

With option 2, your total salary is the sum of a geometric sequence with $a = 2,000,000$, $r = 1.045$ and $n = 7$.
$$S_7 = 2,000,000\left(\frac{1 - (1.045)^7}{1 - 1.045}\right) = \$16,038,304$$

With option 3, your total salary is the sum of an arithmetic sequence with $a = 2,000,000$, $d = 95,000$ and $n = 7$, so
$$S_7 = \frac{7}{2}(4,000,000 + 6(95,000)) = \$15,995,000$$

So option 2 provides the most moeny over the 7-year period, and option 1 provides the least.

69. The total number grains required would be the sum of a geometric sequence with $a = 1$, $r = 2$, and $n = 64$.

$$S_{64} = 1\left(\frac{1 - 2^{64}}{1 - 2}\right) = 1.845 \times 10^{19} \text{ grains}$$

71.

x	$\dfrac{1}{1 - x}$	n
0.1	1.111	3
0.25	1.333	4
0.5	2.000	8
0.75	4.00	21
0.9	10	66

We want the error to be less than .01. So, we solve the following inequality for n.

$$\left|\frac{1 - x^n}{1 - x} - \frac{1}{1 - x}\right| < 0.01$$

$$\left|\frac{-x^n}{1 - x}\right| < 0.01$$

$$\left|\frac{x^n}{1 - x}\right| < 0.01$$

Therefore, the expansion requires 3 terms. Similar calculations are used for $x = 0.25$, $x = 0.5$, $x = 0.75$, and $x = 0.9$.

73. Both options are geometric sequences.

A: After the 5th year, your salary would be $a_5 = \$20{,}000(1.06)^4 = \$25{,}250$

Your salary over the 5 years would be $S_5 = \$20{,}000\left(\dfrac{1 - 1.06^5}{1 - 1.06}\right) = \$112{,}472$

B: After the 5th year, your salary would be $a_5 = \$22{,}000(1.03)^4 = \$24{,}761$

Your salary over the 5 years would be $S_5 = \$22{,}000\left(\dfrac{1 - 1.03^5}{1 - 1.03}\right) = \$116{,}801$

Option B is better than A.

12.4 Mathematical Induction

1. (I) $n = 1$: $2 \cdot 1 = 2$ and $1(1 + 1) = 2$

(II)   If $2 + 4 + 6 + \cdots + 2k = k(k + 1)$,

then $2 + 4 + 6 + \cdots + 2k + 2(k + 1)$

$$= [2 + 4 + 6 + \cdots + 2k] + 2(k + 1) = k(k + 1) + 2(k + 1)$$
$$= k^2 + k + 2k + 2 = k^2 + 3k + 2 = (k + 1)(k + 2)$$

3. (I) $n = 1$: $1 + 2 = 3$ and $\frac{1}{2}(1)(1 + 5) = \frac{1}{2}(6) = 3$

(II) If $3 + 4 + 5 + \cdots + (k + 2) = \frac{1}{2}k(k + 5)$,

then $3 + 4 + 5 + \cdots + (k + 2) + [(k + 1) + 2]$
$= [3 + 4 + 5 + \cdots + (k + 2)] + (k + 3)$
$= \frac{1}{2}k(k + 5) + k + 3 = \frac{1}{2}k^2 + \frac{5}{2}k + k + 3$
$= \frac{1}{2}k^2 + \frac{7}{2}k + 3 = \frac{1}{2}(k^2 + 7k + 6) = \frac{1}{2}(k + 1)(k + 6)$

5. (I) $n = 1$: $3 \cdot 1 - 1 = 2$ and $\frac{1}{2}(1)[3(1) + 1] = \frac{1}{2}(4) = 2$

(II) If $2 + 5 + 8 + \cdots + (3k - 1) = \frac{1}{2}k(3k + 1)$,

then $2 + 5 + 8 + \cdots + (3k - 1) + [3(k + 1) - 1]$
$= [2 + 5 + 8 + \cdots + (3k - 1)] + 3k + 2$
$= \frac{1}{2}k(3k + 1) + (3k + 2) = \frac{3}{2}k^2 + \frac{1}{2}k + 3k + 2$
$= \frac{3}{2}k^2 + \frac{7}{2}k + 2 = \frac{1}{2}(3k^2 + 7k + 4) = \frac{1}{2}(k + 1)(3k + 4)$

7. (I) $n = 1$: $2^{1-1} = 1$ and $2^1 - 1 = 1$

(II) If $1 + 2 + 2^2 + \cdots + 2^{k-1} = 2^k - 1$,
then $1 + 2 + 2^2 + \cdots + 2^{k-1} + 2^{(k+1)} - 1$
$= (1 + 2 + 2^2 + \cdots + 2^{k-1}) + 2^k = 2k - 1 + 2^k = 2(2^k) - 1$
$= 2^{k+1} - 1$

9. (I) $n = 1$: $4^{1-1} = 1$ and $\frac{1}{3}(4^1 - 1) = \frac{1}{3}(3) = 1$

(II) If $1 + 4 + 4^2 + \cdots + 4^{k-1} = \frac{1}{3}(4^k - 1)$,

then $1 + 4 + 4^2 + \cdots + 4^{k-1} + 4^{k+1)-1}$
$= (1 + 4 + 4^2 + \cdots + 4^{(k-1)}) + 4^k$
$= \frac{1}{3}(4^k - 1) + 4^k = \frac{1}{3}[4^k - 1 + 3(4^k)] = \frac{1}{3}[4(4^k) - 1]$
$= \frac{1}{3}(4^{k+1} - 1)$

11. **(I)** $n = 1$: $\dfrac{1}{1 \cdot 2} = \dfrac{1}{2}$ and $\dfrac{1}{1 + 1} = \dfrac{1}{2}$

(II) If $\dfrac{1}{1 \cdot 2} + \dfrac{1}{2 \cdot 3} + \dfrac{1}{3 \cdot 4} + \cdots + \dfrac{1}{k(k + 1)} = \dfrac{k}{k + 1}$,

then $\dfrac{1}{1 \cdot 2} + \dfrac{1}{2 \cdot 3} + \dfrac{1}{3 \cdot 4} + \cdots + \dfrac{1}{k(k + 1)} + \dfrac{1}{(k + 1)[(k + 1) + 1]}$

$= \left[\dfrac{1}{1 \cdot 2} + \dfrac{1}{2 \cdot 3} + \dfrac{1}{3 \cdot 4} + \cdots + \dfrac{1}{k(k + 1)}\right] + \dfrac{1}{(k + 1)(k + 2)}$

$= \dfrac{k}{k + 1} + \dfrac{1}{(k + 1)(k + 2)} = \dfrac{k(k + 2) + 1}{(k + 1)(k + 2)} = \dfrac{k^2 + 2k + 1}{(k + 1)(k + 2)}$

$= \dfrac{(k + 1)(k + 1)}{(k + 1)(k + 2)} = \dfrac{k + 1}{k + 2}$

13. **(I)** $n = 1$: $1^2 = 1$ and $\dfrac{1}{6} \cdot 1 \cdot 2 \cdot 3 = 1$

(II) If $1^2 + 2^2 + 3^2 + \cdots + k^2 = \dfrac{1}{6}k(k + 1)(2k + 1)$,

then $1^2 + 2^2 + 3^2 + \cdots + k^2 + (k + 1)^2$
$= (1^2 + 2^2 + 3^2 + \cdots + k^2) + (k + 1)^2$

$= \dfrac{1}{6}k(k + 1)(2k + 1) + (k + 1)^2 = \dfrac{1}{6}k(2k^2 + 3k + 1) + (k^2 + 2k + 1)$

$= \dfrac{1}{3}k^3 + \dfrac{1}{2}k^2 + \dfrac{1}{6}k + k^2 + 2k + 1 = \dfrac{1}{3}k^3 + \dfrac{3}{2}k^2 + \dfrac{13}{6}k + 1$

$= \dfrac{1}{6}(2k^3 + 9k^2 + 13k + 6) = \dfrac{1}{6}(k + 1)(2k^2 + 7k + 6)$

$= \dfrac{1}{6}(k + 1)(k + 2)(2k + 3)$

15. **(I)** $n = 1$: $5 - 1 = 4$ and $\dfrac{1}{2}(9 - 1) = \dfrac{1}{2} \cdot 8 = 4$

(II) If $4 + 3 + 2 + \cdots + (5 - k) = \dfrac{1}{2}k(9 - k)$,

then $4 + 3 + 2 + \cdots + (5 - k) = [5 - (k + 1)] = [4 + 3 + 2 + \cdots + (5 - k)] + [5 - (k + 1)]$

$= \dfrac{1}{2}k(9 - k) + 4 - k = \dfrac{9}{2}k - \dfrac{1}{2}k^2 + 4 - k = -\dfrac{1}{2}k^2 + \dfrac{7}{2}k + 4$

$= \dfrac{1}{2}(-k^2 + 7k + 8) = \dfrac{1}{2}(k + 1)(8 - k) = \dfrac{1}{2}(k + 1)[9 - (k + 1)]$

17. **(I)** $n = 1$: $1 \cdot (1 + 1) = 2$ and $\dfrac{1}{3} \cdot 1 \cdot 2 \cdot 3 = 2$

(II) If $1 \cdot 2 + 2 \cdot 3 + 3 \cdot 4 + \cdots + k(k + 1) = \dfrac{1}{3}k(k + 1)(k + 2)$,

then $1 \cdot 2 + 2 \cdot 3 + 3 \cdot 4 + \cdots + k(k + 1) + (k + 1)(k + 1 + 1)$
$= [1 \cdot 2 + 2 \cdot 3 + 3 \cdot 4 + \cdots + k(k + 1)] + (k + 1)(k + 2)$

$= \dfrac{1}{3}k(k + 1)(k + 2) + (k + 1)(k + 2) = (k + 1)(k + 2)\left[\dfrac{1}{3}k + 1\right]$

$= (k + 1)(k + 2)\dfrac{1}{3}(k + 3) = \dfrac{1}{3}(k + 1)(k + 2)(k + 3)$

19.   (I) $n = 1$: $1^2 + 1 = 2$ is divisible by 2.

     (II)   If $k^2 + k$ is divisible by 2,

         then $(k + 1)^2 + (k + 1) = k^2 + 2k + 1 + k + 1 = (k^2 + k) + (2k + 2)$

         Since $k^2 + k$ is divisible by 2 and $2k + 2$ is divisible by 2, then $(k + 1)^2 + k + 1$ is divisible by 2.

21.   (I) $n = 1$: $1^2 - 1 + 2 = 2$ is divisible by 2.

     (II)   If $k^2 - k + 2$ is divisible by 2,

         then $(k + 1)^2 - (k + 1) + 2$

$$= k^2 + 2k + 1 - k - 1 + 2 = (k^2 - k + 2) + 2k + 1 - 1$$
$$= (k^2 - k + 2) + 2k$$

         Since $k^2 - k + 2$ is divisible by 2 and $2k$ is divisible by 2, then $(k + 1)^2 - (k + 1) + 2$ is divisible by 2.

23.   (I) $n = 1$: If $x > 1$, then $x^1 = x > 1$.

     (II)   Assume, for any natural number k, that if $x > 1$, then $x^k > 1$. Show that if $x^k > 1$, then $x^{k+1} > 1$:

$$x^{k+1} = x^k \cdot x > 1 \cdot x = x > 1$$
$$\uparrow$$
$$x^k > 1$$

25.   (I) $n = 1$: $a - b$ is a factor of $a^1 - b^1 = a - b$.

     (II)   If $a - b$ is a factor of $a^k - b^k$, show that $a - b$ is a factor of $a^{k+1} - b^{k+1} = a(a^k - b^k) + b^k(a - b)$. Since $a - b$ is a factor of $a^k - b^k$ and $a - b$ is a factor of $a - b$, then $a - b$ is a factor of $a^{k+1} - b^{k+1}$.

27. $n = 1$: $1^2 - 1 + 41 = 41$ is a prime number.

     $n = 41$: $41^2 - 41 + 41 = 1681 = 41^2$ is not prime.

29.   (I) $n = 1$: $ar^{1-1} = a \cdot 1 = a$ and $a \cdot \dfrac{1 - r^1}{1 - r} = a$ because $r \neq 1$.

     (II)   If $a + ar + ar + ar^2 + \cdots + ar^{k-1} = a \cdot \left(\dfrac{1 - r^k}{1 - r} \right)$

         then, $a + ar + ar^2 + \cdots + ar^{k-1} + ar^{(k+1)-1}$

$$= (a + ar + ar^2 + \cdots + ar^{k-1}) + ar^k$$
$$= a \cdot \left(\frac{1 - r^k}{1 - r} \right) + ar^k = \frac{a(1 - r^k) + ar^k(1 - r)}{1 - r}$$
$$= \frac{a - ar^k + ar^k - ar^{k+1}}{1 - r} = a \cdot \left(\frac{1 - r^{k+1}}{1 - r} \right)$$

31.   (I) $n = 3$: The sum of the angles of a triangle is $(3 - 2) \cdot 180° = 180°$.

     (II)   Assume that for any integer k the sum of the angles of a convex polygon of k sides is $(k - 2) \cdot 180°$. A convex polygon of $k + 1$ sides consists of a convex polygon k sides plus a triangle. See the illustration. The sum of the angles is $(k - 2)180° + 180° = (k - 1)180°$. Since Conditions I and II have been met, the result follows.

k sides

$k + 1$ sides

12.5 The Binomial Theorem

1. $\begin{pmatrix} 5 \\ 3 \end{pmatrix} = \dfrac{5!}{3!2!} = \dfrac{5 \cdot 4}{2} = 10$

3. $\begin{pmatrix} 7 \\ 5 \end{pmatrix} = \dfrac{7!}{5!2!} = \dfrac{7 \cdot 6}{2} = 21$

5. $\begin{pmatrix} 50 \\ 49 \end{pmatrix} = \dfrac{50!}{49!1!} = 50$

7. $\begin{pmatrix} 1000 \\ 1000 \end{pmatrix} = 1$

9. $\begin{pmatrix} 55 \\ 23 \end{pmatrix} = 1.866 \times 10^{15}$

11. $\begin{pmatrix} 47 \\ 25 \end{pmatrix} = 1.483 \times 10^{13}$

13. $(x + 1)^5 = \begin{pmatrix} 5 \\ 0 \end{pmatrix} x^5 + \begin{pmatrix} 5 \\ 1 \end{pmatrix} x^4 + \begin{pmatrix} 5 \\ 2 \end{pmatrix} x^3 + \begin{pmatrix} 5 \\ 3 \end{pmatrix} x^2 + \begin{pmatrix} 5 \\ 4 \end{pmatrix} x + \begin{pmatrix} 5 \\ 5 \end{pmatrix}$
$= x^5 + 5x^4 + 10x^3 + 10x^2 + 5x + 1$

15. $(x - 2)^6 = \begin{pmatrix} 6 \\ 0 \end{pmatrix} x^6 + \begin{pmatrix} 6 \\ 1 \end{pmatrix}(-2)x^5 + \begin{pmatrix} 6 \\ 2 \end{pmatrix}(-2)^2 x^4 + \begin{pmatrix} 6 \\ 3 \end{pmatrix}(-2)^3 x^3 + \begin{pmatrix} 6 \\ 4 \end{pmatrix}(-2)^4 x^2$
$+ \begin{pmatrix} 6 \\ 5 \end{pmatrix}(-2)^5 x + \begin{pmatrix} 6 \\ 6 \end{pmatrix}(-2)^6$
$= x^6 + 6 \cdot -2x^5 + 15 \cdot 4x^4 + 20 \cdot -8x^3 + 15 \cdot 16x^2 + 6 \cdot -32x + 64$
$= x^6 - 12x^5 + 60x^4 - 160x^3 + 240x^2 - 192x + 64$

17. $(3x + 1)^4 = \begin{pmatrix} 4 \\ 0 \end{pmatrix}(3x)^4 + \begin{pmatrix} 4 \\ 1 \end{pmatrix}(3x)^3 + \begin{pmatrix} 4 \\ 2 \end{pmatrix}(3x)^2 + \begin{pmatrix} 4 \\ 3 \end{pmatrix}(3x) + \begin{pmatrix} 4 \\ 4 \end{pmatrix}$
$= 81x^4 + 4 \cdot 27x^3 + 6 \cdot 9x^2 + 4 \cdot 3x + 1 = 81x^4 + 108x^3 + 54x^2 + 12x + 1$

19. $(x^2 + y^2)^5 = \begin{pmatrix} 5 \\ 0 \end{pmatrix}(x^2)^5 + \begin{pmatrix} 5 \\ 1 \end{pmatrix}(y^2)(x^2)^4 + \begin{pmatrix} 5 \\ 2 \end{pmatrix}(y^2)^2(x^2)^3 + \begin{pmatrix} 5 \\ 3 \end{pmatrix}(y^2)^3(x^2)^2$
$+ \begin{pmatrix} 5 \\ 4 \end{pmatrix}(y^2)^4(x^2) + \begin{pmatrix} 5 \\ 5 \end{pmatrix}(y^2)^5$
$= x^{10} + 5y^2x^8 + 10y^4x^6 + 10y^6x^4 + 5y^8x^2 + y^{10}$

21. $(\sqrt{x} + \sqrt{2})^6 = \begin{pmatrix} 6 \\ 0 \end{pmatrix}(\sqrt{x})^6 + \begin{pmatrix} 6 \\ 1 \end{pmatrix}\sqrt{2}(\sqrt{x})^5 + \begin{pmatrix} 6 \\ 2 \end{pmatrix}(\sqrt{2})^2(\sqrt{x})^4 + \begin{pmatrix} 6 \\ 3 \end{pmatrix}(\sqrt{2})^3(\sqrt{x})^3$
$+ \begin{pmatrix} 6 \\ 4 \end{pmatrix}(\sqrt{2})^4(\sqrt{x})^2 + \begin{pmatrix} 6 \\ 5 \end{pmatrix}(\sqrt{2})^5(\sqrt{x}) + \begin{pmatrix} 6 \\ 6 \end{pmatrix}(\sqrt{2})^6$
$= x^3 + 6\sqrt{2}(\sqrt{x})^5 + 15(\sqrt{2})^2(\sqrt{x})^4 + 20(\sqrt{2})^3(\sqrt{x})^3 + 15(\sqrt{2})^4(\sqrt{x})^2 + 6(\sqrt{2})^5\sqrt{x} + (\sqrt{2})^6$
$= x^3 + 6\sqrt{2}x^{5/2} + 30x^2 + 40\sqrt{2}x^{3/2} + 60x + 24\sqrt{2}x^{1/2} + 8$

23. $(ax + by)^5 = \begin{pmatrix} 5 \\ 0 \end{pmatrix}(ax)^5 + \begin{pmatrix} 5 \\ 1 \end{pmatrix}(by)(ax)^4 + \begin{pmatrix} 5 \\ 2 \end{pmatrix}(by)^2(ax)^3 + \begin{pmatrix} 5 \\ 3 \end{pmatrix}(by)^3(ax)^2$
$+ \begin{pmatrix} 5 \\ 4 \end{pmatrix}(by)^4(ax) + \begin{pmatrix} 5 \\ 5 \end{pmatrix}(by)^5$
$= (ax)^5 + 5by(ax)^4 + 10(by)^2(ax)^3 + 10(by)^3(ax)^2 + 5(by)^4(ax) + (by)^5$

25. $n = 10$, $a = 3$, $x = x$, and $j = 6$

$$\binom{10}{10-6}3^{10-6}x^6 = \binom{10}{4}3^4 \cdot x^6 = \frac{10!}{4!6!} \cdot 81 \cdot x^6 = \frac{10 \cdot 9 \cdot 8 \cdot 7}{4 \cdot 3 \cdot 2} \cdot 81x^6 = 17,010x^6$$

The coefficient of x^6 is 17,010.

27. $n = 12$, $a = -1$, $x = 2x$, and $j = 7$

$$\binom{12}{12-7}(-1)^{12-7}(2x)^7 = \binom{12}{5}(-1)^5 \cdot 128 \cdot x^7 = \frac{-12!}{5!7!} \cdot 128 \cdot x^7 = -101,376x^7$$

The coefficient of x^7 is $-101,376$.

29. $n = 9$, $a = 3$, $x = 2x$, and $j = 7$

$$\binom{9}{9-7}(3)^{9-7}(2x)^7 = \binom{9}{2}3^2 \cdot 128 \cdot x^7 = \frac{9!}{2!7!} \cdot 9 \cdot 128 \cdot x^7 = 41,472x^7$$

The coefficient of x^7 is 41,472.

31. The fifth term contains x^3. $n = 7$, $a = 3$, $x = x$, and $j = 3$

$$\binom{7}{7-3}(3)^{7-3}x^3 = \binom{7}{4}3^4 \cdot x^3 = \frac{7!}{4!3!} \cdot 81 \cdot x^3 = 2835x^3$$

33. The third term contains x^7. $n = 9$, $a = -2$, $x = 3x$, and $j = 7$

$$\binom{9}{9-7}(-2)^{9-7}(3x)^7 = \binom{9}{2}(-2)^2(3x)^7 = \frac{9!}{2!7!} \cdot 4 \cdot 2187 \cdot x^7 = 314,928x^7$$

35. $\left(x^2 + \dfrac{1}{x}\right)^{12}$

Constant term $= \dbinom{12}{12-j}\left(\dfrac{1}{x}\right)^{12-j}(x^2)^j$ where $2j = 12 - j$ or $j = 4$

Constant term $= \dbinom{12}{8}\left(\dfrac{1}{x}\right)^8(x^2)^4 = \dbinom{12}{8} = \dfrac{12!}{8!4!} = \dfrac{12 \cdot 11 \cdot 10 \cdot 9}{4 \cdot 3 \cdot 2} = 495$

37. $\left(x - \dfrac{2}{\sqrt{x}}\right)^{10}$

x^4 term $= \dbinom{10}{10-j}\left(\dfrac{-2}{\sqrt{10}}\right)^{10-j}x^j$ where $j - \dfrac{10-j}{2} = 4$ or $\dfrac{3}{2}j = 9$ or $j = 6$

x^4 term $= \dbinom{10}{4}\left(\dfrac{-2}{\sqrt{10}}\right)^4 x^6 = 16\dbinom{10}{6}\left(\dfrac{x^6}{x^2}\right) = \dfrac{16 \cdot 10 \cdot 9 \cdot 8 \cdot 7}{4 \cdot 3 \cdot 2} = 3360$

39. $(1.001)^5 = (1 + 10^{-3})^5 = 1 + \dbinom{5}{1} \cdot 10^{-3} + \dbinom{5}{2} \cdot \left(10^{-3}\right)^2 + \dbinom{5}{3}\left(10^{-3}\right)^3 + \cdots$

$$= 1 + .005 + 10(.000001) + 10(.000000001) + \cdots$$

$$= 1 + .005 + .00001 + \cdots$$

$$\approx 1.00501 \text{ correct to five decimal places}$$

41. $\dbinom{n}{n} = 1$

$$\dfrac{n!}{n!(n-n)!} = \dfrac{n!}{n!0!} = \dfrac{n!}{n! \cdot 1} = \dfrac{n!}{n!} = 1$$

43.
$$\binom{n}{0} + \binom{n}{1} + \cdots + \binom{n}{n} = 2^n$$

$$\binom{n}{0} + \binom{n}{1} + \cdots + \binom{n}{n} = (1 + 1)^n$$

$$= \binom{n}{0}1^n + \binom{n}{1}(1)^1(1)^{n-1} + \binom{n}{2}(1)^2(1)^{n-2} + \cdots + \binom{n}{n}(1)^n(1)^{n-n}$$

$$= \binom{n}{0} + \binom{n}{1} + \cdots + \binom{n}{n}$$

45.
$$\binom{5}{0}\left(\frac{1}{4}\right)^5 + \binom{5}{1}\left(\frac{1}{4}\right)^4\left(\frac{3}{4}\right) + \binom{5}{2}\left(\frac{1}{4}\right)^3\left(\frac{3}{4}\right)^2 + \binom{5}{3}\left(\frac{1}{4}\right)^2\left(\frac{3}{4}\right)^3 + \binom{5}{4}\left(\frac{1}{4}\right)\left(\frac{3}{4}\right)^4 + \binom{5}{5}\left(\frac{3}{4}\right)^5$$

$$= 1 \cdot \frac{1}{4^5} + 5 \cdot \frac{1}{4^4} \cdot \frac{3}{4} + 10 \cdot \frac{1}{4^3} \cdot \frac{9}{4^2} + 10 \cdot \frac{1}{4^2} + \frac{27}{4^3} + 5 \cdot \frac{1}{4} \cdot \frac{81}{4^4} + \frac{243}{4^5}$$

$$= \frac{1 + 15 + 90 + 270 + 405 + 243}{4^5} = \frac{1024}{4^5} = \frac{1024}{1024} = 1$$

12.6 Sets and Counting

In Problems 1–9, we have A = {1, 3, 5, 7, 9}, B = { 1, 5, 6, 7}, and C = {1, 2, 4, 6, 8, 9}:

1. $A \cup B = \{1, 3, 5, 7, 9\} \cup \{1, 5, 6, 7\} = \{1, 3, 5, 6, 7, 9\}$

3. $A \cap B = \{1, 3, 5, 7, 9\} \cap \{1, 5, 6, 7\} = \{1, 5, 7\}$

5. $(A \cup B) \cap C = (\{1, 3, 5, 7, 9\} \cup \{1, 5, 6, 7\}) \cap \{1, 2, 4, 6, 8, 9\}$
$$= \{1, 3, 5, 6, 7, 9\} \cap \{1, 2, 4, 6, 8, 9\}$$
$$= \{1, 6, 9\}$$

7. $(A \cap B) \cup C = (\{ 1, 3, 5, 7, 9\} \cap \{1, 5, 6, 7\}) \cup \{1, 2, 4, 6, 8, 9\}$
$$= \{1, 5, 7\} \cup \{1, 2, 4, 6, 8, 9\}$$
$$= \{1, 2, 4, 5, 6, 7, 8, 9\}$$

9. $(A \cup C) \cap (B \cup C) = (\{1, 3, 5, 7, 9\} \cup \{1, 2, 4, 6, 8, 9\}) \cap (\{1, 5, 6, 7\} \cup \{1, 2, 4, 6, 8, 9\})$
$$= \{1, 2, 3, 4, 5, 6, 7, 8, 9\} \cap \{1, 2, 4, 5, 6, 7, 8, 9\}$$
$$= \{1, 2, 4, 5, 6, 7, 8, 9\}$$

In Problems 11–19, we have U = Universal set = {0, 1, 2, 3, 4, 5, 6, 7, 8, 9}, A = {1, 3, 4, 5, 9},
B = {2, 4, 6, 7, 8}, and C = {1, 3, 4, 6}:

11. $A' = \{1, 3, 4, 5, 9\}' = \{0, 2, 6, 7, 8\}$

13. $(A \cap B)' = (\{1, 3, 4, 5, 9\} \cap \{2, 4, 6, 7, 8\})' = \{4\}'$
$$= \{0, 1, 2, 3, 5, 6, 7, 8, 9\}$$

15. $A' \cup B' = \{1, 3, 4, 5, 9\}' \cup \{2, 4, 6, 7, 8\}'$
$$= \{0, 2, 6, 7, 8\} \cup \{0, 1, 3, 5, 9\}$$
$$= \{0, 1, 2, 3, 5, 6, 7, 8, 9\}$$

17. $(A \cap C')' = (\{1, 3, 4, 5, 9\} \cap \{1, 3, 4, 6\}')'$
$$= (\{1, 3, 4, 5, 9\} \cap \{0, 2, 5, 7, 8, 9\})'$$
$$= \{5, 9\}'$$
$$= \{0, 1, 2, 3, 4, 6, 7, 8\}$$

19. $(A \cup B \cup C)' = (\{1, 3, 4, 5, 9\} \cup \{2, 4, 6, 7, 8\} \cup \{1, 3, 4, 6\})'$
$$= \{1, 2, 3, 4, 5, 6, 7, 8, 9\}' = \{0\}$$

21. Subsets of $\{a, b, c, d\}$ with

zero elements:	$\varnothing$
one element:	$\{a\}, \{b\}, \{c\}, \{d\}$
two elements:	$\{a, b\}, \{a, c\}, \{a, d\}, \{b, c\}, \{b, d\}, \{c, d\}$
three elements:	$\{a, b, c\}, \{a, b, d\}, \{a, c, d\}, \{b, c, d\}$
four elements:	$\{a, b, c, d\}$

23. For $n(A) = 15$, $n(B) = 20$, $n(A \cap B) = 10$, we have by (1) that
$$n(A \cup B) = n(A) + n(B) - n(A \cap B) = 15 + 20 - 10 = 25$$

25. For $n(A \cup B) = 50$, $n(A \cap B) = 10$, $n(B) = 20$, we have that
$$n(A \cup B) = n(A) + n(B) - n(A \cap B)$$
$$50 = n(A) + 20 - 10$$
$$40 = n(A)$$

27. From figure, $n(A) = 15 + 3 + 5 + 2 = 25$

29. From figure, $n(A \text{ or } B)$
$$= n(A \cup B) = n(A) + n(B) - n(A \cap B)$$
$$= 25 + 20 - 8 = 37$$

31. From figure, $n(A \text{ but not } C)$
$$= n(A) - n(A \cap C) = 25 - 7 = 18$$

33. From figure, $n(A \text{ and } B \text{ and } C)$
$$= n(A \cap B \cap C) = 5$$

35. Let $A = \{\text{those who will purchase a major appliance}\}$

$B = \{\text{those who will buy a car}\}$

Note that $n(A) = 200$, $n(B) = 150$, $n(A \cap B) = 25$

Using (1), $n(A \cup B) = n(A) + n(B) - n(A \cap B) = 200 + 150 - 25 = 325$

Since a total of 500 were asked the number that will purchase neither, $n((A \cup B)') = 500 - 325 = 175$. The number who will purchase only a car is:
$$n(B \text{ but not } A) = n(B \cap A')n(B) - n(A \cap B) = 150 - 25 = 125.$$

This problem can be solved by Venn Diagram as well.

37. We fill in a Venn Diagram, reading the list from the bottom up:
    - (a) 15
    - (b) 15
    - (c) 15
    - (d) 25
    - (e) 40

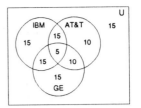

12.7 Permutations and Combinations

Use $P(n, r) = \dfrac{n!}{(n - r)!}$ for Problems 1–7:

1. $P(6, 2) = \dfrac{6!}{(6 - 2)!} = \dfrac{6!}{4!} = \dfrac{6 \cdot 5 \cdot 4!}{4!} = 30$

3. $P(5, 5) = \dfrac{5!}{(5 - 5)!} = \dfrac{5!}{0!} = \dfrac{5 \cdot 4 \cdot 3 \cdot 2 \cdot 1}{1} = 120$

5. $P(8, 0) = \dfrac{8!}{(8 - 0)!} = \dfrac{8!}{8!} = 1$ 7. $P(8, 3) = \dfrac{8!}{(8 - 3)!} = \dfrac{8!}{5!} = \dfrac{8 \cdot 7 \cdot 6 \cdot 5!}{5!} = 336$

In Problems 9–15, we use $C = \dfrac{n!}{(n - r)!r!}$:

9. $C(8, 2) = \dfrac{8!}{(8 - 2)!2!} = \dfrac{8!}{6!2!} = \dfrac{8 \cdot 7 \cdot 6!}{2 \cdot 1 \cdot 6!} = 28$

11. $C(6, 4) = \dfrac{6!}{(6 - 4)!4!} = \dfrac{6!}{2!4!} = \dfrac{6 \cdot 5 \cdot 4!}{2 \cdot 1 \cdot 4!} = 15$

13. $C(15, 15) = \dfrac{15!}{(15 - 15)!15!} = \dfrac{15!}{0!15!} = \dfrac{15!}{1 \cdot 15!} = \dfrac{15!}{15!} = 1$

15. $C(26, 13) = \dfrac{26!}{(26 - 13)!13!} = \dfrac{26!}{13!13!}$

    $= \dfrac{26 \cdot 25 \cdot 24 \cdot 23 \cdot 22 \cdot 21 \cdot 20 \cdot 19 \cdot 18 \cdot 17 \cdot 16 \cdot 15 \cdot 14 \cdot 13!}{13 \cdot 12 \cdot 11 \cdot 10 \cdot 9 \cdot 8 \cdot 7 \cdot 6 \cdot 5 \cdot 4 \cdot 3 \cdot 2 \cdot 1 \cdot 13!}$

    $= 10,400,600$

17. {abc, abd, abe, acb, acd, ace, adb, adc, ade, aeb, aec, aed, bac, bad, bae, bca, bcd, bce, bda, bde, bea, bec, bed, cab, cad, cae, cba, cbd, cde, cda, cdb, cde, cea, ceb, ced, dab, dac, dae, dba, dbc, dbe, dca, dcb, dce, dea, deb, dec, eab, eac, ead, eba, ebc, ebd, eca, ecb, ecd, eda, edb, edc}

    $P(5, 3) = \dfrac{5!}{(5 - 3)!} = \dfrac{5!}{2!} = \dfrac{5 \cdot 4 \cdot 3 \cdot 2 \cdot 1}{2 \cdot 1} = 60.$

19. $\{123, 124, 132, 134, 142, 143, 213, 214, 231, 234, 241, 243, 312, 314, 321, 324, 341, 342, 412, 413, 421, 423, 431, 432\}$

$$P(4, 3) = \frac{4!}{(4-3)!} = \frac{4!}{1!} = 4 \cdot 3 \cdot 2 \cdot 1 = 24$$

21. Combinations of a, b, c, d, e taken 3 at a time are:

$(abc), (abd), (abe), (acd), (ace), (ade),$
$(bcd), (bce), (bde), (cde)$

Thus, $C(5, 3) = 10$.

23. Combinations of 1, 2, 3, 4 taken 3 at a time are:

$(123), (124), (134), (234)$

Thus, $C(4, 3) = 4$

25. There are 5 choices and 3 choices or $5 \cdot 3 = 15$ combinations.

27. There are $4 \cdot 4 = 16$ ways since there are four choices for each of two positions.

29. For a 3-digit number using 0 and 1, we have two choices in each of three positions or $2 \cdot 2 \cdot 2 = 8$ numbers.

31. To line 4 people up, we have 4 choices for the first position, 3 for the second, 2 for the third, and only 1 for the fourth position; i.e., $4 \cdot 3 \cdot 2 \cdot 1 = 4! = 24$ arrangements.

33. Since no letter can be repeated, we have 5 choices for the first position, 4 choices for the second, and 3 choices for the third; i.e., $5 \cdot 4 \cdot 3 = 60$ codes.

35. We have 5 letters and use each only once so that there are 5 choices for the first position, 4 for the second, 3 for the third, 2 for the second, and only 1 for the last position, or $5 \cdot 3 \cdot 4 \cdot 2 \cdot 1 = 5! = 120$ arrangements.

37. To form a committee of 4 from a total of 7 students is given by:

$$C(7, 4) = \frac{7}{(7-4)!4!} = \frac{7!}{3!4!} = \frac{7 \cdot 6 \cdot 5 \cdot 4!}{3 \cdot 2 \cdot 1 \cdot 4!} = 35$$

39. There are 2 possibilities for each of the ten questions; therefore, there are $2^{10} = 1024$ different possible arrangements.

41. There are 9 possible choices for the first digit. There are 10 possible choices for the remaining 3 digits. Therefore, there are $9 \cdot 10 \cdot 10 \cdot 10 = 9000$ possible four-digit numbers.

43. There are 5 positions available for the first book, 4 positions available for the second book, 3 for the third, 2 for the fourth and 1 for the fifth. Therefore, there are $5 \cdot 4 \cdot 3 \cdot 2 \cdot 1 = 120$ different arrangements.

45. $\begin{pmatrix} 8 \\ 1 \end{pmatrix} \begin{pmatrix} 15 \\ 1 \end{pmatrix} \begin{pmatrix} 4 \\ 1 \end{pmatrix} = 8 \cdot 15 \cdot 4 = 480$ different portfolios.

47. Choosing 2 boys from 4 can be done $C(4, 2)$ ways and choosing 3 girls from 8 can be done $C(8, 3)$ ways, giving a total of

$$C(4, 2) \times C(8, 3) = \frac{4!}{(4-2)!2!} \cdot \frac{8!}{(8-3)!3!} = \frac{4!}{2!2!} \cdot \frac{8!}{5!3!}$$

$$= \frac{4 \cdot 3!}{2 \cdot 1 \cdot 2 \cdot 1} \cdot \frac{8 \cdot 7 \cdot 6 \cdot 5!}{3!5!} = 336$$

There are 336 ways of choosing a committee.

49. The committee is formed as follows:

$$\underbrace{\text{Administrators}}_{\text{2 from 4}} \times \underbrace{\text{faculty}}_{\text{3 from 8}} \times \underbrace{\text{students}}_{\text{5 from 20}}$$

$$C(4,\ 2) \times \quad C(8,\ 3) \quad \times\ C(20,\ 5)$$

$$\frac{4!}{(4-2)!2!} \times \frac{8!}{(8-3)!3!} \times \frac{20!}{(20-5)!15!}$$

$$\frac{4!}{2!2!} \times \frac{8!}{5!3!} \times \frac{20!}{15!5!}$$

$$\frac{4 \cdot 3!}{2 \cdot 1 \cdot 2 \cdot 1} \times \frac{8 \cdot 7 \cdot 6 \cdot 5!}{3!5!} \times \frac{20 \cdot 19 \cdot 18 \cdot 17 \cdot 16 \cdot 15!}{5 \cdot 4 \cdot 3 \cdot 2 \cdot 1 \cdot 15!}$$

= 5,209,344 different committees

51. There are 9 choices for the first position, 8 for the second, 7 for the third, etc., or
$9 \cdot 8 \cdot 7 \cdot 6 \cdot 5 \cdot 4 \cdot 3 \cdot 2 \cdot 1 = 9! = 362,880$ possible batting orders.

53. $\dfrac{9!}{2!2!} = 90,720$ different words

55. $C(100, 22)$ ways to form the first committee. After choosing the first committee, there are $100 - 22 = 78$ senators left. Thus, there are $C(78, 13)$ ways to form the second committee. There are $C(65, 10)$ ways to form the third, $C(55, 5)$ ways to form the fourth, $C(50, 16)$ ways to form the fifth, $C(34, 17)$ ways to form the sixth, and $C(17, 17)$ ways to form the seventh committee.

Total number of committees $= C(100, 22) \cdot C(78, 13) \cdot C(65, 10) \cdot C(55, 5) \cdot C(50, 16) \cdot C(34, 17)$
$$\cdot\ C(17, 17)$$
$$= 1.157 \times 10^{76}$$

57. We have 6 players and we must choose 2 of them. Therefore, there are $C(6, 2) = 15$ different teams possible.

59. (a) $C(7, 2) \cdot C(3, 1) = 21 \cdot 3 = 63$ (b) $C(7, 3) = 35$ (c) $C(3, 3) = 1$

12.8 Probability

1. The sample space, all of the logical possibilities that can occur, is $S = \{HH, HT, TH, TT\}$
No one outcome is more likely than another, so that using $P(E) = \dfrac{n(E)}{n(S)}$

The probabilities are $P(HH) = \dfrac{1}{4}$, $P(HT) = \dfrac{1}{4}$, $P(TH) = \dfrac{1}{4}$, and $P(TT) = \dfrac{1}{4}$

3. The sample space, generated by combining each of the outcomes of the two coins as in Problems 1 and 2 with each face of the die, is:
$S = \{HH1, HH2, HH3, HH4, HH5, HH6, HT1, HT2, HT3, HT4, HT5, HT6, TH1, TH2, TH3,$
$TH4, TH5, TH6, TT1, TT2, TT3, TT4, TT5, TT6\}$

There are 24 equally likely outcomes, and the probability of each is $\dfrac{1}{24}$.

5. The sample space for three coins builds on that for two coins:
$$S = \{HHH, HHT, HTH, HTT, THH, THT, TTH, TTT\}.$$

Since there are 8 equally likely outcomes, the probability of each is $\dfrac{1}{8}$.

In Problems 7–11, we use the figures in the text. Note that the sample space for Spinner I is $S_1 = \{1, 2, 3, 4\}$, and the sample space for Spinner II is $S_2 = \{Yellow, Red, Green\}$, and the sample space for Spinner III is $S_3 = \{Forward, Backward\}$.

7. The sample space for spinning Spinner I and then Spinner II is:
$$S = \{1 \text{ Yellow, } 1 \text{ Red, } 1 \text{ Green, } 2 \text{ Yellow, } 2 \text{ Red, } 2 \text{ Green, } 3 \text{ Yellow, } 3 \text{ Red, } 3 \text{ Green,}$$
$$4 \text{ Yellow, } 4 \text{ Red, } 4 \text{ Green}\},$$

or 12 equally likely outcomes, and the probability of each one is $\dfrac{1}{12}$. The probability of getting a 2 or a 4, followed by Red is $P(2 \text{ Red}) + P(4 \text{ Red}) = \dfrac{1}{12} + \dfrac{1}{12} = \dfrac{1}{6}$

9. The sample space for spinning Spinner I, then Spinner II, and then Spinner III is
$$S = \{1 \text{ Yellow Forward, } 1 \text{ Yellow Backward, } 1 \text{ Red Forward, } 1 \text{ Red Backward,}$$
$$1 \text{ Green Forward, } 1 \text{ Green Backward, } 2 \text{ Yellow Forward, } 2 \text{ Yellow Backward,}$$
$$2 \text{ Red Forward, } 2 \text{ Red Backward, } 2 \text{ Green Forward, } 2 \text{ Green Backward,}$$
$$3 \text{ Yellow Forward, } 3 \text{ Yellow Backward, } 3 \text{ Red Forward, } 3 \text{ Red Backward,}$$
$$3 \text{ Green Forward, } 3 \text{ Green Backward, } 4 \text{ Yellow Forward, } 4 \text{ Yellow Backward,}$$
$$4 \text{ Red Forward, } 4 \text{ Red Backward, } 4 \text{ Green Forward, } 4 \text{ Green Backward}\}$$

or 24 equally likely outcomes, and the probability of each is $\dfrac{1}{24}$.

The probability of getting a 1, followed by Red or Green, followed by Backward is:
$$P(1 \text{ Red Backward}) + P(1 \text{ Green Backward}) = \dfrac{1}{24} + \dfrac{1}{24} = \dfrac{1}{12}$$

11. The sample space for spinning Spinner I twice and then Spinner II is:
$$S = \{11 \text{ Yellow, } 11 \text{ Red, } 11 \text{ Green, } 12 \text{ Yellow, } 12 \text{ Red, } 12 \text{ Green, } 13 \text{ Yellow, } 13 \text{ Red,}$$
$$13 \text{ Green, } 14 \text{ Yellow, } 14 \text{ Red, } 14 \text{ Green, } 21 \text{ Yellow, } 21 \text{ Red, } 21 \text{ Green, } 22 \text{ Yellow,}$$
$$22 \text{ Red, } 22 \text{ Green, } 23 \text{ Yellow, } 23 \text{ Red, } 23 \text{ Green, } 24 \text{ Yellow, } 24 \text{ Red, } 24 \text{ Green,}$$
$$31 \text{ Yellow, } 31 \text{ Red, } 31 \text{ Green, } 32 \text{ Yellow, } 32 \text{ Red, } 32 \text{ Green, } 33 \text{ Yellow, } 33 \text{ Red,}$$
$$33 \text{ Green, } 34 \text{ Yellow, } 34 \text{ Red, } 34 \text{ Green, } 41 \text{ Yellow, } 41 \text{ Red, } 41 \text{ Green, } 42 \text{ Yellow,}$$
$$42 \text{ Red, } 42 \text{ Green, } 43 \text{ Yellow, } 43 \text{ Red, } 43 \text{ Green, } 44 \text{ Yellow, } 44 \text{ Red, } 44 \text{ Green}\}$$

or 48 equally likely outcomes, and the probability of each is $\dfrac{1}{48}$.

The probability of getting a 2, followed by a 2 or a 4, followed by a Red or Green is:
$$P(2\ 2 \text{ Red}) + P(2\ 2 \text{ Green}) + P(2\ 4 \text{ Red}) + P(2\ 4 \text{ Green}) = \dfrac{1}{48} + \dfrac{1}{48} + \dfrac{1}{48} + \dfrac{1}{48} = \dfrac{1}{12}$$

13. A, B, C, F   15. B   17. $P(\text{heads}) = \dfrac{4}{5}; P(\text{tails}) = \dfrac{1}{5}$

19. $P(1) = P(3) = P(5) = \dfrac{2}{9}; P(2) = P(4) = P(6) = \dfrac{1}{9}$

21. $P(A \cup B) = P(A) + P(B)$
       $= 0.30 + 0.40 = 0.70$

23. $P(A \cup B) = P(A) + P(B) - P(A \cap B)$
       $= 0.30 + 0.40 - 0.15 = 0.55$

In Problems 25 and 27 a golf ball is selected from a container with 9 white balls, 8 green balls, and 3 orange balls.

25. The sample sapce, the total number of golf balls, contains 20 equally likely outcomes There are 9 white balls. Thus, the probability of getting a white ball is:

$$P(E) = \frac{n(E)}{n(S)} = \frac{n(\text{white balls})}{n(\text{golf balls present})} = \frac{9}{20}$$

27. The sample space, the total number of golf balls, contains 20 equally likely outcomes. There are 9 white balls and 8 green ones. Thus, the probability of getting a white or a green one is

$$P(E) = \frac{n(E)}{n(S)} = \frac{n(\text{white or green or balls})}{n(\text{golf balls present})} = \frac{17}{20}$$

29. There are 36 different possible rolls. The number of ways to get a 6 is 5 and the number of ways to get an 8 is 5. Therefore,

$$P(6 \text{ or } 8) = P(6) + P(8) = \frac{n(6) + n(8)}{36} = \frac{5 + 5}{36} = \frac{10}{36} = \frac{5}{18}$$

In Problems 31 and 33, we are using the table given in the text. The sample space is 100 households in each case.

31. From the table there are 30 households total out of 100 which have an income in excess of $30,000. Thus, the probability that a household has an annual income in excess of $30,000 is

$$P(E) = \frac{n(E)}{n(S)} = \frac{n(\text{in excess of } \$30,000)}{n(\text{total households })} = \frac{30}{100} = \frac{3}{10}$$

33. From the table there are 40 households total out of 100 which have an income less than $20,000. Thus, the probability that a household has an annual income less than $20,000 is

$$P(E) = \frac{n(E)}{n(S)} = \frac{n(\text{less than } \$20,000)}{n(\text{total households })} = \frac{40}{100} = \frac{2}{5}$$

35. (a) $P(1 \text{ or } 2) = P(1) + P(2) = 0.24 + 0.33 = 0.57$

   (b) $P(1 \text{ or more}) = P(1) + P(2) + P(3) + P(4 \text{ or more})$
   $$= 0.24 + 0.33 + 0.21 + 0.17$$
   $$= 0.95$$

   (c) $P(3 \text{ or fewer}) = P(0) + P(1) + P(2) + P(3)$
   $$= 0.05 + 0.24 + 0.331 + 0.21$$
   $$= 0.83$$

   (d) $P(3 \text{ or more}) = P(3) + P(4 \text{ or more})$
   $$= 0.21 + 0.17$$
   $$= 0.38$$

   (e) $P(\text{less than } 2) = P(0) + P(1)$
   $$= 0.05 + 0.24$$
   $$= 0.29$$

   (f) $P(\text{less than } 1) = P(0) = 0.05$

   (g) $P(1,2, \text{ or } 3) = P(1) + P(2) + P(3)$
   $$= 0.24 + 0.33 + 0.21$$
   $$= 0.78$$

   (h) $P(2 \text{ or more}) = P(2) + P(3) + P(4 \text{ or more})$
   $$= 0.33 + 0.21 + 0.17$$
   $$= 0.71$$

37. The sample space for picking 5 numbers from a total of 10 numbers contains
$$P(10, 5) = \frac{10!}{(10 - 5)!} = \frac{10!}{5!} = \frac{10 \cdot 9 \cdot 8 \cdot 7 \cdot 6 \cdot 5!}{5!} = 30{,}240$$
possible outcomes. Only one of these outcomes is the desired event. Thus, the probability of winning is:
$$P(E) = \frac{n(E)}{n(S)} = \frac{n(\text{drawn number})}{n(\text{total possible outcomes})} = \frac{1}{30{,}240} = 0.000033068$$

39. (a) $P(3 \text{ heads}) = C(5, 3) \cdot \left(\dfrac{1}{2}\right)^5 = \dfrac{10}{32}$

(b) $P(0 \text{ heads}) = C(5, 0) \cdot \left(\dfrac{1}{2}\right)^5 = \dfrac{1}{32}$

41. (a) $P(\text{Sum} = 7 \text{ three times}) = P(\text{sum} = 7) \cdot P(\text{sum} = 7) \cdot P(\text{sum} = 7)$
$$= \frac{1}{6} \cdot \frac{1}{6} \cdot \frac{1}{6} = \frac{1}{216} = .00463$$

(b) $P(\text{sum} = 7 \text{ or } 11 \text{ at least twice}) = P(\text{sum} = 7 \text{ or } 11) \cdot P(\text{sum} = 7 \text{ or } 11) \cdot P(\text{sum} \neq 7 \text{ or } 11)$
$+ P(\text{sum} = 7 \text{ or } 11) \cdot P(\text{sum} = 7 \text{ or } 11) \cdot P(\text{sum} = 7 \text{ or } 11)$
$$= \frac{8}{36} \cdot \frac{8}{36} \cdot \frac{28}{36} + \frac{8}{36} \cdot \frac{8}{36} \cdot \frac{8}{36} = .049$$

43. $P(\text{all five defective}) = \dfrac{1}{C(30, 5)} = 7.02 \times 10^{-6}$

$P(\text{at least two defective})$
$$= P(2 \text{ def}) + P(3 \text{ def}) + P(4 \text{ def}) + P(5 \text{ def})$$
$$= \frac{C(5, 2) \cdot C(25, 3)}{C(30, 5)} + \frac{C(5, 3) \cdot C(25, 2)}{C(30, 5)} + \frac{C(5, 4) \cdot C(25, 1)}{C(30, 5)} + \frac{(C(5, 5) \cdot C(25, 0)}{(30, 5)}$$
$$= 0.183$$

45. $P(\text{one of the 5 coins is the one valued at more than \$10,000}) = \dfrac{C(49, 4) \times C(1, 1)}{C(50, 5)} = 0.1$

12 Chapter Review

In Problems 1-8, U = Universal set = {1, 2, 3, 4, 5, 6, 7, 8, 9}, A = {1, 3, 5, 7}, B = {3, 5, 6, 7, 8}, C = {2, 3, 7, 8, 9}.

1. $5! = 5 \cdot 4 \cdot 3 \cdot 2 \cdot 1 = 120$

3. $\dbinom{5}{2} = \dfrac{5!}{2!3!} = \dfrac{5 \cdot 4}{2} = 10$

5. $P(8, 3) = \dfrac{8!}{(8 - 3)!} = \dfrac{8!}{5!} = \dfrac{8 \cdot 7 \cdot 6 \cdot 5!}{5!} = 336$

7. $C(8, 3) = \dfrac{8!}{(8 - 3)!3!} = \dfrac{8!}{5!3!} = \dfrac{8 \cdot 7 \cdot 6 \cdot 5!}{3 \cdot 2 \cdot 1 \cdot 5!} = 56$

9.
$$a_1 = (-1)^1 \frac{1+3}{1+2} = \frac{-4}{3}$$
$$a_2 = (-1)^2 \frac{2+3}{2+2} = \frac{5}{4}$$
$$a_3 = (-1)^3 \frac{3+3}{3+2} = \frac{-6}{5}$$
$$a_4 = (-1)^4 \frac{4+3}{4+2} = \frac{7}{6}$$
$$a_5 = (-1)^5 \frac{5+3}{5+2} = \frac{-8}{7}$$

11.
$$a_1 = \frac{2^1}{1^2} = 2$$
$$a_2 = \frac{2^2}{2^2} = 1$$
$$a_3 = \frac{2^3}{3^2} = \frac{8}{9}$$
$$a_4 = \frac{2^4}{4^2} = 1$$
$$a_5 = \frac{2^5}{5^2} = \frac{32}{25}$$

13.
$$a_1 = 3$$
$$a_2 = \frac{2}{3}(3) = 2$$
$$a_3 = \frac{2}{3}(2) = \frac{4}{3}$$
$$a_4 = \frac{2}{3}\left(\frac{4}{3}\right) = \frac{8}{9}$$
$$a_5 = \frac{2}{3}\left(\frac{8}{9}\right) = \frac{16}{27}$$

15.
$$a_1 = 2$$
$$a_2 = 2 - 2 = 0$$
$$a_3 = 2 - 0 = 2$$
$$a_4 = 2 - 2 = 0$$
$$a_7 = 2 - 0 = 2$$

17. Arithmetic
$$a_{n+1} - a_n = d$$
$$n + 6 - (n + 5) = d = 1$$
$$a = 6, \; a_n = (n + 5)$$
$$S_n = \frac{n}{2}[6 + (n + 5)] = \frac{n}{2}(n + 11)$$

19. Neither

21. Geometric
$$r = \frac{a_{n+1}}{a_n}$$
$$r = \frac{64}{8} = 8$$
$$S_r = \sum_{k=1}^{n} 2^{3k} = 8 + 64 + 512 + \ldots + 2^{3n}$$
$$= 8\left(\frac{1 - 8^n}{1 - 8}\right) = \frac{-8}{7}(1 - 8^n)$$
$$= \frac{-8}{7} + \frac{8^{n+1}}{7} = \frac{8}{7}(8^n - 1)$$

23. Arithmetic
$$d = 4 - 0$$
$$d = 4$$
$$a = 0, \; a_n = 4(n - 1)$$
$$S_n = \frac{n}{2}[0 + 4(n-1)]$$
$$= \frac{n}{2}[4(n - 1)]$$
$$= 2n(n - 1)$$

25. Geometric
$$r = \frac{\frac{3}{2}}{3} = \frac{1}{2}$$
$$S_n = \sum_{k=1}^{n} 3\left(\frac{1}{2}\right)^{k-1} = 3 + \frac{3}{2} + \frac{3}{4} + \ldots + 3\left(\frac{1}{2}\right)^{n-1}$$
$$= 3\left(\frac{1 - \left(\frac{1}{2}\right)^n}{1 - \frac{1}{2}}\right) = 3\frac{\left(1 - \left(\frac{1}{2}\right)^n\right)}{\frac{1}{2}} = 6\left[1 - \left(\frac{1}{2}\right)^n\right]$$

27. Neither

29. Arithmetic: $d = 4, \; a_1 = 3$
$$a_9 = 3 + 8(4)$$
$$a_9 = 35$$

31. Geometric: $r = \frac{1}{10}, \; a_1 = 1$
$$a_{11} = \left(\frac{1}{10}\right)^{10}(1)$$
$$a_{11} = \left(\frac{1}{10}\right)^{10}$$

33. Arithmetic: $d = \sqrt{2}$,

$a_1 = \sqrt{2}$

$a_9 = \sqrt{2} + 8\sqrt{2}$

$a_9 = 9\sqrt{2}$

35. $a_7 = a + 6d = 31$

$a_{20} = a + 19d = 96$

$\phantom{a_{20} = a + } 13d = 65$

$\phantom{a_{20} = a + 13} = 5$

$a = 31 - 6d$

$a = 31 - 6(5)$

$a = 1$

$a_n = 1 + (n - 1)5$

$a_n = 1 + 5n - 5$

$a_n = 5n - 4$

37. $a_{10} = a + 9d = 0$

$a_{18} = a + 17d = 8$

$\phantom{a_{18} = a + 1} 8d = 8$

$\phantom{a_{18} = a + 1} d = 1$

$a = -9d$

$a = -9$

$a_n = -9 + (n - 1)1$

$a_n = -9 + n - 1$

$a_n = n - 10$

39. $a = 3, r = \dfrac{1}{3}$

$$3 + 1 + \frac{1}{3} + \frac{1}{9} + \cdots = \frac{3}{1 - \frac{1}{3}} = \frac{9}{2}$$

41. $a = 2, r = \dfrac{-1}{2}$

$$2 - 1 + \frac{1}{2} - \frac{1}{4} + \cdots = \frac{2}{1 + \frac{1}{2}} = \frac{4}{3}$$

43. $a = 4, r = \dfrac{1}{2}$

$$\sum_{k=1}^{\infty} 4\left(\frac{1}{2}\right)^{k-1} = \frac{4}{1 - \frac{1}{2}} = 8$$

45. (I) $n = 1$: $3 \cdot 1 = 3$ and $\dfrac{3 \cdot 1}{2}(2) = 3$

(II) If $\quad 3 + 6 + 9 + \cdots + 3k = \dfrac{3k}{2}(k + 1)$,

then $\quad 3 + 6 + 9 + \cdots + 3k + 3(k + 1)$

$$= (3 + 6 + 9 + \cdots + 3k) + (3k + 3) = \frac{3k^2}{2}(k + 1) + (3k + 3)$$

$$= \frac{3k^2}{2} + \frac{3k}{2} + 3k + 3 = \frac{3k^2}{2} + \frac{3k}{2} + \frac{6k}{2} + \frac{6}{2}$$

$$= \frac{3k^2}{2} + \frac{9k}{2} + \frac{6}{2} = \frac{3}{2}(k^2 + 3k + 2) = \frac{3}{2}(k + 1)(k + 2)$$

47. (I) $n = 1$: $2 \cdot 3^{1-1} = 2$ and $3^1 - 1 = 2$

(II) If $\quad 2 + 6 + 18 + \cdots + 2 \cdot 3^{k-1} = 3^k - 1$,

then $\quad 2 + 6 + 18 + \cdots + 2 \cdot 3^{k-1} + 2 \cdot 3^{(k+1)-1}$

$$= (2 + 6 + 18 + \cdots + 2 \cdot 3^{k-1}) + 2 \cdot 3^k = 3^k - 1 + 2 \cdot 3^k$$

$$= 3^k(1 + 2) - 1 = 3 \cdot 3^k - 1 = 3^{k+1} - 1$$

49. (I) $n = 1$: $1^2 = 1$ and $\dfrac{1}{2}(6 - 3 - 1) = \dfrac{1}{2}(2) = 1$

(II) If $\quad 1^2 + 4^2 + 7^2 + \cdots + (3k - 2)^2 = \dfrac{1}{2}k(6k^2 - 3k - 1)$

then $1^2 + 4^2 + 7^2 + \cdots + (3k - 2)^2 + [3(k + 1) - 2]^2$

$$= [1^2 + 4^2 + 7^2 + \cdots + (3k - 2)^2] + (3k + 1)^2$$

$$= \frac{1}{2}k(6k^2 - 3k - 1) + (3k + 1)^2 = \frac{1}{2}(6k^3 + 15k^2 + 11k + 2)$$

$$= \frac{1}{2}(k + 1)\left[6k^2 + 12k - 3k + 6 - 3 - 1\right]$$

$$= \frac{1}{2}(k + 1)\left[6k^2 + 12k + 6 - 3k - 3 - 1\right]$$

$$= \frac{1}{2}(k + 1)\left[6(k^2 + 2k + 1) - 3(k + 1) - 1\right]$$

$$= \frac{1}{2}(k + 1)\left[6(k + 1)^2 - 3(k + 1) - 1\right]$$

51. $(x + 2)^4 = \binom{4}{0}x^4 + \binom{4}{1}2 \cdot x^3 + \binom{4}{2}2^2 x^2 + \binom{4}{3}2^3 \cdot x + \binom{4}{4}2^4$

$\qquad = x^4 + 4 \cdot 2x^3 + 6 \cdot 4x^2 + 4 \cdot 8x + 16$

$\qquad = x^4 + 8x^3 + 24x^2 + 32x + 16$

53. $(2x + 3)^5 = \binom{5}{0}(2x)^5 + \binom{5}{1}3(2x)^4 + \binom{5}{2}3^2(2x)^3 + \binom{5}{3}3^3(2x^2) + \binom{5}{4}3^4(2x) + \binom{5}{5}3^5$

$\qquad = 32x^5 + 5 \cdot 3 \cdot 16x^4 + 10 \cdot 9 \cdot 8x^3 + 10 \cdot 27 \cdot 4x^2 + 5 \cdot 81 \cdot 2x + 243$

$\qquad = 32x^5 + 240x^4 + 720x^3 + 1080x^2 + 810x + 243$

55. $n = 9, a = 2, x = x, \text{ and } j = 7$

$\binom{9}{9 - 7}2^{9-7}x^7 = \binom{9}{2} \cdot 2^2 x^7$

$\qquad\qquad = \frac{9!}{2!7!} \cdot 4 \cdot x^7$

$\qquad\qquad = \frac{9 \cdot 8}{2} \cdot 4 \cdot x^7$

$\qquad\qquad = 144x^7$

The coefficient of x^7 is 144.

57. $n = 7, a = 1, x = 2x, \text{ and } j = 2$

$\binom{7}{7 - 2}1^{7-2}(2x)^2 = \binom{7}{5}1^5 \cdot 4x^2$

$\qquad\qquad = \frac{7!}{5!2!} \cdot 1 \cdot 4 \cdot x^2$

$\qquad\qquad = \frac{7 \cdot 6}{2} \cdot 1 \cdot 4x^2$

$\qquad\qquad = 84x^2$

The coefficient of x^2 is 84.

In Problems 59–65, U = Universal set = {1, 2, 3, 4, 5, 6, 7, 8, 9}, A = {1, 3, 5, 7}, B = {3 5, 6, 7, 8},
C = {2, 3, 7, 8, 9}.

59. $A \cup B = \{1, 3, 5, 7\} \cup \{3, 5, 6, 7, 8\}$
$\qquad = \{1, 3, 5, 6, 7, 8\}$

61. $A \cap C = \{1, 3, 5, 7\} \cap \{2, 3, 7, 8, 9\}$
$\qquad = \{3, 7\}$

63. $A' \cup B' = \{1, 3, 5, 7\}' \cup \{3, 5, 6, 7, 8\}'$
$\qquad = \{2, 4, 6, 8, 9\} \cup \{1, 2, 4, 9\}$
$\qquad = \{1, 2, 4, 6, 8, 9\}$

65. $(B \cap C)' = (\{3, 5, 6, 7, 8\} \cap \{2, 3, 7, 8, 9\})'$
$\qquad = (\{3, 7, 8\})' = \{1, 2, 4, 5, 6, 9\}$

67. For $n(A) = 8, n(B) = 12, \text{ and } n(A \cap B) = 3,$
$n(A \cup B) = n(A) + n(B) - n(A \cap B)$
$\qquad = 8 + 12 - 3 = 17$

69. From the figure $n(A) = 20 + 2 + 6 + 1 = 29$

71. From the figure , $n(A \text{ and } C) = n(A \cap C)$
$\qquad = 6 + 1 = 7$

73. From the figure, $n(\text{neither in } A \text{ nor in } C)$
$\qquad = n((A \cup C)') = 25$

75. We have 2 choices of material, 3 choices of color, and 10 choices of size, or $2 \times 3 \times 10 = 60$ choices for a complete assortment.

77. There are two possible outcomes for each game, or
$$2 \times 2 \times 2 \times 2 \times 2 \times 2 \times 2 = 2^7 = 128 \text{ outcomes for 7 games.}$$

79. Some order is significant; we have
$$P(9, 4) = \frac{9!}{(9 - 4)!} = \frac{9!}{5!} = \frac{9 \cdot 8 \cdot 7 \cdot 6 \cdot 5!}{5!} = 3024 \text{ ways to seat 4 people in 9 seats.}$$

81. Choosing 4 runners from 8 where order is not significant gives
$$C(8, 4) = \frac{8!}{(8 - 4)!4!} = \frac{8!}{4!4!} = \frac{8 \cdot 7 \cdot 6 \cdot 5 \cdot 4!}{4 \cdot 3 \cdot 2 \cdot 1 \cdot 4!} = 70 \text{ ways a squad can be chosen.}$$

83. We have 14 teams to be taken 2 at a time, or
$$C(14, 2) = \frac{14!}{(14 - 2)!2!} = \frac{14!}{12!2!} = \frac{14 \cdot 13 \cdot 12!}{2 \cdot 1 \cdot 12!} = 91 \text{ways to pair 14 teams.}$$

85. There are $8 \cdot 10 \cdot 10 \cdot 10 \cdot 10 \cdot 2 = 1{,}600{,}000$ possible phone numbers.

87. There are $24 \cdot 9 \cdot 10 \cdot 10 \cdot 10 = 216{,}000$ possible license plates possible.

89. $\dfrac{7!}{2!2!} = 1260$ different words.

91. (a) $\dbinom{9}{4} \cdot \dbinom{9}{3} \cdot \dbinom{9}{2} = 126 \cdot 84 \cdot 36 = 381{,}024$

    (b) $\dbinom{9}{4} \cdot \dbinom{5}{3} \cdot \dbinom{2}{2} = 126 \cdot 10 \cdot 1 = 1{,}260$

93. The sample space contains $3 + 6 + 11 = 20$ elements. The probability of choosing a 40-watt bulb is:
$$P(E) = \frac{n(E)}{n(S)} = \frac{3}{20}$$
The probability of not choosing a 75 watt bulb is:
$$P(E) = \frac{n(E)}{n(S)} = \frac{3 + 6}{20} = \frac{9}{20}$$

95. The sample space for dealing out 4 cards contains $P(4, 4) = 4! = 24$ possible outcomes. Only one of these outcomes fulfills the desired event of spelling out ROSE so that the probability of this event is:
$$P(E) = \frac{n(E)}{n(S)} = \frac{1}{24}$$

97. (a) $P(\text{all 3 are Merlot}) = \dfrac{5}{12} \cdot \dfrac{4}{11} \cdot \dfrac{3}{10} = 0.045$

    (b) $P(\text{2 are Merlot}) = \dfrac{5}{12} \cdot \dfrac{4}{11} \cdot \dfrac{7}{10} + \dfrac{7}{12} \cdot \dfrac{5}{11} \cdot \dfrac{4}{10} + \dfrac{5}{12} \cdot \dfrac{7}{11} \cdot \dfrac{4}{10} = 0.318$

    (c) $P(\text{none Merlot}) = P(\text{all Cabernet}) = \dfrac{7}{12} \cdot \dfrac{6}{11} \cdot \dfrac{5}{10} = 0.159$

99. The bottom step requires 80 bricks and each successive step requires 3 less bricks than the prior step. This is an arithmetic sequence with $a_1 = 80$ and $d = -3$.
    (a) $a_{25} = a + (n - 1)d = 80 + (24)(-3) = 8$ bricks

(b) The total number of bricks required is:

$$S = 80 + 77 + \cdots + 8$$

This is the sum of an arithmetic sequence. The number of terms to be added is 25.

$$S = \frac{25}{2}(80 + 8) = 1100$$

Thus, 1100 bricks will be needed.

101. This is an example of a geometric series, with $r = \dfrac{3}{4}$.

(a) After striking the third time, the height is $20\left(\dfrac{3}{4}\right)^3 = \dfrac{135}{16}$ feet.

(b) After striking the nth time, the height is $20\left(\dfrac{3}{4}\right)^n$ feet.

(c) If the height is 6 inches (0.5 feet), then

$$0.5 = 20\left(\frac{3}{4}\right)^n$$

$$0.025 = \left(\frac{3}{4}\right)^n$$

$$\log 0.025 = n \log\left(\frac{3}{4}\right)$$

$$n = \frac{\log 0.025}{\log\left(\dfrac{3}{4}\right)} = 12.82$$

The height is less than 6 inches after the 13th strike.

(d) $\text{Distance} = 20 + 20\left(\dfrac{3}{4}\right) + 20\left(\dfrac{3}{4}\right) + 20\left(\dfrac{3}{4}\right)^2 + 20\left(\dfrac{3}{4}\right)^2 + \cdots + 20\left(\dfrac{3}{4}\right)^n + 20\left(\dfrac{3}{4}\right)^n + \cdots$

↑   ↑
Ball   Ball
Going   Coming
Up   Down

$$= 20 + 40\left(\frac{3}{4}\right) + 40\left(\frac{3}{4}\right)^2 + \cdots + 40\left(\frac{3}{4}\right)^n + \cdots$$

$$= 20 + 40\left(\frac{3}{4}\right)\left[1 + \left(\frac{3}{4}\right) + \cdots\right] = 20 + \frac{40\left(\dfrac{3}{4}\right)}{1 - \dfrac{3}{4}}$$

$$= 20 + 120 = 140 \text{ feet}$$

103. (a) $P(\text{tune-up or brake}) = P(\text{tune-up}) + P(\text{brake}) - P(\text{tune-up and brake})$

$$= 0.6 + 0.1 - 0.02$$
$$= 0.68$$

(b) $P(\text{tune-up and brake}') = P(\text{tune-up}) - P(\text{tune-up and brake})$

$$= 0.6 - 0.02$$
$$= 0.58$$

(c) $P(\text{tune-up and brake})' = 1 - P(\text{tune-up or brake})$

$$= 1 - 0.68$$
$$= 0.32$$

APPENDIX

Section 1 Topics from Algebra and Geometry

1. $\frac{1}{2} > 0$ since $\frac{1}{2} - 0 = \frac{1}{2}$ is positive.

3. $-1 > -2$ since $-1 - (-2) = -1 + 2 = 1$ is positive.

5. $\pi > 3.14$ since $\pi \approx 3.1416$ and $3.1416 - 3.1400 = .0016$ is positive, so then $\pi - 3.14$ is positive.

7. $\frac{1}{2} = 0.5$

9. $\frac{2}{3} < 0.67$ since $\frac{2}{3} = 0.666\ldots \approx 0.667$ and $0.670 - 0.667 = 0.003$ is positive, so that $0.67 - \frac{2}{3}$ is positive.

11.
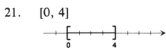

13. $x > 0$

15. $x < 2$

17. $x \leq 1$

19. $2 < x < 5$

21. $[0, 4]$

23. $[4, 6)$

25. $2 \leq x \leq 5$

27. $x \geq 4$

29. $|x + y| = |2 + (-3)| = |2 - 3| = |-1| = 1$

31. $|x| + |y| = |2| + |-3| = 2 + 3 = 5$

33. $3^0 = 1$

35. $4^{-2} = \frac{1}{4^2} = \frac{1}{16}$

37. $\left(\frac{2}{3}\right)^2 = \frac{2^2}{3^2} = \frac{4}{9}$

39. $3^{-6} \cdot 3^4 = 3^{-6+4} = 3^{-2} = \frac{1}{3^2} = \frac{1}{9}$

41. $\left(\frac{2}{3}\right)^{-2} = \dfrac{1}{\left(\frac{2}{3}\right)^2} = \dfrac{1}{\frac{2^2}{3^2}} = \frac{3^2}{2^2} = \frac{9}{4}$

43. $\dfrac{2^3 \cdot 3^2}{2 \cdot 3^{-2}} = \frac{2^3}{2} \cdot \frac{3^2}{3^{-2}} = 2^{3-1} \cdot 3^{2-(-2)} = 2^2 \cdot 3^4 = 4 \cdot 81 = 324$

45. $9^{3/2} = \sqrt{9^3} = \left(\sqrt{9}\right)^3 = 3^3 = 27$

47. $(-8)^{4/3} = \left(\sqrt[3]{-8}\right)^4 = (-2)^4 = 16$

49. $\sqrt{32} = \sqrt{16 \cdot 2} = \sqrt{16} \cdot \sqrt{2} = 4\sqrt{2}$

51. $\sqrt[3]{\dfrac{-8}{27}} = \dfrac{-2}{3}$

53. $x^0 y^2 = 1 \cdot y^2 = y^2$

55. $x^{-2}y = \dfrac{1}{x^2}y = \dfrac{y}{x^2}$

57. $\dfrac{x^{-2}y^3}{xy^4} = \dfrac{x^{-2}}{x} \cdot \dfrac{y^3}{y^4} = x^{-2}y^{3-4} = x^{-3}y^{-1} = \dfrac{1}{x^3}\dfrac{1}{y} = \dfrac{1}{x^3 y}$

59. $\left(\dfrac{4x}{5y}\right)^{-2} = \dfrac{1}{\left(\dfrac{4x}{5y}\right)^2} = \dfrac{1}{\dfrac{(4x)^2}{(5y)^2}} = \dfrac{(5y)^2}{(4x)^2} = \dfrac{5^2 \cdot y^2}{4^2 \cdot x^2} = \dfrac{25y^2}{16x^2}$

61. $\dfrac{x^{-1}y^{-2}z}{x^2yz^3} = \dfrac{x^{-1}}{x^2}\dfrac{y^{-2}}{y}\dfrac{z}{z^3} = x^{-1-2}y^{-2-1}z^{1-3} = x^{-3}y^{-3}z^{-2} = \dfrac{1}{x^3}\dfrac{1}{y^3}\dfrac{1}{z^2} = \dfrac{1}{x^3 y^3 z^2}$

63. $\dfrac{(-2)^3 x^4 (yz)^2}{3^2 xy^3 z^4} = \dfrac{-8}{9}\dfrac{x^4}{x}\dfrac{y^2}{y^3}\dfrac{z^2}{z^4} = \dfrac{-8}{9}x^{4-1}y^{2-3}z^{2-4} = \dfrac{-8}{9}x^3 y^{-1}z^{-2} = \dfrac{-8}{9}x^3\dfrac{1}{y}\dfrac{1}{z^2} = \dfrac{-8x^3}{9yz^2}$

65. $\dfrac{x^{-2}}{x^{-2}+y^{-2}} = \dfrac{\dfrac{1}{x^2}}{\dfrac{1}{x^2}+\dfrac{1}{y^2}} = \dfrac{\dfrac{1}{x^2}}{\dfrac{y^2+x^2}{x^2 y^2}} = \dfrac{x^2 y^2}{x^2(y^2+x^2)} = \dfrac{x^2}{x^2} \cdot \dfrac{y^2}{y^2+x^2}$

$= x^{2-2}\dfrac{y^2}{y^2+x^2} = x^0 \dfrac{y^2}{y^2+x^2} = \dfrac{y^2}{x^2+y^2}$

67. $\left(\dfrac{3x^{-1}}{4y^{-1}}\right)^{-2} = \dfrac{1}{\left(\dfrac{3x^{-1}}{4y^{-1}}\right)^2} = \dfrac{1}{\dfrac{(3x^{-1})^2}{(4y^{-1})^2}} = \dfrac{(4y^{-1})^2}{(3x^{-1})^2} = \dfrac{4^2 y^{-2}}{3^2 x^{-2}} = \dfrac{16}{9}\dfrac{\dfrac{1}{y^2}}{\dfrac{1}{x^2}} = \dfrac{16}{9}\dfrac{x^2}{y^2} = \dfrac{16x^2}{9y^2}$

For legs a and b of a right triangle in Problems 69–78, we use $c^2 = a^2 + b^2$ to find the hypotenuse c:

69. For $a = 5$ and $b = 12$, $c^2 = a^2 + b^2$
$$= 5^2 + 12^2$$
$$= 25 + 144$$
$$c^2 = 169$$
then $c = 13$

71. For $a = 10$ and $b = 24$, $c^2 = 10^2 + 24^2$
$$= 100 + 576$$
$$c^2 = 676$$
then $c = 26$

73. For $a = 7$ and $b = 24$, $c^2 = 7^2 + 24^2$
$$= 49 + 576$$
$$c^2 = 625$$
then $c = 25$

75. For $a = 3$ and $c = 5$, $c^2 = a^2 + b^2$
$$5^2 = 3^2 + b^2$$
$$b^2 = 5^2 - 3^2$$
$$b^2 = 25 - 9$$
$$b^2 = 16$$
then $b = 4$

77. For $b = 7$ and $c = 25$, $c^2 = a^2 + b^2$
$$a^2 = c^2 - b^2$$
$$a^2 = 25^2 - 7^2$$
$$= 625 - 49$$
$$a^2 = 576$$
then $a = 24$

In Problems 79–83, we will test whether the given triangles are right triangles by using $c^2 = a^2 + b^2$. The hypotenuse must be the longest side in any case:

79. For sides 3, 4, 5, let $c = 5$: $c^2 = a^2 + b^2$
$$5^2 = 3^2 + 4^2$$
$$25 = 9 + 16$$
$$25 = 25$$
The given triangle is a right triangle with hypotenuse of length 5.

81. For sides 4, 5, and 6, let $c = 6$: but $6^2 \neq 4^2 + 5^2$
since $36 \neq 16 + 25 = 41$
The triangle is not a right triangle.

83. For sides 7, 24, 25, let $c = 25$: $25^2 = 7^2 + 24^2$
$$625 = 49 + 576$$
$$625 = 625$$
The triangle is a right triangle with hypotenuse of length 25.

85. The diagonal of the rectangle forms two right triangles. The length, 8 inches, is the base of the triangle, and the width, 5 inches, is the altitude or height of the triangle. The diagonal is the hypotenuse, so
$$c^2 = a^2 + b^2$$
$$c^2 = 5^2 + 8^2$$
$$= 89$$
$$c = \sqrt{89} \approx 9.4 \text{ inches}$$

87.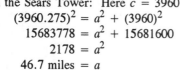
$$c^2 = a^2 + b^2$$
$$c^2 = 50^2 + 30^2$$
$$c^2 = 3400$$
$$c \approx 58.3 \text{ feet}$$

The guy wire needs to be 58.3 feet.

89. We have the triangle with side 3960 miles and hypotenuse
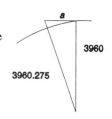
$3960 + \dfrac{1454}{5280} = 3960.275$ miles. Let side a be the distance in miles one can see
from the Sears Tower: Here $c = 3960.3$ and $b = 3960$:
$$(3960.275)^2 = a^2 + (3960)^2$$
$$15683778 = a^2 + 15681600$$
$$2178 = a^2$$
$$46.7 \text{ miles} = a$$

91. We have the triangle with hypotenuse c representing the distance

$3960 + \dfrac{6}{5280} = 3960.0011$ miles. Let a be the distance in miles of the ship from

shore that a 6-foot-tall person can see.

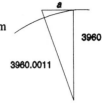

$$c^2 = a^2 + b^2$$
$$(3960.0011)^2 = a^2 + (3960)^2$$
$$15681609 = a^2 + 15681600$$
$$9 = a^2$$
$$3 \text{ miles} = a$$

93. Since $a \le b$, then $a - b \le 0$
$$(a - b)c \le 0 \cdot c$$
$$ca - cb \le 0$$
$$ca \le cb \text{ or } ac \le bc$$
Thus, if $a \le b$ and $c > 0$, then $ac \le bc$

95. Since $a < b$, $a \cdot \dfrac{1}{2} < b \cdot \dfrac{1}{2}$
$$\dfrac{a}{2} < \dfrac{b}{2}$$
$$\dfrac{a}{2} + \dfrac{a}{2} < \dfrac{b}{2} + \dfrac{a}{2}$$
$$a < \dfrac{a + b}{2}$$

Also, $\dfrac{a}{2} < \dfrac{b}{2}$
$$\dfrac{a}{2} + \dfrac{b}{2} < \dfrac{b}{2} + \dfrac{b}{2}$$
$$\dfrac{a + b}{2} < b$$

Thus, $a < \dfrac{a + b}{2} < b$

97. No; no   99. 1   101. 3.15 or 3.16

103. The light can be seen on the horizon 23.3 miles distant. Planes flying at 10,000 feet can see it 99 miles away. The ship would need to be 185.8 feet tall for the data to be correct.

Section 2 Polynomials and Rational Expressions

1. $(10x^5 - 8x^2) + (3x^3 - 2x^2 + 6)$
$10x^5 + 3x^3 + (-8x^2 - 2x^2) + 6$
$10x^5 + 3x^3 - 10x^2 + 6$

3. $(x + a)^2 - x^2 = x^2 + 2xa + a^2 - x^2$ by (3a)
$= (x^2 - x^2) + 2xa + a^2 = 2ax + a^2$

5. $(x + 8)(2x + 1) = 2x^2 + 17x + 8$ by (4b)

7. $(x^2 + x - 1)(x^2 - x + 1)$
$= x^2(x^2 - x + 1) + x(x^2 - x + 1) - 1(x^2 - x + 1)$ by distributive law
$= x^2 \cdot x^2 + x^2 \cdot (-x) + x^2 \cdot 1 + x \cdot x^2 + x \cdot (-x)$
$\quad + x \cdot 1 - 1 \cdot x^2 - 1 \cdot (-x) - 1 \cdot 1$ by distributive law
$= x^4 - x^3 + x^2 + x^3 - x^2 + x - x^2 + x - 1$
$= x^4 - x^2 + 2x - 1$

9. $(x + 1)^3 - (x - 1)^3 = (x^3 + 3x^2 + 3x + 1) - (x^3 - 3x^2 + 3x - 1)$ by (5a) and (5b)
$= x^3 + 3x^2 + 3x + 1 - x^3 + 3x^2 - 3x + 1 = 6x^2 + 2$

$$
\begin{array}{r}
4x^2 - 3x + 1 \\
\hline
\end{array}
$$

11. $x)\overline{4x^3 - 3x^2 + x + 1}$

$\quad\quad\quad\underline{4x^3}$

$\quad\quad\quad\quad\quad -3x^2$

$\quad\quad\quad\quad\quad\underline{-3x^2}$

$\quad\quad\quad\quad\quad\quad\quad x$

$\quad\quad\quad\quad\quad\quad\quad\underline{x}$

$\quad\quad\quad\quad\quad\quad\quad\quad 1$

The quotient is $4x^2 - 3x + 1$; the remainder is 1.

Check: $x(4x^2 - 3x + 1) + 1 = 4x^3 - 3x + x + 1$

13. $x + 2)\overline{4x^3 - 3x^2 + x + 1}$ with quotient $4x^2 - 11x + 23$

$\quad\quad\quad\underline{4x^3 + 8x^2}$

$\quad\quad\quad\quad\quad -11x^2 + x$

$\quad\quad\quad\quad\quad\underline{-11x^2 - 22x}$

$\quad\quad\quad\quad\quad\quad\quad 23x + 1$

$\quad\quad\quad\quad\quad\quad\quad\underline{23x + 46}$

$\quad\quad\quad\quad\quad\quad\quad\quad -45$

The quotient is $4x^2 - 11x + 23$; the remainder is -45.

Check: $(x + 2)(4x^2 - 11x + 23) + (-45) = 4x^3 - 11x^2 + 23x + 8x^2 - 22x + 46 + (-45)$

$\quad\quad\quad\quad\quad\quad\quad\quad\quad\quad\quad\quad = 4x^3 - 3x^2 + x + 1$

15. $x - 4)\overline{4x^3 - 3x^2 + x + 1}$ with quotient $4x^2 + 13x + 53$

$\quad\quad\quad\underline{4x^3 - 16x^2}$

$\quad\quad\quad\quad\quad 13x^2 + x$

$\quad\quad\quad\quad\quad\underline{13x^2 - 52x}$

$\quad\quad\quad\quad\quad\quad\quad 53x + 1$

$\quad\quad\quad\quad\quad\quad\quad\underline{53x - 212}$

$\quad\quad\quad\quad\quad\quad\quad\quad 213$

The quotient is $4x^2 + 13x + 53$; the remainder is 213.

Check: $(x - 4)(4x^2 + 13x + 53) + 213 = 4x^3 + 13x^2 + 53x - 16x^2 - 52x - 212 + 213$

$\quad\quad\quad\quad\quad\quad\quad\quad\quad\quad\quad = 4x^3 - 3x^2 + x + 1$

17. $x^2)\overline{4x^3 - 3x^2 + x + 1}$ with quotient $4x - 3$

$\quad\quad\quad\underline{4x^3}$

$\quad\quad\quad\quad\quad -3x^2$

$\quad\quad\quad\quad\quad\underline{-3x^2}$

$\quad\quad\quad\quad\quad\quad\quad x + 1$

The quotient is $4x - 3$; the remainder is $x + 1$.

Check: $x^2(4x - 3) + (x + 1) = 4x^3 - 3x^2 + x + 1$

19.
$$\require{enclose}
\begin{array}{r}
4x - 3 \\
x^2 + 2 \enclose{longdiv}{4x^3 - 3x^2 + x + 1} \\
\underline{4x^3 \qquad\quad + 8x} \\
-3x^2 - 7x + 1 \\
\underline{-3x^2 \qquad - 6} \\
-7x + 7
\end{array}$$

The quotient is $4x - 3$; the remainder is $-7x + 7$.

Check: $(x^2 + 2)(4x - 3) + (-7x + 7) = 4x^3 - 3x + 8x - 6 - 7x + 7 = 4x^3 - 3x^2 + x + 16$

21.
$$x^2 - 2x - 15 = (x \quad)(x \quad) = \begin{cases} (x \quad 1)(x \quad 15) \\ (x \quad 3)(x \quad 5) \end{cases}$$

$$(\text{signs must alternate}) = \begin{cases} (x + 1)(x - 15) \\ (x - 1)(x + 15) \\ (x + 3)(x - 5) \\ (x - 3)(x + 5) \end{cases}$$

$$= \begin{cases} x^2 - 14x - 15 \\ x^2 + 14x - 15 \\ x^2 - 2x - 15 \\ x^2 + 2x - 15 \end{cases}$$

Thus, $x^2 - 2x - 15 = (x - 5)(x + 3)$

23. $ax^2 - 4a^2x - 45a^3 = a(x^2 - 4ax - 45a^2)$

where $x^2 - 4ax - 45a^2 = (x \quad a)(x \quad a)$

$$= \begin{cases} (x \quad a)(x \quad 45a) \\ (x \quad 3a)(x \quad 15a) \\ (x \quad 5a)(x \quad 9a) \end{cases}$$

$$(\text{signs must alternate}) = \begin{cases} (x + a)(x - 45a) \\ (x - a)(x + 45a) \\ (x + 3a)(x - 15a) \\ (x - 3a)(x + 15a) \\ (x + 5a)(x - 9a) \\ (x - 5a)(x + 9a) \end{cases}$$

$$= \begin{cases} x^2 - 44a - 45a^2 \\ x^2 + 44a - 45a^2 \\ x^2 - 12a - 45a^2 \\ x^2 + 12a - 45a^2 \\ x^2 - 4a - 45a^2 \\ x^2 + 4a - 45a^2 \end{cases}$$

Thus, $x^2 - 4ax - 45a^2 = (x + 5a)(x - 9a)$ and $ax^2 - 4a^2x - 45a^3 = a(x - 9a)(x + 5a)$

25. $x^2 - 27 = x^3 - 3^3 = (x - 3)(x^2 + 3x + 9)$

27. $3x^2 + 4x + 1 = (3x + 1)(x + 1)$

29. $x^7 - x^5 = x^5(x^2 - 1) = x^5(x - 1)(x + 1)$

31. $\dfrac{3x - 6}{5x} \cdot \dfrac{x^2 - x - 6}{x^2 - 4} = \dfrac{3(x - 2)}{5x} \cdot \dfrac{(x - 3)(x + 2)}{(x + 2)(x - 2)} = \dfrac{3(x - 3)}{5x}$

33. $\dfrac{4x^2 - 1}{x^2 - 16} \cdot \dfrac{x^2 - 4x}{2x + 1} = \dfrac{(2x + 1)(2x - 1)}{(x + 4)(x - 4)} \cdot \dfrac{x(x - 4)}{(2x + 1)} = \dfrac{x(2x - 1)}{x + 4}$

35.

$$\frac{x}{x^2 - 7x + 6} - \frac{x}{x^2 - 2x - 24} = \frac{x}{(x-1)(x-6)} - \frac{x}{(x+4)(x-6)} = \frac{x(x+4) - x(x-1)}{(x-1)(x+4)(x-6)}$$

$$= \frac{x^2 + 4x - (x^2 - x)}{(x-1)(x+4)(x-6)} = \frac{5x}{(x-6)(x-1)(x+4)}$$

37. $\dfrac{4}{x^2 - 4} - \dfrac{2}{x^2 + x - 6} = \dfrac{4}{(x+2)(x-2)} - \dfrac{2}{(x+3)(x-2)} = \dfrac{4(x+3) - 2(x+2)}{(x+2)(x-2)(x+3)}$

$$= \frac{4x + 12 - (2x+4)}{(x+2)(x-2)(x+3)} = \frac{2x+8}{(x+2)(x-2)(x+3)} = \frac{2(x+4)}{(x-2)(x+2)(x+3)}$$

39. $\dfrac{x - \dfrac{1}{x}}{x + \dfrac{1}{x}} = \dfrac{\dfrac{x^2 - 1}{x}}{\dfrac{x^2 + 1}{x}} = \dfrac{x^2 - 1}{x} \cdot \dfrac{x}{x^2 + 1} = \dfrac{x^2 - 1}{x^2 + 1} = \dfrac{(x-1)(x+1)}{x^2 + 1}$

41. $\dfrac{3 - \dfrac{x^2}{x+1}}{1 + \dfrac{x}{x^2 - 1}} = \dfrac{\dfrac{3(x+1)}{x+1} - \dfrac{x^2}{x+1}}{\dfrac{x^2 - 1}{x^2 + 1} + \dfrac{x}{x^2 - 1}} = \dfrac{\dfrac{3x + 3 - x^2}{x+1}}{\dfrac{x^2 - 1 + x}{x^2 - 1}} = \dfrac{\dfrac{-x^2 + 3x + 3}{x+1}}{\dfrac{x^2 + x - 1}{(x+1)(x-1)}}$

$$= \frac{-x^2 + 3x + 3}{x+1} \cdot \frac{(x+1)(x-1)}{x^2 + x - 1} = \frac{(x-1)(-x^2 + 3x + 3)}{x^2 + x - 1}$$

Section 3 Radicals; Rational Exponents

1. $\sqrt{8} = \sqrt{4 \cdot 2} = \sqrt{4}\sqrt{2} = 2\sqrt{2}$

3. $\sqrt[3]{16x^4} = \sqrt[3]{8x^3 \cdot 2x} = \sqrt[3]{8x^3}\sqrt[3]{2x} = 2x\sqrt[3]{2x}$

5. $\sqrt[3]{\sqrt{x^6}} = \sqrt[3]{\sqrt{(x^3)^2}} = \sqrt[3]{x^3} = x$

7. $\sqrt{\dfrac{32x^3}{9x}} = \sqrt{\dfrac{16x^2}{9} \cdot 2} = \dfrac{\sqrt{16x^2}}{9}\sqrt{2} = \dfrac{4}{3}x\sqrt{2}$

9. $\sqrt[4]{x^{12}y^8} = \sqrt[4]{(x^3)^4(y^2)^4} = x^3y^2$

11. $\sqrt[4]{\dfrac{x^9y^7}{xy^3}} = \sqrt[4]{x^8y^4} = \sqrt[4]{(x^2)^4y^4} = x^2y$

13. $\sqrt{36x} = \sqrt{36}\sqrt{x} = 6\sqrt{x}$

15. $\sqrt{3x^2}\sqrt{12x} = \sqrt{36x^3} = \sqrt{36x^2}\sqrt{x} = 6x\sqrt{x}$

17. $\left(\sqrt{5}\sqrt[3]{9}\right)^2 = \sqrt{5^2}\sqrt[3]{9^2} = 5\sqrt[3]{81} = 5\sqrt[3]{3^4} = 5\sqrt[3]{3^3 \cdot 3} = 15\sqrt[3]{3}$

19. $\sqrt{\dfrac{2x-3}{2x^4 + 3x^3}}\sqrt{\dfrac{x}{4x^2 - 9}} = \sqrt{\dfrac{x(2x-3)}{(2x^4 + 3x^3)(4x^2 - 9)}} = \sqrt{\dfrac{x(2x-3)}{x^3(2x+3)(2x+3)(2x-3)}}$

$$= \sqrt{\frac{1}{x^2(2x+3)^2}} = \frac{\sqrt{1}}{\sqrt{x^2(2x+3)^2}} = \frac{1}{x(2x+3)}$$

21. $\left(3\sqrt{6}\right)\left(2\sqrt{2}\right) = 6\sqrt{12} = 6\sqrt{4 \cdot 3} = 6\sqrt{4}\sqrt{3} = 12\sqrt{3}$

23. $\left(\sqrt{3} + 3\right)\left(\sqrt{3} - 1\right) = \left(\sqrt{3}\right)^2 + 2\sqrt{3} - 3 = 3 + 2\sqrt{3} - 3 = 2\sqrt{3}$

25. $\left(\sqrt{x} - 1\right)^2 = \left(\sqrt{x}\right)^2 - 2\sqrt{x} + 1 = x - 2\sqrt{x} + 1$

27. $\dfrac{1}{\sqrt{2}} = \dfrac{1}{\sqrt{2}} \cdot \dfrac{\sqrt{2}}{\sqrt{2}} = \dfrac{\sqrt{2}}{\left(\sqrt{2}\right)^2} = \dfrac{\sqrt{2}}{2}$

29. $\dfrac{-\sqrt{3}}{\sqrt{5}} = \dfrac{-\sqrt{3}}{\sqrt{5}} \cdot \dfrac{\sqrt{5}}{\sqrt{5}} = \dfrac{-\sqrt{3}\sqrt{5}}{\left(\sqrt{5}\right)^2} = \dfrac{-\sqrt{15}}{5}$

31. $\dfrac{\sqrt{3}}{5 - \sqrt{2}} = \dfrac{\sqrt{3}}{5 - \sqrt{2}} \cdot \dfrac{5 + \sqrt{2}}{5 + \sqrt{2}} = \dfrac{\sqrt{3}\left(5 + \sqrt{2}\right)}{5^2 - \left(\sqrt{2}\right)^2} = \dfrac{\sqrt{3}\left(5 + \sqrt{2}\right)}{25 - 2} = \dfrac{\sqrt{3}\left(5 + \sqrt{2}\right)}{23}$

33. $\dfrac{2 - \sqrt{5}}{2 + 3\sqrt{5}} = \dfrac{2 - \sqrt{5}}{2 + 3\sqrt{5}} \cdot \dfrac{2 - 3\sqrt{5}}{2 - 3\sqrt{5}} = \dfrac{\left(2 - \sqrt{5}\right)\left(2 - 3\sqrt{5}\right)}{2^2 - \left(3\sqrt{5}\right)^2} = \dfrac{4 - 8\sqrt{5} + 3\left(\sqrt{5}\right)^2}{4 - 9 \cdot 5}$

$$= \dfrac{4 - 8\sqrt{5} + 15}{4 - 45} = \dfrac{19 - 8\sqrt{5}}{-41} = \dfrac{-19 + 8\sqrt{5}}{41}$$

35. $\dfrac{\sqrt{x + h} - \sqrt{x}}{\sqrt{x + h} + \sqrt{x}} = \dfrac{\sqrt{x + h} - \sqrt{x}}{\sqrt{x + h} + \sqrt{x}} \cdot \dfrac{\sqrt{x + h} - \sqrt{x}}{\sqrt{x + h} - \sqrt{x}} = \dfrac{\left(\sqrt{x + h} - \sqrt{x}\right)^2}{\left(\sqrt{x + h}\right)^2 - \left(\sqrt{x}\right)^2}$

$$= \dfrac{\left(\sqrt{x + h}\right)^2 - 2\sqrt{x + h}\sqrt{x} + \left(\sqrt{x}\right)^2}{x + h - x} = \dfrac{x + h - 2\sqrt{x(x + h)} + x}{h}$$

$$= \dfrac{2x + h - 2\sqrt{x(x + h)}}{h}$$

37. $8^{2/3} = \left(\sqrt[3]{8}\right)^2 = 2^2 = 4$

39. $(-27)^{1/3} = \sqrt[3]{-27} = -3$

41. $16^{3/2} = \left(\sqrt{16}\right)^3 = 4^3 = 64$

43. $9^{-3/2} = \left(\sqrt{9}\right)^{-3} = 3^{-3} = \dfrac{1}{3^3} = \dfrac{1}{27}$

45. $\left[\dfrac{9}{8}\right]^{3/2} = \left[\sqrt{\dfrac{9}{8}}\right]^3 = \left[\dfrac{\sqrt{9}}{\sqrt{8}}\right]^3 = \left[\dfrac{3}{2\sqrt{2}}\right]^3 = \dfrac{3^3}{\left(2\sqrt{2}\right)^3} = \dfrac{27}{8\left(\sqrt{2}\right)^3} = \dfrac{27}{16\sqrt{2}} = \dfrac{27}{16\sqrt{2}} \cdot \dfrac{\sqrt{2}}{\sqrt{2}} = \dfrac{27\sqrt{2}}{32}$

47. $\left[\dfrac{8}{9}\right]^{-3/2} = \left[\sqrt{\dfrac{8}{9}}\right]^{-3} = \left[\dfrac{\sqrt{8}}{\sqrt{9}}\right]^{-3} = \left[\dfrac{2\sqrt{2}}{3}\right]^{-3} = \dfrac{1}{\left[\dfrac{2\sqrt{2}}{3}\right]^3} = \dfrac{1}{\dfrac{\left(2\sqrt{2}\right)^3}{3^3}} = \dfrac{3^3}{\left(2\sqrt{2}\right)^3}$

$$= \dfrac{27}{16\sqrt{2}} \cdot \dfrac{\sqrt{2}}{\sqrt{2}} = \dfrac{27\sqrt{2}}{32}$$

49. $x^{5/4}x^{2/3}x^{-1/2} = x^{(5/4+2/3-1/2)} = x^{(15/12)+(8/12)-(6/12)} = x^{17/12}$

51. $(x^3y^6)^{2/3} = x^{6/3}y^{12/3} = x^2y^4$

53. $(x^2y)^{1/3}(xy^2)^{2/3} = x^{2/3}y^{1/3} \cdot x^{2/3}y^{4/3} = x^{(2/3)+2/3}y^{(1/3)+(4/3)} = y^{4/3}y^{5/3}$

55. $(16x^2y^{-1/3})^{3/4} = 16^{3/4}x^{6/4}y^{-3/12} = (2^4)^{3/4}x^{3/2}y^{-1/4} = 2^3 \, x^{3/2} \dfrac{1}{y^{1/4}} = \dfrac{8x^{3/2}}{y^{1/4}}$

57. $\dfrac{x}{(1+x)^{1/2}} + 2(1+x)^{1/2} = \dfrac{x + 2(1+x)^{1/2}(1+x)^{1/2}}{(1+x)^{1/2}} = \dfrac{x + 2(1+x)}{(1+x)^{1/2}} = \dfrac{3x+2}{(1+x)^{1/2}}$

59. $\dfrac{\sqrt{1+x} - x \cdot \dfrac{1}{2\sqrt{1+x}}}{1+x} = \dfrac{\sqrt{1+x} - \dfrac{x}{2\sqrt{1+x}}}{1+x} = \dfrac{\dfrac{\sqrt{1+x}\left(2\sqrt{1+x}\right) - x}{2\sqrt{1+x}}}{\dfrac{1+x}{1}}$

$= \dfrac{2(1+x) - x}{\rule{2cm}{0.4pt}}$

61. $\dfrac{(x+4)^{1/2} - 2x(x+4)^{-1/2}}{x+4} = \dfrac{(x+4)^{1/2} - \dfrac{2x}{(x+4)^{1/2}}}{x+4} = \dfrac{\dfrac{(x+4)(x+4)^{1/2} - 2x}{(x+4)^{1/2}}}{x+4}$

$= \dfrac{\dfrac{x+4-2x}{(x+4)^{1/2}}}{\dfrac{x+4}{1}} = \dfrac{4-x}{(x+4)^{3/2}}$

63. $(x+1)^{3/2} + x \cdot \dfrac{3}{2}(x+1)^{1/2} = (x+1)^{1/2}\left[(x+1)^{2/2} + \dfrac{3x}{2}\right] = (x+1)^{1/2}\left[x+1+\dfrac{3}{2}x\right]$

$= (x+1)^{1/2}\left[\dfrac{3}{2}x+1\right] = \dfrac{1}{2}(5x+2)(x+1)^{1/2}$

65. $6x^{1/2}(x^2+x) - 8x^{3/2} - 8x^{1/2} = 2x^{1/2}(3(x^2+x) - 4x^{2/2} - 4) = 2x^{1/2}(3x^2 + 3x - 4x - 4)$

$= 2x^{1/2}(3x^2 - x - 4) = 2x^{1/2}(3x-4)(x+1)$

Section 4 Solving Equations

1.

$$6 - x = 2x + 9$$
$$(6-x) - 6 = (2x+9) - 6$$
$$-x = 2x + 3$$
$$-x - 2x = 2x + 3 - 2x$$
$$-3x = 3$$
$$\dfrac{-3x}{-3} = \dfrac{3}{-3}$$
$$x = -1$$

Check:

$$6 - x = 2x + 9$$
$$6 - (-1) \overset{?}{=} 2(-1) + 9$$
$$6 + 1 \overset{?}{=} -2 + 9$$
$$7 = 7$$

$2(1$

3.
$$2(3 + 2x) = 3(x - 4)$$
$$6 + 4x = 3x - 12$$
$$(6 + 4x) - 6 = (3x - 12) - 6$$
$$4x = 3x - 18$$
$$4x - 3x = 3x - 18 - 3x$$
$$x = -18$$

Check:
$$2(3 + 2x) = 3(x - 4)$$
$$2(3 + 2)(-18)) \stackrel{?}{=} 3(-18 - 4)$$
$$2(3 - 36 \stackrel{?}{=} 3(-22)$$
$$2(-33) \stackrel{?}{=} -66$$
$$-66 = -66$$

5.
$$8x - (2x + 1) = 3x - 10$$
$$8x - 2x - 1 = 3x - 10$$
$$6x - 1 = 3x - 10$$
$$(6x - 1) + 1 = (3x - 10 + 1$$
$$6x = 3x - 9$$
$$6x - 3x = (3x - 9) - 3x$$
$$3x = -9$$
$$\frac{3x}{3} = \frac{-9}{3}$$
$$x = -3$$

Check:
$$8x - (2x + 1) = 3x - 10$$
$$8(-3) - (2(-3) + 1) \stackrel{?}{=} 3(-3) - 10$$
$$-24 - (-6 + 1) \stackrel{?}{=} -9 - 10$$
$$-24 - (-5) \stackrel{?}{=} -19$$
$$-24 + 5 \stackrel{?}{=} -19$$
$$-19 = -19$$

7.
$$\frac{1}{2}x - 4 = \frac{3}{4}x$$
$$4\left[\frac{1}{2}x - 4\right] = 4\left[\frac{3}{4}x\right]$$
$$2x - 16 = 3x$$
$$(2x - 16) + 16 = 3x + 16$$
$$2x = 3x + 16$$
$$2x - 3x = 3x + 16 - 3x$$
$$-x = 16$$
$$\frac{-x}{-1} = \frac{16}{-1}$$
$$x = -16$$

Check:
$$\frac{1}{2}x - 4 = \frac{3}{4}x$$
$$\frac{1}{2}(-16) - 4 \stackrel{?}{=} \frac{3}{4}(-16)$$
$$-8 - 4 \stackrel{?}{=} -12$$
$$-12 = -12$$

9.
$$0.9t = 0.4 + 0.1t$$
$$0.9t - 0.1t = (0.4 + 0.1t) - 0.1t$$
$$0.8t = 0.4$$
$$\frac{0.8t}{0.8} = \frac{0.4}{0.8}$$
$$t = 0.5$$

Check:
$$0.9t = 0.4 + 0.1t$$
$$0.9(0.5) \stackrel{?}{=} 0.4 + 0.01(0.5)$$
$$0.45 \stackrel{?}{=} 0.4 + 0.05$$
$$0.45 = 0.45$$

11.
$$\frac{2}{y} + \frac{4}{y} = 3$$
$$y\left[\frac{2}{y} + \frac{4}{y}\right] = [3]$$
$$2 + 4 = 3y$$
$$6 = 3y$$
$$\frac{6}{3} = \frac{3y}{3}$$
$$2 = y$$
$$y = 2$$

Check:
$$\frac{2}{y} + \frac{4}{y} = 3$$
$$\frac{2}{2} + \frac{4}{2} \stackrel{?}{=} 3$$
$$1 + 2 \stackrel{?}{=} 3$$
$$3 = 3$$

13.

$$(x + 7)(x - 1) = (x + 1)^2$$
$$x^2 + 6x - 7 = x^2 + 2x + 1$$
$$(x^2 + 6x - 7) - x^2 = (x^2 + 2x + 1) - x^2$$
$$6x - 7 = 2x + 1$$
$$(6x - 7) + 7 = (2x + 1) + 7$$
$$6x = 2x + 8$$
$$6x - 2x = (2x + 8) - 2x$$
$$4x = 8$$
$$\frac{4x}{4} = \frac{8}{4}$$
$$x = 2$$

Check:
$$x + 7)(x - 1) = (x + 1)^2$$
$$(2 + 7)(2 - 1) \overset{?}{=} (2 + 1)^2$$
$$(9)(1) \overset{?}{=} 3^2$$
$$9 = 9$$

15.

$$x(2x - 3) = (2x + 1)(x - 4)$$
$$2x^2 - 3x = 2x^2 - 7x - 4$$
$$(2x - 3x) - 2x^2 = (2x^2 - 7x - 4) - 2x^2$$
$$-3x = -7x - 4$$
$$-3x + 7x = (-7x - 4) + 7x$$
$$4x = -4$$
$$\frac{4x}{4} = \frac{-4}{4}$$
$$x = -1$$

Check:
$$x(2x - 3) = (2x + 1)(x - 4)$$
$$-1(2(-1) - 3) \overset{?}{=} (2(-1) + 1)(-1 - 4)$$
$$-1(-2 - 3) \overset{?}{=} (-2 + 1)(-5)$$
$$-1(-5) \overset{?}{=} (-1)(-5)$$
$$5 = 5$$

17.

$$z(z^2 + 1) = 3 + z^3$$
$$z^3 + z = 3 + z^3$$
$$(z^3 + z) - z^3 = (3 + z^3) - z^3$$
$$z = 3$$

Check:
$$z(z^2 + 1) = 3 + z^3$$
$$3(3^2 + 1) \overset{?}{=} 3 + 3^3$$
$$3(9 + 1) \overset{?}{=} 30$$
$$3(10) \overset{?}{=} 30$$
$$30 = 30$$

19. $\dfrac{x}{x - 3} + 3 = \dfrac{3}{x - 3}$

Note that $x - 3$ cannot equal zero so $x = 3$ is **NOT** in the domain of the variable.

$$(x - 3)\left[\frac{x}{x - 3} + 3\right] = \frac{3}{x - 3}(x - 3)$$
$$x + 3(x - 3) = 3$$
$$x + 3x - 9 = 3$$
$$4x = 12$$
$$x = 3$$

But $x = 3$ is not in the domain of the variable. Hence, the equation has no solution.

21.
$$x^2 = 9x$$
$$x^2 - 9x = 0$$
$$x(x - 9) = 0$$
$$x = 0, x = 9$$
The solution set is $\{0, 9\}$.

23.
$$t^3 - 9t^2 = 0$$
$$t^2(t - 9) = 0$$
$$t^2 = 0, t - 9 = 0$$
$$t = 0, \qquad t = 9$$
The solution set is $\{0, 9\}$.

25. $\dfrac{2x}{x^2 - 4} = \dfrac{4}{x^2 - 4} - \dfrac{1}{x + 2}$

Note that $x^2 - 4$ cannot equal zero and $x + 2$ cannot equal zero. Therefore, $x = -2$ and $x = 2$ are not in the domain of the variable.

$$(x^2 - 4)\frac{2x}{x^2 - 4} = (x^2 - 4)\left[\frac{4}{x^2 - 4} - \frac{1}{x + 2}\right]$$

$$2x = 4 - \frac{x^2 - 4}{x + 2}$$

$$2x = 4 - \frac{(x - 2)(x + 2)}{x + 2}$$

$$2x = 4 - (x - 2)$$
$$2x = 4 - x + 2$$
$$2x = 6 - x$$
$$3x = 6$$
$$x = 2$$

But $x = 2$ is not in the domain of the variable. The equation has no solution.

27.

$$\frac{x}{x + 2} = \frac{1}{2}$$

$$2(x + 2)\left[\frac{x}{x + 2}\right] = 2(x + 2)\left[\frac{1}{2}\right]$$

$$2x = x + 2$$
$$2x - x = (x + 2) - x$$
$$x = 2$$

Check:

$$\frac{x}{x + 2} = \frac{1}{2}$$

$$\frac{2}{2 + 2} \overset{?}{=} \frac{1}{2}$$

$$\frac{2}{4} \overset{?}{=} \frac{1}{2}$$

$$\frac{1}{2} = \frac{1}{2}$$

29.

$$\frac{3}{2x - 3} = \frac{2}{x + 5}$$

$$(2x - 3)(x + 5)\left[\frac{3}{2x - 3}\right] = (2x - 3)(x + 5)\left[\frac{2}{x + 5}\right]$$

$$3(x + 5) = 2(2x - 3)$$
$$(3x + 15) = 4x - 6$$
$$(3x + 15) - 15 = (4x - 6) - 15$$
$$3x = 4x - 21$$
$$3x - 4x = (4x - 21) - 4x$$
$$-x = -21$$
$$\frac{-x}{-1} = \frac{-21}{-1}$$
$$x = 21$$

Check:

$$\frac{3}{2x - 3} = \frac{2}{x + 5}$$

$$\frac{3}{2(21) - 3} \overset{?}{=} \frac{2}{21 + 5}$$

$$\frac{3}{42 - 3} = \frac{2}{26}$$

$$\frac{3}{39} \overset{?}{=} \frac{1}{13}$$

$$\frac{1}{13} = \frac{1}{13}$$

31.

$$(x + 2)(3x) = (x + 2)(6)$$
$$3x^2 + 6x = 6x + 12$$
$$(3x + 6x) - 6x = (6x + 12) - 6x$$
$$3x^2 = 12$$

$$\frac{3x^2}{3} = \frac{12}{3}$$
$$x^2 = 4$$
$$x = 2 \text{ or } x = -2$$
$$\{-2, 2\}$$

Check:
$$(x + 2)(3x) = (x + 2)(6)$$
$$(2 + 2)(3(2)) \overset{?}{=} (2 + 2)(6)$$
$$(4)(6) \overset{?}{=} (4)(6)$$
$$24 = 24$$

Alternate Solution:

$$(x + 2)(3x) = (x + 2)(6)$$

$$\frac{(x + 2)(3x)}{x + 2} = \frac{(x + 2)(6)}{x + 2}$$

$$3x = 6$$
$$x = 2$$

Check:
$$(x + 2)(3x) = (x + 2)(6)$$
$$(-2 + 2)(3(-2)) \overset{?}{=} (-2 + 2)(6)$$
$$0 = 0$$

33.

$$\frac{6t + 7}{4t - 1} = \frac{3t + 8}{2t - 4}$$

$$(4t - 1)(2t - 4)\left[\frac{6t + 7}{4t - 1}\right] = (4t - 1)(2t - 4)\left[\frac{3t + 8}{2t - 4}\right]$$

$$(2t - 4)(6t + 7) = (4t - 1)(3t + 8)$$

$$12t^2 - 10t - 28 = 12t^2 + 29t - 8$$

$$(12t^2 - 10t - 28) - 12t^2 = (12t^2 + 29t - 8) - 12t^2$$

$$-10t - 28 = 29t - 8$$

$$(-10t - 28) + 28 = (29t - 8) + 28$$

$$-10t = 29t + 20$$

$$-10t - 29t = 29t + 20 - 29t$$

$$-39t = 20$$

$$\frac{-39t}{-39} = \frac{20}{-39}$$

$$t = \frac{-20}{39}$$

Check:

$$\frac{6t + 7}{4t - 1} = \frac{3t + 8}{2t - 4}$$

$$\frac{6\left[\dfrac{-20}{39}\right] + 7}{4\left[\dfrac{-20}{39}\right] - 1} \overset{?}{=} \frac{3\left[\dfrac{-20}{39}\right] + 8}{2\left[\dfrac{-20}{39}\right] - 4}$$

$$\frac{\dfrac{-120}{39} + 7}{\dfrac{-80}{39} - 1} \overset{?}{=} \frac{\dfrac{-60}{39} + 8}{\dfrac{-40}{39} - 4}$$

$$\frac{\dfrac{-120 + 273}{39}}{\dfrac{-80 - 39}{39}} \overset{?}{=} \frac{\dfrac{-60 + 312}{39}}{\dfrac{-40 - 156}{39}}$$

$$\frac{\dfrac{153}{39}}{\dfrac{-119}{39}} \overset{?}{=} \frac{\dfrac{252}{39}}{\dfrac{-196}{39}}$$

$$\frac{153}{39} \cdot \frac{39}{-119} = \frac{252}{39} \cdot \frac{39}{-196}$$

$$\frac{153}{119} - \frac{252}{196} = -\frac{252}{39} \cdot \frac{39}{-196}$$

$$-\frac{153}{119} = -\frac{252}{196}$$

$$-\frac{9 \cdot 17}{7 \cdot 17} = -\frac{4 \cdot 7 \cdot 9}{4 \cdot 7 \cdot 9}$$

$$-\frac{9}{7} = -\frac{9}{7}$$

35.

$$\frac{2}{x - 2} = \frac{3}{x + 5} + \frac{10}{(x + 5)(x - 2)}$$

$$(x + 5)(x - 2)\left[\frac{2}{x - 2}\right] = (x + 5)(x - 2)\left[\frac{3}{x + 5} + \frac{10}{(x + 5)(x - 2)}\right]$$

$$2(x + 5) = 3(x - 2) + 10$$

$$2x + 10 = 3x - 6 + 10$$

$$2x + 10 = 3x + 4$$

$$(2x + 10) - 10 = (3x + 4) - 10$$

$$2x = 3x - 6$$

$$2x - 3x = (3x - 6) - 3x$$

$$-x = -6$$

$$\frac{-x}{-1} = \frac{-6}{-1}$$

$$x = 6$$

Check:
$$\frac{2}{x-2} = \frac{3}{x+5} + \frac{10}{(x+5)(x-2)}$$

$$\frac{2}{6-2} \stackrel{?}{=} \frac{3}{6+5} + \frac{10}{(6+5)(6-2)}$$

$$\frac{2}{4} \stackrel{?}{=} \frac{3}{11} + \frac{10}{(11)(4)}$$

$$\frac{1}{2} \stackrel{?}{=} \frac{12}{44} + \frac{10}{44}$$

$$\frac{1}{2} \stackrel{?}{=} \frac{22}{44}$$

$$\frac{1}{2} = \frac{1}{2}$$

Section 5 Completing the Square

1. $x^2 - 4x + \underline{\ ?\ }$ 4 should be added to complete the square.

3. $x^2 + \frac{1}{2}x + \underline{\ ?\ }$ $\frac{1}{16}$ should be added to complete the square.

5. $x^2 - \frac{2}{3}x + \underline{\ ?\ }$ $\frac{1}{9}$ should be added to complete the square.

7. $x^2 + y^2 - 4x + 4y - 1 = 0$
    $x^2 - 4x + \underline{\ \ } + y^2 + 4y + \underline{\ \ } = 1$
    $x^2 - 4x + 4 + y^2 + 4y + 4 = 1 + 4 + 4$
    $(x - 2)^2 + (y + 2)^2 = 9$

9. $x^2 + y^2 + 6x - 2y + 1 = 0$
    $x^2 + 6x + \underline{\ \ } + y^2 - 2y + \underline{\ \ } = -1$
    $x^2 + 6x + 9 + y^2 - 2y + 1 = -1 + 9 + 1$
    $(x + 3)^2 + (y - 1)^2 = 9$

11. $x^2 + y^2 + x - y - \frac{1}{2} = 0$

    $x^2 + x + \underline{\ \ } + y^2 - y + \underline{\ \ } = \frac{1}{2}$

    $x^2 + x + \frac{1}{4} + y^2 - y + \frac{1}{4} = \frac{1}{2} + \frac{1}{4} + \frac{1}{4}$

    $\left(x + \frac{1}{2}\right)^2 + \left(y - \frac{1}{2}\right)^2 = 1$

13. $x^2 + 4x - 21 = 0$
                $x^2 + 4x + 4 = 21 + 4$
                  $(x + 2)^2 = 25$

                    $x + 2 = \pm\sqrt{25}$
                    $x + 2 = \pm 5$
                        $x = -2 \pm 5$
    $x = -2 + 5 = 3$ or $x = -2 - 5 = -7$
    The solution set is $\{-7, 3\}$.

15.

$$x^2 - \frac{1}{2}x = \frac{3}{16}$$

$$x^2 - \frac{1}{2}x + \frac{1}{16} = \frac{3}{16} + \frac{1}{16}$$

$$\left(x - \frac{1}{4}\right)^2 = \frac{1}{4}$$

$$x - \frac{1}{4} = \pm\sqrt{\frac{1}{4}}$$

$$x - \frac{1}{4} = \pm\frac{1}{2}$$

$$x = \frac{1}{4} \pm \frac{1}{2}$$

$$x = \frac{1}{4} + \frac{1}{2} = \frac{3}{4} \quad \text{or} \quad x = \frac{1}{4} - \frac{1}{2} = \frac{-1}{4}$$

The solution set is $\left\{\dfrac{-1}{4}, \dfrac{3}{4}\right\}$.

17.

$$3x^2 + x - \frac{1}{2} = 0$$

$$x^2 + \frac{1}{3}x = \frac{1}{6}$$

$$x^2 + \frac{1}{3}x + \frac{1}{36} = \frac{1}{6} + \frac{1}{36}$$

$$\left(x + \frac{1}{6}\right)^2 = \frac{7}{36}$$

$$x + \frac{1}{6} = \pm\sqrt{\frac{7}{36}}$$

$$x = \frac{-1}{6} \pm \frac{\sqrt{7}}{6}$$

$$x = \frac{-1 - \sqrt{7}}{6} \quad \text{or} \quad x = \frac{-1 + \sqrt{7}}{6}$$

The solution set is $\left\{\dfrac{-1 - \sqrt{7}}{6}, \dfrac{-1 + \sqrt{7}}{6}\right\}$.

Section 6 Synthetic Division

1.

```
2| 1  -1   2   4        ← f(x) = x³ - x² + 2x + 4
        2   2   8
   1   1   4  12        ← Remainder = 12
```

Quotient $= x^2 + x + 47$

3.

```
3| 3   2  -1    3       ← f(x) = 3x³ + 2x² - x + 3
       3   9  33   96
   3  11  32   99       ← Remainder = 99
```

Quotient $= 3x^2 + 11x + 32$

5.
$$
\begin{array}{r|rrrrrr}
-3 & 1 & 0 & -4 & 0 & 1 & 0 \\
  &   & -3 & 9 & -15 & 45 & -138 \\
\hline
  & 1 & -3 & 5 & -15 & 46 & -138
\end{array}
$$
← $f(x) = x^5 - 4x^3 + x$

← Remainder = -138

Quotient = $x^4 - 3x^3 + 5x^2 - 15x + 46$

7.
$$
\begin{array}{r|rrrrrrr}
1 & 4 & 0 & -3 & 0 & 1 & 0 & 5 \\
  &   & 4 & 4 & 1 & 1 & 2 & 2 \\
\hline
  & 4 & 4 & 1 & 1 & 2 & 2 & 7
\end{array}
$$
← Remainder = 7

Quotient = $4x^5 + 4x^4 + x^3 + x^2 + 2x + 2$

9.
$$
\begin{array}{r|rrrr}
-1.1 & 0.1 & 0 & 0.2 & 0 \\
     &   & -0.11 & 0.121 & -0.3531 \\
\hline
     & 0.1 & -0.11 & 0.321 & -0.3531
\end{array}
$$
← Remainder = -0.3531

Quotient = $0.1x^2 - 0.11x + 0.321$

11.
$$
\begin{array}{r|rrrrrr}
1 & 1 & 0 & 0 & 0 & 0 & -1 \\
  &   & 1 & 1 & 1 & 1 & 1 \\
\hline
  & 1 & 1 & 1 & 1 & 1 & 0
\end{array}
$$
← $f(x) = x^5 - 1$

← Remainder = 0

Quotient = $x^4 + x^3 + x^2 + x + 1$

In Problems 13–22, we use the following facts: the remainder in synthetic division when f(x) is divided by x − c, is f(c); and x − c is a factor of f(x) only if f(c) = 0.

13. We divide by $x - c = x - 2$:
$$
\begin{array}{r|rrrr}
2 & 4 & -3 & -8 & 4 \\
  &   & 8 & 10 & 4 \\
\hline
  & 4 & 5 & 2 & 8
\end{array}
$$
Remainder = 8 ≠ 0; therefore, $x - 2$ is *not* a factor of $f(x)$.

15. $x - c = x - 2$
$$
\begin{array}{r|rrrr}
2 & 3 & -6 & 0 & -5 & 10 \\
  &   & 6 & 0 & 0 & -10 \\
\hline
  & 3 & 0 & 0 & -5 & \boxed{0}
\end{array}
$$
The remainder = 0; therefore, $x - 2$ *is* a factor of $f(x)$.

17. $x - c = x - (-3) = x + 3$
$$
\begin{array}{r|rrrrrrr}
-3 & 3 & 0 & 0 & 82 & 0 & 0 & 27 \\
   &   & -9 & 27 & -81 & -3 & 9 & -27 \\
\hline
   & 3 & -9 & 27 & 1 & -3 & 9 & \boxed{0}
\end{array}
$$
Remainder = 0; therefore, $x + 3$ is a factor.

19. $x - c = x - (-4) = x + 4$
$$
\begin{array}{r|rrrrrrr}
-4 & 4 & 0 & -64 & 0 & 1 & 0 & -15 \\
   &   & -16 & 64 & 0 & 0 & -4 & 16 \\
\hline
   & 4 & -16 & 0 & 0 & 1 & -4 & \boxed{1}
\end{array}
$$
$x + 4$ is not a factor, since the remainder = 1 ≠ 0.

21. $x - c = x - \dfrac{1}{2}$

$$\dfrac{1}{2} \bigg) \begin{array}{rrrrr} 2 & -1 & 0 & 2 & -1 \\ & 1 & 0 & 0 & 1 \\ \hline 2 & 0 & 0 & 2 & \boxed{0} \end{array}$$

Since the remainder $= 0$; therefore, $x - \dfrac{1}{2}$ *is* a factor of $f(x)$.

For Problems 23–28, recall that the remainder that results when $f(x)$ is divided by $(x - c)$ is precisely $f(c)$.

23. Divide $f(x) = 5x^4 - 3x^2 + 1$ by $x - c = x - 2$

$$2 \bigg) \begin{array}{rrrrr} 5 & 0 & -3 & 0 & 1 \\ & 10 & 20 & 34 & 68 \\ \hline 5 & 10 & 17 & 34 & \boxed{69} \end{array} \quad \leftarrow f(c) = f(2) = 69$$

25. Divide $f(x) = 4x^5 - 3x^3 + 2x - 1$ by $x - c = x + 1$

$$-1 \bigg) \begin{array}{rrrrrr} 4 & 0 & -3 & 0 & 2 & -1 \\ & -4 & 4 & -1 & 1 & -3 \\ \hline 4 & -4 & 1 & -1 & 3 & \boxed{-4} \end{array} \quad \leftarrow f(c) = f(-1) = -4$$

27. Divide $f(x) = 9x^{17} - 8x^{10} + 9x^8 + 5$ by $x - c = x - 1$

$$1 \bigg) \begin{array}{rrrrrrrrrrrrrrrrr} 9 & 0 & 0 & 0 & 0 & 0 & 0 & -8 & 0 & 9 & 0 & 0 & 0 & 0 & 0 & 0 & 0 & 5 \\ & 9 & 9 & 9 & 9 & 9 & 9 & 9 & 1 & 1 & 10 & 10 & 10 & 10 & 10 & 10 & 10 & 10 \\ \hline 9 & 9 & 9 & 9 & 9 & 9 & 9 & 1 & 1 & 10 & 10 & 10 & 10 & 10 & 10 & 10 & 10 & \boxed{15} \end{array} \quad \leftarrow f(1)$$

Note: In this case, it would be much easier to find $f(1)$ by substitution:
$$f(1) = 9(1)^{17} - 8(1)^{10} + 9(1)^8 + 5 = 9 - 8 + 9 + 5 = 15$$

TRIGONOMETRIC FUNCTIONS

Chapter 6

6.1 Angles and Their Measure

1.
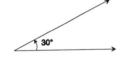

3.
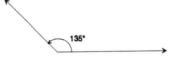

5.
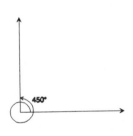

7.
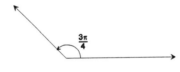

9.

11.
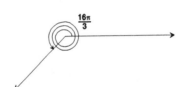

13. $30° = 30 \cdot 1 \text{ degree} = 30 \cdot \dfrac{\pi}{180} \text{ radian} = \dfrac{\pi}{6} \text{ radians}$

15. $240° = 240 \cdot 1 \text{ degree} = 240 \cdot \dfrac{\pi}{180} \text{ radian} = \dfrac{4\pi}{3} \text{ radians}$

17. $-60° = -60 \cdot 1 \text{ degree} = -60 \cdot \dfrac{\pi}{180} \text{ radian} = -\dfrac{\pi}{3} \text{ radians}$

19. $180° = 180 \cdot 1 \text{ degree} = 180 \cdot \dfrac{\pi}{180} \text{ radian} = \pi \text{ radians}$

21. $135° = 135 \cdot 1 \text{ degree} = 135 \cdot \dfrac{\pi}{180} \text{ radian} = \dfrac{3\pi}{4} \text{ radians}$

23. $\dfrac{\pi}{3} \text{ radian} = \dfrac{\pi}{3} \cdot 1 \text{ radian} = \dfrac{\pi}{3} \cdot \dfrac{180}{\pi} \text{ degrees} = 60°$

Section 6.1 Angles and Their Measure

649

25. $\frac{-5\pi}{4}$ radian $= \frac{-5\pi}{4} \cdot 1$ radian $= \frac{-5\pi}{4} \cdot \frac{180}{\pi}$ degrees $= -225°$

27. $\frac{\pi}{2}$ radians $= \frac{\pi}{2} \cdot 1$ radian $= \frac{\pi}{2} \cdot \frac{180}{\pi}$ degrees $= 90°$

29. $\frac{\pi}{12}$ radian $= \frac{\pi}{12} \cdot 1$ radian $= \frac{\pi}{12} \cdot \frac{180}{\pi}$ degrees $= 15°$

31. $\frac{2\pi}{3}$ radian $= \frac{2\pi}{3} \cdot 1$ radian $= \frac{2\pi}{3} \cdot \frac{180}{\pi}$ degrees $= 120°$

33. $r = 10$ meters

$\theta = \frac{1}{2}$ radian

$s = ?$

Use $s = r\theta$, $s = 10\left[\frac{1}{2}\right]$

$= 5$ meters

35. $\theta = \frac{1}{3}$ radian

$s = 2$ feet

$\theta = ?$

Use $s = r\theta$, $2 = r\left[\frac{1}{3}\right]$

$6 = r$

$r = 6$ feet

37. $r = 5$ miles

$s = 3$ miles

$\theta = ?$

Use $s = r\theta$ or $\theta = \frac{s}{r}$

$\theta = \frac{3}{5} = 0.6$ radian

39. $r = 2$ inches

$\theta = 30°$

$s = ?$

Convert $30°$ to $\frac{\pi}{6}$ radian.

Use $s = r\theta$

$s = 2\left[\frac{\pi}{6}\right] = \frac{\pi}{3}$ inches ≈ 1.047 inches

41. $17° = 17 \cdot 1$ degree $= 17 \cdot \frac{\pi}{180}$ radians $= \frac{17\pi}{180}$ radians $= 0.30$

43. $-40° = -40 \cdot 1$ degree $= -40 \cdot \frac{\pi}{180}$ radians $= \frac{-2\pi}{9}$ radians $= -0.70$

45. $125° = 125 \cdot 1$ degree $= 125 \cdot \frac{\pi}{180}$ radians $= \frac{25\pi}{36}$ radians $= 2.18$

47. $340° = 340 \cdot 1$ degree $= 340 \cdot \frac{\pi}{180}$ radians $= \frac{17\pi}{9}$ radians $= 5.93$

49. 3.14 radians $= 3.14 \cdot 1$ radian $= 3.14 \cdot \frac{180}{\pi}$ degrees $= 179.91°$

51. 10.25 radians $= 10.25 \cdot 1$ radian $= 10.25 \cdot \frac{180}{\pi}$ degrees $= 587.28°$

53. 2 radians $= 2 \cdot 1$ radian $= 2 \cdot \frac{180}{\pi}$ degrees $= 114.59°$

55. 6.32 radians = 6.32 · 1 radian = 6.32 · $\frac{180}{\pi}$ degrees = 362.11°

57. $40°10'25'' = \left[40 + 10 \cdot \frac{1}{60} + 25 \cdot \frac{1}{60} \cdot \frac{1}{60}\right]° = (40 + 0.16667 + 0.00694)° = 40.1736$
$$= 40.17°$$

59. $1°2'3'' = \left[1 + 2 \cdot \frac{1}{60} + 3 \cdot \frac{1}{60} \cdot \frac{1}{60}\right]° = (1 + 0.03333 + 0.00083)° = 1.03416 = 1.03°$

61. $9°9'9'' = \left[9 + 9 \cdot \frac{1}{60} + 9 \cdot \frac{1}{60} \cdot \frac{1}{60}\right] = (9 + 0.15 + 0.0025) = 9.15°$

63. $40.32° = ?$
$0.32° = (0.32)(1°) = (0.32)(60') = 19.2'$
$0.2' = (0.2)(1') = (0.2)(60'') = 12''$
$40.32° = 40° + 0.32° = 40° + 19.2' = 40° + 19' + 0.2' = 40° + 19' + 12'' = 40°19'12''$

65. $18.255° = ?$
$0.255° = (0.255)(1 \text{ degree}) = (0.255)(60') = 15.3'$
$0.3' = (0.3)(1') = (0.3)(60'') = 18''$
$18.255° = 18° + 0.255° = 18° + 15.3' = 18° + 15' + 0.3' = 18° + 15' + 18'' = 18°15'18''$

67. $19.99° = ?$
$0.99° = (0.99)(1 \text{ degree}) = (0.99)(60') = 59.4'$
$0.4' = (0.4)(1') = (0.4)(60') = 24''$
$19.99° = 19° + 0.99° = 19° + 59.4' = 19° + 59' + 0.4' = 19° + 59' + 24'' = 19°59'24''$

69. $s = r\theta$ $\theta = 90° = \frac{\pi}{2}$ in 15 minutes

$s = (6)\left[\frac{\pi}{2}\right]$; $s = 3\pi$ inches ≈ 9.4248 in.

$\theta = \frac{25}{60} = \frac{5}{12} \cdot \frac{360}{1} = 150° = \frac{5}{6}\pi$ in 25 minutes

$s = 6\left[\frac{5\pi}{6}\right]$; $s = 5\pi$ inches ≈ 15.7080 in.

71. $r = 5$ cm $t = 20$ sec $\theta = \frac{1}{3}$ rad $\omega = \frac{\theta}{t}$

$\omega = \frac{\frac{1}{3}}{20} = \frac{1}{3} \cdot \frac{1}{20} = \frac{1}{60}$ rad/sec

$v = \frac{s}{t}$ where $s = r\theta$; $s = 5\left[\frac{1}{3}\right]$

$v = \frac{\frac{5}{3}}{20} = \frac{5}{3} \cdot \frac{1}{20} = \frac{1}{12}$ cm/sec

73. $d = 26$ inches; $v = 35$ mi/hr
$\frac{35 \text{ mi}}{\text{hr}} \cdot \frac{5280 \text{ ft}}{\text{mi}} \cdot \frac{12 \text{ in}}{\text{ft}} \cdot \frac{\text{rev}}{\pi(26 \text{ in})} \cdot \frac{1 \text{ hr}}{60 \text{ min}} = 452.5$ rpm

75. $s = r\theta;\ r = 18$ inches

$$\theta = \frac{1}{3} \cdot 2 = \frac{2\pi}{3}$$

$$s = 18 \cdot \frac{2\pi}{3} = 12\pi \approx 37.7 \text{ inches}$$

77. $v = r\omega = 2.39 \times 10^5 \cdot \dfrac{1 \text{ rev}}{27.3 \text{ days}} \cdot \dfrac{1 \text{ day}}{24 \text{ hrs}} \cdot \dfrac{2\pi \text{ rad}}{\text{rev}} = 2292$ mph

79. Find distance (d) traveled by 2″ pulley in 3 revolutions.
$$d = \pi D \cdot N = \pi \cdot 2 \cdot 3 = 6\pi''$$

This distance is the same traveled by 8″ pulley. Solve for N.
$$6\pi = \pi \cdot 8 \cdot N$$
$$N = \frac{6}{8} = \frac{3}{4} \text{ rpm}$$

81. Find linear speed v using $v = r\omega$

$$v = r\omega = 4' \left[\frac{2\pi \text{ rad}}{\text{rev}} \right] \left[\frac{10 \text{ rev}}{\text{min}} \right] \left[\frac{1 \text{ mile}}{5280 \text{ ft}} \right] \left[\frac{60 \text{ min}}{\text{hr}} \right] = 2.86 \text{ mph}$$

83. The linear speed is 9.55 mi/hr. Since the diameter is 8.5 feet, the radius of the wheel is
$$\frac{8.5 \text{ feet}}{2} = 4.25 \text{ feet. Thus,}$$

$$v = r \cdot \omega$$

$$\omega = \frac{v}{r} = \frac{9.55 \text{ mi/hr}}{4.25 \text{ feet}} = \frac{50424 \text{ feet/hr}}{4.25 \text{ feet}} = 11864.47 \text{ radians/hr}$$

One revolution is 2π radians. So,

$$\omega = 11864.47 \frac{\text{radians}}{\text{hr}} \cdot \frac{1 \text{ revolution}}{2\pi} \cdot \frac{1 \text{ hr}}{60 \text{ minutes}} = 31.47 \text{ rev/min}$$

85. The earth makes one full rotation in 24 hours. The distance, s, traveled in 24 hours is the circumference of the earth. The circumference of the earth is 2π (3960 miles). Therefore, the linear velocity a person must travel to keep up with the sun is:

$$v = \frac{s}{t} = \frac{2\pi(3960 \text{ miles})}{24 \text{ hours}} \approx 1037 \text{ mph}$$

6.2 Trigonometric Functions: Unit Circle Approach

1. $(-3, 4)$

For $(a, b) = (-3, 4)$, we find $a = -3$, $b = 4$, and $r = \sqrt{a^2 + b^2} = \sqrt{9 + 16} = \sqrt{25} = 5$. Thus,

$$\sin \theta = \frac{b}{r} = \frac{4}{5} \qquad \csc \theta = \frac{r}{b} = \frac{5}{4}$$

$$\cos \theta = \frac{a}{r} = -\frac{3}{5} \qquad \sec \theta = \frac{r}{a} = -\frac{5}{3}$$

$$\tan \theta = \frac{b}{a} = -\frac{4}{3} \qquad \cot \theta = \frac{a}{b} = -\frac{3}{4}$$

3. $(2, -3)$

For $(a, b) = (2, -3)$, we find $a = 2$, $b = -3$, and $r = \sqrt{a^2 + b^2} = \sqrt{4 + 9} = \sqrt{13}$. Thus,

$$\sin \theta = \frac{b}{r} = \frac{-3}{\sqrt{13}} \cdot \frac{\sqrt{13}}{\sqrt{13}} = -\frac{3\sqrt{13}}{13} \qquad \csc \theta = \frac{r}{b} = -\frac{\sqrt{13}}{3}$$

$$\cos \theta = \frac{a}{r} = \frac{2}{\sqrt{13}} \cdot \frac{\sqrt{13}}{\sqrt{13}} = \frac{2\sqrt{13}}{13} \qquad \sec \theta = \frac{r}{a} = \frac{\sqrt{13}}{2}$$

$$\tan \theta = \frac{b}{a} = -\frac{3}{2} \qquad \cot \theta = \frac{a}{b} = -\frac{2}{3}$$

5. $(-2, -2)$

For $(a, b) = (2, -2)$, we find $a = -2$, $b = -2$, and $r = \sqrt{a^2 + b^2} = \sqrt{4 + 4} = \sqrt{8} = 2\sqrt{2}$. Thus,

$$\sin \theta = \frac{b}{r} = \frac{-2}{2\sqrt{2}} = -\frac{1}{\sqrt{2}} \cdot \frac{\sqrt{2}}{\sqrt{2}} = -\frac{\sqrt{2}}{2} \qquad \csc \theta = \frac{r}{b} = \frac{2\sqrt{2}}{-2} = -\sqrt{2}$$

$$\cos \theta = \frac{a}{r} = \frac{-2}{2\sqrt{2}} = -\frac{1}{\sqrt{2}} \cdot \frac{\sqrt{2}}{\sqrt{2}} = -\frac{\sqrt{2}}{2} \qquad \sec \theta = \frac{r}{a} = \frac{2\sqrt{2}}{-2} = -\sqrt{2}$$

$$\tan \theta = \frac{b}{a} = \frac{-2}{-2} = 1 \qquad \cot \theta = \frac{a}{b} = \frac{-2}{-2} = 1$$

7. $(-3, -2)$

For $(a, b) = (-3, -2)$, we find $a = -3$, $b = -2$, and $r = \sqrt{a^2 + b^2} = \sqrt{9 + 4} = \sqrt{13}$. Thus,

$$\sin \theta = \frac{b}{r} = \frac{-2}{\sqrt{13}} = -\frac{2}{\sqrt{13}} \cdot \frac{\sqrt{13}}{\sqrt{13}} = -\frac{2\sqrt{13}}{13} \qquad \csc \theta = \frac{r}{b} = \frac{\sqrt{13}}{-2} = -\frac{\sqrt{13}}{2}$$

$$\cos \theta = \frac{a}{r} = \frac{-3}{\sqrt{13}} = -\frac{3}{\sqrt{13}} \cdot \frac{\sqrt{13}}{\sqrt{13}} = -\frac{3\sqrt{13}}{13} \qquad \sec \theta = \frac{r}{a} = \frac{\sqrt{13}}{-3} = -\frac{\sqrt{13}}{3}$$

$$\tan \theta = \frac{b}{a} = \frac{-2}{-3} = \frac{2}{3} \qquad \cot \theta = \frac{a}{b} = \frac{-3}{-2} = \frac{3}{2}$$

9. $\left[\dfrac{1}{3}, \dfrac{-1}{4}\right]$

For $(a, b) = \left[\dfrac{1}{3}, \dfrac{-1}{4}\right]$, we find $a = \dfrac{1}{3}$, $b = \dfrac{-1}{4}$, and $r = \sqrt{\dfrac{1}{9} + \dfrac{1}{16}} = \sqrt{\dfrac{25}{144}} = \dfrac{5}{12}$.

$$\sin \theta = \frac{\frac{-1}{4}}{\frac{5}{12}} = \frac{-1}{4} \cdot \frac{12}{5} = \frac{-3}{5} \qquad \csc \theta = \frac{\frac{5}{12}}{\frac{-1}{4}} = \frac{5}{12} \cdot \frac{-4}{1} = \frac{-5}{3}$$

$$\cos \theta = \frac{\frac{1}{3}}{\frac{5}{12}} = \frac{1}{3} \cdot \frac{12}{5} = \frac{4}{5} \qquad \sec \theta = \frac{\frac{5}{12}}{\frac{1}{3}} = \frac{5}{12} \cdot \frac{3}{1} = \frac{5}{4}$$

$$\tan \theta = \frac{\frac{-1}{4}}{\frac{1}{3}} = \frac{-1}{4} \cdot \frac{3}{1} = \frac{-3}{4} \qquad \cot \theta = \frac{\frac{1}{3}}{\frac{-1}{4}} = \frac{1}{3} \cdot \frac{-4}{1} = \frac{-4}{3}$$

11. $\sin 45° + \cos 60° = \dfrac{\sqrt{2}}{2} + \dfrac{1}{2} = \dfrac{1}{2}\left(\sqrt{2} + 1\right)$ 13. $\sin 90° + \tan 45° = 1 + 1 = 2$

15. $\sin 45° \cos 45° = \dfrac{\sqrt{2}}{2} \cdot \dfrac{\sqrt{2}}{2} = \dfrac{2}{4} = \dfrac{1}{2}$ 17. $\csc 45° \tan 60° = \sqrt{2} \cdot \sqrt{3} = \sqrt{6}$

19. $4 \sin 90° - 3 \tan 180° = 4(1) = 3(0) = 4$

21. $2 \sin \dfrac{\pi}{3} - 3 \tan \dfrac{\pi}{6} = 2\left[\dfrac{\sqrt{3}}{2}\right] - 3\left[\dfrac{\sqrt{3}}{3}\right] = 0$ 23. $\sin \dfrac{\pi}{4} - \cos \dfrac{\pi}{4} = \dfrac{\sqrt{2}}{2} - \dfrac{\sqrt{2}}{2} = 0$

25. $2 \sec \dfrac{\pi}{4} + 4 \cot \dfrac{\pi}{3} = 2\sqrt{2} + 4\left[\dfrac{\sqrt{3}}{3}\right]$ 27. $\tan \pi - \cos 0 = 0 - 1 = -1$

29. $\csc \dfrac{\pi}{2} + \cot \dfrac{\pi}{2} = 1 + 0 = 1$

31.   The point $P = \left[\dfrac{-1}{2}, \dfrac{\sqrt{3}}{2}\right]$ lies on the terminal side of $\theta = \dfrac{2\pi}{3}$. Thus,

$$\sin \frac{2\pi}{3} = \frac{\sqrt{3}}{2} \qquad\qquad \csc \frac{2\pi}{3} = \frac{1}{\frac{\sqrt{3}}{2}} = \frac{2\sqrt{3}}{3}$$

$$\cos \frac{2\pi}{3} = \frac{-1}{2} \qquad\qquad \sec \frac{2\pi}{3} = \frac{1}{\frac{-1}{2}} = -2$$

$$\tan \frac{2\pi}{3} = \frac{\frac{\sqrt{3}}{2}}{\frac{-1}{2}} = -\sqrt{3} \qquad\qquad \cot \frac{2\pi}{3} = \frac{\frac{-1}{2}}{\frac{\sqrt{3}}{2}} = \frac{-\sqrt{3}}{3}$$

33.   The point $P = \left[\dfrac{-\sqrt{3}}{2}, \dfrac{1}{2}\right]$ lies on the terminal side of $\theta = 150° = \dfrac{5\pi}{6}$. Thus,

$$\sin 150° = \frac{1}{2} \qquad\qquad \csc 150° = \frac{1}{\frac{1}{2}} = 2$$

$$\cos 150° = \frac{-\sqrt{3}}{2} \qquad\qquad \sec 150° = \frac{1}{\frac{-\sqrt{3}}{2}} = \frac{-2\sqrt{3}}{3}$$

$$\tan 150° = \frac{\frac{1}{2}}{\frac{-\sqrt{3}}{2}} = \frac{-\sqrt{3}}{3} \qquad\qquad \cot 150° = \frac{\frac{-\sqrt{3}}{2}}{\frac{1}{2}} = -\sqrt{3}$$

35. The point $P = \left[\dfrac{\sqrt{3}}{2}, \dfrac{-1}{2}\right]$ lies on the terminal side of $\theta = \dfrac{-\pi}{6}$. Thus,

$$\sin \frac{\pi}{6} = \frac{-1}{2}$$

$$\csc \frac{-\pi}{6} = \frac{1}{\dfrac{-1}{2}} = -2$$

$$\cos \frac{-\pi}{6} = \frac{\sqrt{3}}{2}$$

$$\sec \frac{-\pi}{6} = \frac{1}{\dfrac{\sqrt{3}}{2}} = \frac{2\sqrt{3}}{3}$$

$$\tan \frac{-\pi}{6} = \frac{\dfrac{-1}{2}}{\dfrac{\sqrt{3}}{2}} = \frac{-\sqrt{3}}{3}$$

$$\cot \frac{-\pi}{6} = \frac{\dfrac{\sqrt{3}}{2}}{\dfrac{-1}{2}} = -\sqrt{3}$$

37. The point $P = \left[\dfrac{-\sqrt{2}}{2}, \dfrac{-\sqrt{2}}{2}\right]$ lies on the terminal side of $\theta = 225° = \dfrac{5\pi}{4}$. Thus,

$$\sin 225° = \frac{-\sqrt{2}}{2}$$

$$\csc 225° = \frac{1}{\dfrac{-\sqrt{2}}{2}} = -\sqrt{2}$$

$$\cos 225° = \frac{-\sqrt{2}}{2}$$

$$\sec 225° = \frac{1}{\dfrac{-\sqrt{2}}{2}} = -\sqrt{2}$$

$$\tan 225° = \frac{\dfrac{-\sqrt{2}}{2}}{\dfrac{-\sqrt{2}}{2}} = 1$$

$$\cot 225° = \frac{\dfrac{-\sqrt{2}}{2}}{\dfrac{-\sqrt{2}}{2}} = 1$$

39. Since $\dfrac{5\pi}{2} = 2\pi + \dfrac{\pi}{2}$ the point $P = (0, 1)$ lies on the terminal side of $\theta = \dfrac{5\pi}{2}$. Thus,

$$\sin \frac{5\pi}{2} = 1 \qquad\qquad \csc \frac{5\pi}{2} = 1$$

$$\cos \frac{5\pi}{2} = 0 \qquad\qquad \sec \frac{5\pi}{2} \text{ is not defined}$$

$$\tan \frac{5\pi}{2} \text{ is not defined} \qquad\qquad \cot \frac{5\pi}{2} = 0$$

41. The point $(-1, 0)$ lies on the terminal side of $\theta = -180°$. Thus,

$$\sin (-180°) = 0 \qquad\qquad \csc (-180°) \text{ is not defined}$$
$$\cos (-180°) = -1 \qquad\qquad \sec (-180°) = -1$$
$$\tan (-180°) = 0 \qquad\qquad \cot (-180°) \text{ is not defined}$$

43.

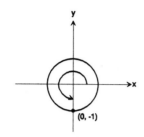

$$\sin \frac{3\pi}{2} = \frac{-1}{1} = -1 \qquad \csc \frac{3\pi}{2} = -1$$

$$\cos \frac{3\pi}{2} = \frac{0}{1} = 0 \qquad \sec \frac{3\pi}{2} = \text{undefined}$$

$$\tan \frac{3\pi}{2} = \frac{1}{0} = \text{undefined} \qquad \cot \frac{3\pi}{2} = \frac{0}{1} = 0$$

45.

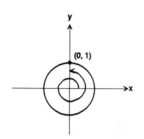

$$\sin 450° = \frac{1}{1} = 1 \qquad \csc 450° = \frac{1}{1} = 1$$

$$\cos 450° = \frac{0}{1} = 0 \qquad \sec 450° = \frac{1}{0} = \text{undefined}$$

$$\tan 450° = \frac{1}{0} = \text{undefined} \qquad \cot 450° = \frac{0}{1} = 0$$

47. Using your calculator, $\sin 28° \approx 0.4694716 \approx 0.47$.

49. Using your calculator, $\tan 21° \approx 0.383864 \approx 0.38$.

51. Using your calculator, $\sec 41° = \dfrac{1}{\cos 41°} \approx 1.325012993 \approx 1.33$

53. Using your calculator, $\cot 70° = \dfrac{1}{\tan 70°} \approx 0.3639702 \approx 0.36$

55. Set the mode to receive radians. Then, $\sin \dfrac{\pi}{10} \approx 0.309017 \approx 0.31$.

57. Set the mode to receive radians. Then, $\tan \dfrac{5\pi}{12} \approx 3.7320508 \approx 3.73$.

59. Set the mode to receive radians. Then, $\sec \dfrac{\pi}{12} = \dfrac{1}{\cos\left(\dfrac{\pi}{12}\right)} \approx 1.0352762 \approx 1.04$.

61. Set the mode to receive radians. Then, $\cot\left(\dfrac{\pi}{18}\right) = \dfrac{1}{\tan\left(\dfrac{\pi}{18}\right)} \approx 5.6712819 \approx 5.67$.

63. Set the mode to receive radians. Then, $\sin 1 \approx 0.841471 \approx 0.84$.

65. Use the (regular) degree mode. Then, $\sin 1° \approx 0.0174524 \approx 0.02$.

67. Use the degree mode. Then, $\cos 21.5° \approx 0.9304176 \approx 0.93$.

69. Set the mode to receive radians. Then, $\tan 0.3 \approx 0.3093362 \approx 0.31$.

71. $\sin 60° = \dfrac{\sqrt{3}}{2}$ **73.** $\sin \dfrac{60°}{2} = \sin 30° = \dfrac{1}{2}$

75. $(\sin 60°)^2 = \left[\dfrac{\sqrt{3}}{2}\right]^2 = \dfrac{3}{4}$ **77.** $\sin (2 \cdot 60°) = \sin 120° = \dfrac{\sqrt{3}}{2}$

79. $2 \sin 60° = 2\left[\dfrac{\sqrt{3}}{2}\right] = \sqrt{3}$ **81.** $\dfrac{\sin 60°}{2} = \dfrac{\dfrac{\sqrt{3}}{2}}{2} = \dfrac{\sqrt{3}}{4}$

83. $\sin 45° + \sin 135° + \sin 225° + \sin 315° = \dfrac{\sqrt{2}}{2} + \dfrac{\sqrt{2}}{2} - \dfrac{\sqrt{2}}{2} - \dfrac{\sqrt{2}}{2} = 0$

85. If $\sin \theta = 0.1$, then $\sin (\theta + \pi) = -0.1$ **87.** If $\tan \theta = 3$, then $\tan (\theta + \pi) = 3$

89. If $\sin \theta = \dfrac{1}{5}$, then $\csc \theta = \dfrac{1}{\dfrac{1}{5}} = 5$

91. Using the formula $R = \dfrac{v_0^2 \sin 2\theta}{g}$ and $g \approx 32.2$ ft/sec^2, and given $\theta = 45°$ and $v_0 = 100$ ft/sec, we get:

$R = \dfrac{(100)^2 \sin 2(45°)}{32.2}$

$R = \dfrac{(10{,}000)(\sin 90°)}{32.2}$, using calculator

$R = \dfrac{(10{,}000)(1)}{32.2} \approx 310.559$

$R \approx 310.56$ feet

Using the formula $H = \dfrac{v_0^2 \sin^2 \theta}{2g}$ and $g \approx 32.2$ ft/sec^2, and given $\theta = 45°$ and $v_0 = 100$ ft/sec, we get:

$H = \dfrac{(100)^2(\sin 45°)^2}{2(32.2)}$, using calculator or table $\sin 45° = 0.7071$

$H = \dfrac{(10{,}000)(0{,}7071)^2}{(64.4)}$, using calculator

$H \approx 77.638262$

$H \approx 77.64$ feet

93. Using the formula $R = \dfrac{v_0^2 \sin 2\theta}{g}$ and $g \approx 9.8$ m/sec^2, and given $\theta = 25°$ and $v_0 = 500$ m/sec, we get:

$R = \dfrac{(500)^2 \sin 2(25°)}{9.8}$

$R = \dfrac{(250{,}000)(\sin 50°)}{9.8}$

$R = \dfrac{(250{,}000)(0.7660444)}{9.8}$

$R \approx 19{,}541.95$

$R \approx 19{,}542$ meters

Using the formula $H = \dfrac{v_0^2 \sin^2 \theta}{2g}$ and $g \approx 9.8$ m/sec^2, and given $\theta = 25°$ and $v_0 = 500$ m/sec, we get:

$$H = \dfrac{(500)^2 \sin^2(25°)}{2(9.8)}$$
$$H \approx 2278 \text{ m}$$

95. We use the formula, $t = \sqrt{\dfrac{2a}{g \sin \theta \cos \theta}}$, where $g \approx 32$ ft/sec/sec and $a = 10$. Then

(a) $t = \sqrt{\dfrac{20}{32 \sin 30° \cos 30°}} = \sqrt{\dfrac{20}{32\left[\dfrac{1}{2}\right]\left[\dfrac{\sqrt{3}}{2}\right]}} = \sqrt{\dfrac{20}{8\sqrt{3}}} = \sqrt{\dfrac{5}{2\sqrt{3}}} \approx 1.2$ sec

(b) $t = \sqrt{\dfrac{20}{32 \sin 45° \cos 45°}} = \sqrt{\dfrac{20}{32\left[\dfrac{\sqrt{2}}{2}\right]\left[\dfrac{\sqrt{2}}{2}\right]}} = \sqrt{\dfrac{20}{16}} = \sqrt{\dfrac{5}{4}} \approx 1.12$ sec

(c) $t = \sqrt{\dfrac{20}{32 \sin 60° \cos 60°}} = \sqrt{\dfrac{20}{32\left[\dfrac{\sqrt{3}}{2}\right]\left[\dfrac{1}{2}\right]}} = \sqrt{\dfrac{5}{2\sqrt{3}}} \approx 1.2$ sec

97. (a) $T(30°) = 1 + \dfrac{2}{3 \sin 30°} - \dfrac{1}{4 \tan 30°} = 1 + \dfrac{2}{3\left[\dfrac{1}{2}\right]} - \dfrac{1}{4\left[\dfrac{1}{\sqrt{3}}\right]}$

$$= 1 + \dfrac{4}{3} - \dfrac{\sqrt{3}}{4} \approx 1.9 \text{ hours}$$

The time Sally is on the paved road is given by $P(\theta) = \dfrac{8 - \dfrac{2}{\tan \theta}}{8}$.

Thus, $P(30°) = \dfrac{8 - \dfrac{2}{\tan 30°}}{8} \approx 0.57$ hours.

(b) $T(45°) = 1 + \dfrac{2}{3 \sin 45°} - \dfrac{1}{4 \tan 45°} = 1 + \dfrac{2}{3\left[\dfrac{1}{\sqrt{2}}\right]} - \dfrac{1}{4(1)}$

$$= 1 + \dfrac{2\sqrt{2}}{3} - \dfrac{1}{4} \approx 1.69 \text{ hours}$$

$P(45°) = \dfrac{8 - \dfrac{2}{\tan 45°}}{8} \approx 0.75$ hours

(c) $T(60°) = 1 + \dfrac{2}{3 \sin 60°} - \dfrac{1}{4 \tan 60°} = 1 + \dfrac{2}{3\left[\dfrac{\sqrt{3}}{2}\right]} - \dfrac{1}{4\left[\dfrac{\sqrt{3}}{1}\right]}$

$$= 1 + \dfrac{4}{3\sqrt{3}} - \dfrac{1}{4\sqrt{3}} \approx 1.63 \text{ hours}$$

$$P(60°) = \frac{8 - \dfrac{2}{\tan 60°}}{8} \approx 0.86 \text{ hours}$$

(d) $T(90°) = 1 + \dfrac{2}{3 \sin 90°} - \dfrac{1}{4 \tan 90°} = 1 + \dfrac{2}{3(1)} - \dfrac{\cos \dfrac{\pi}{2}}{4 \sin \dfrac{\pi}{2}} = 1 + \dfrac{2}{3} \approx 1.67 \text{ hours}$

Sally walks directly to the path from the house, takes the path, then walks directly to the house from the path.

99. (a) $R = \dfrac{(32)^2 \sqrt{2}}{32}(\sin(2 \cdot 60) - \cos(2 \cdot 60) - 1) = 32\sqrt{2}\,(0.866 - (-0.5) - 1)$

$$= 32\sqrt{2}\,(0.366) \approx 16.6 \text{ ft} \approx 16.6 \text{ ft}$$

(b)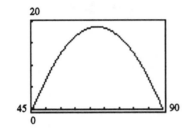

(c) R is largest when $\theta = 67.50°$.

6.3 Properties of the Trigonometric Functions

1. $\sin 405° = \sin(360° + 45°) = \sin 45° = \dfrac{\sqrt{2}}{2}$

3. $\tan 405° = \tan(180° + 180° + 45°) = \tan 45° = 1$

5. $\csc 450° = \csc(360° + 90°) = \csc 90° = 1$

7. $\cot 390° = \cot(180° + 180° + 30°) = \cot 30° = \sqrt{3}$

9. $\cos \dfrac{33\pi}{4} = \cos\left[\dfrac{\pi}{4} + \dfrac{32\pi}{4}\right] = \cos\left[\dfrac{\pi}{4} + 8\pi\right] = \cos\left[\dfrac{\pi}{4} + 2\pi \cdot 4\right] = \cos \dfrac{\pi}{4} = \dfrac{\sqrt{2}}{2}$

11. $\tan 21\pi = \tan(\pi + 20\pi) = \tan(\pi + \pi \cdot 20) = \tan \pi = 0$

13. $\sec \dfrac{17\pi}{4} = \sec\left[\dfrac{\pi}{4} + \dfrac{16\pi}{4}\right] = \sec\left[\dfrac{\pi}{4} + 4\pi\right] = \sec\left[\dfrac{\pi}{4} + 2\pi \cdot 2\right] = \sec\dfrac{\pi}{4} = \sqrt{2}$

15. $\tan \dfrac{19\pi}{6} = \tan\left[\dfrac{\pi}{6} + \dfrac{18\pi}{6}\right] = \tan\left[\dfrac{\pi}{6} + 3\pi\right] = \tan\left[\dfrac{\pi}{6} + \pi \cdot 3\right] = \tan\dfrac{\pi}{6} = \dfrac{\sqrt{3}}{3}$

17. Using Table 5, we find that $\sin \theta > 0$ for points P in quadrant I and II, and $\cos \theta < 0$ for points P in quadrants II and III. Both conditions are satisfied only if P lies in quadrant II.

19. Since $\sin \theta < 0$ for points P in quadrants III and IV, and $\tan \theta < 0$ for points P in quadrants II and IV, P lies in quadrant IV.

21. Since $\cos \theta > 0$ for points P in quadrants I and IV, and $\tan \theta < 0$ for points P in quadrants II and IV, P lies in quadrant IV.

23. Since $\sec \theta < 0$ for points P quadrants II and III, and $\sin \theta > 0$ for points P in quadrants I and II, P lies in quadrant II.

25. $\sin \theta = \dfrac{2}{\sqrt{5}}$, $\cos \theta = \dfrac{1}{\sqrt{5}}$

Based on the fundamental identities,

$$\tan \theta = \frac{\sin \theta}{\cos \theta} = \frac{\frac{2}{\sqrt{5}}}{\frac{1}{\sqrt{5}}} = 2 \qquad\qquad \sec \theta = \frac{1}{\cos \theta} = \frac{1}{\frac{1}{\sqrt{5}}} = \sqrt{5}$$

$$\csc \theta = \frac{1}{\sin \theta} = \frac{1}{\frac{2}{\sqrt{5}}} = \frac{\sqrt{5}}{2} \qquad\qquad \cot \theta = \frac{1}{\tan \theta} = \frac{1}{2}$$

27. $\sin \theta = \dfrac{1}{2}$, $\cos \theta = \dfrac{\sqrt{3}}{2}$

Based on the fundamental identities,

$$\tan \theta = \frac{\sin \theta}{\cos \theta} = \frac{\frac{1}{2}}{\frac{\sqrt{3}}{2}} = \frac{1}{\sqrt{3}} = \frac{\sqrt{3}}{3} \qquad \sec \theta = \frac{1}{\cos \theta} = \frac{1}{\frac{\sqrt{3}}{2}} = \frac{2}{\sqrt{3}} = \frac{2\sqrt{3}}{3}$$

$$\csc \theta = \frac{1}{\sin \theta} = \frac{1}{\frac{1}{2}} = 2 \qquad\qquad \cot \theta = \frac{1}{\tan \theta} = \frac{1}{\frac{1}{\sqrt{3}}} = \sqrt{3}$$

29. $\sin \theta = -\dfrac{1}{3}$, $\cos \theta = \dfrac{2\sqrt{2}}{3}$

Based on the fundamental identities,

$$\tan \theta = \frac{\sin \theta}{\cos \theta} = \frac{\frac{-1}{3}}{\frac{2\sqrt{2}}{3}} = \frac{-1}{2\sqrt{2}} = \frac{-\sqrt{2}}{4} \qquad \sec \theta = \frac{1}{\cos \theta} = \frac{1}{\frac{2\sqrt{2}}{3}} = \frac{3}{2\sqrt{2}} = \frac{3\sqrt{2}}{4}$$

$$\csc \theta = \frac{1}{\sin \theta} = \frac{1}{\frac{-1}{3}} = -3 \qquad\qquad \cot \theta = \frac{1}{\tan \theta} = \frac{1}{\frac{-1}{2\sqrt{2}}} = -2\sqrt{2}$$

31. $\sin \theta = 0.2588$, $\cos \theta = 0.9659$

Based on the fundamental identities,

$$\tan \theta = \frac{\sin \theta}{\cos \theta} = \frac{0.2588}{0.9659} \approx 0.2679 \qquad \sec \theta = \frac{1}{\cos \theta} = \frac{1}{0.9659} \approx 1.0353$$

$$\csc \theta = \frac{1}{\sin \theta} = \frac{1}{0.2588} \approx 3.8640 \qquad \cot \theta = \frac{\cos \theta}{\sin \theta} = \frac{0.9659}{0.2588} \approx 3.7322$$

33. $\sin \theta = \dfrac{12}{13}$, $90° < \theta < 180°$

First, we solve for $\cos \theta$:

$$\sin^2 \theta + \cos^2 \theta = 1$$
$$\cos^2 = 1 - \sin^2 \theta$$
$$\cos \theta = \pm\sqrt{1 - \sin^2 \theta}$$

Because $90° < \theta < 180°$, $\cos \theta < 0$.

$$\cos \theta = -\sqrt{1 - \sin^2 \theta} = -\sqrt{1 - \dfrac{144}{169}} = -\sqrt{\dfrac{25}{169}} = \dfrac{-5}{13}$$

$$\tan \theta = \dfrac{\sin \theta}{\cos \theta} = \dfrac{\dfrac{13}{13}}{\dfrac{-5}{13}} = \dfrac{-12}{5} \qquad\qquad \sec \theta = \dfrac{1}{\cos \theta} = \dfrac{1}{\dfrac{-5}{13}} = \dfrac{-13}{5}$$

$$\csc \theta = \dfrac{1}{\sin \theta} = \dfrac{1}{\dfrac{12}{13}} = \dfrac{13}{12} \qquad\qquad \cot \theta = \dfrac{1}{\tan \theta} = \dfrac{1}{\dfrac{-12}{5}} = \dfrac{-5}{12}$$

35. $\cos \theta = \dfrac{-4}{5}$, $\pi < \theta < \dfrac{3\pi}{2}$

$$\sin^2 \theta = 1 - \cos^2 \theta$$
$$\sin \theta = \pm\sqrt{1 - \cos^2 \theta}$$

Because $\pi < \theta < \dfrac{3\pi}{2}$, $\sin \theta < 0$.

$$\sin \theta = -\sqrt{1 - \cos^2 \theta} = -\sqrt{1 - \dfrac{16}{25}} = -\sqrt{\dfrac{9}{25}} = \dfrac{-3}{5}$$

$$\tan \theta = \dfrac{\sin \theta}{\cos \theta} = \dfrac{\dfrac{-3}{5}}{\dfrac{-4}{5}} = \dfrac{3}{4} \qquad\qquad \csc \theta = \dfrac{1}{\sin \theta} = \dfrac{1}{\dfrac{-3}{5}} = \dfrac{-5}{3}$$

$$\cot \theta = \dfrac{1}{\tan \theta} = \dfrac{1}{\dfrac{3}{4}} = \dfrac{4}{3} \qquad\qquad \sec \theta = \dfrac{1}{\cos \theta} = \dfrac{1}{\dfrac{-4}{5}} = \dfrac{-5}{4}$$

37. $\sin \theta = \dfrac{5}{13}$, $\cos \theta < 0$

First, we solve for $\cos \theta$:

$$\sin^2 t + \cos^2 \theta = 1$$
$$\cos^2 \theta = 1 - \sin^2 \theta$$
$$\cos \theta = \pm\sqrt{1 - \sin^2 \theta}$$

Because $\cos \theta < 0$, we use the minus sign:

$$\cos \theta = -\sqrt{1 - \sin^2 \theta} = -\sqrt{1 - \left(\dfrac{5}{13}\right)^2} = -\sqrt{1 - \dfrac{25}{169}} = -\sqrt{\dfrac{144}{169}} = \dfrac{-12}{13}$$

$$\tan \theta = \dfrac{\sin \theta}{\cos \theta} = \dfrac{\dfrac{5}{13}}{\dfrac{-12}{13}} = \dfrac{-5}{12} \qquad\qquad \sec \theta = \dfrac{1}{\cos \theta} = \dfrac{1}{\dfrac{-12}{13}} = \dfrac{-13}{12}$$

$$\csc \theta = \dfrac{1}{\sin \theta} = \dfrac{1}{\dfrac{5}{13}} = \dfrac{13}{5} \qquad\qquad \cot \theta = \dfrac{1}{\tan \theta} = \dfrac{1}{\dfrac{-5}{12}} = \dfrac{-12}{5}$$

39. $\cos \theta = \dfrac{-1}{3}$, $\csc \theta > 0$

$\sin^2 \theta = 1 - \cos^2 \theta$

$\sin \theta = \pm\sqrt{1 - \cos^2 \theta}$

Because $\csc \theta > 0$, and $\sin \theta = \dfrac{1}{\csc \theta}$, it follows that $\sin \theta > 0$.

$$\sin \theta = \sqrt{1 - \cos^2 \theta} = \sqrt{1 - \left[\dfrac{-1}{3}\right]^2} = \sqrt{1 - \dfrac{1}{9}} = \sqrt{\dfrac{8}{9}} = \dfrac{2\sqrt{2}}{3}$$

$$\tan \theta = \dfrac{\sin \theta}{\cos \theta} = \dfrac{\dfrac{2\sqrt{2}}{3}}{\dfrac{-1}{3}} = -2\sqrt{2} \qquad\qquad \csc \theta = \dfrac{1}{\sin \theta} = \dfrac{1}{\dfrac{2\sqrt{2}}{3}} = \dfrac{3\sqrt{2}}{4}$$

$$\cot \theta = \dfrac{1}{\tan \theta} = \dfrac{1}{-2\sqrt{2}} = \dfrac{-\sqrt{2}}{4} \qquad\qquad \sec \theta = \dfrac{1}{\cos \theta} = \dfrac{1}{\dfrac{-1}{3}} = -3$$

41. $\sin \theta = \dfrac{2}{3}$, $\tan \theta < 0$

$\cos^2 \theta = 1 - \sin^2 \theta$

Because $\tan \theta = \dfrac{\sin \theta}{\cos \theta} < 0$ and $\sin \theta > 0$, it follows that $\cos \theta < 0$.

Therefore, we use the minus sign:

$$\cos \theta = -\sqrt{1 - \sin^2 \theta} = -\sqrt{1 - \left[\dfrac{2}{3}\right]^2} = -\sqrt{1 - \dfrac{4}{9}} = -\sqrt{\dfrac{5}{9}} = \dfrac{-\sqrt{5}}{3}$$

$$\csc \theta = \dfrac{1}{\sin \theta} = \dfrac{1}{\dfrac{2}{3}} = \dfrac{3}{2} \qquad\qquad \tan \theta = \dfrac{\sin \theta}{\cos \theta} = \dfrac{\dfrac{2}{3}}{\dfrac{-\sqrt{5}}{3}} = \dfrac{-2\sqrt{5}}{5}$$

$$\sec \theta = \dfrac{1}{\cos \theta} = \dfrac{1}{\dfrac{-\sqrt{5}}{3}} = \dfrac{-3\sqrt{5}}{5} \qquad\qquad \cot \theta = \dfrac{1}{\tan \theta} = \dfrac{1}{\dfrac{-2}{\sqrt{5}}} = \dfrac{-\sqrt{5}}{2}$$

43. $\sec \theta = 2$, $\sin \theta < 0$

Because $\sec \theta = \dfrac{1}{\cos \theta}$ and $\sec \theta = 2$, then $\cos \theta = \dfrac{1}{2}$.

$\cos \theta = \dfrac{1}{2}$

$\sin^2 \theta = 1 - \cos^2 \theta$

$\sin \theta = \pm\sqrt{1 - \cos^2 \theta}$

Because $\sin \theta < 0$, we use the minus sign:

$$\sin \theta = -\sqrt{1 - \cos^2 \theta} = -\sqrt{1 - \left[\dfrac{1}{2}\right]^2} = -\sqrt{1 - \dfrac{1}{4}} = -\sqrt{\dfrac{3}{4}} = \dfrac{-\sqrt{3}}{2}$$

$$\csc \theta = \frac{1}{\sin \theta} = \frac{1}{\frac{-\sqrt{3}}{2}} = \frac{-2\sqrt{3}}{3}$$

$$\tan \theta = \frac{\sin \theta}{\cos \theta} = \frac{\frac{-\sqrt{3}}{2}}{\frac{1}{2}} = -\sqrt{3} \qquad\qquad \cot \theta = \frac{1}{\tan \theta} = \frac{1}{-\sqrt{3}} = \frac{-\sqrt{3}}{3}$$

45. $\tan \theta = \dfrac{3}{4}$, $\sin \theta < 0$

$$\cot \theta = \frac{1}{\tan \theta} = \frac{1}{\frac{3}{4}} = \frac{4}{3}$$

Because $\tan \theta = \dfrac{\sin \theta}{\cos \theta} = \dfrac{3}{4} > 0$ and $\sin \theta < 0$, it follows that $\cos \theta < 0$.

We know that $\tan^2 \theta + 1 = \sec^2 \theta$.

$$\sec \theta = \pm\sqrt{\tan^2 \theta + 1}$$

Because $\cos \theta = \dfrac{1}{\sec \theta} < 0$, it follows that $\sec \theta < 0$. Therefore, we use the minus sign:

$$\sec \theta = -\sqrt{\tan^2 \theta + 1} = -\sqrt{\left(\frac{3}{4}\right)^2 + 1} = \sqrt{\frac{9}{16} + 1} = -\sqrt{\frac{25}{16}} = \frac{-5}{4}$$

$$\cos \theta = \frac{1}{\sec \theta} = \frac{1}{\frac{-5}{4}} = \frac{-4}{5}$$

$$\sin \theta = -\sqrt{1 - \left(\frac{-4}{5}\right)^2} = -\sqrt{\frac{9}{25}} = \frac{-3}{5} \qquad\qquad \csc \theta = \frac{1}{\sin \theta} = \frac{1}{\frac{-3}{5}} = \frac{-5}{3}$$

47. $\tan \theta = \dfrac{-1}{3}$, $\sin \theta > 0$

Because $\tan \theta = \dfrac{\sin \theta}{\cos \theta} < 0$ and $\sin \theta > 0$, it follows that $\cos \theta < 0$.

$$\sec^2 \theta = \tan^2 \theta + 1$$

$$\sec \theta = \pm\sqrt{\tan^2 \theta + 1}$$

Because $\cos \theta = \dfrac{1}{\sec \theta}$, it follows that $\sec \theta < 0$. Therefore, we use the minus sign.

$$\sec \theta = \pm\sqrt{-\tan^2 \theta + 1} = -\sqrt{\left(\frac{-1}{3}\right)^2 + 1} = -\sqrt{\frac{1}{9} + 1} = \frac{-\sqrt{10}}{3}$$

$$\cos \theta = \frac{1}{\sec \theta} = \frac{1}{\frac{-\sqrt{10}}{3}} = \frac{-3\sqrt{10}}{10}$$

$$\sin \theta = \sqrt{1 - \left(\frac{-3}{\sqrt{10}}\right)^2} = \sqrt{1 - \frac{9}{10}} = \frac{1}{\sqrt{10}} = \frac{\sqrt{10}}{10}$$

$$\csc \theta = \frac{1}{\sin \theta} = \frac{1}{\frac{\sqrt{10}}{10}} = \sqrt{10} \qquad\qquad \cot \theta = \frac{1}{\tan \theta} = \frac{1}{\frac{-1}{3}} = -3$$

49. $\sin(-60°) = -\sin 60° = \dfrac{-\sqrt{3}}{2}$

51. $\tan(-30°) = -\tan 30° = \dfrac{-\sqrt{3}}{3}$

53. $\sec(-60°) = \sec 60° = 2$

55. $\sin(-90°) = -\sin 90° = -1$

57. $\tan\left[\dfrac{-\pi}{4}\right] = -\tan \dfrac{\pi}{4} = -1$

59. $\cos\left[\dfrac{-\pi}{4}\right] = \cos \dfrac{\pi}{4} = \dfrac{\sqrt{2}}{2}$

61. $\tan(-\pi) = -\tan \pi = 0$

63. $\csc\left[\dfrac{-\pi}{4}\right] = -\csc \dfrac{\pi}{4} = -\sqrt{2}$

65. $\sec\left[\dfrac{-\pi}{6}\right] = \sec \dfrac{\pi}{6} = \dfrac{2\sqrt{3}}{3}$

67. $\sin(-\pi) + \cos 5\pi = -\sin \pi + \cos(\pi + 4\pi)$
$$= 0 + \cos \pi = -1$$

69. $\sec(-\pi) + \csc\left[-\dfrac{\pi}{2}\right] = \sec \pi - \csc \dfrac{\pi}{2} = -1 - 1 = -2$

71. $\sin\left[\dfrac{-9\pi}{4}\right] - \tan\left[\dfrac{-9\pi}{4}\right] = -\sin \dfrac{9\pi}{4} + \tan \dfrac{9\pi}{4} = -\sin\left[\dfrac{\pi}{4} + \dfrac{8\pi}{4}\right] + \tan\left[\dfrac{\pi}{4} + \dfrac{8\pi}{4}\right]$
$$= -\sin \dfrac{\pi}{4} + \tan \dfrac{\pi}{4} = \dfrac{-\sqrt{2}}{2} + 1 = 1 - \dfrac{\sqrt{2}}{2}$$

73. $\sin^2 40° + \cos^2 40° = 1$

75. $\sin 80° \csc 80° = \sin 80° \cdot \dfrac{1}{\sin 80°} = \dfrac{\sin 80°}{\sin 80°} = 1$

77. $\tan 40° - \dfrac{\sin 40°}{\cos 40°} = \tan 40° - \tan 40° = 0$

79. If $\sin \theta = 0.3$, then
$$\sin \theta + \sin(\theta + 2\pi) + \sin(\theta + 4\pi) = \sin \theta + \sin \theta + \sin \theta = 0.3 + 0.3 + 0.3 = 0.9$$

81. If $\tan \theta = 3$, then $\tan \theta + \tan(\theta + \pi) + \tan(\theta + 2\pi) = \tan \theta + \tan \theta + \tan \theta = 3 + 3 + 3 = 9$

83. Since $\tan \theta = \dfrac{500}{1500} = \dfrac{1}{3}$, then $\sin \theta = \dfrac{1}{\sqrt{1 + 9}} = \dfrac{1}{\sqrt{10}}$.

Thus, $T = 5\left[1 - \dfrac{1}{3\left[\dfrac{1}{3}\right]} + \dfrac{1}{\dfrac{1}{\sqrt{10}}}\right] = 5\left(\sqrt{10}\right) \approx 15.8$ min.

85. $f(\theta) = \tan \theta$ is defined for all real numbers except odd multiples of $\dfrac{\pi}{2}$.

87. $f(\theta) = \sec \theta$ is defined for all real numbers except odd multiples of $\dfrac{\pi}{2}$.

89. The value of $\sin k\pi$ where k is any integer is 0.

91. Let $P = (x, y)$ be the point on the unit circle that corresponds to an angle θ.

Consider the equation $\tan \theta = \dfrac{y}{x} = a$. Then $y = ax$. But $x^2 + y^2 = 1$ so that $x^2 + a^2x^2 = 1$.

Thus, $x = \pm\dfrac{1}{\sqrt{1 + a^2}}$ and $y = \pm\dfrac{a}{\sqrt{1 + a^2}}$; that is, for any real number a, there is a point

$P = (x, y)$ on the unit circle for which $\tan \theta = a$. In other words, $-\infty < \tan \theta < +\infty$, and the range of the tangent function is the set of all real numbers.

93. Suppose there is a number p, $0 < p < 2\pi$, for which $\sin(\theta + p) = \sin \theta$ for all θ. If $\theta = 0$, then

$\sin(0 + p) = \sin p = \sin 0 = 0$; so that $p = \pi$. If $\theta = \dfrac{\pi}{2}$, then $\sin\left[\dfrac{\pi}{2} + p\right] = \sin\left[\dfrac{\pi}{2}\right]$. But

$p = \pi$. Thus, $\sin\left[\dfrac{3\pi}{2}\right] = -1 = \sin\left[\dfrac{\pi}{2}\right] = 1$. This is impossible. The smallest positive number p for which $\sin(\theta + p) = \sin \theta$ for all θ is therefore $p = 2\pi$.

95. $\sec \theta = \dfrac{1}{\cos \theta}$; since $\cos \theta$ has period 2π, so does $\sec \theta$.

97. If $P = (a, b)$ is the point on the unit circle corresponding to θ, then $Q = (-a, -b)$ is the point on the unit circle corresponding to $\theta + \pi$. Thus, $\tan(\theta + \pi) = \dfrac{-b}{-a} = \dfrac{b}{a} = \tan \theta$; that is, the period of the tangent function is π.

99. Let $P = (a, b)$ be the point on the unit circle corresponding to θ.

Then $\csc \theta = \dfrac{1}{b} = \dfrac{1}{\sin \theta}$; $\sec \theta = \dfrac{1}{a} = \dfrac{1}{\cos \theta}$; $\cot \theta = \dfrac{a}{b} = \dfrac{1}{\left[\dfrac{b}{a}\right]} = \dfrac{1}{\tan \theta}$

101. $(\sin \theta \cos \phi)^2 + (\sin \theta \sin \phi)^2 + \cos^2 \theta = \sin^2 \theta \cos^2 \phi + \sin^2 \theta \sin^2 \phi + \cos^2 \theta$
$= \sin^2 \theta(\cos^2 \phi + \sin^2 \phi) + \cos^2 \theta = \sin^2 \theta + \cos^2 \theta = 1$

6.4 Right Triangle Trigonometry

1. opposite = 5
    adjacent = 12
    By the Pythagorean Theorem:
    $$5^2 + 12^2 = (\text{hypotenuse})^2$$
    $$25 + 144 = (\text{hypotenuse})^2$$
    $$169 = (\text{hypotenuse})^2$$
    $$13 = \text{hypotenuse}$$

    $\sin \theta = \dfrac{\text{opp}}{\text{hyp}} = \dfrac{5}{13}$ $\qquad \cos \theta = \dfrac{\text{adj}}{\text{hyp}} = \dfrac{12}{13}$ $\qquad \tan \theta = \dfrac{\text{opp}}{\text{adj}} = \dfrac{5}{12}$

    $\csc \theta = \dfrac{\text{hyp}}{\text{opp}} = \dfrac{13}{5}$ $\qquad \sec \theta = \dfrac{\text{hyp}}{\text{adj}} = \dfrac{13}{12}$ $\qquad \cot \theta = \dfrac{\text{adj}}{\text{opp}} = \dfrac{12}{5}$

3.   opposite $= 2$
adjacent $= 3$
By the Pythagorean Theorem:

$$2^2 + 3^2 = (\text{hypotenuse})^2$$
$$4 + 9 = (\text{hypotenuse})^2$$
$$13 = \text{hypotenuse}$$
$$\sqrt{13} = \text{hypotenuse}$$

$\sin \theta = \dfrac{\text{opp}}{\text{hyp}} = \dfrac{2}{\sqrt{13}} \cdot \dfrac{\sqrt{13}}{\sqrt{13}} = \dfrac{2\sqrt{13}}{13}$ $\csc \theta = \dfrac{\text{hyp}}{\text{opp}} = \dfrac{\sqrt{13}}{2}$

$\cos \theta = \dfrac{\text{adj}}{\text{hyp}} = \dfrac{3}{\sqrt{13}} \cdot \dfrac{\sqrt{13}}{\sqrt{13}} = \dfrac{3\sqrt{13}}{13}$ $\sec \theta = \dfrac{\text{hyp}}{\text{adj}} = \dfrac{\sqrt{13}}{3}$

$\tan \theta = \dfrac{\text{opp}}{\text{adj}} = \dfrac{2}{3}$ $\cot \theta = \dfrac{\text{adj}}{\text{opp}} = \dfrac{3}{2}$

5.   adjacent $= 2$
hypotenuse $= 4$
By the Pythagorean Theorem:

$$2^2 + (\text{opp})^2 = 4^2$$
$$4 + (\text{opp})^2 = 16$$
$$(\text{opp})^2 = 12$$
$$\text{opp} = \sqrt{12} = 2\sqrt{3}$$

$\sin \theta = \dfrac{\text{opp}}{\text{hyp}} \dfrac{2\sqrt{3}}{4} = \dfrac{\sqrt{3}}{2}$ $\csc \theta = \dfrac{\text{hyp}}{\text{opp}} = \dfrac{4}{2\sqrt{3}} \cdot \dfrac{\sqrt{3}}{\sqrt{3}} = \dfrac{4\sqrt{3}}{6} = \dfrac{2\sqrt{3}}{3}$

$\cos \theta = \dfrac{\text{adj}}{\text{hyp}} = \dfrac{2}{4} = \dfrac{1}{2}$ $\sec \theta = \dfrac{\text{hyp}}{\text{adj}} = \dfrac{4}{2} = 2$

$\tan \theta = \dfrac{\text{opp}}{\text{adj}} = \dfrac{2\sqrt{3}}{2} = \sqrt{3}$ $\cot \theta = \dfrac{\text{adj}}{\text{opp}} = \dfrac{2}{2\sqrt{3}} \cdot \dfrac{\sqrt{3}}{\sqrt{3}} = \dfrac{\sqrt{3}}{3}$

7.   opposite $= \sqrt{2}$
adjacent $= 1$
By the Pythagorean Theorem:

$$\left(\sqrt{2}\right)^2 + 1^2 = (\text{hypotenuse})^2$$
$$2 + 1 = (\text{hypotenuse})^2$$
$$3 = (\text{hypotenuse})^2$$
$$\sqrt{3} = (\text{hypotenuse})$$

$\sin \theta = \dfrac{\text{opp}}{\text{hyp}} = \dfrac{\sqrt{2}}{\sqrt{3}} \cdot \dfrac{\sqrt{3}}{\sqrt{3}} = \dfrac{\sqrt{6}}{3}$ $\csc \theta = \dfrac{\text{hyp}}{\text{opp}} = \dfrac{\sqrt{3}}{\sqrt{2}} \cdot \dfrac{\sqrt{2}}{\sqrt{2}} = \dfrac{\sqrt{6}}{2}$

$\cos \theta = \dfrac{\text{adj}}{\text{hyp}} = \dfrac{1}{\sqrt{3}} \cdot \dfrac{\sqrt{3}}{\sqrt{3}} = \dfrac{\sqrt{3}}{3}$ $\sec \theta = \dfrac{\text{hyp}}{\text{adj}} = \dfrac{\sqrt{3}}{1} = \sqrt{3}$

$\tan \theta = \dfrac{\text{opp}}{\text{adj}} = \dfrac{\sqrt{2}}{1} = \sqrt{2}$ $\cot \theta = \dfrac{\text{adj}}{\text{opp}} = \dfrac{1}{\sqrt{2}} \cdot \dfrac{\sqrt{2}}{\sqrt{2}} = \dfrac{\sqrt{2}}{2}$

9.   opposite $= 1$

hypotenuse $= \sqrt{5}$
By the Pythagorean Theorem:

$$1^2 + (\text{adjacent})^2 = \left(\sqrt{5}\right)^2$$
$$1 + (\text{adjacent})^2 = 5$$
$$(\text{adjacent})^2 = 4$$
$$\text{adjacent} = 2$$

$\sin\theta = \dfrac{\text{opp}}{\text{hyp}} = \dfrac{1}{\sqrt{5}} \cdot \dfrac{\sqrt{5}}{\sqrt{5}} = \dfrac{\sqrt{5}}{5}$ $\csc\theta = \dfrac{\text{hyp}}{\text{opp}} = \dfrac{\sqrt{5}}{1} = \sqrt{5}$

$\cos\theta = \dfrac{\text{adj}}{\text{hyp}} = \dfrac{2}{\sqrt{5}} \cdot \dfrac{\sqrt{5}}{\sqrt{5}} = \dfrac{2\sqrt{5}}{5}$ $\sec\theta = \dfrac{\text{hyp}}{\text{adj}} = \dfrac{\sqrt{5}}{2}$

$\tan\theta = \dfrac{\text{opp}}{\text{adj}} = \dfrac{1}{2}$ $\cot\theta = \dfrac{\text{adj}}{\text{opp}} = \dfrac{2}{1} = 2$

11.   The reference angle for $-30°$ is
$0° - (-30°) = 30°$.

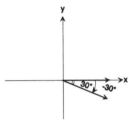

13.   Let α represent the reference angle.
$$120° + \alpha = 180°$$
$$\alpha = 60°$$

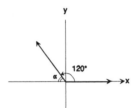

15. $180° + \alpha = 210°$
$\alpha = 30°$

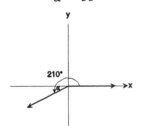

17.   Remember that the reference angle is
the acute angle formed by the terminal
side of θ and the positive or negative
$x-$axis.

Hence, $\pi + \alpha = \dfrac{5\pi}{4}$

$\alpha = \dfrac{5\pi}{4} - \pi = \dfrac{\pi}{4}$

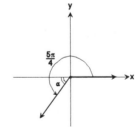

19. $\dfrac{8\pi}{3} + \alpha = 3\pi$

$\alpha = 3\pi - \dfrac{8\pi}{3}$

$= \dfrac{9\pi}{3} - \dfrac{8\pi}{3}$

$\alpha = \dfrac{\pi}{3}$

21. $180° - 135° = \alpha$

$45° = \alpha$

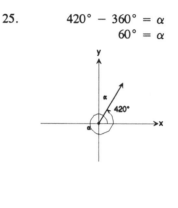

23. $\pi - \dfrac{2\pi}{3} = \alpha$

$\dfrac{\pi}{3} = \alpha$

25. $420° - 360° = \alpha$

$60° = \alpha$

27. sin 150°
The angle 150° is in quadrant II, where the sin θ is positive. The reference angle for 150° is 30°. Thus,

$$\sin 150° = \sin 30° = \dfrac{1}{2}$$

29. cos 315°
The angle 315° is in quadrant IV where the cos θ is positive. The reference angle for 315° is 45°. Thus,

$$\cos 315° = \cos 45° = \dfrac{\sqrt{2}}{2}$$

31. sec 240°
The angle 240° is in quadrant III where sec θ is negative. The reference angle for 240° is 60°. Thus,

$$\sec 240° = -\sec 60° = -2$$

33. cot 330°
The angle 330° is in quadrant IV where cot θ is negative. The reference angle for 330° is 30°. Thus,

$$\cot 330° = -\cot 30° = -\dfrac{\sqrt{3}}{1} = -\sqrt{3}$$

35. $\sin \dfrac{3\pi}{4}$

The angle $\dfrac{3\pi}{4}$ is in quadrant II where sin is positive. The reference angle for $\dfrac{3\pi}{4}$ is $\dfrac{\pi}{4}$. Thus,

$$\sin \dfrac{3\pi}{4} = \sin \dfrac{\pi}{4} = \dfrac{\sqrt{2}}{2}$$

37. $\cot \dfrac{7\pi}{6}$

The angle $\dfrac{7\pi}{6}$ is in quadrant III where $\cot \theta$ is positive. The reference angle for $\dfrac{7\pi}{6}$ is $\dfrac{\pi}{6}$. Thus,

$$\cot \dfrac{7\pi}{6} = \cot \dfrac{\pi}{6} = \sqrt{3}$$

39. $\cos(-60°)$
The angle $(-60°)$ is in quadrant IV where $\cos \theta$ is positive. The reference angle for $-60°$ is $60°$. Thus,

$$\cos(-60°) = \cos 60° = \dfrac{1}{2}$$

41. $\sin\left[-\dfrac{2\pi}{3}\right]$

The angle $-\dfrac{2\pi}{3}$ is in quadrant III where $\sin \theta$ is negative. The reference angle for $\left[-\dfrac{2\pi}{3}\right]$ is $\dfrac{\pi}{3}$. Thus,

$$\sin\left[-\dfrac{2\pi}{3}\right] = -\sin \dfrac{\pi}{3} = -\dfrac{\sqrt{3}}{2}$$

43. $\tan \dfrac{14\pi}{3}$

$$\tan \dfrac{14\pi}{3} = \tan\left[\dfrac{12\pi}{3} + \dfrac{2\pi}{3}\right] = \tan\left[4\pi + \dfrac{2\pi}{3}\right]$$

The angle $\dfrac{14\pi}{3}$ is in quadrant II where $\tan \theta$ is negative. The reference angle for $\dfrac{14\pi}{3}$ is $\dfrac{\pi}{3}$. Thus,

$$\tan \dfrac{14\pi}{3} = -\tan \dfrac{\pi}{3} = -\sqrt{3}$$

45. $\csc(-315°)$
The angle $-315°$ is in quadrant I where $\csc \theta$ is positive. The reference angle for $-315°$ is $45°$. Thus,

$$\csc(-315°) = \csc 45° = \sqrt{2}$$

47. $\sin 38° - \cos 52° = 0$
Since $\sin(90° - 52°) = \cos 52°$ or $\sin 38°$ by the cofunction formula.

49. $\dfrac{\cos 10°}{\sin 80°} = 1$
Since $\cos(90° - 80°) = \sin 80°$ or $\cos 10° = \sin 80°$ by the cofunction formula.

51. $1 - \cos^2 20° - \cos^2 70°$

Since by the fundamental identity,
$$\sin^2 20° + \cos^2 20° = 1, \text{ then}$$
$$\sin^2 20° = 1 - \cos^2 20°$$
Also, $\cos^2(90° - 20°) = \sin^2 20°$ by the cofunction formula so
$$\cos^2 70° = \sin^2 20°$$
By substitution, $1 - \cos^2 20° - \cos^2 70° = (1 - \cos^2 20°) - \cos^2 70°$
$$= \sin^2 20° - \sin^2 20° = 0$$

53. $\tan 20° - \dfrac{\cos 70°}{\cos 20°} = \tan 20° - \dfrac{\sin 20°}{\cos 20°} = \tan 20° - \tan 20° = 0$

55. $\cos 35° \sin 55° + \sin 35° \cos 55° = \cos 35° \cos 35° + \sin 35° \sin 35° = \cos^2 35° + \sin^2 35° = 1$

57. (a) Since $\sin \theta = \cos(90° - \theta)$ (they are cofunctions), $\cos(90° - \theta) = \dfrac{1}{3}$

(b) Find $\cos^2 \theta$. By the fundamental identity,
$$\sin^2 \theta + \cos^2 \theta = 1, \text{ so } \cos^2 \theta = 1 - \sin^2 \theta = 1 - \left[\frac{1}{3}\right]^2 = 1 - \frac{1}{9} = \frac{8}{9}$$

(c) Since $\csc \theta = \dfrac{1}{\sin \theta} = \dfrac{1}{\frac{1}{3}} = 3$

(d) $\sec\left[\dfrac{\pi}{2} - \theta\right] = \csc \theta = 3$ (from part (c))
$$= \csc \theta = 3$$

59. (a) $1 + \tan^2 \theta = \sec^2 \theta$ (fundamental identity)
given $\tan \theta = 4$ so
$$1 + (4)^2 = \sec^2 \theta$$
$$1 + 16 = \sec^2 \theta$$
$$17 = \sec^2 \theta$$

(b) $\cot \theta = \dfrac{1}{\tan \theta}$ (fundamental identity)
given $\tan \theta = 4$ so
$$\cot \theta = \frac{1}{4}$$

(c) $\cot\left[\dfrac{\pi}{2} - \theta\right] = \tan \theta$ (cofunctions)

$\cot\left[\dfrac{\pi}{2} - \theta\right] = 4$ since $\tan \theta = 4$ is given

(d) Using part (b), $\cot \theta = \dfrac{1}{4}$ and
the fundamental identity $1 + \cot^2 \theta = \csc^2$, we have
$$1 + \left[\frac{1}{4}\right]^2 = \csc^2 \theta$$
$$1 + \frac{1}{16} = \csc^2 \theta$$
$$\frac{17}{16} = \csc^2 \theta$$

61. (a) Using $\csc \theta = 4$ (given),

and the fundamental identity $\sin \theta = \dfrac{1}{\csc \theta}$ we have

$$\sin \theta = \frac{1}{4}$$

(b) Using $\csc \theta = 4$ (given) and the fundamental identity
$$1 + \cot^2 \theta = \csc^2, \text{ we have}$$
$$1 + \cot^2 \theta = (4)^2$$
$$1 + \cot^2 \theta = 16$$
$$\cot^2 \theta = 15$$

(c) $\sec(90° - \theta) = \csc \theta$ (cofunctions) and $\csc \theta = 4$ (given) so $\sec(90° - \theta) = 4$

(d) Using part (b), $\cot^2 \theta = 15$ and the fundamental identities

$$\tan \theta = \frac{1}{\cot \theta} \text{ and } 1 + \tan^2 = \sec^2 \theta,$$

we have $\qquad \tan^2 \theta = \dfrac{1}{\cot^2 \theta} = \dfrac{1}{15}$

so $\qquad\qquad 1 + \dfrac{1}{15} = \sec^2 \theta$

$$\frac{16}{15} = \sec^2 \theta$$

63. $\cos\left[\dfrac{\pi}{2} - \theta\right] = \sin \theta$ since they are cofunctions. Thus,

$$\sin \theta + \cos\left[\frac{\pi}{2} - \theta\right] = \sin \theta + \sin \theta = 0.3 + 0.3 = 0.6$$

65. $\sin 1° + \sin 2° + \sin 3° + ... + \sin 358° + \sin 359°$
$= (\sin 1° + \sin 359°) + (\sin 2° + \sin 358°) + \cdots$
$= [\sin 1° + \sin (-1°)] + [\sin 2° + (\sin -2°)] + \cdots + [\sin 179° + \sin (-179°)]$
$\qquad + \sin 90° + \sin 270° + \sin 180°$
$= (\sin 1° - \sin 1°) + (\sin 2° - \sin 2°) + \cdots + (\sin 179° - \sin 179°) + 1 + (-1) + 0$
$= 0 + 0 + \cdots + 0 + 0 + 0 = 0$

67. $\sin \theta = \cos(2\theta + 30°)$
Since $\sin \theta = \cos(90° - \theta)$, then
$$2\theta + 30° = 90° - \theta$$
$$3\theta = 60°$$
$$\theta = 20°$$

69. (a) $T(\theta) = \dfrac{\dfrac{1}{\sin \theta}}{3} + \dfrac{8 - \dfrac{2}{\tan \theta}}{8} + \dfrac{\dfrac{1}{\sin \theta}}{3} = \dfrac{1}{3 \sin \theta} + 1 - \dfrac{1}{4 \tan \theta} + \dfrac{1}{3 \sin \theta}$

$\qquad = \dfrac{2}{3 \sin \theta} + 1 - \dfrac{1}{4 \tan \theta} = 1 + \dfrac{2}{3 \sin \theta} - \dfrac{1}{4 \tan \theta}$

(b)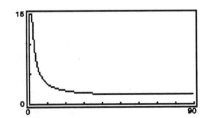

The angle, θ, which results in the least time is 67.97° correct to two decimal places. The least time is 1.62 hours correct to two decimal places. Sally is on the road for 0.9 hours.

71. (a) $T = \dfrac{1500 \text{ feet}}{300 \text{ ft/min}} + \dfrac{500 \text{ feet}}{100 \text{ ft/min}} = 5 \text{ min} + 5 \text{ min} = 10 \text{ minutes}$

(b) $T = \dfrac{500 \text{ feet}}{100 \text{ ft/min}} + \dfrac{1500 \text{ feet}}{100 \text{ ft/min}} = 5 \text{ min} + 15 \text{ min} = 20 \text{ minutes}$

(c) Since $\tan \theta = \dfrac{\text{opposite}}{\text{adjacent}}$, we have $\tan \theta = \dfrac{500}{x}$; therefore, $x = \dfrac{500}{\tan \theta}$.

Since $\sin \theta = \dfrac{\text{opposite}}{\text{hypotenuse}}$, we have $\sin \theta = \dfrac{500}{\text{length of trip in the sand}}$; therefore, length of

trip in the sand $= \dfrac{500}{\sin \theta}$. Thus, the time of the trip is:

$$T = \dfrac{1500 - x}{300} + \dfrac{500/\sin\theta}{100} = \dfrac{1500 - \dfrac{500}{\tan\theta}}{300} + \dfrac{500}{100 \sin \theta}$$

$$T(\theta) = 5 - \dfrac{5}{3 \tan \theta} + \dfrac{5}{\sin \theta} = 5\left[1 - \dfrac{1}{3 \tan \theta} + \dfrac{1}{\sin \theta}\right]$$

(d) From the figure, we see $\tan \theta = \dfrac{500}{500} = 1$ and $\sin \theta = \dfrac{1}{\sqrt{2}}$.

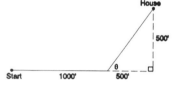

Thus, $T = 5\left[1 - \dfrac{1}{3 \cdot 1} + \dfrac{1}{\dfrac{1}{\sqrt 2}}\right] = 5\left[1 - \dfrac{1}{3} + \sqrt 2\right]$

≈ 10.4 minutes

(e) The time, T, is least when $\theta \approx 70.52°$ correct to two decimal places. To find x, we solve the following

equation: $\tan 70.52 = \dfrac{500}{x}$

Thus, $x = \dfrac{500}{\tan 70.52} \approx 176.9$ feet.

75.

θ	0.5	0.4	0.2	0.1	0.01	0.001	0.0001	0.00001
$\sin \theta$	0.4794	0.3894	0.1987	0.0998	0.0100	0.0010	0.0001	0.00001
$\dfrac{\sin \theta}{\theta}$	0.9589	0.9735	0.9933	0.9983	1.0000	1.0000	1.0000	1.0000

$\dfrac{\sin \theta}{\theta}$ approaches 1 as θ approaches 0

77. (a) $|OA| = |OC| = 1$; Angle OAC = Angle OCA. Thus,

Angle OAC + Angle OAC + $180° - \theta = 180°$

$$2(\text{Angle } OAC) = \theta$$

$$\text{Angle } OAC = \dfrac{\theta}{2}$$

(b) $\sin \theta = \dfrac{|CD|}{|OC|} = |CD|$, since $|OC| = 1$

$\cos \theta = \dfrac{|OD|}{|OC|} = |OD|$

(c) $\quad \tan \dfrac{\theta}{2} = \dfrac{|CD|}{|AD|}$, since angle $OAC = \dfrac{\theta}{2}$ by part (a)

$$= \dfrac{\sin \theta}{1 + |OD|}, \text{ since } |CD| = \sin \theta \text{ by part (b) and}$$

$$|AD| = |AO| + |OD| = 1 + |OD|, \text{ since } |AO| = 1$$

$$= \dfrac{\sin \theta}{1 + \cos \theta}, \text{ since } |OD| = \cos \theta \text{ by part (b)}$$

79. $\quad h = x \cdot \dfrac{h}{x} = x \tan \theta$ and $h = (1 - x)\dfrac{h}{1 - x} = (1 - x)\tan n\theta$

Thus, $\qquad\qquad x \tan \theta = (1 - x)\tan n\theta$

$$x \tan \theta = \tan n\theta - x \tan n\theta$$

$$x(\tan \theta + \tan n\theta) = \tan n\theta$$

$$x = \dfrac{\tan n\theta}{\tan \theta + \tan n\theta}$$

81. (a) $\quad$ Area $\triangle OAC = \dfrac{1}{2}|OC|\,|AC| = \dfrac{1}{2}\dfrac{|OC|}{1} \cdot \dfrac{|AC|}{1} = \dfrac{1}{2}\sin \alpha \cos \alpha$

(b) $\quad$ Area $\triangle OCB = \dfrac{1}{2}|OC|\,|BC| = \dfrac{1}{2}|OB|^2 \dfrac{|BC|}{|OB|} \cdot \dfrac{|OC|}{|OB|} = \dfrac{1}{2}|OB|^2 \sin \beta \cos \beta$

(c) $\quad$ Area $\triangle OAB = \dfrac{1}{2}|BD|\,|OA| = \dfrac{1}{2}|BD| \cdot 1 = \dfrac{1}{2}|OB|\dfrac{|BD|}{|OB|} = \dfrac{1}{2}|OB|\sin(\alpha + \beta)$

(d) $\quad \dfrac{\cos \alpha}{\cos \beta} = \dfrac{\dfrac{|OC|}{|OA|}}{\dfrac{|OC|}{|OB|}} = \dfrac{|OC|}{1} \cdot \dfrac{|OB|}{|OC|} = |OB|$

(e) $\quad$ Area $\triangle OAB = $ Area $\triangle OAC + $ Area $\triangle OCB$

$$\dfrac{1}{2}|OB|\sin(\alpha + \beta) = \dfrac{1}{2}\sin \alpha(\cos \alpha + \dfrac{1}{2}|OB|^2 \sin \beta \cos \beta$$

$$\dfrac{\cos \alpha}{\cos \beta}\sin(\alpha + \beta) = \sin \alpha \cos \alpha + \dfrac{\cos^2 \alpha}{\cos^2 \beta}\sin \beta \cos \beta$$

$$\sin(\alpha + \beta) = \dfrac{\cos \beta}{\cos \alpha}(\sin \alpha \cos \alpha) + \dfrac{\cos \alpha}{\cos \beta}(\sin \beta \cos \beta)$$

$$\sin(\alpha + \beta) = \cos \beta \sin \alpha + \cos \alpha \sin \beta$$

83.
$$\sin \alpha = \frac{\sin \alpha}{\cos \alpha} \cos \alpha = \tan \alpha \cos \alpha = \cos \beta \cos \alpha = \cos \beta \tan \beta = \cos \beta \cdot \frac{\sin \beta}{\cos \beta} = \sin \beta$$

$\sin^2 \alpha + \cos^2 \alpha = 1$, thus

$\sin^2 \alpha + \tan^2 \beta = 1$

$$\sin^2 \alpha + \frac{\sin^2 \beta}{\cos^2 \beta} = 1$$

$$\sin^2 \alpha + \frac{\sin^2 \alpha}{1 - \sin^2 \alpha} = 1$$

$\sin^2 \alpha - \sin^4 \alpha + \sin^2 \alpha = 1 - \sin^2 \alpha$

$\sin^4 \alpha - 3 \sin^2 \alpha + 1 = 0$

$$\sin^2 \alpha = \frac{3 \pm \sqrt{5}}{2}$$

$$\sin^2 \alpha = \frac{3 - \sqrt{5}}{2}$$

$$\sin \alpha = \sqrt{\frac{3 - \sqrt{5}}{2}}$$

6.5   Graphs of the Trigonometric Functions

1. 0

3. The graph of $y = \sin x$ is increasing for $\frac{-\pi}{2} \le x \le \frac{\pi}{2}$.

5. The largest value of $y = \sin x$ is 1.

7. $\sin x = 0$ when $x = 0, \pi, 2\pi$

9. $\sin x = 1$ for $x = \frac{-3\pi}{2}, \frac{\pi}{2}$ if $-2\pi \le x \le 2\pi$; $\sin x = -1$ for $x = \frac{-\pi}{2}, \frac{3\pi}{2}$ if $-2\pi \le x \le 2\pi$

11. 0   **13.** 1

15. $\sec x = 1$ for $x = -2\pi, 0, 2\pi$ if $-2\pi \le x \le 2\pi$
$\sec x = -1$ for $x = -\pi, \pi$ if $-2\pi \le x \le 2\pi$

17. $y = \sec x$ has vertical asymptotes for $x = \frac{-3\pi}{2}, \frac{-\pi}{2}, \frac{\pi}{2}, \frac{3\pi}{2}$ if $-2\pi \le x \le 2\pi$.

19. $y = \tan x$ has vertical asymptotes for $x = \frac{-3\pi}{2}, \frac{-\pi}{2}, \frac{\pi}{2}, \frac{3\pi}{2}$ if $-2\pi \le x \le 2\pi$.

21. B, C, F   **23.** C   **25.** D   **27.** B   **29.** A

31.

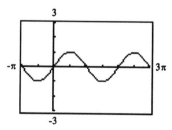
$y = \sin x$

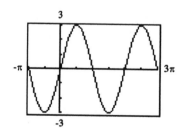
$y = 3 \sin x$

$y = 3 \sin x$ is a vertical stretch by a factor of 3 of the graph of $y = \sin x$.

33.

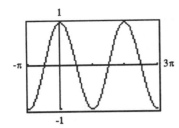

$$y = \cos x$$

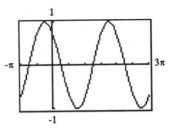

$$y = \cos\left(x + \frac{\pi}{4}\right)$$

$y = \cos\left(x + \dfrac{\pi}{4}\right)$ is a horizontal shift $\dfrac{\pi}{4}$ units left of the graph of $y = \cos x$.

35.

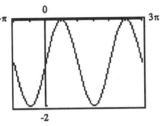

$$y = \sin x - 1$$
$y = \sin x - 1$ is the graph of $y = \sin x$
shifted down one unit.

37.

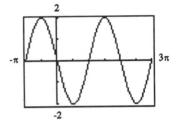

$y = 2 \sin x$ is a vertical stretch by a factor of 2 of the graph of $y = \sin x$. $y = -2 \sin x$ is a reflection of the graph of $y = 2 \sin x$ about the x-axis.

39.

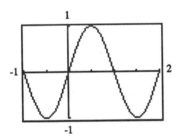

$y = \sin(\pi x)$ is a horizontal compression of $y = \sin x$.

41.

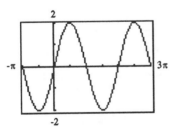

$$y = 2 \sin x$$

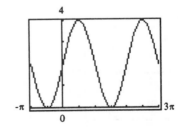

$$y = 2 \sin x + 2$$

$y = 2 \sin x + 2$ is the graph of $y = \sin x$ vertically stretched by a factor of 2 and shifted up two units.

43.

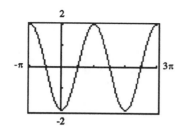

$$y = -2 \cos x$$

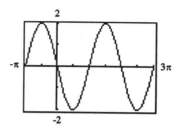

$$y = -2 \cos\left[x - \frac{\pi}{2}\right]$$

$y = -2 \cos\left[x - \dfrac{\pi}{2}\right]$ is the graph of $y = \cos x$ reflected about the x-axis, vertically stretched by a factor of 2, and shifted horizontally $\dfrac{\pi}{2}$ units right.

45.

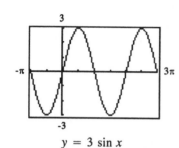

$$y = 3 \sin x$$

$$y = 3 \sin(-x)$$

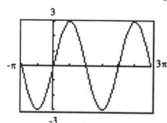

$$y = 3 \sin(\pi - x) = 3 \sin(-(x - \pi))$$

$y = 3 \sin(\pi - x)$ is obtained by vertically stretching $y = \sin x$ by a factor of 3. Then reflect the graph of $y = 3 \sin x$ about the y-axis. Finally, horizontally shift the graph of $y = 3 \sin (-x)$ π units to the right.

47.

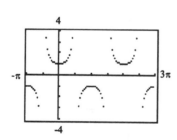

$$y = \sec x = \frac{1}{\cos x}$$

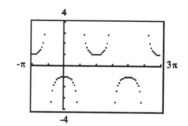

$$y = -\sec x = \frac{-1}{\cos x}$$

The graph of $y = -\sec x$ is the graph of $y = \sec x$ reflected about the x-axis.

49.

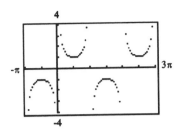

$$y = \sec\left[x - \frac{\pi}{2}\right]$$

The graph of $y = \sec\left[x - \frac{\pi}{2}\right]$ is the graph of $y = \sec x$ shifted horizontally $\frac{\pi}{2}$ units to the right.

51.

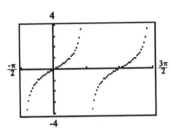

$y = \tan x$

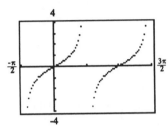

$y = \tan(x - \pi)$

The graph of $y = \tan x$ is the same as the graph of $y = \tan(x - \pi)$.

53.

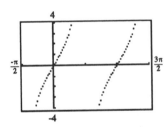

$y = 3\tan x$

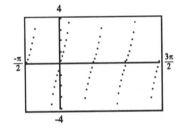

$y = 3\tan(2x)$

The graph of $y = 3\tan(2x)$ is the graph of $y = \tan x$ vertically stretched by a factor of 3 and horizontally compressed by a factor of $\frac{1}{2}$ (i.e., each x-value is multiplied by $\frac{1}{2}$).

55.

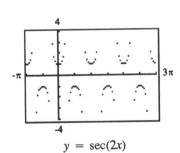

$y = \sec(2x)$

The graph of $y = \sec(2x)$ is the graph of $y = \sec x$ horizontally compressed by a factor of 2.

57.

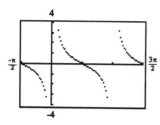

$y = \cot x$

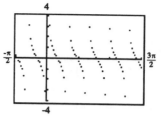

$y = \cot(\pi x)$

The graph of $y = \cot(\pi x)$ is the graph of $y = \cot x$ horizontally compressed by a factor of π.

59.

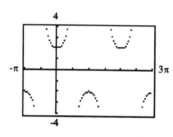

$y = -3 \tan x$

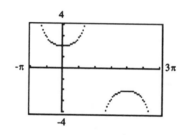

$y = -3 \tan(4x)$

The graph of $y = -3 \tan(4x)$ is obtained by vertically stretching the graph of $y = \tan x$ by a factor of 3, reflecting $y = 3 \tan x$ about the x-axis to obtain $y = -3 \tan x$ and horizontally compressing $y = -3 \tan x$ by a factor of 4.

61.

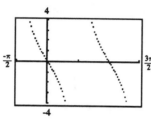

$y = 2 \sec x$

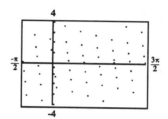

$y = 2 \sec \dfrac{1}{2}x$

The graph of $y = 2 \sec \dfrac{1}{2}x$ is the graph of $y = \sec x$ vertically stretched by a factor of 2 and horizontally compressed by a factor of 2.

63. (a) $L = \dfrac{3}{\cos \theta} + \dfrac{4}{\sin \theta}$

$L = 3 \sec \theta + 4 \csc \theta$

(c) L is least when $\theta = 0.83$.

(b)

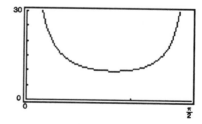

65. The graph of $y = \sin \omega x$ has period $\dfrac{2\pi}{\omega}$.

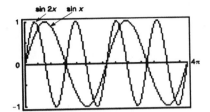

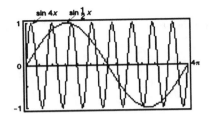

67. It would appear that $\sin x = \cos\left(x - \dfrac{\pi}{2}\right)$ because if the cosine function was shifted to the right $\dfrac{\pi}{2}$ units, it would become the sine function.

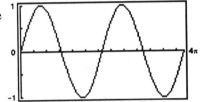

6.6 The Inverse Trigonometric Functions

1. $\sin^{-1} 0$

We seek the angle θ, $-\dfrac{\pi}{2} \leq \theta \leq \dfrac{\pi}{2}$, whose sine equals 0.

$$\sin \theta = 0 \qquad -\frac{\pi}{2} \leq \theta \leq \frac{\pi}{2}$$
$$\theta = 0$$
$$\sin^{-1} 0 = 0$$

3. $\sin^{-1}(-1)$

We seek the angle θ, $-\dfrac{\pi}{2} \leq \theta \leq \dfrac{\pi}{2}$, whose sine equals (-1).

$$\sin \theta = -1 \qquad -\frac{\pi}{2} \leq \theta \leq \frac{\pi}{2}$$
$$\theta = -\frac{\pi}{2}$$
$$\sin^{-1}(-1) = -\frac{\pi}{2}$$

5. $\tan^{-1} 0$

We seek the angle θ, $-\dfrac{\pi}{2} < \theta < \dfrac{\pi}{2}$, whose tangent equals 0.

$$\tan \theta = 0 \qquad -\frac{\pi}{2} < \theta < \frac{\pi}{2}$$
$$\theta = 0$$
$$\tan^{-1} 0 = 0$$

7. $\sin^{-1} \dfrac{\sqrt{2}}{2}$

   We seek the angle θ, $-\dfrac{\pi}{2} \leq \theta \leq \dfrac{\pi}{2}$, whose sine equals $\dfrac{\sqrt{2}}{2}$.

   $$\sin \theta = \dfrac{\sqrt{2}}{2} \qquad -\dfrac{\pi}{2} \leq \theta \leq \dfrac{\pi}{2}$$

   $$\theta = \dfrac{\pi}{4}$$

   $$\sin^{-1} \dfrac{\sqrt{2}}{2} = \dfrac{\pi}{4}$$

9. $\tan^{-1} \sqrt{3}$

   We seek the angle θ, $-\dfrac{\pi}{2} < \theta < \dfrac{\pi}{2}$, whose tangent equals $\sqrt{3}$.

   $$\tan \theta = \sqrt{3} \qquad -\dfrac{\pi}{2} < \theta < \dfrac{\pi}{2}$$

   $$\theta = \dfrac{\pi}{3}$$

   $$\tan^{-1} \sqrt{3} = \dfrac{\pi}{3}$$

11. $\cos^{-1}\left(-\dfrac{\sqrt{3}}{2}\right)$

   We seek the angle θ, $0 \leq \theta \leq \pi$, whose tangent equals $-\dfrac{\sqrt{3}}{2}$.

   $$\cos \theta = \left(\dfrac{-\sqrt{3}}{2}\right) \qquad 0 \leq \theta \leq \pi$$

   $$\theta = \dfrac{5\pi}{6}$$

   $$\cos^{-1}\left(-\dfrac{\sqrt{3}}{2}\right) = \dfrac{5\pi}{6}$$

13. Set the mode of the calculator to radians.
    $\sin^{-1} 0.1 = 0.10$

15. Set the mode of the calculator to radians.
    $\tan^{-1} 5 = 1.37$

17. Set the mode of the calculator to radians.
    $\cos^{-1} \dfrac{7}{8} = 0.51$

19. Set the mode of the calculator to radians.
    $\tan^{-1}(-0.4) = -0.38$

21. Set the mode of the calculator to radians.
    $\sin^{-1}(-0.12) = -0.12$

23. Set the mode of the calculator to radians.
    $\cos^{-1} \dfrac{\sqrt{2}}{3} = 1.08$

25. $\cos\left[\sin^{-1}\dfrac{\sqrt{2}}{2}\right]$

First find the angle θ, $-\dfrac{\pi}{2} \le \theta \le \dfrac{\pi}{2}$, whose sine equals $\dfrac{\sqrt{2}}{2}$.

$$\sin\theta = \dfrac{\sqrt{2}}{2} \qquad -\dfrac{\pi}{2} \le \theta \le \dfrac{\pi}{2}$$

$$\theta = \dfrac{\pi}{4}$$

Now, $\cos\left[\sin^{-1}\dfrac{\sqrt{2}}{2}\right] = \cos\theta = \cos\dfrac{\pi}{4} = \dfrac{\sqrt{2}}{2}$.

27. $\tan\left[\cos^{-1}\left(-\dfrac{\sqrt{3}}{2}\right)\right]$

First find the angle θ, $0 \le \theta \le \pi$, whose cosine equals $-\dfrac{\sqrt{3}}{2}$.

$$\cos\theta = -\dfrac{\sqrt{3}}{2} \qquad 0 \le \theta \le \pi$$

$$\theta = \dfrac{5\pi}{4}$$

Now, $\tan\left[\cos^{-1}\left(-\dfrac{\sqrt{3}}{2}\right)\right] = \tan\theta = \tan\dfrac{5\pi}{6} = -\dfrac{\sqrt{3}}{3}$.

29. $\sec\left[\cos^{-1}\dfrac{1}{2}\right]$

First find the angle θ, $0 \le \theta \le \pi$, whose cosine equals $\dfrac{1}{2}$.

$$\cos\theta = \dfrac{1}{2} \qquad 0 \le \theta \le \pi$$

$$\theta = \dfrac{\pi}{3}$$

Now, $\sec\left[\cos^{-1}\dfrac{1}{2}\right] = \sec\theta = \sec\dfrac{\pi}{3} = 2$.

31. $\csc\left(\tan^{-1}1\right)$

First find the angle θ, $-\dfrac{\pi}{2} < \theta < \dfrac{\pi}{2}$, whose tangent equals 1.

$$\tan\theta = 1 \qquad -\dfrac{\pi}{2} < \theta < \dfrac{\pi}{2}$$

$$\theta = \dfrac{\pi}{4}$$

Now, $\csc\left(\tan^{-1}1\right) = \csc\theta = \cos\dfrac{\pi}{4} = \sqrt{2}$.

33. $\sin\left(\tan^{-1}(-1)\right)$

First find the angle θ, $-\dfrac{\pi}{2} < \theta < \dfrac{\pi}{2}$, whose tangent equals -1.

$$\tan\theta = -1 \qquad\qquad -\dfrac{\pi}{2} < \theta < \dfrac{\pi}{2}$$

$$\theta = -\dfrac{\pi}{4}$$

Now, $\sin\left[\tan^{-1}(-1)\right] = \sin\theta = \sin\left(-\dfrac{\pi}{4}\right) = -\dfrac{\sqrt{2}}{2}.$

35. $\sec\left[\sin^{-1}\left(-\dfrac{1}{2}\right)\right]$

First find the angle θ, $-\dfrac{\pi}{2} \le \theta \le \dfrac{\pi}{2}$, whose sine equals $-\dfrac{1}{2}$.

$$\sin\theta = -\dfrac{1}{2}$$

$$\theta = -\dfrac{\pi}{6}$$

Now, $\sec\left[\sin^{-1}\left(-\dfrac{1}{2}\right)\right] = \sec\theta = \sec\left(-\dfrac{\pi}{6}\right) = \dfrac{2\sqrt{3}}{3}.$

37. $\tan\left[\sin^{-1}\dfrac{1}{3}\right]$

First we know that $\sin\theta = \dfrac{1}{3}$, $-\dfrac{\pi}{2} \le \theta \le \dfrac{\pi}{2}$, so we have:

By the Pythagorean Theorem, the missing side of the triangle is:
$$x^2 + 1 = 9$$
$$x^2 = 8$$
$$x = \pm\sqrt{8}$$
but x is positive in quadrant I, so
$$x = \sqrt{8} = 2\sqrt{2}$$

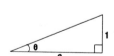

Now, $\tan\left[\sin^{-1}\dfrac{1}{3}\right] = \tan\theta = \dfrac{1}{2\sqrt{2}}$ (using $\dfrac{\text{opp}}{\text{adj}}$ in triangle)

$$\tan\theta = \dfrac{1}{2\sqrt{2}} \cdot \dfrac{\sqrt{2}}{\sqrt{2}} = \dfrac{\sqrt{2}}{4}$$

39. $\sec\left[\tan^{-1}\dfrac{1}{2}\right]$

First we know that $\tan\theta = \dfrac{1}{2}$, $-\dfrac{\pi}{2} < \theta < \dfrac{\pi}{2}$, so we have:

By the Pythagorean Theorem, the hypotenuse is:
$$1^2 + 2^2 = r^2$$
$$1 + 4 = r^2$$
$$5 = r^2$$
$$\sqrt{5} = r$$

Now, $\sec\left[\tan^{-1}\dfrac{1}{2}\right] = \sec\theta = \dfrac{\text{hyp}}{\text{adj}} = \sec\theta = \dfrac{\sqrt{5}}{2}$

41. $\cot\left[\sin^{-1}\left(-\dfrac{\sqrt{2}}{3}\right)\right]$

First draw the angle θ, $-\dfrac{\pi}{2} \le \theta \le \dfrac{\pi}{2}$, whose sine equals $-\dfrac{\sqrt{2}}{3}$

$$\sin\theta = -\frac{\sqrt{2}}{3}$$

By the Pythagorean Theorem, the missing side is:

$$x^2 + \left(-\sqrt{2}\right)^2 = 3^2$$
$$x^2 + 2 = 9$$
$$x^2 = 7$$
$$x = \pm\sqrt{7}\text{, but } x \text{ is positive in quadrant IV}$$
$$x = \sqrt{7}$$

Now, $\cot\left[\sin^{-1}\left(-\dfrac{\sqrt{2}}{3}\right)\right] = \cot\theta = \dfrac{\text{adj}}{\text{opp}}$

$$\cot\theta = \frac{\sqrt{7}}{-\sqrt{2}} = -\frac{\sqrt{7}}{\sqrt{2}}\cdot\frac{\sqrt{2}}{\sqrt{2}} = -\frac{\sqrt{14}}{2}$$

43. $\sin\left[\tan^{-1}(-3)\right]$

First draw the angle θ, $-\dfrac{\pi}{2} < \theta < \dfrac{\pi}{2}$, whose tangent is -3.

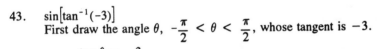

$$\tan\theta = -3$$

By the Pythagorean Theorem, the hypotenuse is:
$$r^2 = 1^2 + (-3)^2$$
$$r^2 = 1 + 9$$
$$r^2 = 10$$
$$r = \sqrt{10}$$

Now, $\sin\left[\tan^{-1}(-3)\right] = \sin\theta = \dfrac{\text{opp}}{\text{hyp}}$

$$\sin\theta = \frac{-3}{\sqrt{10}} = \frac{-3}{\sqrt{10}}\cdot\frac{\sqrt{10}}{\sqrt{10}} = \frac{-3\sqrt{10}}{10}$$

45. $\sec\left[\sin^{-1}\dfrac{2\sqrt{5}}{5}\right]$

First draw the angle θ, $-\dfrac{\pi}{2} \le \theta \le \dfrac{\pi}{2}$, whose sine is $\dfrac{2\sqrt{5}}{5}$

$$\sin\theta = \frac{2\sqrt{5}}{5}$$

By the Pythagorean Theorem, we find the missing side,

$$\left(2\sqrt{5}\right)^2 + x^2 = (5)^2$$
$$20 + x^2 = 25$$
$$x^2 = 5$$
$$x = \pm\sqrt{5}\text{ but } x > 0 \text{ in quadrant I}$$
$$x = \sqrt{5}$$

Now, $\sec\left[\sin^{-1}\dfrac{2\sqrt{5}}{5}\right] = \sec\theta = \dfrac{\text{hyp}}{\text{adj}}$

$$\sec\theta = \dfrac{5}{\sqrt{5}} = \dfrac{5}{\sqrt{5}} \cdot \dfrac{\sqrt{5}}{\sqrt{5}} = \dfrac{5\sqrt{5}}{5} = \sqrt{5}$$

47. Use radian mode on calculator.
$\sin^{-1}(\tan 0.5) = 0.58$

49. Use radian mode on calculator.
$\tan^{-1}(\sin 0.1) = 0.10$

51. Use radian mode on calculator.
$\cos^{-1}(\sin 1) = 0.57$

53. Use radian mode on calculator.
$\sin^{-1}\left[\tan\dfrac{\pi}{8}\right] = 0.43$

55. Use radian mode on calculator.
$\tan^{-1}\left[\sin\dfrac{\pi}{8}\right] = 0.37$

57. $\sec(\tan^{-1}\nu) = \sqrt{1+\nu^2}$

Let $\theta = \tan^{-1}\nu$ Then, $\tan\theta = \nu$, $-\dfrac{\pi}{2} < \theta < \dfrac{\pi}{2}$.
Hence, $\sec\theta > 0$ and $\tan^2\theta + 1 = \sec^2\theta$
$$\nu^2 + 1 = \sec^2\theta$$
$$\sqrt{\nu^2 + 1} = \sec\theta$$
Thus, $\sec(\tan^{-1}\nu) = \sec\theta = \sqrt{\nu^2 + 1} = \sqrt{1 + \nu^2}$

59. Let $\theta = \cos^{-1}\nu$. Then $\cos\theta = \nu$, $0 \le \theta \le \pi$
$$\tan(\cos^{-1}\nu) = \tan\theta = \dfrac{\sin\theta}{\cos\theta} = \dfrac{\sqrt{1-\cos^2\theta}}{\cos\theta} = \dfrac{\sqrt{1-\nu^2}}{\nu}$$

61. Let $\theta = \sin^{-1}\nu$. Then $\sin\theta = \nu$, $\dfrac{-\pi}{2} \le \theta \le \dfrac{\pi}{2}$
$$\cos(\sin^{-1}\nu) = \cos\theta = \sqrt{1-\sin^2\theta} = \sqrt{1-\nu^2}$$

63. Let $\alpha = \sin^{-1}\nu$ and $\beta = \cos^{-1}\nu$

Then $\sin\alpha = \nu = \cos\beta$, so α, β are complementary. Thus, $\alpha + \beta = \dfrac{\pi}{2}$.

65. Let $\alpha = \tan^{-1}\dfrac{1}{\nu}$. Then, $\dfrac{1}{\nu} = \tan\alpha$, $\dfrac{-\pi}{2} < \alpha < \dfrac{\pi}{2}$, $\alpha \ne 0$.

Let $\beta = \tan^{-1}\nu$. Then $\nu = \tan\beta$, $\dfrac{-\pi}{2} < \beta < \dfrac{\pi}{2}$.

Thus, $\tan\alpha\tan\beta = 1$, so that $\tan\alpha = \cot\beta$. Thus, $\alpha + \beta = \dfrac{\pi}{2}$

67. $\sec^{-1} 4$

Let $\nu = \sec^{-1} 4$. Then $\sec \theta = 3, 0 \le \nu \le \pi, \theta \ne \dfrac{\pi}{2}$.

Thus, $\cos \theta = \dfrac{1}{4}$ and $\sec^{-1} 4 = \theta = \cos^{-1} \dfrac{1}{4} \approx 1.32$.

69. $\cot^{-1} 2$

Let $\theta = \cot^{-1} 2$. Then $\cot \theta = 2, 0 < \nu < \pi$.

Thus, $\tan \theta = \dfrac{1}{2}$ and $\cot^{-1} 2 = \theta = \tan^{-1} \dfrac{1}{2} \approx 0.46$.

71. $\csc^{-1} (-3)$

Let $\theta = \csc^{-1} (-3)$. Then $\csc \theta = -3, \dfrac{-\pi}{2} \le \theta \le \dfrac{\pi}{2}, \theta \ne 0$.

Thus, $\sin \theta = \dfrac{-1}{3}$ and $\csc^{-1} (-3) = \theta = \sin^{-1} \left[\dfrac{-1}{3}\right] \approx -0.34$.

73. $\cot^{-1}\left(-\sqrt{5}\right)$

Let $\theta = \cot^{-1}\left(-\sqrt{5}\right)$. Then $\cot \theta = -\sqrt{5}, 0 < \theta < \pi$.

Thus, $\tan \theta = \dfrac{-1}{\sqrt{5}}$ and $\cot^{-1}\left(-\sqrt{5}\right) = \theta = \tan^{-1} \left[\dfrac{-1}{\sqrt{5}}\right] \approx 2.72$

75. Let $\theta = \csc^{-1} \left[-\dfrac{3}{2}\right]$. Then $\csc \theta = -\dfrac{3}{2}$, so $\sin \theta = -\dfrac{2}{3}$ and $\theta = \sin^{-1} \left[-\dfrac{2}{3}\right] \approx -0.73$.

77. Let $\theta = \cot^{-1} \left[-\dfrac{3}{2}\right]$. Then $\cot \theta = -\dfrac{3}{2}, 0 < \theta < \pi$. Thus, $\cos \theta = \dfrac{-3}{\sqrt{13}}, \dfrac{\pi}{2} < \theta < \pi$, and

$\theta = \cos^{-1} \left[\dfrac{-3}{\sqrt{13}}\right] \approx 2.55$.

79. $\dfrac{6.5}{2.5} = \dfrac{26.5 + x}{2.5 + x}$

$(2.5)(26.5) + 2.5x = 6.5x + (6.5)(2.5)$

$4x = 2.5(20)$

$x = 12.5$

$\cos \theta = \dfrac{2.5}{15}$

$\theta = 1.4$ radians $= 80.4°, \alpha = 180 - 80.4° = 99.6° = 1.73$

$s_1 = r_1\alpha = 6.5(1.73) = 11.3$

$s_2 = r_2\alpha = 2.5(1.4) = 3.5$

Length of belt $= 2(11.3 + 24 + 3.5) = 77.6$ inches

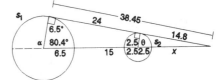

81. $\sin(\sin^{-1}x) = x$

Let $\theta = \sin^{-1} x$

$\sin \theta = x$ where $-\dfrac{\pi}{2} \le \theta \le \dfrac{\pi}{2}$ and

$-1 \le x \le 1$.

Hence, $-1 \le x \le 1$.

83. $\sin^{-1}(\sin x) = x$

Then x is the angle whose sine equals the

$\sin x$, i.e., $\sin x = x, \dfrac{-\pi}{2} \le x \le \dfrac{\pi}{2}$.

This is true only at $x = 0$.

85. $y = \sec^{-1} x$

$\sec y = x$

$\dfrac{1}{\cos y} = x$

$\cos y = \dfrac{1}{x}$

$y = \cos^{-1}\left[\dfrac{1}{x}\right] = \sec^{-1} x$

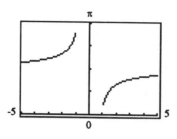

6 Chapter Review

1. $135° = 135 \cdot 1 \text{ degree} = 135 \cdot \dfrac{\pi}{180} \text{ radian} = \dfrac{3\pi}{4} \text{ radians.}$

3. $18° = 18 \cdot 1 \text{ degree} = 18 \cdot \dfrac{\pi}{180} \text{ radian} = \dfrac{\pi}{10} \text{ radians.}$

5. $\dfrac{3\pi}{4} \text{ radian} = \dfrac{3\pi}{4} \cdot 1 \text{ radian} = \dfrac{3\pi}{4} \cdot \dfrac{180}{\pi} \text{ degrees} = 135°$

7. $-\dfrac{5\pi}{2} \text{ radian} = -\dfrac{5\pi}{2} \cdot 1 \text{ radian} = -\dfrac{5\pi}{2} \cdot \dfrac{180}{\pi} \text{ degrees} = -450°$

9. $\tan \dfrac{\pi}{4} - \sin \dfrac{\pi}{6} = 1 - \dfrac{1}{2} = \dfrac{1}{2}$

11. $3 \sin 45° - 4 \tan \dfrac{\pi}{6} = 3\left[\dfrac{\sqrt{2}}{2}\right] - 4\left[\dfrac{\sqrt{3}}{3}\right] = \dfrac{3\sqrt{2}}{2} - \dfrac{4\sqrt{3}}{3}$

13. $6 \cos \dfrac{3\pi}{4} + 2 \tan\left[-\dfrac{\pi}{3}\right]$

Using reference angles: $\cos \dfrac{3\pi}{4} = -\cos \dfrac{\pi}{4} = -\dfrac{\sqrt{2}}{2}$

$\tan\left[-\dfrac{\pi}{3}\right] = -\tan \dfrac{\pi}{3} = -\sqrt{3}$

Hence, $6\left[\dfrac{-\sqrt{2}}{2}\right] + 2\left(-\sqrt{3}\right) = -3\sqrt{2} - 2\sqrt{3}$

15. $\sec\left[-\dfrac{\pi}{3}\right] - \cot\left[-\dfrac{5\pi}{4}\right]$

Using reference angles: $\sec\left[-\dfrac{\pi}{3}\right] = \sec \dfrac{\pi}{3} = 2$

$\cot\left[-\dfrac{5\pi}{4}\right] = -\cot\left[\dfrac{\pi}{4}\right] = -1$

Hence, $\sec\left[-\dfrac{\pi}{3}\right] - \cot\left[-\dfrac{5\pi}{4}\right] = 2 - (-1) = 3$

17. $\tan \pi + \sin \pi = 0 + 0 = 0$ 19. $\cos 180° - \tan(-45°) = -1 - (-1) = -1 + 1 = 0$

21. $\sin^2 20° + \dfrac{1}{\sec^2 20°} = \sin2 \, 20° + \cos^2 20° = 1$

23. $\sec 50° \cdot \cos 50° = \dfrac{1}{\cos 50°} \cdot \cos 50° = 1$

25. $\dfrac{\sin 50°}{\cos 40°} = \dfrac{\sin 50°}{\sin(90° - 40°)} = \dfrac{\sin 50°}{\sin 50°} = 1$ 27. $\dfrac{\sin(-40°)}{\cos 50°} = \dfrac{-\sin 40°}{\sin(90° - 50°)} = \dfrac{-\sin 40°}{\sin 40°} = -1$

29. $\sin 400° \sec(-50°) = \sin(400° - 360°)\sec 50° = \sin 40° \csc(90° - 50°) = \sin 40° \csc 40°$
$$= \sin 40° \cdot \dfrac{1}{\sin 40°} = 1$$

31. $\sin \theta = \dfrac{-4}{5}$, $\cos \theta > 0$

First we solve for $\cos \theta$:
$$\cos^2 \theta = 1 - \sin^2 \theta$$

$$\cos \theta = \sqrt{1 - \sin^2 \theta} = \sqrt{1 - \left(\dfrac{-4}{5}\right)^2} = \sqrt{1 - \dfrac{16}{25}} = \dfrac{3}{5}$$

$$\tan \theta = \dfrac{\sin \theta}{\cos \theta} = \dfrac{\dfrac{-4}{5}}{\dfrac{3}{5}} = \dfrac{-4}{3} \qquad \csc \theta = \dfrac{1}{\sin \theta} = \dfrac{1}{\dfrac{-4}{5}} = \dfrac{-5}{4}$$

$$\sec \theta = \dfrac{1}{\cos \theta} = \dfrac{1}{\dfrac{3}{5}} = \dfrac{5}{3} \qquad \cot \theta = \dfrac{1}{\tan \theta} = \dfrac{1}{\dfrac{-4}{3}} = \dfrac{-3}{4}$$

33. $\tan \theta = \dfrac{12}{5}$, $\sin \theta < 0$

Because $\tan \theta = \dfrac{\sin \theta}{\cos \theta} > 0$ and $\sin \theta < 0$, $\cos \theta < 0$. Since $\cos \theta = \dfrac{1}{\sec \theta} < 0$, $\sec \theta < 0$.
$$\sec^2 \theta = \tan^2 \theta + 1$$

$$\sec \theta = -\sqrt{\tan^2 \theta + 1} = -\sqrt{\left(\dfrac{12}{5}\right)^2 + 1} = -\sqrt{\dfrac{144}{25} + 1} = -\sqrt{\dfrac{169}{25}} = \dfrac{-13}{5}$$

$$\cos \theta = \dfrac{1}{\sec \theta} = \dfrac{1}{\dfrac{-13}{5}} = \dfrac{-5}{13}$$

$$\sin \theta = -\sqrt{1 - \cos^2 \theta} = -\sqrt{1 - \left(\dfrac{-5}{13}\right)^2} = -\sqrt{1 - \dfrac{25}{169}} = -\sqrt{\dfrac{144}{169}} = \dfrac{-12}{13}$$

$$\csc \theta = \dfrac{1}{\sin \theta} = \dfrac{1}{\dfrac{-12}{13}} = \dfrac{-13}{12} \qquad \cot \theta = \dfrac{1}{\tan \theta} = \dfrac{1}{\dfrac{12}{5}} = \dfrac{5}{12}$$

35. $\sec \theta = \dfrac{-5}{4}$, $\tan \theta < 0$
$$\tan^2 \theta + 1 = \sec^2 \theta$$
$$\tan^2 \theta = \sec^2 \theta - 1$$

$$\tan \theta = \pm\sqrt{\sec^2 \theta - 1}$$

Because tan $t < 0$, we use the minus sign:

$$\tan \theta = -\sqrt{\left(\frac{-5}{4}\right)^2 - 1} = -\sqrt{\frac{25}{16} - 1} = -\sqrt{\frac{9}{16}} = \frac{-3}{4}$$

$$\cot \theta = \frac{1}{\tan \theta} = \frac{1}{\dfrac{-3}{4}} = \frac{-4}{3} \qquad\qquad \cos \theta = \frac{1}{\sec \theta} = \frac{1}{\dfrac{-5}{4}} = \frac{-4}{5}$$

$$\sin \theta = \pm\sqrt{1 - \cos^2 \theta}$$

Since $\tan \theta = \dfrac{\sin \theta}{\cos \theta} < 0$ and $\cos \theta < 0$, $\sin \theta > 0$. Therefore, we use the plus sign:

$$\sin \theta = \sqrt{1 - \left(\frac{-4}{5}\right)^2} = \sqrt{1 - \frac{16}{25}} = \sqrt{\frac{9}{25}} = \frac{3}{5} \qquad\qquad \csc \theta = \frac{1}{\sin \theta} = \frac{1}{\dfrac{3}{5}} = \frac{5}{3}$$

37. $\sin \theta = \dfrac{12}{13}$, θ in quadrant II

$\sin \theta = \dfrac{b}{r} = \dfrac{12}{13}$ so $b = 12$ and $r = 13$

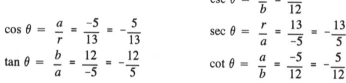

Since $a < 0$ in quadrant II and using $r = \sqrt{a^2 + b^2}$, we get:

$$13 = \sqrt{a^2 + 12^2}$$
$$169 = a^2 + 144$$
$$25 = a^2$$
$$\pm 5 = a$$
$$-5 = a$$

$$\csc \theta = \frac{r}{b} = \frac{13}{12}$$

$$\cos \theta = \frac{a}{r} = \frac{-5}{13} = -\frac{5}{13} \qquad\qquad \sec \theta = \frac{r}{a} = \frac{13}{-5} = -\frac{13}{5}$$

$$\tan \theta = \frac{b}{a} = \frac{12}{-5} = -\frac{12}{5} \qquad\qquad \cot \theta = \frac{a}{b} = \frac{-5}{12} = -\frac{5}{12}$$

39. $\sin \theta = -\dfrac{5}{13}$, $\dfrac{3\pi}{2} < \theta < 2\pi$

$\sin \theta = \dfrac{b}{r} = \dfrac{-5}{13}$, so $b = -5$ and $r = 13$

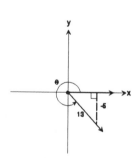

Since $a > 0$ in quadrant IV and using $r = \sqrt{a^2 + b^2}$, we get:

$$\sqrt{a^2 + (-5)^2} = 13$$
$$a^2 + 25 = 169$$
$$a^2 = 144$$
$$a = \pm 12$$
$$a = 12$$

$$\csc \theta = \frac{r}{b} = \frac{13}{-5} = -\frac{13}{5}$$

$$\cos \theta = \frac{a}{r} = \frac{12}{13} \qquad\qquad \sec \theta = \frac{r}{a} = \frac{13}{12}$$

$$\tan \theta = \frac{b}{a} = \frac{-5}{12} = -\frac{5}{12} \qquad\qquad \cot \theta = \frac{a}{b} = \frac{12}{-5} = -\frac{12}{5}$$

41. $\tan \theta = \dfrac{1}{3}$, $180° < \theta < 270°$

$\tan \theta = \dfrac{b}{a}$; but $a < 0$ and $b < 0$ in quadrant III, so

$\tan \theta = \dfrac{b}{a} = \dfrac{1}{3}$, so $a = -3$ and $b = -1$.

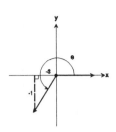

Using $r = \sqrt{a^2 + b^2}$, we find:

$$\sqrt{(-3)^2 + (-1)^2} = r$$
$$\sqrt{9 + 1} = r$$
$$\sqrt{10} = r$$

$\sin \theta = \dfrac{b}{r} = -\dfrac{1}{\sqrt{10}} = -\dfrac{1}{\sqrt{10}} \cdot \dfrac{\sqrt{10}}{\sqrt{10}} = -\dfrac{\sqrt{10}}{10}$ $\csc \theta = \dfrac{r}{b} = \dfrac{\sqrt{10}}{-1} = -\sqrt{10}$

$\cos \theta = \dfrac{a}{r} = \dfrac{-3}{\sqrt{10}} = -\dfrac{3}{\sqrt{10}} \cdot \dfrac{\sqrt{10}}{\sqrt{10}} = -\dfrac{3\sqrt{10}}{10}$ $\sec \theta = \dfrac{r}{a} = \dfrac{\sqrt{10}}{-3} = -\dfrac{\sqrt{10}}{3}$

$\cot \theta = \dfrac{a}{b} = \dfrac{-3}{-1} = 3$

43. $\sec \theta = 3$, $\dfrac{3\pi}{2} < \theta < 2\pi$

$\sec \theta = \dfrac{r}{a} = \dfrac{3}{1}$ so $a = 1$ and $r = 3$

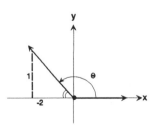

Sin $b < 0$ in quadrant IV and using $\sqrt{a^2 + b^2} = r$, we get:

$$\sqrt{1^2 + b^2} = 3$$
$$1 + b^2 = 9$$
$$b^2 = 8$$
$$b = \pm\sqrt{8}$$
$$b = -2\sqrt{2}$$

$\sin \theta = \dfrac{b}{r} = \dfrac{-2\sqrt{2}}{3}$ $\csc \theta = \dfrac{r}{b} = \dfrac{3}{-2\sqrt{2}} = -\dfrac{3}{2\sqrt{2}} \cdot \dfrac{\sqrt{2}}{\sqrt{2}} = -\dfrac{3\sqrt{2}}{4}$

$\cos \theta = \dfrac{a}{r} = \dfrac{1}{3}$

$\tan \theta = \dfrac{b}{a} = \dfrac{-2\sqrt{2}}{1} = -2\sqrt{2}$ $\cot \theta = \dfrac{a}{b} = \dfrac{1}{-2\sqrt{2}} = -\dfrac{1}{2\sqrt{2}} \cdot \dfrac{\sqrt{2}}{\sqrt{2}} = -\dfrac{\sqrt{2}}{4}$

45. $\cot \theta = -2$, $\dfrac{\pi}{2} < \theta < \pi$

$\cot \theta = \dfrac{a}{b}$, so $a < 0$ and $b > 0$ in quadrant II

$\cot \theta = \dfrac{a}{b} = \dfrac{-2}{1}$, so $a = -2$ and $b = 1$

Using $\sqrt{a^2 + b^2} = r$, we get:

$$\sqrt{(-2)^2 + 1^2} = r$$
$$\sqrt{5} = r$$

$$\sin \theta = \frac{b}{r} = \frac{1}{\sqrt{5}} = \frac{\sqrt{5}}{5} \qquad\qquad \csc \theta = \frac{r}{b} = \sqrt{5}$$

$$\cos \theta = \frac{a}{r} = \frac{-2}{\sqrt{5}} = \frac{-2\sqrt{5}}{5} \qquad \sec \theta = \frac{r}{a} = \frac{\sqrt{5}}{-2} = \frac{-\sqrt{5}}{2}$$

$$\tan \theta = \frac{b}{a} = \frac{1}{-2} = \frac{-1}{2}$$

47.

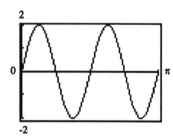

The graph of $y = 2 \sin 4x$ is the graph of $y = \sin x$ vertically stretched by a factor of 2 and horizontally compressed by a factor of 4

(i.e., each x-value is multiplied by $\frac{1}{4}$).

49.

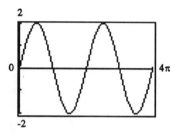

The graph of $y = -2 \cos \left(x + \frac{\pi}{2}\right)$ is the graph of $y = \cos x$ vertically stretched by a factor of 2, reflected about the x-axis and shifted horizontally $\frac{\pi}{2}$ units left.

51.

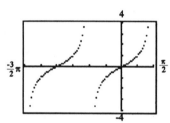

The graph of $y = \tan(x + \pi)$ is the graph of $y = \tan x$ shifted horizontally π units to the left.

53.

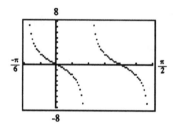

The graph $y = -2 \tan 3x$ is the graph of $y = \tan x$ vertically stretched by a factor of 2, reflected about the x-axis and horizontally compressed by a factor of 3.

55. $\sin^{-1} 1$

We are looking for the angle θ, $-\frac{\pi}{2} \leq \theta \leq \frac{\pi}{2}$, whose sine is 1.

$$\sin \theta = 1 \qquad\qquad -\frac{\pi}{2} \leq \theta \leq \frac{\pi}{2}$$

$$\theta = \frac{\pi}{2}$$

Hence, $\sin^{-1} 1 = \frac{\pi}{2}$

57. $\tan^{-1} 1$

We are looking for the angle θ, $-\dfrac{\pi}{2} < \theta < \dfrac{\pi}{2}$, whose tangent is 1.

$$\tan \theta = 1 \qquad\qquad -\frac{\pi}{2} < \theta < \frac{\pi}{2}$$

$$\theta = \frac{\pi}{4}$$

Hence, $\tan^{-1} 1 = \dfrac{\pi}{4}$

59. $\cos^{-1}\left(-\dfrac{\sqrt{3}}{2}\right)$

We are looking for the angle θ, $0 \le \theta \le \pi$, whose cosine is $-\dfrac{\sqrt{3}}{2}$.

$$\cos \theta = -\frac{\sqrt{3}}{2} \qquad\qquad 0 \le \theta \le \pi$$

$$\theta = \frac{5\pi}{6}$$

Hence, $\cos^{-1}\left(-\dfrac{\sqrt{3}}{2}\right) = \dfrac{5\pi}{6}$

61. $\sin\left(\cos^{-1} \dfrac{\sqrt{2}}{2}\right)$

First find the angle θ, $0 \le \theta \le \pi$, whose cosine equals $\dfrac{\sqrt{2}}{2}$.

$$\cos \theta = \frac{\sqrt{2}}{2}, \, 0 \le \theta \le \pi$$

$$\theta = \frac{\pi}{4}$$

Now, $\sin\left(\cos^{-1} \dfrac{\sqrt{2}}{2}\right) = \sin \theta = \sin \dfrac{\pi}{4} = \dfrac{\sqrt{2}}{2}$

63. $\tan\left[\sin^{-1}\left(-\dfrac{\sqrt{3}}{2}\right)\right]$

First find the angle θ, $-\dfrac{\pi}{2} \le \theta \le \dfrac{\pi}{2}$, whose sine equals $-\dfrac{\sqrt{3}}{2}$.

$$\sin \theta = -\frac{\sqrt{3}}{2}, \, -\frac{\pi}{2} \le \theta \le \frac{\pi}{2}$$

$$\theta = -\frac{\pi}{3}$$

Now, $\tan\left[\sin^{-1}\left(-\dfrac{\sqrt{3}}{2}\right)\right] = \tan \theta = \tan\left(-\dfrac{\pi}{3}\right) = -\sqrt{3}$

65. $\sec\left[\tan^{-1}\dfrac{\sqrt{3}}{3}\right]$

First find the angle θ, $-\dfrac{\pi}{2}<\theta<\dfrac{\pi}{2}$, whose tangent equals $\dfrac{\sqrt{3}}{3}$.

$$\tan\theta = \dfrac{\sqrt{3}}{3} \quad -\dfrac{\pi}{2}<\theta<\dfrac{\pi}{2}$$

$$\theta = \dfrac{\pi}{6}$$

Now, $\sec\left[\tan^{-1}\dfrac{\sqrt{3}}{3}\right] = \sec\theta = \sec\dfrac{\pi}{6} = \dfrac{2\sqrt{3}}{3}$

67. $\sin\left[\tan^{-1}\dfrac{3}{4}\right]$

Let $\theta = \tan^{-1}\dfrac{3}{4}$

Then, $\tan\theta = \dfrac{3}{4} = \dfrac{\text{opp}}{\text{adj}}$, $-\dfrac{\pi}{2}<\theta<\dfrac{\pi}{2}$

The hypotenuse is $r = \sqrt{x^2+y^2}$

$$r = \sqrt{16+9}$$
$$r = \sqrt{25}$$
$$r = 5$$

Thus, $\sin\left[\tan^{-1}\dfrac{3}{4}\right] = \sin\theta = \dfrac{3}{5}$

69. $\tan\left[\sin^{-1}\left(-\dfrac{4}{5}\right)\right]$

Let $\theta = \sin^{-1}\left(-\dfrac{4}{5}\right)$

Then, $\sin\theta = -\dfrac{4}{5} = \dfrac{\text{opp}}{\text{hyp}}$, $-\dfrac{\pi}{2}\le\theta\le\dfrac{\pi}{2}$.

Using the Pythagorean Theorem, we find the missing side:

$$x^2 + (-4)^2 = 5^2$$
$$x^2 + 16 = 25$$
$$x^2 = 9$$
$$x = \pm 3$$
$$x = 3, \; x > 0 \text{ in quadrant IV}$$

Hence, $\tan\left[\sin^{-1}\left(-\dfrac{4}{5}\right)\right] = \tan\theta = -\dfrac{4}{3}$

71. $\theta = 30° \text{ or } \theta = \dfrac{\pi}{6}$

radius = 2 feet

Using $s = r\theta$, we get:

$$s = 2\left(\dfrac{\pi}{6}\right)$$
$$s = \dfrac{\pi}{3}\text{ feet}$$

73. $\nu = 180\text{ mi/hr}$

diameter $= \dfrac{1}{2}$ mi, so $r = \dfrac{1}{4}$ mi

Find angular speed: ω

Using $\nu = r\omega$, we get:

$$180 = \dfrac{1}{4}\omega$$

$720\text{ rad/hr} = \omega$ (Remember that ω is expressed in radians per unit time.)

$$\dfrac{720\text{ rad}}{\text{hr}} \cdot \dfrac{1\text{ rev}}{2\pi\text{ rad}} = \omega$$

$$\dfrac{720}{2\pi}\text{ rev/hr} = \omega$$

$$114.59\text{ rev/hr} = \omega$$